W9-AXW-104

Handbook of Chemical Engineering Calculations

Other McGraw-Hill Chemical Engineering Books of Interest

Austin et al. • Shreve's Chemical Process Industries
Cook, DuMont • Process Drying Practice
Chopey • Environmental Engineering in the Process Plant
Dean • Lange's Handbook of Chemistry
Dillon • Materials Selection for the Chemical Process Industries
Freeman • Hazardous Waste Minimization
Freeman • Standard Handbook of Hazardous Waste Treatment and Disposal
Grant, Grant • Grant & Hackh's Chemical Dictionary
Kister • Distillation Operation
Kister • Distillation Design
McGee • Molecular Engineering
Miller • Flow Measurement Handbook
Perry, Green • Perry's Chemical Engineers' Handbook
Reid et al. • Properties of Gases and Liquids
Reist • Introduction to Aerosol Science
Ryans, Roper • Process Vacuum Systems and Design Operation
Sandler, Luckiewicz • Practical Process Engineering
Satterfield • Heterogenous Catalysis in Practice
Shinskey • Process Control Systems
Shugar, Dean • The Chemist's Ready Reference Handbook
Shugar, Ballinger • The Chemical Technicians' Ready Reference Handbook
Smith, Van Laan • Piping and Pipe Support Systems
Stock • AI in Process Control
Tatterson • Fluid Mixing and Gas Dispersion in Agitated Tanks
Yokell • A Working Guide to Shell-and-Tube Heat Exchangers

Handbook of Chemical Engineering Calculations

NICHOLAS P. CHOPEY, Editor
Executive Editor of Chemical Engineering Magazine
Member, American Institute of Chemical Engineers,
American Society for Engineering Education

Second Edition

McGRAW-HILL, INC.
New York San Francisco Washington, D.C. Auckland
Bogotá Caracas Lisbon London Madrid
Mexico City Milan Montreal New Delhi
San Juan Singapore Sydney Tokyo Toronto

Library of Congress Cataloging-in-Publication Data

Handbook of chemical engineering calculations/Nicholas P. Chopey, editor.—2nd ed.
p. cm.

ISBN 0-07-011021-2
1. Chemical engineering—Mathematics. I. Chopey, Nicholas P.
TP149.H285 1993 93-25590
660′.212—dc20 CIP

4 5 6 7 8 9 0 DOC/DOC 9 9 8 7 6 5

ISBN 0-07-011021-2

The sponsoring editor for this book was Gail F. Nalven, the editing supervisor was Jim Halston, and the production supervisor was Donald F. Schmidt.

Printed and bound by R. R. Donnelley & Sons Company.

This book is printed on acid-free paper.

To Kathleen

Contents

Contributors

Robert M. Baldwin, Ph.D., Professor and Head, Chemical and Petroleum-Refining Engineering Department, Colorado School of Mines, Golden, CO (*Reaction Kinetics and Reactor Design*)

James R. Beckman, Ph.D., Associate Professor, Department of Chemical and Bio Engineering, Arizona State University, Tempe, AZ *(Crystallization)*

Gerald D. Button, Market Manager, Pharmaceuticals, Rohm & Haas Co., Philadelphia, PA (*Ion Exchange,* in *Other Chemical Engineering Calculations*)

David S. Dickey, Ph.D., Manager, Fluid Mixing Technology, Prochem, Division of Robbins & Myers, Inc., Springfield, Ohio *(Liquid Agitation)*

Otto Frank, Supervisor, Process Engineering, Allied Corp., Morristown, NJ *(Distillation)*

James H. Gary, Ph.D., Professor Emeritus, Chemical and Petroleum-Refining Engineering Department, Colorado School of Mines, Golden, CO *(Stoichiometry)*

Michael S. Graboski, Ph.D., Director, Colorado Institute for Fuels and High-Altitude Engine Research, Colorado School of Mines, Golden, CO *(Reaction Kinetics and Reactor Design)*

Avinash Gupta, Ph.D., Senior Principal Chemical Engineer, ABB Lummus Crest, Bloomfield, NJ *(Physical and Chemical Properties)*

Wenfang Leu, Ph.D., Research Scientist, Department of Chemical Engineering, University of Houston, Houston, TX *(Filtration)*

Kenneth J. McNulty, Sc.D., Technical Director of R&D, Koch Engineering Co., Wilmington, Mass. *(Absorption and Stripping)*

Paul E. Minton, Principal Engineer, Union Carbide Corp., South Charleston, WV *(Heat Transfer)*

Edward S. S. Morrison, Senior Staff Engineer, Union Carbide Corp., Houston, TX *(Heat Transfer)*

A. K. S. Murthy, Eng.Sc.D., Head, Fuels Research Group, Allied Corp., Morristown, NJ *(Phase Equilibrium)*

E. Dendy Sloan, Ph.D., Professor, Chemical and Petroleum-Refining Engineering Department, Colorado School of Mines, Golden, CO *(Chemical-Reaction Equilibrium)*

Ross W. Smith, Ph.D., Professor, Department of Chemical and Metallurgical Engineering, Mackay School of Mines, University of Nevada at Reno, Reno, NV *(Size Reduction)*

Louis Theodore, Eng.Sc.D., Professor, Department of Chemical Engineering, Manhattan College, Bronx, N.Y. *(Air-Pollution Control)*

Frank M. Tiller, Ph.D., M. D. Anderson Professor, Department of Chemical Engineering, University of Houston, Houston, TX *(Filtration)*

Frank H. Verhoff, Ph.D., Director of Chemical Engineering Research, Miles Laboratories, Inc., Elkhart, IN *(Extraction and Leaching)*

Preface

This handbook shows how to solve the main process-related problems that crop up often in chemical engineering practice.

The book, now in its second edition, is an outgrowth of the highly successful *Standard Handbook of Engineering Calculations,* which contains separate sections devoted to the various engineering disciplines, including a relatively short section on chemical engineering. The desirability of publishing a similar handbook that focused exclusively on chemical engineering was obvious. Tyler Hicks, editor of the previously mentioned volume, was co-editor of the first edition, which helped to assure continuity in employing the well-accepted approach that the *Standard Handbook* features.

The approach consists of the use of solved, numerical illustrative examples. Except for introductory paragraphs in a few of the chapters where introduction seemed especially appropriate, this entire volume consists of solved examples. In each chapter, these have been chosen so as to bring out the most important problems that arise in the topic covered by that chapter (excepting those problems that are too complex and unwieldy for this approach). In some work-situations, readers will use the solution techniques directly, with a calculator; in other cases, the examples will provide the understanding that is needed for implementing a computer-based solution.

All chapters and all examples included in the first edition have been retained. The main change in this second edition is the adding of two major chapters: one on absorption and stripping, the other on air pollution control. Also, new examples have been added to several of the existing chapters.

All of the chapter on flow of fluids and solids comes from the *Standard Handbook of Engineering Calculations,* and that book was the source for part of the material in the stoichiometry chapter as well as most of the final chapter, "Other Chemical Engineering Calculations." Virtually everything else in this volume was prepared specifically for it. Each section was prepared by one or more experts in the given field, and to a large extent the actual choice of examples was made by the expert(s) in question. Thus, we owe a doubly deep debt of gratitude to our contributors.

Nicholas P. Chopey

Handbook of Chemical Engineering Calculations

1

Physical and Chemical Properties

Avinash Gupta, Ph.D.
Senior Principal Chemical Engineer
ABB Lummus Crest
Bloomfield, NJ

REFERENCES: [1] Reid, Prausnitz, and Sherwood—*The Properties of Gases and Liquids,* McGraw-Hill; [2] Lewis, Randall, and Pitzer—*Thermodynamics,* McGraw-Hill; [3] Prausnitz—*Molecular Thermodynamics of Fluid-Phase Equilibria,* Prentice-Hall; [4] Hougen, Watson, and Ragatz—*Chemical Process Principles,* parts I and II, Wiley; [5] Bretsznajder—*Prediction of Transport and Other Physical Properties of Fluids,* Pergamon; [6] Smith and Van Ness—*Chemical Engineering Thermodynamics,* McGraw-Hill; [7] Perry and Chilton—*Chemical Engineers' Handbook,* McGraw-Hill; [8] The Chemical Rubber Company—*Handbook of Chemistry and Physics;* [9] Bland and Davidson—*Petroleum Processing Handbook,* McGraw-Hill; [10] American Petroleum Institute—*Technical Data Book—Petroleum Refining;* [11] Natural Gas Processors Suppliers Association—*Engineering Data Book;* [12] Dreisbach—*Physical Properties of Chemical Compounds,* vols. 1–3, American Chemical Society; [13] Timmermans—*Physico-Chemical Constants of Binary Systems and Pure Organic Compounds,* Elsevier; [14] Rossini—*Selected Values of Properties of Chemical Compounds,* Thermodynamics Research Center, Texas A&M University; [15] Stull and Prophet—*JANAF Thermochemical Tables,* NSRDS, NBS-37; [16] Gunn and Yamada—*AIChE Journal 17*:1341, 1971; [17] Rackett—*J. Chem. Eng. Data 15*:514, 1970; [18] Yen and Woods—*AIChE Journal 12*:95, 1966; [19] Yuan and Stiel—*Ind. Eng. Chem. Fund. 9*:393, 1970; [20] *J. Chem. Eng. Data 10*:207, 1965; [21] *Chem. Eng. Tech. 26*:679, 1954; [22] *J. Am. Chem. Soc. 77*:3433, 1955; [23] *AIChE Journal 21*:510, 1975; [24] *AIChE Journal 13*:626, 1967; [25] *J. Chem. Phys. 29*:546, 1958; [26] *Ind. Eng. Chem. Process Des. Dev. 10*:576, 1971; [27] *AIChE Journal 15*:615, 1969; [28] Palmer—*Chemical Engineering 82*:80, 1975; [29] Letsou and Stiel—*AIChE Journal* 19:409, 1973; [30] Jossi et al—*AIChE Journal* 8:59, 1962.

1-1 Molar Gas Constant

Calculate the molar gas constant R in the following units:

a. (atm)(cm^3)/(g·mol)(K)

b. (psia)(ft^3)/(lb·mol)(°R)

c. (atm)(ft^3)/(lb·mol)(K)

d. kWh/(lb·mol)(°R)

e. hp·h/(lb·mol)(°R)

f. (kPa)(m^3)/(kg·mol)(K)

g. cal/(g·mol)(K)

Calculation Procedure:

1. *Assume a basis.*

Assume gas is at standard conditions, that is, 1 g·mol gas at 1 atm (101.3 kPa) pressure and 0°C (273 K, or 492°R), occupying a volume of 22.4 L.

2. *Compute the gas constant.*

Apply suitable conversion factors and obtain the gas constant in various units. Use $PV = RT$; that is, $R = PV/T$. Thus,

a. R = (1 atm)[22.4 L/(g·mol)](1000 cm^3/L)/273 K = 82.05 (atm)(cm^3)/(g·mol)(K)

b. R = (14.7 psia)[359 ft^3/(lb·mol)]/492°R = 10.73 (psia)(ft^3)/(lb·mol)(°R)

c. R = (1 atm)[359 ft^3/(lb·mol)]/273 K = 1.315 (atm)(ft^3)/(lb·mol)(K)

d. R = [10.73 (psia)(ft^3)/(lb·mol)(°R)](144 in^2/ft^2)[3.77 $\times$ 10^{-7} kWh/(ft·lbf)] = 5.83 $\times$ 10^{-4} kWh/(lb·mol)(°R)

e. R = [5.83 $\times$ 10^{-4} kWh/(lb·mol)(°R)](1/0.746 hp·h/kWh) = 7.82 $\times$ 10^{-4} hp·h/(lb·mol)(°R)

f. R = (101.325 kPa/atm)[22.4 L/(g·mol)][1000 g·mol/(kg·mol)]/(273 K)(1000 L/m^3) = 8.31(kPa)(m^3)/(kg·mol)(K)

g. R = [7.82 $\times$ 10^{-4} hp·h/(lb·mol)(°R)][6.4162 $\times$ 10^5 cal/(hp·h)][1/453.6 lb·mol/(g·mol)](1.8°R/K) = 1.99 cal/(g·mol)(K)

1-2 Estimation of Critical Temperature from Empirical Correlation

Predict the critical temperature of (a) n-eicosane, (b) 1-butene, and (c) benzene using the empirical correlation of Nokay. The Nokay relation is

$$\log T_c = A + B \log SG + C \log T_b$$

where T_c is critical temperature in kelvins, T_b is normal boiling point in kelvins, and SG is specific gravity of liquid hydrocarbons at 60°F relative to water at the same temperature. As for A, B, and C, they are correlation constants given in Table 1-1.

TABLE 1-1 Correlation Constants for Nokay's Equation

Family of compounds	A	B	C
Alkanes (paraffins)	1.359397	0.436843	0.562244
Cycloalkanes (naphthenes)	0.658122	−0.071646	0.811961
Alkenes (olefins)	1.095340	0.277495	0.655628
Alkynes (acetylenes)	0.746733	0.303809	0.799872
Alkadienes (diolefins)	0.147578	−0.396178	0.994809
Aromatics	1.057019	0.227320	0.669286

Calculation Procedure:

1. Obtain normal boiling point and specific gravity.

Obtain T_b and SG for these three compounds from, for instance, Reid, Prausnitz, and Sherwood [1]. These are (a) for *n*-eicosane ($C_{20}H_{42}$), $T_b = 617$ K and $SG = 0.775$; (b) for 1-butene (C_4H_8), $T_b = 266.9$ K and $SG = 0.595$; and (c) for benzene (C_6H_6), $T_b = 353.3$ K and $SG = 0.885$.

2. Compute critical temperature using appropriate constants from Table 1-1.

Thus (a) for *n*-eicosane:

$$\log T_c = 1.359397 + 0.436843 \log 0.775 + 0.562244 \log 617 = 2.87986$$

so $T_c = 758.3$ K (905°F). (b) For 1-butene:

$$\log T_c = 1.095340 + 0.277495 \log 0.595 + 0.655628 \log 266.9 = 2.62355$$

so $T_c = 420.3$ K (297°F). (c) For benzene:

$$\log T_c = 1.057019 + 0.22732 \log 0.885 + 0.669286 \log 353.3 = 2.75039$$

so $T_c = 562.8$ K (553°F).

Related Calculations: This procedure may be used to estimate the critical temperature of hydrocarbons containing a single family of compounds, as shown in Table 1-1. Tests of the equation on paraffins in the range C_1–C_{20} and various other hydrocarbon families in the range C_3–C_{14} have shown average and maximum deviations of about 6.5 and 35°F (3.6 and 19 K), respectively.

1-3 Critical Properties from Group-Contribution Method

Estimate the critical properties of *p*-xylene and *n*-methyl-2-pyrrolidone using Lydersen's method of group contributions.

Calculation Procedure:

1. Obtain molecular structure, normal boiling point T_b, and molecular weight MW.

From handbooks, for *p*-xylene (C_8H_{10}), MW = 106.16, $T_b = 412.3$ K, and the structure is

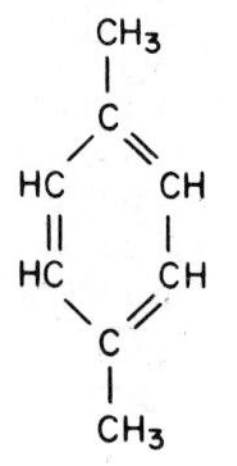

For *n*-methyl-2-pyrrolidone (C_5H_9NO), MW = 99.1, T_b = 475.0 K, and the structure is

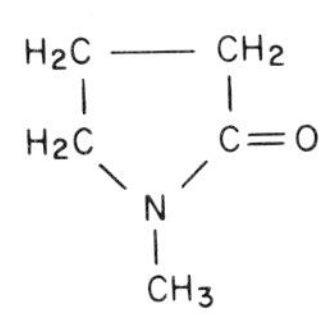

2. *Sum up structural contributions of the individual property increments from Table 1-2.*

The calculations can be set out in the following arrays, in which N stands for the number of groups. For *p*-xylene:

Group type	N	ΔT	ΔP	ΔV	$(N)(\Delta T)$	$(N)(\Delta P)$	$(N)(\Delta V)$
$-CH_3$ (nonring)	2	0.020	0.227	55	0.04	0.454	110
$-\overset{|}{C}=$ (ring)	2	0.011	0.154	36	0.022	0.308	72
$H\overset{|}{C}=$ (ring)	4	0.011	0.154	37	0.044	0.616	148
Total					0.106	1.378	330

For *n*-methyl-2-pyrrolidone:

Group type	N	ΔT	ΔP	ΔV	$(N)(\Delta T)$	$(N)(\Delta P)$	$(N)(\Delta V)$
$-CH_3$ (nonring)	1	0.020	0.227	55	0.020	0.227	55
$-CH_2-$ (ring)	3	0.013	0.184	44.5	0.039	0.552	133.5
$C=O$ (ring)	1	0.033	0.2	50	0.033	0.20	50
$-N-$ (ring)	1	0.007	0.13	32	0.007	0.13	32
Total					0.099	1.109	270.5

3. *Compute the critical properties.*

The formulas are

$$T_c = T_b\{[(0.567) + \Sigma(N)(\Delta T) - [\Sigma(N)(\Delta T)]^2\}^{-1}$$

$$P_c = \text{MW}[0.34 + (N)(\Delta P)]^{-2}$$

$$V_c = [40 + (N)(\Delta V)]$$

$$Z_c = P_c V_c / RT_c$$

where T_c, P_c, V_c, and Z_c are critical temperature, critical pressure, critical volume, and critical compressibility factor, respectively. Thus, for *p*-xylene,

$$T_c = 412.3[0.567 + 0.106 - (0.106)^2]^{-1}$$

$$= 623.0 \text{ K } (661.8°\text{F}) \text{ (literature value is 616.2 K)}$$

$$P_c = 106.16(0.34 + 1.378)^{-2} = 35.97 \text{ atm } (3644 \text{ kPa}) \text{ (literature value is 34.7 atm)}$$

TABLE 1-2 Critical-Property Increments—Lydersen's Structural Contributions

Symbols	ΔT	ΔP	ΔV
	Nonring increments		
$-CH_3$	0.020	0.227	55
$\overset{\vert}{-CH_2}$	0.020	0.227	55
$\underset{\vert}{\overset{\vert}{-CH}}$	0.012	0.210	51
$\underset{\vert}{\overset{\vert}{-C-}}$	0.00	0.210	41
$=CH_2$	0.018	0.198	45
$\overset{\vert}{=CH}$	0.018	0.198	45
$\overset{\vert}{=C-}$	0.0	0.198	36
$=C=$	0.0	0.198	36
$\equiv CH$	0.005	0.153	(36)
$\equiv C-$	0.005	0.153	(36)
	Ring increments		
$-CH_2-$	0.013	0.184	44.5
$\underset{\vert}{\overset{\vert}{-CH}}$	0.012	0.192	46
$\underset{\vert}{\overset{\vert}{-C-}}$	(−0.007)	(0.154)	(31)
$\overset{\vert}{=CH}$	0.011	0.154	37
$\overset{\vert}{=C-}$	0.011	0.154	36
$=C=$	0.011	0.154	36
	Halogen increments		
$-F$	0.018	0.221	18
$-Cl$	0.017	0.320	49
$-Br$	0.010	(0.50)	(70)
$-I$	0.012	(0.83)	(95)
	Oxygen increments		
$-OH$ (alcohols)	0.082	0.06	(18)
$-OH$ (phenols)	0.031	(−0.02)	(3)
$-O-$ (nonring)	0.021	0.16	20
$-O-$ (ring)	(0.014)	(0.12)	(8)
$\overset{\vert}{-C}=O$ (nonring)	0.040	0.29	60

Symbols	ΔT	ΔP	ΔV
Oxygen increments (cont)			
$-\overset{\vert}{C}=O$ (ring)	(0.033)	(0.2)	(50)
$H\overset{\vert}{C}=O$ (aldehyde)	0.048	0.33	73
$-COOH$ (acid)	0.085	(0.4)	80
$-COO-$ (ester)	0.047	0.47	80
$=O$ (except for combinations above)	(0.02)	(0.12)	(11)
Nitrogen increments			
$-NH_2$	0.031	0.095	28
$-\overset{\vert}{N}H$ (nonring)	0.031	0.135	(37)
$-\overset{\vert}{N}H$ (ring)	(0.024)	(0.09)	(27)
$-\overset{\vert}{N}-$ (nonring)	0.014	0.17	(42)
$-\overset{\vert}{N}-$ (ring)	(0.007)	(0.13)	(32)
$-CN$	(0.060)	(0.36)	(80)
$-NO_2$	(0.055)	(0.42)	(78)
Sulfur increments			
$-SH$	0.015	0.27	55
$-S-$ (nonring)	0.015	0.27	55
$-S-$ (ring)	(0.008)	(0.24)	(45)
$=S$	(0.003)	(0.24)	(47)
Miscellaneous			
$-\overset{\vert}{Si}-$	0.03	(0.54)	
$-\overset{\vert}{\underset{\vert}{B}}-$	(0.03)		

Note: There are no increments for hydrogen. All bonds shown as free are connected with atoms other than hydrogen. Values in parentheses are based on too few experimental data to be reliable.

Source: A. L. Lydersen, U. of Wisconsin Eng. Exp. Station, 1955.

$$V_c = 40 + 330$$

$$= 370 \text{ cm}^3/(\text{g}\cdot\text{mol})\ [5.93 \text{ ft}^3/(\text{lb}\cdot\text{mol})]\ [\text{literature value} = 379 \text{ cm}^3/(\text{g}\cdot\text{mol})]$$

And since $R = 82.06\ (\text{cm}^3)(\text{atm})/(\text{g}\cdot\text{mol})(\text{K})$,

$$Z_c = (35.97)(370)/(82.06)(623) = 0.26$$

For *n*-methyl-2-pyrrolidone,

$$T_c = 475[0.567 + 0.099 - (0.099)^2]^{-1} = 723.9 \text{ K } (843°\text{F})$$

$$P_c = 99.1(0.34 + 1.109)^{-2} = 47.2 \text{ atm } (4780 \text{ kPa})$$

$$V_c = 40 + 270.5 = 310.5 \text{ cm}^3/(\text{g}\cdot\text{mol}) \; [4.98 \text{ ft}^3/(\text{lb}\cdot\text{mol})]$$

$$Z_c = (47.2)(310.5)/(82.06)(723.9) = 0.247$$

Related Calculations: Extensive comparisons between experimental critical properties and those estimated by several other methods have shown that the Lydersen group-contribution method is the most accurate. This method is relatively easy to use for both hydrocarbons and organic compounds in general, provided that the structure is known. Unlike Nokay's correlation (see Example 1-2), it can be readily applied to hydrocarbons containing characteristics of more than a single family, such as an aromatic with olefinic side chains. A drawback of the Lydersen method, however, is that it cannot distinguish between isomers of similar structure, such as 2,3-dimethylpentane and 2,4-dimethylpentane.

Based on tests with paraffins in the C_1–C_{20} range and other hydrocarbons in the C_3–C_{14} range, the average deviation from experimental data for critical pressure is 18 lb/in^2 (124 kPa), and the maximum error is around 70 lb/in^2 (483 kPa). In general, the accuracy of the correlation is lower for unsaturated compounds than for saturated ones. As for critical temperature, the typical error is less than 2 percent; it can range up to 5 percent for nonpolar materials of relatively high molecular weight (e.g., 7100). Accuracy of the method when used with multifunctional polar groups is uncertain.

1-4 Redlich-Kwong Equation of State

Estimate the molar volume of isopropyl alcohol vapor at 10 atm (1013 kPa) and 473 K (392°F) using the Redlich-Kwong equation of state. For isopropyl alcohol, use 508.2 K as the critical temperature T_c and 50 atm as the critical pressure P_c. The Redlich-Kwong equation is

$$P = RT/(V - b) - a/T^{0.5}V(V - b)$$

where P is pressure, T is absolute temperature, V is molar volume, R is the gas constant, and a and b are equation-of-state constants given by

$$a = 0.4278R^2T_c^{2.5}/P_c \qquad \text{and} \qquad b = 0.0867RT_c/P_c$$

when the critical temperature is in kelvins, the critical pressure is in atmospheres, and R is taken as 82.05 (atm)(cm^3)/(g·mol)(K).

In an alternate form, the Redlich-Kwong equation is written as

$$Z = 1/(1 - h) - (A/B)[h/(1 + h)]$$

where $h = b/V = BP/Z$, $B = b/RT$, $A/B = a/bRT^{1.5}$, and Z, the compressibility factor, is equal to PV/RT.

Calculation Procedure:

1. Calculate the compressibility factor Z.

Since the equation is not explicit in Z, solve for it by an iterative procedure. For Trial 1, assume that $Z = 0.9$; therefore,

$$h = 0.0867(P/P_c)/Z(T/T_c) = \frac{0.087(10/50)}{(0.9)(473/508.2)} = 0.0208$$

Substituting for the generalized expression for A/B in the Redlich-Kwong equation,

$$Z = \frac{1}{1-h} - \left[\frac{(0.4278R^2T_c^{2.5}/P_c)}{(0.0867RT_c/P_c)(RT^{1.5})}\right]\left(\frac{h}{1+h}\right)$$

$$= \frac{1}{1-h} - (4.9343)(T_c/T)^{1.5}\left(\frac{h}{1+h}\right)$$

$$= \frac{1}{1-0.0208} - \left[(4.9343)\left(\frac{508.2}{473}\right)^{1.5}\right]\left[\frac{0.0208}{1+0.0208}\right]$$

$$= 0.910.$$

For Trial 2, then, assume that $Z = 0.91$; therefore,

$$h = \frac{0.0867(10/50)}{0.91(473/508.2)} = 0.0205$$

and

$$Z = \frac{1}{1-0.0205} - (4.9343)(508.2/473)^{1.5}\frac{0.0205}{1+0.0205} = 0.911$$

which is close enough.

2. Calculate molar volume.

By the definition of Z,

$$V = ZRT/P$$

$$= (0.911)(82.05)(473)/(10)$$

$$= 3535.6 \text{ cm}^3/(\text{g}\cdot\text{mol}) \ [3.536 \text{ m}^3/(\text{kg}\cdot\text{mol}) \text{ or } 56.7 \text{ ft}^3/(\text{lb}\cdot\text{mol})]$$

Related Calculations: This two-constant equation of Redlich-Kwong is extensively used for engineering calculations and enjoys wide popularity. Many modifications of the Redlich-Kwong equations of state, such as those by Wilson, Barnes-King, Soave, and Peng-Robinson, have been made and are discussed in Reid et al. [1]. The constants for the equation of state may be obtained by least-squares fit of the equation to experimental P-V-T data. However, such data are often not available. When this is the case, estimate the constants on the basis of the critical properties, as shown in the example.

1-5 *P-V-T* Properties of a Gas Mixture

A gaseous mixture at 25°C (298 K) and 120 atm (12,162 kPa) contains 3% helium, 40% argon, and 57% ethylene on a mole basis. Compute the volume of the mixture per mole using the following: (a) ideal-gas law, (b) compressibility factor based on pseudoreduced conditions (Kay's method), (c) mean compressibility factor and Dalton's law, (d) van der Waal's equation and Dalton's law, and (e) van der Waal's equation based on averaged constants.

Calculation Procedure:

1. Solve the ideal-gas law for volume.

By definition, $V = RT/P$, where V is volume per mole, T is absolute temperature, R is the gas constant, and P is pressure. Then,

$$V = [82.05\ (\text{cm}^3)(\text{atm})/(\text{g}\cdot\text{mol})(\text{K})]298\ \text{K}/120\ \text{atm} = 203.8\ \text{cm}^3/(\text{g}\cdot\text{mol})$$

2. Calculate the volume using Kay's method.

In this method, V is found from the equation $V = ZRT/P$, where Z, the compressibility factor, is calculated on the basis of pseudocritical constants that are computed as mole-fraction-weighted averages of the critical constants of the pure compounds. Thus, $T'_c = \Sigma Y_i T_{c,i}$ and similarly for P'_c and Z'_c, where the subscript c denotes critical, the prime denotes pseudo, the subscript i pertains to the ith component, and Y is mole fraction. Pure-component critical properties can be obtained from handbooks. The calculations can then be set out as a matrix:

Component, i	Y_i	$T_{c,i}$ (K)	$Y_i T_{c,i}$ (K)	$P_{c,i}$ (atm)	$Y_i P_{c,i}$ (atm)	$Z_{c,i}$	$Y_i Z_{c,i}$
He	0.03	5.2	0.16	2.24	0.07	0.301	0.009
A	0.40	150.7	60.28	48.00	19.20	0.291	0.116
C_2H_4	0.57	283.0	161.31	50.50	28.79	0.276	0.157
Σ =	1.00		221.75		48.06		0.282

Then the reduced temperature $T_r = T/T'_c = 298/221.75 = 1.34$, and the reduced pressure $P_r = P/P'_c = 120/48.06 = 2.50$. Now $Z'_c = 0.282$. Refer to the generalized compressibility plots in Figs. 1-2 and 1-3, which pertain respectively to Z'_c values of 0.27 and 0.29. Figure 1-2 gives a Z of 0.71, and Figure 1-3 gives a Z of 0.69. By linear interpolation, then, Z for the present case is 0.70. Therefore, the mixture volume is given by

$$V = ZRT/P = (0.70)(82.05)(298)/120 = 138.8\ \text{cm}^3/(\text{g}\cdot\text{mol})$$

3. Calculate the volume using the mean compressibility factor and Dalton's law.

Dalton's law states that the total pressure exerted by a gaseous mixture is equal to the sum of the partial pressures. In using this method, assume that the partial pressure of a

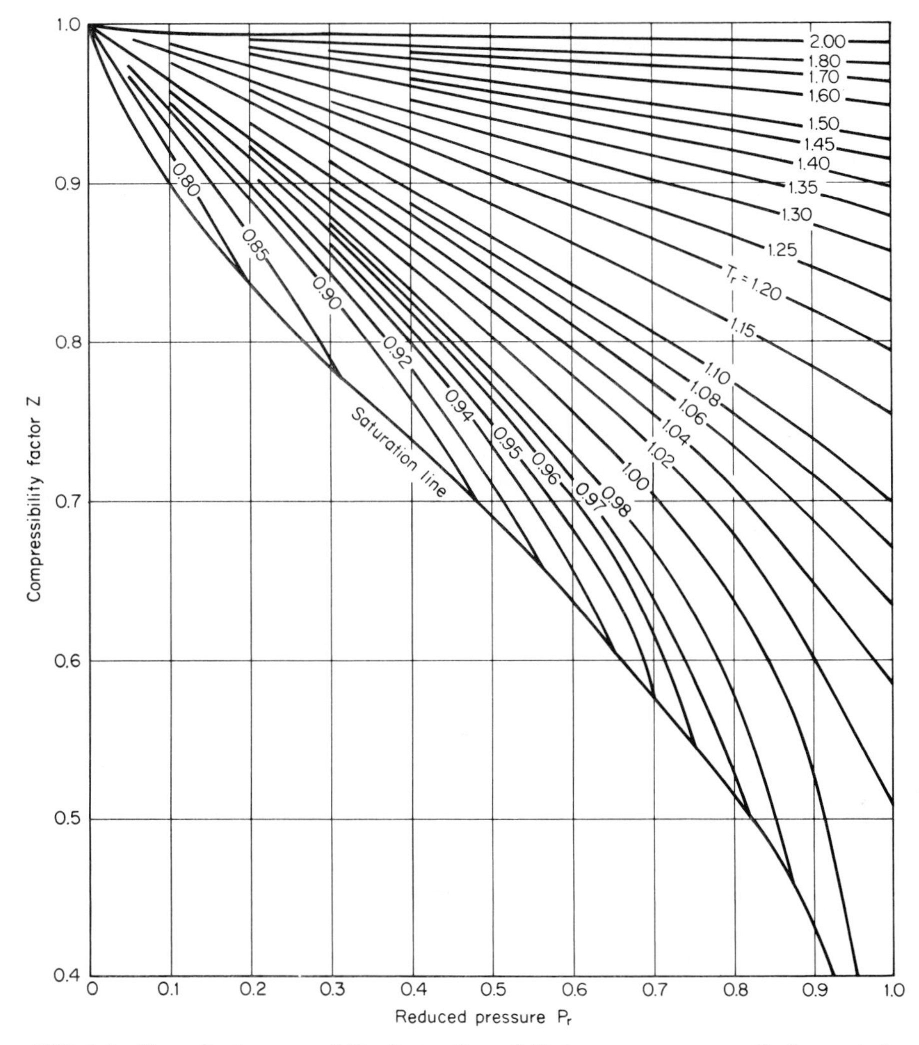

FIG. 1-1 Generalized compressibility factor; Z_c = 0.27; low-pressure range. *(Lydersen et al., University of Wisconsin Engineering Experiment Station, 1955.)*

component of a mixture is equal to the product of its mole fraction and the total pressure. Thus the method consists of calculating the partial pressure for each component, calculating the reduced pressure and reduced temperature, finding the corresponding compressibility factor for each component (from a conventional compressibility-factor chart in a handbook), and then taking the mole-fraction-weighted average of those compressibility factors and using that average value to find V. The calculations can be set out in matrix form, employing the critical properties from the matrix in step 2:

Component (i)	Y_i	Partial pressure ($p_i = PY_i$)	Reduced pressure ($p_i/P_{c,i}$)	Reduced temperature ($T/T_{c,i}$)	Compressibility factor (Z_i)	Z_iY_i
Helium	0.03	3.6	1.61	57.3	1.000	0.030
Argon	0.40	48.0	1.00	1.98	0.998	0.399
Ethylene	0.57	68.4	1.35	1.05	0.368	0.210
Total	1.00	120.0				0.639

Therefore

$$V = ZRT/P = (0.639)(82.05)(298)/120 = 130.2 \text{ cm}^3/(\text{g}\cdot\text{mol})$$

4. Calculate the volume using van der Waal's equation and Dalton's law.

Van der Waal's equation is

$$P = RT/(V - b) - a/V^2$$

where a and b are van der Waal constants, available from handbooks, that pertain to a given substance. The values for helium, argon, and ethylene are as follows (for calculations with pressure in atmospheres, volume in cubic centimeters, and quantity in gram-moles):

Component	van der Waal constant a	b
Helium	0.0341×10^6	23.7
Argon	1.350×10^6	32.3
Ethylene	4.480×10^6	57.2

For a mixture obeying Dalton's law, the equation can be rewritten as

$$P = RT\left[\frac{Y_{\text{He}}}{(V - Y_{\text{He}}b_{\text{He}})} + \frac{Y_{\text{A}}}{(V - Y_{\text{A}}b_{\text{A}})} + \frac{Y_{\text{Eth}}}{(V - Y_{\text{Eth}}b_{\text{Eth}})}\right] - \left(\frac{1}{V^2}\right)(Y_{\text{He}}^2a_{\text{He}} + Y_{\text{A}}^2a_{\text{A}} + Y_{\text{Eth}}^2a_{\text{Eth}})$$

Upon substitution,

$$120 = (82.05)(298)\left[\frac{0.03}{V - (0.03)(23.7)} + \frac{0.40}{V - (0.4)(32.3)} + \frac{0.57}{V - (0.57)(57.2)}\right] - \frac{1}{V^2}[(0.0341)(10^6)(0.03^2) + (1.35)(10^6)(0.4^2) + (4.48)(10^6)(0.57^2)]$$

Solving for volume by trial and error,

$$V = 150.9 \text{ cm}^3/(\text{g}\cdot\text{mol}) \; [2.42 \text{ ft}^3/(\text{lb}\cdot\text{mol})]$$

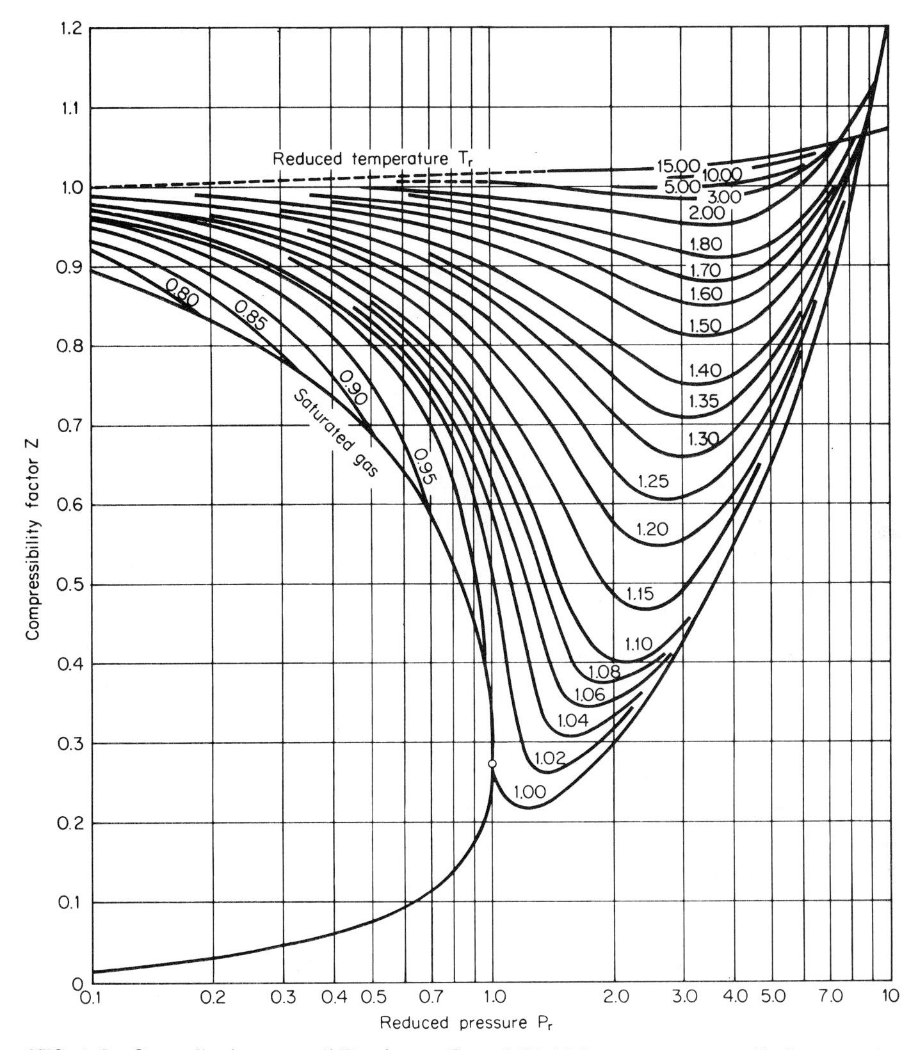

FIG. 1-2 Generalized compressibility factor; $Z_c = 0.27$; high-pressure range. *(Lydersen et al., University of Wisconsin Engineering Experiment Station, 1955.)*

5. Calculate the volume using van der Waal's equation with averaged constants.

In this method it is convenient to rearrange the van der Waal equation into the form

$$V^3 - (b_{avg} + RT/P)V^2 + a_{avg}V/P - a_{avg}b_{avg}/P = 0$$

For a_{avg}, take the expression $[\Sigma Y_i(a_i)^{0.5}]^2$; for b_{avg}, use the straightforward mole-fraction-weighted linear average $\Sigma Y_i b_i$. Thus, taking the values of a_i and b_i from the matrix in step 4,

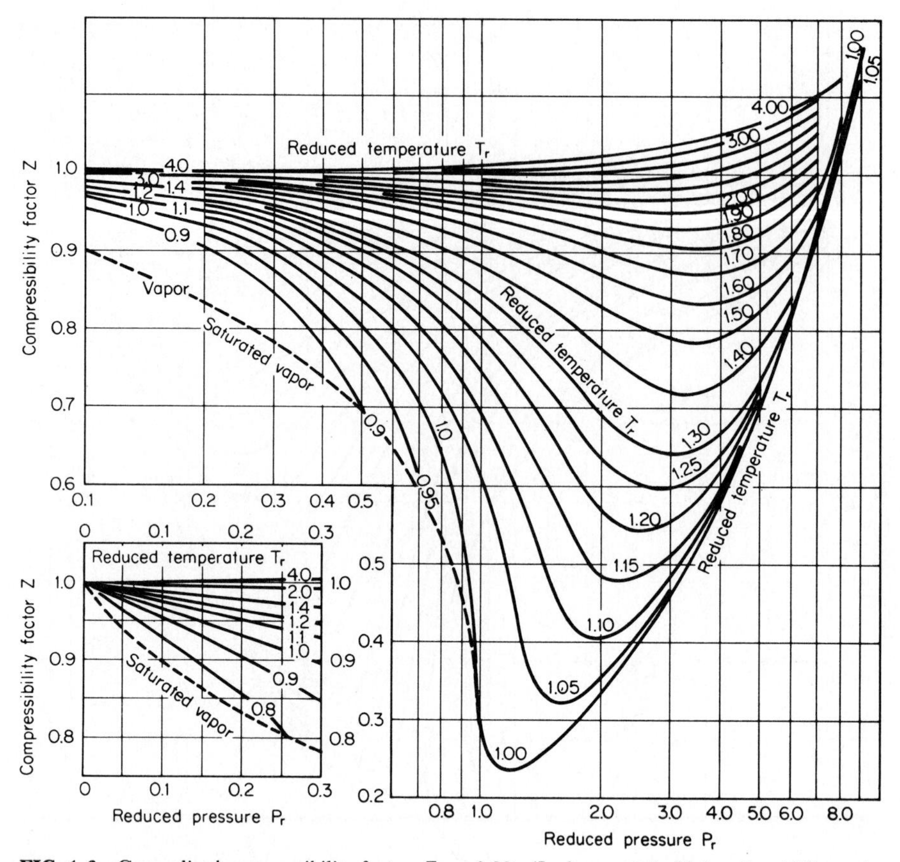

FIG. 1-3 Generalized compressibility factor; $Z_c = 0.29$. *(Lydersen et al., University of Wisconsin Engineering Experiment Station, 1955.)*

$$a_{avg} = [(0.03)(0.0341 \times 10^6)^{0.5} + (0.40)(1.350 \times 10^6)^{0.5} + (0.57)(4.48 \times 10^6)^{0.5}]^2$$

$$= 2.81 \times 10^6$$

$$b_{avg} = (0.03)(23.7) + (0.4)(32.3) + (0.57)(57.2)$$

$$= 46.23$$

Upon substitution,

$$V^3 - [46.23 + (82.05)(298)/120]V^2 + (2.81 \times 10^6)V/120 - (2.81 \times 10^6)(46.23)/120 = 0$$

Trial-and-error solution gives

$$V = 137 \text{ cm}^3/(\text{g}\cdot\text{mol}) \ [2.20 \text{ ft}^3/(\text{lb}\cdot\text{mol})]$$

Related Calculations: This illustration outlines various simple techniques for estimating P-V-T properties of gaseous mixtures. Obtain the compressibility factor from the generalized corresponding-state correlation, as shown in step 2.

The ideal-gas law is a simplistic model that is applicable to simple molecules at low pressure and high temperature. As for Kay's method, which in general is superior to the others, it is basically suitable for nonpolar/nonpolar mixtures and some polar/polar mixtures, but not for nonpolar/polar ones. Its average error ranges from about 1 percent at low pressures to 5 percent at high pressures and to as much as 10 percent when near the critical pressure.

For a quick estimate one may compute the pseudocritical parameters for the mixture using Kay's mole-fraction-averaging mixing rule and obtain the compressibility factor from the generalized corresponding-state correlation as shown in step 2.

1-6 Density of a Gas Mixture

Calculate the density of a natural gas mixture containing 32.1% methane, 41.2% ethane, 17.5% propane, and 9.2% nitrogen (mole basis) at 500 psig (3,550 kPa) and 250°F (394 K).

Calculation Procedure:

1. Obtain the compressibility factor for the mixture.

Employ Kay's method, as described in step 2 of Example 1-5. Thus Z is found to be 0.933.

2. Calculate the mole-fraction-weighted average molecular weight for the mixture.

The molecular weights of methane, ethane, propane, and nitrogen are 16, 30, 44, and 28, respectively. Therefore, average molecular weight $M' = (0.321)(16) + (0.412)(30) + (0.175)(44) + (0.092)(28) = 27.8$ lb/mol.

3. Compute the density of the mixture.

Use the formula

$$\rho = M'P/ZRT$$

where ρ is density, P is pressure, R is the gas constant, and T is absolute temperature. Thus,

$$\rho = (27.8)(500 + 14.7)/(0.933)(10.73)(250 + 460)$$
$$= 2.013 \text{ lb/ft}^3 \ (32.2 \text{ kg/m}^3)$$

Related Calculations: Use of the corresponding-states three-parameter graphic correlation developed by Lydersen, Greenkorn, and Hougen (Figs. 1-1, 1-2, and 1-3) gives fairly good results for predicting the gas-phase density of nonpolar pure components and their mixtures. Errors are within 4 to 5 percent. Consequently, this generalized correlation can be used to perform related calculations, except in the regions near the critical point. For improved accuracy in estimating P-V-T properties of pure components and their mixtures, use the Soave-modified Redlich-Kwong equation or the Lee-Kesler form of the Bendict-Webb-Rubin (B-W-R) generalized equation. For hydrocarbons, either of the two are accurate to within 2 to 3 percent, except near the critical point; for nonhydrocarbons, the Lee-Kesler modification of the B-W-R equation is recommended, the error probably being within a few percent except for polar molecules near the critical point. However, these equations are fairly complex and therefore not suitable for hand calculation. For a general discussion of various corresponding-state and analytical equations of state, see Reid et al. [1].

1-7 Estimation of Liquid Density

Estimate the density of saturated liquid ammonia at 37°C (310 K, or 99°F) using (a) the Gunn-Yamada generalized correlation, and (b) the Rackett equation. The Gunn-Yamada correlation [16] is

$$V/V_{\mathrm{Sc}} = V_r^{(0)}(1 - \omega\Gamma)$$

where V is the liquid molar specific volume in cubic centimeters per gram-mole; ω is the acentric factor; Γ is as defined below; V_{Sc} is a scaling parameter equal to $(RT_c/P_c)(0.2920 - 0.0967\omega)$, where R is the gas constant, P is pressure, and the subscript c denotes a critical property; and $V_r^{(0)}$ is a function whose value depends on the reduced temperature T/T_c:

$$V_r^{(0)} = 0.33593 - 0.33953(T/T_c) + 1.51941(T/T_c)^2 - 2.02512(T/T_c)^3 + 1.11422(T/T_c)^4 \qquad \text{for } 0.2 \leq T/T_c \leq 0.8$$

or

$$V_r^{(0)} = 1.0 + 1.3(1 - T/T_c)^{0.5} \log (1 - T/T_c) - 0.50879(1 - T/T_c) - 0.91534(1 - T_r)^2 \qquad \text{for } 0.8 \leq T/T_c \leq 1.0$$

and

$$\Gamma = 0.29607 - 0.09045\,(T/T_c) - 0.04842\,(T/T_c)^2 \qquad \text{for } 0.2 \leq T/T_c \leq 1.0$$

The Rackett equation [17] is

$$V_{\mathrm{sat\ liq}} = V_c Z_c^{\,(1 - T/T_c)^{0.2857}}$$

where $V_{\mathrm{sat\ liq}}$ is the molar specific volume for saturated liquid, V_c is the critical molar volume, and Z_c is the critical compressibility factor. Use these values for ammonia: T_c = 405.6 K, P_c = 111.3 atm, Z_c = 0.242, V_c = 72.5 cm^3/(g·mol), and ω = 0.250.

Calculation Procedure:

1. Compute saturated-liquid density using the Gunn-Yamada equation.

$$V_{Sc} = (82.05)(405.6)[0.2920 - (0.0967)(0.250)]/111.3$$
$$= 80.08 \text{ cm}^3/(\text{g}\cdot\text{mol})$$

and the reduced temperature is given by

$$T/T_c = (37 + 273)/405.6$$
$$= 0.764$$

Therefore,

$$V_r^{(0)} = 0.33593 - (0.33953)(0.764) + (1.51941)(0.764)^2 - (2.02512)(0.764)^3$$
$$+ (1.11422)(0.764)^4$$
$$= 0.4399$$
$$\Gamma = 0.29607 - 0.09045\,(0.764) - 0.04842\,(0.764)^2 = 0.1987$$

and the saturated liquid volume is given by

$$V = (0.4399)(80.08)[1 - (0.250)(0.1987)]$$
$$= 33.48 \text{ cm}^3/(\text{g}\cdot\text{mol})$$

Finally, letting M equal the molecular weight, the density of liquid ammonia is found to be

$$\rho = M/V = 17/33.48 = 0.508 \text{ g/cm}^3 \text{ (31.69 lb/ft}^3\text{)}$$

(The experimental value is 0.5834 g/cm^3, so the error is 12.9 percent.)

2. Compute saturated-liquid density using the Rackett equation.

$$V_{sat} = (72.5)(0.242)^{(1-0.764)0.2857} = 28.34 \text{ cm}^3/(\text{g}\cdot\text{mol})$$

So

$$\rho = 17/28.34 = 0.5999 \text{ g/cm}^3 \text{ (37.45 lb/ft}^3\text{)} \qquad \text{(error} = 2.8 \text{ percent)}$$

Related Calculations: Both the Gunn-Yamada and Rackett equations are limited to saturated liquids. At or below a T_r of 0.99, the Gunn-Yamada equation appears to be quite accurate for nonpolar as well as slightly polar compounds. With either equation, the errors for nonpolar compounds are generally within 1 percent. The correlation of Yen and Woods [18] is more general, being applicable to compressed as well as saturated liquids.

1-8 Estimation of Ideal-Gas Heat Capacity

Estimate the ideal-gas heat capacity C_p° of 2-methyl-1,3-butadiene and *n*-methyl-2-pyrrolidone at 527°C (800 K, or 980°F) using the group-contribution method of Rihani and Doraiswamy. The Rihani-Doraiswamy method is based on the equation

$$C_p^\circ = \sum_i N_i a_i + \sum_i N_i b_i T + \sum_i N_i c_i T^2 + \sum_i N_i d_i T^3$$

where N_i is the number of groups of type i, T is the temperature in kelvins, and a_i, b_i, c_i, and d_i are the additive group parameters given in Table 1-3.

Calculation Procedure:

1. ***Obtain the molecular structure from a handbook, and list the number and type of groups.***

For 2-methyl-1,3-butadiene, the structure is

$$\begin{array}{c} H_2C{=}CH{-}C{=}CH_2 \\ \qquad\quad | \\ \qquad\quad CH_3 \end{array}$$

and the groups are

$$-CH_3 \qquad {}^{H}\!\!\searrow\!\!\nearrow C{=}CH_2 \qquad \text{and} \qquad \searrow\!\!\nearrow C{=}CH_2$$

For *n*-methyl-2-pyrrolidone, the structure is

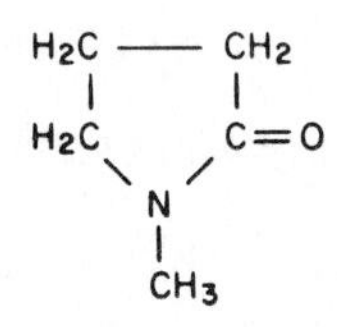

and the groups are

$$-CH_3 \qquad -\overset{|}{C}H_2 \qquad -\overset{|}{C}{=}O \qquad \searrow\!\!\nearrow N-$$

and a 5-membered (pentene) ring.

2. ***Sum up the group contributions for each compound.***

Obtain the values of a, b, c, and d from Table 1-3, and set out the calculations in a matrix:

	N	a	$b \times 10^2$	$c \times 10^4$	$d \times 10^6$
2-Methyl-1,3-butadiene:					
$-CH_3$	1	0.6087	2.1433	−0.0852	0.01135
\ HC=CH_2	1	0.2773	3.4580	−0.1918	0.004130
\ C=CH_2 /	1	−0.4173	3.8857	−0.2783	0.007364
$\sum$ (N)(group parameter)		0.4687	9.4870	−0.5553	0.02284
n-Methyl-2-pyrrolidone:					
5-membered (pentene) ring	1	−6.8813	0.7818	−0.0345	0.000591
$-CH_3$	1	0.6087	2.1433	−0.0852	0.01135
| $-CH_2$	3	0.3945	2.1363	−0.1197	0.002596
\ −C=O	1	1.0016	2.0763	−0.1636	0.004494
\ N− /	1	−3.4677	2.9433	−0.2673	0.007828
$\sum$ (N)(group parameter)		−7.5552	14.3536	−0.9097	0.026859

3. Compute the ideal-gas heat capacity for each compound.

Refer to the equation in the statement of the problem. Now, $T = 527 + 273 = 800$ K. Then, for 2-methyl-1,3-butadiene,

$$C_p^\circ = 0.4687 + (9.4870 \times 10^{-2})(800) + (-0.5553 \times 10^{-4})(800)^2 + (0.02284 \times 10^{-6})(800)^3$$
$$= 52.52 \text{ cal/(g·mol)(K) [52.52 Btu/(lb·mol)(°F)]}$$

And for *n*-methyl-2-pyrrolidone,

$$C_p^\circ = -7.5552 + (14.3536 \times 10^{-2})(800) + (-0.9097 \times 10^{-4})(800)^2 + (0.02686 \times 10^{-6})(800)^3$$
$$= 62.81 \text{ cal/(g·mol)(K) [62.81 Btu/(lb·mol)(°F)]}$$

Related Calculations: The Rihani-Doraiswamy method is applicable to a large variety of compounds, including heterocyclics; however, it is not applicable to acetylenics. It predicts to within 2 to 3 percent accuracy. Accuracy levels are somewhat less when predicting at temperatures below about 300 K (80°F). Good accuracy is obtainable using the methods of Benson [25] and of Thinh [26].

TABLE 1-3 Group Contributions to Ideal-Gas Heat Capacity

Symbol	Coefficients			
	a	$b \times 10^2$	$c \times 10^4$	$d \times 10^6$
Aliphatic hydrocarbon groups				
$-CH_3$	0.6087	2.1433	−0.0852	0.01135
$-CH_2-$ (with one further bond, |)	0.3945	2.1363	−0.1197	0.002596
$=CH_2$	0.5266	1.8357	−0.0954	0.001950
$-\overset{\vert}{\underset{\vert}{C}}-H$	−3.5232	3.4158	−0.2816	0.008015
$-\overset{\vert}{\underset{\vert}{C}}-$	−5.8307	4.4541	−0.4208	0.012630
$H\backslash C{=}CH_2 /$	0.2773	3.4580	−0.1918	0.004130
$\backslash C{=}CH_2 /$	−0.4173	3.8857	−0.2783	0.007364
$H\backslash C{=}C /H$, with one free bond on each C (both free bonds below; cis)	−3.1210	3.8060	−0.2359	0.005504
$H\backslash C{=}C \backslash H$, with one free bond on each C (one above, one below; trans)	0.9377	2.9904	−0.1749	0.003918
$\backslash C{=}C /H$, with two free bonds on the first C and one on the second	−1.4714	3.3842	−0.2371	0.006063
$\backslash C{=}C /$, with two free bonds on each C	0.4736	3.5183	−0.3150	0.009205
$H\backslash C{=}C{=}CH_2 /$	2.2400	4.2896	−0.2566	0.005908
$\backslash C{=}C{=}CH_2 /$	2.6308	4.1658	−0.2845	0.007277
$H\backslash C{=}C{=}C /H$, with one free bond on each terminal C	−3.1249	6.6843	−0.5766	0.017430
$\equiv CH$	2.8443	1.0172	−0.0690	0.001866
$-C\equiv$	−4.2315	7.8689	−0.2973	0.00993
Aromatic hydrocarbon groups				
HC (two aromatic bonds)	−1.4572	1.9147	−0.1233	0.002985

TABLE 1-3 Group Contributions to Ideal-Gas Heat Capacity (*continued*)

Symbol	Coefficients			
	a	$b \times 10^2$	$c \times 10^4$	$d \times 10^6$
Aromatic hydrocarbon groups				
$-C\langle$ (aromatic bonds ↗ ↘)	−1.3883	1.5159	−0.1069	0.002659
$\leftrightarrow C\langle$ (aromatic bonds ↗ ↘)	0.1219	1.2170	−0.0855	0.002122
Oxygen-containing groups				
$-OH$	6.5128	−0.1347	0.0414	−0.001623
$-O-$	2.8461	−0.0100	0.0454	−0.002728
$-C(H)=O$	3.5184	0.9437	0.0614	−0.006978
$>C=O$	1.0016	2.0763	−0.1636	0.004494
$-C(=O)-O-H$	1.4055	3.4632	−0.2557	0.006886
$-C(=O)O-$	2.7350	1.0751	0.0667	−0.009230
$O\langle$ (aromatic bonds ↗ ↘)	−3.7344	1.3727	−0.1265	0.003789
Nitrogen-containing groups				
$-C\equiv N$	4.5104	0.5461	0.0269	−0.003790
$-N\equiv C$	5.0860	0.3492	0.0259	−0.002436
$-NH_2$	4.1783	0.7378	0.0679	−0.007310
$>NH$	−1.2530	2.1932	−0.1604	0.004237
$>N-$	−3.4677	2.9433	−0.2673	0.007828
$N\langle$ (aromatic bonds ↗ ↘)	2.4458	0.3436	0.0171	−0.002719
$-NO_2$	1.0898	2.6401	−0.1871	0.004750
Sulfur-containing groups				
$-SH$	2.5597	1.3347	−0.1189	0.003820
$-S-$	4.2256	0.1127	−0.0026	−0.000072

TABLE 1-3 Group Contributions to Ideal-Gas Heat Capacity (*continued*)

Symbol	Coefficients			
	a	$b \times 10^2$	$c \times 10^4$	$d \times 10^6$
↙S↘ (with two arrows)	4.0824	−0.0301	0.0731	−0.006081
$-SO_3H$	6.9218	2.4735	0.1776	−0.022445
Halogen-containing groups				
−F	1.4382	0.3452	−0.0106	−0.000034
−Cl	3.0660	0.2122	−0.0128	0.000276
−Br	2.7605	0.4731	−0.0455	0.001420
−I	3.2651	0.4901	−0.0539	0.001782
Contributions due to ring formation (for cyclics only)				
Three-membered ring	−3.5320	−0.0300	0.0747	−0.005514
Four-membered ring	−8.6550	1.0780	0.0425	−0.000250
Five-membered ring:				
c-Pentane	−12.2850	1.8609	−0.1037	0.002145
c-Pentene	−6.8813	0.7818	−0.0345	0.000591
Six-membered ring:				
c-Hexane	−13.3923	2.1392	−0.0429	−0.001865
c-Hexene	−8.0238	2.2239	−0.1915	0.005473

1-9 Heat Capacity of Real Gases

Calculate the heat capacity C_p of ethane vapor at 400 K (260°F) and 50 atm (5065 kPa). Also estimate the heat-capacity ratio C_p/C_v at these conditions. The ideal-gas heat capacity for ethane is given by

$$C_p^\circ = 2.247 + (38.201 \times 10^{-3})T - (11.049 \times 10^{-6})T^2$$

where C_p° is in cal/(g·mol)(K), and T is in kelvins. For ethane, critical temperature $T_c = 305.4$ K and critical pressure $P_c = 48.2$ atm.

Calculation Procedure:

1. Compute reduced temperature T_r and reduced pressure P_r.

Thus $T_r = T/T_c = 400/305.4 = 1.310$, and $P_r = P/P_c = 50/48.2 = 1.04$.

2. Obtain ΔC_p from Fig. 1-4.

Thus $\Delta C_p = C_p - C_p^\circ = 3$ cal/(g·mol)(K) at $T_r = 1.31$ and $P_r = 1.04$.

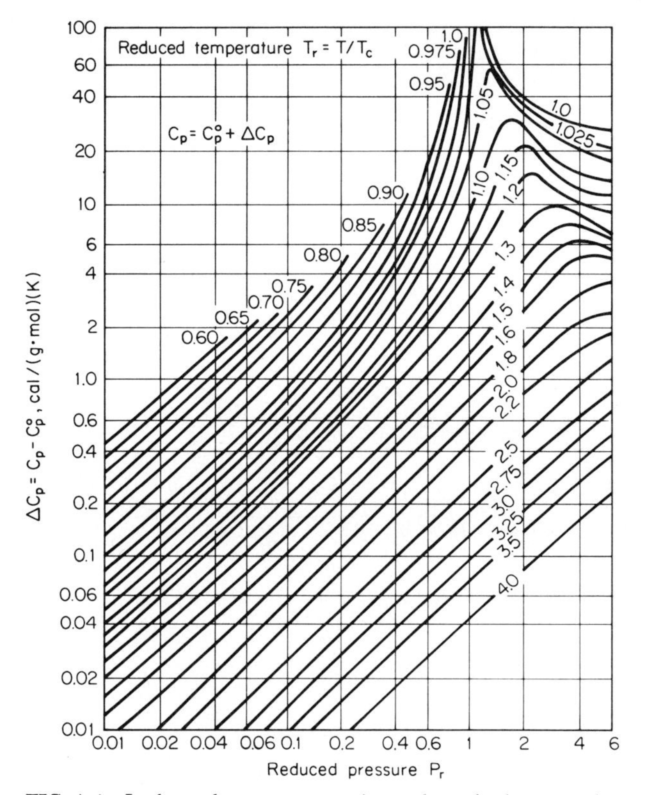

FIG. 1-4 Isothermal pressure correction to the molar heat capacity of gases. *(Perry and Chilton—Chemical Engineers' Handbook, McGraw-Hill, 1973.)*

3. Calculate ideal-gas heat capacity.

$$C_p^o = 2.247 + (38.201 \times 10^{-3})(400) - (11.049 \times 10^{-6})(400^2)$$
$$= 15.76 \text{ cal/(g·mol)(K)}$$

4. Compute real-gas heat capacity.

$$C_p = \Delta C_p + C_p^o = 3 + 15.76 = 18.76 \text{ cal/(g·mol)(K) [18.76 Btu/(lb·mol)(°F)]}$$

5. Estimate heat-capacity ratio.

From Fig. 1-5, $C_p - C_v = 4$ at $T_r = 1.31$ and $P_r = 1.04$. So the real-gas heat-capacity ratio is

$$\frac{C_p}{C_v} = \frac{C_p}{C_p - (C_p - C_v)} = \frac{18.76}{18.76 - 4} = 1.27$$

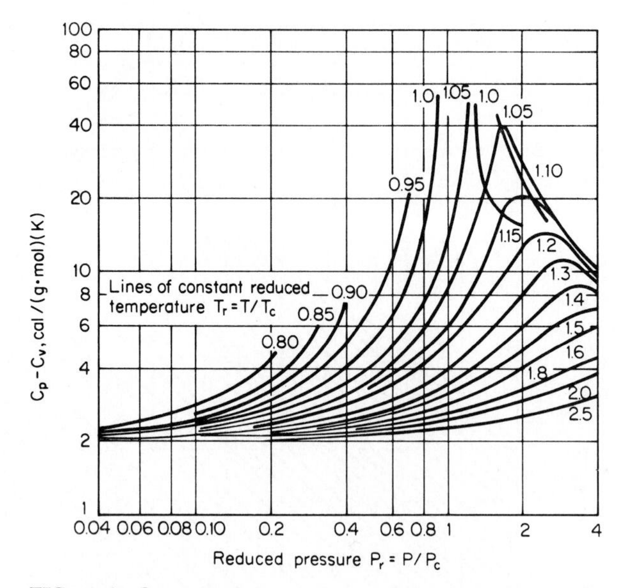

FIG. 1-5 Generalized heat-capacity differences, $C_p - C_v$. *(Perry and Chilton—Chemical Engineers' Handbook, McGraw-Hill, 1973.)*

Note that the ideal-gas heat-capacity ratio is

$$\frac{C_p^\circ}{C_v^\circ} = \frac{C_p^\circ}{(C_p^\circ - R)} = 15.76/(15.76 - 1.987) = 1.144$$

Related Calculations: This graphic correlation may be used to estimate the heat-capacity ratio of any nonpolar or slightly polar gas. The accuracy of the correlation is poor for highly polar gases and (as is true for correlations in general) near the critical region. For polar gases, the Lee-Kesler method [27] is suggested.

1-10 Liquid Heat Capacity—Generalized Correlation

Estimate the saturated-liquid heat capacity of (a) *n*-octane and (b) ethyl mercaptan at 27°C (80.6°F) using the Yuan-Stiel corresponding-states correlation [19], given as

$$C_{\sigma,L} - C_p^\circ = (\Delta C_{\sigma,L})^{(0)} + \omega(\Delta C_{\sigma,L})^{(1)}$$

for nonpolar liquids, or

$$C_{\sigma,L} - C_p^\circ = (\Delta C_{\sigma,L})^{(0p)} + \omega(\Delta C_{\sigma,L})^{(1p)} + X(\Delta C_{\sigma,L})^{(2p)} + X^2(\Delta C_{\sigma,L})^{(3p)} + \omega^2(\Delta C_{\sigma,L})^{(4p)} + X\omega(\Delta C_{\sigma,L})^{(5p)}$$

for polar liquids, where $C_{\sigma,L}$ is saturated-liquid heat capacity and C_p° is ideal-gas heat capacity, both in calories per gram-mole kelvin; ω is the Pitzer acentric factor; the $\Delta C_{\sigma,L}$

TABLE 1-4 Yuan and Stiel Deviation Functions for Saturated-Liquid Heat Capacity

Reduced temperature	$(\Delta C_\sigma)^{(0)}$	$(\Delta C_\sigma)^{(1)}$	$(\Delta C_\sigma)^{(0p)}$	$(\Delta C_\sigma)^{(1p)}$	$(\Delta C_\sigma)^{(2p)}$	$(\Delta C_\sigma)^{(3p)} \times 10^{-2}$	$(\Delta C_\sigma)^{(4p)}$	$(\Delta C_\sigma)^{(5p)}$
0.96	14.87	37.0						
0.94	12.27	29.2	12.30	29.2	−126	*	*	*
0.92	10.60	27.2	10.68	27.4	−123	*	*	*
0.90	9.46	26.1	9.54	25.9	−121	*	*	*
0.88	8.61	25.4	8.67	24.9	−117.5	*	*	*
0.86	7.93	24.8	8.00	24.2	−115	*	*	*
0.84	7.45	24.2	7.60	23.5	−112.5	*	*	*
0.82	7.10	23.7	7.26	23.0	−110	*	*	*
0.80	6.81	23.3	7.07	22.6	−108	*	*	*
0.78	6.57	22.8	6.80	22.2	−107	*	*	*
0.76	6.38	22.5	6.62	21.9	−106	*	*	*
0.74	6.23	22.2	6.41	22.5	−105	−0.69	−4.22	−29.5
0.72	6.11	21.9	6.08	23.6	−107	0.15	−7.20	−30.0
0.70	6.01	21.7	6.01	24.5	−110	1.31	−10.9	−29.1
0.68	5.91	21.6	5.94	25.7	−113	2.36	−15.2	−22.8
0.66	5.83	21.8	5.79	27.2	−118	3.06	−20.0	−7.94
0.64	5.74	22.2	5.57	29.3	−124	3.24	−25.1	14.8
0.62	5.64	22.8	5.33	31.8	−132	2.87	−30.5	43.0
0.60	5.54	23.5	5.12	34.5	−141	1.94	−36.3	73.1
0.58	5.42	24.5	4.92	37.6	−151	0.505	−42.5	102
0.56	5.30	25.6	4.69	41.1	−161	−1.37	−49.2	128
0.54	5.17	26.9	4.33	45.5	−172	−3.58	−56.3	149
0.52	5.03	28.4	3.74	50.9	−184	−6.02	−64.0	165
0.50	4.88	30.0	2.87	57.5	−198	−8.56	−72.1	179
0.48	4.73	31.7	1.76	65.0	−213	−11.1	−80.6	192
0.46	4.58	33.5	0.68	72.6	−229	−13.3	−89.4	206
0.44	4.42	35.4	0.19	78.5	−244	−15.0	−98.2	221
0.42	4.26	37.4						
0.40	4.08	39.4						

*Data not available for $(\Delta C_\sigma)^{(3p)}$ to $(\Delta C_\sigma)^{(5p)}$ above $T_r = 0.74$; assume zero.

Source: R. C. Reid, J. M. Prausnitz, and T. K. Sherwood, *Properties of Gases and Liquids,* McGraw-Hill, New York, 1977.

terms are deviation functions for saturated-liquid heat capacity (given in Table 1-4); and X is the Stiel polarity factor (from Table 1-5).

For n-octane, $T_c = 568.8$ K, $\omega = 0.394$, $X = 0$ (nonpolar liquid), and

$$C_p^\circ = -1.456 + (1.842 \times 10^{-1})T - (1.002 \times 10^{-4})T^2 + (2.115 \times 10^{-8})T^3$$

where T is in kelvins.

For ethyl mercaptan, $T_c = 499$ K, $\omega = 0.190$, and $X = 0.004$ (slightly polar), and

$$C_p^\circ = 3.564 + (5.615 \times 10^{-2})T - (3.239 \times 10^{-5})T^2 + (7.552 \times 10^{-9})T^3$$

where T is in kelvins.

Calculation Procedure:

1. Estimate the deviation functions.

For n-octane, $T_r = (273 + 27)/568.8 = 0.527$. From Table 1-4, using linear interpolation and the nonpolar terms,

$$(\Delta C_{\sigma,L})^{(0)} = 5.08 \qquad \text{and} \qquad (\Delta C_{\sigma,L})^{(1)} = 27.9$$

TABLE 1-5 Stiel Polarity Factors of Some Polar Materials

Material	Polarity factor	Material	Polarity factor
Methanol	0.037	Water	0.023
Ethanol	0.0	Hydrogen chloride	0.008
n-Propanol	−0.057	Acetone	0.013
Isopropanol	−0.053	Methyl fluoride	0.012
n-Butanol	−0.07	Ethylene oxide	0.012
Dimethylether	0.002	Methyl acetate	0.005
Methyl chloride	0.007	Ethyl mercaptan	0.004
Ethyl chloride	0.005	Diethyl ether	−0.003
Ammonia	0.013		

Source: R. C. Reid, J. M. Prausnitz, and T. K. Sherwood, *Properties of Gases and Liquids,* McGraw-Hill, New York, 1977.

For ethyl mercaptan, $T_r = (273 + 27)/499 = 0.60$. From Table 1-4, for polar liquids,

$$(\Delta C_{\sigma,L})^{(0p)} = 5.12 \qquad (\Delta C_{\sigma,L})^{(1p)} = 34.5 \qquad (\Delta C_{\sigma,L})^{(2p)} = -141$$

$$(\Delta C_{\sigma,L})^{(3p)} = 0.0194 \qquad (\Delta C_{\sigma,L})^{(4p)} = -36.3 \qquad \text{and} \qquad (\Delta C_{\sigma,L})^{(5p)} = 73.1$$

2. *Compute ideal-gas heat capacity.*

For *n*-octane,

$$C_p^\circ = -1.456 + (1.842 \times 10^{-1})(300) - (1.002 \times 10^{-4})(300^2) + (2.115 \times 10^{-8})(300^3) = 45.36 \text{ cal/(g·mol)(K)}$$

And for ethyl mercaptan,

$$C_p^\circ = 3.564 + (5.615 \times 10^{-2})(300) - (3.239 \times 10^{-5})(300^2) + (7.552 \times 10^{-9})(300^3) = 17.7 \text{ cal/(g·mol)(K)}$$

3. *Compute saturated-liquid heat capacity.*

For *n*-octane,

$$C_{\sigma,L} = 5.08 + (0.394)(27.9) + 45.36 = 61.43 \text{ cal/(g·mol)(K)}$$

The experimental value is 60 cal/(g·mol)(K), so the error is 2.4 percent.

For ethyl mercaptan,

$$C_{\sigma,L} = 5.12 + (0.19)(34.5) + (0.004)(-141) + (0.004^2)(1.94)(10^{-2}) + (0.190^2)(-36.3) + (0.004)(0.19)(73.1) + 17.7$$
$$= 27.6 \text{ cal/(g·mol)(K) [27.6 Btu/(lb·mol)(°F)]}$$

The experimental value is 28.2 cal/(g·mol)(K), so the error is 2.1 percent.

1-11 Enthalpy Difference for Ideal Gas

Compute the ideal-gas enthalpy change for *p*-xylene between 289 and 811 K (61 and 1000°F), assuming that the ideal-gas heat-capacity equation is (with T in kelvins)

$$C_p^\circ = -7.388 + (14.9722 \times 10^{-2})T - (0.8774 \times 10^{-4})T^2 + (0.019528 \times 10^{-6})T^3$$

Calculation Procedure:

1. Compute the ideal-gas enthalpy difference.

The ideal-gas enthalpy difference $(H_2^\circ - H_1^\circ)$ is obtained by integrating the C_p° equation between two temperature intervals:

$$\begin{aligned}(H_2^\circ - H_1^\circ) &= \int_{T1}^{T2} C_p^\circ \, dt \\ &= \int_{T1}^{T2} [-7.388 + (14.9772)(10^{-2})T \\ &\quad - (0.8774)(10^{-4})T^2 + (0.019528)(10^{-6})T^3] \, dT \\ &= (-7.388)(811 - 289) + (14.9772 \times 10^{-2})(811^2 - 289^2)/2 \\ &\quad - (0.8774 \times 10^{-4})(811^3 - 289^3)/3 \\ &\quad + (0.019528 \times 10^{-6})(811^4 - 289^4)/4 \\ &= 26{,}327 \text{ cal/(g}\cdot\text{mol) [47,400 Btu/(lb}\cdot\text{mol)]}\end{aligned}$$

The literature value is 26,284 cal/(g·mol).

Related Calculations: Apply this procedure to compute enthalpy difference for any ideal gas. In absence of the ideal-gas heat-capacity equation, estimate C_p° using the Rihani-Doraiswamy group-contribution method, Example 1-8.

1-12 Estimation of Heat of Vaporization

Estimate the enthalpy of vaporization of acetone at the normal boiling point using the following relations, and compare your results with the experimental value of 7230 cal/(g·mol).

1. Clapeyron equation and compressibility factor [20]:

$$\Delta H_{v,b} = (RT_c \, \Delta Z_v T_{b,r} \ln P_c)/(1 - T_{b,r})$$

2. Chen method [21]:

$$\Delta H_{v,b} = RT_c T_{b,r} \left(\frac{3.978 T_{b,r} - 3.938 + 1.555 \ln P_c}{1.07 - T_{b,r}} \right)$$

3. Riedel method [22]:

$$\Delta H_{v,b} = 1.093RT_c \left[T_{b,r} \frac{(\ln P_c - 1)}{0.930 - T_{b,r}} \right]$$

4. Pitzer correlation:

$$\Delta H_{v,b} = RT_c[7.08(1 - T_{b,r})^{0.354} + 10.95\omega(1 - T_{b,r})^{0.456}]$$

where $\Delta H_{v,b}$ = enthalpy of vaporization at the normal boiling point in cal/(g·mol)
T_c = critical temperature in kelvins
P_c = critical pressure in atmospheres
ω = Pitzer acentric factor
R = gas constant = 1.987 cal/(g·mol)(K)
$T_{b,r}$ = T_b/T_c, reduced temperature at the normal boiling point T_b
ΔZ_v = $Z_v - Z_L$, the difference in the compressibility factor between the saturated vapor and saturated liquid at the normal boiling point, given in Table 1-6.

Also estimate the heat of vaporization of water at 300°C (572°F) by applying the Watson correlation:

$$\frac{\Delta H_2}{\Delta H_1} = \left(\frac{1 - T_{r,2}}{1 - T_{r,1}}\right)^{0.38}$$

where ΔH_1 and ΔH_2 are the heats of vaporization at reduced temperatures of $T_{r,1}$ and $T_{r,2}$, respectively.

Data for acetone are T_b = 329.7 K, T_c = 508.7 K, P_c = 46.6 atm, and ω = 0.309. Data for water are T_b = 373 K, T_c = 647.3 K, and $\Delta H_{v,b}$ = 9708.3 cal/(g·mol).

Calculation Procedure:

1. Calculate reduced temperature $T_{b,r}$ and reduced pressure P_r for the acetone at normal-boiling-point conditions (1 atm) and obtain ΔZ_v.

Thus,

$$T_{b,r} = \frac{329.7}{508.7} = 0.648 \quad \text{and} \quad P_r = \frac{1}{46.6} = 0.0215$$

From Table 1-6, by extrapolation, ΔZ_v = 0.966.

2. Compute the heat of vaporization of acetone using the Clapeyron equation.

From the preceding equation,

$$\Delta H_{v,b} = (1.987)(508.7)(0.966)(0.648)(\ln 46.6)/(1 - 0.648)$$
$$= 6905 \text{ cal/(g·mol) [12,430 Btu/(lb·mol)]}$$

Percent error is 100(7230 − 6905)/7230, or 4.5 percent.

TABLE 1-6 Values of ΔZ_v as a Function of Reduced Pressure

P_r	$Z_v - Z_L$	P_r	$Z_v - Z_L$	P_r	$Z_v - Z_L$
0	1.0	0.25	0.769	0.80	0.382
0.01	0.983	0.30	0.738	0.85	0.335
0.02	0.968	0.35	0.708	0.90	0.280
0.03	0.954	0.40	0.677	0.92	0.256
0.04	0.942	0.45	0.646	0.94	0.226
0.05	0.930	0.50	0.612	0.95	0.210
0.06	0.919	0.55	0.578	0.96	0.192
0.08	0.899	0.60	0.542	0.97	0.170
0.10	0.880	0.65	0.506	0.98	0.142
0.15	0.838	0.70	0.467	0.99	0.106
0.20	0.802	0.75	0.426	1.00	0.000

3. Compute the heat of vaporization using the Chen method.

Thus,

$$\Delta H_{v,b} = (1.987)(508.7)(0.648)\left[\frac{(3.978)(0.648) - (3.938) + (1.555)\ln 46.6}{1.07 - 0.648}\right]$$

$$= 7160 \text{ cal/(g·mol) [12,890 Btu/(lb·mol)]}$$

Error is 1.0 percent.

4. Compute the heat of vaporization using the Riedel method.

Thus,

$$\Delta H_{v,b} = (1.093)(1.987)(508.7)\left\{0.648\left[\frac{\ln(46.6) - 1}{0.930 - 0.648}\right]\right\}$$

$$= 7214 \text{ cal/(g·mol) [12,985 Btu/(lb·mol)]}$$

Error is 0.2 percent.

5. Compute the heat of vaporization using the Pitzer correlation.

Thus,

$$\Delta H_{v,b} = (1.987)(508.7)[7.08(1 - 0.648)^{0.354} + (10.95)(0.309)(1 - 0.648)^{0.456}]$$

$$= 7069 \text{ cal/(g·mol) [12,720 Btu/(lb·mol)]}$$

Error is 2.2 percent.

6. Compute the heat of vaporization of the water.

Now, $T_{r,1} = (100 + 273)/647.3 = 0.576$ and $T_{r,2} = (300 + 273)/647.3 = 0.885$, where the subscript 1 refers to water at its normal boiling point and the subscript 2 refers to water at 300°C. In addition, $\Delta H_{v,b}$ (= ΔH_1) is given above as 9708.3 cal/(g·mol). Then, from the Watson correlation,

$$\Delta H_v \text{ (at 300°C)} = 9708.3\left(\frac{1-0.885}{1-0.576}\right)^{0.38}$$

$$= 5913 \text{ cal/(g·mol) [10,640 Btu/(lb·mol)]}$$

The value given in the steam tables is 5949 cal/(g·mol), so the error is 0.6 percent.

Related Calculations: This illustration shows several techniques for estimating enthalpies of vaporization for pure liquids. The Clapeyron equation is inherently accurate, especially if ΔZ_v is obtained from reliable P-V-T correlations. The other three techniques yield approximately the same error when averaged over many types of fluids and over large temperature ranges. They are quite satisfactory for engineering calculations. A comparison of calculated and experimental results for 89 compounds has shown average errors of 1.8 and 1.7 percent for the Riedel and Chen methods, respectively. For estimating ΔH_v at any other temperature from a single value at a given temperature, use the Watson correlation. Such a value is normally available at some reference temperature.

1-13 Prediction of Vapor Pressure

Estimate the vapor pressure of 1-butene at 100°C (212°F) using the vapor-pressure correlation of Lee and Kesler [23]. Also compute the vapor pressure of ethanol at 50°C (122°F) from the Thek-Stiel generalized correlation [24]. The Lee-Kesler equation is

$$(\ln P_r^*) = (\ln P_r^*)^{(0)} + \omega(\ln P_r^*)^{(1)}$$

at constant T_r, and the Thek-Stiel correlation for polar and hydrogen-bonded molecules is

$$\ln P_r^* = \frac{\Delta H_{vb}}{RT_c(1-T_{b,r})^{0.375}} \times \left(1.14893 - 0.11719T_r - 0.03174T_r^2 - \frac{1}{T_r} - 0.375 \ln T_r\right) + \left[1.042\alpha_c - \frac{0.46284H_{vb}}{RT_c(1-T_{b,r})^{0.375}}\right]\left[\frac{(T_r)^A - 1}{A} + 0.040\left(\frac{1}{T_r} - 1\right)\right]$$

where $P_r^* = P^*/P_c$, reduced vapor pressure, P^* = vapor pressure at T_r, P_c = critical pressure, ω = acentric factor, $T_r = T/T_c$, reduced temperature, $(\ln P_r^*)^{(0)}$ and $(\ln P_r^*)^{(1)}$ are correlation functions given in Table 1-7, ΔH_{vb} = heat of vaporization at normal boiling point T_b, α_c = a constant obtained from the Thek-Stiel equation from conditions P^* = 1 atm at $T = T_b$,

$$A = \left[5.2691 + \frac{2.0753\,\Delta H_{vb}}{RT_c(1-T_{b,r})^{0.375}} - \frac{3.1738T_{b,r}\ln P_c}{1-T_{b,r}}\right]$$

and $T_{b,r}$ is the reduced normal boiling point.

TABLE 1-7 Correlation Terms for the Lee-Kesler Vapor-Pressure Equation

T_r	$-\ln (P_r^*)^{(0)}$	$-\ln (P_r^*)^{(1)}$	T_r	$-\ln (P_r^*)^{(0)}$	$-\ln (P_r^*)^{(1)}$
1.00	0.000	0.000	0.60	3.568	3.992
0.98	0.118	0.098	0.58	3.876	4.440
0.96	0.238	0.198	0.56	4.207	4.937
0.94	0.362	0.303	0.54	4.564	5.487
0.92	0.489	0.412	0.52	4.951	6.098
0.90	0.621	0.528	0.50	5.370	6.778
0.88	0.757	0.650	0.48	5.826	7.537
0.86	0.899	0.781	0.46	6.324	8.386
0.84	1.046	0.922	0.44	6.869	9.338
0.82	1.200	1.073	0.42	7.470	10.410
0.80	1.362	1.237	0.40	8.133	11.621
0.78	1.531	1.415	0.38	8.869	12.995
0.76	1.708	1.608	0.36	9.691	14.560
0.74	1.896	1.819	0.34	10.613	16.354
0.72	2.093	2.050	0.32	11.656	18.421
0.70	2.303	2.303	0.30	12.843	20.820
0.68	2.525	2.579			
0.66	2.761	2.883			
0.64	3.012	3.218			
0.62	3.280	3.586			

Source: R. C. Reid, J. M. Prausnitz, and T. K. Sherwood, *Properties of Gases and Liquids,* McGraw-Hill, New York, 1977.

Calculation Procedure:

1. Obtain critical properties and other necessary basic constants from Reid, Prausnitz, and Sherwood [1].

For 1-butene, $T_c = 419.6$ K, $P_c = 39.7$ atm, and $\omega = 0.187$. For ethanol, $T_b = 351.5$ K, $T_c = 516.2$ K, $P_c = 63$ atm, and $\Delta H_{vb} = 9260$ cal/(g·mol).

2. Obtain correlation terms in the Lee-Kesler equation.

From Table 1-7, interpolating linearily at $T_r = (273 + 100)/419.6 = 0.889$, $(\ln P_r^*)^{(0)} = -0.698$ and $(\ln P_r^*)^{(1)} = -0.595$.

3. Compute the vapor pressure of 1-butene.

Using the Lee-Kesler equation, $(\ln P_r^*) = -0.698 + (0.187)(-0.595) = -0.8093$, so

$$P_r^* = 0.4452 \quad \text{and} \quad P^* = (0.4452)(39.7) = 17.67 \text{ atm (1790 kPa)}$$

The experimental value is 17.7 atm, so the error is only 0.2 percent.

4. Compute the constant α_c for ethanol.

Now, when $T_{b,r} = 351.5/516.2 = 0.681$, $P^* = 1$ atm. So, in the Thek-Stiel equation, the A term is

$$5.2691 + \left[\frac{(2.0753)(9260)}{(1.987)(516.2)(1 - 0.681)^{0.375}} - \frac{(3.1738)(0.681)(\ln 63)}{(1 - 0.681)}\right] = 5.956$$

at those conditions. Substituting into the full Thek-Stiel equation,

$$\ln \frac{1}{63} = \frac{9260}{(1.987)(516.2)(1 - 0.681)^{0.375}}$$

$$\times [1.14893 - (0.11719)(0.681) - (0.03174)(0.681)^2 - 1/0.681 - (0.375) \ln 0.681]$$

$$+ \left[1.042\alpha_c - \frac{(0.46284)(9260)}{(1.987)(516.2)(1 - 0.681)^{0.375}}\right]$$

$$\times \left[\frac{(0.681)^{5.956} - 1}{5.956} + (0.040)(1/0.681 - 1)\right]$$

Solving for α_c, we find it to be 9.078.

5. *Compute the vapor pressure of ethanol.*

Now, $T_r = (273 + 50)/516.2 = 0.626$. Substituting into the Thek-Stiel correlation,

$$\ln P_r^* = \frac{9260}{(1.987)(516.2)(1 - 0.681)^{0.375}} \left[1.14893 - 0.11719(0.626)\right.$$

$$\left. - 0.03174(0.626)^2 - \frac{1}{0.626} - 0.375 \ln 0.626\right]$$

$$+ \left[1.042(9.078) - \frac{(0.46284)(9260)}{(1.987)(516.2)(1 - 0.681)^{0.375}}\right]$$

$$\times \left[\frac{(0.626)^{5.956} - 1}{5.956} + 0.040\left(\frac{1}{0.626} - 1\right)\right]$$

$$= -5.37717$$

Therefore, $P_r^* = 0.00462$, so $P^* = (0.00462)(63) = 0.2911$ atm (29.5 kPa). The experimental value is 0.291 atm, so the error in this case is negligible.

Related Calculations: For nonpolar liquids, use the Lee-Kesler generalized correlation. For polar liquids and those having a tendency to form hydrogen bonds, the Lee-Kesler equation does not give satisfactory results. For predicting vapor pressure of those types of compounds, use the Thek-Stiel correlation. This method, however, requires heat of vaporization at the normal boiling point, besides critical constants. If heat of vaporization at the normal boiling point is not available, estimate using the Pitzer correlation discussed in Example 1-12. If a heat-of-vaporization value at any other temperature is available, use the Watson correlation (Example 1-12) to obtain the value at the normal boiling point.

The Lee-Kesler and Thek-Stiel equations each can yield vapor pressures whose accuracy is within ±1 percent. The Antoine equations, based on a correlation with three constants, is less accurate; in some cases, the accuracy is within ±4 or 5 percent. For an example using the Antoine equation, see Example 3-1.

1-14 Enthalpy Estimation—Generalized Method

Calculate (a) enthalpy H_V of ethane vapor at 1000 psia (6900 kPa) and 190°F (360 K), (b) enthalpy H_L of liquid ethane at 50°F (283 K) and 450 psia (3100 kPa). Use generalized enthalpy departure charts (Figs. 1-6 through 1-9) to estimate enthalpy values, and base the calculations relative to H = 0 for saturated liquid ethane at −200°F. The basic constants for ethane are molecular weight MW = 30.07, critical temperature T_c = 550°R, critical pressure P_c = 709.8 psia, and critical compressibility factor Z_c = 0.284. Ideal-gas enthalpy $H°$ (relative to saturated liquid ethane at −200°F) at 190°F = 383 Btu/lb, and at 50°F = 318 Btu/lb.

Calculation Procedure:

1. Compute reduced temperature T_r and reduced pressure P_r.

a. For the vapor, $T_r = (190 + 459.7)/550 = 1.18$, and $P_r = 1000/709.8 = 1.41$.

b. For the liquid, $T_r = (50 + 459.7)/550 = 0.927$, and $P_r = 450/709.8 = 0.634$.

2. Obtain the enthalpy departure function.

a. From Fig. 1-9, for $Z_c = 0.29$, $T_r = 1.18$, $P_r = 1.41$, $(H° - H)/T_c = 2.73$ Btu/(lb·mol)(°R); and from Fig. 1-8, for $Z_c = 0.27$, $T_r = 1.18$, $P_r = 1.41$, $(H° - H)/T_c = 2.80$ Btu/(lb·mol)(°R). Interpolating linearly for $Z_c = 0.284$, $(H° - H)/T_c = 2.75$.

b. From Fig. 1-9, for $Z_c = 0.29$, $P_r = 0.634$, and $T_r = 0.927$, in the liquid region, $(H° - H)/T_c = 7.83$ Btu/(lb·mol)(°R); and from Fig. 1-8, for $Z_c = 0.27$, $P_r = 0.634$, and $T_r = 0.927$, in the liquid region, $(H° - H)/T_c = 9.4$ Btu/(lb·mol)(°R). Interpolating for $Z_c = 0.284$, $(H° - H)/T_c = 8.3$.

3. Compute enthalpy of vapor and liquid.

a. Enthalpy of ethane vapor at 190°F and 1000 psia:

$$H_v = H°_{190°F} - \left(\frac{H° - H}{T_c}\right)\left(\frac{T_c}{MW}\right) = 383 - (2.75)(550)/(30.07)$$

$$= 332.7 \text{ Btu/lb } (773{,}800 \text{ J/kg})$$

b. Enthalpy of ethane liquid at 50°F and 450 psia:

$$H_L = H°_{50°F} - \left(\frac{H° - H}{T_c}\right)\left(\frac{T_c}{MW}\right) = 318 - (8.3)(550)/(30.07)$$

$$= 166 \text{ Btu/lb } (386{,}100 \text{ J/kg})$$

Related Calculations: This procedure may be used to estimate the enthalpy of any liquid or vapor for nonpolar or slightly polar compounds. Interpolation is required if Z_c values lie between 0.23, 0.25, 0.27, and 0.29. However, extrapolation to Z_c values less than 0.23 or higher than 0.29 should not be made, because serious errors may result. When estimating enthalpy departures for mixtures, estimate mixture pseudocritical prop-

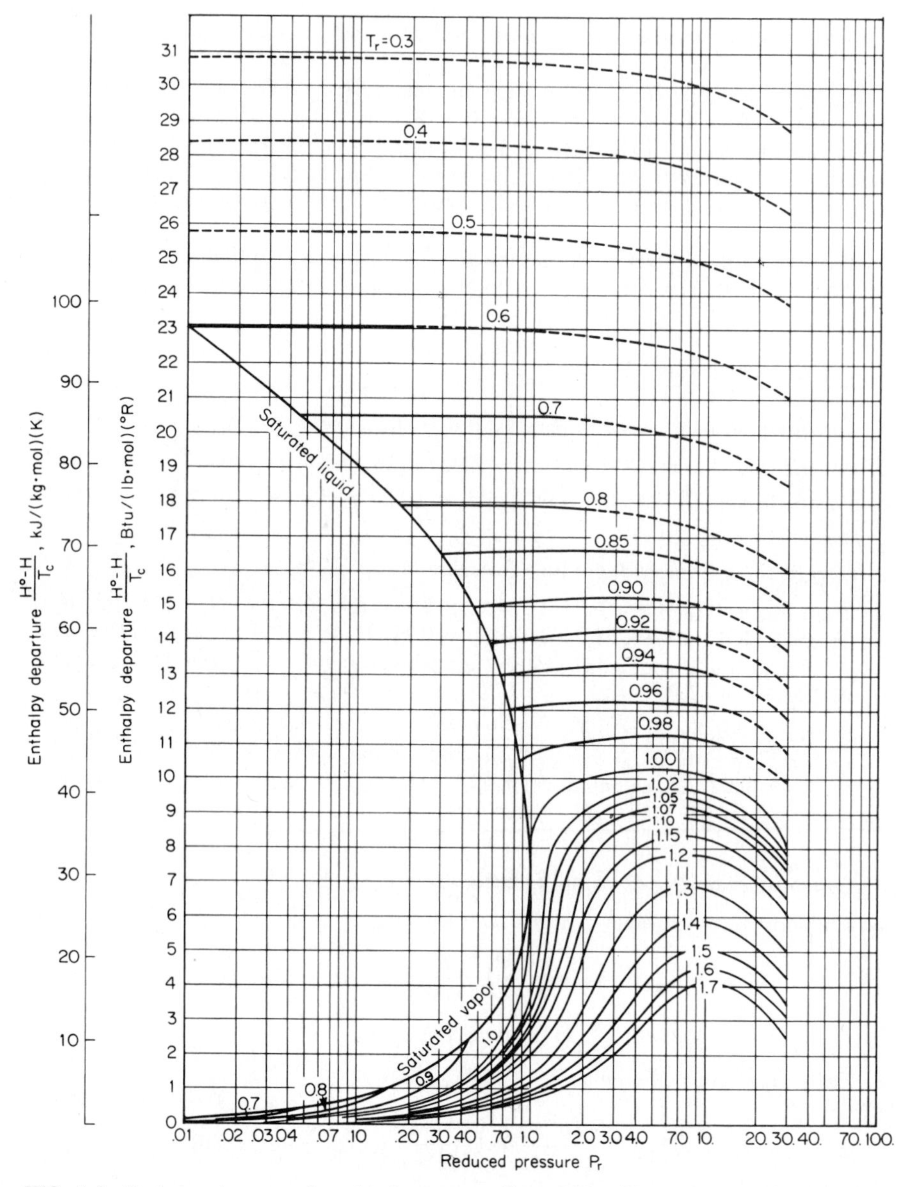

FIG. 1-6 Enthalpy departure from ideal-gas state; $Z_c = 0.23$. *(Yen and Alexander—AICHE Journal 11:334, 1965.)*

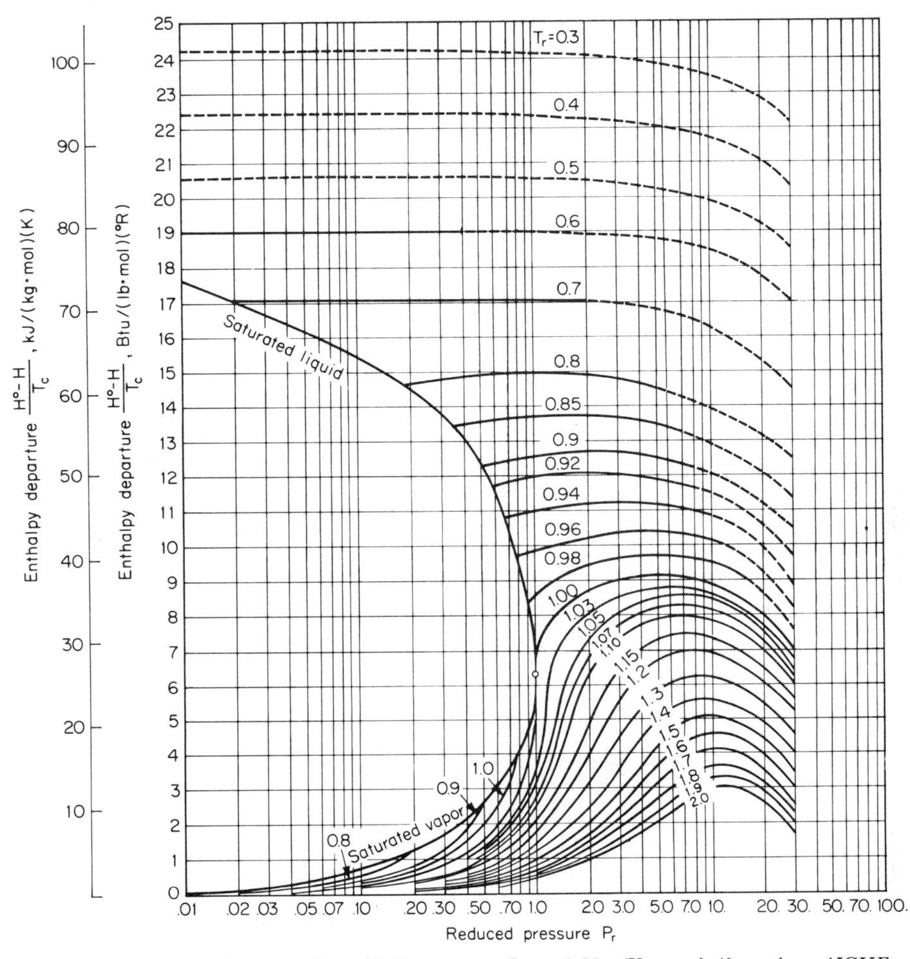

FIG. 1-7 Enthalpy departure from ideal-gas state; $Z_c = 0.25$. *(Yen and Alexander—AICHE Journal 11:334, 1965.)*

erties by taking mole-fraction-weighted averages. Do not use this correlation for gases having a low critical temperature, such as hydrogen, helium, or neon.

1-15 Entropy Involving a Phase Change

Calculate the molar entropies of fusion and vaporization for benzene. Having a molecular weight of 78.1, benzene melts at 5.5°C with a heat of fusion of 2350 cal/(g·mol). Its normal boiling point is 80.1°C, and its heat of vaporization at that temperature is 94.1 cal/g.

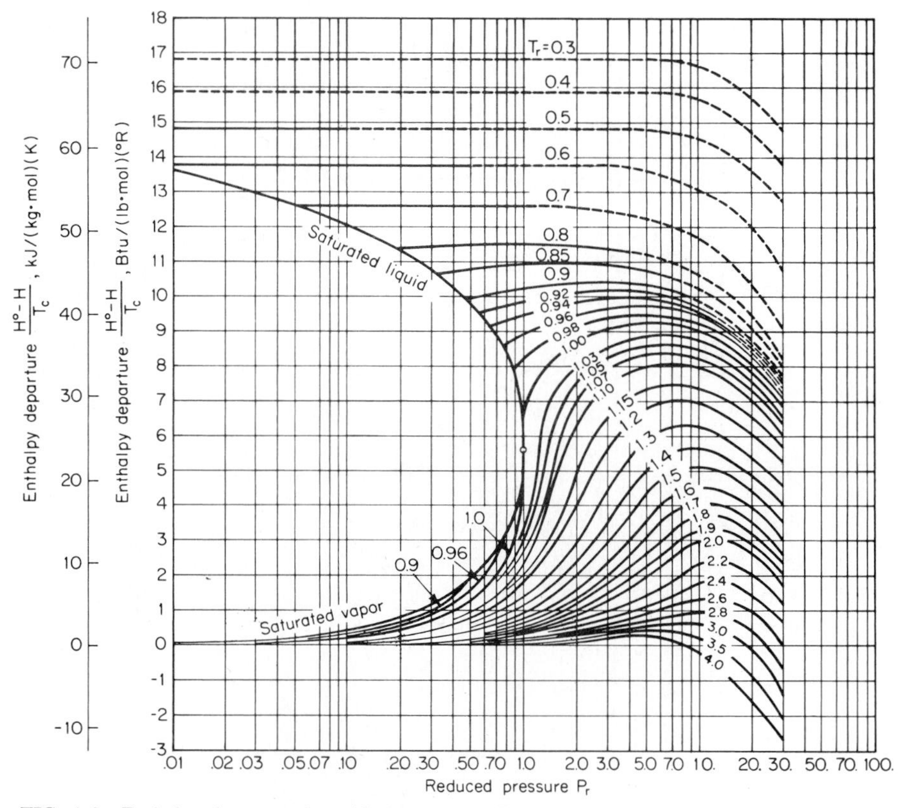

FIG. 1-8 Enthalpy departure from ideal-gas state; $Z_c = 0.27$. *(Yen and Alexander—AICHE Journal 11:334, 1965.)*

Calculation Procedure:

1. Calculate the entropy of fusion ΔS_{fusion}.

By definition, $\Delta S_{\text{fusion}} = \Delta H_{\text{fusion}}/T_{\text{fusion}}$, where the numerator is the heat of fusion and the denominator is the melting point in absolute temperature. Thus

$$\Delta S_{\text{fusion}} = \frac{2350}{(5.5 + 273)} = 8.44 \text{ cal/(g·mol)(K) [8.44 Btu/(lb·mol)(°F)]}$$

2. Calculate the entropy of vaporization ΔS_{vap}.

By definition, $\Delta S_{\text{vap}} = \Delta H_{\text{vap}}/T_{\text{vap}}$, where the numerator is the heat of vaporization and the denominator is the absolute temperature at which the vaporization takes place. Since the heat of vaporization is given on a weight basis, it must be multiplied by the molecular weight to obtain the final result on a molar basis. Thus

$$\Delta S_{\text{vap}} = \frac{(94.1)(78.1)}{(80.1 + 273)} = 20.81 \text{ cal/(g·mol)(K) [20.81 Btu/(lb·mol)(°F)]}$$

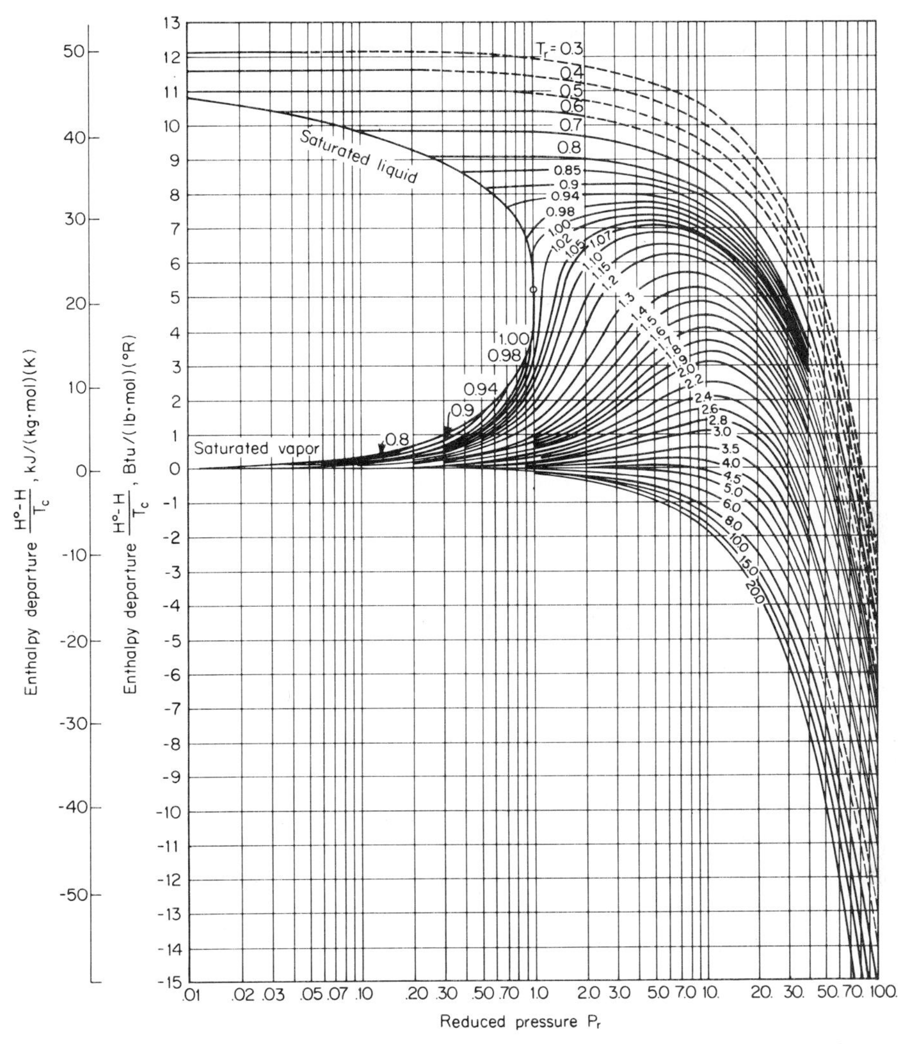

FIG. 1-9 Enthalpy departure from ideal-gas state; $Z_c = 0.29$. *(Garcia-Rangel and Yen—Paper presented at 159th National Meeting of American Chemical Society, Houston, Tex., 1970.)*

Related Calculations: This procedure can be used to obtain the entropy of phase change for any compound. If heat-of-vaporization data are not available, the molar entropy of vaporization for nonpolar liquids can be estimated via an empirical equation of Kistyakowsky:

$$\Delta S_{vap} = 8.75 + 4.571 \log T_b$$

where T_b is the normal boiling point in kelvins and the answer is in calories per gram-mole per kelvin. For benzene, the calculated value is 20.4, which is in close agreement with the value found in step 2.

1-16 Absolute Entropy from Heat Capacities

Calculate the absolute entropy of liquid *n*-hexanol at 20°C (68°F) and 1 atm (101.3 kPa) from these heat-capacity data:

Temperature, K	Phase	Heat capacity, cal/(g·mol)(K)
18.3	Crystal	1.695
27.1	Crystal	3.819
49.9	Crystal	8.670
76.5	Crystal	15.80
136.8	Crystal	24.71
180.9	Crystal	29.77
229.6	Liquid	46.75
260.7	Liquid	50.00
290.0	Liquid	55.56

The melting point of *n*-hexanol is −47.2°C (225.8 K), and its enthalpy of fusion is 3676 cal/(g·mol). The heat capacity of crystalline *n*-hexanol $(C_p)_{\text{crystal}}$ at temperatures below 18.3 K may be estimated using the Debye-Einstein equation:

$$(C_p)_{\text{crystal}} = aT^3$$

where a is an empirical constant and T is the temperature in kelvins. The absolute entropy may be obtained from

$$S^\circ_{\text{liq,20°C}} = \underbrace{\int_0^{18.3} (C_p/T)dT}_{\text{(A)}} + \underbrace{\int_{18.3}^{225.8} (C_p/T)dT}_{\text{(B)}} + \underbrace{\Delta H_{\text{fusion}}/225.8}_{\text{(C)}} + \underbrace{\int_{225.8}^{(273+20)} (C_p/T)dT}_{\text{(D)}} \qquad (1\text{-}1)$$

where $S^\circ_{\text{liq,20°C}}$ = absolute entropy of liquid at 20°C (293 K)
(A) = absolute entropy of crystalline *n*-hexanol at 18.3 K, from the Debye-Einstein equation
(B) = entropy change between 18.3 K and fusion temperature, 225.8 K
(C) = entropy change due to phase transformation (melting)
(D) = entropy change of liquid *n*-hexanol from melting point to the desired temperature (293 K)

Calculation Procedure:

1. *Estimate absolute entropy of crystalline n-hexanol.*

Since no experimental data are available below 18.3 K, estimate the entropy change below this temperature using the Debye-Einstein equation. Use the crystal entropy value of 1.695 cal/(g·mol)(°K) at 18.3 K to evaluate the coefficient a. Hence $a = 1.695/18.3^3 = 0.2766 \times 10^{-3}$. The "**A**" term in Eq. 1-1 therefore is

$$\int_0^{18.3} [(0.2766 \times 10^{-3})T^3/T]dT = (0.2766 \times 10^{-3})18.3^3/3 = 0.565 \text{ cal/(g·mol)(K)}$$

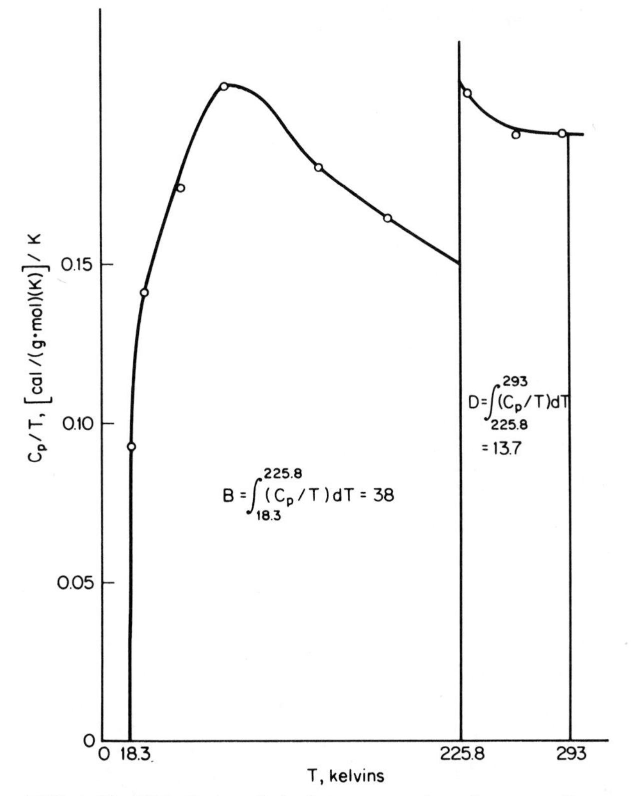

FIG. 1-10 Calculation of absolute entropy from heat-capacity data (Example 1-16).

2. Compute entropy change between 18.3 K and fusion temperature.

Plot the given experimental data on the crystal heat capacity versus temperature in kelvins and evaluate the integral (the "**B**" term in Eq. 1-1) graphically. See Fig. 1-10. Thus,

$$\int_{18.3}^{225.8} (C_p/T)\,dT = 38.0 \text{ cal/(g·mol)(K)}$$

3. Compute the entropy change due to phase transformation.

In this case,

$$\Delta H_{\text{fusion}}/225.8 = 3676/225.8 = 16.28 \text{ cal/(g·mol)(K)}$$

This is the "**C**" term in Eq. 1-1.

4. Compute the entropy change between 225.8 K and 293 K.

Plot the given experimental data on the liquid heat capacity versus temperature in kelvins and evaluate the integral (the "**D**" term in Eq. 1-1) graphically. See Fig. 1-10. Thus,

$$\int_{225.8}^{293} (C_p/T)dT = 13.7 \text{ cal/(g·mol)(K)}$$

5. Calculate the absolute entropy of liquid n-hexanol.

The entropy value is obtained from the summation of the four terms; i.e., $S^\circ_{\text{liq}}, 20°\text{C}$ = 0.565 + 38.0 + 16.28 + 13.7 = 68.5 cal/(g·mol)(K) [68.5 Btu/(lb·mol)(°F)].

Related Calculations: This general procedure may be used to calculate entropy values from heat-capacity data. However, in many situations involving practical computations, the entropy changes rather than absolute values are required. In such situations, the "**A**" term in Eq. 1-1 may not be needed. Entropy changes associated with phase changes, such as melting and vaporization, can be evaluated from the $\Delta H/T$ term (see Example 1-15).

1-17 Expansion under Isentropic Conditions

Calculate the work of isentropic expansion when 1000 lb·mol/h of ethylene gas at 1500 psig (10,450 kPa) and 104°F (313 K) is expanded in a turbine to a discharge pressure of 150 psig (1135 kPa). The ideal-gas heat capacity of ethylene is

$$C_p^\circ = 0.944 + (3.735 \times 10^{-2})T - (1.993 \times 10^{-5})T^2$$

where C_p° is in British thermal units per pound-mole per degree Rankine, and T is temperature in degrees Rankine. Critical temperature T_c is 282.4 K (508.3°R), critical pressure P_c is 49.7 atm, and critical compressibility factor Z_c is 0.276.

Calculation Procedure:

1. Estimate degree of liquefaction, if any.

Since the expansion will result in cooling, the possibility of liquefaction must be considered. If it does occur, the final temperature will be the saturation temperature T_{sat} corresponding to 150 psig. On the assumption that liquefaction has occurred from the expansion, the reduced saturation temperature T_{rs} at the reduced pressure P_r of (150 + 14.7)/(49.7)(14.7), or 0.23, is 0.81 from Fig. 1-1. Therefore, T_{sat} = (0.81)(508.3) or 412°R. Use the following equation to estimate the mole fraction of ethylene liquefied:

$$\Delta S = 0 = (S_1^\circ - S_1) + \int_{T_1}^{T_{\text{sat}}} (C_p^\circ/T)dT - R \ln (P_2/P_1) - (S_2^\circ - S_2)_{SG} - x\Delta S_{\text{vap}}$$

where $(S_1^\circ - S_1)$ and $(S_2^\circ - S_2)_{SG}$ are the entropy departure functions for gas at inlet conditions and saturation conditions, respectively; ΔS_{vap} is the entropy of vaporization at the saturation temperature; x equals moles of ethylene liquefied per mole of ethylene entering the turbine; and P_1 and P_2 are the inlet and exhaust pressures. Obtain entropy

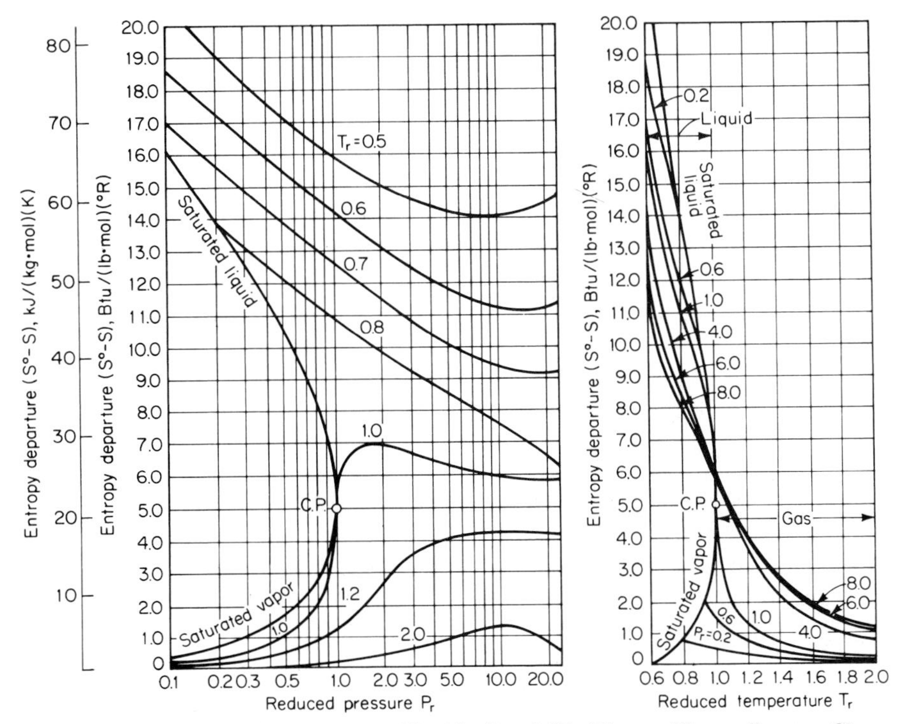

FIG. 1-11 Entropy departure of gases and liquids; $Z_c = 0.27$. *(Hougen, Watson, Ragatz—Chemical Process Principles, Part II, Wiley, 1959.)*

departure functions from Fig. 1-11 at inlet and exhaust conditions. Thus, at inlet conditions, $T_{r_1} = (104 + 460)/508.3 = 1.11$, and $P_{r_1} = (1500 + 14.7)/(14.7)(49.7) = 2.07$, so $(S_1^\circ - S_1) = 5.0$ Btu/(lb·mol)(°R). At outlet conditions, $T_{r2} = 0.81$, and $P_{r2} = 0.23$, so $(S_2^\circ - S_2)_{SG} = 1.0$ Btu/(lb·mol)(°R) and $(S_2^\circ - S_2)_{SL} = 13.7$ Btu/(lb·mol)(°R). The difference between these last two values is the entropy of vaporization at the saturation temperature ΔS_{vap}.

Substituting in the entropy equation, $\Delta S = 0 = 5.0 + 0.944 \ln (412/564) + (3.735 \times 10^{-2})(412 - 564) - (1/2)(1.993 \times 10^{-5})(412^2 - 564^2) - 1.987 \ln (164.7/1514.7) - 1.0 - x(13.7 - 1.0)$. Upon solving, $x = 0.31$. Since the value of x is between 0 and 1, the assumption that liquefaction occurs is valid.

2. *Compute the work of isentropic expansion.*

The work of isentropic expansion is obtained from enthalpy balance equation:

$$\text{Work} = -\Delta H$$

$$= -\left\{ T_c[(H_1^\circ - H_1)/T_c] + \int_{T_1}^{T_{\text{sat}}} C_p^\circ dT \right.$$

$$\left. - T_c[(H_2^\circ - H_2)_{SG}/T_c] - x\Delta H_{\text{vap}} \right\}$$

where $(H_1^\circ - H_1)/T_c$ and $(H_2^\circ - H_2)_{SG}/T_c$ are enthalpy departure functions for gas at inlet conditions and saturation conditions, respectively; and ΔH_{vap} is enthalpy of vaporization at the saturation temperature ($= T\,\Delta S_{\text{vap}}$).

The enthalpy departure functions, obtained as in Example 1-14, are

$$(H_1^\circ - H_1)/T_c = 6.1 \text{ Btu/(lb·mol)(°R)}$$

$$(H_2^\circ - H_2)_{SG}/T_c = 0.9 \text{ Btu/(lb·mol)(°R)}$$

Substituting these in the enthalpy balance equation,

$$\begin{aligned}\text{Work} &= -\Delta H \\ &= -[(508.3)(6.1) + (412 - 564)(0.944) + (1/2)(3.735)(10^{-2})(412^2 \\ &\quad - 564^2) - (1/3)(1.993)(10^{-5})(412^3 - 564^3) - (508.3)(0.9) \\ &\quad - (0.31)(412)(13.7 - 1.0)] \\ &= 1165.4 \text{ Btu/(lb·mol)}\end{aligned}$$

The power from the turbine equals [1165.4 Btu/(lb·mol)](1000 lb·mol/h)(0.000393 hp·h/Btu) = 458 hp (342 kW).

Related Calculations: When specific thermodynamic charts, namely, enthalpy-temperature, entropy-temperature, and enthalpy-entropy, are not available for a particular system, use the generalized enthalpy and entropy charts to perform expander-compressor calculations, as shown in this example.

1-18 Calculation of Fugacities

Calculate fugacity of (a) methane gas at 50°C (122°F) and 60 atm (6080 kPa), (b) benzene vapor at 400°C (752°F) and 75 atm (7600 kPa), (c) liquid benzene at 428°F (493 K) and 2000 psia (13,800 kPa), and (d) each component in a mixture of 20% methane, 40% ethane, and 40% propane at 100°F (310 K) and 300 psia (2070 kPa) assuming ideal-mixture behavior. The experimental pressure-volume data for benzene vapor at 400°C (752°F) from very low pressures up to about 75 atm are represented by

$$Z = \frac{PV}{RT} = 1 - 0.0046P$$

where Z is the compressibility factor and P is the pressure in atmospheres.

Calculation Procedure:

1. Obtain the critical-property data.

From any standard reference, the critical-property data are

Compound	T_c, K	P_c, atm	P_c, kPa	Z_c
Benzene	562.6	48.6	4924	0.274
Methane	190.7	45.8	4641	0.290
Ethane	305.4	48.2	4884	0.285
Propane	369.9	42.0	4256	0.277

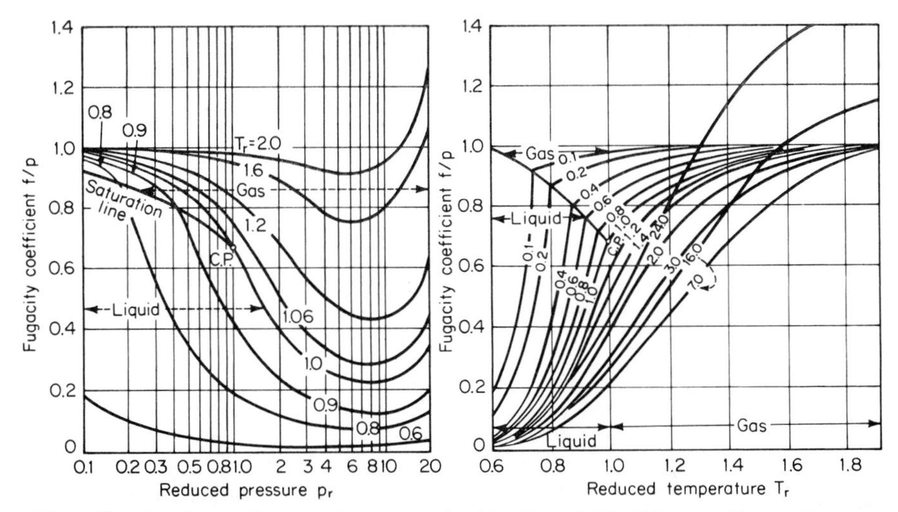

FIG. 1-12 Fugacity coefficients of gases and liquids; $Z_c = 0.27$. *(Hougen, Watson, Ragatz—Chemical Process Principles, Part II, Wiley, 1959.)*

2. Calculate the fugacity of the methane gas.

Now, reduced temperature $T_r = T/T_c = (50 + 273)/190.7 = 1.69$, and reduced pressure $P_r = P/P_c = 60/45.8 = 1.31$. From the generalized fugacity-coefficient chart (Fig. 1-12), the fugacity coefficient f/P at the reduced parameters is 0.94. (Ignore the fact that Z_c differs slightly from the standard value of 0.27). Therefore, the fugacity $f = (0.94)(60) = 56.4$ atm (5713 kPa).

3. Calculate the fugacity of the benzene vapor using the P-V-T relationship given in the statement of the problem.

Start with the equation

$$d \ln (f/P)_T = [(Z - 1) d \ln P]_T$$

Upon substituting the given P-V-T relationship,

$$d \ln (f/P) = (1 - 0.0046P)(d \ln P) - d \ln P = -0.0046P(d \ln P)$$

Since $dP/P = d \ln P$, $d \ln (f/P) = -0.0046\, dP$. Upon integration,

$$\ln (f_2/P_2) - \ln (f_1/P_1) = -0.0046(P_2 - P_1)$$

From the definition of fugacity, f/P approaches 1.0 as P approaches 0; therefore, $\ln (f_1/P_1)$ approaches ln 1 or 0 as P_1 approaches 0. Hence $\ln (f_2/P_2) = -0.0046P_2$, and $f_2 = P_2 \exp(-0.0046P_2)$. So, when $P_2 = 75$ atm, fugacity $f = 75 \exp[(-0.0046)(75)] = 53.1$ atm (5380 kPa).

4. Calculate the fugacity of the benzene vapor from the generalized correlation.

Now $T_r = (400 + 273)/562.6 = 1.20$, and $P_r = 75/48.6 = 1.54$. Ignoring the slight difference in Z_c from Fig. 1-12, $f/P = 0.78$. So fugacity $f = (0.78)(75) = 58.5$ atm (5926 kPa).

5. Calculate the fugacity of the liquid benzene from the generalized correlation.

Now $T_r = (428 + 460)/[(562.6)(1.8)] = 0.88$, and $P_r = 2000/[(14.69)(48.6)] = 2.80$. From Fig. 1-12, $f/P = 0.2$. So fugacity $f = (0.2)(2000) = 400$ psia (2760 kPa).

6. Calculate the fugacity of each component in the mixture.

The calculations for this step can be set out as follows:

	Methane	Ethane	Propane
Mole fraction, Y_i	0.2	0.4	0.4
Reduced temperature, $T_r = T/T_c$	1.63	1.02	0.84
Reduced pressure, $P_r = P/P_c$	0.45	0.42	0.49
Fugacity coefficient, f_i°/P (from Fig. 1-12)	0.98	0.89	0.56
Fugacity of pure component, $f_i^\circ = (f_i^\circ/P)(300)$, psia (kPa)	294 (2027)	267 (1841)	168 (1158)
Fugacity of component in the mixture, $f_i = Y_i f_i^\circ$, (ideal mixture), psia (kPa)	58.8 (405)	106.8 (736)	67.2 (463)

Related Calculations: If experimental *P-V-T* data are available, either as an analytical expression or as tabular values, the fugacity coefficient may be calculated by integrating the data (numerically or otherwise) as shown in step 3 above. However, if such data are not available, use the generalized fugacity coefficient chart to estimate fugacity values. Refer to Hougen, Watson, and Ragatz [4] for deviation-correction terms for values of Z_c above and below the standard value of 0.27.

The generalized-correlation method used in this example is fast and adequate for calculations requiring typical engineering accuracy. Fugacities can also be calculated by thermodynamically rigorous methods based on equations of state. Although these are cumbersome for hand calculation, they are commonly used for estimating vapor-phase nonidealities and making phase-equilibrium calculations. (Examples are given in Sec. 3, on phase equilibrium; in particular, see Examples 3-2 and 3-3.)

In the present example involving a mixture (part d), ideal behavior was assumed. For handling nonideal gaseous mixtures, volumetric data are required, preferably in the form of an equation of state at the temperature under consideration and as a function of composition and density, from zero density (lower integration limit) to the density of interest. These computations often require trial-and-error solutions and consequently are tedious for hand calculation.

1-19 Activity Coefficients from the Scatchard-Hildebrand Equation

Experimental vapor-liquid-equilibrium data for benzene(1)/n-heptane(2) system at 80°C (176°F) are given in Table 1-8. Calculate the vapor compositions in equilibrium with the corresponding liquid compositions, using the Scatchard-Hildebrand regular-solution model for the liquid-phase activity coefficient, and compare the calculated results

TABLE 1-8 Vapor-Liquid Equilibrium Data for Benzene(1)/n-Heptane(2) System at 80°C

Mole fraction of benzene in liquid phase, x_1	0.000	0.0861	0.2004	0.3842	0.5824	0.7842	0.8972	1.000
Mole fraction of benzene in vapor phase, y_1	0.000	0.1729	0.3473	0.5464	0.7009	0.8384	0.9149	1.000
Total pressure, P								
mmHg:	427.8	476.25	534.38	613.53	679.74	729.77	748.46	757.60
kPa:	57.0	63.49	71.24	81.80	90.62	97.29	99.79	101.01

with the experimentally determined composition. Ignore the nonideality in the vapor phase. Also calculate the solubility parameters for benzene and n-heptane using heat-of-vaporization data.

The following data are available on the two components:

Compound	Vapor pressure at 80°C, mmHg	Normal boiling point, °C	Heat of vaporization at normal boiling point, cal/(g·mol)	Solubility parameter, (cal/cm³)$^{1/2}$	Liquid molar volume at 25°C, [cm³/(g·mol)]	Critical temperature T_c, K
Benzene	757.6	80.3	7352	9.16	89.4	562.1
n-heptane	427.8	98.6	7576	7.43	147.5	540.2

The Scatchard-Hildebrand regular-solution model expresses the liquid activity coefficients γ_i in a binary mixture as

$$\ln \gamma_1^L = \frac{V_1^L \phi_2^2}{RT} (\delta_1 - \delta_2)^2$$

and

$$\ln \gamma_2^L = \frac{V_2^L \phi_1^2}{RT} (\delta_1 - \delta_2)^2$$

and the activity coefficient of liquid component i in a multicomponent mixture as

$$\ln \gamma_i^L = \frac{V_i^L}{RT} (\delta_i - \bar{\delta})^2$$

where V_i^L = liquid molar volume of component i at 25°C
R = gas constant
T = system temperature
ϕ_i = molar volume fraction of component i at 25°C = $(x_i V_i^L)/(\Sigma x_i V_i^L)$
δ_i = solubility parameter of component i
$\bar{\delta}$ = a molar volume fraction average of δ_i = $\Sigma \phi_i \delta_i$

The solubility parameter is defined as

$$\delta_i = \left(\frac{\Delta H_i^V - 298.15R}{V_i^L}\right)^{1/2}$$

where ΔH_i^V = heat of vaporization of component i from saturated liquid to the ideal-gas state at 25°C, cal/(g·mol)

R = gas constant, 1.987 cal/(g·mol)(K)

V_i^L = liquid molar volume of component i at 25°C, cm^3/(g·mol)

Calculation Procedure:

1. Compute the liquid-phase activity coefficients.

Using the given values of liquid molar volumes and solubility parameters, the activity coefficients are calculated for each of the eight liquid compositions in Table 1-8. For instance, when $x_1 = 0.0861$ and $x_2 = 1 - x_1 = 0.914$, then $\phi_1 = (0.0861)(89.4)/[(0.0861)(89.4) + (0.914)(147.5)] = 0.0540$ and $\phi_2 = 1 - \phi_1 = 0.9460$. Therefore,

$$\ln \gamma_1 = \frac{(89.4)(0.9460)^2}{(1.987)(273 + 80)}(9.16 - 7.43)^2 = 0.341$$

so $\gamma_1 = 1.407$, and

$$\ln \gamma_2 = \frac{(147.5)(0.054)^2}{(1.987)(273 + 80)}(9.16 - 7.43)^2 = 0.002$$

so $\gamma_2 = 1.002$.

The activity coefficients for other liquid compositions are calculated in a similar fashion and are given in Table 1-9.

2. Compute the vapor-phase mole fractions.

Assuming ideal vapor-phase behavior, $y_1 = x_1P_1^\circ\gamma_1/P$ and $y_2 = x_2P_2^\circ\gamma_2/P = 1 - y_1$, where P_i° is the vapor pressure of component i. From Table 1-8, when $x_1 = 0.0861$, $P = 476.25$ mmHg. Therefore, the *calculated* vapor-phase mole fraction of benzene is $y_1 = (0.0861)(757.6)(1.407)/476.25 = 0.1927$. The mole fraction of n-heptane is $(1 - 0.1927)$ or 0.8073. The vapor compositions in equilibrium with other liquid compositions are calculated in a similar fashion and are tabulated in Table 1-9. The last column in the table shows the deviation of the calculated composition from that determined experimentally.

3. Estimate heats of vaporization at 25°C.

Use the Watson equation:

$$\Delta H_{\text{vap},25^\circ\text{C}} = \Delta H_{\text{vap},TNBP}\left[\frac{1 - (273 + 25)/T_c}{1 - T_{NBP}/T_c}\right]^{0.38}$$

where ΔH_{vap} is heat of vaporization, and T_{NBP} is normal boiling-point temperature.

TABLE 1-9 Experimental and Calculated Vapor-Liquid Equilibrium Data for the Benzene(1)/n-Heptane(2) System at 80°C

x_1	y_1^{exp}	γ_1	γ_2	y_1^{calc}	Percent deviation $(y_1^{calc} - y_1^{exp})(100)/y_1^{exp}$
0.0000	0.0000	1.4644	1.0000	0.0000	—
0.0861	0.1729	1.4070	1.002	0.1927	11.5
0.2004	0.3473	1.3330	1.0110	0.3787	9.0
0.3842	0.5464	1.2224	1.0485	0.5799	6.1
0.5824	0.7009	1.1185	1.1412	0.7260	3.6
0.7842	0.8384	1.0379	1.3467	0.8450	0.8
0.8972	0.9194	1.0097	1.5607	0.9170	− 0.3
1.0000	1.0000	1.0000	1.8764	1.0000	0.00
					(Overall = 5.1%)

Note: Superscript ^{exp} means experimental; superscript ^{calc} means calculated.

Thus, for benzene,

$$\Delta H_{vap,25°C} = 7353 \left[\frac{1 - 298/562.1}{1 - (273 + 80.3)/562.1} \right]^{0.38} = 8039 \text{ cal/(g·mol)}$$

And for n-heptane,

$$\Delta H_{vap,25°C} = 7576 \left[\frac{1 - 298/540.2}{1 - (273 + 98.6)/540.2} \right]^{0.38} = 8694 \text{ cal/(g·mol)}$$

4. Compute the solubility parameters.

By definition,

$$\delta_{benzene} = \left[\frac{8039 - (298.15)(1.987)}{89.4} \right]^{1/2} = 9.13 \text{ (cal/cm}^3)^{1/2}$$

and

$$\delta_{n\text{-heptane}} = \left[\frac{8694 - (298.15)(1.987)}{147.5} \right]^{1/2} = 7.41 \text{ (cal/cm}^3)^{1/2}$$

These calculated values are reasonably close to the true values given in the statement of the problem.

Related Calculations: The regular-solution model of Scatchard and Hildebrand gives a fair representation of activity coefficients for many solutions containing nonpolar components. This procedure is suggested for estimating vapor-liquid equilibria if experimental data are not available. The solubility parameters and liquid molar volumes used as characteristic constants may be obtained from Table 1-10. For substances not listed there, the solubility parameters may be calculated from heat of vaporization and liquid molar volume data as shown in step 4.

For moderately nonideal liquid mixtures involving similar types of compounds, the method gives activity coefficients within ±10 percent. However, extension of the corre-

TABLE 1-10 Selected Values of Solubility Parameters at 25°C

Formula	Substance	Liquid molar volume, cm^3	Molar heat of vaporization at 25°C, kcal	Solubility parameter, $(cal/cm^3)^{1/2}$
	Aliphatic hydrocarbons			
C_5H_{12}	*n*-Pentane	116	6.40	7.1
	2-Methyl butane (isopentane)	117	6.03	6.8
	2,2-Dimethyl propane (neopentane)	122	5.35	6.2
C_6H_{14}	*n*-Hexane	132	7.57	7.3
C_7H_{16}	*n*-Heptane	148	8.75	7.4
C_8H_{18}	*n*-Octane	164	9.92	7.5
	2,2,4-Trimethylpentane ("isooctane")	166	8.40	6.9
$C_{16}H_{34}$	*n*-Hexadecane	294	19.38	8.0
C_5H_{10}	Cyclopentane	95	6.85	8.1
C_6H_{12}	Cyclohexane	109	7.91	8.2
C_7H_{14}	Methylcyclohexane	128	8.46	7.8
C_6H_{12}	1-Hexene	126	7.34	7.3
C_8H_{16}	1-Octene	158	9.70	7.6
C_6H_{10}	1,5-Hexadiene	118	7.6	7.7
	Aromatic hydrocarbons			
C_6H_6	Benzene	89	8.10	9.2
C_7H_8	Toluene	107	9.08	8.9
C_8H_{10}	Ethylbenzene	123	10.10	8.8
	o-Xylene	121	10.38	9.0
	m-Xylene	123	10.20	8.8
	p-Xylene	124	10.13	8.8
C_9H_{12}	*n*-Propyl benzene	140	11.05	8.6
	Mesitylene	140	11.35	8.8
C_8H_8	Styrene	116	10.5	9.3
$C_{10}H_8$	Naphthalene	123		9.9
$C_{14}H_{10}$	Anthracene	(150)		9.9
$C_{14}H_{10}$	Phenanthrene	158		9.8
	Fluorocarbons			
C_6F_{14}	Perfluoro-*n*-hexane	205	7.75	5.9
C_7F_{16}	Perfluoro-*n*-heptane (pure)	226	8.69	6.0
	Perfluoroheptane (mixture)			5.85
C_6F_{12}	Perfluorocyclohexane	170	6.9	6.1
C_7F_{14}	Perfluoro (methylcyclohexane)	196	7.9	6.1
	Other fluorochemicals			
$(C_4F_9)_3N$	Perfluoro tributylamine	360	13.0	5.9
$C_4Cl_2F_6$	Dichlorohexafluorocyclobutane	142		7.1
$C_4Cl_3F_7$	2,2,3-Trichloroheptafluorobutane	165	8.51	6.9
$C_2Cl_3F_3$	1,1,2-Trichloro, 1,2,2-trifluoroethane	120	6.57	7.1
$C_7F_{15}H$	Pentadecafluoroheptane	215	9.01	6.3

TABLE 1-10 Selected Values of Solubility Parameters at 25°C (*continued*)

Formula	Substance	Liquid molar volume, cm^3	Molar heat of vaporization at 25°C, kcal	Solubility parameter, $(cal/cm^3)^{1/2}$
	Other aliphatic halogen compounds			
CH_2Cl_2	Methylene chloride	64	6.84	9.8
$CHCl_3$	Chloroform	81	7.41	9.2
CCl_4	Carbon tetrachloride	97	7.83	8.6
$CHBr_3$	Bromoform	88	10.3	10.5
CH_3I	Methyl iodide	63	6.7	9.9
CH_2I_2	Methylene iodide	81		11.8
C_2H_5Cl	Ethyl chloride	74	5.7	8.3
C_2H_5Br	Ethyl bromide	75	6.5	8.9
C_2H_5I	Ethyl iodide	81	7.7	9.4
$C_2H_4Cl_2$	1,2-Dichloroethane (ethylene chloride)	79	8.3	9.9
$C_2H_4Cl_2$	1,1-Dichloroethane (ethylidene chloride)	85	7.7	9.1
$C_2H_4Br_2$	1,2-Dibromoethane	90	9.9	10.2
$C_2H_3Cl_3$	1,1,1-Trichloroethane	100	7.8	8.5

Source: J. Hildebrand, J. Prausnitz, and R. Scott, *Regular and Related Solutions,* (c) 1970 by Litton Educational Publishing Inc. Reprinted with permission of Van Nostrand Co.

lation to hydrogen-bonding compounds and highly nonideal mixtures can lead to larger errors.

When experimental equilibrium data on nonideal mixtures are not available, methods such as those based on Derr and Deal's analytical solution of groups (ASOG) [28] or the UNIFAC correlation (discussed in Example 3-4) may be used. Activity-coefficient estimation methods are also available in various thermodynamic-data packages, such as Chemshare. Further discussion may be found in Prausnitz [3] and in Reid, Prausnitz, and Sherwood [1].

The following two examples show how to use experimental equilibrium data to obtain the *equation* coefficients (as opposed to the *activity* coefficients themselves) for activity-coefficient correlation equations. Use of these correlations to calculate the activity coefficients and make phase-equilibrium calculations is discussed in Sec. 3.

1-20 Activity-Coefficient-Correlation Equations and Liquid-Liquid Equilibrium Data

Calculate the coefficients of Van Laar equations and the three-suffix Redlich-Kister equations from experimental solubility data at 70°C (158°F, or 343 K) for the water(1)/trichloroethylene(2) system. The Van Laar equations are

$$\log \gamma_1 = \frac{A_{1,2}}{\left(1 + \dfrac{A_{1,2}X_1}{A_{2,1}X_2}\right)^2} \qquad \log \gamma_2 = \frac{A_{2,1}}{\left(1 + \dfrac{A_{2,1}X_2}{A_{1,2}X_1}\right)^2}$$

The Redlich-Kister equation is

$$g^E/RT = X_1X_2\,[B_{1,2} + C_{1,2}(2X_1-1)]$$

From the relation $\ln \gamma_i = [\delta(g^E/RT)/\delta X_i]$

$$\ln \gamma_i = X_1X_2\,[B_{1,2} + C_{1,2}(X_1 - X_2)] +$$

$$X_2\,[B_{1,2}(X_2 - X_1) + C_{1,2}(6X_1X_2 - 1)]$$

In these equations, γ_1 and γ_2 are activity coefficients of components 1 and 2, respectively, X_1 and X_2 are mole fractions of components 1 and 2, respectively, $A_{1,2}$ and $A_{2,1}$ are Van Laar coefficients, $B_{1,2}$ and $C_{1,2}$ are Redlich-Kister coefficients, R is the gas constant, and T is the absolute system temperature.

The solubility data at 70°C in terms of mole fraction are: mole fraction of water in water-rich phase $X_1^W = 0.9998$; mole fraction of trichloroethylene in water-rich phase $X_2^W = 0.0001848$; mole fraction of water in trichloroethylene-rich phase $X_1^O = 0.007463$; and mole fraction of trichloroethylene in trichloroethylene-rich phase $X_2^O = 0.9925$.

Calculation Procedure:

1. Estimate the Van Laar constants.

In the water-rich phase,

$$\log \gamma_1^W = \frac{A_{1,2}}{\left[1 + \dfrac{(A_{1,2})(0.9998)}{(A_{2,1})(0.0001848)}\right]^2} = \frac{A_{1,2}}{[1 + (5410.2)(A_{1,2})/A_{2,1}]^2}$$

$$\log \gamma_2^W = \frac{A_{2,1}}{\left[1 + \dfrac{(A_{2,1})(0.0001848)}{(A_{1,2})(0.9998)}\right]^2} = \frac{A_{2,1}}{[1 + (0.0001848)(A_{2,1})/A_{1,2}]^2}$$

In the trichloroethylene-rich phase,

$$\log \gamma_1^O = \frac{A_{1,2}}{\left[1 + \dfrac{(A_{1,2})(0.007463)}{(A_{2,1})(0.9925)}\right]^2} = \frac{A_{1,2}}{[1 + (0.007520)(A_{1,2})/A_{2,1}]^2}$$

$$\log \gamma_2^O = \frac{A_{2,1}}{\left[1 + \dfrac{(A_{2,1})(0.9925)}{(A_{1,2})(0.007463)}\right]^2} = \frac{A_{2,1}}{[1 + (132.99)(A_{2,1})/A_{1,2}]^2}$$

Now, at liquid-liquid equilibria, the partial pressure of component i is the same both in water-rich and organic-rich phases, and partial pressure $\overline{P}_i = P_i^* \gamma_i X_i$, where P_i^* is the vapor pressure of component i. Hence $\gamma_1^O/\gamma_1^W = K_1 = X_1^W/X_1^O = 0.9998/0.007463 = 134$, and $\gamma_2^O/\gamma_2^W = K_2 = X_2^W/X_2^O = 0.0001848/0.9925 = 0.0001862$.

From the Van Laar equation,

$$\log \gamma_1^O - \log \gamma_1^W = \log K_1$$

$$= \frac{A_{1,2}}{[1 + (0.007520)(A_{1,2})/A_{2,1}]^2} - \frac{A_{1,2}}{[1 + (5410.2)(A_{1,2})/A_{2,1}]^2}$$

Substituting for K_1 and rearranging,

$$A_{1,2}^2 + 0.0246A_{2,1}^2 + 132.9788A_{1,2}A_{2,1} - 62.5187A_{1,2}^2A_{2,1} = 0 \qquad (1\text{-}2)$$

Similarly,

$$\log \gamma_2^O - \log \gamma_2^W = \log K_2$$

$$= \frac{A_{2,1}}{[1 + (132.99)(A_{2,1})/A_{1,2}]^2} - \frac{A_{2,1}}{[1 + (0.0001848)(A_{2,1})/A_{1,2}]^2}$$

Substituting for K_2 and rearranging,

$$A_{1,2}^2 + 0.02456A_{2,1}^2 + 132.8999A_{1,2}A_{2,1} + 35.6549A_{1,2}A_{2,1}^2 = 0 \qquad (1\text{-}3)$$

Solving Eqs. 1-2 and 1-3 simultaneously for the Van Laar constants, $A_{1,2} = 2.116$ and $A_{2,1} = -3.710$.

2. ***Estimate the Redlich-Kister coefficients.***

$$\log (\gamma_1^O/\gamma_1^W) = \log K_1$$
$$= B_{1,2}(X_2^{2,O} - X_2^{2,W}) + C_{1,2}[(X_2^{2,O})(4X_1^O - 1) - (X_2^{2,W})(4X_1^W - 1)] \qquad (1\text{-}4)$$

$$\log (\gamma_2^O/\gamma_2^W) = \log K_2$$
$$= B_{1,2}(X_1^{2,O} - X_1^{2,W}) + C_{1,2}[(X_1^{2,O})(1 - 4X_2^O) - (X_1^{2,W})(1 - 4X_2^W)] \qquad (1\text{-}5)$$

Substituting numerical values and simultaneously solving Eqs. 1-4 and 1-5 for the Redlich-Kister constants, $B_{1,2} = 2.931$ and $C_{1,2} = 0.799$.

Related Calculations: This illustration outlines the procedure for obtaining coefficients of a liquid-phase activity-coefficient model from mutual solubility data of partially miscible systems. Use of such models to calculate activity coefficients and to make phase-equilibrium calculations is discussed in Sec. 3. This leads to estimates of phase compositions in liquid-liquid systems from limited experimental data. At ordinary temperature and pressure, it is simple to obtain experimentally the composition of two coexisting phases, and the technical literature is rich in experimental results for a large variety of binary and ternary systems near 25°C (77°F) and atmospheric pressure. Example 1-21 shows how to apply the same procedure with vapor-liquid equilibrium data.

TABLE 1-11 Vapor-Liquid Equilibria for Ethanol(1)/Toluene(2) System at 55°C

P, mmHg (1)	X_1^{exp} (2)	Y_1^{exp} (3)	Y_1P, mmHg (4)	Y_2P, mmHg (5)	Y_1, ideal (6)	$\ln \gamma_1^{exp}$ (7)	$\ln \gamma_2^{exp}$ (8)	$\left(\frac{G^E}{X_1X_2RT}\right)^{exp}$ (9)
114.7	0.0000	0.0000	0.0000	114.7	0.0000	(2.441)*	0.0000	(2.441)*
144.2	0.0120	0.2127	30.7	113.5	0.0233	2.213	0.002	2.407
194.6	0.0400	0.4280	83.3	111.3	0.0575	2.010	0.011	2.369
243.0	0.1000	0.5567	135.3	107.7	0.1151	1.577	0.043	2.182
294.5	0.4000	0.6699	197.3	97.2	0.3798	0.568	0.345	1.809
308.2 (Azeotrope)	0.7490	0.7490	230.8	77.4	0.6795	0.097	0.988	1.706
305.7	0.8400	0.7994	244.4	61.3	0.7683	0.040	1.206	1.686
295.2	0.9400	0.8976	265.0	30.2	0.8903	0.008	1.480	1.708
279.6	1.000	1.0000	279.6	0.0000	1.0000	0.000	(1.711)*	(1.711)*

Note: Superscript ^{exp} means experimental.
*Values determined graphically by extrapolation.

1-21 Activity-Coefficient-Correlation Equations and Vapor-Liquid Equilibrium Data

From the isothermal vapor-liquid equilibrium data for the ethanol(1)/toluene(2) system given in Table 1-11, calculate (a) vapor composition, assuming that the liquid phase and the vapor phase obey Raoult's and Dalton's laws, respectively, (b) the values of the infinite-dilution activity coefficients, γ_1^∞ and γ_2^∞, (c) Van Laar parameters using data at the azeotropic point as well as from the infinite-dilution activity coefficients, and (d) Wilson parameters using data at the azeotropic point as well as from the infinite-dilution activity coefficients.

For a binary system, the Van Laar equations are

$$\log \gamma_1 = \frac{A_{1,2}}{\left(1 + \dfrac{A_{1,2}X_1}{A_{2,1}X_2}\right)^2} \quad \text{and} \quad \log \gamma_2 = \frac{A_{2,1}}{\left(1 + \dfrac{A_{2,1}X_2}{A_{1,2}X_1}\right)^2}$$

or

$$A_{1,2} = \log \gamma_1 \left(1 + \frac{X_2 \log \gamma_2}{X_1 \log \gamma_1}\right)^2 \quad \text{and} \quad A_{2,1} = \log \gamma_2 \left(1 + \frac{X_1 \log \gamma_1}{X_2 \log \gamma_2}\right)^2$$

The Wilson equations are

$$G^E/RT = -X_1 \ln (X_1 + X_2G_{1,2}) - X_2 \ln (X_2 + X_1G_{1,2})$$

$$\ln \gamma_1 = -\ln (X_1 + X_2G_{1,2}) + X_2 \left[\frac{G_{1,2}}{(X_1 + X_2G_{1,2})} - \frac{G_{2,1}}{(X_2 + X_1G_{2,1})}\right]$$

$$\ln \gamma_2 = -\ln (X_2 + X_1G_{2,1}) - X_1 \left[\frac{G_{1,2}}{(X_1 + X_2G_{1,2})} - \frac{G_{2,1}}{(X_2 + X_1G_{2,1})}\right]$$

$$G^E/X_1X_2RT = \ln \gamma_1/X_2 + \ln \gamma_2/X_1$$

In these equations, γ_1 and γ_2 are activity coefficients of components 1 and 2, respectively, G^E is Gibbs molar excess free energy, $A_{1,2}$ and $A_{2,1}$ are Van Laar parameters, $G_{1,2}$ and $G_{2,1}$ are Wilson parameters, that is,

$$G_{i,j} = \frac{V_j}{V_i} e^{-a_{i,j}/RT} \qquad i \neq j$$

$a_{i,j}$ are Wilson constants ($a_{i,j} \neq a_{j,i}$ and $G_{i,j} \neq G_{j,i}$), X_1 and X_2 are liquid-phase mole fractions of components 1 and 2, respectively, Y_1 and Y_2 are vapor-phase mole fractions of components 1 and 2, respectively, R is the gas constant, and T is the absolute system temperature.

Calculation Procedure:

1. Compute vapor compositions, ignoring liquid- and vapor-phase nonidealities.

By Raoult's and Dalton's laws, $Y_1 = X_1P_1^O/P$, $Y_2 = X_2P_2^O/P$ or $Y_2 = 1 - Y_1$, where P^O is vapor pressure and P is total system pressure. Now, $P_1^O = 279.6$ mmHg (the value of pressure corresponding to $X_1 = Y_1 = 1$), and $P_2^O = 114.7$ mmHg (the value of pressure corresponding to $X_2 = Y_2 = 1$).

The vapor compositions calculated using the preceding equations are shown in col. 6 of Table 1-11.

2. Compute logarithms of the activity coefficients from experimental X-Y data.

Assuming an ideal vapor phase, the activity coefficients are given as $\gamma_1 = PY_1/X_1P_1^O$ and $\gamma_2 = PY_2/X_2P_2^O$. The natural logarithms of the activity coefficients calculated using the preceding equations are shown in cols. 7 and 8 of Table 1-11.

3. Compute infinite-dilution activity coefficients.

First, calculate $G^E/X_1X_2RT = \ln \gamma_1/X_2 + \ln \gamma_2/X_1$. The function G^E/X_1X_2RT calculated using the preceding equation is tabulated in col. 9 of Table 1-11.

Next, obtain the infinite-dilution activity coefficients graphically by extrapolating G^E/X_1X_2RT values to $X_1 = 0$ and $X_2 = 0$ (Fig. 1-13). Thus, $\ln \gamma_1^\infty = 2.441$ and $\ln \gamma_2^\infty = 1.711$, so $\gamma_1^\infty = 11.48$ and $\gamma_2^\infty = 5.53$.

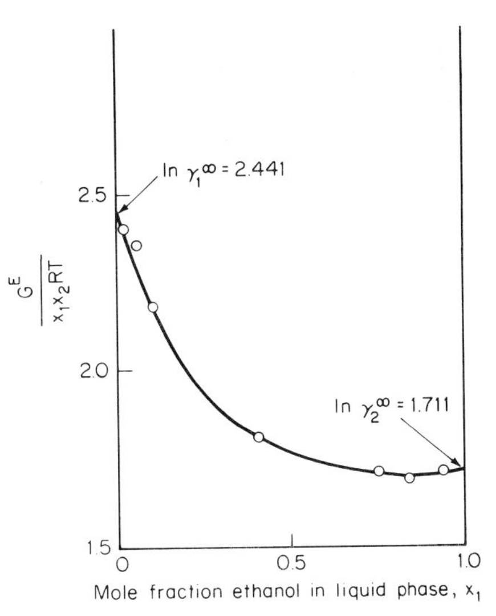

FIG. 1-13 Graphic solution for infinite-dilution activity coefficients (Example 1-21).

4. Calculate Van Laar parameters from azeotropic data.

At the azeotropic point, $X_1 = 0.7490$, $X_2 = 1 - X_1 = 0.251$, and $\ln \gamma_1 = 0.097$ and $\ln \gamma_2 = 0.988$ (see Table 1-11). Therefore,

$$A_{1,2} = \frac{0.097}{2.303}\left[1 + \frac{(0.251)(0.988)(2.303)}{(0.749)(0.097)(2.303)}\right]^2$$
$$= 0.820 \qquad (\ln \gamma_i = 2.303 \log \gamma_i)$$

and

$$A_{2,1} = \frac{0.988}{2.303}\left[1 + \frac{(0.749)(0.097)(2.303)}{(0.251)(0.988)(2.303)}\right]^2$$
$$= 0.717$$

5. Calculate Van Laar parameters from the infinite-dilution activity coefficients.

By definition, $A_{1,2} = \log \gamma_1^\infty$ and $A_{2,1} = \log \gamma_2^\infty$. Therefore, $A_{1,2} = 2.441/2.303 = 1.059$, and $A_{2,1} = 1.711/2.303 = 0.743$.

6. Calculate Wilson parameters from azeotropic data.

From the azeotropic data,

$$0.097 = -\ln(0.749 + 0.251G_{1,2}) + 0.251\left[\frac{G_{1,2}}{(0.749 + 0.251G_{1,2})} - \frac{G_{2,1}}{(0.251 + 0.749G_{2,1})}\right]$$

and

$$0.988 = -\ln(0.251 + 0.749G_{2,1}) - 0.749\left[\frac{G_{1,2}}{(0.749 + 0.251G_{1,2})} - \frac{G_{2,1}}{(0.251 + 0.749G_{2,1})}\right]$$

Solving for $G_{1,2}$ and $G_{2,1}$ by trial and error gives $G_{1,2} = 0.1260$ and $G_{2,1} = 0.4429$.

7. Calculate Wilson parameters from the infinite-dilution activity coefficients.

The equations are $\ln \gamma_1^\infty = 2.441 = -\ln G_{1,2} + 1 - G_{2,1}$, and $\ln \gamma_2^\infty = 1.711 = -\ln G_{2,1} + 1 - G_{1,2}$. Solving by trial and error, $G_{1,2} = 0.1555$ and $G_{2,1} = 0.4209$.

Related Calculations: These calculations show how to use vapor-liquid equilibrium data to obtain parameters for activity-coefficient correlations such as those of Van Laar and Wilson. (Use of liquid-liquid equilibrium data for the same purpose is shown in Example 1-20.) If the system forms an azeotrope, the parameters can be obtained from a single measurement of the azeotropic pressure and the composition of the constant boiling mixture. If the activity coefficients at infinite dilution are available, the two parameters for the Van Laar equation are given directly, and the two in the case of the Wilson equation can be solved for as shown in the example.

In principle, the parameters can be evaluated from minimal experimental data. If vapor-liquid equilibrium data at a series of compositions are available, the parameters in a given excess-free-energy model can be found by numerical regression techniques. The goodness of fit in each case depends on the suitability of the form of the equation. If a plot of G^E/X_1X_2RT versus X_1 is nearly linear, use the Margules equation (see Sec. 3). If a plot of X_1X_2RT/G^E is linear, then use the Van Laar equation. If neither plot approaches linearity, apply the Wilson equation or some other model with more than two parameters.

The use of activity-coefficient-correlation equations to calculate activity coefficients and make phase-equilibrium calculations is discussed in Sec. 3. For a detailed discussion, see Prausnitz [3].

1-22 Convergence-Pressure Vapor-Liquid Equilibrium *K* Values

In a natural-gas processing plant (Fig. 1-14), a 1000 lb·mol/h (453.6 kg·mol/h) stream containing 5 mol % nitrogen, 65% methane, and 30% ethane is compressed from 80 psia (552 kPa) at 70°F (294 K) to 310 psia (2137 kPa) at 260°F (400 K) and subsequently cooled in a heat-exchanger chilling train (a cooling-water heat exchanger and two refrigeration heat exchangers) to partially liquefy the feed stream. The liquid is disengaged from the vapor phase in a separating drum at 300 psia (2070 kPa) and −100°F (200 K) and pumped to another part of the process. The vapor from the drum is directed to a 15-tray absorption column.

In the column, the vapor is countercurrently contacted with 250 mol/h of liquid propane, which absorbs from the vapor feed 85 percent of its ethane content, 9 percent of its methane, and a negligible amount of nitrogen. The pressure of the bottom stream of liquid propane with its absorbed constituents is raised to 500 psia (3450 kPa), and the stream is heated to 50°F (283 K) in a heat exchanger before being directed to another section of the process. The vapor, leaving the overhead tower at −60°F (222 K), consists of unabsorbed constituents plus some vaporized propane from the liquid-propane absorbing medium.

a. In which heat exchanger of the chilling train does condensation start? Assume that the convergence pressure is 800 psia (5520 kPa).

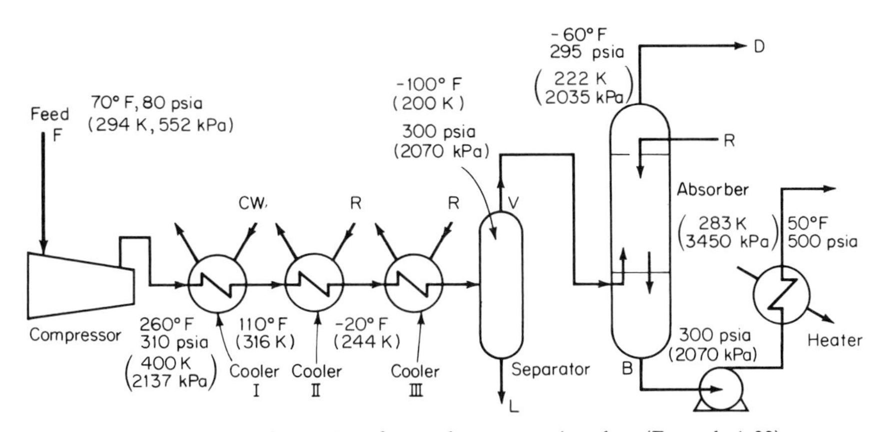

FIG. 1-14 Flow diagram for portion of natural-gas-processing plant (Example 1-22).

b. How much propane is in the vapor leaving the overhead of the column? Assume again that the convergence pressure is 800 psia (5520 kPa).

c. Does any vaporization take place in the heat exchanger that heats the bottoms stream from the column?

Use convergence-pressure vapor-liquid equilibrium K-value charts (Figs. 1-15 through 1-26).

Calculation Procedure:

1. Calculate dew-point equilibrium for the feed.

A vapor is at its dew-point temperature when the first drop of liquid forms upon cooling the vapor at constant pressure and the composition of the vapor remaining is the same as that of the initial vapor mixture. At dew-point conditions, $Y_i = N_i = K_iX_i$, or $X_i = N_i/K_i$, and $\Sigma N_i/K_i = 1.0$, where Y_i is the mole fraction of component i in the vapor phase, X_i is the mole fraction of component i in the liquid phase, N_i is the mole fraction of component i in the original mixture, and K_i is the vapor-liquid equilibrium K value.

Calculate the dew-point temperature of the mixture at a pressure of 310 psia and an assumed convergence pressure of 800 psia. At various assumed temperatures and at a pressure of 310 psia, the K values for methane, nitrogen, and ethane, as obtained from Figs. 1-15, 1-18, and 1-21, are listed in Table 1-12, as are the corresponding values of N_i/K_i. At the dew-point temperature, the latter will add up to 1.0. It can be seen from the table that the dew point lies between −60 and −50°F (222 and 227 K). Therefore, the condensation will take place in the last heat exchanger in the train, because that one lowers the stream temperature from −20°F (244 K) to −100°F (200 K).

2. Estimate the compositions of the liquid and vapor phases leaving the separator.

At the separator conditions, −100°F and 310 psia, estimate the mole ratio of vapor to liquid V/L from the relationship

$$L_i = F_i/[1 + (V/L)K_i]$$

where L_i is moles of component i in the liquid phase, F_i is moles of component i in the feed (given in the statement of the problem), and K_i is the vapor-liquid equilibrium K

TABLE 1-12 Calculation of Dew Point for Feed Gas in Example 1-22 (convergence pressure = 800 psia)

		Assumed temperature							
	Mole fraction	0°F		−50°F		−100°F		−60°F	
Component i	N_i	K_i	N_i/K_i	K_i	N_i/K_i	K_i	N_i/K_i	K_i	N_i/K_i
Methane	0.65	3.7	0.176	3.1	0.210	2.6	0.250	3.0	0.217
Nitrogen	0.05	18.0	0.003	16.0	0.003	12.0	0.004	15.0	0.003
Ethane	0.30	0.8	0.375	0.45	0.667	0.17	1.765	0.35	0.857
Total	1.00		0.554		0.880		2.019		1.077

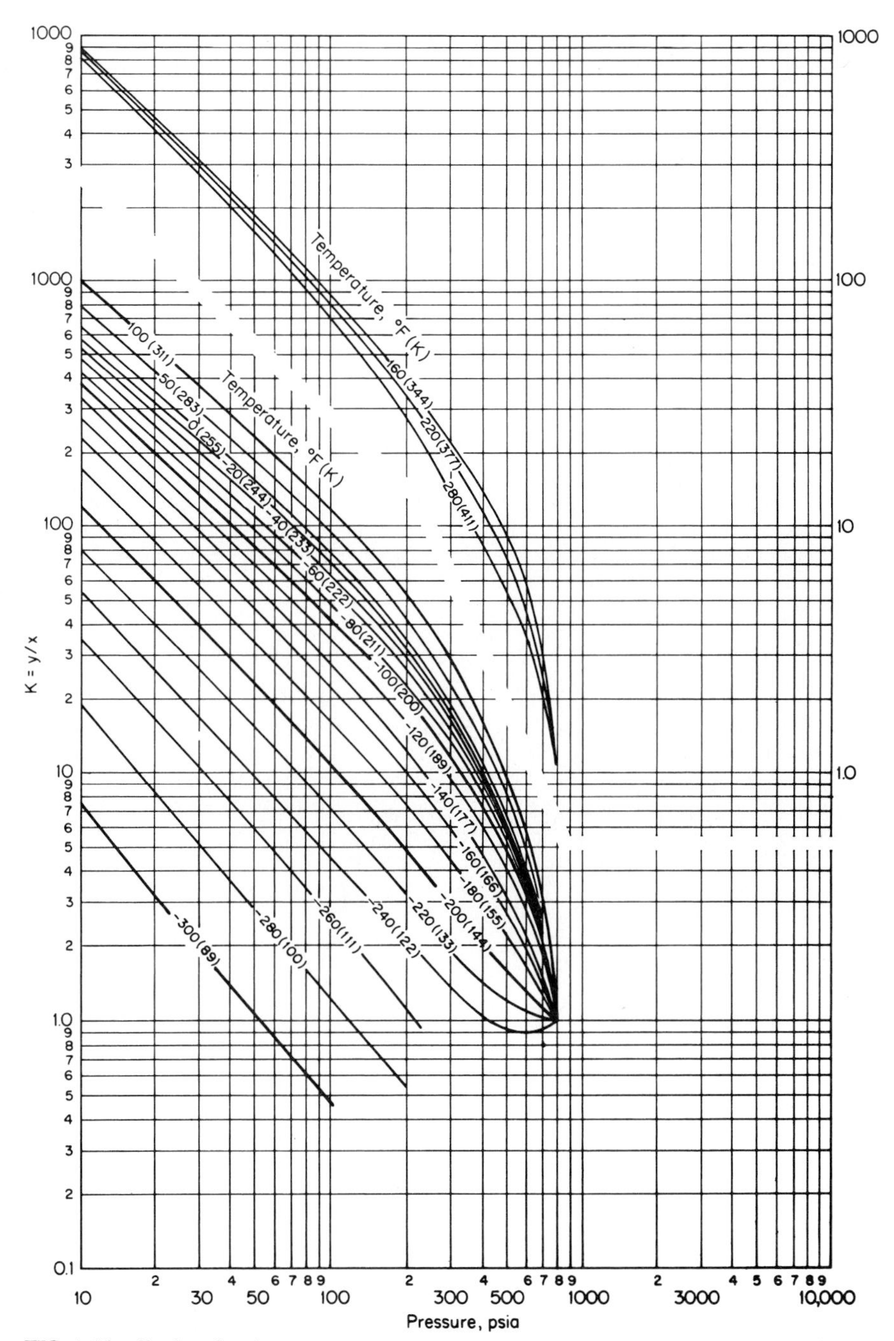

FIG. 1-15 *K* values for nitrogen; convergence pressure = 800 psia. (Note: 1 psi = 6.895 kPa.) *(Courtesy of NGPA.)*

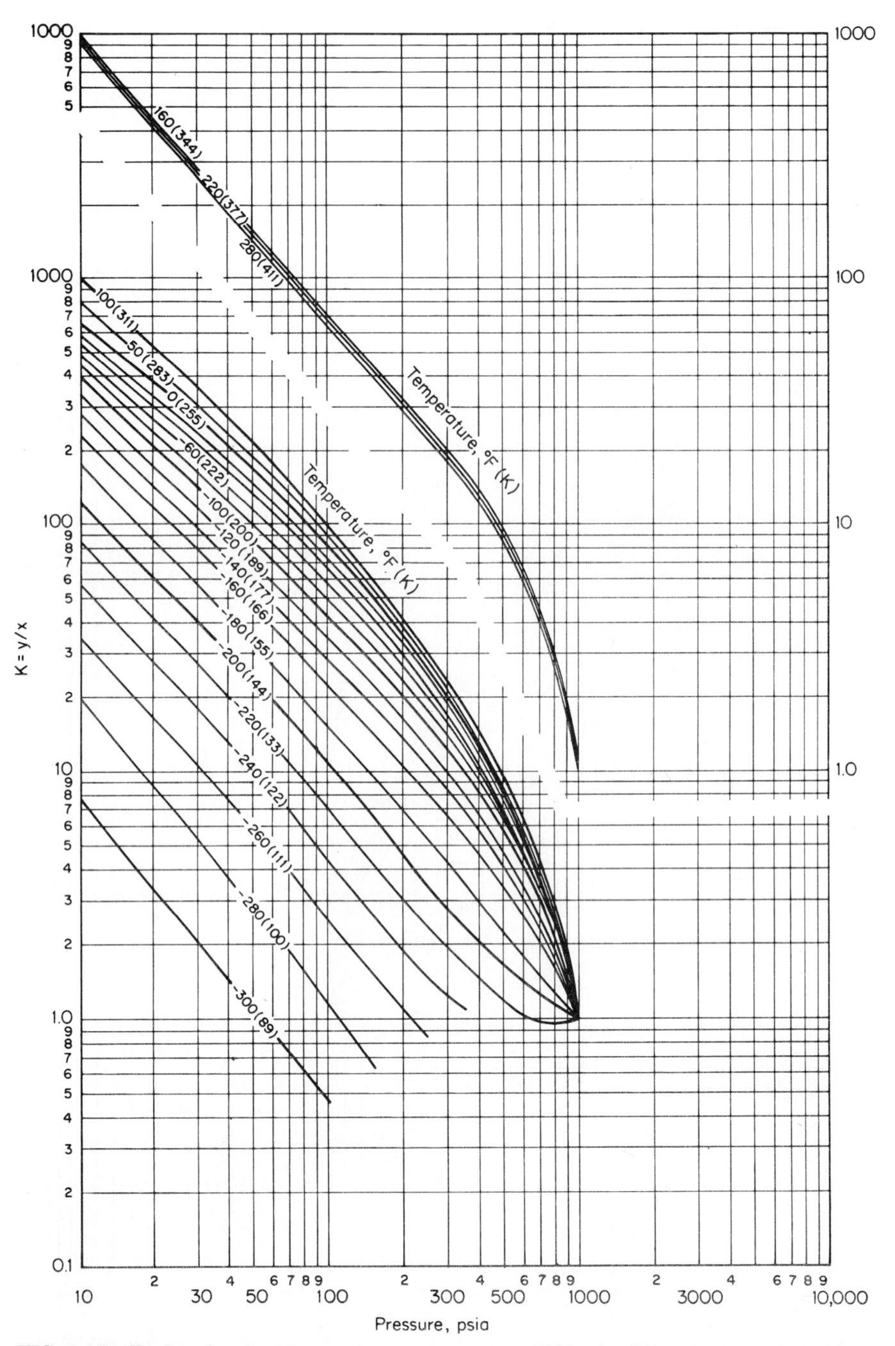

FIG. 1-16 *K* values for nitrogen; convergence pressure = 1000 psia. (Note: 1 psi = 6.895 kPa.) *(Courtesy of NGPA.)*

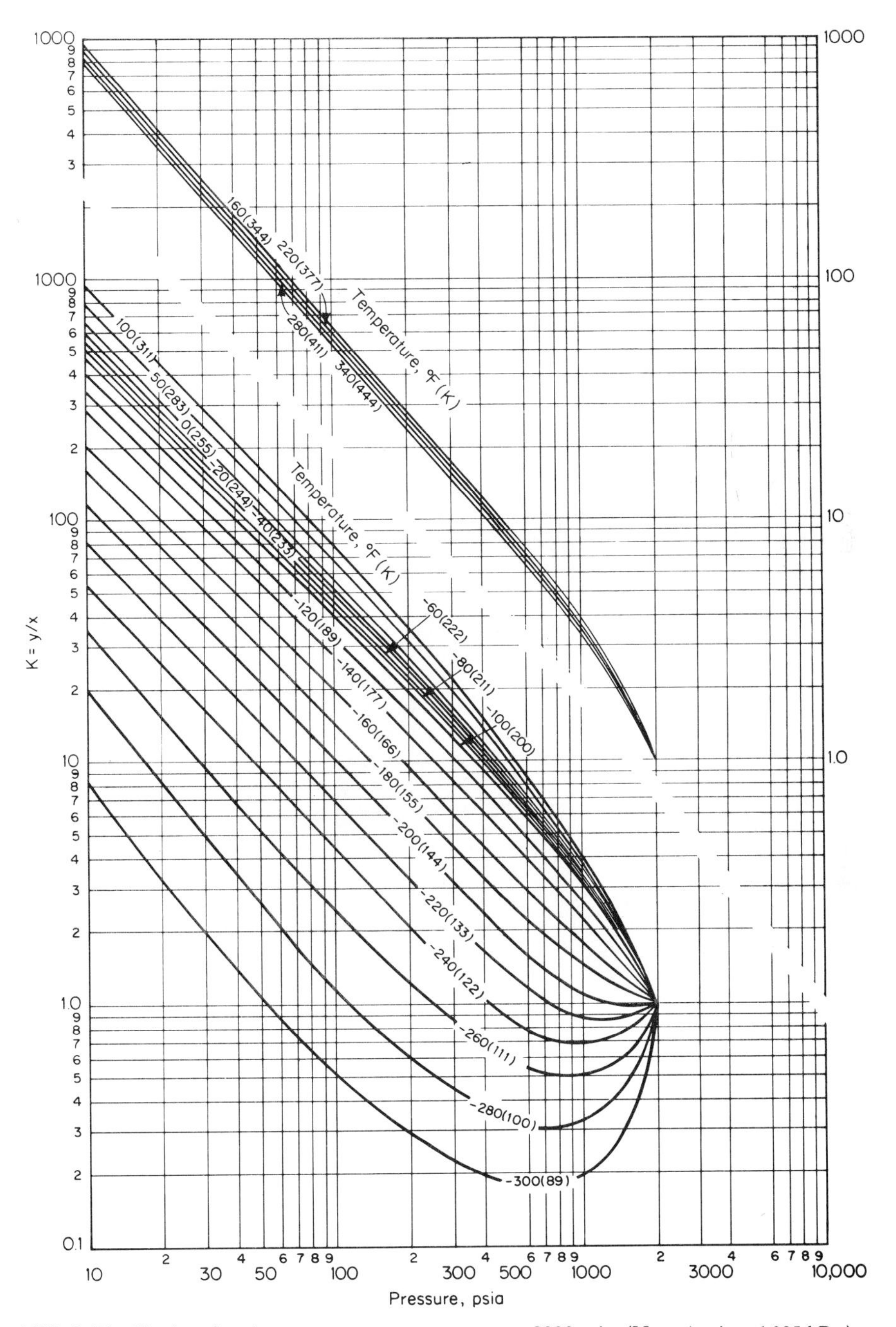

FIG. 1-17 *K* values for nitrogen; convergence pressure = 2000 psia. (Note: 1 psi = 6.895 kPa.) *(Courtesy of NGPA.)*

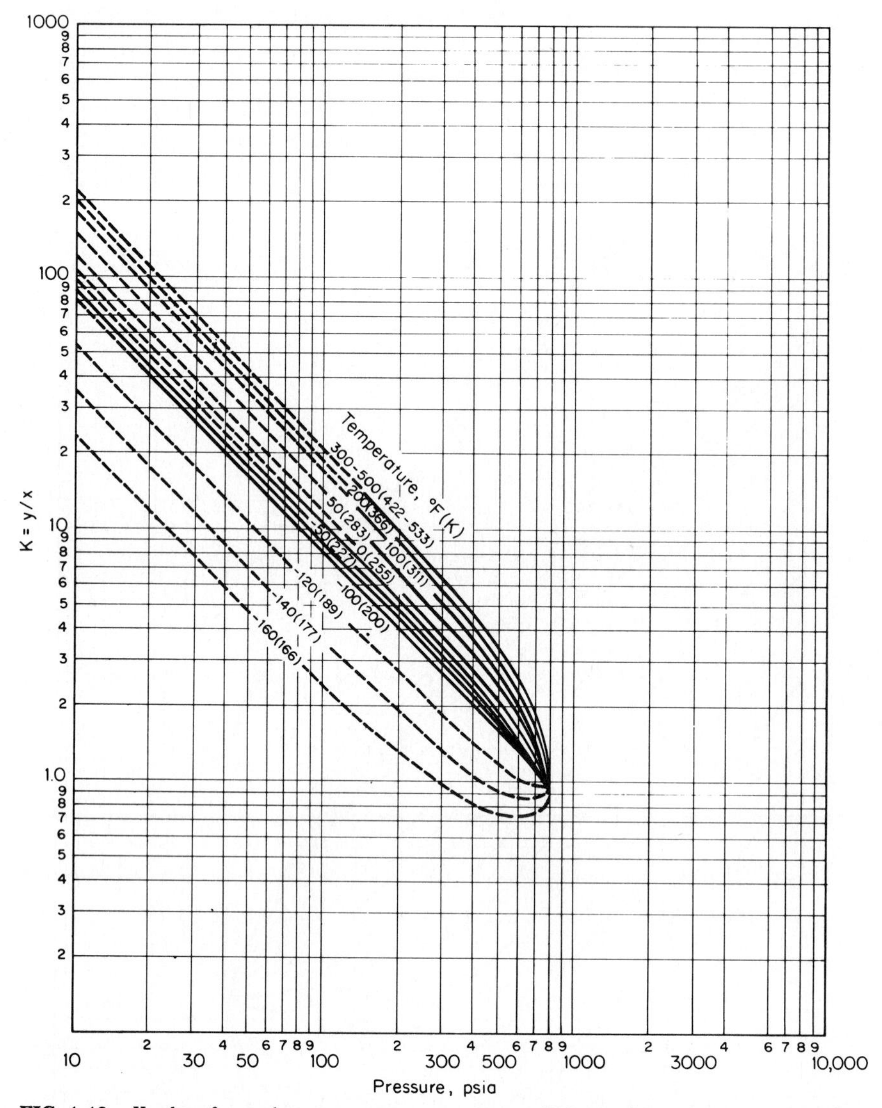

FIG. 1-18 *K* values for methane; convergence pressure = 800 psia. (Note: 1 psi = 6.895 kPa.) *(Courtesy of NGPA.)*

value for that component, obtainable from Figs. 1-15, 1-18, and 1-21. Use trial and error for finding V/L: Using the known F_i and K_i, assume a value of V/L and calculate the corresponding L_i; add up these L_i to obtain the total moles per hour of liquid, and subtract the total from 1000 mol/h (given in the statement of the problem) to obtain the moles per hour of vapor; take the ratio of the two, and compare it with the assumed V/L ratio. This trial-and-error procedure, leading to a V/L of 2.2, is summarized in Table 1-13. Thus the composition of the liquid phase leaving the separator (i.e., the L_i corresponding to a V/L of 2.2) appears as the next-to-last column in that table. The vapor composition

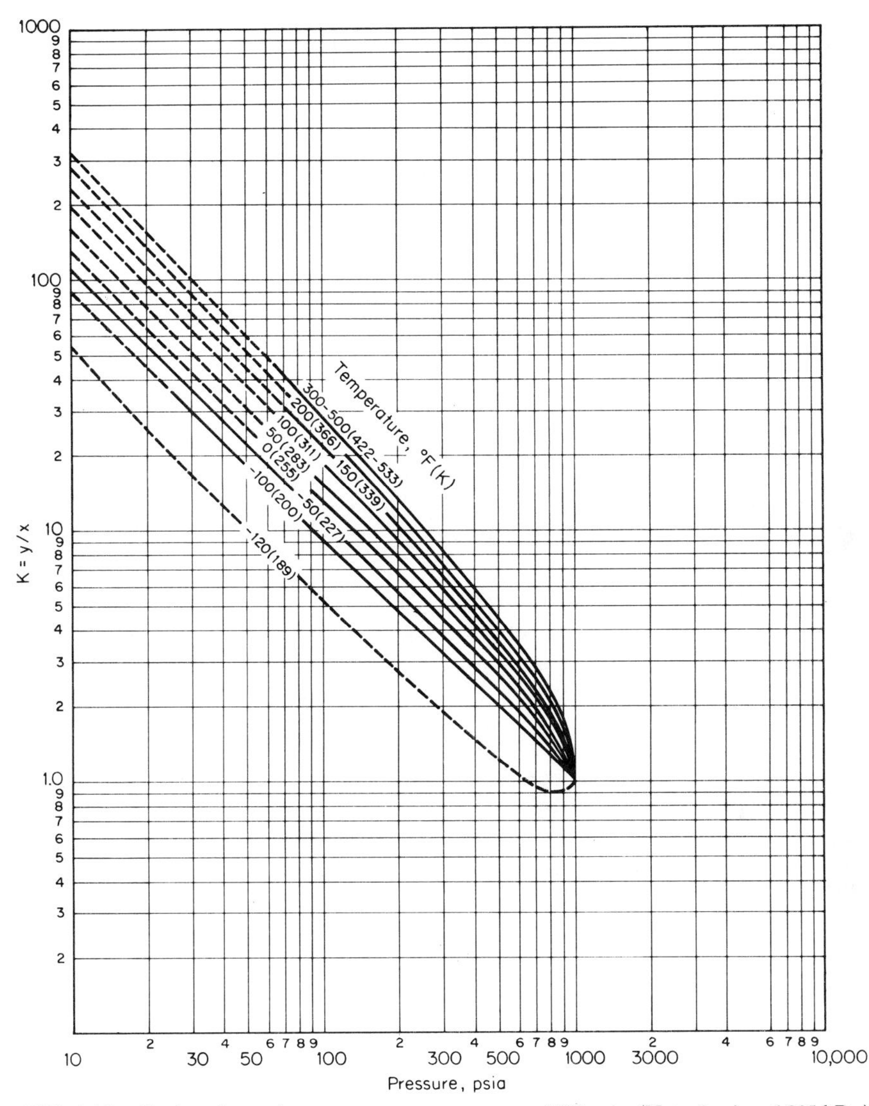

FIG. 1-19 *K* values for methane; convergence pressure = 1000 psia. (Note: 1 psi = 6.985 kPa.) *(Courtesy of NGPA.)*

(expressed as moles per hour of each component), given in the last column, is found by subtracting the moles per hour in the liquid phase from the moles per hour of feed.

3. Estimate the amount of propane in the vapor leaving the column overhead.

The vapor stream leaving the column overhead may be assumed to be at its dew point; in other words, the dew point of the overhead mixture is −60°F and 295 psia. At the dew point, $\Sigma(Y_i/K_i) = 1.0$. Designate Y_i, the mole fraction of component i in the column

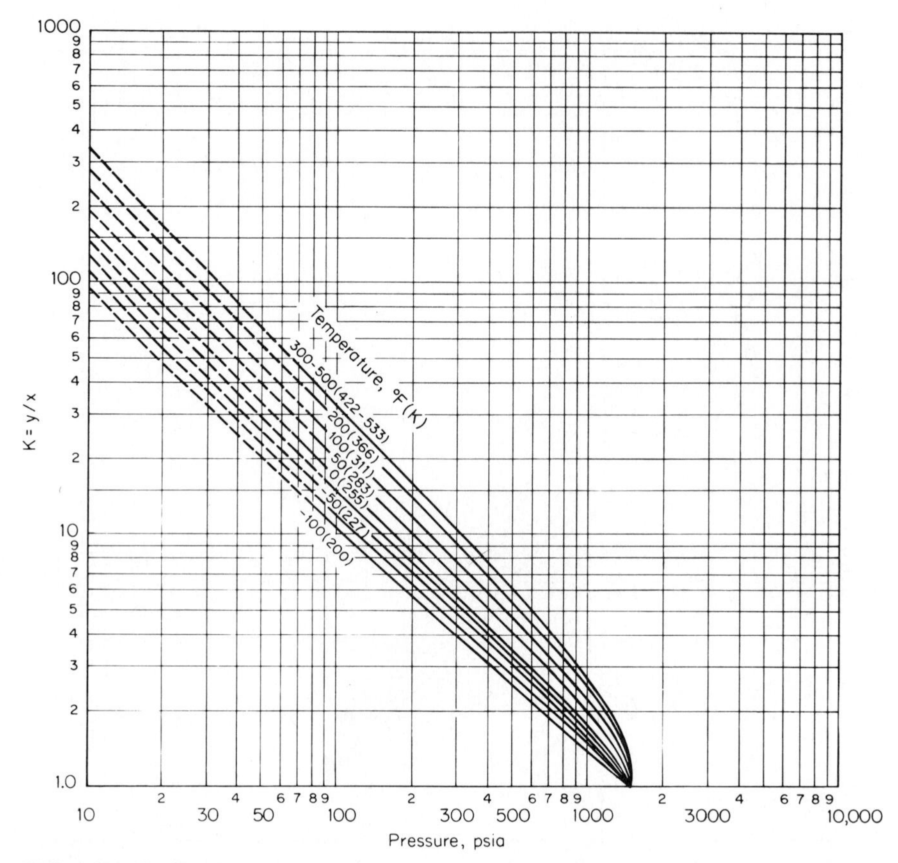

FIG. 1-20 *K* values for methane; convergence pressure = 1500 psia. (Note: 1 psi = 6.895 kPa.) *(Courtesy of NGPA.)*

TABLE 1-13 Equilibrium Flash Calculations for Example 1-22 (convergence pressure = 800 psia; feed rate = 1000 mol/h)

Component i	F_i, lb·mol/h	K_i at −100°F, 300 psia	L_i, lb·mol/h, at assumed V/L ratio of 1	2	2.5	2.3	2.2	V_i, lb·mol/h, at V/L = 2.2
Methane	650	2.6	180.6	104.8	86.7	93.1	96.7	553.3
Nitrogen	50	12.0	3.8	2.0	1.6	1.7	1.8	48.2
Ethane	300	0.18	254.2	220.6	206.9	212.2	214.9	85.1
Total	1000		438.6	327.4	295.2	307	313.4	686.6
$V = 1000 - L$			561.4	672.6	704.8	693	686.6	
Calculated V/L			1.280	2.054	2.388	2.257	2.191 (close enough)	

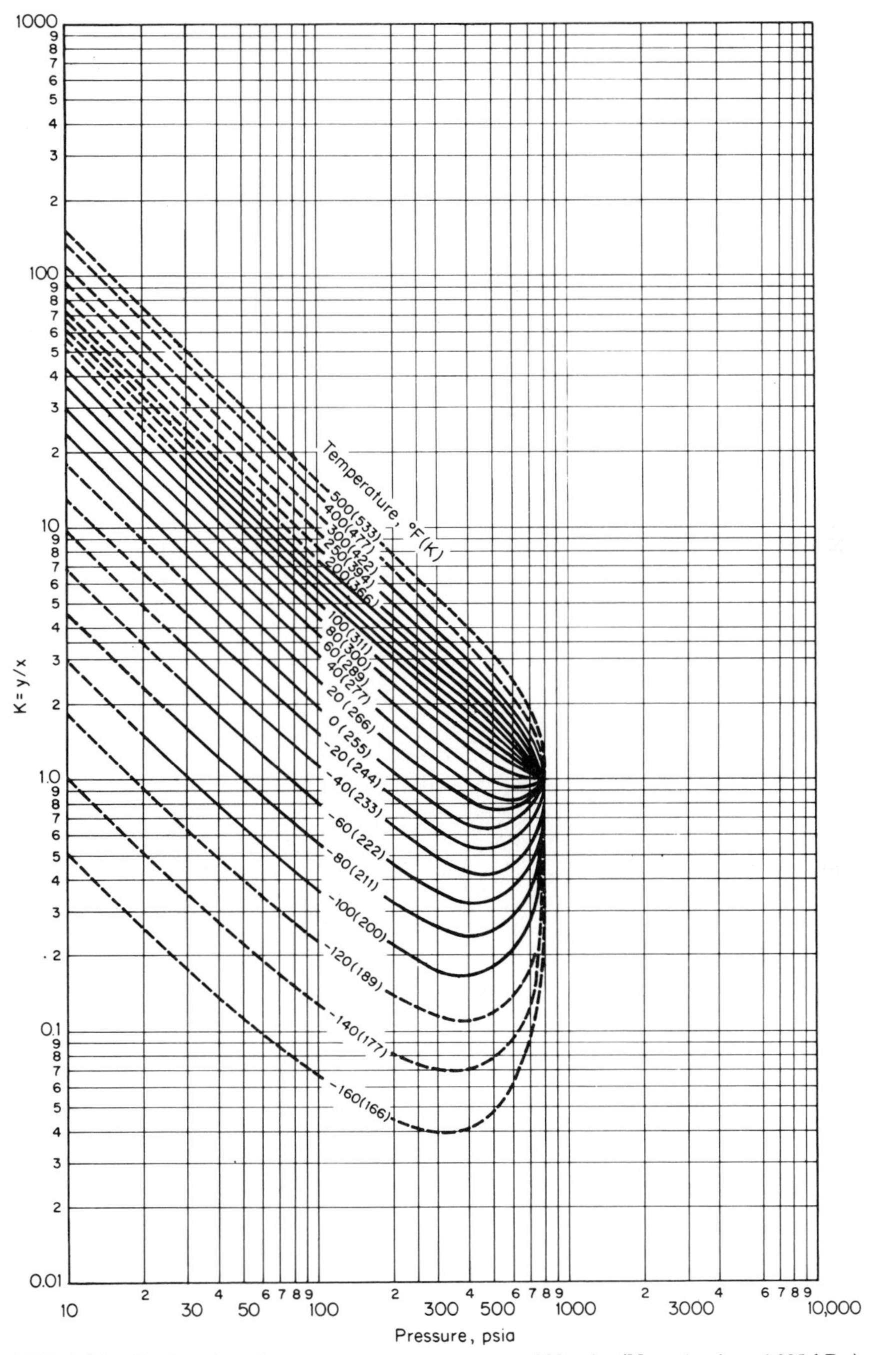

FIG. 1-21 *K* values for ethane; convergence pressure = 800 psia. (Note: 1 psi = 6.895 kPa.) *(Courtesy of NGPA.)*

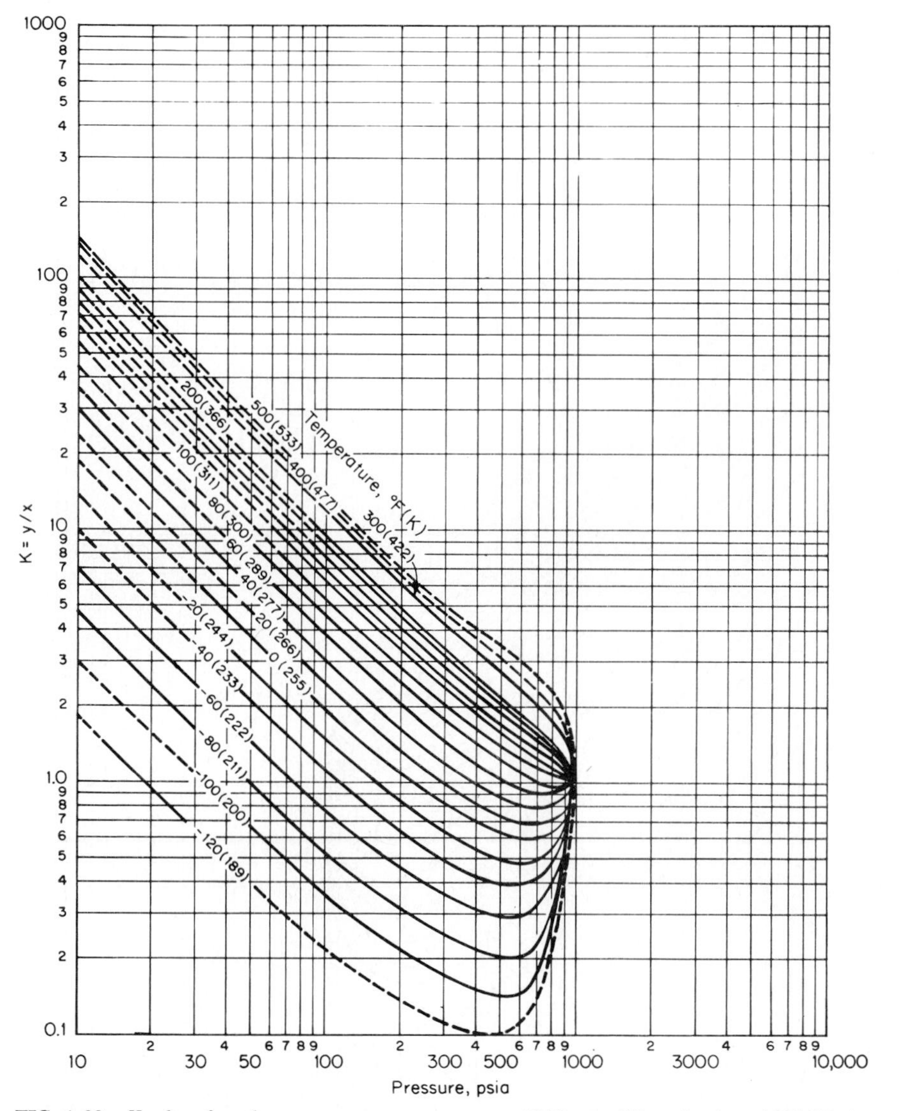

FIG. 1-22 *K* values for ethane; convergence pressure = 1000 psia. (Note: 1 psi = 6.895 kPa.) *(Courtesy of NGPA.)*

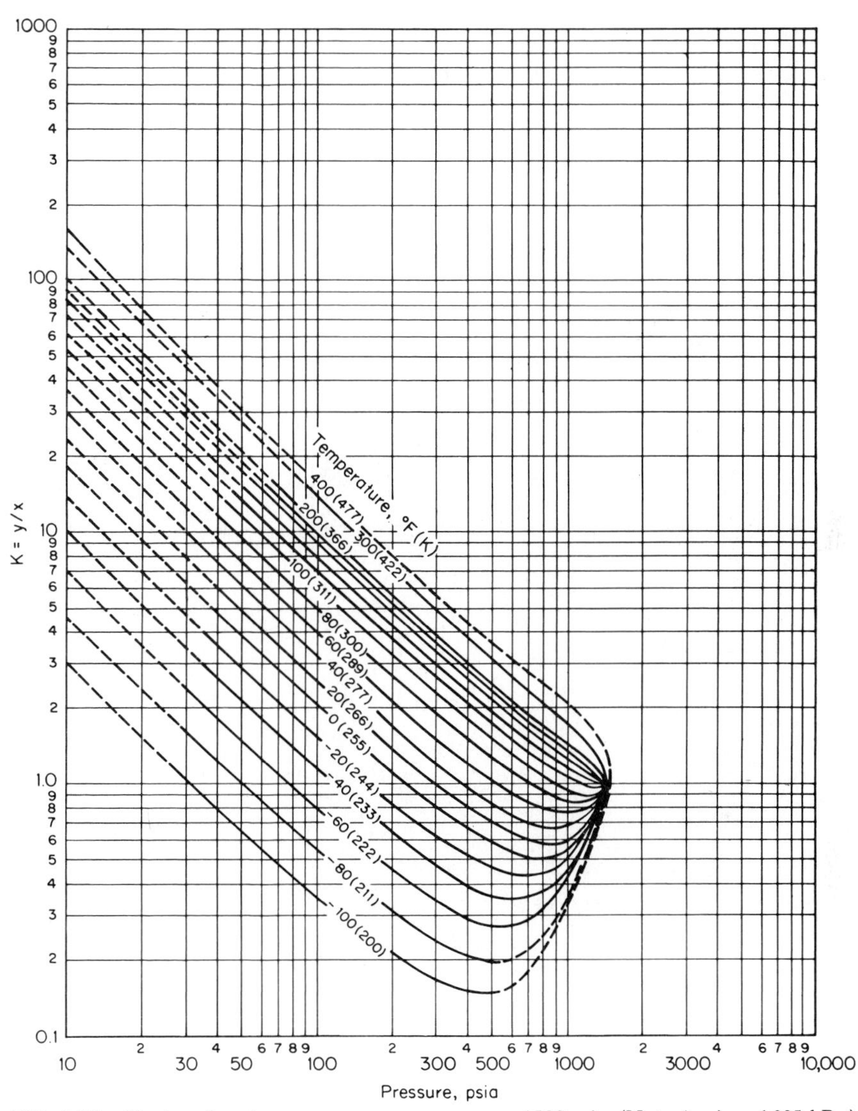

FIG. 1-23 *K* values for ethane; convergence pressure = 1500 psia. (Note: 1 psi = 6.895 kPa.) *(Courtesy of NGPA.)*

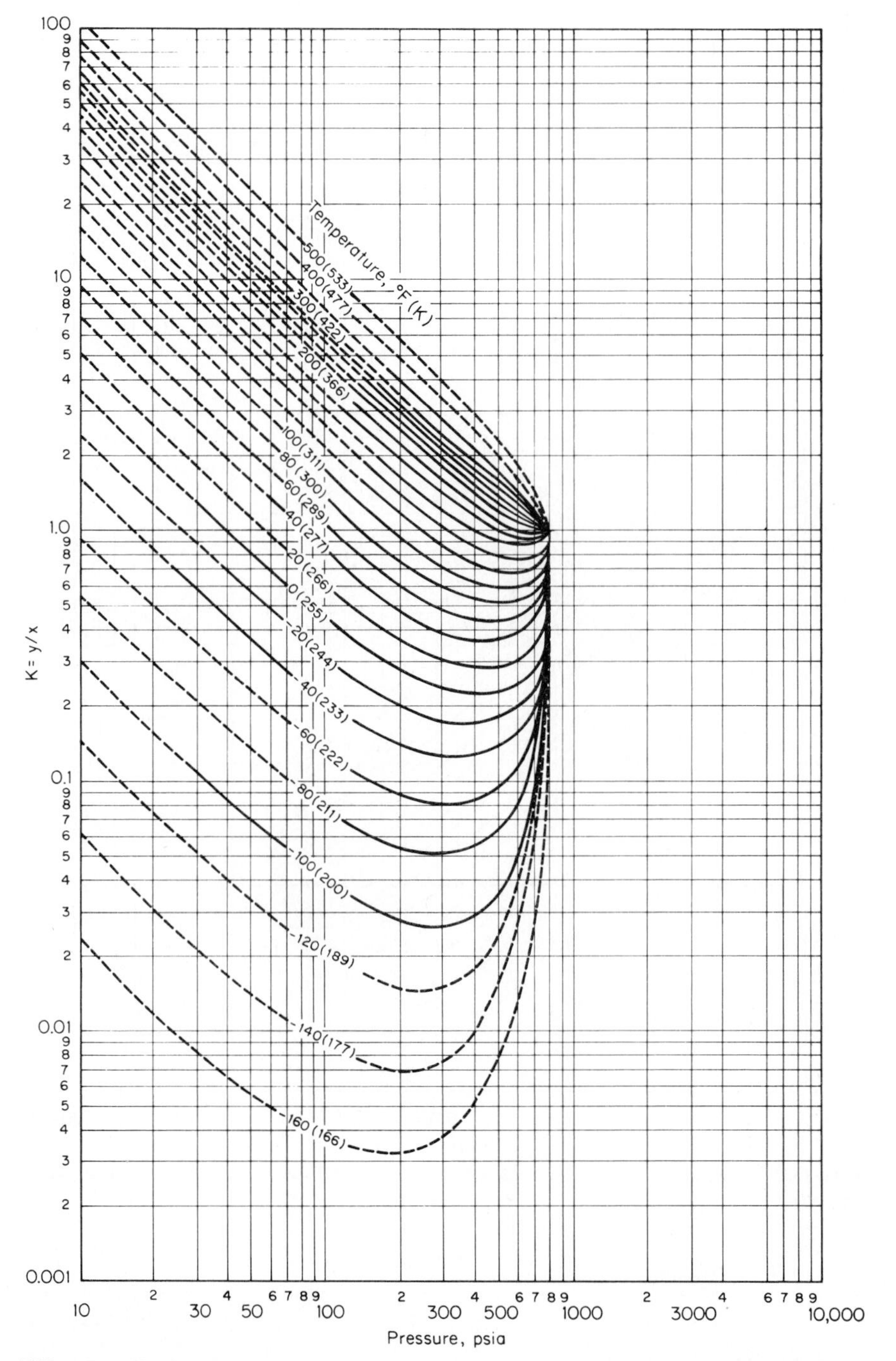

FIG. 1-24 *K* values for propane; convergence pressure = 800 psia. (Note: 1 psi = 6.895 kPa.) *(Courtesy of NGPA.)*

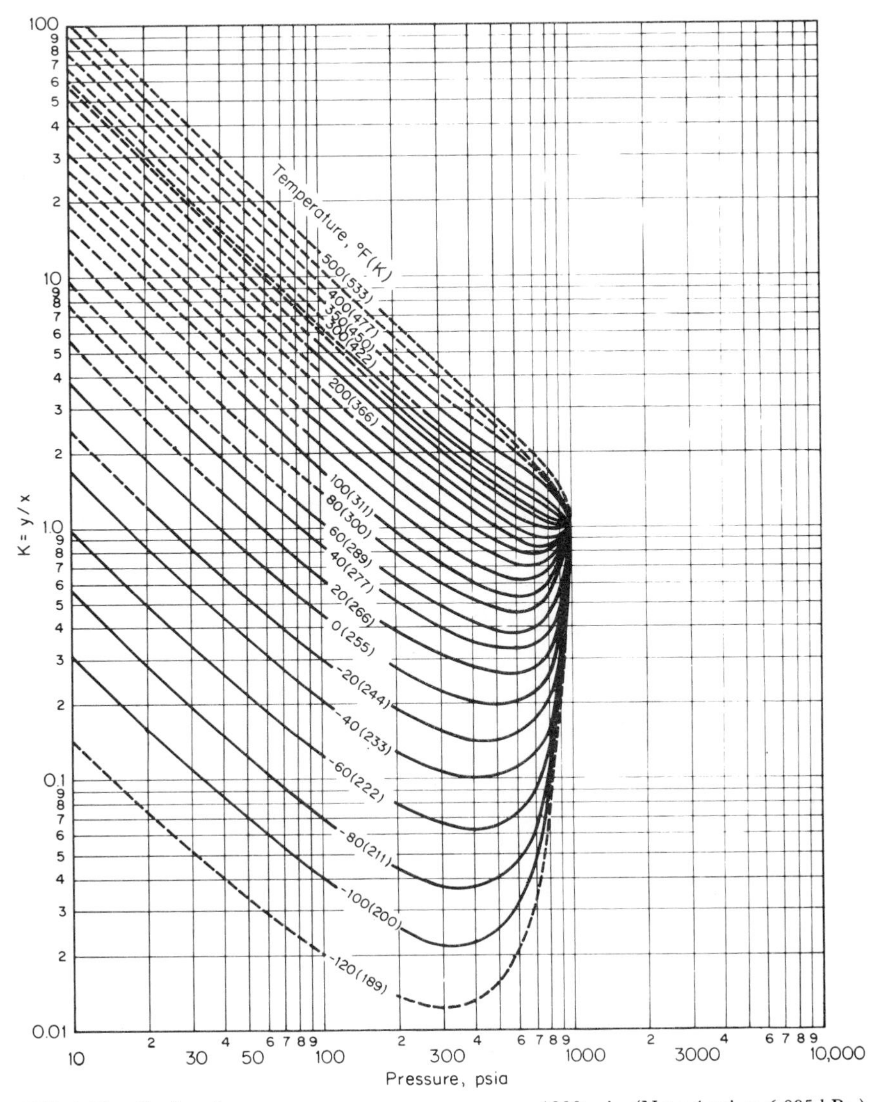

FIG. 1-25 *K* values for propane; convergence pressure = 1000 psia. (Note: 1 psi = 6.895 kPa.) *(Courtesy of NGPA.)*

overhead, as D_i/D, where D_i is the flow rate of component i in the overhead stream and D is the flow rate of the total overhead stream. Then the dew-point equation can be rearranged into $\Sigma(D_i/K_i) = D$. Then the material balance and distribution of components in the vapor leaving the overhead can be summarized as in Table 1-14. (From a graph not shown, K_i for propane at the dew point and 800 psia convergence pressure is

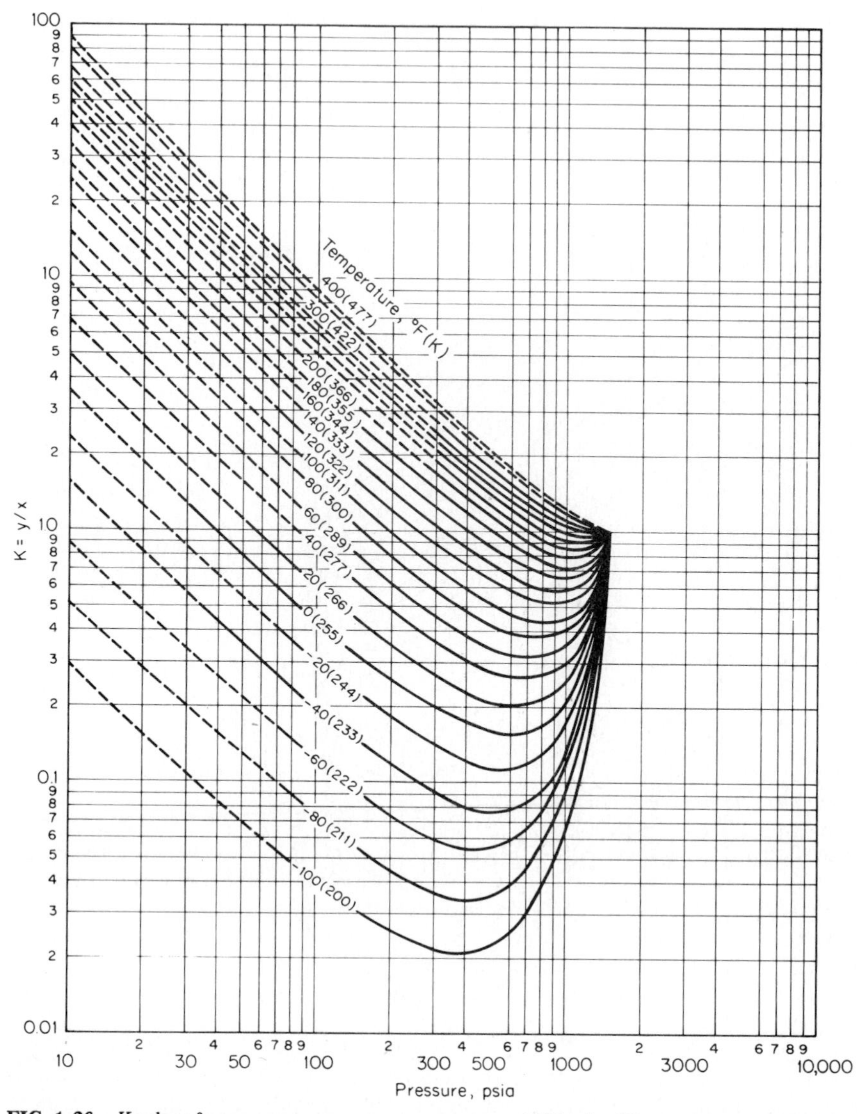

FIG. 1-26 *K* values for propane; convergence pressure = 1500 psia. (Note: 1 psi = 6.895 kPa.) *(Courtesy of NGPA.)*

0.084.) Since $\Sigma D_i = D$, the following equation can be written and solved for D_{pr}, the flow rate of propane:

$$206.6 + 11.9D_{pr} = 564.5 + D_{pr}$$

Thus D_{pr} is found to be 32.8 lb·mol/h (14.9 kg·mol/h). (The amount of propane leaving the column bottom, needed for the next step in the calculation procedure, is 250 − 32.8, or 217.2 lb·mol/h.)

TABLE 1-14 Overhead from Absorption Column in Example 1-22 (convergence pressure = 800 psia)

Component i	Amount in feed to absorption column F_i, lb·mol/h	Fraction absorbed	Quantity absorbed L_i, lb·mol/h	Quantity in overhead D_i, lb·mol/h	K_i at −60°F, 295 psia	D_i/K_i
Nitrogen	48.2	0	0	48.2	15	3.21
Methane	553.3	0.09	49.8	503.5	3	167.8
Ethane	85.1	0.85	72.3	12.8	0.36	35.6
Propane	0	—	—	D_{pr}	0.084	$11.9D_{pr}$
Total	686.6			$(564.5 + D_{pr})$		$(206.6 + 11.9D_{pr})$

4. Compute the convergence pressure at the tower bottom.

The convergence pressure is the pressure at which the vapor-liquid K values of all components in the mixture converge to a value of $K = 1.0$. The concept of convergence pressure is used empirically to account for the effect of composition. Convergence pressure can be determined by the critical locus of the system: For a binary mixture, the convergence pressure is the pressure corresponding to the system temperature read from the binary critical locus. Critical loci of many hydrocarbon binaries are given in Fig. 1-27. This figure forms the basis for determining the convergence pressure for use with the K-value charts (Figs. 1-15 through 1-26).

Strictly speaking, the convergence pressure of a binary mixture equals the critical pressure of the mixture only if the system temperature coincides with the mixture critical temperature. For multicomponent mixtures, furthermore, the convergence pressure depends on both the temperature and the liquid composition of mixture. For convenience, a multicomponent mixture is treated as a pseudobinary mixture in this K-value approach. The pseudobinary mixture consists of a light component, which is the lightest component present in not less than 0.001 mol fraction in the liquid, and a pseudoheavy component that represents the remaining heavy components. The critical temperature and the critical pressure of the pseudoheavy component are defined as $T_{c,\text{heavy}} = \Sigma W_i T_{c,i}$ and $P_{c,\text{heavy}} = \Sigma W_i P_{c,i}$, where W_i is the weight fraction of component i in the liquid phase on a lightest-component-free basis, and $T_{c,i}$ and $P_{c,i}$ are critical temperature and critical pressure of component i, respectively. The pseudocritical constants computed by following this procedure (outlined in Table 1-15) are $T_{c,\text{heavy}} = 186.7°\text{F}$ and $P_{c,\text{heavy}} = 633.1$ psia. Locate points $(T_{c,\text{light}}, P_{c,\text{light}})$ and $(T_{c,\text{heavy}}, P_{c,\text{heavy}})$ of the pseudobinary mixture on Fig. 1-27, and construct the critical locus by interpolating between the adjacent loci as shown by the dotted line. Read off convergence-pressure values corresponding to the system temperature from this critical locus. Thus, at 50 and 75°F (283 and 297 K), the convergence pressures are about 1400 and 1300 psia (9653 and 8964 kPa), respectively.

5. Obtain convergence-pressure K values.

Table 1-16 lists K values of each component from the appropriate charts corresponding to given temperature and pressure at two convergence pressures, namely, 1000 and 1500 psia. Values are interpolated linearly for convergence pressures of 1400 psia (corresponding to 50°F) and 1300 psia (corresponding to 75°F).

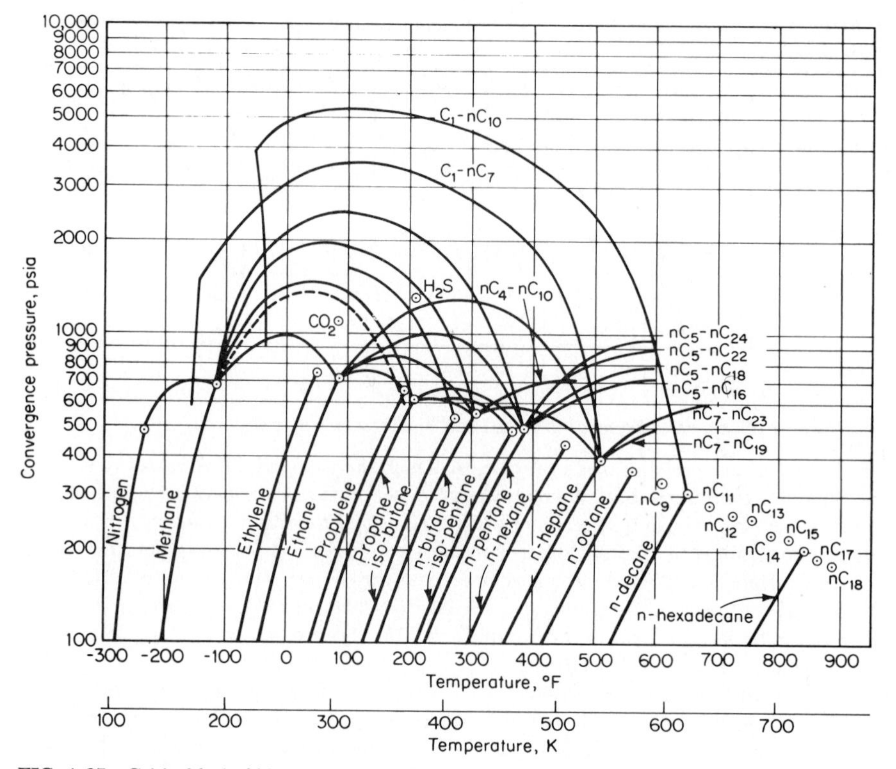

FIG. 1-27 Critical loci of binary mixtures. (Note: 1 psi = 6.895 kPa.) *(Courtesy of NGPA.)*

6. *Estimate bubble-point temperature of tower bottom stream after it leaves the pump and heat exchanger.*

At the bubble point, $\Sigma K_i X_i = 1.0$, where X_i is the mole fraction of component i in the liquid mixture. Designate X_i as B_i/B, where B_i is the flow rate of component i and B is the flow rate of the total bottoms stream. Then the bubble-point equation can be rearranged into $\Sigma K_i B_i = B$. Using the K_i values that correspond to temperatures of 50 and 75°F (from the previous step), we find $\Sigma K_i B_i$ at 50°F to be 331.1 and $\Sigma K_i B_i$ at 75°F to be 367.3. However, at the bubble point, $\Sigma K_i B_i$ should be 339.3 mol/h, i.e., the amount of the total bottom stream. By linear interpolation, therefore, the bubble-point tempera-

TABLE 1-15 Pseudocritical Constants for Tower Bottoms in Example 1-22

Component	Flow rate B_i, mol/h	Molecular weight, lb/(lb·mol)	Flow rate, lb/h	Critical temperature $T_{c,i}$, °F	Critical pressure $P_{c,i}$, psia	Weight fraction W_i	$W_i T_{c,i}$	$W_i P_{c,i}$
Methane	49.8	—	—	—	—	—	—	—
Ethane	72.3	30.07	2174	90.3	709.8	0.169	15.3	120.0
Propane	217.2	49.09	10,662	206.3	617.4	0.831	171.4	513.1
Total	339.3		12,836			1.000	186.7	633.1

TABLE 1-16 Convergence-Pressure K Values for Example 1-22

	K values					
	50°F, 500 psia			75°F, 500 psia		
Component	Conv. press. = 1000 psia	Conv. press. = 1500 psia	Conv. press.* = 1400 psia	Conv. press. = 1000 psia	Conv. press. = 1500 psia	Conv. press.* = 1300 psia
CH_4	2.9	4.1	3.86	3.05	4.4	3.86
C_2H_6	0.95	0.93	0.93	1.18	1.2	1.19
C_3H_8	0.36	0.32	0.33	0.46	0.38	0.41

*Values interpolated linearly.

ture is 56°F (286 K). Since the temperature of the stream leaving the heat exchanger is 50°F (286 K), no vaporization will take place in that exchanger.

Related Calculations: The convergence-pressure K-value charts provide a useful and rapid graphical approach for phase-equilibrium calculations. The Natural Gas Processors Suppliers Association has published a very extensive set of charts showing the vapor-liquid equilibrium K values of each of the components methane to n-decane as functions of pressure, temperature, and convergence pressure. These charts are widely used in the petroleum industry. The procedure shown in this illustration can be used to perform similar calculations. See Examples 3-10 and 3-11 for straightforward calculation of dew points and bubble points, respectively.

1-23 Heat of Formation from Elements

Calculate the values of standard heat of formation ΔH_f° and standard free energy of formation ΔG_f° of 2-methyl propene (isobutene) from the elements at 400 K (260°F):

$$4C(s) + 4H_2(g) \rightarrow C_4H_8(g)$$

The standard heat of formation of a compound relative to its elements, all in their standard state of unit activities, is expressed as

$$\Delta H_{f,T}^\circ = [(H_T^\circ - H_0^\circ) + \Delta H_{f,0}^\circ]_{\text{compound}} - [\Sigma(H_T^\circ - H_0^\circ)]_{\text{elements}}$$

where $\Delta H_{f,T}^\circ$ = standard heat of formation at temperature T
H_T° = enthalpy of compound or element at temperature T
H_0° = enthalpy of compound or element at 0 K
$\Delta H_{f,0}^\circ$ = standard heat of formation at 0 K

The standard free energy of formation of a compound at a temperature T from the elements at the same temperature is expressed as

$$\left(\frac{\Delta G_f^\circ}{T}\right)_T = \left[\left(\frac{G_T^\circ - H_0^\circ}{T}\right) + \frac{\Delta H_{f,0}^\circ}{T}\right]_{\text{compound}} - \left[\sum \frac{G_T^\circ - H_0^\circ}{T}\right]_{\text{elements}}$$

TABLE 1-17 Enthalpy above 0 K

	State	298.16 K	400 K	500 K	600 K	800 K	1000 K	1500 K
		$(H_T^\circ - H_0^\circ)$, kg·cal/(g·mol)						
Methane	*g*	2.397	3.323	4.365	5.549	8.321	11.560	21.130
Ethane	*g*	2.856	4.296	6.010	8.016	12.760	18.280	34.500
Ethene (ethylene)	*g*	2.525	3.711	5.117	6.732	10.480	14.760	27.100
Ethyne (acetylene)	*g*	2.3915	3.5412	4.7910	6.127	8.999	12.090	20.541
Propane	*g*	3.512	5.556	8.040	10.930	17.760	25.670	48.650
Propene (propylene)	*g*	3.237	4.990	7.076	9.492	15.150	21.690	40.570
n-Butane	*g*	4.645	7.340	10.595	14.376	23.264	33.540	63.270
2-Methylpropane (isobutane)	*g*	4.276	6.964	10.250	14.070	23.010	33.310	63.050
1-Butene	*g*	4.112	6.484	9.350	12.650	20.370	29.250	54.840
cis-2-Butene	*g*	3.981	6.144	8.839	12.010	19.510	28.230	53.620
trans-2-Butene	*g*	4.190	6.582	9.422	12.690	20.350	29.190	54.710
2-Methylpropene (isobutene)	*g*	4.082	6.522	9.414	12.750	20.490	29.370	55.000
n-Pentane	*g*	5.629	8.952	12.970	17.628	28.568	41.190	77.625
2-Methylbutane (isopentane)	*g*	5.295	8.596	12.620	17.300	28.300	41.010	77.740
2,2-Dimethylpropane (neopentane)	*g*	5.030	8.428	12.570	17.390	28.640	41.510	78.420
n-Hexane	*g*	6.622	10.580	15.360	20.892	33.880	48.850	92.010
2-Methylpentane	*g*	6.097	10.080	14.950	20.520	33.600	48.700	
3-Methylpentane	*g*	6.622	10.580	15.360	20.880	33.840	48.800	
2,2-Dimethylbutane	*g*	5.912	9.880	14.750	20.340	33.520	48.600	
2,3-Dimethylbutane	*g*	5.916	9.833	14.610	20.170	33.230	48.240	
Graphite	*s*	0.25156	0.5028	0.8210	1.1982	2.0816	3.0750	5.814
Hydrogen, H_2	*g*	2.0238	2.7310	3.4295	4.1295	5.5374	6.9658	10.6942
Water, H_2O	*g*	2.3677	3.1940	4.0255	4.8822	6.6896	8.6080	13.848
CO	*g*	2.0726	2.7836	3.4900	4.2096	5.7000	7.2570	11.3580
CO_2	*g*	2.2381	3.1948	4.2230	5.3226	7.6896	10.2220	17.004
O_2	*g*	2.0698	2.7924	3.5240	4.2792	5.8560	7.4970	11.7765
N_2	*g*	2.07227	2.7824	3.4850	4.1980	5.6686	7.2025	11.2536
NO	*g*	2.1942	2.9208	3.6440	4.3812	5.9096	7.5060	11.6940

Note: H_T° = molal enthalpy of the substance in its standard state, at temperature T.
H_0° = molal enthalpy of the substance in its standard state, at 0 K.
Source: A. Hougen, K. M. Watson, and R. A. Ragatz, *Chemical Process Principles,* part III, Wiley, New York, 1959.

where ΔG_f° = standard free energy of formation at temperature T
G_T° = standard free energy of a compound or element at temperature T

Free-energy functions and enthalpy functions are given in Tables 1-17 and 1-18.

Calculation Procedure:

1. *Tabulate free-energy and enthalpy functions.*

From Tables 1-17 and 1-18:

Energy function	$C_4H_8(g)$	$C(s)$	$H_2(g)$
$(G_T^\circ - H_0^\circ)/T$, cal/(g·mol)(K)	−60.90	−0.824	−26.422
$\Delta H_{f,0}^\circ$, cal/(g·mol)	980	0	0
$(H_T^\circ - H_0^\circ)_{400\ K}$, cal/(g·mol)	6522	502.8	2731

TABLE 1-18 Free-Energy Function and Standard Heat of Formation at 0 K

	State	$-(G^\circ_T - H^\circ_0)/T$							
		298.16 K	400 K	500 K	600 K	800 K	1000 K	1500 K	$(\Delta H^\circ_f)_0$
Methane	*g*	36.46	38.86	40.75	42.39	45.21	47.65	52.84	−15.987
Ethane	*g*	45.27	48.24	50.77	53.08	57.29	61.11	69.46	−16.517
Ethene (ethylene)	*g*	43.98	46.61	48.74	50.70	54.19	57.29	63.94	+14.522
Ethyne (acetylene)	*g*	39.976	42.451	44.508	46.313	49.400	52.005	57.231	+54.329
Propane	*g*	52.73	56.48	59.81	62.93	68.74	74.10	85.86	−19.482
Propene (propylene)	*g*	52.95	56.39	59.32	62.05	67.04	71.57	81.43	+8.468
n-Butane	*g*	58.54	63.51	67.91	72.01	70.63	86.60	101.95	−23.67
2-Methylpropane (isobutane)	*g*	56.08	60.72	64.95	68.95	76.45	83.38	98.64	−25.30
1-Butene	*g*	59.25	63.64	67.52	71.14	77.82	83.93	97.27	+4.96
cis-2-Butene	*g*	58.67	62.89	66.51	69.94	76.30	82.17	95.12	+3.48
trans-2-Butene	*g*	56.80	61.31	65.19	68.84	75.53	81.62	94.91	+2.24
2-Methylpropene (isobutene)	*g*	56.47	60.90	64.77	68.42	75.15	81.29	94.66	+0.98
n-Pentane	*g*	64.52	70.57	75.94	80.96	90.31	98.87	117.72	−27.23
2-Methylbutane (isopentane)	*g*	64.36	70.67	75.28	80.21	89.44	97.96	116.78	−28.81
2,2-Dimethylpropane (neopentane)	*g*	56.36	61.93	67.04	71.96	81.27	89.90	108.91	−31.30
n-Hexane	*g*	70.62	77.75	84.11	90.06	101.14	111.31	133.64	−30.91
2-Methylpentane	*g*	70.50	77.2	83.3	89.1	100.1	110.3	132.5	−32.08
3-Methylpentane	*g*	68.56	75.69	82.05	88.0	99.08	109.3	131.6	−31.97
2,2-Dimethylbutane	*g*	65.79	72.3	78.3	84.1	95.0	105.1	127.4	−34.65
2,3-Dimethylbutane	*g*	67.58	74.06	80.05	85.77	96.54	106.57	128.70	−32.73
Graphite	*s*	0.5172	0.824	1.146	1.477	2.138	2.771	4.181	0
Hydrogen, H_2	*g*	24.423	26.422	27.950	29.203	31.186	32.738	35.590	0
H_2O	*g*	37.165	39.505	41.293	42.766	45.128	47.010	50.598	−57.107
CO	*g*	40.350	42.393	43.947	45.222	47.254	48.860	51.864	−27.2019
CO_2	*g*	43.555	45.828	47.667	49.238	51.895	54.109	58.481	−93.9686
O_2	*g*	42.061	44.112	45.675	46.968	49.044	50.697	53.808	0
N_2	*g*	38.817	40.861	42.415	43.688	45.711	47.306	50.284	0
NO	*g*	42.980	45.134	46.760	48.090	50.202	51.864	54.964	+21.477

Note: G°_T = molal free energy of the substance in its standard state, at temperature T, g·cal/(g·mol).
H°_0 = molal enthalpy of the substance in its standard state, at 0 K, g·cal/(g·mol).
$(\Delta H^\circ_f)_0$ = standard modal heat of formation at 0 K, kg·cal/(g·mol).
Source: O. A. Hougen, K. M. Watson, and R. A. Ragatz, *Chemical Process Principles,* part III, Wiley, New York, 1959.

2. *Calculate standard heat of formation.*

Thus, $\Delta H^\circ_{f,400\text{ K}}$ = (6522 + 980) − [(4)(502.8) + (4)(2731)] = −5433.2 cal/(g·mol)[−9779.8 Btu/(lb·mol)].

3. *Calculate standard free energy of formation.*

Thus, $\Delta G^\circ_f/T$ = (−60.90 + 980/400) − [(4)(−0.824) + (4)(−26.422)] = 50.534 cal/(g·mol)(K). So, ΔG°_f = (50.534)(400) = 20,213.6 cal/(g·mol) [36,384 Btu/(lb·mol)].

Related Calculations: Use this procedure to calculate standard heats and free energies of formation of any compound relative to its elements. The functions $(H^\circ_T - H^\circ_0)$, $(G^\circ_T - H^\circ_0)/T$, and $\Delta H^\circ_{f,0}/T$ not listed in Tables 1-17 and 1-18 may be found in other sources, such as Stull and Prophet [15].

1-24 Standard Heat of Reaction, Standard Free-Energy Change, and Equilibrium Constant

Calculate the standard heat of reaction ΔH°_T, the standard free-energy change ΔG°_T, and the reaction equilibrium constant K_T for the water-gas shift reaction at 1000 K (1340°F):

$$CO(g) + H_2O(g) = CO_2(g) + H_2(g)$$

The standard heat of reaction is expressed as

$$\Delta H^\circ_T = \Sigma[(H^\circ_T - H^\circ_0) + \Delta H^\circ_{f,0}]_{products} - \Sigma[(H^\circ_T - H^\circ_0) + \Delta H^\circ_{f,0}]_{reactants}$$

The standard free-energy change is expressed as

$$\Delta G^\circ/T = \Sigma[(G^\circ_T - H^\circ_0)/T + \Delta H^\circ_{f,0}/T]_{products} - \Sigma[(G^\circ_T - H^\circ_0)/T + \Delta H^\circ_{f,0}/T]_{reactants}$$

The equilibrium constant is expressed as $K = e^{-\Delta G^\circ/RT}$, where the terms are as defined in Example 1-23.

Calculation Procedure:

1. Tabulate free-energy and enthalpy functions.

From Tables 1-17 and 1-18:

Energy function	$CO_2(g)$	$H_2(g)$	$CO(g)$	$H_2O(g)$
$(G^\circ_T - H^\circ_0)/T$, cal/(g·mol)(K)	−54.109	−32.738	−48.860	−47.010
$\Delta H^\circ_{f,0}$, cal/(g·mol)	−93,968.6	0	−27,201.9	−57,107
$(H^\circ_T - H^\circ_0)$, cal/(g·mol)	10,222	6965.8	7257	8608

2. Calculate standard heat of reaction.

Thus, $\Delta H^\circ_{1000\ K}$ = [(10,222.0) + (−93,968.6) + (6,965.8) + (0)] − [(7,257.0) + (−27,201.9) + (8,608) + (−57,107)] = −8336.9 cal/(g·mol) [−15,006.4 Btu/(lb·mol)].

3. Calculate standard free-energy change.

Thus, $\Delta G^\circ/1000$ = [−54.109 + (−93,968.6/1000) + (−32.738) + 0/1000] − [−48.860 + (−27201.9/1000 + (−47.010) + (−57,107.0/1000)] = −0.638 cal/(g·mol)(K). Therefore, $\Delta G^\circ_{1000\ K}$ = −638.0 cal/(g·mol) [−1148.4 Btu/(lb·mol)].

4. Calculate reaction equilibrium constant.

Thus, $K_{1000\ K} = e^{638/(1.987)(1000)} = 1.379.$

Related Calculations: Use this procedure to calculate heats of reaction and standard free-energy changes for reactions that involve components listed in Tables 1-17 and 1-18.

Heat of reaction, free-energy changes, and reaction equilibrium constants are discussed in more detail in Sec. 4 in the context of chemical-reaction equilibrium.

1-25 Standard Heat of Reaction from Heat of Formation—Aqueous Solutions

Calculate the standard heat of reaction $\Delta H°$ for the following acid-base-neutralization reaction at standard conditions [25°C, 1 atm (101.3 kPa)]:

$$2NaOH(aq) + H_2SO_4(l) = Na_2SO_4(aq) + 2H_2O(l)$$

Calculation Procedure:

1. Calculate the heat of reaction.

The symbol (*aq*) implies that the sodium hydroxide and sodium sulfate are in infinitely dilute solution. Therefore, the heat of solution must be included in the calculations. Data on both heat of formation and heat of solution at the standard conditions (25°C and 1 atm) are available in Table 1-19. Since the answer sought is also to be at standard conditions, there is no need to adjust for differences in temperature (or pressure), and the equation to be used is simply

$$\Delta H° = \Sigma(\Delta H_F^\circ)_{\text{products}} + \Sigma(\Delta H_s^\circ)_{\text{dissolved products}} - \Sigma(\Delta H_F^\circ)_{\text{reactants}} - \Sigma(\Delta H_s^\circ)_{\text{dissolved reactants}}$$

where ΔH_F° is standard heat of formation, and ΔH_s° is standard integral heat of solution at infinite dilution.

Thus, from Table 1-19, and taking into account that there are 2 mol each of water and sodium hydroxide,

$$\begin{aligned}\Delta H° &= [-330{,}900 + 2(-68{,}317)] + (-560) \\ &\quad - [2(-101{,}990) + (-193{,}910)] - 2(-10{,}246) \\ &= -49{,}712 \text{ cal/(g·mol) } [-89{,}482 \text{ Btu/(lb·mol)}]\end{aligned}$$

The reaction is thus exothermic. To maintain the products at 25°C, it will be necessary to remove 49,712 cal of heat per gram-mole of sodium sulfate produced.

Related Calculations: Use this general procedure to calculate heats of reaction for other aqueous-phase reactions. Calculation of heat of reaction from standard heat of reaction is covered in Sec. 4 in the context of chemical-reaction equilibrium; see in particular Example 4-1.

TABLE 1-19 Standard Heats of Formation and Standard Integral Heats of Solution at Infinite Dilution (25°C, 1 atm)

Compound	Formula	State	ΔH_f°, cal/(g·mol)	ΔH_s°, cal/(g·mol)
Ammonia	NH_3	*g*	−11,040	−8,280
		l	−16,060	−3,260
Ammonium nitrate	NH_4NO_3	*s*	−87,270	6,160
Ammonium sulfate	$(NH_4)_2SO_4$	*s*	−281,860	1,480
Calcium carbide	CaC_2	*s*	−15,000	
Calcium carbonate	$CaCO_3$	*s*	−288,450	
Calcium chloride	$CaCl_2$	*s*	−190,000	−19,820
Calcium hydroxide	$Ca(OH)_2$	*s*	−235,800	−3,880
Calcium oxide	CaO	*s*	−151,900	−19,400
Carbon (graphite)	C	*s*	0	
Amorphous, in coke	C	*s*	2,600	
Carbon dioxide	CO_2	*g*	−94,051.8	−4,640
Carbon disulfide	CS_2	*g*	27,550	
		l	21,000	
Carbon monoxide	CO	*g*	−26,416	
Carbon tetrachloride	CCl_4	*g*	−25,500	
		l	−33,340	
Copper sulfate	$CuSO_4$	*s*	−181,000	−17,510
Hydrochloric acid	HCl	*g*	−22,063	−17,960
Hydrogen sulfide	H_2S	*g*	−4,815	−4,580
Iron oxide	Fe_3O_4	*s*	−267,000	
Iron sulfate	$FeSO_4$	*s*	−220,500	−15,500
Nitric acid	HNO_3	*l*	−41,404	−7,968
Potassium chloride	KCl	*s*	−104,175	4,115
Potassium hydroxide	KOH	*s*	−101,780	−13,220
Potassium nitrate	KNO_3	*s*	−117,760	8,350
Potassium sulfate	K_2SO_4	*s*	−342,660	5,680
Sodium carbonate	Na_2CO_3	*s*	−270,300	−5,600
	$Na_2CO_3 \cdot 10H_2O$	*s*	−975,600	16,500
Sodium chloride	NaCl	*s*	−98,232	930
Sodium hydroxide	NaOH	*s*	−101,990	−10,246
Sodium nitrate	$NaNO_3$	*s*	−101,540	−5,111
Sodium sulfate	Na_2SO_4	*s*	−330,900	−560
	$Na_2SO_4 \cdot 10H_2O$	*s*	−1,033,480	18,850
Sulfur dioxide	SO_3	*g*	−70,960	9,900
Sulfur trioxide	SO_3	*g*	−94,450	−54,130
Sulfuric acid	H_2SO_4	*l*	−193,910	−22,990
Water	H_2O	*g*	−57,798	
		l	−68,317	
Zinc sulfate	$ZnSO_4$	*s*	−233,880	−19,450

Source: F. D. Rossini et al., *Selected Values of Chemical Thermodynamic Properties,* National Bureau of Standards, Circular 500, 1952.

1-26 Standard Heat of Reaction from Heat of Combustion

Calculate the standard heat of reaction ΔH° of the following reaction using heat-of-combustion data:

$$CH_3OH(l) + CH_3COOH(l) = CH_3OOCCH_3(l) + H_2O(l)$$

(Methanol) (Acetic acid) (Methyl acetate) (Water)

Calculation Procedure:

1. Obtain heats of combustion.

Data on heats of combustion for both organic and inorganic compounds are given in numerous reference works. Thus:

Compound	State	$\Delta H_{combustion}$, kcal/(g·mol)
CH_3OH	Liquid	−173.65
CH_3COOH	Liquid	−208.34
CH_3OOCCH_3	Liquid	−538.76
H_2O	Liquid	0

2. Calculate $\Delta H°$.

The heat of reaction is the difference between the heats of combustion of the reactants and of the products:

$$\Sigma\Delta H^\circ_{combustion,\ reactants} - \Sigma\Delta H^\circ_{combustion,\ products}$$

So, $\Delta H° = (-173.65) + (-208.34) - (-538.76) = 156.77$ kcal/(g·mol).

Related Calculations: For a reaction between organic compounds, the basic thermochemical data are generally available in the form of standard heats of combustion. Use the preceding procedure for calculating the standard heats of reaction when organic compounds are involved, using the standard heats of combustion directly instead of standard heats of formation. Heat-of-formation data must be used when organic and inorganic compounds both appear in the reaction.

Calculation of heat of reaction from standard heat of reaction is covered in Sec. 4 in the context of chemical-reaction equilibrium; see in particular Example 4-1.

1-27 Standard Heat of Formation from Heat of Combustion

Calculate the standard heats of formation of benzene(l), methanol(l), aniline(l), methyl chloride(g), and ethyl mercaptan(l) using heat-of-combustion data, knowledge of the combustion products, and the equations in Table 1-20.

Calculation Procedure:

1. Obtain data on the heats of combustion, note the corresponding final combustion products, and select the appropriate equation in Table 1-20.

From standard reference sources, the standard heats of combustion and the final products are as follows. The appropriate equation, A or B, is selected based on what the final combustion products are, i.e., where they are within col. 1 of Table 1-20.

Compound (and number of constituent atoms)	Combustion products	Heat of combustion, kcal/(g·mol)	Equation to be used
Benzene, $C_6H_6(l)$	$CO_2(g)$, $H_2O(l)$	780.98	A or B
Methanol, $CH_4O(l)$	$CO_2(g)$, $H_2O(l)$	173.65	A or B
Aniline, $C_6H_7N(l)$	$CO_2(g)$, $H_2O(l)$, $N_2(g)$	812.0	B
Methyl chloride, $CH_3Cl(g)$	$CO_2(g)$, $H_2O(l)$, $HCl(aq)$	182.81	A
Ethyl mercaptan, $C_2H_6S(l)$	$CO_2(g)$, $H_2O(l)$, $SO_2(g)$	448.0	B

2. ***Calculate the heats of formation and compare with the literature values.***

This step can be set out in matrix form, as follows:

		Heat of formation, cal/(g·mol)	
Compound	Equation	Calculated value	Literature value
Benzene(l)	A	$= -(-780.98)(10^3) - (94{,}051.8)(6) - (34{,}158.7)(6) = 11{,}717$	11,630
Methanol(l)	A	$= -(-173.65)(10^3) - (94{,}051.8)(1) - (34{,}158.7)(4) = -57{,}036$	−57,040
Aniline(l)	B	$= -(-812.0)(10^3) - (94{,}051.8)(6) - (34{,}158.7)(7) = 8{,}578$	8,440
Methyl chloride(l)	A	$= -(-182.81)(10^3) - (94{,}051.8)(1) - (34{,}158.7)(3) - (5{,}864.3)(1) = -19{,}582$	−19,580
Ethyl mercaptan(l)	B	$= -(-448.0)(10^3) - (94{,}051.8)(2) - (34{,}158.7)(6) - (70{,}960)(1) = -16{,}015.8$	−16,000

Related Calculations: This approach may be used to find the heat of formation of a compound all of whose constituent atoms are among the following: carbon, hydrogen, bromine, chlorine, fluorine, iodine, nitrogen, oxygen, and sulfur.

1-28 Heat of Absorption from Solubility Data

Estimate the heat of absorption of carbon dioxide in water at 15°C (59°F, or 288 K) from these solubility data:

Temperature, °C	0	5	10	15	20
Henry constant, atm/mol-fraction	728	876	1040	1220	1420

The formula is

$$\Delta H_{abs} = (\bar{h}_i - H_i) = R[d \ln H/d(1/T)]$$

where ΔH_{abs} is heat of absorption, $\bar{h}_i$ is partial molar enthalpy of component i at infinite dilution in the liquid at a given temperature and pressure, H_i is molar enthalpy of pure gas i at the given temperature and pressure, H is the Henry constant, the partial pressure of the gas divided by its solubility, R is the gas constant, and T is absolute temperature.

TABLE 1-20 Heat of Formation from Heat of Combustion

Final products of the combustion (1)	Equation for heat of formation H_f°, cal/(g·mol) (2)
$CO_2(g)$, $H_2O(l)$, $Br_2(g)$, $HCl(aq)$, $I(s)$, $HNO_3(aq)$, $H_2SO_4(aq)$	$H_f^\circ = -\Delta H_c - 94{,}051.8a - 34{,}158.7b + 3670c - 5864.3d - 44{,}501e - 15{,}213.3g - 148{,}582.6i$ (A)
$CO_2(g)$, $H_2O(l)$, $Br_2(g)$, $HF(aq)$, $I(s)$, $N_2(g)$, $SO_2(g)$	$H_f^\circ = -\Delta H_c - 94{,}051.8a - 34{,}158.7b + 3670c - - 44{,}501e - 70{,}960i$ (B)

Note: In the above equations, ΔH_c = heat of combustion corresponding to the final products in col. 1, a = atoms of carbon in the compound, b = atoms of hydrogen, c = atoms of bromine, d = atoms of chlorine, e = atoms of fluorine, g = atoms of nitrogen, and i = atoms of sulfur.

Calculation Procedure:

1. Determine d ln H/d(1/T).

This is the slope of the plot of ln H against $1/T$. A logarithmic plot based on the data given in the statement of the problem is shown in Fig. 1-28. The required slope is found to be -2.672×10^3 K.

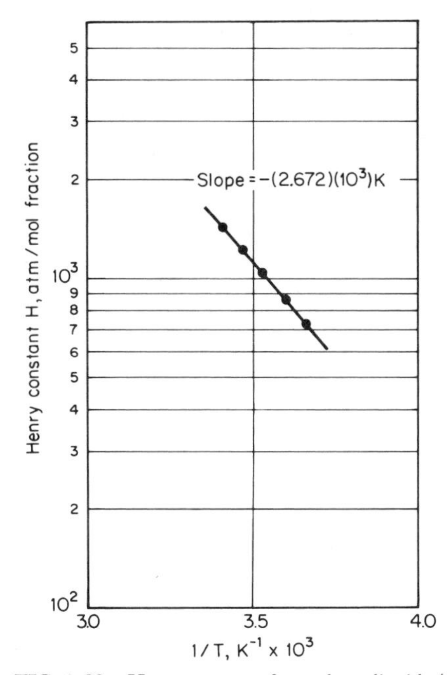

FIG. 1-28 Henry constant for carbon-dioxide/water system at 15°C (Example 1-28). Note: 1 atm = 101.3 kPa.

2. Calculate the heat of absorption.

Substitute directly into the formula. Thus,

$$\Delta H_{abs} = R[d \ln H/d(1/T)] = 1.987(-2.672 \times 10^3)$$
$$= -5.31 \times 10^3 \text{ cal/(g·mol) [9.56 Btu/(lb·mol)]}$$

Related Calculations: This procedure may be used for calculating the heats of solution from low-pressure solubility data where Henry's law in its simple form is obeyed.

1-29 Estimation of Liquid Viscosity at High Temperatures

Use the Letsou-Stiel high-temperature generalized correlation to estimate the viscosity of liquid benzene at 227°C (500 K, 440°F).

The correlation is expressed as

$$\mu_L\psi = (\mu_L\psi)^{(0)} + \omega(\mu_L\psi)^{(1)}$$

where $(\mu_L\psi)^{(0)} = 0.015174 - 0.02135T_R + 0.0075T_R^2$
$(\mu_L\psi)^{(1)} = 0.042552 - 0.07674T_R + 0.0340T_R^2$
μ_L = liquid viscosity at the reduced temperature, in centipoise (cP)

$$\psi = \frac{T_c^{1/6}}{(M^{1/2}P_c^{2/3})}$$

ω = Pitzer accentric factor
T_c = critical temperature, K
P_c = critical pressure, atm
M = molecular weight

Use these values for benzene: $T_c = 562.6$ K, $P_c = 48.6$ atm, $M = 78.1$, and $\omega = 0.212$.

Calculation Procedure:

1. Calculate the correlating parameters for the Letsou-Stiel method.

By use of the equations outlined above,

$$\psi = T_c^{1/6}/(M^{1/2}P_c^{2/3}) = (562.6)^{1/6}/(78.1)^{1/2}(48.6)^{2/3} = 0.0244$$
$$T_R = T/T_c = (227 + 273)/562.6 = 0.889$$
$$(\mu_L\psi)^{(0)} = 0.015174 - (0.02135)(0.889) + (0.0075)(0.889)^2 = 0.0021$$
$$(\mu_L\psi)^{(1)} = 0.042552 - (0.07674)(0.889) + (0.0340)(0.889)^2 = 0.0012$$

2. Calculate the viscosity of liquid benzene.

Thus, upon dividing through by ψ, we have the equation

$$\mu_L = (1/\psi)[(\mu_L\psi)^{(0)} + \omega(\mu_L\psi)^{(1)}]$$
$$= (1/0.024)[(0.0021 + (0.212)(0.0012)] = 0.0981 \text{ cP}$$

Related Calculations: The Letsou-Stiel correlation is a fairly accurate method for estimating viscosities of liquids at relatively high temperatures, $T_R = 0.75$ or higher. It has been tested on a large number of compounds and is reported to fit most materials up to a reduced temperature of about 0.9, with an average error of ±3 percent.

1-30 Viscosity of Nonpolar and Polar Gases at High Pressure

Use the Jossi-Stiel-Thodos generalized correlation to estimate the vapor viscosity of (1) methane (a nonpolar gas) at 500 psig (35 atm abs) and 250°F (394 K), and (2) ammonia (a polar gas) at 340°F (444.4 K) and 1980 psig (135.8 atm abs). Experimentally determined viscosities for those two gases at low pressure and the same temperatures are 140 μP for methane and 158 μP for ammonia.

The Jossi-Stiel-Thodos correlation is summarized in these equations:

For nonpolar gases,

$$[(\mu - \mu^0)\psi + 1]^{0.25} = 1.0230 + 0.2336\rho_R + 0.58533\rho_R^2 - 0.040758\rho_R^3 + 0.093324\rho_R^4$$

For polar gases,

$$(\mu - \mu^0)\psi = 1.656\rho_R^{1.111} \qquad \text{if } \rho_R \le 0.1$$

$$(\mu - \mu^0)\psi = 0.0607(9.045\rho_R + 0.63)^{1.739} \qquad \text{if } 0.1 \le \rho_R \le 0.9$$

$$\log\{4 - \log[(\mu - \mu^0)\psi]\} = 0.6439 - 0.1005\rho_R - D \qquad \text{if } 0.9 \le \rho_R < 2.6$$

where $D = 0$ if $0.9 \le \rho_R \le 2.2$, or
$D = (0.000475)(\rho_R^3 - 10.65)^2 \qquad \text{if } 2.2 < \rho_R \le 2.6$
μ = viscosity of high-pressure (i.e., dense) gas, μP
μ^0 = viscosity of gas at low pressure, μP
ρ_R = reduced gas density, ρ/ρ_c (which equals V_c/V)
ρ = density
ρ_c = critical density
V = specific volume
V_c = critical specific volume
$\psi = T_c^{1/6}/(M^{1/2}P_c^{2/3})$
T_c = critical temperature, K
P_c = critical pressure, atm
M = molecular weight

Use these values for methane: $T_c = 190.6$ K; $P_c = 45.4$ atm; $V_c = 99.0$ cm^3/(g · mol); $Z_c = 0.288$; $M = 16.04$. And use these values for ammonia: $T_c = 405.6$ K; $P_c = 111.3$ atm; $V_c = 72.5$ cm^3/(g · mol); $Z_c = 0.242$; $M = 17.03$.

Calculation Procedure:

1. Calculate the nonideal compressibility factor for the methane and the ammonia.

Use the generalized correlations shown in Figs. 1-1 through 1-3 in this chapter. For methane, $T_R = 394/190.6 = 2.07$, and $P_R = 35/45.4 = 0.771$. From the statement of the problem, $Z_c = 0.288$. Thus, by interpolation from Fig. 1-2 (for $Z_c = 0.27$) and Fig. 1-3 (for $Z_c = 0.29$), Z is found to be 0.98.

Similarly for ammonia, $T_R = 444.4/405.6 = 1.10$, $P_R = 135.8/111.3 = 1.22$, and $Z_c = 0.242$. By extrapolation from Figs. 1-2 and 1-3, $Z = 0.65$.

2. *Calculate the reduced density for the high-pressure gas.*

For methane, $\rho_R = \rho/\rho_c = \rho V_c = PV_c/ZRT = (35)(99.0)/(0.98)(82.07)(394) = 0.109$.
Similarly for ammonia, $\rho_R = PV_c/ZRT = (135.8)(72.5)/(0.65)(82.07)(444.4) = 0.415$.

3. *Calculate the parameter ψ.*

For methane, $\psi = T_c^{1/6}/(M^{1/2}P_c^{2/3}) = (190.6)^{1/6}/(16.04)^{1/2}(45.4)^{2/3} = 0.0471$
And for ammonia, $\psi = (405.6)^{1/6}/(17.03)^{1/2}(111.3)^{2/3} = 0.0285$.

4. *Calculate the viscosity for the high-pressure methane, using the nonpolar equation.*

Thus, $[(\mu - \mu^0)\psi + 1]^{0.25} = 1.023 + (0.2336)(0.109) + (0.58533)(0.109)^2 - (0.040758)(0.109)^3 + (0.093324)(0.109)^4 = 1.0550$. And accordingly,

$$\mu - \mu^0 + [(1.0550)^{1/0.25} - 1]/\psi = 140 + [(1.0550)^{1/0.25} - 1]/0.0471 = 145.07\ \mu\text{P}$$

This correlation thus indicates that the pressure effect raises the viscosity of the methane by about 4 percent.

5. *Calculate the viscosity for the high-pressure ammonia, using the appropriate polar equation.*

Because the reduced density (0.415) lies between 0.1 and 0.9, the appropriate equation is

$$(\mu - \mu^0)\psi = 0.0607(9.045\rho_R + 0.63)^{1.739}$$

Thus, $(\mu - \mu^0)\psi = 0.0607[(9.045)(0.415) + 0.63]^{1.739} = 0.7930$, so $\mu = \mu_0 + 0.7930/\psi = 158 + 0.7930/0.0285 = 185.8\ \mu\text{P}$. This correlation thus predicts that the pressure effect raises the viscosity of the ammonia by about 18 percent.

1-31 Thermal Conductivity of Gases

Use the modified Euken correlation to estimate the thermal conductivity of nitric oxide (NO) vapor at 300°C (573 K, 572°F) and 1 atm (101.3 kPa). At that temperature, the viscosity of nitric oxide is 32.7×10^{-5} P.

The modified Euken correlation is

$$\lambda M/\mu = 3.52 + 1.32C_v^\circ = 3.52 + 1.32C_p^\circ/\gamma$$

where M = molecular weight
μ = viscosity, P
C_p°, C_v° = ideal-gas heat capacity at constant pressure and constant volume, respectively, cal/(g · mol)K
γ = heat capacity ratio, equal to C_p°/C_v°
λ = thermal conductivity, cal/(cm)(s)(K)

Use these values and relationships for nitric oxide: $M = 30.01$, $C_p^\circ = 6.461 + 2.358 \times 10^{-3}\,T$ (where T is in kelvins), and $C_v^\circ = C_p^\circ - R = C_p^\circ - 1.987$.

Calculation Procedure:

1. Calculate C_v°.

Thus, $C_v^\circ = C_p^\circ - R = 6.461 + (2.358 \times 10^{-3})(300 + 273) - 1.987 = 5.826$ cal/(g · mol)(K).

2. Calculate the thermal conductivity.

Thus, upon rearranging the Eucken equation,

$$\begin{aligned}\lambda &= (\mu/M)(3.52 + 1.32C_v^\circ)\\ &= [(32.7 \times 10^{-5})/30.01][3.52 + (1.32)(5.826)]\\ &= 1.22 \times 10^{-4} \text{ cal/(cm)(s)(K)}\end{aligned}$$

The experimentally determined value is reported as 1.07×10^{-4} cal/(cm)(s)(K). The error is thus 14 percent.

1-32 Thermal Conductivity of Liquids

Estimate the thermal conductivity of carbon tetrachloride at 10°C (283 K, 50°F) using the Robbins and Kingrea correlation:

$$\lambda_L = [(88 - 4.94\text{H})(0.001)/\Delta S][(0.55/T_R)^N][C_p\rho^{4/3}]$$

where λ_L = liquid thermal conductivity, cal/(cm)(s)(K)
T_R = reduced temperature, equal to T/T_c where T_c is critical temperature
C_p = molal heat capacity of liquid, cal/(g · mol)(K)
ρ = molal liquid density, (g · mol)/cm³
$\Delta S = (\Delta H_{vb}/T_b) + R \ln (273/T_b)$, cal/(g · mol)(K)
ΔH_{vb} = molal heat of vaporization at normal boiling point, cal/(g · mol)
T_b = normal boiling point, K
H = empirical parameter whose value depends on molecular structure and is obtained from Table 1-21
N = empirical parameter whose value depends on liquid density at 20°C (It equals 0 if the density is greater than 1.0 g/cm³, and 1 otherwise.)

Use these values for carbon tetrachloride: $T_c = 556.4$ K, molecular weight = 153.8, liquid density = 1.58 g/cm³, molal heat capacity = 31.37 cal/(g · mol)(K), $T_b = 349.7$ K, $\Delta H_{vb} = 7170$ cal/(g · mol)

Calculation Procedure:

***1. Obtain the structural constant H* from Table 1-21.**

Since carbon tetrachloride is an unbranched hydrocarbon with three chlorine substitutions, H has a value of 3.

2. Calculate ΔS.

Thus, $\Delta S = (\Delta H_{vb}/T_b) + R[\ln(273/T_b)] = 7170/349.7 + 1.987[\ln(273/349.7)] = 20.0$ cal/(g · mol)(K).

3. Calculate the thermal conductivity.

Thus,

$$\lambda_L = [(88 - 4.94H)(0.001)/\Delta S][(0.55/T_R)^N][C_p\rho^{4/3}]$$
$$= \{[88 - (4.94)(3)][0.001]/[20.0]\}\{[0.55/(283/556.4)]^0\}\{[31.37][1.58/153.8]^{4/3}\}$$
$$= 2.564 \times 10^{-6} \text{ cal/(cm)(s)(K)}$$

The experimentally determined value is reported as 2.510×10^{-6} cal/(cm)(s)(K). The error is thus 2.2 percent.

TABLE 1-21 *H*-factors for Robbins-Kingrea Correlation

Functional group	Number of groups	*H*
Unbranched hydrocarbons:		
Paraffins		0
Olefins		0
Rings		0
CH_3 branches	1	1
	2	2
	3	3
C_2H_5 branches	1	2
i-C_3H_7 branches	1	2
C_4H_9 branches	1	2
F substitutions	1	1
	2	2
Cl substitutions	1	1
	2	2
	3 or 4	3
Br substitutions	1	4
	2	6
I substitutions	1	5
OH substitutions	1 (iso)	1
	1 (normal)	−1
	2	0
	1 (tertiary)	5
Oxygen substitutions:		
$-\overset{\vert}{C}{=}O$ (ketones, aldehydes)		0
$-\overset{O}{\overset{\Vert}{C}}-O-$ (acids, esters)		0
—O— (ethers)		2
NH_2 substitutions	1	1

Source: Reprinted by permission from "Estimate Thermal Conductivity," *Hydrocarbon Processing,* May 1962,

2

Stoichiometry*

James H. Gary, Ph.D.

Professor Emeritus
Chemical and Petroleum-Refining
Engineering Department
Colorado School of Mines
Golden, CO

The first law of thermodynamics is the basis for material- or energy-balance calculations. Usually there is no significant transformation of mass to energy, and for a material balance, the first law can be reduced to the form

*Examples 2-8, 2-9, and 2-10 are taken from T. G. Hicks, *Standard Handbook of Engineering Calculations*, McGraw-Hill Book Co., Inc.

Mass in = mass out + accumulation

A similar equation can be used to express the energy balance

Energy in (above datum) + energy generated = energy out (above datum)

Energy balances differ from mass balances in that the total mass is known but the total energy of a component is difficult to express. Consequently, the heat energy of a material is usually expressed relative to its standard state at a given temperature. For example, the heat content, or enthalpy, of steam is expressed relative to liquid water at 273 K (32°F) at a pressure equal to its own vapor pressure.

Regardless of how complicated a material-balance system may appear, the use of a systematic approach can resolve it into a number of independent equations equal to the number of unknowns. One suitable stepwise approach is (1) state the problem, (2) list available data, (3) draw a sketch of the system, (4) define the system boundaries, (5) establish the bases for the system parameters, (6) write component material balances, (7) write an overall material balance, (8) solve the equations, and (9) make another mass balance as a check.

2-1 Material Balance—No Chemical Reactions Involved

A slurry containing 25 percent by weight of solids is fed into a filter. The filter cake contains 90 percent solids and the filtrate contains 1 percent solids. Make a material balance around the filter for a slurry feed rate of 2000 kg/h (4400 lb/h). For that feed rate, what are the corresponding flow rates for the cake and the filtrate?

Calculation Procedure:

1. Sketch the system, showing the available data, indicating the unknowns, defining the system boundary, and establishing the basis for the calculations.

When no chemical reactions are involved, the balances are based on the masses of individual chemical compounds appearing in more than one incoming or outgoing stream. Components appearing in only one incoming and one outgoing stream can be lumped together as though they are one component to simplify calculations and increase precision. A convenient unit of mass is selected, usually the kilogram or pound, and all components are expressed in that unit.

As the basis for a continuous process, always choose a unit of time or a consistent set of flow rates per unit of time. For batch processes, the appropriate basis is 1 batch. In the present (continuous) process, let the basis be 1 h. Let C be the mass flow rate of filter cake and F the mass flow rate of filtrate, in kilograms per hour. Figure 2-1 is a sketch of the system.

2. Set up and solve the material-balance equations.

This is a steady-state operation, so accumulation equals zero and the amount of mass in equals the amount of mass out (per unit of time). Since there are two unknowns, C and F, two independent equations must be written. One will be an overall balance; the other can be either a liquid balance (the option chosen in this example) or a solids balance.

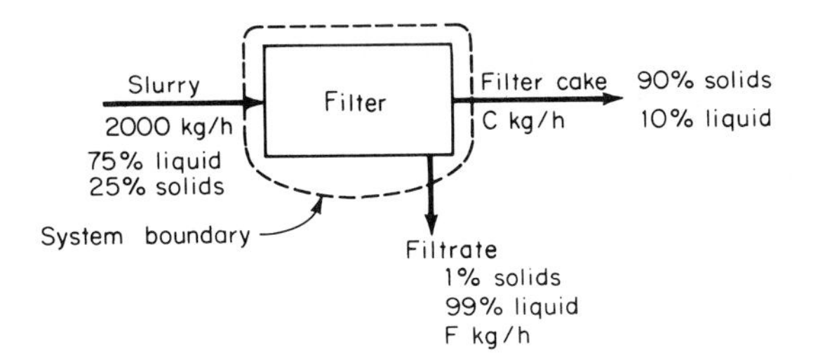

FIG. 2-1 Material balance for filter (Example 2-1).

Overall balance: Filtrate out + cake out = slurry in, or $F + C = 2000$ kg/h (4400 lb/h).

Liquid balance: Liquid in filtrate + liquid in cake = liquid in slurry, or (wt fraction liquid in filtrate)(mass of filtrate) + (wt fraction liquid in cake)(mass of cake) = (wt fraction liquid in slurry)(mass of slurry), or $(1.0 - 0.01)F + (1.0 - 0.90)C = (1.0 - 0.25)(2000)$.

Simultaneous solving of the two equations, $F + C = 2000$ and $0.99F + 0.1C = 1500$, gives F to be 1460.7 kg/h (3214 lb/h) of filtrate and C to be 539.3 kg/h (1186 lb/h) of cake.

3. Check the results.

It is convenient to check the answers by substituting them into the equation not used above, namely, the solids balance. Thus, solid in filtrate + solid in cake = solid in slurry, or $0.01(1460.7) + 0.9(539.3) = 0.25(2000)$. The answers, therefore, are correct.

2-2 Material Balance—Chemical Reactions Involved

Natural gas consisting of 95% methane and 5% nitrogen by volume is burned in a furnace with 15% excess air. How much air at 289 K (61°F) and 101.3 kPa (14.7 psia) is required if the fuel consumption is 10 m^3/s (353 ft^3/s) measured at 289 K and 101.3 kPa? Make an overall material balance and calculate the quantity and composition of the flue gas.

Calculation Procedure:

1. Sketch the system, setting out the available data, indicating the unknowns, defining the system boundary, and establishing the basis for the calculations.

For a process involving chemical reactions, the usual procedure is to express the compositions of the streams entering and leaving the process in molar concentrations. The balances are made in terms of the largest components remaining unchanged in the reactions. These can be expressed as atoms (S), ions (SO_4^{2-}), molecules (O_2), or other suitable units.

Whenever the reactants involved are not present in the proper stoichiometric ratios, the limiting reactant should be determined and the excess quantities of the other reactants calculated. Unconsumed reactants and inert materials exit with the products in their original form.

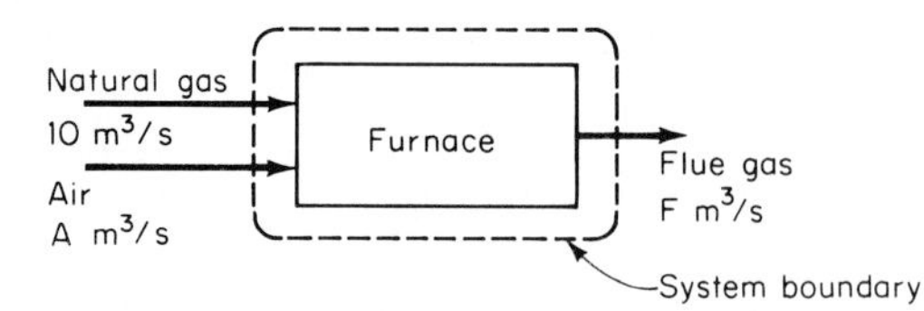

FIG. 2-2 Material balance for furnace (Example 2-2).

By convention, the amount of excess reactant in a reaction is always defined on the basis of the reaction going to 100 percent completion for the limiting reactant. The degree of completion is not a factor in determining or specifying the excess of reactants. For example, if methane is burned with 10 percent excess air, the volume of air needed to burn the methane is calculated as though there is total combustion of methane to carbon dioxide and water.

In the present problem, let the basis be 1 s. Let A and F be the volumetric flow rates for air and flue gas, in cubic meters per second. Figure 2-2 is a sketch of the system. The data are as follows:

Natural gas at 289 K and 101.3 kPa = 10 m^3/s
  95% CH_4 MW = 16
  5% N_2 MW = 28

Air at 289 K and 101.3 kPa = A m^3/s
  21% O_2 MW = 32
  79% N_2 MW = 28

R = 8.314 kJ/(kg·mol)(K) (ideal-gas constant)

2. *Convert the natural gas flow rate to kilogram-moles per second.*

At the conditions of this problem, the ideal-gas law can be used. Thus, $n = PV/RT$, where n is number of moles, P is pressure, V is volume, R is the gas constant, and T is absolute temperature.

For CH_4, n = (101.3 kPa)[(10 m^3/s)(0.95)]/[8.314 kJ/(kg·mol)(K)](289 K) = 0.40 kg·mol/s. For N_2, since the volumetric composition is 95% CH_4 and 5% N_2, n = (0.05/0.95)(0.40) = 0.02 kg·mol/s.

3. *Determine the amount of oxygen required and the airflow rate.*

The combustion reaction for CH_4 is $CH_4 + 2O_2 \rightarrow CO_2 + 2H_2O$. Thus 0.40 mol/s of CH_4 requires 2(0.40), or 0.80, mol/s of O_2 for stoichiometric combustion. Since 15 percent excess air is specified, the number of moles of oxygen in the air is (1.15)(0.80), or 0.92, kg·mol/s. The amount of nitrogen in with the air is [(0.79 mol N_2/mol air)/(0.21 mol O_2/mol air)](0.92 kg·mol/s O_2) = 3.46 kg·mol/s. Total moles in the incoming air are 0.92 + 3.46 = 4.38 kg·mol/s. Finally, using the ideal-gas law to convert to volumetric flow rate, $V = nRT/P$ = (4.38)(8.314)(289)/101.3 = 103.9 m^3/s (3671 ft^3/s) of air.

4. Set up the material balance and calculate the composition and quantity of the flue gas.

Convert to a mass basis because it is always true (unless there is a conversion between mass and energy) that from a mass standpoint the input equals the output plus the accumulation. In the present problem, there is no accumulation. The output (the flue gas) includes nitrogen from the air and from the natural gas, plus the 15 percent excess oxygen, plus the reaction products, namely, 0.40 mol/s CO_2 and 2(0.40) = 0.80 mol/s water.

Select 1 s as the basis. Then the inputs and outputs are as follows:

Component	Kilogram-moles	×	Kilograms per mole	=	Kilograms
			Inputs		
Natural gas:					
CH_4	0.40		16		6.40
N_2	0.02		28		0.56
Air:					
N_2	3.46		28		96.88
O_2	0.92		32		29.44
Total					133.28
			Output		
Flue gas:					
N_2	(3.46 + 0.02)		28		97.44
O_2	(0.92 − 0.80)		32		3.84
CO_2	0.40		44		17.60
H_2O	0.80		18		14.40
Total					133.28

The accumulation is zero. The overall material balance is 133.28 = 133.28 + 0. The total quantity of flue gas, therefore, is 133.28 kg/s (293 lb/s).

Related Calculations: The composition of the flue gas as given above is by weight. If desired, the composition by volume (which, indeed, is the more usual basis for expressing gas composition) can readily be obtained by calculating the moles per second of nitrogen (molar and volumetric compositions are equal to each other).

For more complex chemical reactions, it may be necessary to make mass balances for each molecular or atomic species rather than for the compounds.

2-3 Material Balance—Incomplete Data on Composition or Flow Rate

Vinegar with a strength of 4.63% (by weight) acetic acid is pumped into a vat to which 1000 kg (2200 lb) of 36.0% acetic acid is added. The resulting mixture contains 8.50% acid. How much of this 8.50% acid solution is in the vat?

Calculation Procedure:

1. List the available data, establish a basis for the calculations, and assign letters for the unknown quantities.

In many situations, such as in this example, some streams entering or leaving a process may have incomplete data to express their compositions or flow rates. The usual procedure is to write material balances as in the preceding examples but to assign letters to represent the unknown quantities. There must be one independent material balance written for each unknown in order to have a unique solution.

The present problem is a batch situation, so let the basis be 1 batch. There are two inputs: an unknown quantity of vinegar having a known composition (4.63% acetic acid) and a known amount of added acetic acid [1000 kg (2200 lb)] of known composition (36.0% acid). There is one output: a final batch of unknown quantity but known composition (8.50% acid). Let T represent the kilograms of input vinegar and V the size of the final batch in kilograms.

2. Set up and solve the material-balance equations.

Two independent material balances can be set up, one for acetic acid or for water and the other for the overall system.

Acetic acid: Input = output, or $0.0463T + 0.360(1000 \text{ kg}) = 0.0850V$

Water: Input = output, or $(1 - 0.0463)T + (1 - 0.360)(1000 \text{ kg}) = (1 - 0.0850)V$

Overall: Input = output, or $T + 1000 = V$

Use the overall balance and one of the others, say, the one for acetic acid. By substitution, then, $0.0463T + 0.360(1000) = 0.0850(T + 1000)$; so T is found to be 7106 kg vinegar, and $V = T + 1000 = 8106$ kg (17,833 lb) solution in the vat.

3. Check the results.

It is convenient to make the check by substituting into the equation not used above, namely, the water balance:

$$(1 - 0.0463)(7106) + (1 - 0.360)(1000) = (1 - 0.0850)(8106)$$

Thus, the results check.

2-4 Use of a Tie Element in Material-Balance Calculations

The spent catalyst from a catalytic-cracking reactor is taken to the regenerator for reactivation. Coke deposited on the catalyst in the reactor is removed by burning with air, and the flue gas is vented. The coke is a mixture of carbon and high-molecular-weight tars considered to be hydrocarbons. For the following conditions, calculate the weight percent of hydrogen in the coke. Assume that the coke on the regenerated catalyst has the same composition as the coke on the spent catalyst:

Carbon on spent catalyst	1.50 wt %
Carbon on regenerated catalyst	0.80 wt %
Air from blower	150,000 kg/h (330,000 lb/h)
Hydrocarbon feed to reactor	300,000 kg/h (660,000 lb/h)

Flue gas analysis (dry basis):

CO_2	12.0 vol %
CO	6.0 vol %
O_2	0.7 vol %
N_2	81.3 vol %
	100.0

Assume that all oxygen not reported in flue gas analysis reacted with hydrogen in the coke to form water. All oxygen is reported as O_2 equivalent. Assume that air is 79.02% nitrogen and 20.98% oxygen.

Calculation Procedure:

1. Select a basis and a tie component, and write out the relevant equations involved.

Select 100 kg·mol dry flue gas as the basis. Since nitrogen passes through the system unreacted, select it as the tie component. That is, the other components of the system can be referred to nitrogen as a basis, thus simplifying the calculations.

Since dry flue gas is the basis, containing CO_2 and CO, the relevant reactions and the quantities per 100 mol flue gas are

$$\underset{12\text{ kg·mol}}{C} + \underset{12\text{ kg·mol}}{O_2} \rightarrow \underset{12\text{ kg·mol}}{CO_2}$$

$$\underset{6\text{ kg·mol}}{C} + \underset{3\text{ kg·mol}}{\tfrac{1}{2}O_2} \rightarrow \underset{6\text{ kg·mol}}{CO}$$

2. Calculate the amount of oxygen in the entering air.

The total moles of nitrogen in the entering air must equal the total moles in the flue gas, namely, 81.3 kg·mol. Since air is 79.02% nitrogen and 20.98% oxygen, the oxygen amounts to (20.98/79.02)(81.3), or 21.59 kg·mol.

3. Calculate the amount of oxygen that leaves the system as water.

The number of kilogram-moles of oxygen in the regenerator exit gases should be the same as the number in the entering air, that is, 21.59. Therefore, the oxygen not accounted for in the dry analysis of the flue gas is the oxygen converted to water. The dry analysis accounts for 12 mol oxygen as CO_2, 3 mol as CO, and 0.7 mol as unreacted oxygen. Thus the water leaving the system in the (wet) flue gas accounts for (21.59 − 12 − 3 − 0.7) = 5.89 mol oxygen.

4. Calculate the weight percent of hydrogen in the coke.

Since 2 mol water is produced per mole of oxygen reacted, the amount of water in the wet flue gas is 2(5.89) = 11.78 mol. This amount of water contains 11.78 mol hydrogen or (11.78)[2.016 kg/(kg·mol)] = 23.75 kg hydrogen.

Now, the amount of carbon associated with this is the 12 mol that reacted to form CO_2 plus the 6 mol that reacted to form CO or (12 + 6)[12.011 kg/(kg·mol)] = 216.20 kg carbon. Therefore, the weight percent of hydrogen in the coke is (100)(23.75)/(23.75 + 216.0) = 9.91 percent.

2-5 Material Balance—Chemical Reaction and a Recycle Stream Involved

In the feed-preparation section of an ammonia plant, hydrogen is produced from methane by a combination steam-reforming/partial-oxidation process. Enough air is used in partial oxidation to give a 3:1 hydrogen-nitrogen molar ratio in the feed to the ammonia unit. The hydrogen-nitrogen mixture is heated to reaction temperature and fed into a fixed-bed reactor where 20 percent conversion of reactants to ammonia is obtained per pass. After leaving the reactor, the mixture is cooled and the ammonia removed by condensation. The unreacted hydrogen-nitrogen mixture is recycled and mixed with fresh feed. On the basis of 100 kg·mol/h (220 lb·mol/h) of fresh feed, make a material balance and determine the ammonia-production and recycle rates.

Calculation Procedure:

1. Sketch the system, showing the available data, indicating the unknowns, and establishing the system boundary.

Since one of the answers sought is the recycle rate, the system boundary must be selected in such a way as to be crossed by the recycle stream.

Since the feed is a 3:1 ratio, the 100 mol/h will consist of 25 mol/h of nitrogen and 75 mol/h of hydrogen. Let x equal the moles per hour of ammonia produced and y the moles per hour of recycle. The sketch and the system boundary are shown in Figure 2-3.

2. Determine the amount of ammonia produced.

Set out the ammonia-production reaction, namely, $N_2 + 3H_2 \rightarrow 2NH_3$. Thus 4 mol hydrogen-nitrogen mixture in a 3:1 ratio (as is the case for the feed in this example) will yield 2 mol ammonia. Since the system boundary is drawn in such a way that the exiting and reentering recycle streams offset each other algebraically, the net output from the system, consisting of liquid ammonia, must equal the net input, consisting of fresh feed. The amount x of ammonia produced per hour thus can be determined by straightforward stoichiometry: x = [100 kg·mol/h ($H_2 + N_2$)][(2 mol NH_3)/4 mol ($H_2 + N_2$)], or x = 50 kg·mol/h (110 lb·mol/h) NH_3.

3. Determine the recycle rate.

The total feed to the heater and reactor consists of (100 + y) kg·mol/h. Twenty percent of this feed is converted to ammonia, and during that conversion, 2 mol ammonia is pro-

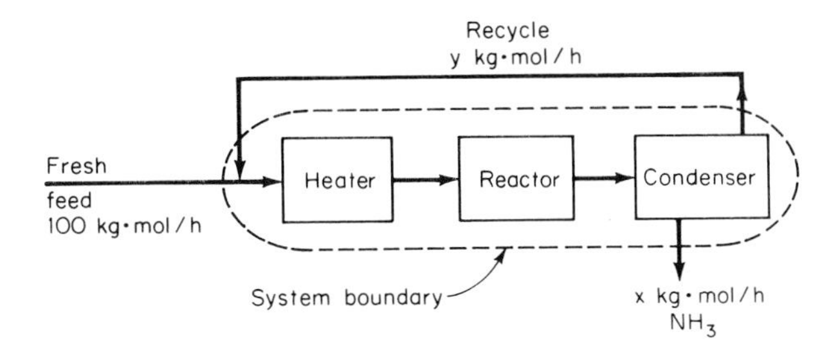

FIG. 2-3 Material balance for ammonia plant with no purge stream (Example 2-5).

duced per 4 mol feed (inasmuch as it consists of 3:1 hydrogen-nitrogen mixture). Therefore, the amount of ammonia produced equals $[0.20(100 + y)][(2 \text{ mol } NH_3)/4 \text{ mol } (H_2 + N_2)]$. Ammonia production is 50 kg·mol/h, so solving this equation for y gives a recycle rate of 400 kg·mol/h (880 lb·mol/h).

4. Check the results.

A convenient way to check is to set up an overall mass balance. Since there is no accumulation, the input must equal the output. Input = (25 kg·mol/h N_2)(28 kg/mol) + (75 kg·mol/h H_2)(2 kg/mol) = 850 kg/h, and output = (50 kg·mol/h NH_3)(17 kg/mol) = 850 kg/h. The results thus check.

Related Calculations: It is also possible to calculate the recycle rate in the preceding example by making a material balance around the reactor-condenser system.

The ratio of the quantity of material recycled to the quantity of fresh feed is called the "recycle ratio." In the preceding problem, the recycle ratio is 400/100, or 4.

2-6 Material Balance—Chemical Reaction, Recycle Stream, and Purge Stream Involved

In the preceding example for producing ammonia, the amount of air fed is set by the stoichiometric ratio of hydrogen to nitrogen for the ammonia feed stream. In addition to nitrogen and oxygen, the air contains inert gases, principally argon, that gradually build up in the recycle stream until the process is affected adversely. It has been determined that the concentration of argon in the reactor must be no greater than 4 mol argon per 100 mol hydrogen-nitrogen mixture. Using the capacities given in the preceding example, calculate the amount of the recycle stream that must be vented to meet the concentration requirement. The fresh feed contains 0.31 mol argon per 100 mol hydrogen-nitrogen mixture. Also calculate the amount of ammonia produced.

Calculation Procedure:

1. Select the basis for calculation and sketch the system, showing the available data and indicating the unknowns.

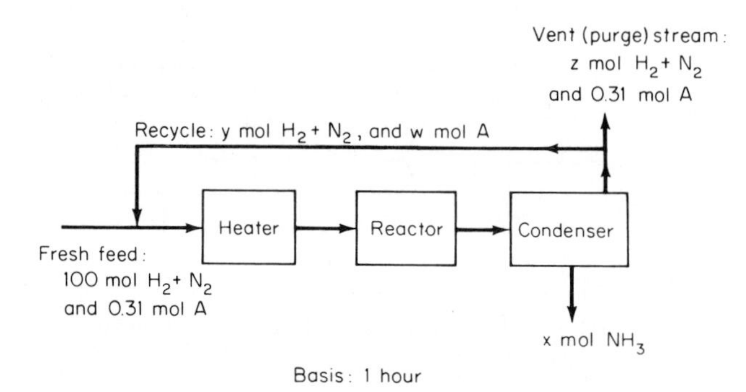

FIG. 2-4 Material balance for ammonia plant with purge stream (Example 2-6).

For ease of comparison with the preceding example, let the basis be 100.31 kg·mol/h total fresh feed, consisting of 100 mol ($H_2 + N_2$) and 0.31 mol argon (A). Let x equal the moles of NH_3 produced per hour, y the moles of $H_2 + N_2$ recycled per hour, w the moles of A recycled per hour, and z the moles of $H_2 + N_2$ purged per hour. The sketch is Fig. 2-4.

2. *Calculate the amount of recycle stream that must be vented.*

As noted in the preceding example, the conversion per pass through the reactor is 20 percent. Therefore, for every 100 mol ($H_2 + N_2$) entering the heater-reactor-condenser train, 20 mol will react to form ammonia and 80 mol will leave the condenser to be recycled or purged. All the argon will leave with this recycle-and-purge stream. Since the maximum allowable argon level in the reactor input is 4 mol argon per 100 mol ($H_2 + N_2$), there will be 4 mol argon per 80 mol ($H_2 + N_2$) in the recycle-and-purge stream.

Under steady-state operating conditions, the argon purged must equal the argon entering in the fresh feed. The moles of argon in the purge equals $(4/80)z = 0.05z$. Therefore, $0.05z = 0.31$, so $z = 6.2$ mol/h ($H_2 + N_2$) purged. The total purge stream consists of 6.2 mol/h ($H_2 + N_2$) plus 0.31 mol/h argon.

3. *Calculate the amount of $H_2 + N_2$ recycled.*

The moles of $H_2 + N_2$ in the feed to the reactor is 100 mol fresh feed plus y mol recycle. Of this, 80 percent is to be either purged or recycled; that is, $0.80(100 + y) = y + z = y + 6.2$. Therefore $y = 369$ mol/h ($H_2 + N_2$) recycled.

Although not needed for the solution of this problem, the amount of argon in the recycle can be calculated as a matter of interest. Total argon entering the reactor is $0.31 + w$ mol; then, according to the argon limitation stipulated, $0.31 + w = 0.04(100 + 369)$, so $w = 18.45$ mol/h.

4. *Calculate the amount of ammonia produced.*

Of the $100 + y$ mol ($H_2 + N_2$) (in the stoichiometric 3:1 ratio) entering the reactor, 20 percent is converted to ammonia. The reaction is $N_2 + 3H_2 \rightarrow 2NH_3$, so 4 mol reactants yields 2 mol ammonia. Therefore, total ammonia production is 0.20(100 + y)(2/4) = 0.20(100 + 369)(2/4) = 46.9 mol/h.

5. *Check the results.*

It is convenient to check by making an overall mass balance. The input per hour consists of 100 mol (H_2 + N_2) in a 3:1 ratio plus 0.31 mol A; that is, (75 mol H_2)(2 kg/mol) + (25 mol N_2)(28 kg/mol) + (0.31 mol A)(40 kg/mol) = 862.4 kg. The output per hour consists of 46.9 mol ammonia plus a vent-stream mixture of 6.2 mol (H_2 + N_2) (in a 3:1 ratio) and 0.31 mol A; that is, (46.9 mol NH_3)(17 kg/mol) + (3/4)[6.2 mol (H_2 + N_2)](2 kg H_2/mol) + (1/4)(6.2)(28 kg N_2/mol) + (0.31 mol A)(40 kg/mol) = 862.4 kg.

There is no accumulation in the system, so input should equal output. Since 862.4 kg = 862.4 kg, the results thus check.

2-7 Use of Energy Balance with Material Balance

A particular crude oil is heated to 510 K (458°F) and charged at 10 L/h (0.01 m^3/h, or 2.6 gal/h) to the flash zone of a laboratory distillation tower. The flash zone is at an absolute pressure of 110 kPa (16 psi). Determine the percent vaporized and the amounts of the overhead and bottoms streams. Assume that the vapor and liquid are in equilibrium.

Calculation Procedure:

1. *Select the approach to be employed.*

In this problem there is not enough information available to employ a purely material-balance approach. Instead, use an energy balance as well. Such an approach is especially appropriate in cases such as this one in which some of the components undergo a phase change.

From the American Petroleum Institute's (API) *Technical Data Book—Petroleum Refining*, specific heats, specific gravities, latent heats of vaporization, and percent vaporization can be obtained, for a given oil, as a function of flash-zone temperature (percent vaporization and flash-zone temperature are functionally related because the flash vaporization takes place adiabatically). This suggests a trial-and-error procedure: Assume a flash-zone temperature and the associated percent vaporization; then make an energy balance to check the assumptions. Finally, complete the material balance.

2. *Assume a flash-zone temperature and percent vaporization, and obtain the data for the system at those conditions.*

Assume, for a first guess, that 30 percent (by volume) of the feed is vaporized. The *API Data Book* indicates that for this oil, the corresponding flash-zone temperature is 483 K (410°F); the fraction vaporized has a latent heat of vaporization of 291 kJ/kg (125 Btu/lb) and a density of 0.750 kg/L (750 kg/m^3, or 47.0 lb/ft^3) and a specific heat of 2.89 kJ/(kg)(K) [0.69 Btu/(lb)(°F)]. The unvaporized portion has a density of 0.892 kg/L

(892 kg/m^3, or 55.8 lb/ft^3) and a specific heat of 2.68 kJ/(kg)(K) [0.64 Btu/(lb)(°F)]. In addition, the feed has a density of 0.850 kg/L (850 kg/m^3, or 53.1 lb/ft^3) and a specific heat of 2.85 kJ/(kg)(K) [0.68 Btu/(lb)(°F)].

3. Make an energy balance.

For convenience, use the flash temperature, 483 K, as the datum temperature. The energy brought into the system by the feed, consisting of sensible-heat energy with reference to the datum temperature, must equal the energy in the vapor stream (its latent heat plus its sensible heat) plus the energy in the bottoms stream (its sensible heat). However, since the flash temperature is the datum, and since both the vapor and the bottoms streams are at the datum temperature, neither of those product streams has a sensible-heat term associated with it. Thus the energy balance on the basis of 1 h (10 L) is as follows:

$$(10\ \text{L})(0.850\ \text{kg/L})[2.85\ \text{kJ/(kg)(K)}](510\ \text{K} - 483\ \text{K})$$
$$= (3\ \text{L})(0.750\ \text{kg/L})\{(291\ \text{kJ/kg}) + [2.89\ \text{kJ/(kg)(K)}](483\ \text{K} - 483\ \text{K})\}$$
$$+ (7\ \text{L})(0.892\ \text{kg/L})[2.68\ \text{kJ/(kg)(K)}](483\ \text{K} - 483\ \text{K})$$

Or, 654 = 655 + 0. Since this is within the limits of accuracy, the assumption of 30 percent vaporized is correct.

4. Make the material balance to determine the amount in the overhead and bottoms streams.

On the basis of 1 h, the mass in is (10 L)(0.850 kg/L) = 8.5 kg. The mass out consists of the mass that becomes vaporized (the overhead) plus the mass that remains unvaporized (the bottoms). The overhead is (3 L)(0.750 kg/L) = 2.25 kg (4.96 lb). The bottoms stream is (7 L)(0.892 kg/L) = 6.24 kg (13.76 lb). Thus, 8.5 kg = (2.25 + 6.24) kg. The material balance is consistent, within the limits of accuracy.

Related Calculations: The problem can be worked in similar fashion using values from enthalpy tables. In this case, the datum temperature is below the flash-zone temperature; therefore, sensible heat in the two exiting streams must be taken into account.

2-8 Combustion of Coal Fuel in a Furnace

A coal has the following ultimate analysis: C = 0.8339, H_2 = 0.0456, O_2 = 0.0505, N_2 = 0.0103, S = 0.0064, ash = 0.0533, total = 1.000. This coal is burned in a steam-boiler furnace. Determine the weight of air required for theoretically perfect combustion, the weight of gas formed per pound of coal burned, and the volume of flue gas at the boiler exit temperature of 600°F (589 K) per pound of coal burned; the air required with 20 percent excess air and the volume of gas formed with this excess; and the CO_2 percentage in the flue gas on a dry and wet basis.

Calculation Procedure:

1. Compute the weight of oxygen required per pound of coal.

To find the weight of oxygen required for theoretically perfect combustion of coal, set up the following tabulation, based on the ultimate analysis of the coal:

Element	×	Molecular-weight ratio	=	Pounds O_2 required
C; 0.8339	×	32/12	=	2.2237
H_2; 0.0456	×	16/2	=	0.3648
O_2; 0.0505; decreases external O_2 required			=	− 0.0505
N_2; 0.0103; is inert in combustion and is ignored				
S; 0.0064	×	32/32	=	0.0064
Ash; 0.0533; is inert in combustion and is ignored				
Total 1.0000				
Pounds external O_2 per lb fuel			=	2.5444

Note that of the total oxygen needed for combustion, 0.0505 lb is furnished by the fuel itself and is assumed to reduce the total external oxygen required by the amount of oxygen present in the fuel. The molecular-weight ratio is obtained from the equation for the chemical reaction of the element with oxygen in combustion. Thus, for carbon, $C + O_2 \rightarrow CO_2$, or $12 + 32 = 44$, where 12 and 32 are the molecular weights of C and O_2, respectively.

2. Compute the weight of air required for perfect combustion.

Air at sea level is a mechanical mixture of various gases, principally 23.2% oxygen and 76.8% nitrogen by weight. The nitrogen associated with the 2.5444 lb of oxygen required per pound of coal burned in this furnace is the product of the ratio of the nitrogen and oxygen weights in the air and 2.5444, or $(2.5444)(0.768/0.232) = 8.4219$ lb. Then the weight of air required for perfect combustion of 1 lb coal = sum of nitrogen and oxygen required = $8.4219 + 2.5444 = 10.9663$ lb of air per pound of coal burned.

3. Compute the weight of the products of combustion.

Find the products of combustion by addition:

Fuel constituents	+	Oxygen	→	Products of combustion
C; 0.8339	+	2.2237	→	$CO_2 = 3.0576$ lb
H; 0.0456	+	0.3648	→	$H_2O = 0.4104$
O_2; 0.0505; this is *not* a product of combustion				
N_2; 0.0103; inert but passes through furnace				= 0.0103
S; 0.0064	+	0.0064	→	SO_2 = 0.0128
Outside nitrogen from step 2				= N_2 = 8.4219
Pounds of flue gas per pound of coal burned				= 11.9130

4. Convert the flue-gas weight to volume.

Use Avogadro's law, which states that under the same conditions of pressure and temperature, 1 mol (the molecular weight of a gas expressed in pounds) of any gas will occupy the same volume.

At 14.7 psia and 32°F, 1 mol of any gas occupies 359 ft^3. The volume per pound of any gas at these conditions can be found by dividing 359 by the molecular weight of the gas and correcting for the gas temperature by multiplying the volume by the ratio of the absolute flue-gas temperature and the atmospheric temperature. To change the weight analysis (step 3) of the products of combustion to volumetric analysis, set up the calculation thus:

Products	Weight, lb	Molecular weight	Temperature correction		Volume at 600°F, ft^3
CO_2	3.0576	44	(359/44)(3.0576)(2.15)	=	53.8
H_2O	0.4104	18	(359/18)(0.4104)(2.15)	=	17.6
Total N_2	8.4322	28	(359/28)(8.4322)(2.15)	=	233.0
SO_2	0.0128	64	(359/64)(0.0128)(2.15)	=	0.17
Cubic feet of flue gas per pound of coal burned				=	304.57

In this calculation, the temperature correction factor 2.15 = (absolute flue-gas temperature)/(absolute atmospheric temperature), R = (600 + 460)/(32 + 460). The total weight of N_2 in the flue gas is the sum of the N_2 in the combustion air and the fuel, or 8.4219 + 0.0103 = 8.4322 lb. This value is used in computing the flue-gas volume.

5. *Compute the CO_2 content of the flue gas.*

The volume of CO_2 in the products of combustion at 600°F is 53.8 ft^3 as computed in step 4, and the total volume of the combustion products is 304.57 ft^3. Therefore, the percent CO_2 on a wet basis (i.e., including the moisture in the combustion products) = ft^3 CO_2/total ft^3 = 53.8/304.57 = 0.1765, or 17.65 percent.

The percent CO_2 on a dry, or Orsat, basis is found in the same manner except that the weight of H_2O in the products of combustion, 17.6 lb from step 4, is subtracted from the total gas weight. Or, percent CO_2, dry, or Orsat, basis = (53.8)/(304.57 − 17.6) = 0.1875, or 18.75 percent.

6. *Compute the air required with the stated excess flow.*

With 20 percent excess air, the airflow required = (0.20 + 1.00)(airflow with no excess) = 1.20(10.9663) = 13.1596 lb of air per pound of coal burned. The airflow with no excess is obtained from step 2.

7. *Compute the weight of the products of combustion.*

The excess air passes through the furnace without taking part in the combustion and increases the weight of the products of combustion per pound of coal burned. Therefore, the weight of the products of combustion is the sum of the weight of the combustion products without the excess air and the product of (percent excess air)(air for perfect combustion, lb); or using the weights from steps 3 and 2, respectively, = 11.9130 + (0.20)(10.9663) = 14.1063 lb of gas per pound of coal burned with 20 percent excess air.

8. *Compute the volume of the combustion products and the percent CO_2.*

The volume of the excess air in the products of combustion is obtained by converting from the weight analysis to the volumetric analysis and correcting for temperature as in step

4, using the air weight from step 2 for perfect combustion and the excess-air percentage, or (10.9663)(0.20)(359/28.95)(2.15) = 58.5 ft^3 (1.66 m^3). In this calculation, the value 28.95 is the molecular weight of air. The total volume of the products of combustion is the sum of the column for perfect combustion, step 4, and the excess-air volume, above, or 304.57 + 58.5 = 363.07 ft^3 (10.27 m^3).

Using the procedure in step 5, the percent CO_2, wet basis, = 53.8/363.07 = 14.8 percent. The percent CO_2, dry basis, = 53.8/(363.07 − 17.6) = 15.6 percent.

Related Calculations: Use the method given here when making combustion calculations for any type of coal—bituminous, semibituminous, lignite, anthracite, cannel, or coking—from any coal field in the world used in any type of furnace—boiler, heater, process, or waste-heat. When the air used for combustion contains moisture, as is usually true, this moisture is added to the combustion-formed moisture appearing in the products of combustion. Thus, for 80°F (300 K) air of 60 percent relative humidity, the moisture content is 0.013 lb per pound of dry air. This amount appears in the products of combustion for each pound of air used and is a commonly assumed standard in combustion calculations.

2-9 Combustion of Fuel Oil in a Furnace

A fuel oil has the following ultimate analysis: C = 0.8543, H_2 = 0.1131, O_2 = 0.0270, N_2 = 0.0022, S = 0.0034, total = 1.0000. This fuel oil is burned in a steam-boiler furnace. Determine the weight of air required for theoretically perfect combustion, the weight of gas formed per pound of oil burned, and the volume of flue gas at the boiler exit temperature of 600°F (589 K) per pound of oil burned; the air required with 20 percent excess air and the volume of gas formed with this excess; and the CO_2 percentage in the flue gas on a dry and wet basis.

Calculation Procedure:

1. Compute the weight of oxygen required per pound of oil.

The same general steps as given in the previous Calculation Procedure will be followed. Consult that procedure for a complete explanation of each step. Using the molecular weight of each element, the following table can be set up:

Element	×	Molecular-weight ratio	=	Pounds O_2 required
C; 0.8543	×	32/12	=	2.2781
H_2; 0.1131	×	16/2	=	0.9048
O_2; 0.0270; decreases external O_2 required			=	−0.0270
N_2; 0.0022; is inert in combustion and is ignored				
S; 0.0034	×	32/32	=	0.0034
Total 1.0000				
Pounds of external O_2 per pound fuel			=	3.1593

2. *Compute the weight of air required for perfect combustion.*

The weight of nitrogen associated with the required oxygen = (3.1593)(0.768/0.232) = 10.4583 lb. The weight of air required = 10.4583 + 3.1593 = 13.6176 lb per pound of oil burned.

3. *Compute the weight of the products of combustion.*

As before:

Fuel constituents + oxygen =			Products of combustion
C; 0.8543 + 2.2781	=	3.1324	CO_2
H_2; 0.1131 + 0.9148	=	1.0179	H_2O
O_2; 0.270; *not* a product of combustion			
N_2; 0.0022; inert but passes through furnace	=	0.0022	N_2
S; 0.0034 + 0.0034	=	0.0068	SO_2
Outside N_2 from step 2	=	10.458	N_2
Pounds of flue gas per pound of oil burned	=	14.6173	

4. *Convert the flue-gas weight to volume.*

As before:

Products	Weight, lb	Molecular weight	Temperature correction		Volume at 600°F, ft^3
CO_2	3.1324	44	(359/44)(3.1324)(2.15)	=	55.0
H_2O	1.0179	18	(359/18)(1.0179)(2.15)	=	43.5
N_2 (total)	10.460	28	(359/28)(10.460)(2.15)	=	288.5
SO_2	0.0068	64	(359/64)(0.0068)(2.15)	=	0.82
Cubic feet of flue gas per pound of oil burned				=	387.82

In this calculation, the temperature correction factor 2.15 = (absolute flue-gas temperature)/(absolute atmospheric temperature) = (600 + 460)/(32 + 460). The total weight of N_2 in the flue gas is the sum of the N_2 in the combustion air and the fuel, or 10.4583 + 0.0022 = 10.4605 lb.

5. *Compute the CO_2 content of the flue gas.*

The CO_2, wet basis, = 55.0/387.82 = 0.142, or 14.2 percent. The CO_2, dry basis, = 55.0/(387.2 − 43.5) = 0.160, or 16.0 percent.

6. *Compute the air required with stated excess flow.*

The pounds of air per pound of oil with 20 percent excess air = (1.20)(13.6176) = 16.3411 lb air per pound of oil burned.

7. Compute the weight of the products of combustion.

The weight of the products of combustion = product weight for perfect combustion, lb + (percent excess air)(air for perfect combustion, lb) = 14.6173 + (0.20)(13.6176) = 17.3408 lb flue gas per pound of oil burned with 20 percent excess air.

8. Compute the volume of the combustion products and the percent CO_2.

The volume of excess air in the products of combustion is found by converting from the weight to the volumetric analysis and correcting for temperature as in step 4, using the air weight from step 2 for perfect combustion and the excess-air percentage, or (13.6176)(0.20)(359/28.95)(2.15) = 72.7 ft^3 (2.06 m^3). Add this to the volume of the products of combustion found in step 4, or 387.82 + 72.70 = 460.52 ft^3 (13.03 m^3).

Using the procedure in step 5, the percent CO_2, wet basis, = 55.0/460.52 = 0.1192, 11.92 percent. The percent CO_2, dry basis, = 55.0/(460.52 − 43.5) = 0.1318, or 13.18 percent.

Related Calculations: Use the method given here when making combustion calculations for any type of fuel oil—paraffin-base, asphalt-base, Bunker C, No. 2, 3, 4, or 5—from any source, domestic or foreign, in any type of furnace—boiler, heater, process, or waste-heat. When the air used for combustion contains moisture, as is usually true, this moisture is added to the combustion-formed moisture appearing in the products of combustion. Thus, for 80°F air of 60 percent relative humidity, the moisture content is 0.013 lb per pound of dry air. This amount appears in the products of combustion for each pound of air used and is a commonly assumed standard in combustion calculations.

2-10 Combustion of Natural Gas in a Furnace

A natural gas has the following volumetric analysis at 60°F: CO_2 = 0.004, CH_4 = 0.921, C_2H_6 = 0.041, N_2 = 0.034, total = 1.000. This natural gas is burned in a steam-boiler furnace. Determine the weight of air required for theoretically perfect combustion, the weight of gas formed per pound of natural gas burned, and the volume of the flue gas at the boiler exit temperature of 650°F per pound of natural gas burned; the air required with 20 percent excess air and the volume of gas formed with this excess; and the CO_2 percentage in the flue gas on a dry and wet basis.

Calculation Procedure:

1. Compute the weight of oxygen required per pound of gas.

The same general steps as given in the previous Calculation Procedures will be followed, except that they will be altered to make allowances for the differences between natural gas and coal.

The composition of the gas is given on a volumetric basis, which is the usual way of expressing a fuel-gas analysis. To use the volumetric-analysis data in combustion calculations, they must be converted to a weight basis. This is done by dividing the weight of each component by the total weight of the gas. A volume of 1 ft^3 of the gas is used for this computation. Find the weight of each component and the total weight of 1 ft^3 as follows, using the properties of the combustion elements and compounds given in Table 2-1:

Component	Percent by volume	Density, lb/ft^3	Component weight, lb = col. 2 × col. 3
CO_2	0.004	0.1161	0.0004644
CH_4	0.921	0.0423	0.0389583
C_2H_6	0.041	0.0792	0.0032472
N_2	0.034	0.0739	0.0025026
Total	1.000		0.0451725 lb/ft^3

$$\text{Percent } CO_2 = 0.0004644/0.0451725 = 0.01026, \text{ or } 1.03 \text{ percent}$$

$$\text{Percent } CH_4 \text{ by weight} = 0.0389583/0.0451725 = 0.8625, \text{ or } 86.25 \text{ percent}$$

$$\text{Percent } C_2H_6 \text{ by weight} = 0.0032472/0.0451725 = 0.0718, \text{ or } 7.18 \text{ percent}$$

$$\text{Percent } N_2 \text{ by weight} = 0.0025026/0.0451725 = 0.0554, \text{ or } 5.54 \text{ percent}$$

The sum of the weight percentages = 1.03 + 86.25 + 7.18 + 5.54 = 100.00. This sum checks the accuracy of the weight calculation, because the sum of the weights of the component parts should equal 100 percent.

Next, find the oxygen required for combustion. Since both the CO_2 and N_2 are inert, they do not take part in the combustion; they pass through the furnace unchanged. Using the molecular weights of the remaining compounds in the gas and the weight percentages, we have:

Compound	×	Molecular-weight ratio	=	Pounds O_2 required
CH_4; 0.8625	×	64/16	=	3.4500
C_2H_6; 0.0718	×	112/30	=	0.2920
Pounds of external O_2 required per pound fuel			=	3.7420

In this calculation, the molecular-weight ratio is obtained from the equation for the combustion chemical reaction, or $CH_4 + 2O_2 = CO_2 + 2H_2O$, that is, 16 + 64 = 44 + 36, and $C_2H_6 + \frac{7}{2}O_2 = 2CO_2 + 3H_2O$, that is, 30 + 112 = 88 + 54.

2. ***Compute the weight of air required for perfect combustion.***

The weight of nitrogen associated with the required oxygen = (3.742)(0.768/0.232) = 12.39 lb. The weight of air required = 12.39 + 3.742 = 16.132 lb per pound of gas burned.

3. ***Compute the weight of the products of combustion.***

Use the following relation:

Fuel constituents	+	Oxygen	=	Products of combustion
CO_2; 0.0103; inert but passes through the furnace			=	0.010300
CH_4; 0.8625	+	3.45	=	4.312500
C_2H_6; 0.003247	+	0.2920	=	0.032447
N_2; 0.0554; inert but passes through the furnace			=	0.055400
Outside N_2 from step 2			=	12.390000
Pounds of flue gas per pound of natural gas burned			=	16.800347

4. *Convert the flue-gas weight to volume.*

The products of complete combustion of any fuel that does not contain sulfur are CO_2, H_2O, and N_2. Using the combustion equation in step 1, compute the products of combustion thus: $CH_4 + 2O_2 = CO_2 + H_2O$; $16 + 64 = 44 + 36$; or the CH_4 burns to CO_2 in the ratio of 1 part CH_4 to 44/16 parts CO_2. Since, from step 1, there is 0.03896 lb CH_4 per ft^3 natural gas, this forms $(0.03896)(44/16) = 0.1069$ lb CO_2. Likewise, for C_2H_6, $(0.003247)(88/30) = 0.00952$ lb. The total CO_2 in the combustion products $= 0.00464 + 0.1069 + 0.00952 = 0.11688$ lb, where the first quantity is the CO_2 in the fuel.

Using a similar procedure for the H_2O formed in the products of combustion by CH_4, $(0.03896)(36/16) = 0.0875$ lb. For C_2H_6, $(0.003247)(54/30) = 0.005816$ lb. The total H_2O in the combustion products $= 0.0875 + 0.005816 = 0.093316$ lb.

Step 2 shows that 12.39 lb N_2 is required per pound of fuel. Since 1 ft^3 of the fuel weighs 0.04517 lb, the volume of gas that weighs 1 lb is $1/0.04517 = 22.1$ ft^3. Therefore, the weight of N_2 per cubic foot of fuel burned $= 12.39/22.1 = 0.560$ lb. This, plus the

TABLE 2-1 Properties of Combustion Elements

			At 14.7 psia, 60°F		Nature		Heat value, Btu		
Element or compound	Formula	Molecular weight	Weight, lb/ft^3	Volume, ft^3/lb	Gas or solid	Combustible	Per pound	Per ft^3 at 14.7 psia, 60°F	Per molecule
Carbon	C	12	—	—	S	Yes	14,540	—	174,500
Hydrogen	H_2	2.02*	0.0053	188	G	Yes	61,000	325	123,100
Sulfur	S	32	—	—	S	Yes	4,050	—	129,600
Carbon monoxide	CO	28	0.0739	13.54	G	Yes	4,380	323	122,400
Methane	CH_4	16	0.0423	23.69	G	Yes	24,000	1,012	384,000
Acetylene	C_2H_2	26	0.0686	14.58	G	Yes	21,500	1,483	562,000
Ethylene	C_2H_4	28	0.0739	13.54	G	Yes	22,200	1,641	622,400
Ethane	C_2H_6	30	0.0792	12.63	G	Yes	22,300	1,762	668,300
Oxygen	O_2	32	0.0844	11.84	G				
Nitrogen	N_2	28	0.0739	13.52	G				
Air†	—	29	0.0765	13.07	G				
Carbon dioxide	CO_2	44	0.1161	8.61	G				
Water	H_2O	18	0.0475	21.06	G				

*For most practical purposes, the value of 2 is sufficient.
†The molecular weight of 29 is merely the weighted average of the molecular weight of the constituents.
Source: P. W. Swain and L. N. Rowley, "Library of Practical Power Engineering" (collection of articles published in *Power*).

weight of N_2 in the fuel, step 1, is 0.560 + 0.0025 = 0.5625 lb N_2 in the products of combustion.

Next, find the total weight of the products of combustion by taking the sum of the CO_2, H_2O, and N_2 weights, or 0.11688 + 0.09332 + 0.5625 = 0.7727 lb. Now convert each weight to cubic feet at 650°F, the temperature of the combustion products, or:

Products	Weight, lb	Molecular weight	Temperature correction		Volume at 650°F, ft^3
CO_2	0.11688	44	(379/44)(0.11688)(2.255)	=	2.265
H_2O	0.09332	18	(379/18)(0.09332)(2.255)	=	4.425
N_2 (total)	0.5625	28	(379/28)(0.5625)(2.255)	=	17.190
Cubic feet of flue gas per cubic foot of natural-gas fuel				=	23.880

In this calculation, the value of 379 is used in the molecular-weight ratio because at 60°F and 14.7 psia the volume of 1 lb of any gas = 379/gas molecular weight. The fuel gas used is initially at 60°F and 14.7 psia. The ratio 2.255 = (650 + 460)/(32 + 460).

5. Compute the CO_2 content of the flue gas.

The CO_2, wet basis, = 2.265/23.88 = 0.0947, or 9.47 percent. The CO_2, dry basis, = 2.265/(23.88 − 4.425) = 0.1164, or 11.64 percent.

6. Compute the air required with the stated excess flow.

The pounds of air per pound of natural gas with 20 percent excess air = (1.20)(16.132) = 19.3584 lb air per pound of natural gas, or 19.3584/22.1 = 0.875 lb of air per cubic foot of natural gas (14.02 kg/m^3). See step 4 for an explanation of the value 22.1.

7. Compute the weight of the products of combustion.

Weight of the products of combustion = product weight for perfect combustion, lb + (percent excess air)(air for perfect combustion, lb) = 16.80 + (0.20)(16.132) = 20.03 lb.

8. Compute the volume of the combustion products and the percent CO_2.

The volume of excess air in the products of combustion is found by converting from the weight to the volumetric analysis and correcting for temperature as in step 4, using the air weight from step 2 for perfect combustion and the excess-air percentage, or (16.132/22.1)(0.20)(379/28.95)(2.255) = 4.31 ft^3. Add this to the volume of the products of combustion found in step 4, or 23.88 + 4.31 = 28.19 ft^3 (0.80 m^3).

Using the procedure in step 5, the percent CO_2, wet basis, = 2.265/28.19 = 0.0804, or 8.04 percent. The percent CO_2, dry basis, = 2.265/(28.19 − 4.425) = 0.0953, or 9.53 percent.

Related Calculations: Use the method given here when making combustion calculations for any type of gas used as a fuel—natural gas, blast-furnace gas, coke-oven gas, producer gas, water gas, sewer gas—from any source, domestic or foreign, in any type of furnace—boiler, heater, process, or waste-heat. When the air used for combustion contains moisture, as is usually true, this moisture is added to the combustion-formed moisture appearing in the products of combustion. Thus, for 80°F (300 K) air of 60 percent relative humidity, the moisture content is 0.013 lb per pound of dry air. This amount appears in the products of combustion for each pound of air used and is a commonly assumed standard in combustion calculations.

3

Phase Equilibrium

A.K.S. Murthy, Eng. Sc.D.

Head, Fuels Research Group
Allied Corporation
Morristown, NJ

REFERENCES: VAPOR PRESSURE OF PURE COMPOUNDS [1] Boublik, Fried, and Hala—*The Vapour Pressures of Pure Substances,* Elsevier; [2] Riddick and Bunger—*Organic Solvents,* 3d ed., vol. 2, Wiley-Interscience; [3] Wichterle and Linek—*Antoine Vapor Pressure Con-*

stants of Pure Compounds, Academia; [4] Zwolinski and Wilhoit—*Vapor Pressures and Heats of Vaporization of Hydrocarbons and Related Compounds,* Thermodynamics Research Center, Texas A&M University; [5] Zwolinski et al.—*Selected Values of Properties of Hydrocarbons and Related Compounds,* API Research Project 44, Thermodynamics Research Center, Texas A&M University; [6] Ohe—*Computer Aided Data Book of Vapor Pressure,* Data Book Publishing Co.; [7] Stull—*Ind. Eng. Chem. 39*:517, 1947. EQUATIONS OF STATE [8] Redlich and Kwong—*Chem. Rev. 44*:233, 1949; [9] Wohl—*Z. Phys. Chem. B2*:77, 1929; [10] Benedict, Webb, and Rubin—*Chem. Eng. Prog. 47*:419, 1951; *J. Chem. Physics 8*:334, 1940; *J. Chem. Physics 10*:747, 1942; [11] Soave—*Chem. Engr. Sci. 27*:1197, 1972; [12] Peng and Robinson—*Ind. Eng. Chem. Fund. 15*:59, 1976. FUGACITY OF PURE LIQUID [13] Pitzer and Curl—*J. Am. Chem. Soc. 79*:2369, 1957; [14] Chao and Seader—*AIChE Journal 7*:598, 1961. ACTIVITY COEFFICIENT [15] Wohl—*Trans. Am. Inst. Chem. Engrs. 42*:217, 1946; [16] Wilson—*J. Am. Chem. Soc. 86*:127, 1964; [17] Renon and Prausnitz—*AIChE Journal 14*:135, 1968; [18] Abrams and Prausnitz—*AIChE Journal 21*:116, 1975; [19] Fredenslund, Gmehling, and Rasmussen—*Vapor-Liquid Equilibria Using UNIFAC,* Elsevier; [20] Herington—*J. Inst. Petrol. 37*:457, 1951; [21] Van Ness, Byer, and Gibbs—*AIChE Journal 19*:238, 1973. VLE DATA [22] Chu, Wang, Levy, and Paul—*Vapor-Liquid Equilibrium Data,* Edwards; [23] Gmehling and Onken—*Vapor-Liquid Equilibrium Data Collection,* Chemistry Data Series, DECHEMA; [24] Hala, Wichterle, Polak, and Boublik—*Vapor-Liquid Equilibrium Data at Normal Pressures,* Pergamon Press; [25] Hirata, Ohe, and Nagahama—*Computer Aided Data Book of Vapor-Liquid Equilibria,* Kodansha/Elsevier; [26] Horsley—*Azeotropic Data,* American Chemical Society; [27] Wichterle, Linek, and Hala—*Vapor-Liquid Equilibrium Data Bibliography,* Elsevier; *Supplement I,* Elsevier, 1976. LIQUID-LIQUID EQUILIBRIUM [28] Francis—*Liquid-Liquid Equilibrium,* Interscience; [29] Francis—*Handbook for Components in Solvent Extraction,* Gordon & Breach; [30] Seidel—*Solubilities of Organic Compounds,* 3d ed., Van Nostrand; *Supplement,* Van Nostrand, 1952; [31] Stephen and Stephen—*The Solubilities of Inorganic and Organic Compounds,* Pergamon; [32] Sorenson and Arlt—*Liquid-Liquid Equilibrium Data Collection,* Chemistry Data Series, DECHEMA; [33] Sorensen et al.—*Fluid Phase Equilibria 2*:297, 1979; *Fluid Phase Equilibria 3*:47, 1979. GENERAL [34] Prausnitz—*Molecular Thermodynamics of Fluid-Phase Equilibria,* Prentice-Hall; [35] Hunter—*Ind. Eng. Chem. Fund. 6*:461, 1967; [36] Himmelblau—*Applied Nonlinear Programming,* McGraw-Hill.

3-1 Vapor-Liquid Equilibrium Ratios for Ideal-Solution Behavior

Assuming ideal-system behavior, calculate the K values and relative volatility for the benzene-toluene system at 373 K (212°F) and 101.3 kPa (1 atm).

Calculation Procedure:

1. Determine the relevant vapor-pressure data.

Design calculations involving vapor-liquid equilibrium (VLE), such as distillation, absorption, or stripping, are usually based on vapor-liquid equilibrium ratios, or K values. For the ith species, K_i is defined as $K_i = y_i/x_i$, where y_i is the mole fraction of that species in the vapor phase and x_i is its mole fraction in the liquid phase. Sometimes the design calculations are based on relative volatility $\alpha_{i,j}$, which equals K_i/K_j, the subscripts i and j referring to two different species. In general, K values depend on temperature and pressure and the compositions of both phases.

When a system obeys Raoult's law and Dalton's law, it is known as an "ideal system" (see Related Calculations for guidelines). Then $K_i = P_i^\circ/P$, and $\alpha_{i,j} = P_i^\circ/P_j^\circ$, where P_i° is the vapor pressure of the (pure) ith component at the system temperature and P is the total pressure.

One way to obtain the necessary vapor-pressure data for benzene and toluene is to employ the Antoine equation:

$$\log_{10} P^\circ = A - B/(t + C)$$

(See also Sec. 1.) When P° is in millimeters of mercury and t is temperature in degrees Celsius, the following are the values for the constants A, B, and C:

	A	B	C
Benzene	6.90565	1211.033	220.790
Toluene	6.95464	1424.255	219.482

Then, at 373 K (i.e., 100°C),

$$\log_{10} P^\circ_{\text{benzene}} = 6.90565 - 1211.033/(100 + 220.790) = 3.1305$$

and P°_{benzene} therefore equals 1350.5 mmHg (180.05 kPa). Similarly,

$$\log_{10} P^\circ_{\text{toluene}} = 6.95464 - 1424.255/(100 + 219.482) = 2.4966$$

so P°_{toluene} equals 313.8 mmHg (41.84 kPa).

2. *Divide vapor pressures by total pressure to obtain K values.*

Total pressure P is 1 atm, or 760 mmHg. Therefore, $K_{\text{benzene}} = 1350.5/760 = 1.777$, and $K_{\text{toluene}} = 313.8/760 = 0.413$.

3. *Calculate relative volatility of benzene with respect to toluene.*

Divide the vapor pressure of benzene by that of toluene. Thus, $\alpha_{\text{benzene-toluene}} = 1350.5/313.8 = 4.304$.

Related Calculations: Many systems deviate from the ideal solution behavior in either or both phases, so the K values given by $K_i = P_i^\circ/P$ are not adequate. The rigorous thermodynamic definition of K is

$$K_i = \gamma_i f_i^\circ/\phi_i P$$

where γ_i is the activity coefficient of the ith component in the liquid phase, f_i° is the fugacity of pure liquid i at system temperature T and pressure P, and ϕ_i is the fugacity coefficient of the ith species in the vapor phase.

In this definition, the activity coefficient takes account of nonideal liquid-phase behavior; for an ideal liquid solution, the coefficient for each species equals 1. Similarly, the fugacity coefficient represents deviation of the vapor phase from ideal gas behavior and is equal to 1 for each species when the gas obeys the ideal gas law. Finally, the fugacity takes the place of vapor pressure when the pure vapor fails to show ideal gas behavior, either because of high pressure or as a result of vapor-phase association or dissociation. Methods for calculating all three of these follow.

The vapor-phase fugacity coefficient can be neglected when the system pressure is low [e.g., less than 100 psi (689.5 kPa), generally] and the system temperature is not below a reduced temperature of 0.8. The pure-liquid fugacity is essentially equal to the vapor pressure at system temperatures up to a reduced temperature of 0.7. Unfortunately, however, many molecules (among them hydrogen fluoride and some organic acids) associate in the vapor phase and behave nonideally even under the preceding conditions. There is no widely recognized listing of all such compounds. As for nonideality in the liquid phase, perhaps the most important cause of it is hydrogen bonding. For general rules of thumb for predicting hydrogen bonding, see R. H. Ewell, J. M. Harrison, and L. Berg, *Ind. Eng. Chem. 36(10)*:871, 1944.

3-2 Fugacity of Pure Liquid

Calculate the fugacity of liquid hydrogen chloride at 40°F (277.4 K) and 200 psia (1379 kPa). (The role of fugacity in phase equilibrium is discussed under Related Calculations in Example 3-1).

Calculations Procedure:

1. Calculate the compressibility factor.

For components whose critical temperature is greater than the system temperature,

$$f^\circ = \nu P^\circ \exp\left[V(P - P^\circ)/RT\right]$$

where f° is the pure-liquid fugacity, ν is the fugacity coefficient for pure vapor at the system temperature, P° is the vapor pressure at that temperature, V is the liquid molar volume, P is the system pressure, and T is the absolute temperature.

Thermodynamically, the fugacity coefficient is given by

$$\ln \nu = \int_0^{P^\circ} \frac{Z - 1}{P}\, dP$$

where Z is the compressibility factor. This integral has been evaluated for several equations of state. For instance, for the Redlich-Kwong equation, which is very popular in engineering design and is employed here, the relationship is

$$\ln \nu = (Z - 1) - \ln(Z - BP^\circ) - \frac{A^2}{B}\ln\left(1 + \frac{BP^\circ}{Z}\right)$$

where $A^2 = 0.4278/T_r^{2.5}P_c$
$B = 0.0867/T_rP_c$
T_r = reduced temperature (T/T_c)
T_c = critical temperature
P_c = critical pressure

The compressibility factor Z is calculated by solving the following cubic equation (whose symbols are as defined above):

$$Z^3 - Z^2 + [A^2P^\circ - BP^\circ(1 + BP^\circ)]Z + (A^2P^\circ)(BP^\circ) = 0$$

The critical temperature of HCl is 584°R, so the reduced temperature at 40°F is (460 + 40)/584 or 0.85616. The critical pressure of HCl is 1206.9 psia. Then, $A^2 = 0.4278/[0.85616^{2.5}(1206.9)] = 5.226 \times 10^{-4}$, and $B = 0.0867/(0.85616 \times 1206.9) = 8.391 \times 10^{-5}$. The vapor pressure P° of HCl at 40°F, is 423.3 psia, so $A^2P^\circ = 0.2212$, $BP^\circ = 0.03552$, $A^2P^\circ - BP^\circ(1 + BP^\circ) = 0.1844$, and $(A^2P)(BP^\circ) = 7.857 \times 10^{-3}$. The cubic equation thus becomes $Z^3 - Z^2 + 0.1844Z + 7.856 \times 10^{-3} = 0$.

This equation can be solved straightforwardly or by trial and error. The largest real root is the compressibility factor for the vapor; in this case, $Z = 0.73431$.

2. ***Calculate the fugacity coefficient.***

Using the preceding relationship based on the Redlich-Kwong equation,

$$\ln \nu = (0.73431 - 1) - \ln(0.73431 - 0.03552) - \frac{5.226 \times 10^{-4}}{8.391 \times 10^{-5}} \ln\left(1 + \frac{0.03552}{0.73431}\right)$$

$$= -0.2657 + 0.3584 - 0.2942 = -0.2015$$

Therefore, $\nu = 0.8175$.

3. ***Calculate the fugacity.***

The density of saturated vapor at 40°F is 55 lb/ft^3, and the molecular weight of HCl is 36.46. Therefore, the liquid molar volume V is 36.46/55 = 0.663 ft^3/(lb·mol). The gas constant R is 10.73 (psia)(ft^3)/(lb·mol)(°R). Therefore,

$$\exp[V(P - P^\circ)/RT] = \exp\left[\frac{0.663(200 - 423.3)}{10.73(460 + 40)}\right] = 0.9728$$

Finally, using the equation for fugacity at the beginning of this problem, $f^\circ = 0.8175 \times 423.3 \times 0.9728 = 336.6$ psi (2321 kPa).

The exponential term, 0.9728 in this equation, is known as the "Poynting correction." It is greater than unity if system pressure is greater than the vapor pressure. The fugacity coefficient ν is always less than unity. Depending on the magnitudes of ν and the Poynting correction, fugacity of pure liquid f° can thus be greater or less than the vapor pressure.

Related Calculations: This procedure is valid only for those components whose critical temperature is above the system temperature. When the system temperature is instead above the critical temperature, generalized fugacity-coefficient graphs can be used. However, such an approach introduces the concept of hypothetical liquids. When accurate results are needed, experimental measurements should be made. The Henry constant, which can be experimentally determined, is simply $\gamma^\infty f^\circ$, where γ^∞ is the activity coefficient at infinite dilution (see Example 3-8).

Use of generalized fugacity coefficients (e.g., see Example 1-18) eliminates some computational steps. However, the equation-of-state method used here is easier to program on a programmable calculator or computer. It is completely analytical, and use of an equation of state permits the computation of all the thermodynamic properties in a consistent manner.

3-3 Nonideal Gas-Phase Mixtures

Calculate the fugacity coefficients of the components in a gas mixture containing 80% HCl and 20% dichloromethane (DCM) at 40°F (277.4 K) and 200 psia (1379 kPa).

Calculation Procedure:

1. Calculate the compressibility factor for the mixture.

In a manner similar to that used in the previous problem, an expression for the fugacity coefficient in vapor mixtures can be derived from any equation of state applicable to such mixtures. If the Redlich-Kwong equation of state is used, the expression is

$$\ln \phi_i = (Z - 1)\frac{B_i}{B} - \ln (Z - BP) - \frac{A^2}{B}\left(\frac{2A_i}{A} - \frac{B_i}{B}\right) \ln \left(1 + \frac{BP}{Z}\right)$$

where ϕ_i = the fugacity coefficient of the ith component in the vapor
$A_i^2 = 0.4278/T_{r,i}^{2.5}P_{c,i}$
$B_i = 0.0867/T_{r,i}P_{c,i}$
$A = \Sigma y_i A_i$
$B = \Sigma y_i B_i$
$T_{r,i}$ = the reduced temperature of component i ($T/T_{c,i}$)
$T_{c,i}$ = the critical temperature of component i
$P_{c,i}$ = the critical pressure of component i
y_i = the mole fraction of component i in the vapor mixture.

The compressibility factor Z is calculated by solving the cubic equation:

$$Z^3 - Z^2 + [A^2P - BP(1 + BP)]Z + (A^2P)(BP) = 0$$

The relevant numerical inputs are:

	HCl	DCM
Critical temperature $T_{c,i}$, °R	584	933
Critical pressure $P_{c,i}$, psia	1206.9	893
Reduced temperature $T_{r,i}$ at 40°F	0.8562	0.5359
A_i, equal to $(0.4278/T_{r,i}^{2.5}P_{c,i})^{1/2}$	0.02286	0.04774
B_i, equal to $0.0867/T_{r,i}P_{c,i}$	8.390×10^{-5}	1.812×10^{-4}
Mole fraction y_i	0.8	0.2

Then, $A = y_1A_1 + y_2A_2 = 0.0278$, $B = y_1B_1 + y_2B_2 = 1.0336 \times 10^{-4}$, $A^2P = (0.0278)^2(200) = 0.1546$, $BP = (1.0336 \times 10^{-4})(200) = 0.02067$, $A^2P - BP(1 + BP) = 0.1335$, and $(A^2P)(BP) = 0.003195$. The cubic equation becomes $Z^3 - Z^2 + 0.1335Z + 0.003195 = 0$.

This equation can be solved straightforwardly or by trial and error. The largest real root is the compressibility factor; in this case, $Z = 0.8357$.

2. Calculate the fugacity coefficients.

Substituting into the preceding relationship based on the Redlich-Kwong equation,

$$\begin{aligned}\ln \phi_i &= (0.8357 - 1)B_i/1.0336 \times 10^{-4} - \ln (0.8357 - 0.02067) \\ &\quad - (0.0278^2/1.0336 \times 10^{-4})[(2A_i/0.0278) \\ &\quad - (B_i/1.0336 \times 10^{-4})][\ln (1 + 0.02067/0.8357)] \\ &= -1589.6B_i + 0.2045 - 0.1827(71.94A_i - 9674.9B_i)\end{aligned}$$

Letting subscript 1 correspond to HCl and subscript 2 to DCM, the equation yields $\ln \phi_1 = -0.08095$ and $\ln \phi_2 = -0.3905$. Therefore, $\phi_1 = 0.9222$ and $\phi_2 = 0.6767$.

Related Calculations: Certain compounds, such as acetic acid and hydrogen fluoride, are known to form dimers, trimers, or other oligomers by association in the vapor phase. Simple equations of state are not adequate for representing the nonideality in systems containing such compounds. Unfortunately, there is no widely recognized listing of all such compounds.

3-4 Nonideal Liquid Mixtures

Calculate the activity coefficients of chloroform and acetone at 0°C in a solution containing 50 mol % of each component, using the Wilson-equation model. The Wilson constants for the system (with subscript 1 pertaining to chloroform and subscript 2 to acetone) are

$$\lambda_{1,2} - \lambda_{1,1} = -332.23 \text{ cal/(g}\cdot\text{mol)} \; [-1390 \text{ kJ/(kg}\cdot\text{mol)}]$$

$$\lambda_{1,2} - \lambda_{2,2} = -72.20 \text{ cal/(g}\cdot\text{mol)} \; [-302.1 \text{ kJ/(kg}\cdot\text{mol)}]$$

Calculation Procedure:

1. Compute the $G_{i,j}$ parameters for the Wilson equation.

General engineering practice is to establish liquid-phase nonideality through experimental measurement of vapor-liquid equilibrium. Models with adjustable parameters exist for adequately representing most nonideal-solution behavior. Because of these models, the amount of experimental information needed is not excessive (see Example 3-9, which shows procedures for calculating such parameters from experimental data).

One such model is the Wilson-equation model, which is applicable to multicomponent systems while having the attraction of entailing only parameters that can be calculated from binary data alone. Another attraction is that the Wilson constants are approximately independent of temperature. This model is

$$\ln \gamma_i = 1 - \ln \sum_{j=1}^{N} x_j G_{j,i} - \sum_{j=1}^{N} \left(x_j G_{i,j} \bigg/ \sum_k x_k G_{k,j} \right)$$

where γ_i is the activity coefficient of the ith component, x_i is the mole fraction of that component, and $G_{i,j} = V_i/V_j \exp [-(\lambda_{i,j} - \lambda_{j,j})/RT]$, with V_i being the liquid molar volume of the ith component and $(\lambda_{i,j} - \lambda_{j,j})$ being the Wilson constants. Note that $\lambda_{i,j} = \lambda_{j,i}$, but that $G_{i,j}$ is not equal to $G_{j,i}$.

For a binary system, $N = 2$ and $(x_i + x_j) = 1.0$, and the model becomes

$$\ln \gamma_1 = -\ln (x_1 + x_2 G_{2,1}) + x_2 \left[\frac{G_{2,1}}{x_1 + x_2 G_{2,1}} - \frac{G_{1,2}}{x_2 + x_1 G_{1,2}} \right]$$

and

$$\ln \gamma_2 = -\ln (x_2 + x_1 G_{1,2}) + x_1 \left[\frac{G_{1,2}}{x_2 + x_1 G_{1,2}} - \frac{G_{2,1}}{x_1 + x_2 G_{2,1}} \right]$$

where

$$G_{1,2} = \frac{V_1}{V_2} \exp \left(- \frac{\lambda_{1,2} - \lambda_{2,2}}{RT} \right) \quad \text{and} \quad G_{2,1} = \frac{V_2}{V_1} \exp \left(- \frac{\lambda_{1,2} - \lambda_{1,1}}{RT} \right)$$

At 0°C, $V_1 = 71.48$ cc/(g·mol) and $V_2 = 78.22$ cc/(g·mol). Therefore, for the chloroform/acetone system,

$$G_{1,2} = \frac{71.48}{78.22} \exp \left(- \frac{(-72.20)}{(1.9872)(273.15)} \right) = 1.0438$$

and

$$G_{2,1} = \frac{78.22}{71.48} \exp \left(- \frac{(-332.23)}{(1.9872)(273.15)} \right) = 2.0181$$

2. *Calculate the activity coefficients.*

The mole fractions x_1 and x_2 each equal 0.5. Therefore, $x_1 + x_2 G_{2,1} = 1.509$, and $x_2 + x_1 G_{1,2} = 1.022$. Substituting into the Wilson model,

$$\ln \gamma_1 = -\ln 1.509 + 0.5 \left(\frac{2.0181}{1.509} - \frac{1.0438}{1.022} \right) = -0.2534$$

Therefore, $\gamma_1 = 0.7761$. And,

$$\ln \gamma_2 = -\ln 1.022 + 0.5 \left(\frac{1.0438}{1.022} - \frac{2.0181}{1.509} \right) = -0.1798$$

Therefore, $\gamma_2 = 0.8355$.

This system, where the activity coefficients are less than unity, is an example of negative deviation from ideal behavior.

Related Calculations: When the Wilson model is used for systems with more than two components, it is important to remember that the $G_{i,j}$ summations must be made over every possible pair of components (also, remember that $G_{i,j} \neq G_{j,i}$).

A limitation on the Wilson equation is that it is not applicable to systems having more than one liquid phase. The NRTL model, which is similar to the Wilson equation, may be used for systems forming two liquid phases.

An older model for predicting liquid-phase activity coefficients is that of Van Laar, in which (for two components) $\ln \gamma_1 = A_{1,2}/(1 + A_{1,2}x_1/A_{2,1}x_2)^2$, with the A's being constants to be determined from experimental data. This model can handle systems having more than one liquid phase. Another older model is that of Margules, available in "two-suffix" and "three-suffix" versions. These are (for two components), respectively, $\ln \gamma_1 = Ax_2^2$ and $\ln \gamma_1 = x_2^2[A_{1,2} + 2(A_{2,1} - A_{1,2})x_1]$. Ternary versions of these older models are available, but extending the Van Laar and Margules correlations to multicomponent systems is in general rather awkward.

Fredenslund and coworkers have developed the UNIFAC correlation, which is satisfactory for those systems covered by the extensive amount of experimental data in their work. For details, see A. Fredenslund, J. Gmehling, and P. Rasumssen, *Vapor-Liquid Equilibria Using UNIFAC,* Elsevier Scientific Publishing Co., Amsterdam, 1977.

3-5 *K* Value for Ideal Liquid Phase, Nonideal Vapor Phase

Assuming the liquid phase but not the vapor phase to be ideal, calculate the K values for HCl and dichloromethane (DCM) in a system at 200 psia (1,379 kPa) and 40°F (277.4 K) and whose vapor composition is 80 mol % HCl. Also calculate the relative volatility. Compare the calculated values with those which would exist if the system showed ideal behavior.

Calculation Procedure:

1. Set out the relevant form of the thermodynamic definition of K value.

Refer to the rigorous definition of K value given under Related Calculations in Example 3-1. When the liquid phase is ideal, then $\gamma_i = 1$. Thus the relevant form is

$$K_i = f_i^\circ/\phi_i P$$

where f_i° is the fugacity of pure liquid i at system temperature and pressure, ϕ_i is the fugacity coefficient of the ith species in the vapor phase, and P is the system pressure. Let subscript 1 pertain to HCl and subscript 2 pertain to DCM.

2. Determine the pure-liquid fugacities and the vapor-phase fugacity coefficients.

The fugacity of HCl at 200 psia and 40°F was calculated in Example 3-2 to be 336.6 psi. The same procedure can be employed to find f_2°, the pure-liquid fugacity of DCM.

With reference to that example, $A^2 = 0.04774$ and $B_2 = 1.812 \times 10^{-4}$; the vapor pressure of DCM at 40°F is 3.28 psi. The resulting cubic equation for compressibility factor yields a Z of 0.993; the Redlich-Kwong relationship yields a ν_2 of 0.993. The liquid molar volume of DCM at 40°F is 1.05 ft³/(lb·mol), so the Poynting correction is calculated to be 1.039, and the resulting value for f_2° emerges as 3.38 psi (23.3 kPa). As for the fugacity coefficients, they were calculated in Example 3-3 to be $\phi_1 = 0.9222$ and $\phi_2 = 0.6767$.

3. Calculate the K values and relative volatility.

Using the equation in step 1, $K_1 = 336.6/(0.9222)(200) = 1.82$, and $K_2 = 3.38/(0.6767)(200) = 0.025$. From the definition in Example 3-1, relative volatility $\alpha_{1,2} = K_1/K_2 = 1.82/0.025 = 72.8$.

4. Compare these results with those which would prevail if the system were ideal.

As indicated in Example 3-1, the K value based on ideal behavior is simply the vapor pressure of the component divided by the system pressure. Vapor pressures of HCl and DCM at 40°F are, respectively, 423.3 and 3.28 psia. Thus $K_1^{\text{ideal}} = 423.3/200 = 2.12$, $K_2^{\text{ideal}} = 3.28/200 = 0.016$, and $\alpha_{1,2}^{\text{ideal}} = 2.12/0.016 = 133$.

It may be seen that under the system conditions that prevail in this case, the actual relative volatility is considerably lower than it would be if the system were ideal.

3-6 *K* Value for Ideal Vapor Phase, Nonideal Liquid Phase

Assuming the vapor phase but not the liquid phase to be ideal, calculate the K values for ethanol and water in an 80% ethanol solution at 500 mmHg (66.7 kPa) and 70°C (158°F, 343 K). Also calculate the relative volatility, and compare the calculated values with those which would exist if the system showed ideal behavior.

Calculation Procedure:

1. Set out the relevant form of the thermodynamic definition of K value.

At this low system pressure, the vapor-phase nonideality is negligible. Since neither component has a very high vapor pressure at the system temperature, and since the differences between the vapor pressures and the system pressure are relatively small, the pure-liquid fugacities can be taken to be essentially the same as the vapor pressures.

Refer to the rigorous definition of K value given under Related Calculations in Example 3-1. Taking into account the assumptions in the preceding paragraph, this definition simplifies into $K_i = \gamma_i P_i/P$, where γ_i is the activity coefficient of the ith component in the liquid phase, P_i is the vapor pressure of that component, and P is the system pressure.

2. Determine the activity coefficients and the vapor pressures.

The activity coefficients can be calculated from the Wilson-equation model, as discussed in Example 3-4, or by one of the other methods discussed in the same example under

Related Calculations. For instance, the experimentally determined Van Laar constants for the ethanol/water system are $A_{1,2} = 1.75$ and $A_{2,1} = 0.91$, where subscripts 1 and 2 refer to ethanol and water, respectively. So the Van Laar model discussed there becomes $\ln \gamma_1 = 1.75/[1 + (1.75)(0.08)/(0.91)(0.2)]^2$, and (by interchanging subscripts) $\ln \gamma_2 = 0.91/[1 + (0.91)(0.2)/(1.75)(0.8)]^2$. Accordingly, $\gamma_1 = 1.02$, and $\gamma_2 = 2.04$.

From tables, the vapor pressures of ethanol and water at 70°C are 542 and 233 mmHg, respectively.

3. *Calculate the K values and relative volatility.*

Using the equation in step 1, $K_1 = (1.02)(542)/500 = 1.11$, and $K_2 = (2.04)(233)/500 = 0.95$.

From the definition in Example 3-1, relative volatility $\alpha_{1,2} = K_1/K_2 = 1.11/0.95 = 1.17$.

4. *Compare these results with those which would prevail if the system were ideal.*

As indicated in that example, the K value based on ideal behavior is simply the vapor pressure of the component divided by the system pressure. Thus $K_1^{\text{ideal}} = 542/500 = 1.084$, $K_2^{\text{ideal}} = 233/500 = 0.466$, and $\alpha_{1,2}^{\text{ideal}} = 1.084/0.466 = 2.326$. Under the system conditions that prevail in this case, actual relative volatility is considerably lower than the ideal-system value.

3-7 Thermodynamic Consistency of Experimental Vapor-Liquid Equilibrium Data

Vapor-liquid equilibrium data for the ethanol/water system (subscripts 1 and 2, respectively) at 70°C (158°F, 343 K) are given in the three left columns of Table 3-1. Check to see if the data are thermodynamically consistent.

TABLE 3-1 Ethanol/Water System at 70°C (Examples 3-7 and 3-8)

Experimental data			Calculated results				
Pressure P, mmHg	Mole fraction ethanol in liquid x_1	Mole fraction ethanol in vapor y_1	γ_1, $Py_1/P_1^\circ x_1$	γ_2, $Py_2/P_2^\circ x_2$	$\ln (\gamma_1/\gamma_2)$	$\ln \gamma_1/x_2^2$	$-\ln \gamma_2/x_1^2$
362.5	0.062	0.374	4.034	1.038	1.357	1.585	—
399.0	0.095	0.439	3.402	1.062	1.165	1.495	—
424.0	0.131	0.482	2.878	1.085	0.976	1.400	—
450.9	0.194	0.524	2.247	1.143	0.676	1.246	—
468.0	0.252	0.552	1.891	1.203	0.452	1.139	—
485.5	0.334	0.583	1.564	1.305	0.181	1.008	—
497.6	0.401	0.611	1.399	1.387	0.009	0.936	—
525.9	0.593	0.691	1.131	1.714	−0.416	—	−1.532
534.3	0.680	0.739	1.071	1.870	−0.557	—	−1.354
542.7	0.793	0.816	1.030	2.070	−0.698	—	−1.157
543.1	0.810	0.826	1.022	2.135	−0.737	—	−1.156
544.5	0.943	0.941	1.002	2.419	−0.881	—	−0.993
544.5	0.947	0.945	1.002	2.425	−0.883	—	−0.988

Calculation Procedure:

1. *Select the criterion to be used for thermodynamic consistency.*

Deviations from thermodynamic consistency arise as a result of experimental errors. Impurities in the samples used for vapor-liquid equilibrium measurements are often the source of error. A complete set of vapor-liquid equilibrium data includes temperature T, pressure P, liquid composition x_i, and vapor composition y_i. Usual practice is to convert these data into activity coefficients by the following equation, which is a rearranged form of the equation that rigorously defines K values (i.e., defines the ratio y_i/x_i under Related Calculations in Example 3-1):

$$\gamma_i = \phi_i P y_i / f_i^\circ x_i$$

The fugacity coefficients ϕ_i are estimated using procedures described in Example 3-3. Fugacity f_i° of pure liquid is calculated using procedures described in Example 3-2. Assuming that ϕ_i and f_i° are correctly calculated, the γ_i obtained using the preceding equation must obey the Gibbs-Duhem equation. The term "thermodynamically consistent data" is used to refer to data that obey that equation.

At constant temperature, the Gibbs-Duhem equation can be rearranged to give the approximate equality

$$\int_0^1 \ln \frac{\gamma_1}{\gamma_2}\, dx_1 = 0$$

(This would be an exact equality if both temperature and pressure were constant, but that would be inconsistent with the concept of vapor-liquid equilibrium.) In other words, the net area under the curve ln (γ_1/γ_2) versus x_1 should be zero. This means that the area above and below the x axis must be equal. Since real data entail changes in system pressure and are subject to experimental errors and errors in estimating f_i° and ϕ_i, the preceding requirement cannot be expected to be exactly satisfied. However, the deviation should not be more than a few percent of the total absolute area.

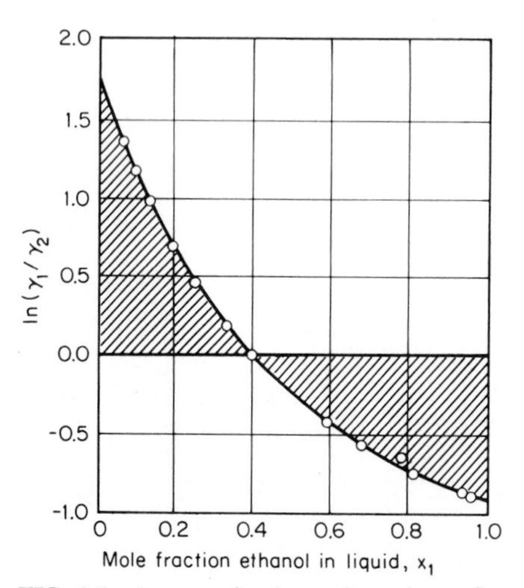

FIG. 3-1 Area test for thermodynamic consistency of data on ethanol(1)/water(2) system (Example 3-7).

2. *Determine the system activity coefficients, and plot the natural logarithm of their ratio against liquid-phase ethanol content.*

As in the previous example, and for the reasons discussed there, the system pressures and vapor pressures in the present example are such that the vapor pressure may be used for f° and $\phi_i = 1$. Activity coefficients calculated from the experimental data, as well as ln (γ_1/γ_2) values, are also given in Table 3-1. A plot of ln (γ_1/γ_2) versus x_1 is shown in Fig. 3-1.

3. Determine and evaluate the net area under the plot.

By measurement, the area above the x axis is 0.295, and the area below it is 0.325. Net area is $0.295 - 0.325 = 0.03$. The total absolute area is $0.295 + 0.325 = 0.62$. Net area as a percent of total area is thus 5 percent. This figure is small enough that the data can be assumed to be thermodynamically consistent.

Related Calculations: When vapor-liquid equilibrium data are taken under isobaric rather than isothermal conditions, as is often the case, the right-hand side of the preceding Gibbs-Duhem equation cannot as readily be taken to approximate zero. Instead, the equation should be taken as

$$\int_0^1 \ln\left(\frac{\gamma_1}{\gamma_2}\right) dx = \int_{x=0}^{x=1} \left(\frac{\Delta H}{RT^2}\right) dT$$

where ΔH is the heat of mixing.

For some systems, the integral on the right-hand side may indeed be neglected, i.e., set equal to zero. These include systems consisting of chemically similar components with low values for the heat of mixing and ones in which the boiling points of pure components are close together. In general, however, this will not be the case.

Since the enthalpy of mixing necessary for the evaluation of the integral is often not available, the integral can instead be estimated as $1.5(\Delta T_{max})/T_{min}$. Here, T_{min} is the lowest-boiling temperature in the isobaric system. Usually, this will be the boiling temperature of the lower-boiling component. In cases of low-boiling azeotropes, it is the boiling temperature of the azeotrope. Moreover, ΔT_{max} is the maximum difference of boiling points in the total composition range of the isobaric system. Usually, this will be the boiling-point difference of the pure components. For azeotropic systems, ΔT_{max} is the difference between the boiling temperature of the azeotrope and the component boiling most distant from it.

The preceding test for thermodynamic data treats the data set as a whole. It does not determine whether individual data points are consistent. A point-consistency test has been proposed, but it is too cumbersome for manual calculations.

3-8 Estimating Infinite-Dilution Activity Coefficients

Estimate the infinite-dilution activity coefficients of ethanol and water at 70°C (158°F, 343 K) using the data given in the left three columns of Table 3-1.

Calculation Procedure:

1. Determine the system activity coefficients, calculate the ratios ln γ_1/x_2^2, −ln γ_2/x_1^2, and ln (γ_1/γ_2), and plot them against x_1.

In a binary system, the infinite-dilution activity coefficients are defined as

$$\gamma_1^\infty = \lim_{x_1 \to 0} \gamma_1 \qquad \text{and} \qquad \gamma_2^\infty = \lim_{x_1 \to 1} \gamma_2$$

They are generally calculated for two reasons: First, activity coefficients in the very dilute range are experimentally difficult to measure but are commonly needed when designing separation systems, and estimation of infinite-dilution activity coefficients is needed for meaningful extrapolation of whatever experimental data are available. Second, these coefficients are useful when estimating the parameters that are required in several mathematical models used for determining activity coefficients.

One way to estimate infinite-dilution activity coefficients is related to the preceding example on thermodynamic consistency; in addition, it takes advantage of the fact that the plot of $\ln \gamma_1/x_2^2$ versus x_1 is considerably more linear than is $\ln \gamma_1$ versus x_1.

The first step is to calculate $\ln \gamma_1/x_2^2$, $-\ln \gamma_2/x_1^2$, and $\ln (\gamma_1/\gamma_2)$. The results are shown in Table 3-1 and plotted as a function of x_1 in Fig. 3-2.

2. ***Extrapolate the two curves relating to γ_1 so that they converge on a common point at $x_1 = 0$ and the two curves relating to γ_2 so that they converge at $x_1 = 1$. Determine these two points of convergence and find their antilogarithms.***

The extrapolations are carried out in Fig. 3-2. The point of convergence along $x_1 = 0$ that corresponds to $\ln \gamma_1^\infty$ is 1.75, so γ_1^∞ is 5.75. The point of convergence along $x_1 = 1$ that corresponds to $-\ln \gamma_2^\infty$ is -0.91, so γ_2^∞ is 2.48.

3-9 Estimating the Parameters for the Wilson-Equation Model for Activity Coefficients

Calculate the Wilson constants for the ethanol/water system using the infinite-dilution activity coefficients calculated in the preceding example: $\gamma_1^\infty = 5.75$ and $\gamma_2^\infty = 2.48$, with subscript 1 pertaining to ethanol.

Calculation Procedure:

1. ***Rearrange the Wilson-equation model so that its constants can readily be calculated from infinite-dilution activity coefficients.***

In Example 3-4, given Wilson constants were employed in the Wilson-equation model to calculate the activity coefficients for the two components of a binary nonideal liquid mixture. The present example, in essence, reverses the procedure; it employs known activity coefficients (at infinite dilution) in order to calculate Wilson constants, so that these can be employed to determine activity coefficients in other situations concerning the same two components.

As shown in the Example 3-4, $G_{1,2} = (V_1/V_2) \exp [-(\lambda_{1,2} - \lambda_{2,2})RT]$. This can be rearranged into $\lambda_{1,2} - \lambda_{2,2} = -RT \ln (G_{1,2}V_2/V_1)$. Similarly, $\lambda_{1,2} - \lambda_{1,1} = -RT \ln (G_{2,1}V_1/V_2)$. Moreover, at infinite dilution, when x_1 and x_2, respectively, equal zero, the Wilson-equation model becomes $G_{1,2} + \ln G_{2,1} = 1 - \ln \gamma_1^\infty$, and $G_{2,1} + \ln G_{1,2} = 1 - \ln \gamma_2^\infty$.

Thus the problem consists of first calculating $G_{1,2}$ and $G_{2,1}$ from the known values of γ_1^∞ and γ_2^∞ and then employing the calculated G values to determine the Wilson constants, $\lambda_{1,2} - \lambda_{1,1}$ and $\lambda_{1,2} - \lambda_{2,2}$.

2. ***Calculate the G values.***

This calculation, involving a transcendental function, must be made by trial and error. Since $\ln \gamma_1^\infty = \ln 5.75 = 1.75$ and $\ln \gamma_2^\infty = \ln 2.48 = 0.91$, the preceding G equations become $G_{1,2} + \ln G_{2,1} = 1 - 1.75 = -0.75$, and $G_{2,1} + \ln G_{1,2} = 1 - 0.91 = 0.09$. The procedure consists of guessing a value for $G_{1,2}$, then using this value in the first equation to calculate a $G_{2,1}$, then using this $G_{2,1}$ in the second equation to calculate a $G_{1,2}$ (designated $G^*_{1,2}$), and repeating these steps until the difference between (guessed) $G_{1,2}$ and (calculated) $G^*_{1,2}$ becomes sufficiently small.

Start with $G_{1,2}$ guessed to be 1.0. Then $G_{2,1} = \exp(-0.75 - 1) = 0.1738$, $G^*_{1,2} = \exp(0.09 - 0.1738) = 0.9196$, and $G^*_{1,2} - G_{1,2} = 0.9196 - 1 = -0.0804$. Next, guess $G_{1,2}$ to be 0.9. Then $G_{2,1} = 0.192$, $G^*_{1,2} = 0.903$, and $G^*_{1,2} - G_{1,2} = 0.003$. Next, guess $G_{1,2}$ to be 0.9036 (by linear interpolation). Then $G_{2,1} = 0.1914$, $G^*_{1,2} = 0.9036$, and $G^*_{1,2} - G_{1,2} = 0$. Hence the solution is $G_{1,2} = 0.9036$ and $G_{2,1} = 0.1914$.

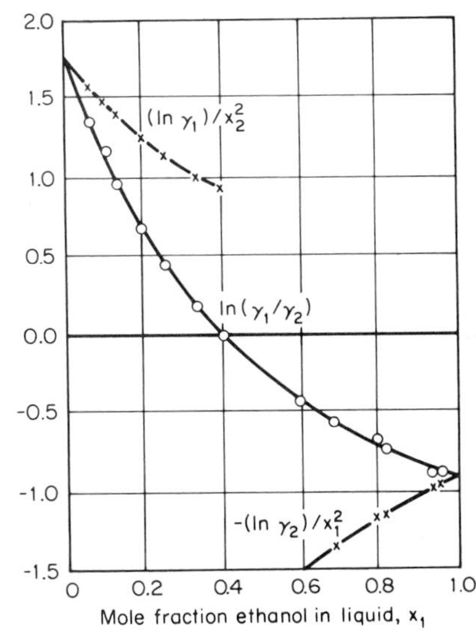

FIG. 3-2 Determining infinite-dilution activity coefficients by extrapolation (Example 3-8).

3. Calculate the Wilson constants.

At 70°C (343 K), the temperature at which the activity coefficients were determined, $V_1 = 62.3$ cc/(g·mol), and $V_2 = 18.5$ cc/(g·mol). Then $\lambda_{1,2} - \lambda_{2,2} = -RT \ln (G_{1,2}V_1/V_2) = -(1.987)343 \ln [0.9036(18.5)/62.3] = 897$ cal/(g·mol) [3724 kJ/(kg·mol)], and $\lambda_{1,2} - \lambda_{1,1} = -(1.987)343 \ln [0.1914(62.3)/18.5] = 299$ cal/(g·mol) [1251 kJ/(kg·mol)].

Related Calculations: The constants for the binary Margules and Van Laar models for predicting activity coefficients (see Related Calculations under Example 3-4) are simply the natural logarithms of the infinite-dilution activity coefficients: $A_{1,2} = \ln \gamma_1^\infty$ and $A_{2,1} = \ln \gamma_2^\infty$.

The Wilson, Margules, and Van Laar procedures described in this example are suitable for manual calculation. However, to take full advantage of all the information available from whatever vapor-liquid equilibrium data are at hand, statistical procedures for estimating the parameters should instead be employed, with the aid of digital computers. These procedures fall within the domain of nonlinear regression analysis.

In such an analysis, one selects a suitable objective function and then varies the parameters so as to maximize or minimize the function. Theoretically, the objective function should be derived using the statistical principles of maximum-likelihood estimation. In practice, however, it is satisfactory to use a weighted-least-squares analysis, as follows:

Usually, the liquid composition and the temperature are assumed to be error-free independent variables. Experimental values of vapor composition and total pressure are employed to calculate vapor-phase fugacity coefficients and pure-liquid fugacities. Then one uses trial and error in guessing parameters in the activity-coefficient model to be determined so as to arrive at parameter values that minimize the weighted sums of the squared differences between (1) experimental values of total pressure and of vapor mole fractions and (2) the values of pressure and mole fraction that are calculated on the basis of the model-generated activity coefficients. For several algorithms for minimizing the sum of squares, see D. M. Himmelblau, *Applied Nonlinear Programming,* McGraw-Hill, New York, 1972.

3-10 Calculating Dew Point When Liquid Phase Is Ideal

Calculate the dew point of a vapor system containing 80 mol % benzene and 20 mol % toluene at 1000 mmHg (133.3 kPa).

Calculation Procedure:

1. Select a temperature, and test its suitability by trial and error.

The dew point of a system at pressure P whose vapor composition is given by mole fractions y_i is that temperature at which there is onset of condensation. Mathematically, it is that temperature at which

$$\sum_{i=1}^{N} \frac{y_i}{K_i} = 1$$

where the K_i are vapor-liquid equilibrium ratios as defined in Example 3-1.

When the liquid phase is ideal, K_i depends only on the temperature, the pressure, and the vapor composition. The procedure for determining the dew point in such a case is to (1) guess a temperature; (2) calculate the K_i, which equal $f_i^\circ/\phi_i P$, where f_i° is the fugacity of pure liquid i at the system temperature and pressure, ϕ_i is the fugacity coefficient of the ith species in the vapor phase, and P is the system pressure; and (3) check if the preceding dew-point equation is satisfied. If it is not, repeat the procedure with a different guess.

For this system, f_i° may be assumed to be the same as the vapor pressure. The vapor phase can be assumed to be ideal, that is, $\phi_i = 1$. (For a discussion of the grounds for these assumptions, see Example 3-6).

The boiling points of benzene and toluene at 1000 mmHg are first calculated (for instance, by using the Antoine equation, as discussed in Example 3-1). They are 89°C and 141°C, respectively. As a first guess at the dew-point temperature, try a linear interpolation of these boiling points: $T = (0.8)(89) + (0.2)(141)$, which approximately equals 100. Let subscript 1 refer to benzene; subscript 2 to toluene.

At 100°C, $P_1^\circ = 1350$ mmHg and $P_2^\circ = 314$ mmHg (see Example 3-1). Therefore, $K_1 = 1350/1000 = 1.35$ and $K_2 = 314/1000 = 0.314$. Moreover, the dew-point equation becomes $(y_1/K_1) + (y_2/K_2) = (0.8)/(1.35) + (0.2)/(0.314) = 1.230$.

Since this sum is greater than 1, the K values are too low. So the vapor pressures and, accordingly, the assumed temperature are too low.

2. Repeat the trial-and-error procedure with the aid of graphic interpolation until the dew-point temperature is found.

As the next estimate of the dew point, try 110°C. At this temperature, $K_1 = 1.756$ and $K_2 = 0.428$, as found via the procedure outlined in the previous step. Then $(y_1/K_1) + (y_2/K_2) = 0.923$. Since the sum is less than 1, the assumed temperature is too high.

Next plot $(y_1/K_1) + (y_2/K_2)$ against temperature, as shown in Fig. 3-3. Linear interpolation of the first two trial values suggests 107.5°C as the next guess. At this temperature, $K_1 = 1.647$ and $K_2 = 0.397$, and $(y_1/K_1) + (y_2/K_2) = 0.989$.

The second and third guess points on Fig. 3-3 suggest that the next guess be 107.1°C. At this temperature, $K_1 = 1.630$ and $K_2 = 0.392$, and $(y_1/K_1) + (y_2/K_2) = 1.001$. Thus, the dew point can be taken as 107.1°C (380.3 K or 224.8°F).

As a matter of interest, the liquid that condenses at the dew point has the composition $x_i = y_i/K_i$, where the x_i are the liquid-phase mole fractions. In this case, $x_1 = 0.8/1.630 = 0.49$ and $x_2 = 0.2/0.392 = 0.51$ $(= 1 - 0.49)$.

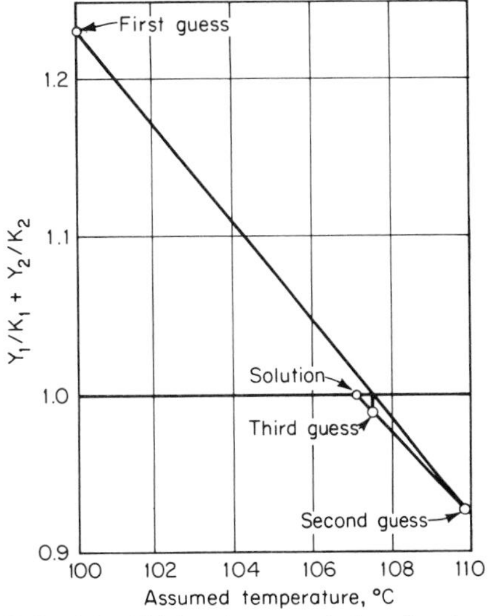

FIG. 3-3 Trial-and-error solution for dew point (Example 3-10).

Related Calculations: When the liquid phase is nonideal, the K values depend on the liquid composition, in this case the dew composition, which is not known. The procedure then is to guess both the liquid composition and the temperature and to check not only whether the dew-point equation is satisfied, but also whether the values of x_i calculated by the equations $x_i = y_i/K_i$ are the same as the guessed values. (When there is no other basis for the initial guess of the dew composition, an estimate of it may be obtained by dew-point calculations presuming an ideal liquid phase, as outlined above.) Such nonideal liquid-phase dew-point calculations are usually carried out using a digital computer. A logic diagram for dew-point determination is shown in Fig. 3-4.

3-11 Calculating Bubble Point When Vapor Phase Is Ideal

Calculate the bubble point of a liquid system containing 80 mol % ethanol and 20 mol % water at 500 mmHg (66.7 kPa).

Calculation Procedure:

1. Select a temperature, and test its suitability by trial and error.

The bubble point of a system at pressure P and whose liquid composition is given by mole fractions x_i is that temperature at which there is onset of vaporization. Mathematically, it is the temperature such that

$$\sum_{i=1}^{N} K_i x_i = 1$$

where the K_i are vapor-liquid equilibrium ratios.

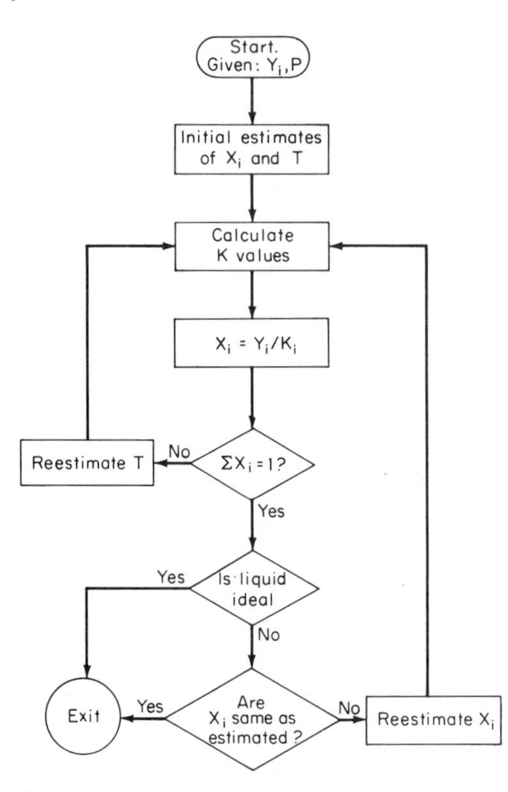

FIG. 3-4 Logic diagram for determining dew point (Example 3-10).

When the vapor phase is ideal, the K_i are independent of the vapor composition. In such a case, the procedure for bubble-point determination is to (1) guess a temperature; (2) calculate the K_i, which equal $\gamma_i f_i^\circ/P$, where γ_i is the activity coefficient of the ith component in the liquid phase, f_i° is the fugacity of pure liquid i at system temperature and pressure, and P is the system pressure; and (3) check if the preceding bubble-point equation is satisfied. If it is not, repeat the procedure with a different guess.

For this system, f_i° may be assumed to be the same as the vapor pressure (for a discussion of the grounds for this assumption, see Example 3-6). Activity coefficients can be calculated using the Wilson, Margules, or Van Laar equations (see Example 3-4).

From vapor-pressure calculations, the boiling points of ethanol and water at 500 mmHg are 68°C and 89°C, respectively. As a first guess at the bubble-point temperature, try a linear interpolation of the boiling points: $T = (0.8)(68) + (0.2)(89)$, which is approximately equal to 72. Let subscript 1 refer to ethanol and subscript 2 to water. At 72°C, activity coefficients calculated via the Van Laar equation are 1.02 for ethanol and 2.04 for water. From tables or calculation, the vapor pressures are 589 and 254 mmHg. Then $K_1 = (1.02)(589)/500 = 1.202$ and $K_2 = (2.04)(254)/500 = 1.036$, and the bubble-point equation becomes $K_1x_1 + K_2x_2 = (1.202)(0.8) + (1.036)(0.2) = 1.169$.

Since this is greater than 1, the K values are too high. This means that the vapor pressures and, accordingly, the assumed temperature are too high.

2. ***Repeat the trial-and-error procedure until the bubble-point temperature is found.***

As a second guess, try 67°C. Assume that because the temperature difference is small, the activity coefficients remain the same as in step 1. At 67°C, the vapor pressures are 478 and 205 mmHg. Then, $K_1 = 0.9746$ and $K_2 = 0.8344$, as found by the procedure outlined in the previous step. Then $K_1x_1 + K_2x_2 = 0.9466$. Since this sum is less than 1, the assumed temperature is too low.

As a next estimate, try 68°C. At this temperature, $K_1 = 1.039$ and $K_2 = 0.891$, and $K_1x_1 + K_2x_2 = 1.009$.

Finally, at 68.3°C, $K_1 = 1.030$ and $K_2 = 0.883$, and $K_1x_1 + K_2x_2 = 1.0006$. This is very close to 1.0, so 68.3°C (341.5 K or 154.9°F) can be taken as the bubble point.

As a matter of interest, the vapor that is generated at the bubble point has the composition $y_i = K_ix_i$, where the y_i are the vapor-phase mole fractions. In this case, $y_1 = (1.030)(0.8) = 0.82$ and $y_2 = (0.883)(0.2) = 0.18$ $(= 1.0 - 0.82)$.

Related Calculations: When the vapor phase is nonideal, the K values depend on the vapor composition, in this case the bubble composition, which is unknown. The procedure then is to guess both the vapor composition and the temperature and to check not only whether the bubble-point equation is satisfied, but also whether the values of y_i given by the equations $y_i = K_ix_i$ are the same as the guess values. (When there is no other basis for guessing the bubble composition, an estimate based on calculations that presume an ideal vapor phase may be used, as outlined above.) As in the case of nonideal dew points, nonideal bubble points are calculated using a digital computer. Figure 3-5 shows a logic diagram for computing bubble points.

3-12 Binary Phase Diagrams for Vapor-Liquid Equilibrium

A two-phase binary mixture at 100°C (212°F or 373 K) and 133.3 kPa (1.32 atm or 1000 mmHg) has an overall composition of 68 mol % benzene and 32 mol % toluene. Determine the mole fraction benzene in the liquid phase and in the vapor phase.

Calculation Procedure:

1. ***Obtain (or plot from data) a phase diagram for the benzene/toluene system.***

Vapor-liquid equilibrium behavior of binary systems can be represented by a temperature-composition diagram at a given constant pressure (such as Fig. 3-6) or by a pressure-composition diagram at a given constant temperature (such as Fig. 3-7).

Curve *ABC* in each figure represents the states of saturated-liquid mixtures; it is called the "bubble-point curve" because it is the locus of bubble points in the temperature-composition diagram. Curve *ADC* represents the states of saturated vapor; it is called the "dew-point curve" because it is the locus of the dew points. The bubble- and dew-point

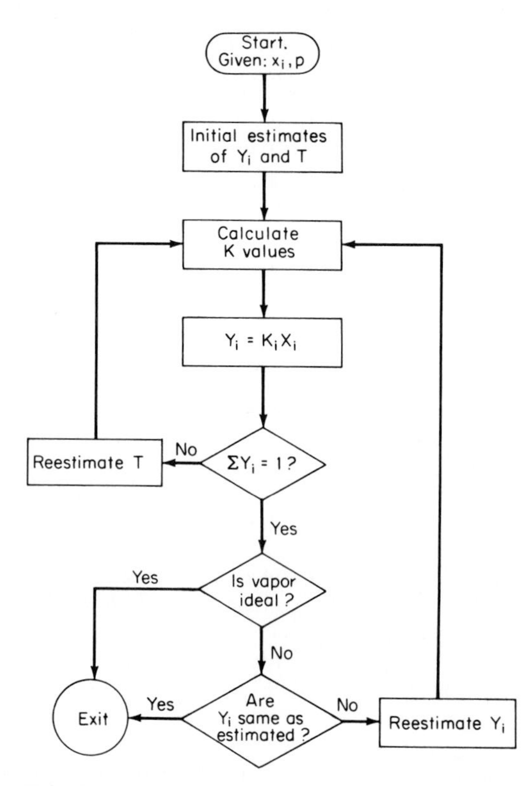

FIG. 3-5 Logic diagram for determining bubble point (Example 3-11).

curves converge at the two ends, which represent the saturation points of the two pure components. Thus in Fig. 3-6, point *A* corresponds to the boiling point of toluene at 133.3 kPa, and point *C* corresponds to the boiling point of benzene. Similarly, in Fig. 3-7, point *A* corresponds to the vapor pressure of toluene at 100°C, and point *C* corresponds to the vapor pressure of benzene.

The regions below *ABC* in Fig. 3-6 and above *ABC* in Fig. 3-7 represent subcooled liquid; no vapor is present. The regions above *ADC* in Fig. 3-6 and below *ADC* in Fig. 3-7 represent superheated vapor; no liquid is present. The area between the curves is the region where both liquid and vapor phases coexist.

2. *Determine vapor and liquid compositions directly from the diagram.*

If a system has an overall composition and temperature or pressure such that it is in the two-phase region, such as the conditions at point *E* in either diagram, it will split into a vapor phase whose composition is given by point *D* and a liquid phase whose composition is given by point *B*, where line *BD* is the horizontal (constant-temperature or constant-pressure) line passing through *E*.

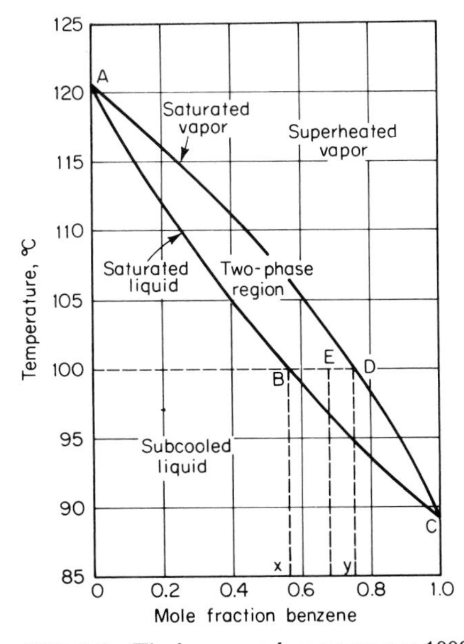

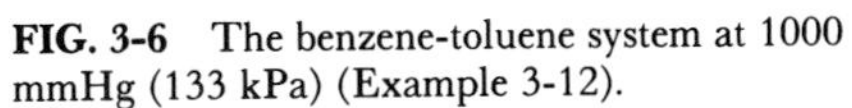

FIG. 3-6 The benzene-toluene system at 1000 mmHg (133 kPa) (Example 3-12).

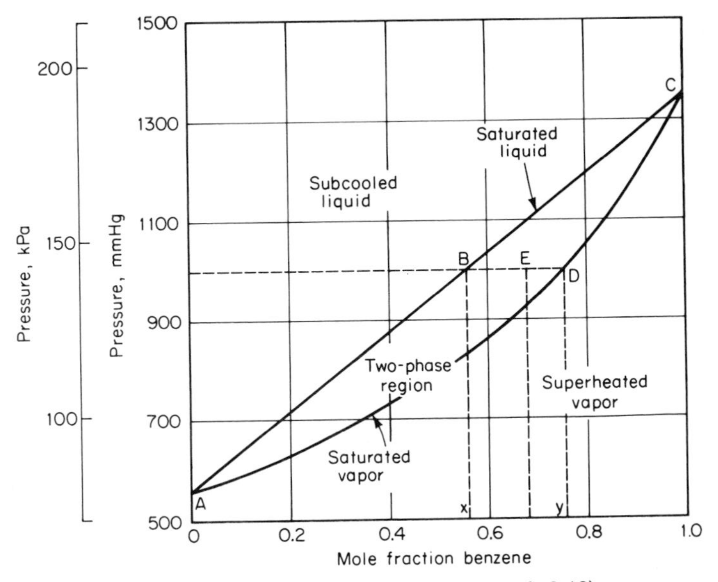

FIG. 3-7 The benzene-toluene system at 100°C (Example 3-12).

Related Calculations: The relative amounts of vapor and liquid present at equilibrium are given by (moles of liquid)/(moles of vapor) = (length of line *ED*)/(length of line *BE*).

Figures 3-6 and 3-7 have shapes that are characteristic for ideal systems. Certain nonideal systems deviate so much from these as to form maxima or minima at an intermediate composition rather than at one end or the other of the diagram. Thus the dew-point and bubble-point curves meet at this intermediate composition as well as at the ends. Such a composition is called an "azeotropic composition."

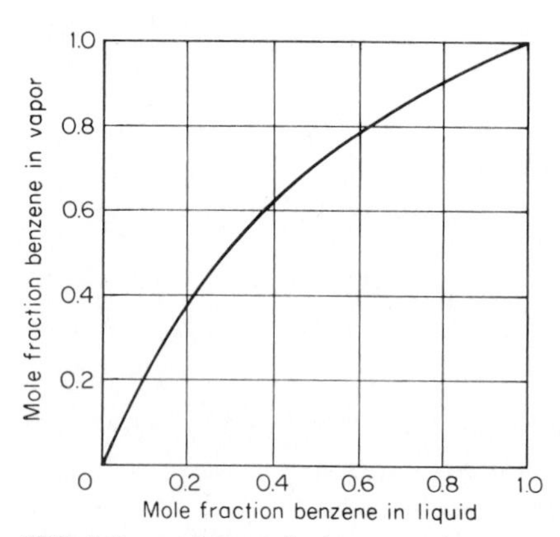

FIG. 3-8 *xy* diagram for benzene-toluene system at 1 atm (Example 3-12).

The example discussed here pertains to binary systems. By contrast, multicomponent vapor-liquid equilibrium behavior cannot easily be represented on diagrams and instead is usually calculated at a given state by using the procedures described in the preceding two examples.

Phase equilibrium is important in design of distillation columns. Such design is commonly based on use of an *xy* diagram, a plot of equilibrium vapor composition *y* versus liquid composition *x* for a given binary system at a given pressure. An *xy* diagram can be prepared from a temperature-composition or a pressure-composition diagram, such as Figs. 3-6 and 3-7. Select a value of *x*; find the bubble-point condition (temperature on Fig. 3-6, for instance) for that value on curve *ABC*; then move horizontally to curve *ADC* in order to find the equilibrium value of *y*; and then plot that *y* as ordinate versus that *x* as abscissa. Figure 3-8 shows the *xy* diagram for the benzene/toluene system.

3-13 Multicomponent Liquid-Liquid Equilibrium

Develop a design method for calculating multicomponent liquid-liquid equilibrium.

Calculation Procedure:

1. *Establish what equations must be satisfied.*

Ideal solutions or solutions exhibiting negative deviation from ideal behavior cannot form two liquid phases. Instead, strong positive deviation is necessary for two or more liquid phases to exist together.

Calculations of multicomponent liquid-liquid equilibrium are needed in the design of liquid (solvent) extraction systems. Since these operations take place considerably below the bubble point, it is not necessary to consider the equilibrium-vapor phase. The equations to be solved are:

Material balances:

$$\phi x_i^{(1)} + (1 - \phi)x_i^{(2)} = Z_i$$

Equilibrium relationships:

$$\gamma_i^{(1)}x_i^{(1)} = \gamma_i^{(2)}x_i^{(2)}$$

Constraints:

$$\sum_{i=1}^{N} x_i^{(1)} = \sum_{i=1}^{N} x_i^{(2)} = 1$$

In these equations, ϕ is the mole fraction of phase 1 (i.e., the moles of phase 1 per total moles), Z_i is the mole fraction of the ith species in the total system, $x_i^{(1)}$ is the mole fraction of the ith species in phase 1, $x_i^{(2)}$ is the mole fraction of the ith species in phase 2, $\gamma_i^{(1)}$ is the activity coefficient of the ith species in phase 1, and $\gamma_i^{(2)}$ is its activity coefficient in phase 2.

2. *Set out the calculation procedure.*

The preceding equations are rearranged to facilitate the solution according to the following procedure: (1) guess the liquid compositions of both phases, (2) calculate the activity coefficients in both phases at the solution temperature, (3) solve the following nonlinear equation for ϕ:

$$\begin{aligned} F(\phi) &= \sum_{i=1}^{N} [x_i^{(1)} - x_i^{(2)}] \\ &= \sum_{i=1}^{N} \frac{z_i[1 - \gamma_i^{(1)}/\gamma_i^{(2)}]}{\phi + (1 - \phi)\gamma_i^{(1)}/\gamma_i^{(2)}} \\ &= 0 \end{aligned}$$

(4) calculate the liquid compositions from

$$x_i^{(1)} = \frac{Z_i}{\phi + (1 - \phi)\gamma_i^{(1)}/\gamma_i^{(2)}}$$

and

$$x_i^{(2)} = x_i^{(1)}\gamma_i^{(1)}/\gamma_i^{(2)}$$

and (5) if the calculated values in step 4 are not the same as the guessed values from step 1, repeat the procedure with different guesses.

Any nonlinear equation-solving techniques can be used in the preceding procedure, which is usually carried out with a digital computer.

Related Calculations: Graphic representation of liquid-liquid equilibrium is convenient only for binary systems and isothermal ternary systems. Detailed discussion of such diagrams appears in A. W. Francis, *Liquid-Liquid Equilibrium,* Interscience, New York, 1963. Thermodynamic correlations of liquid-liquid systems using available models for liquid-phase nonideality are not always satisfactory, especially when one is trying to extrapolate outside the range of the data.

Of interest in crystallization calculations is solid-liquid equilibrium. When the solid phase is a pure component, the following thermodynamic relationship holds:

$$\ln \frac{1}{\gamma_i x_i} = \frac{\Delta H_i}{RT}\left(1 - \frac{T}{T_i^*}\right) - \frac{\Delta C_i}{R}\left[\left(\frac{T_i^*}{T} - 1\right) - \ln\left(\frac{T_i^*}{T}\right)\right]$$

where x_i is the saturation mole fraction of the ith component in equilibrium with pure solid, γ_i is the activity coefficient of the ith species with reference to pure solid at absolute temperature T, R is the universal gas constant, T_i^* is the triple-point temperature of the component, ΔH_i is its heat of fusion at T_i^*, and ΔC_i is the difference between the specific heats of pure liquid and solid. For many substances, the second term on the right-hand side of the equation is negligible and can be omitted. With such a simplification, the equation can be rewritten as

$$x_i = \frac{\exp\left[\dfrac{\Delta H_i}{RT_i^*}\left(1 - \dfrac{T_i^*}{T}\right)\right]}{\gamma_i}$$

Often the melting point and the heat of fusion at the melting point are used as estimates of T_i^* and ΔH_i. It should be noted that the latter equation is nonlinear, since γ_i on the right-hand side is a function of x_i. Hence the determination of x_i calls for an iterative numerical procedure, such as the Newton-Raphson or the secant methods.

3-14 Calculations for Isothermal Flashing

A mixture containing 50 mol % benzene and 50 mol % toluene exists at 1 atm (101.3 kPa) and 100°C (212°F or 373 K). Calculate the compositions and relative amounts of the vapor and liquid phases.

Calculation Procedure:

1. Calculate the mole fraction in the vapor phase.

When the components of the system are soluble in each other in the liquid state (as is the case for benzene and toluene), so that there is only one liquid phase, then the moles of vapor V per total moles can be calculated from the equation

$$\sum_{i=1}^{N} [Z_i(1 - K_i)/(1 - V + VK_i)] = 0$$

where Z_i is the mole fraction of the ith component in the total system and K_i is the K value of the ith component, as discussed in the first example in this section.

Since the pressure is not high, ideal-system K values can be used, and these are independent of composition. From the above-mentioned example, $K_1 = 1.777$ and $K_2 = 0.413$, where the subscript 1 refers to benzene and subscript 2 to toluene.

For a binary system, as in this case, the preceding summation equation becomes $Z_1(1 - K_1)/(1 - V + VK_1) + Z_2(1 - K_2)/(1 - V + VK_2) = 0$, or $0.5(1 - 1.777)/(1 - V + 1.777V) + 0.5(1 - 0.413)/(1 - V + 0.413V) = 0$. By algebra, $V = 0.208$, so 20.8 percent of the system will be in the vapor phase.

2. *Calculate the liquid composition.*

The liquid composition can be calculated from the equation $x_i = Z_i/(1 - V + VK_i)$. For benzene, $x_1 = 0.5/[1 - 0.208 + (0.208)(1.777)] = 0.430$, and for toluene, $x_2 = 0.5/[1 - 0.208 + (0.208)(0.413)] = 0.570$ (or, of course, for a binary mixture, $x_2 = 1 - x_1 = 0.570$).

3. *Calculate the vapor composition.*

The vapor composition can be calculated directly from the defining equation for K values in rearranged form: $y_i = K_i x_i$. For benzene, $y_1 = K_1 x_1 = 1.777(0.430) = 0.764$, and for toluene, $y_2 = K_2 x_2 = 0.413(0.570) = 0.236$ (or, of course, for a binary mixture, $y_2 = 1 - y_1 = 0.236$).

Related Calculations: If the system is such that two liquid phases form, then the defining equation for V above must be replaced by a pair of equations in which the unknowns are V and ϕ, the latter being the moles of liquid phase 1 per total liquid moles:

$$\sum_{i=1}^{N} Z_i[1 - K_i^{(1)}]/[\phi(1 - V) + (1 - \phi)(1 - V)K_i^{(1)}/K_i^{(2)} + VK_i^{(1)}] = 0, \text{ and } \sum_{i=1}^{N} Z_i[1 - K_i^{(1)}/K_i^{(2)}]/[\phi(1 - V) + (1 - \phi)(1 - V)K_i^{(1)}/K_i^{(2)} + VK_i^{(1)}] = 0$$

In addition, the equation for liquid composition becomes $x_i^{(1)} = Z_i/[\phi(1 - V) + (1 - \phi)(1 - V)K_i^{(1)}/K_i^{(2)} + VK_i^{(1)}]$. In these three equations, the superscripts 1 and 2 refer to the two liquid phases.

In systems where more than one liquid phase exists, the K values are not independent of composition. In such a case, it is necessary to use a trial-and-error procedure, guessing the composition of all the phases, calculating a K value for the equilibrium of each liquid phase with the vapor phase, solving the equations for V and ϕ, and then calculating the compositions of the phases by the equations for the x_i and y_i and seeing if these calculations agree with the guesses. This procedure should be carried out using a computer.

In the case of adiabatic rather than isothermal flashing, when the total enthalpy of the system rather than its temperature is specified, the equations associated with isothermal flash are solved jointly with an enthalpy-balance equation, treating the temperature as another variable. The general enthalpy balance is

$$H_F = VH_V[1 - V][\phi H_L^{(1)} + (1 - \phi)H_L^{(2)}], \text{ where } H_V = \sum_{i=1}^{N} h_{v,i}y_i + \Delta h_d \text{ and } H_L^{(k)} = \sum_{i=1}^{N} h_{L,i}x_i^{(k)} + \Delta h_m^{(k)}$$

where $h_{v,i}$ is ideal-gas enthalpy per mole of pure ith component, $h_{L,i}$ is the enthalpy per mole of pure liquid of ith component, Δh_d is enthalpy departure from ideal-gas state per mole of vapor of composition given by the y_i, and $\Delta h_m^{(k)}$ is the heat of mixing per mole of liquid phase having the composition given by the $x_i^{(k)}$, k taking the values 1 and 2 if there are two liquid phases or the value 1 if there is only one liquid phase. Very often the heat of mixing is negligible. A convenient procedure for adiabatic flash is to perform isothermal flashes at a series of temperatures to find the temperature at which the enthalpy balance is satisfied.

4

Chemical-Reaction Equilibrium

E. Dendy Sloan, Ph.D.

Professor
Chemical and Petroleum-Refining
Engineering Department
Colorado School of Mines
Golden, CO

REFERENCES: [1] Balzhiser, Samuels, and Eliassen—*Chemical Engineering Thermodynamics,* Prentice-Hall; [2] Benson—*Thermochemical Kinetics,* Wiley; [3] Bett, Rowlinson, and Saville—*Thermodynamics for Chemical Engineers,* MIT Press; [4] Denbigh—*Principles of Chemical Equilibrium,* Cambridge Univ. Press; [5] Hougen, Watson, and Ragatz—*Chemical Process Principles,* part II, Wiley; [6] Modell and Reid—*Thermodynamics and Its Applications,* Prentice-Hall; [7] Perry and Chilton—*Chemical Engineers' Handbook,* McGraw-Hill; [8] Prausnitz—*Molecular Thermodynamics of Fluid Phase Equilibria,* Prentice-Hall; [9] Reid, Prausnitz, and Sherwood—

The Properties of Gases and Liquids, McGraw-Hill; [10] Sandler—*Chemical Engineering Thermodynamics,* Wiley; [11] Smith and Van Ness—*Introduction to Chemical Engineering Thermodynamics,* McGraw-Hill; [12] Soave—*Chem. Eng. Sci. 27*:1197, 1972; [13] Stull and Prophet—*JANAF Thermochemical Tables,* NSRDS-NBS 37, 1971.

Chemical engineers make reaction-equilibria calculations to find the potential yield of a given reaction, as a function of temperature, pressure, and initial composition. In addition, the heat of reaction is often obtained as an integral part of the calculations. Because the calculations are made from thermodynamic properties that are accessible for most common compounds, the feasibility of a large number of reactions can be determined without laboratory study. Normally, thermodynamic feasibility should be determined before obtaining kinetic data in the laboratory, and if a given set of reaction conditions does not yield a favorable final mixture of products and reactants, the reaction-equilibria principles indicate in what direction one or more of the conditions should be changed.

The basic sequence of equilibria calculations consists of (1) determining the standard-state Gibbs free-energy change ΔG° for the reaction under study at the temperature of interest (since this step commonly includes calculation of the standard enthalpy change of reaction ΔH°, an example of that calculation is included below), (2) determining the equilibrium constant K from ΔG°, (3) relating the equilibria compositions to K, and (4) evaluating how this equilibrium composition changes as a function of temperature and pressure. A frequent complication is the need to deal with simultaneous or heterogeneous reactions or both.

4-1 Heat of Reaction

Evaluate the heat of reaction of $CO(g) + 2H_2(g) \rightarrow CH_3OH(g)$ at 600 K (620°F) and 10.13 MPa (100 atm).

Calculation Procedure:

1. Calculate the heat of reaction at 298 K, ΔH°_{298}.

Calculating heat of reaction is a multistep process. One starts with standard heats of formation at 298 K, calculates the standard heat of reaction, and then adjusts for actual system temperature and pressure.

The heat of reaction at 298 K is commonly referred to as the "standard heat of reaction." It can be calculated readily from the standard heats of formation ($\Delta H^\circ_{f\,298}$) of the reaction components; these standard heats of formation are widely tabulated. Thus, Perry and Chilton [7] show $\Delta H^\circ_{f\,298}$ for CH_3OH and CO to be −48.08 and −26.416 kcal/(g·mol), respectively (the heat of formation of hydrogen is zero, as is the case for all other elements). The standard heat of reaction is the sum of the heats of formation of the reaction products minus the heats of formation of the reactants. In this case,

$$\Delta H^\circ_{298} = -48.08 - [-26.416 + 2(0)]$$

$$= -21.664 \text{ kcal/(g·mol)}$$

2. *Calculate the heat-capacity constants.*

The heat capacity per mole of a given substance C_p can be expressed as a function of absolute temperature T by equations of the form $C_p = a + bT + cT^2 + dT^3$. Values of the constants are tabulated in the literature. Thus Sandler [10] shows the following values for the substances in this example:

	CH_3OH	CO	H_2
a	4.55	6.726	6.952
b (× 10^2)	2.186	0.04001	−0.04576
c (× 10^5)	−0.291	0.1283	0.09563
d (× 10^9)	−1.92	−0.5307	−0.2079

In this step, we are determining the constants for the equation $\Delta C_p = \Delta a + \Delta bT + \Delta cT^2 + \Delta dT^3$, where Δa (for instance) equals $\Sigma a_{\text{products}} - \Sigma a_{\text{reactants}}$. Thus, for the reaction in this example, $\Delta a = 4.55 - 6.726 - 2(6.952) = -16.08$; $\Delta b = [2.186 - 0.04001 + 2(0.04576)] \times 10^{-2} = 2.2375 \times 10^{-2}$; $\Delta c = [-0.291 - 0.1283 - 2(0.09563)] \times 10^{-5} = -0.61056 \times 10^{-5}$; and $\Delta d = [-1.92 + 0.5307 + 2(0.2079)] \times 10^{-9} = -0.9735 \times 10^{-9}$. The resulting ΔC_p is in cal/(g·mol)(K) (rather than in kcal).

3. *Calculate the standard heat of reaction at 600 K, ΔH°_{600}.*

This step combines the results of the previous two, via the equation

$$\Delta H^\circ_T = \Delta H^\circ_{298} + \int_{298}^{T} [\Delta a + \Delta bT + \Delta cT^2 + \Delta dT^3]dT$$

Integration gives the equation

$$\Delta H_T = \Delta H^\circ_{298} + \Delta a(T - 298) + \frac{\Delta b}{2}(T^2 - 298^2) + \frac{\Delta c}{3}(T^3 - 298^3) + \frac{\Delta d}{4}(T^4 - 298^4)$$

Therefore,

$$\Delta H^\circ_{600} = -21{,}664 - 16.08(600 - 298) + \left(\frac{2.2375 \times 10^{-2}}{2}\right)(600^2 - 298^2) - \left(\frac{0.61056 \times 10^{-5}}{3}\right)(600^3 - 298^3) - \left(\frac{0.9735 \times 10^{-9}}{4}\right)(600^4 - 298^4)$$

$$= -2.39 \times 10^4 \text{ cal/(g·mol)} \; [-1.00 \times 10^8 \text{ J/(kg·mol) or } -4.30 \times 10^4 \text{ Btu/(lb·mol)}]$$

4. Calculate the true heat of reaction by correcting ΔH°_{600} for the effect of pressure.

The standard heats of formation, from which ΔH°_{600} was calculated, are based on a pressure of 1 atm. The actual pressure in this case is 100 atm. The correction equation is $\Delta H_{600,100\text{ atm}} = \Delta H^\circ_{600} - \Delta H^1$, where the correction factor ΔH^1 is in turn defined as follows: $\Delta H^1 = \Delta H^\circ + \omega \Delta H'$. In the latter equation, ΔH° and $\Delta H'$ are parameters whose values depend on the reduced temperature and reduced pressure, while ω is the acentric factor. Generalized correlations that enable the calculation of ΔH° and $\Delta H'$ can be found, for instance, in Smith and Van Ness [11], and ω is tabulated in the literature. Analogously to the approach in step 2, the calculation is based on the sum of the values for the reaction products minus the sum of the values for the reactants.

For the present example, the values are as follows:

Component	ω	ΔH°	$\Delta H'$	ΔH^1, cal/(g·mol)
CH_3OH	0.556	1222	305.5	1392
CO	0.041	0	0	0
H_2	0	0	0	0

Therefore, ΔH^1 for this example is 1392. In turn,

$$\Delta H_{600,100\text{ atm}} = -2.39 \times 10^4 - 1392$$

$$= -2.529 \times 10^4 \text{ cal/(g·mol)} \; [-1.058 \times 10^8 \text{ J/(kg·mol) or } - 4.55 \text{ Btu/(lb·mol)}]$$

Related Calculations: When reactants enter at a different temperature from the exiting products, the enthalpy changes that are due to this temperature difference should be computed independently and added to $\Delta H^\circ_{298\text{ K}}$, along with the pressure deviations of enthalpy.

When heat capacities of enthalpies of formation are not tabulated, a group-contribution method such as that in the following example may be used.

4-2 Heat of Formation for Uncommon Compounds

Assuming that tabulated heat-of-formation data are not available for ethanol, n-propanol, and n-butanol, use the group-contribution approach of Benson to estimate these data at 298 K.

Calculation Procedure:

1. Draw structural formulas for the chemical compounds.

Take the formulas from a chemical dictionary, chemistry textbook, or similar source. For this example, they are

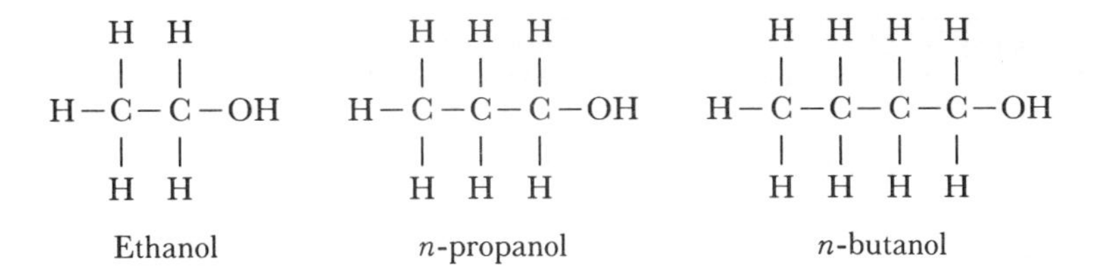

2. *Identify all groups in each formula.*

A group is defined here as a polyvalent atom (ligancy $\geq$ 2) in a molecule together with all of its ligands. Thus the ethanol molecule has one each of three kinds of groups: (1) a carbon atom linked to one other carbon atom and to three hydrogen atoms, the notation for this group being C—(C)(H)$_3$; (2) a carbon atom linked to one other carbon atom and to an oxygen atom and to two hydrogen atoms, which can be expressed as C—(C)(O)(H)$_2$; and (3) an oxygen atom linked to a carbon atom and to a hydrogen atom, which can be expressed as O—(C)(H).

The *n*-propanol molecule has one each of four kinds of groups: C—(C)(H)$_3$, C—(C)$_2$(H)$_2$, C—(C)(O)(H)$_2$, and O—(C)(H). And the *n*-butanol molecule differs from the *n*-propanol molecule only in that it has two (rather than one) of the C—(C)$_2$(H)$_2$ groups.

3. *For each molecule, add up the partial heats of formation for its constituent groups.*

The partial heats of formation can be found in Benson [2]. Adding these up for a given molecule gives an estimate of the heat of formation for that molecule.

For ethanol, the result is as follows:

C—(C)(H)$_3$	−10.20 kcal/(g·mol)
C—(C)(O)(H)$_2$	−8.1
O—(C)(H)	−37.9
ΔH°_{f298} =	−56.2 kcal/(g·mol) [−2.35 × 10^8 J/(kg·mol) or −1.012 × 10^5 Btu/(lb·mol)]

For *n*-propanol:

C—(C)(H)$_3$	−10.20
C—(C)$_2$(H)$_2$	−4.93
C—(C)(O)(H)$_2$	−8.1
O—(C)(H)	−37.9
ΔH°_{f298} =	−61.13 kcal/(g·mol) [−2.556 × 10^8 J/(kg·mol) or −1.10 × 10^5 Btu/(lb·mol)]

And for *n*-butanol:

$C-(C)(H)_3$	-10.20
$2[C-(C)_2(H)_2] = 2(-4.93) =$	-9.86
$C-(C)(O)(H)_2$	-8.1
$O-(C)(H)$	-37.9
$\Delta H^\circ_{f298} =$	-66.06 kcal/(g·mol) [-2.762×10^8 J/(kg·mol) or -1.189×10^5 Btu/(lb·mol)]

Related Calculations: As an indicator of the degree of accuracy of this method, the observed values for ethanol, n-propanol, and n-butanol are -2.35×10^8, -2.559×10^8, and -2.83×10^8 J/(kg·mol), respectively. Deviations may be as high as $\pm 1.25 \times 10^7$ J/(kg·mol). Benson's method may also be used to estimate ideal-gas heat capacities.

4-3 Standard Gibbs Free-Energy Change of Reaction

Calculate the standard Gibbs free-energy change ΔG° for the reaction $CH_3CH_2OH \rightarrow CH_2{=}CH_2 + H_2O$ at 443 K (338°F).

Calculation Procedure:

1. ***Calculate ΔH°_{298} for the reaction, and set out an equation for ΔH°_T as a function of temperature T.***

This step directly employs the procedures developed in the preceding example. Thus ΔH°_{298} is found to be 1.0728×10^4 cal/(g·mol), and the required equation is

$$\Delta H^\circ_T = \Delta H^\circ_{298} + \Delta a(T - 298) + (\Delta b/2)(T^2 - 298^2) + (\Delta c/3)(T^3 - 298^3) + (\Delta d/4)(T^4 - 298^4)$$

The calculated numerical values of the heat-capacity constants are $\Delta a = 3.894$, $\Delta b = -0.01225$, $\Delta c = 0.7381 \times 10^{-5}$, and $\Delta d = -1.4287 \times 10^{-9}$. Therefore, after arithmetic partial simplification, the equation becomes (with the answer emerging in calories per gram-mole)

$$\Delta H^\circ_T = 1.0049 \times 10^4 + \Delta a T + (\Delta b/2)T^2 + (\Delta c/3)T^3 + (\Delta d/4)T^4$$

2. ***Calculate ΔG°_{298} for the reaction.***

This can be determined in a straightforward manner from tabulated data on standard Gibbs free energies of formation at 298 K, ΔG°_{f298}. The answer is the sum of the values for the reaction products minus the sum of the values for the reactants. Thus Perry and Chilton [7] show ΔG°_{f298} for ethanol, ethylene, and water to be, respectively, -40.23, 16.282, and -54.635 kcal/(g·mol). Accordingly, $\Delta G^\circ_{298} = (16.282 - 54.635) - (-40.23) = 1.877$ kcal/(g·mol) [7.854×10^6 J/(kg·mol) or 3.3786×10^3 Btu/(lb·mol)].

3. Calculate ΔG° at the reaction temperature, 443 K.

The equation used is

$$\Delta G^\circ = \Delta H_0 - \Delta a T \ln T - (\Delta b/2)T^2 - (\Delta c/6)T^3 - (\Delta d/12)T^4 - IRT$$

where ΔH_0 is the integration constant that is determined for the ΔH°_T equation in step 1 (that is, 1.0049×10^4), R is the universal gas constant, the heat-capacity constants (Δa, etc.) are as determined in step 1, and I is a constant yet to be determined.

To determine I, substitute into the equation the (known) values pertaining to ΔG° at 298 K. Thus 1.877×10^3 cal/(g·mol) $= 1.0049 \times 10^4 - (3.894)(298) \ln 298 - \frac{1}{2}(-0.01225)(298)^2 - (\frac{1}{6})(0.7381 \times 10^{-5})(298)^3 - (\frac{1}{12})(-1.4287 \times 10^{-9})(298)^4 - (1.987)(298)(I)$. Solving for I, it is found to be 3.503.

Finally, insert this value into the equation and solve for ΔG° at 443 K. Thus $\Delta G^\circ = 1.0049 \times 10^4 - (3.894)(443) \ln 443 - \frac{1}{2}(-0.01225)(443)^2 - (\frac{1}{6})(0.7381 \times 10^{-5})(443)^3 - (\frac{1}{12})(-1.4287 \times 10^{-9})(443)^4 - (1.987)(443)(3.503) = -2.446 \times 10^3$ cal/(g·mol) [or -1.0237×10^7 J/(kg·mol) or -4.4019 Btu/(lb·mol)].

Related Calculations: ΔG°_{298} may also be calculated via the relationship $\Delta G^\circ_{298} = \Delta H^\circ_{298} - T\,\Delta S^\circ_{298}$, where S is entropy. Values for S can be found in the literature, or ΔS°_{298} can be estimated by Benson's group-contribution method (see example for estimating ΔH°_{f298} by that method above). However, the deviations for entropy values calculated by Benson's method are high compared with those for ΔH°_{f298}, so when such group methods are used in the course of calculating ΔG°_{298}, only qualitative feasibility studies are possible, based on these rules-of-thumb: (1) if ΔG°_T is less than zero, the reaction is clearly feasible; (2) if the value is less than 4.184×10^7 J/(kg·mol) [or 1.8×10^4 Btu/(lb·mol)], the reaction should be studied further; and (3) if the value is greater than 4.184×10^7 J/(kg·mol), the reaction is feasible only under exceptional circumstances. (Thus, in the present example, the reaction appears unfavorable at 298 K, but much more promising at 443 K.)

The Gibbs free-energy change of reaction may be related to the heat of reaction by $d(\Delta G^\circ/RT)/dT = -\Delta H^\circ/RT^2$. And if ΔH° is constant, or if changes in ΔG° are needed over a small temperature range, then a given ΔG° value at a known temperature T_1 may be used to obtain ΔG° at a different temperature T_v via the relationship, $\Delta G_v/RT_v = \Delta G^\circ/RT_1 - (\Delta H^\circ/R)[(1/T_v) - (1/T_1)]$.

4-4 Estimation of Equilibrium Compositions

Estimate the equilibrium composition of the reaction *n*-pentane → neopentane, at 500 K (440°F) and 10.13 MPa (100 atm), if the system initially contains 1 g·mol *n*-pentane. Ignore other isomerization reactions.

Calculation Procedure:

1. Calculate ΔG° at 500 K.

This step is carried out as outlined in the previous example. Accordingly, ΔG°_{500} is found to be 330.54 cal/(g·mol) [or 13.83×10^6 J/(kg·mol) or 594.97 Btu/(lb·mol)].

2. *Determine the equilibrium constant K.*

This step consists of solving the equation $K = \exp(-\Delta G°/RT)$. Thus $K = \exp[-330.54/(1.987)(500)] = 0.717$.

3. *Express mole fractions in terms of the reaction-progress variable x.*

The reaction-progress variable x is a measure of the extent to which a reaction has taken place. In the present example, x would be 0 if the "equilibrium" mixture consisted solely of n-pentane; it would be 1 if the mixture consisted solely of neopentane (i.e., if all the n-pentane had reacted). It is defined as $\int_{n_{o,i}}^{n_{f,i}} dn/\nu_i$, where n is number of moles of component i, subscripts o and f refer to initial and final states, respectively, and ν_i is the stoichiometric number for component i, which is equal to its stoichiometric coefficient in that reaction but with the convention that reactants are given a minus sign. In the present example, $\nu_{n\text{-pentane}} = -1$ and $\nu_{\text{neopentane}} = 1$, and $n_{o,n\text{-pentane}} = 1$ and $n_{o,\text{neopentane}} = 0$. Since a given system (such as the system in this example) is characterized by a particular value of the reaction-progress variable, the integrals described above (an integral can be written for each component) must equal each other for that system.

In the present case,

$$\int_1^{n_{f,n\text{-pentane}}} dn_{n\text{-pentane}}/-1 = \int_0^{n_{f,\text{neopentane}}} dn_{\text{neopentane}}/1 = x$$

Carrying out the two integrations and solving for the two n_f values in terms of x, we find $n_{f,n\text{-pentane}} = (1 - x)$, and $n_{f,\text{neopentane}} = x$. Total moles at equilibrium equals $(1 - x) + x = 1$. Thus the mole fractions y_i at equilibrium are $(1 - x)/1 = (1 - x)$ for n-pentane and $x/1 = x$ for neopentane. [This procedure is valid for any system, however complicated. In the present simple example, it is intuitively clear that at equilibrium, the two n_f values must be $(1 - x)$ and x.]

4. *Estimate the final composition by using the Lewis-Randall rule.*

By definition, $K = \Pi \hat{a}_i^{\nu_i}$, where the $\hat{a}_i$ are the activities of the components within the mixture, and the ν_i are the stoichiometric numbers as defined above. Then, using 101.3 kPa (1 atm) as the standard-state fugacity, $K = (\hat{f}/f°)_{\text{neopentane}}/(\hat{f}/f°)_{n\text{-pentane}} = (y\hat{\phi}P)_{\text{neopentane}}/(y\hat{\phi}P)_{n\text{-pentane}}$, where f is fugacity, y is mole fraction, ϕ is fugacity coefficient, P is system pressure, the caret symbol ^ denotes the value of f or ϕ in solution, and the superscript ° denotes standard state, normally taken 1 atm.

The Lewis-Randall rule assumes that the fugacity of a component in solution is directly proportional to its mole fraction. Under this assumption, $\hat{\phi}_i = \phi_i$. Therefore (and noting that P cancels in numerator and denominator), $K = (y\phi)_{\text{neopentane}}/(y\phi)_{n\text{-pentane}}$. The ϕ can be calculated by methods outlined in Sec. 3 (or by use of Pitzer correlations); they are found to be 0.505 for neopentane and 0.378 for n-pentane. Therefore, substituting into the equation for K, $0.717 = 0.505y_{\text{neopentane}}/0.378y_{n\text{-pentane}} = 0.505x/0.378(1 - x)$. Solving for x, it is found to be 0.349. Therefore, the equilibrium composition consists of 0.349 mole fraction neopentane and $(1 - 0.349) = 0.651$ mole fraction n-pentane.

Related Calculations: If the standard heat of reaction $\Delta H°$ is known at a given T and K is known at that temperature, a reasonable estimate for K (designated K_1) at a not-too-distant temperature T_1 can be obtained via the equation $\ln (K/K_1) = (-\Delta H°/R)[(1/T) - (1/T_1)]$, where the heat of reaction is assumed to be constant over the temperature range of interest.

Occasionally, laboratory-measured values of K are available. These can be used to determine either or both constants of integration (ΔH_0 and I) in the calculations for ΔG.

As for the calculation of equilibrium composition, the Lewis-Randall-rule assumption in step 4 is a simplification. The rigorous calculation instead requires use of $\hat{\phi}$, the values of fugacity coefficient as they actually exist in the reaction mixture. This entails calculation from an equation of state. Initially, one would hope to use the virial equation, with *initial estimates* of mole fractions made as in step 4; however, in the present example, the pseudoreduced conditions are beyond the range of convergence of the virial equation. If the modified Soave-Redlich-Kwong equation is used instead, with a calculator program, to calculate the $\hat{\phi}$, the resulting mole fractions are found to be 0.362 for neopentane and 0.638 for n-pentane. Thus the Lewis-Randall-rule simplification in this case leads to errors of 3.6 and 2.0 percent, respectively. The degree of accuracy required should dictate whether the simplified approach is sufficient.

4-5 Activities Based on Mixed Standard States

For the reaction $2A \rightarrow B + C$ at 573 K (572°F), a calculation based on standard states of ideal gas at 101.32 kPa (1 atm) for A and B and pure liquid at its vapor pressure of 202.65 kPa (2 atm) for C produces a $\Delta G°/RT$ of -5, with G in calories per gram-mole. Calculate the equilibrium constant based on ratios of final mole fractions at 303.97 kPa (3 atm), assuming that all three components are ideal gases.

Calculation Procedure:

1. Select the solution method to be employed.

This problem may be solved either by setting out the equilibrium-constant equation in terms of the mixed (two) standard states or by first converting the $\Delta G°$ to a common standard state. Both methods are shown here.

2. Solve the problem by using mixed standard states.

a. Calculate the equilibrium constant K in terms of activities based on the given mixed standard states. By definition, $K = \exp(-\Delta G°/RT) = \exp -(-5) = 148.41$.

b. Relate the K to compositions. By definition (see previous problem), $K = a_C a_B/a_A^2 = (f_C/f_C^\circ)(f_B/f_B^\circ)/(f_A/f_A^\circ)^2$. In the present case, the standard-state fugacities are $f_A^\circ = f_B^\circ = 1$ atm, and $f_C^\circ = 2$ atm. Since the gases are assumed to be ideal, $f_i = y_i P$, where y is mole fraction and P is the system pressure. Therefore

$$K = 148.41 = [y_C(3\text{ atm})/(2\text{ atm})][y_B(3\text{ atm})/(1\text{ atm})]/[y_A(3\text{ atm})/(1\text{ atm})]^2$$

c. Define K in terms of mole fractions and solve for its numerical value. In terms of mole fractions for this reaction, $K = y_C y_B/y_A^2$. Rearranging the final equation in the preceding step,

$$y_C y_B/y_A^2 = 148.41(2/3)(1/3)(3/1)^2 = 296.82 = K$$

3. Solve the problem again, this time using uniform standard states.

a. Correct ΔG° to 1 atm for all the components. For this reaction, $\Delta G^\circ = G_C^\circ + G_B^\circ - 2G_A^\circ$. Now, G_B° and G_A° are already based on 1 atm. To correct G_C° to 1 atm, use the relationship $\Delta G = RT \ln (f_2/f_1)$, where R is the gas constant, T is absolute temperature, and f is fugacity. Since we wish to convert from a basis of 2 atm to one of 1 atm, $f_1 = 2$ atm and $f_2 = 1$ atm; therefore,

$$\Delta G = (1.987)(573)[\ln (1/2)] = -789.2 \text{ cal/(g·mol)}$$

Thus $G_{C,1\text{ atm}}^\circ = G_{C,2\text{ atm}}^\circ - 789.2$. Therefore, to correct ΔG° to 1 atm for all components, 789.2 cal/(g·mol) must be subtracted from the ΔG° as given in the statement of the problem; that is, from $-5RT$. Thus $\Delta G^\circ = (-5)(1.987)(573) - 789.2 = -6481.96$, and $\Delta G^\circ/RT$ corrected to 1 atm becomes -5.693.

b. Solve for K, and relate it to compositions and express it in terms of mole fractions. By definition, $K = \exp(-\Delta G^\circ/RT) = \exp -(-5.693) = 296.82$. And as in step 2b, $K = (y_C P/f_C^\circ)(y_B P/f_B^\circ)/(y_A P/f_A^\circ)^2 = (3y_C/1)(3y_B/1)/(3y_A/1)^2$. Therefore, $y_C y_B / y_A^2 = K = 296.82$, the same result as via the method based on mixed standard states.

Related Calculations: As this example illustrates, one should know the standard state chosen for the Gibbs free energy of formation for each compound (usually, ideal gas at 1 atm) when considering the relation of mole fractions to K values.

When a solid phase occurs in a reaction, it is often pure; in such a case, its activity may be expressed by

$$a_i = f_i/f_i^\circ = \exp[(v_s/RT)(P - P^\circ)]$$

where v_s is the molar volume of the solid and P° its vapor pressure. This activity is usually close to unity.

If a reaction takes place in the liquid phase, the activity may be expressed by $a_i = \gamma_i x_i$, where γ_i is the activity coefficient for component i and x_i is its mole fraction.

4-6 Effects of Temperature and Pressure on Equilibrium Composition

The reaction $H_2O \rightarrow H_2 + \frac{1}{2}O_2$ appears unlikely at 298 K (77°F) and 101.3 kPa (1 atm) because its ΔG° is 2.286×10^8 J/(kg·mol) (see Related Calculations under Example 4-3). If one wishes to increase the equilibrium conversion to hydrogen and oxygen, (a) should the temperature be increased or decreased, and (b) should the pressure be increased or decreased?

Calculation Procedure:

1. Calculate ΔH°, the heat of reaction.

This is done as outlined in Example 4-1. (Actually, since 298 K is a standard temperature, ΔH° for the present example can be read directly from tables.) The value is found to be 2.42×10^8 J/(kg·mol) [1.04×10^5 Btu/(lb·mol)].

2. *Calculate the algebraic sum of the stoichiometric numbers for the reaction $\Sigma\nu_i$.*

As discussed in Example 4-4, stoichiometric numbers for the components in a given reaction are numerically equal to the stoichiometric coefficients, but with the convention that the reactants get minus signs. In the present case, the stoichiometric numbers for H_2O, H_2, and O_2 are, respectively, -1, $+1$, and $+½$. Their sum is $+½$.

3. *Apply rules that pertain to partial derivatives of the reaction-progress variable.*

As discussed in Example 4-4, the reaction-progress variable x is a measure of the extent to which a reaction has taken place. In the present case, the desired increase in equilibrium conversion to hydrogen and oxygen implies an increase in x.

It can be shown that $(\partial x/\partial T)_P$ = (a positive number)($\Delta H°$) for all reactions. In the present example, $\Delta H°$ is positive (i.e., the reaction is endothermic). Therefore, an increase in temperature will increase x and, thus, the conversion to hydrogen and oxygen.

It can also be shown that $(\partial x/\partial P)_T$ = (a negative number)($\Sigma\nu_i$) for all reactions. Since $\Sigma\nu_i$ is positive, an increase in pressure will decrease x. So, a *decrease* in pressure will favor the conversion to hydrogen and oxygen.

Related Calculations: While nothing is said above about kinetics, increasing the temperature very frequently changes the reaction rate favorably. Accordingly, in some exothermic-reaction situations, it may be worthwhile to sacrifice some degree of equilibrium conversion in favor of shorter reactor residence time by raising reaction temperature. Similarly, a pressure change may have an effect on kinetics that is contrary to its effect on equilibrium.

The reaction in this example illustrates another point. It happens that (at 1 atm) water will not appreciably dissociate into hydrogen and oxygen unless the temperature is raised about 1500 K (2240°F); but at such temperatures, molecules of H_2 and O_2 may dissociate into atomic H and O and enter into unexpected reactions. The engineer should keep such possibilities in mind when dealing with extreme conditions.

4-7 Equilibrium Composition for Simultaneous Known Chemical Reactions

Given an initial mixture of 1 g·mol each of CO and H_2, estimate the equilibrium composition that will result from the following set of simultaneous gas-phase reactions at 900 K (1160°F) and 101.3 kPa (1 atm).

$$CO + 3H_2 \rightarrow CH_4 + H_2O \tag{4-1}$$

$$CO + H_2O \rightarrow CO_2 + H_2 \tag{4-2}$$

$$CO_2 + 4H_2 \rightarrow CH_4 + 2H_2O \tag{4-3}$$

$$4CO + 2H_2O \rightarrow CH_4 + 3CO_2 \tag{4-4}$$

Calculation Procedure:

1. Find the independent chemical reactions.

Write equations for the formation of each compound present:

$$C + \tfrac{1}{2}O_2 \rightarrow CO \tag{4-5}$$

$$C + 2H_2 \rightarrow CH_4 \tag{4-6}$$

$$C + O_2 \rightarrow CO_2 \tag{4-7}$$

$$H_2 + \tfrac{1}{2}O_2 \rightarrow H_2O \tag{4-8}$$

Next, algebraically combine these equations to eliminate all elements not present as elements in the system. For instance, eliminate O_2 by combining Eqs. 4-5 and 4-7

$$2CO \rightarrow CO_2 + C \tag{4-9}$$

and Eqs. 4-5 and 4-8

$$C + H_2O \rightarrow CO + H_2 \tag{4-10}$$

Then eliminate C by combining Eqs. 4-6 and 4-9

$$2CO + 2H_2 \rightarrow CO_2 + CH_4 \tag{4-11}$$

and Eqs. 4-6 and 4-10

$$CO + 3H_2 \rightarrow H_2O + CH_4 \tag{4-12}$$

The four initial equations are thus reduced into two independent equations, namely, Eqs. 4-11 and 4-12, for which we must find simultaneous equilibria. All components present in the original four equations are contained in Eqs. 4-11 and 4-12.

2. Calculate $\Delta G°$ and K for each independent reaction.

This may be done as in the relevant examples earlier in this section, with determination of $\Delta G°$ as a function of temperature. An easier route, however, is to use the standard Gibbs free-energy change of formation ΔG_f° for each compound at the temperature of interest in the relationship

$$\Delta G° = \Sigma \Delta G^\circ_{f,\text{products}} - \Sigma \Delta G^\circ_{f,\text{reactants}}$$

Reid et al. [9] give the following values of ΔG_f° at 900 K: for CH_4, 2029 cal/(g·mol); for CO, −45,744 cal/(g·mol); for CO_2, −94,596 cal/(g·mol); and for H_2O, −47,352 cal/(g·mol) (since H_2 is an element, its ΔG_f° is zero).

For Eq. 4-11, $\Delta G° = 2029 + (-94{,}596) - 2(-45{,}744) = -1079$ cal/(g·mol). For Eq. 4-12, $\Delta G° = 2029 + (-47{,}352) - (-45{,}744) = 421$ cal/(g·mol).

Finally, since the equilibrium constant K is equal to $\exp(-\Delta G°/RT)$, the value of K for Eq. 4-11 is $\exp[-(-1079)/(1.987)(900)] = 1.8283$, and the value of K for Eq. 4-12 is $\exp[-421/(1.987)(900)] = 0.7902$.

3. *Express the equilibrium mole fractions in terms of the reaction-progress variables.*

See Example 4-4 for a discussion of the reaction-progress variable. Let x_K and x_L be the reaction-progress variables for Reactions 4-11 and 4-12, respectively. Since Reaction 4-11 has 1 mol CO_2 on the product side, the number of moles of CO_2 at equilibrium is x_K. Since Reactions 4-11 and 4-12 each have 1 mol CH_4 on the product side, the number of moles of CH_4 at equilibrium is $x_K + x_L$. Since Reaction 4-11 has 2 mol CO on the reactant side and Reaction 4-12 has 1 mol, the number of moles of CO at equilibrium is $1 - 2x - x$ (because 1 mol CO was originally present). This approach allows us to express the equilibrium mole fractions as follows:

Component	Number of moles: Initially	Number of moles: At equilibrium	Equilibrium mole fraction y
CO_2	0	x_K	$x_K/2(1 - x_K - x_L)$
CH_4	0	$x_K + x_L$	$(x_K + x_L)/2(1 - x_K - x_L)$
CO	1	$1 - 2x_K - x_L$	$(1 - 2x_K - x_L)/2(1 - x_K - x_L)$
H_2	1	$1 - 2x_K - 3x_L$	$(1 - 2x_K - 3x_L)/2(1 - x_K - x_L)$
H_2O	0	x_L	$x_L/2(1 - x_K - x_L)$
		$2 - 2x_K - 2x_L$	

4. *Relate equilibrium mole fractions to the equilibrium constants.*

By definition, $K = \Pi \hat{a}_i^{\nu_i}$, where the $\hat{a}_i$ are the activities of the components with the mixture, and the ν_i are the stoichiometric numbers for the reaction (see Example 4-4). The present example is at relatively large reduced temperatures and relatively low reduced pressures, so the activities can be represented by the equilibrium mole fractions y_i. For Reaction 4-11, $K_K = y_{CO_2} y_{CH_4}/(y_{CO})^2(y_{H_2})^2$. Substituting the value for K from step 2 and the values for the y_i from the last column of the table in step 3 and algebraically simplifying,

$$1.8283 = \frac{x_K(x_K + x_L)[2(1 - x_K - x_L)]^2}{(1 - 2x_K - x_L)^2(1 - 2x_K - 3x_L)^2} \tag{4-13}$$

And for Reaction 4-12, $K_L = y_{H_2O} y_{CH_4}/y_{CO}(y_{H_2})^3$, or

$$0.7902 = \frac{x_L(x_K + x_L)[2(1 - x_K - x_L)]^2}{(1 - 2x_K - x_L)(1 - 2x_K - 3x_L)^3} \tag{4-14}$$

5. *Solve for the equilibrium conditions.*

Equations 4-13 and 4-14 in step 4 must be solved simultaneously. These are nonlinear and have more than one set of solutions; however, this complication can be eased by imposing two restrictions to ensure that no more CO or H_2 is used than the amount of each that is available (1 g·mol). Thus,

$$2x_K + x_L \leq 1 \tag{4-15}$$

$$2x_K + 3x_L \leq 1 \tag{4-16}$$

Even with Restrictions 4-15 and 4-16, the solution to Eqs. 4-13 and 4-14 requires a sophisticated mathematical technique or multiple trial and error calculations, as done most easily on the computer. At the solution, $x_K = 0.189038$ and $x_L = 0.0632143$; therefore, $y_{CO_2} = 0.1264$, $y_{CH_4} = 0.1687$, $y_{CO} = 0.37360$, $y_{H_2} = 0.28906$, and $y_{H_2O} = 0.04227$.

Related Calculations: If the gas is not ideal, the fugacity coefficients ϕ_i will not be unity, so the activities cannot be represented by the mole fractions. If the pressure is sufficient for a nonideal solution to exist in the gas phase, $\hat{\phi}_i$ will be a function of y_i, the solution to the problem. In this case, the y_i value obtained for the solution with $\hat{\phi}_i = 1$ should be used for the next iteration and so on until convergence. Alternatively, one could initially solve the problem using the Lewis-Randall rule for $\hat{\phi}_i(\hat{\phi}_i = \phi_i)$; the y_i obtained in that solution could be substituted back into $\hat{\phi}_i$ for the next estimate. Many times this is done most easily by computer.

4-8 Equilibrium Composition for Simultaneous Unspecified Chemical Reactions

Given an initial mixture of 1 g·mol CO and 1 g·mol H_2, determine the equilibrium composition of a final system at 900 K (1160°F) and 101.3 kPa (1 atm) that contains CO, CO_2, CH_4, H_2, and H_2O.

Calculation Procedure:

1. Determine the number of gram-atoms of each atom present in the system.

Here 1 g·mol CO contains 1 g·atom each of C and O, and 1 g·mol of H_2 contains 2 g·atoms of H. Let A_k equal the number of gram-atoms of element k present in the system. Then $A_C = 1$, $A_O = 1$, and $A_H = 2$.

2. Determine the number of gram-atoms of each element present per gram-mole of each substance.

Let $a_{i,k}$ equal the number of gram-atoms of element k per gram-mole of substance i. Then the following matrix can be set up:

i	$a_{i,C}$	$a_{i,O}$	$a_{i,H}$
CO	1	1	0
CO_2	1	2	0
CH_4	1	0	4
H_2	0	0	2
H_2O	0	1	2

3. Determine the Gibbs free energy of formation ΔG_f° for each compound at 900 K.

See step 2 of the previous problem. From Reid et al. [9], the values are: for CH_4, 2029 cal/(g·mol); for CO, −45,744 cal/(g·mol); for CO_2, −94,596 cal/(g·mol); for H_2O, −47,352 cal/(g·mol); and for H_2, zero.

4. Write equations for minimization of total Gibbs free energy.

This step employs the method of Lagrange undetermined multipliers for minimization under constraint; for a discussion of this method, refer to mathematics handbooks. As for its application to minimization of total Gibbs free energy, see Perry and Chilton [7] and Smith and Van Ness [11].

Write the following equation for each substance i:

$$\Delta G_f^\circ + RT \ln (y_i \hat{\phi}_i P / f_i^\circ) + \sum_k (\lambda_k a_{i,k}) = 0$$

where R is the gas constant, T is temperature, y is mole fraction, $\hat{\phi}$ is the fugacity coefficient, P is pressure, f° is standard-state fugacity, and λ_k is the Lagrange undetermined multiplier for element k within substance i. Since T is high and P is low, gas ideality is assumed; $\hat{\phi}_i = 1.0$. Set f° at 1 atm. Then the five equations are

For CO: $-45{,}744 + RT \ln y_{CO} + \lambda_C + \lambda_O = 0$

For CO_2: $-94{,}596 + RT \ln y_{CO_2} + \lambda_C + 2\lambda_O = 0$

For CH_4: $2029 + RT \ln y_{CH_4} + \lambda_C + 4\lambda_H = 0$

For H_2: $RT \ln y_{H_2} + 2\lambda_H = 0$

For H_2O: $-47{,}352 + RT \ln y_{H_2O} + 2\lambda_H + \lambda_O = 0$

[In all cases, $RT = (1.987)(900)$.]

5. Write material-balance and mole-fraction equations.

A material-balance equation can be written for each element, based on the values found in steps 1 and 2.

$$\sum_i y_i a_{i,k} = A_k / n_T$$

where n_T is the total number of moles in the system. These three equations are

For O: $y_{CO} + 2y_{CO_2} + y_{H_2O} = 1/n_T$

For C: $y_{CO} + y_{CO_2} + y_{CH_4} = 1/n_T$

For H: $4y_{CH_4} + 2y_{H_2} + 2y_{H_2O} = 2/n_T$

In addition, the requirement that the mole fractions sum to unity yields $y_{CO} + y_{CO_2} + y_{H_4} + y_{H_2} + y_{H_2O} = 1$.

6. Solve the nine equations from steps 4 and 5 simultaneously, to find y_{CO}, y_{CO_2}, y_{CH_4}, y_{H_2}, y_{H_2O}, λ_C, λ_O, λ_H, and n_T.

This step should, of course, be done on a computer. The Lagrange multipliers have no physical significance and should be eliminated from the solution scheme. The equilibrium composition is thus found to be as follows: $y_{CO_2} = 0.122$; $y_{CH_4} = 0.166$; $y_{CO} = 0.378$; $y_{H_2} = 0.290$; and $y_{H_2O} = 0.044$.

Note that these results closely compare with those found in the previous example, which is based on the same set of reaction conditions.

4-9 Heterogeneous Chemical Reactions

Estimate the composition of the liquid and vapor phases when n-butane isomerizes at 311 K (100°F). Assume that the reaction occurs in the vapor phase.

Calculation Procedure:

1. Determine the number of degrees of freedom for the system.

Use the phase rule $F = C - P + 2 - r$, where F is degrees of freedom, C is number of components, P is number of phases, and r is the number of independent reactions. In this case, C is 2, P is 2, and r is 1 (namely, $n\text{-}C_4H_{10} \rightarrow iso\text{-}C_4H_{10}$); therefore, $F = 1$. This means that we can choose either temperature or pressure alone to specify the system; when the temperature is given (311 K in this case), the system pressure is thereby established.

2. Calculate the equilibrium constant at the given temperature.

See Example 4-4. K is found to be 2.24.

3. Relate the equilibrium constant to compositions of the two phases.

Let subscripts 1 and 2 refer to $n\text{-}C_4H_{10}$ and $iso\text{-}C_4H_{10}$, respectively. Then $K = a_2/a_1 = (\hat{\phi}_2 y_2 P/f_2^\circ)/(\hat{\phi}_1 y_1 P/f_1^\circ)$, where a is activity, ϕ is fugacity coefficient, y is mole fraction, P is system pressure, f is fugacity, the caret symbol ˆ denotes the value of ϕ in solution, and the superscript ° denotes the standard state. If f_1° and f_2° are both selected to be 1 atm (101.3 kPa), the expression simplifies to

$$K = \hat{\phi}_2 y_2 P/\hat{\phi}_1 y_1 P \tag{4-17}$$

4. Relate the gas phase to the liquid phase.

Using phase-equilibrium relationships (see Sec. 3), the following equation can be set out for each of the two components:

$$y_i \hat{\phi}_i P = x_i \gamma_i P_i^{\text{sat}} \phi_i^{\text{sat}}$$

where x_i is liquid-phase mole fraction, γ_i is activity coefficient, P_i^{sat} is vapor pressure, and ϕ_i^{sat} is fugacity of pure i in the vapor at the system temperature.

The two butane isomers form an ideal solution in the liquid phase at the system temperature, the molecules being quite similar. Therefore, both activity coefficients can be taken to be unity.

Via relationships discussed in Sec. 3, ϕ_i^{sat} is found to be 0.91 and ϕ_2^{sat} to be 0.89. And from vapor pressure data, P_1^{sat} is 3.53 atm and P_2^{sat} is 4.95 atm.

5. Solve for chemical and phase equilibria simultaneously.

Write the equation in step 4 for each component and substitute into Eq. 4-17 while taking into account that $x_1 + x_2 = 1$. This gives

$$K = x_2 P_2^{\text{sat}} \phi_2^{\text{sat}} / (1 - x_2) P_1^{\text{sat}} \phi_1^{\text{sat}}$$

Or, substituting the known numerical values,

$$2.24 = x_2(4.95)(0.89)/(1 - x_2)(3.53)(0.91)$$

Thus the liquid composition is found to be $x_2 = 0.62$ and $x_1 = 0.38$.

Finding the vapor composition y_1 and y_2 first requires trial-and-error solution for P and ϕ_i in the equations

$$y_1\phi_1 P = x_1 P_1^{sat}\phi_1^{sat} \qquad \text{and} \qquad y_2\phi_2 P = x_2 P_2^{sat}\phi_2^{sat}$$

Initially assuming that $\phi_1 = \phi_2 = 1$ and noting that $y_1 + y_2 = 1$, we can write a combined expression

$$\begin{aligned} P &= x_1 P_1^{sat}\phi_1^{sat} + x_2 P_2^{sat}\phi_2^{sat} \\ &= (0.38)(3.53)(0.91) + (0.62)(4.95)(0.89) \\ &= 3.95 \end{aligned}$$

Therefore, $P = 3.95$ atm.

Using this value of P to estimate the ϕ_i values (see Sec. 3), we obtain $\phi_1 = 0.903$ and $\phi_2 = 0.913$. Then we substitute these and solve for a second-round estimate of P in

$$\begin{aligned} (y_1 + y_2) = 1 &= (x_1 P_1^{sat}\phi_1^{sat}/\phi_1 P) + (x_2 P_2^{sat}\phi_2^{sat}/\phi_2 P) \qquad (4\text{-}18) \\ &= (0.38)(3.53)(0.91)/0.903P + (0.62)(4.95)(0.89)/0.913P \end{aligned}$$

Therefore, $P = 4.344$ atm in this second-round estimate.

Now, use $P = 4.344$ to correct the estimates of the ϕ_i values, to obtain $\phi_1 = 0.894$ and $\phi_2 = 0.904$.

Substituting in Eq. (4-18) yields $P = 4.38$ atm, which converges with the previous estimates. Finally, from the relationship $y_i = x_i P_i^{sat}\phi_i^{sat}/\phi_i P$, $y_1 = (0.38)(3.53)(0.91)/(0.894)(4.38) = 0.31$, and $y_2 = (0.62)(4.95)(0.89)/(0.904)(4.38) = 0.69$.

Thus the liquid composition is $x_1 = 0.38$ and $x_2 = 0.62$, and the vapor composition is $y_1 = 0.31$ and $y_2 = 0.69$. The system pressure at equilibrium is 4.38 atm (443.7 kPa).

Related Calculations: Most frequently, liquid ideality cannot be assumed, and the engineer must use activity coefficients for the liquid phase. Activity coefficients are strong functions of liquid composition and temperature, so these calculations become trial-and-error, most easily done by computer.

In addition, there are frequently more than two nonideal-liquid components, which requires the use of a multicomponent activity-coefficient equation; see Prausnitz [8]. Often the parameters of the activity-coefficient equation have not been experimentally determined and must be estimated by a group-contribution technique; see Reid et al. [9].

5

Reaction Kinetics and Reactor Design

R. M. Baldwin, Ph.D.

Professor and Head
Chemical and Petroleum-Refining Engineering Department
Colorado School of Mines
Golden, CO

M. S. Graboski, Ph.D.

Director
Colorado Institute for Fuels and High-Altitude Engine Research
Colorado School of Mines
Golden, CO

5-1 Determining a Rate Expression by Integral Analysis of Batch-Reactor Data

Saponification of ethyl acetate with sodium hydroxide, that is,

$$CH_3COOC_2H_5 + NaOH \rightarrow CH_3COONa + C_2H_5OH$$

has been investigated at 298 K (77°F) in a well-stirred isothermal batch reactor. The following data were collected:

Time, min	5	9	13	20	25	33	37
Concentration of NaOH, g·mol/L	0.00755	0.00633	0.00541	0.00434	0.00385	0.0032	0.00296

The run began with equimolar (0.1 g·mol/L) amounts of sodium hydroxide and ethyl acetate as the reactants. Calculate the overall order of the reaction and the value of the reaction rate constant at 298 K, and write the rate expression for the reaction.

Calculation Procedure:

1. ***Assume a functional form for the rate expression.***

The rate expression for this reaction may be given by the following equation, which relates the rate of disappearance r of sodium hydroxide to concentrations of reactants and products:

$$-r_{NaOH} = k_1[CH_3COOC_2H_5]^a[NaOH]^b - k_{-1}[CH_3COONa]^c[C_2H_5OH]^d$$

In this expression, a, b, c, and d are the unknown reaction orders, k_1 and k_{-1} are the forward and reverse rate constants, and the bracketed formulas denote the concentrations of the compounds. Integral analysis of the data requires an assumption as to the functional form of the reaction rate expression (e.g., zero-order, first-order, second-order with regard to a given reactant), which is then inserted into the appropriate reactor material balance. Literature values indicate that the equilibrium constant for this reaction is very large ($k_1/k_{-1} \rightarrow \infty$). As an initial guess, the reaction may be considered to be first-order in both reactants and irreversible. Thus $-r_{NaOH} = k[EtAc]^1[NaOH]^1$, where k is the reaction rate constant. Since the reactants are present in equimolar ratio initially, $[EtAc] = [NaOH]$ throughout the run (if the initial guess is correct), and the rate expression can therefore be expressed as $-r_{NaOH} = k[NaOH]^2$.

2. ***Insert the rate expression into batch-reactor material balance.***

A transient material balance for the NaOH on an isothermal batch reactor becomes (NaOH in) − (NaOH out) + (net NaOH generation) = NaOH accumulation. For this system,

$$+r_{NaOH}V = \frac{dN_{NaOH}}{dt}$$

where r = rate of generation
V = reactor volume
N = number of moles
t = time

Rearranging, referring to Step 1, and noting that concentration may be given by N/V leads to the expression

$$\frac{-d[\text{NaOH}]}{dt} = k[\text{NaOH}]^2$$

3. Solve for concentration-versus-time profile.

The preceding expression may be separated and integrated to give a concentration-versus-time profile that may be tested against the experimental data. Integrating,

$$-\int_{[\text{NaOH}]_0}^{[\text{NaOH}]_t} \frac{d[\text{NaOH}]}{[\text{NaOH}]^2} = k \int_0^t dt$$

Carrying out the indicated integrations and simplifying leads to the expression

$$\frac{1}{[\text{NaOH}]_0} - \frac{1}{[\text{NaOH}]} = -kt$$

4. Plot the data.

The assumed model (first-order in both reactants and irreversible) predicts that if the data are plotted as $(1/[\text{NaOH}]_0 - 1/[\text{NaOH}])$ versus time, a straight line passing through the origin should be obtained, and the slope of this line will be the reaction rate constant $-k$. A plot of the experimental data according to this model is shown in Fig. 5-1. As may be seen, the data fit the assumed model quite well. The reaction rate constant, as determined by measuring the slope, is found to be

$$k = 6.42 \text{ L/(g}\cdot\text{mol)(min)}$$

Thus the rate of saponification of ethyl acetate with sodium hydroxide may be adequately modeled by a rate expression of the form

$$-r_{\text{NaOH}} = 6.42[\text{NaOH}]^{1.0}[\text{CH}_3\text{COOC}_2\text{H}_5]^{1.0}$$

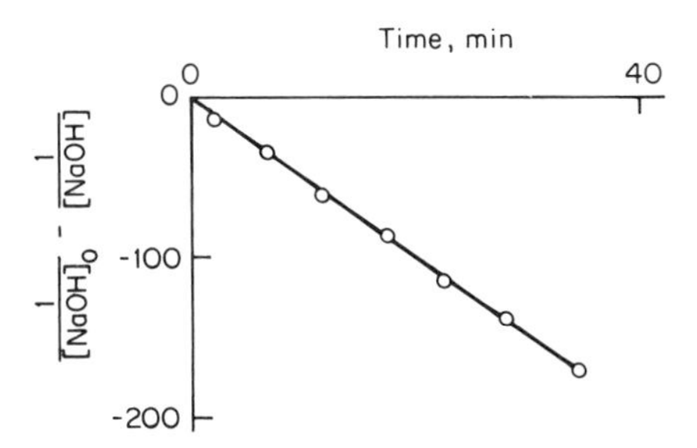

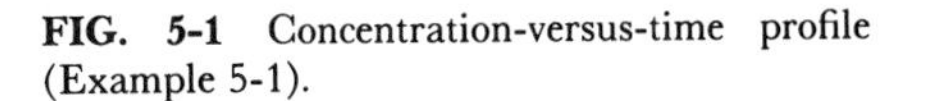
FIG. 5-1 Concentration-versus-time profile (Example 5-1).

Related Calculations: (1) Integral analysis may be used on data from any reactor from which integral reaction rate data have been obtained. The preceding procedure applies equally well to data from an integral tubular-flow reactor, if the tube-flow material balance

$$+r_i = \frac{dC_i}{d\tau}$$

is used rather than the batch-reactor material balance. Here C_i is the concentration of species i, in moles per volume, and τ is the residence time. (2) The form of the material balance to be tested depends on the reacting system under investigation and the data available for testing. The preceding analysis was applied because the system was of constant density and the data were in the form of concentration versus time. Data are often presented as fractional conversion (X_A) versus time, where

$$X_A = \frac{C_{A0} - C_A}{C_{A0}}$$

5-2 Determining a Rate Expression by Differential Analysis of Batch-Reactor Data

Determine an appropriate rate expression for the gas-phase reaction $A \rightarrow 2B$ utilizing the following data from a constant-volume batch reactor:

Time, h	Total pressure, kPa (atm)
0	132.74 (1.31)
0.5	151.99 (1.5)
1	167.19 (1.65)
1.5	178.33 (1.76)
2	186.44 (1.84)
2.5	192.52 (1.90)
3	197.58 (1.95)
3.5	201.64 (1.99)
4	205.18 (2.025)
5	210.76 (2.08)
6	214.81 (2.12)
7	217.85 (2.15)
8	220.38 (2.175)

The reaction mixture consists of 76.94% A with 23.06% inerts at 101.325 kPa (1 atm) and 287 K (57.2°F). The reaction is initiated by dropping the reactor into a constant-temperature bath at 373 K (212°F). Equilibrium calculations have shown the reaction to be essentially irreversible in this temperature range.

Calculation Procedure:

1. *Propose a generalized rate expression for testing the data.*

Analysis of rate data by the differential method involves utilizing the entire reaction-rate expression to find reaction order and the rate constant. Since the data have been obtained from a batch reactor, a general rate expression of the following form may be used:

$$\frac{dC_A}{dt} = -kC_A^{\alpha}$$

where k and α are the reaction rate constant and reaction order to be determined.

2. *Convert the rate expression to units of pressure.*

Since the data are in the form of total pressure versus time, the rate expression to be tested must also be in the form of total pressure versus time. Assuming ideal-gas behavior, $PV = nRT$, and therefore,

$$C_i = \frac{n_i}{V} = \frac{P_i}{RT}$$

where P_i = partial pressure of species i. Thus the rate expression becomes

$$\frac{1}{RT}\frac{dP_A}{dt} = -\left(\frac{1}{RT}\right)^{\alpha} kP_A^{\alpha}$$

Now, partial pressure of species A must be related to total-system pressure. This may be done easily by a general mole balance on the system, resulting in the following relationships:

a. *For any reactant,*

$$P_R = P_{R0} - \frac{r}{\Delta n}(\pi - \pi_0)$$

b. *For any product,*

$$P_S = P_{S0} + \frac{s}{\Delta n}(\pi - \pi_0)$$

where P_{R0} and P_{S0} are initial partial pressures of reactant R and product S, r and s are molar stoichiometric coefficients on R and P, π is total pressure, π_0 is initial total pressure, and Δn is net change in number of moles, equaling total moles of products minus total moles of reactants.

In the present case, r for the reactant A equals 1, s for the product B equals 2, and Δn equals $(2 - 1) = 1$. Using the data, and the relationship between partial pressure and total pressure for a reactant, the form of the rate expression to be tested may be derived:

$$P_A = P_{A0} - \frac{1}{2-1}(\pi - \pi_0)$$

From the data, $\pi_0 = 132.74$ kPa (1.31 atm), so $P_{A0} = (132.74)(0.7694) = 102.13$ kPa (1.0 atm). Therefore, $P_A = 102.13 - (1/1)(\pi - 132.74) = 234.87 - \pi$, with P_A and π in kilopascals, and $dP_A/dt = -d\pi/dt$. Thus the rate expression becomes $d\pi/dt = k'(234.87 - \pi)^\alpha$, where $k' = k(RT)^{1-\alpha}$.

3. *Linearize the rate expression by taking logs, and plot the data.*

The proposed rate expression may be linearized by taking logs, resulting in the following expression:

$$\ln\left(\frac{d\pi}{dt}\right) = \ln k' + \alpha \ln(234.87 - \pi)$$

This expression indicates that if $\ln(d\pi/dt)$ is plotted against $\ln(234.87 - \pi)$, a straight line should result with slope α and y intercept $\ln k'$. Thus, to complete the rate-data analysis, the derivative $d\pi/dt$ must be evaluated.

Three methods are commonly used to estimate this quantity: (1) slopes from a plot of π versus t, (2) equal-area graphic differentiation, or (3) Taylor series expansion. For details on these, see a mathematics handbook. The derivatives as found by equal-area graphic differentiation and other pertinent data are shown in the following table:

Time, h	$234.87 - \pi$, kPa	$d\pi/dt$
0	102.13	44.5
0.5	82.88	34
1	67.68	26
1.5	56.54	19.5
2	48.43	15
2.5	42.35	11
3	37.29	9
3.5	33.23	7.5
4	29.69	6.5
5	24.11	4.5
6	20.06	3.5
7	17.02	2.5
8	14.49	1.5

Plotting $\ln(d\pi/dt)$ versus $\ln(234.87 - \pi)$ yields an essentially straight line with a slope of 1.7 and an intercept of 0.0165. Thus, an appropriate rate expression for this reaction is given by

$$d\pi/dt = 0.0165(234.87 - \pi)^{1.7}$$

or

$$-dP_A/dt = 0.0165P_A^{1.7}$$

Related Calculations: The rate expression derived above may be converted back to concentration units by noting that $C_A = P_A/RT$ and using $T = 373$ K, $R = 0.0821$ (L)(atm)/(g·mol)(K).

5-3 Finding Required Volume for an Adiabatic Continuous-Flow Stirred-Tank Reactor

Determine the volume required for an adiabatic mixed-flow reactor processing 56.64 L/min (2 ft³/min or 0.05664 m³/min) of a liquid feed containing reactant R and inerts I flowing at a rate of 0.67 g·mol/min and 0.33 g·mol/min, respectively. In the reactor, R is isomerized to S and T (90 percent fractional conversion of R) by the following elementary reaction: $R \xrightarrow{k_1} S + T$. Feed enters the reactor at 300 K (80.6°F). Data on the system are as follows:

Heat Capacities

$$R = 7 \text{ cal/(g·mol)(°C)}$$
$$S = T = 4 \text{ cal/(g·mol)(°C)}$$
$$I = 8 \text{ cal/(g·mol)(°C)}$$

Reaction Rate Constant at 298 K

$$k_1 = 0.12 \text{ h}^{-1}$$

Activation Energy

$$25{,}000 \text{ cal/(g·mol)}$$

Heat of Reaction at 273 K

$$\Delta H_R = -333 \text{ cal/(g·mol) of } R$$

Calculation Procedure:

1. Write the material- and energy-balance expressions for the reactor.

This problem must be solved by simultaneous solution of the material- and energy-balance relationships that describe the reacting system. Since the reactor is well insulated and an exothermic reaction is taking place, the fluid in the reactor will heat up, causing the reaction to take place at some temperature other than where the reaction rate constant and heat of reaction are known.

Assuming a constant-density reacting system, a constant volumetric flow rate through the reactor, and steady-state operation, a material balance on species R gives the expression

$$\nu C_{R0} - \nu C_R + V r_R = 0$$

where C_{R0} = concentration of R in feed
C_R = concentration of R in products
ν = volumetric flow rate
V = reactor volume
r_R = net rate of formation of R

This equation can be rearranged into

$$\frac{C_{R0} - C_R}{-r_R} = \tau$$

where τ = residence time (V/ν).

A first-law energy balance on the continuous-flow stirred-tank reactor gives the expression

$$Q = \sum_{i=1}^{n} F_{i0}\hat{C}_{p,i}(T - T_{i0}) + X[\Delta H^\circ + \Delta\hat{C}_p(T - T^\circ)]$$

where Q = rate of heat exchange with surroundings
F_i = outlet molar flow rate of species i
F_{i0} = inlet molar flow rate of species i
$\hat{C}_{p,i}$ = mean heat capacity of species i
T_{i0} = inlet (feed) temperature of species i
T = reactor operating temperature
X = molar conversion rate of species R ($= F_{R0} - F_R$)
ΔH° = heat of reaction at T°
T° = reference temperature for heat-of-reaction data
$\Delta\hat{C}_p = \Sigma v_i C_{p,i\text{products}} - \Sigma v_i C_{p,i\text{reactants}}$
v_i = molar stoichiometric coefficient

2. *Calculate the operating temperature in the reactor.*

Application of the energy balance shown above allows the reaction mass temperature to be calculated, since all quantities in the expression are known except T. Pertinent calculations and parameters are as follows:

$Q = 0$ (adiabatic)

$F_{R0} = 0.67$ mol/min

$F_{I0} = 0.33$ mol/min

$T_{R0} = T_{I0} = 300$ K

$X = X_R F_{R0} = (0.90)(0.67) = 0.603$ g·mol R per minute

$\Delta\hat{C}_p = 4 + 4 - 7 = 1$ cal/(g·mol)(°C)

Substituting into the energy balance,

$$0 = (0.67)(7)(T - 300) + (0.33)(8)(T - 300) + 0.603[-333 + (1)(T - 273)]$$

Solving for reactor temperature,

$$T = 323.4 \text{ K } (122.7°\text{F})$$

3. *Calculate the reaction rate constant at the reactor operating temperature.*

Since the temperature in the reactor is not 25°C (where the value for the reaction rate constant is known), the rate constant must be estimated at the reactor temperature. The Arrhenius form of the rate constant may be used to obtain this estimate:

$$k = A \exp\left[-E_A/(RT)\right]$$

where A = preexponential factor
E_A = activation energy
R = universal gas constant
T = absolute temperature

Dividing this Arrhenius equation for $T = T$ by the Arrhenius equation for $T = 298$ K (25°C) and noting that $(1/T) - (1/298) = (298 - T)/298T$, the following expression can be derived:

$$k_T = k_{298} \exp\left[\frac{-25{,}000}{1.987}\left(\frac{298 - T}{298T}\right)\right]$$

Now $k_{298} = 0.12\ \text{h}^{-1}$, so when $T = 323.4$ K (the reactor operating temperature), the reaction rate constant becomes $k_{323.4} = 3.31\ \text{h}^{-1}$.

4. *Solve for reactor volume using the material-balance expression.*

The material balance for the continuous-flow stirred-tank reactor may now be used to calculate the reactor volume required for the isomerization. Inserting the first-order rate expression into the material balance,

$$\tau = \frac{C_{R0} - C_R}{kC_R} = \frac{V}{\nu}$$

To apply this material balance, it is first necessary to calculate the inlet and outlet concentrations of species R. This may be easily accomplished from the given data and the relationships

$$C_i = \frac{F_i}{\nu} \qquad F_i = F_{i0}(1 - X_i)$$

Since ν = constant, $C_i = C_{i0}(1 - X_i)$. In these relationships, C_i is the concentration of species i, F_i is the molar flow rate of species i, ν is volumetric flow rate, and X_i is fractional conversion of species i.

For this example,

$$C_{R0} = \frac{F_{R0}}{v} = \frac{0.67}{56.64} = 0.012 \text{ g·mol/L}$$

and

$$C_R = 0.012(1 - 0.9) = 0.0012 \text{ g·mol/L}$$

Thus the reactor volume may now be directly calculated after converting the volumetric flow rate to an hourly basis:

$$V = \frac{0.012 - 0.0012}{(3.31)(0.0012)}(56.64)(60) = 9240 \text{ L } (326 \text{ ft}^3 \text{ or } 9.24 \text{ m}^3)$$

Related Calculations: (1) Since the reaction is irreversible, equilibrium considerations do not enter into the calculations. For reversible reactions, the ultimate extent of the reaction should always be checked first, using the procedures outlined in the preceding section. If equilibrium calculations show that the required conversion cannot be attained, then either the conditions of the reaction (e.g., temperature) must be changed or the design is not feasible. Higher temperatures should be investigated to increase ultimate conversions for endothermic reactions, while lower temperatures will favor higher conversions for exothermic reactions.

(2) The simultaneous solving of the material- and energy-balance expressions may yield more than one solution. This is especially true for exothermic reactions occurring in continuous stirred-tank reactors. The existence of other feasible solutions may be determined by plotting the energy-balance and material-balance expressions on molar conversion rate versus temperature coordinates, as shown in Fig. 5-2. In the figure, points *A* and *C* represent stable operating points for the reactor, while point *B* is the metastable, or "ignition," point, where stable operation is difficult. This is due to the relative slopes of the material- and energy-balance lines at *B*. Small positive temperature excursions away from *B* result in "ignition" of the reaction mass because the rate of heat generation is greater than that of heat removal, and the reactor restabilizes at point *A*. Similarly, a

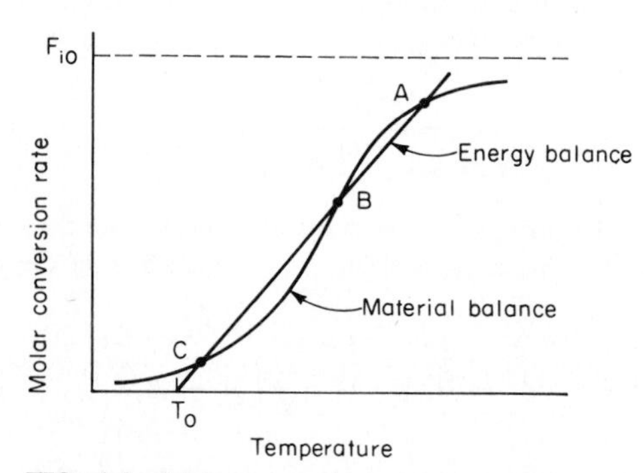

FIG. 5-2 Molar conversion rate versus temperature (Example 5-3).

small negative temperature deviation from B causes the reaction mass to "quench," and the reactor restabilizes at point C.

5-4 Calculating the Size of an Isothermal Plug-Flow Reactor

Laboratory experiments on the irreversible, homogeneous gas-phase reaction $2A + B = 2C$ have shown the reaction rate constant to be 1×10^5 (g·mol/L)$^{-2}$ s^{-1} at 500°C (932°F). Analysis of isothermal data from this reaction has indicated that a rate expression of the form $-r_A = kC_AC_B^2$ provides an adequate representation for the data at 500°C and 101.325 kPa (1 atm) total pressure. Calculate the volume of an isothermal, isobaric plug-flow reactor that would be required to process 6 L/s (0.212 ft^3/s) of a feed gas containing 25% A, 25% B, and 50% inerts by volume if a fractional conversion of 90% is required for component A.

Calculation Procedure:

1. *Develop a plug-flow-reactor design equation from the material balance.*

To properly size a reactor for this reaction and feedstock, a relationship between reactor volume, conversion rate of feed, and rate of reaction is needed. This relationship is provided by the material balance on the plug-flow reactor.

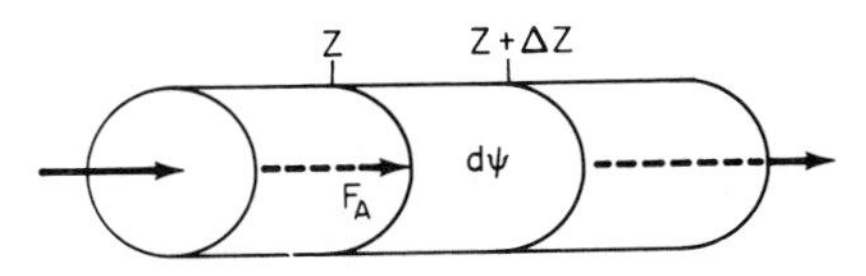

FIG. 5-3 Differential volume element for plug-flow reactor (Example 5-4).

For a single ideal reactor, a component material balance on a differential volume element $d\psi$ (see Fig. 5-3) becomes, for species A

$$F_{AZ} - F_{AZ+\Delta Z} + r_A\Delta\psi = 0 \qquad \text{or} \qquad \frac{dF_A}{d\psi} = r_A$$

where ψ is a reactor-size parameter (volume, mass of catalyst, etc.), F_A is molar flow rate of A, and r_A is the rate of generation of A per unit of volume. Upon rearranging and integrating,

$$\int_{F_{A0}}^{F_A} \frac{dF_A}{r_A} = \int_0^{\psi} d\psi = \psi$$

In terms of total molar conversion rate of species A, designated X, $F_A = F_{A0} - X$, and $dF_A = -dX$, where F_{A0} is molar flow rate of A in feed. Thus the material balance for this homogeneous reaction becomes

$$\psi = V = -\int_0^X \frac{dX}{r_A}$$

2. Relate molar flow rates of products and reactants to conversion rate of A.

To use the mass balance to solve for reactor size V, it is necessary to relate the rate of generation $+r_A$ of A to the molar conversion rate X of A. This is easily done through the rate expression. For this reaction

$$+r_A = -kC_AC_B^2$$

where C_A and C_B are molar concentrations of A and B. For this gas-phase reaction, the concentrations of A and B may be calculated from the ideal-gas law, so

$$C_A = n_A/V = \frac{p_A}{RT} = \frac{p_T y_A}{RT}$$

where p_A = partial pressure of A
V = volume
n_A = number of moles of A
p_T = total system pressure
R = universal gas constant
T = absolute temperature
y_A = mole fraction of A

The rate expression then becomes

$$+r_A = -k\left(\frac{p_T}{RT}\right)^3 y_A y_B^2$$

Mole fractions for components A and B are computed from a component-by-component mole balance on the reactor, noting that ½ mol B and 1 mol C are involved per mole of A reacted. Thus,

$$\begin{aligned} F_A &= F_{A0} - X \\ F_B &= F_{B0} - X/2 \\ F_C &= F_{C0} + X \\ F_I &= F_{I0} \\ \hline F_T &= F_{T0} - X/2 \end{aligned}$$

In this balance, F_A, F_B, and F_C are the molar flow rates of species A, B, and C at any position in the reactor, F_{A0}, F_{B0}, and F_{C0} are the molar flow rates of A, B, and C in the feed, and F_I and F_{I0} are the molar flow rates of inerts in product and in the feed.

Now, by definition,

$$y_A = \frac{F_A}{F_T} = \frac{F_{A0} - X}{F_{T0} - X/2} \quad \text{and} \quad y_B = \frac{F_B}{F_T} = \frac{F_{B0} - X/2}{F_{T0} - X/2}$$

and the rate expression in terms of molar conversion rate of A therefore becomes:

$$+r_A = -k\left(\frac{p_T}{RT}\right)^3 \left[\frac{(F_{A0} - X)(F_{B0} - X/2)^2}{(F_{T0} - X/2)^3}\right]$$

hydrogen and produces 1 mol ethane, the batch reactor will contain at any given time t the following mixture (with subscript 0 referring to the amount initially present):

$$N_{C_2H_2} = N_{C_2H_2,0} - X$$

$$N_{H_2} = N_{H_2,0} - 2X$$

$$N_{C_2H_6} = X$$

The total number of moles is thus the sum of these three equations, or $N_{T,0} - 2X$. The design equation thus becomes

$$\frac{-dX}{dt} = \frac{-k}{V^2}(N_{H_2,0} - 2X)^2(N_{C_2H_2,0} - X)$$

2. *Separate and integrate the batch-reactor design equation.*

The reaction time required for any given initial composition and conversion of reactants may now be calculated directly from the preceding expression, once this expression has been integrated and solved for $t = f(X)$. The appropriate solution method is as follows:

a. *Separate the variables.*

$$\int_0^x \frac{dX}{(N_{H_2,0} - 2X)^2(N_{C_2H_2,0} - X)} = \frac{k}{V^2}\int_0^t dt$$

b. *Integrate.*

$$\int_0^x \frac{dX}{(N_{H_2,0} - 2X)^2(N_{C_2H_2,0} - X)} = \frac{1}{(-N_{H_2,0} + 2N_{C_2H_2,0})}\,\frac{1}{(N_{H_2,0} - 2X)} + \frac{-1}{(-N_{H_2,0} + 2N_{C_2H_2,0})}\ln\left(\frac{N_{C_2H_2,0} - X}{N_{H_2,0} - 2X}\right) = \frac{k}{V^2}t$$

c. *Evaluate between limits.* Evaluation of the integral between 0 and x for the left-hand side and between 0 and t for the right-hand side gives the following expression for reaction time as a function of total molar conversion:

$$\frac{1}{(N_{H_2,0} - 2X)(2N_{C_2H_2,0} - N_{H_2,0})} - \frac{1}{N_{H_2,0}(2N_{C_2H_2,0} - N_{H_2,0})} + \frac{1}{(2N_{C_2H_2,0} - N_{H_2,0})^2}\ln\left[\frac{N_{C_2H_2,0}/N_{H_2,0}}{(N_{C_2H_2,0} - X)/(N_{H_2,0} - 2X)}\right] = \frac{k}{V^2}t$$

3. *Solve for reaction time.*

The total molar conversion of acetylene at 90% fractional conversion may be found by direct application of the definitions for fractional and total molar conversion. Thus, let $X_{C_2H_2}$ stand for the fractional conversion of acetylene, which is defined as

$$\frac{N_{C_2H_2,0} - N_{C_2H_2}}{N_{C_2H_2,0}}$$

According to the problem statement, this equals 0.9. Since $N_{C_2H_2,0} = 0.001$ g·mol, $N_{C_2H_2} = 0.001 - (0.9)(0.001) = 0.0001$ g·mol. Similarly, from step 1, X equals total molar conversion of acetylene, defined as $N_{C_2H_2,0} - N_{C_2H_2}$. It is therefore equal to $(0.001 - 0.0001) = 0.0009$ g·mol.

This value for total molar conversion, along with the initial moles of C_2H_2 and H_2, now allows reaction time to be calculated. Since the initial mixture is 75% hydrogen and 25% acetylene, $N_{H_2,0} = 3(0.001) = 0.003$. Substituting into the design equation from step 2,

$$\frac{1}{(0.003 - 0.0018)(0.002 - 0.003)} - \frac{1}{0.003(0.002 - 0.003)}$$

$$+ \frac{1}{(0.002 - 0.003)^2} \ln \left[\frac{(0.001/0.003)}{(0.001 - 0.0009)/(0.003 - 0.0018)} \right] = \frac{10^5}{1^2} t$$

Solving, $t = 8.86$ min. Thus approximately 9 min is needed to convert 90% of the acetylene originally charged to the 1-L reactor to ethane.

Related Calculations: (1) The stoichiometry of the chemical reaction strongly influences the final form of the integrated design expression. Different rate expressions will lead to different functional relationships between time and total molar conversion. The preceding example was specific for an irreversible reaction, second-order in hydrogen and first-order in acetylene. If the rate expression had been simply first-order in acetylene (as might be the case with excess hydrogen), the integration of the design expression would yield

$$-\ln \left(\frac{N_{C_2H_2} - X}{N_{C_2H_2,0}} \right) = kt$$

Similarly, other rate expressions yield different forms for the time, molar conversion relationship.

(2) In many cases, analytical integration of the design equation is difficult. The integral may still be evaluated, however, by (a) numerical integration methods, such as Euler or Runge-Kutta, or (b) graphic evaluation of $\int_0^x f(X)dX$ by plotting $f(X)$ versus X and finding the area under the curve.

5-6 Calculating Reaction Rates from Continuous-Flow Stirred-Tank Reactor Data

In a study of the nitration of toluene by mixed acids, the following data were obtained in a continuous-flow stirred-tank reactor. It had been previously determined that the reactor was well mixed; the composition within the reactor and in the exit stream can be considered equal. In addition, it had been determined that mass-transfer effects were not limiting the process rate. Thus the rate measured is the true kinetic rate of reaction. Calculate that rate.

Reactor data:	
Mixed-acid feed rate, g/h	325.3
Toluene feed rate, g/h	91.3
Acid-phase leaving, g/h	301.4
Organic-phase leaving, g/h	117.3
Temperature, °C (°F)	36.1 (97.0)
Stirrer speed, r/min	1520
Reactor volume, cm^3	635
Volume fraction of acid phase in reactor	0.67
Organic-phase composition:	
Mononitrotoluene, mol %	68.1
Toluene, mol %	31.1
Sulfuric acid, mol %	0.8
Density at 25°C, g/cm^3	1.0710
Feed acid-phase composition:	
H_2SO_4, mol %	29.68
HNO_3, mol %	9.45
H_2O, mol %	60.87
Density at 25°C, g/cm^3	1.639
Spent acid composition:	
H_2SO_4, mol %	29.3
HNO_3, mol %	0.3
H_2O, mol %	70.4
Density, g/cm^3	1.603

1. Check the elemental material balances.

For reference, the molecular weights involved are as follows:

Component	Molecular weight
Toluene	92
Mononitrotoluene	137
Sulfuric acid	98
Nitric acid	63
Water	18

The feed consists of 0.0913 kg/h of toluene and 0.3253 kg/h of mixed acid. The latter stream can be considered as follows:

Acid component	Mol % (given)	Kilograms per 100 mol feed	Wt %	kg/h
H_2SO_4	29.68	2908.6	63.24	0.2057
HNO_3	9.45	595.4	12.94	0.0421
H_2O	60.87	1095.7	23.82	0.0775
		4599.7	100.00	0.3253

The output consists of 0.1173 kg/h of the organic phase and 0.3014 kg/h of spent acid.

These two streams can be considered as follows:

Organic-phase component	Mol % (given)	Kilograms per 100 mol product	Wt %	kg/h
Toluene	31.1	2,861.2	23.32	0.0274
Mononitrotoluene	68.1	9,329.7	76.04	0.0892
Sulfuric acid	0.8	78.4	0.64	0.0007
		12,269.3	100.00	0.1173

Spent-acid-phase component	Mol % (given)	Kilograms per 100 mol acid	Wt %	kg/h
H_2SO_4	29.3	2871.4	69.07	0.2082
HNO_3	0.3	18.9	0.45	0.0014
H_2O	70.4	1267.2	30.48	0.0918
	100.0	4157.5	100.00	0.3014

The elemental material balances can then be checked. For carbon:

Component	kg/h of C in	kg/h of C out
Toluene	0.0834	0.0250
Mononitrotoluene	0.0000	0.0547
	0.0834	0.0797

Percent difference in C = (100)(0.0834 − 0.0797)/0.0834 = 4.44%

For hydrogen:

Component	kg/h of H in	kg/h of H out
Mononitrotoluene	0.0000	0.0046
Toluene	0.0079	0.0024
H_2SO_4	0.0042	0.0043
HNO_3	0.0007	0.0000
H_2O	0.0086	0.0102
	0.0214	0.0215

Percent difference in H = (100)(0.0215 − 0.0214)/0.0214 = 0.5%

For oxygen:

Component	kg/h of O in	kg/h of O out
Mononitrotoluene	0.0000	0.0208
H_2SO_4	0.1343	0.1364
HNO_3	0.0321	0.0011
H_2O	0.0689	0.0816
	0.2353	0.2399

Percent difference in O = (100)(0.2399 − 0.2353)/0.2353 = 2%

The elemental balances for C, H, and O suggest that the run is reasonably consistent (because the percent differences between feed and product are small), so there is no need to make material balances for the other elements.

2. *Employ the material-balance data to determine the reaction rate.*

In a continuous-flow stirred-tank reactor, the material balance for component A is as follows at steady state:

$$\text{Rate of input of } A - \text{rate of output of } A + \text{rate of generation of } A = 0$$

Thus,

$$(F_{A0} - F_A) + r_A V = 0$$

where F_{A0}, F_A = inlet and outlet molar flow rates
r_A = rate of reaction per unit volume of organic phase
V = volume of the organic phase

Rearranging the material balance,

$$-r_A = \frac{F_{A0} - F_A}{V} = \frac{W_{A0} - W_A}{MV}$$

where W_{A0} and W_A are mass flow rates, and M is the molecular weight of A.

The reaction stoichiometry is as follows:

$$C_6H_5CH_3 + HNO_3 \xrightarrow[H_2O]{H_2SO_4} C_6H_4(CH_3)(NO_2) + H_2O$$

Since the reaction involves 1 mol of each component, $r(\text{toluene}) = r(HNO_3) = -r(\text{mononitrotoluene}) = -r(H_2O)$.

From the reactor data given in the statement of the problem, the organic-phase volume is 0.33(0.635 L) = 0.210 L. Thus, by material balance, the computed rates are

Component	r, kg·mol/(h)(L)
Toluene	0.00331
HNO_3	0.00308
Mononitrotoluene	0.00310
Water	0.00370

These can be averaged to give the mean rate of reaction: 0.00330 ± 0.00021 kg·mol/(h)(L). For a reactant (e.g., toluene), a minus sign should be placed in front of it.

Related Calculations: In this example, it was stated at the outset that mass-transfer effects were not limiting the process rate. In the general case, however, it is important to calculate the effect of mass-transfer resistance on the reaction rate.

Consider, for instance, the gasification of porous carbon pellets in a fixed-bed reactor using steam and oxygen; the reaction rate could be affected both by external-film mass transfer and by pore-diffusion mass transfer.

The first of these pertains to the stagnant film separating the particle surface from the bulk gas. At steady state, the rate of transport to the surface is given by the standard mass-transfer expression

$$W = k_m A_p C(Y_B - Y_S) = k_m A_p (C_B - C_S)$$

where W = transfer rate, in moles per time per weight of solid
k_m = mass-transfer coefficient, in length per time
A_p = external surface area per weight of solid
Y_B = bulk-gas concentration, in mole fraction units
Y_S = concentration of gas adjacent to surface, in mole fraction units
C = total gas concentration, in moles per volume
C_B = concentration of component in the bulk, in moles per volume
C_S = concentration of component adjacent to surface, in moles per volume.

The mass-transfer coefficient k_m is a weak function of absolute temperature and velocity. The total concentration C is given approximately by the ideal-gas law $C = P/(RT)$, where P is absolute pressure, R is the gas constant, and T is absolute temperature.

In fixed-bed operation, Satterfield[1] recommends correlations for mass (and heat) transfer coefficients based on the Colburn j factor, defined as follows:

$$j = \frac{k_m}{(\rho^* V)} Sc^{2/3}$$

where j = Colburn j factor, dimensionless
k_m = mass-transfer coefficient, in moles per unit of time per unit area of particle surface
ρ^* = molar density, in moles per volume
V = superficial velocity, based on empty reactor tube
Sc = Schmidt number, $\mu/\rho D$, dimensionless
μ = viscosity
ρ = mass density
D = diffusivity through the film

The j factor depends on the external bed porosity ϵ and the Reynolds number, $Re = D_p V\rho/\mu$, where D_p is the particle diameter, as follows:

$$\epsilon j = \frac{0.357}{Re^{0.359}} \qquad 3 \leq Re \leq 2000$$

The appropriate particle diameter is given as

$$D_p = \frac{6V_{ex}}{S_{ex}}$$

where V_{ex} = volume of particle
S_{ex} = surface area of particle

External mass transfer reduces the concentration of reactant gas close to the particle surface and thus reduces the overall process rate. Thus, consider gasification to be a first-order reaction. Then at steady state, the rate of gasification equals the rate of mass trans-

[1]Satterfield, C., *Mass Transfer in Heterogeneous Catalysis,* MIT Press (1970).

fer. For a nonporous solid, the surface reaction (whose rate constant is k^*) consumes the diffusing reactant:

$$k^* C_S = k_m A_p (C_B - C_S)$$

Solving for the surface concentration yields

$$C_S = \frac{k_m A_p C_B}{k^* + k_m A_p}$$

Now, the process rate is given by

$$-r_c = k^* C_S = \frac{k^* k_m A_p C_B}{k^* + k_m A_p}$$

So, if the mass-transfer rate constant k_m is large in comparison to k^*, the rate reduces to $-r_c = k^* C_B$; that is, the true kinetic rate is based directly on the bulk concentration.

Next, consider the effect of pore diffusion on the reaction rate. The gasification reaction occurs principally within the particle. Except at very high temperatures, reactants must diffuse into the pore to the reacting surface. The average reaction rate within the particle may be related to the rate based on the surface concentration in terms of an effectiveness factor η defined as

$$\eta = \frac{r_{\text{avg}}}{r_{\text{surface}}}$$

The effectiveness factor is a function of a dimensionless group termed the "Thiele modulus," which depends on the diffusivity in the pore, the rate constant for reaction, pore dimension, and external surface concentration C_S.

The effectiveness factor for a wide range of reaction kinetic models differs little from that of the first-order case. For an isothermal particle, the first-order reaction effectiveness factor is given as

$$\eta = \frac{\tanh \phi}{\phi}$$

where ϕ is the Thiele modulus, that is,

$$\phi = L_p \left(\frac{k C_S^{m-1}}{V_p D} \right)^{1/2}$$

and where L_p = effective pore length, cm = $R/3$ for spheres (R = particle radius)
k = reaction rate constant
C_S = external surface concentration, in moles per cubic centimeter
m = reaction order (equal to 1 for first-order)
V_p = pore volume, in cubic centimeters per gram
D = diffusivity

When porous solids are being used as catalysts or as reactants, the rate constant k^* in the global equation is replaced by ηk^*. Consequently, this equation applies to porous solids as well as nonporous solids.

When diffusion is fast relative to surface kinetics, $\phi \to 0$, $\eta \to 1$, and $r_{avg} = r_{surface}$. Under these conditions, all the pore area is accessible and effective for reaction. When $\phi \to \infty$, that is, when diffusion is slow relative to kinetics, the reaction occurs exclusively at the particle external surface; reactant gas does not penetrate into the pores.

External mass transport generally becomes dominant at temperatures higher than that at which pore diffusion limits the gasification rate. For small particles, smaller than 20 mesh, mass-transfer limitations generally are not important because these particles have external surface areas that are large compared with their unit volume. Furthermore, mass-transfer coefficients are greater in fluid-bed operations owing to the motion of the solid particles. Thus in fluid-bed operations, external mass-transfer limitation in the temperature region below about 900 to 1100°C is never important. For fixed-bed operation, however, mass transfer to large particles can be important.

5-7 Total Surface Area, Active Surface Area, Porosity, and Mean Pore Radius of a Catalyst

A catalyst composed of 10 wt % nickel on γ-alumina is used to promote the catalytic methanation reaction

$$CO + 3H_2 \rightarrow CH_4 + H_2O$$

The important properties of the catalyst for characterization purposes are the total surface area, dispersion of metallic nickel, pore volume, and mean pore radius. Determine each of these, using the experimental data given in the respective calculation steps.

Calculation Procedure:

1. Calculate the total surface area.

The weight gain of the catalyst due to the physical adsorption of nitrogen under various nitrogen pressures is a function of, and thus an indicator of, total surface area. The first three columns of Table 5-1 show the weight of adsorbed nitrogen and the corresponding pressure for experimental runs conducted at the atmospheric boiling point of liquid nitrogen.

The most common way of analyzing such data is by using the so-called BET equation. For multilayer adsorption, this equation can be set out in the form

$$\frac{P}{W(P^* - P)} = \frac{1}{W_m C} + \frac{C - 1}{C W_m} \frac{P}{P^*}$$

where W is weight adsorbed per gram of catalyst at pressure P, P^* is the vapor pressure of the adsorbent, C is a parameter related to the heat of adsorption, and W_m is the weight for monolayer coverage of the solid.

The last-named variable is the one of interest in the present case, because it represents the weight of adsorbed nitrogen that just covers the entire surface of the catalyst, internal and external. (Because the catalyst is highly porous, most of the area is pore wall and is internal to the solid.)

Now, from the form of the equation, a plot of $P/[W(P^* - P)]$ against P/P^* should

TABLE 5-1 BET Calculations for Prototype Catalyst (Example 5-7)

W, mg/g	P, mmHg	P, kPa	$\frac{P}{P^* - P)W} \times 10^3$	$P/P^* \times 10^2$
13	6.25	0.83	0.637	0.82
17	15.6	2.08	1.233	2.05
20	25.0	3.33	1.700	3.29
22	34.4	4.59	2.155	4.53
25	56.3	7.50	3.200	7.41
28	84.4	11.2	4.462	11.11
32	163.0	21.7	8.532	21.45

Note: Let $y = P/(P^* - P)W$, $X = P/P^*$, $y = Sx + I$. By least squares,

$D = (\Sigma x)^2 - n\Sigma x^2$
$S = (\Sigma y \Sigma x - n\Sigma xy)/D$
$I = (\Sigma x \Sigma xy - \Sigma y \Sigma x^2)/D$
$\Sigma y = 21.92 \times 10^{-3}$
$\Sigma xy = 2.747 \times 10^{-3}$
$\Sigma x^2 = 0.06747$
$\Sigma x = 0.5066$
$n = 7$
$S = 0.037 \text{ mg}^{-1}$
$I = 4.05 \times 10^{-4} \text{ mg}^{-1}$

yield a straight line. Let S and I, respectively, stand for the slope $(C - 1)/(CW_m)$ and the intercept $1/(W_mC)$ of that line. Then, by algebraic rearrangement, $W_m = 1/(S + I)$. Since the tests were conducted at the atmospheric boiling point, P^* was essentially 760 mmHg (101.3 kPa).

Values for $P/[W(P^* - P)]$ and P/P^* appear in the fourth and fifth columns of Table 5-1. Application of ordinary least-squares regression analysis to the resulting plot (not shown) shows S to be 0.0377 mg^{-1} and I to be 4.05×10^{-4} mg^{-1}. Therefore, W_m = 26.26 mg nitrogen per gram of catalyst. This equals $0.02626/28 = 9.38 \times 10^{-4}$ g·mol nitrogen. Employing Avogadro's number, this is $(9.38 \times 10^{-4})(6.023 \times 10^{23}) = 5.65 \times 10^{20}$ nitrogen molecules. Finally, the nitrogen molecule can be taken to have a surface area of 15.7×10^{-20} m^2. Therefore, the surface area of the catalyst (in intimate contact with the nitrogen monolayer) can be estimated to be $(5.65 \times 10^{20})(15.7 \times 10^{-20})$, or about 89 m^2/g.

2. *Estimate the dispersion of the nickel.*

Hydrogen dissociatively adsorbs on nickel whereas it does not interact with the catalyst support and is not significantly adsorbed within the nickel crystal lattice. Therefore, the amount of uptake of hydrogen by the catalyst is a measure of how well the nickel has been dispersed when deposited on the support.

At several pressures up to atmospheric, uptake of hydrogen proved to be constant, at 0.256 mg/g of catalyst, suggesting that the exposed nickel sites were saturated.

Thus, per gram of catalyst, the atoms of hydrogen adsorbed equals $(0.256 \times 10^{-3}$ g)(½ mol/g)(2 atoms/molecule)$(6.023 \times 10^{23}$ molecules/mol) $= 1.542 \times 10^{20}$ atoms. This can also be taken as the number of surface nickel atoms. Now, since the catalyst consists of 10% nickel, the total number of nickel atoms per gram of catalyst equals (0.1 g)(1/58.71 mol/g)$(6.023 \times 10^{23}$ atoms/mol) $= 1.0259 \times 10^{21}$ atoms. Therefore, the degree of dispersion of the nickel equals (surface nickel atoms)/(total nickel atoms) =

$(1.542 \times 10^{20})/(1.0259 \times 10^{21}) = 0.15$. Thus only 15 percent of the nickel deposited has been dispersed and is available for catalysis.

3. *Calculate the porosity and the mean pore radius.*

The particle porosity may be readily determined by a helium pycnometer and a mercury porosimeter. In the pycnometer, the solid skeletal volume V_S is obtained. The skeletal density ρ_S is found from the sample weight W_S:

$$\rho_S = \frac{W}{V_S}$$

The total sample volume V_A, including pores, can be determined by mercury displacement at atmospheric pressure, since mercury will not enter the pores under these conditions. The apparent density ρ_A, then, is

$$\rho_A = \frac{W}{V_A}$$

And the porosity ϵ of the solid is given by

$$\epsilon = 1 - \rho_A/\rho_S$$

For the catalyst in question, the apparent density is 1.3 kg/dm^3 and the skeletal density is 3.0 kg/dm^3. The porosity is therefore $\epsilon = 1 - 1.3/3 = 0.57$.

The pore volume V_p equals the porosity divided by the apparent density: $V_p = \epsilon/\rho_A = 0.57/1.3 = 0.44$ dm^3/kg $= 0.44$ cm^3/g. Assuming cylindrical pores of uniform length and radius,

$$\frac{V_p}{A_p} = \frac{n\pi R_p^2 L_p}{n 2\pi R_p L_p}$$

where n = number of pores
A_p = pore surface area (calculated in step 1)
R_p = pore radius
L_p = pore length

Therefore,

$$R_p = \frac{2V_p}{A_p} = \frac{2(0.44 \text{ cm}^3/\text{g})}{89 \text{ m}^2/\text{g}} = 99 \times 10^{-10} \text{ m } (99 \text{ Å})$$

5-8 Sizing and Design of a System of Stirred-Tank Reactors

It is proposed to process 3 m^3/h of a reaction mixture in either one or two (in series) continuous-flow stirred-tank reactors. The reaction is $A + 2B \rightarrow C$. At 50°C, the kinetic rate expression is as follows:

$$-r_A = k_1 C_A C_B/(1 + k_2 C_A)$$

where $k_1 = 0.1$ and $k_2 = 0.6$, with concentrations in kilogram-moles per cubic meter and rates in kilogram-moles per cubic meter per hour.

The mixture specific gravity is constant and equal to 1.2 kg/dm^3. The molecular weight of the feed is 40. The feed contains 10 mol % A, 20% B, and 70% inert solvent S. The liquid viscosity is 0.8 mPa·s (cp) at reaction temperature.

Determine the reactor volume required for one reactor and that for two equal-sized reactors in series for 80 percent conversion of A. And if the capital cost of a continuous-flow stirred-tank reactor unit is given by $200{,}000(V/100)^{0.6}$ (where V is reactor volume in m^3), the life is 20 years with no salvage value, and power costs 3 cents per kilowatt-hour, determine which system has the economic advantage. Assume that overhead, personnel, and other operating costs, except agitation, are constant. The operating year is 340 days. Each reactor is baffled (with a baffle width to tank diameter of 1/12) and equipped with an impeller whose diameter is one-third the tank diameter. The impeller is a six-bladed turbine having a width-to-diameter ratio of 1/5. The impeller is located at one-third the liquid depth from the bottom. The tank liquid-depth-to-diameter ratio is unity.

Calculation Procedure:

1. Develop the necessary material-balance expressions for a single reactor, and find its volume.

For a single reactor, the mass-balance design equation is

$$V = \frac{F_{A,0} - F_A}{-r_A} = \frac{X}{-r_A}$$

where V is the volume of material within the reactor, $F_{A,0}$ is inlet molar flow rate of species A, F_A is its exit molar flow rate, and X is moles of A reacted per unit of time. As noted, $F_A = (F_{A,0} - X)$, and since the reaction of 1 mol A involves 2 mol B and yields 1 mol C, and since the inlet is 10% A and 20% B, $F_B = F_{B,0} - 2X = 2(F_{A,0} - X)$ and $F_C = X$. Because the solvent is inert, $F_S = F_{S,0}$.

Now,

$$-r_A = \frac{K_1 C_A C_B}{1 + K_2 C_A} = \frac{K_1(F_A/\nu)(F_B/\nu)}{1 + K_2(F_A/\nu)} = \frac{2K_1(F_A/\nu)^2}{1 + K_2(F_A/\nu)}$$

where ν is the volumetric flow rate at the outlet (which equals the inlet volumetric rate, since the system is of constant density).

Since 80% conversion of A is specified, $X/F_{A,0} = 0.8$. And since the total inlet molar flow rate $F_{T,0}$ is 10% A,

$$F_{A,0} = 0.1F_{T,0} = 0.1\left(\frac{3\ \text{m}^3}{\text{h}} \times \frac{1.2\ \text{kg}}{\text{dm}^3} \times \left(\frac{10\ \text{dm}}{\text{m}}\right)^3 \times \frac{1\ \text{kg}\cdot\text{mol}}{40\ \text{kg}}\right) = 9\ \text{kg}\cdot\text{mol/h}$$

Therefore, the molar conversion rate X is 0.8(9) = 7.2 kg·mol/h. The outlet concentration of A is given by

$$C_A = F_A/\nu = (F_{A,0} - X)/\nu = \frac{9 - 7.2}{3}\frac{(\text{kg}\cdot\text{mol/h})}{(\text{m}^3/\text{h})} = 0.600\ \text{kg}\cdot\text{mol/m}^3$$

Thus the rate at the outlet conditions is given by

$$-r_A = \frac{(2)(0.1)(0.6)^2}{1 + (0.6)(0.6)} = 0.053 \text{ kg}\cdot\text{mol}/(\text{m}^3)(\text{h})$$

Finally, the volume is

$$V = X/-r_A = \frac{7.2 \text{ kg}\cdot\text{mol/h}}{0.053 \text{ kg}\cdot\text{mol}/(\text{m}^3)(\text{h})} = 136 \text{ m}^3 \ (4803 \text{ ft}^3)$$

2. *Develop the equations for two reactors in series, and find their volume.*

For a pair of reactors in series, define X_1 and X_2 as the moles of A reacted per unit of time in reactors 1 and 2, respectively. Then,

$$V = X_1/-r_{A1} = X_2/-r_{A2} \quad \text{and} \quad X_1 + X_2 = 7.2 \text{ kg}\cdot\text{mol/h}$$

Therefore,

$$\frac{X_1}{7.2 - X_1} = \frac{-r_{A1}}{-r_{A2}}$$

By material balance,

$$-r_{A1} = \frac{2K_1(F_{A,0} - X_1)^2/\nu^2}{1 + K_2(F_{A,0} - X_1)/\nu} \quad \text{and} \quad -r_{A2} = \frac{2K_1(F_{A,0} - X_1 - X_2)^2/\nu^2}{1 + K_2(F_{A,0} - X_1 - X_2)/\nu}$$

Now, the rate $-r_{A2}$ is the same as the rate for the single reactor, since $X_1 + X_2 =$ overall conversion; therefore, $-r_{A2} = 0.053$ kg·mol/(m³)(h). Accordingly,

$$\frac{0.053X_1}{7.2 - X_1} = \frac{2K_1(F_{A,0} - X_1)^2/\nu^2}{1 + K_2(F_{A,0} - X_1)/\nu} = \frac{0.022(9 - X_1)^2}{1 + 0.2(9 - X_1)}$$

Solution of this cubic equation (by, for example, Newton's method) gives $X_1 = 5.44$. Therefore, the volume for each reactor is as follows:

$$V = \frac{X_2}{-r_{A2}} = \frac{7.2 - X_1}{-r_{A2}} = \frac{1.76}{0.053} = 33.2 \text{ m}^3 \ (1173 \text{ ft}^3)$$

3. *Conduct an economic analysis and decide between the one- and two-reactor systems.*

The two costs to be considered are depreciation of capital and power cost for agitation. Using the volumes, 20-year life with no salvage, and the straight-line depreciation method:

System	Capital cost	Annual depreciation expense
One reactor, 136 m³	$240,520	$12,026
Two reactors, each 33.2 m³	206,415	10,321

For normal mixing, the Pfaudler agitation-index (γ) number for this low-viscosity fluid is 2 ft^2/s^3. Most stirrers are designed for impeller Reynolds numbers of 1000 or greater. For the impeller specified, the power number ψ_n is 0.6 at high Reynolds numbers.[1]

The required impeller diameter D_i may be calculated from the given data. With the liquid height H_L equal to the tank diameter D_t,

$$V = \frac{\pi D_t^2}{4}(D_t) \qquad D_i = \tfrac{1}{3}D_t$$

For the two cases, the impeller diameters are thus found to be 1.86 m (6.10 ft) and 1.16 m (3.81 ft), respectively. In terms of the Pfaudler index, for low-viscosity liquids,

$$3\gamma = \frac{4n^3D_i^2}{\pi}\psi_n\left(\frac{D_i}{D_t}\right)^2\frac{D_i}{H_L}$$

where n is the mixer revolutions per minute. Solving this equation for n, and noting that $\gamma = 2\ ft^2/s^3 = 0.186\ m^2/s^3$, the mixer revolutions per minute is given as follows:

$$n = 60\ s/m\left[\frac{3\gamma\pi}{4D_i^2\psi_n}\left(\frac{D_t}{D_i}\right)^3\right]^{1/3} = \frac{162.1}{(D_i)^{2/3}}\ r/min$$

Substituting the impeller diameter in meters gives 107.2 r/min for the large tank and 146.8 r/min for each small tank. The Reynolds numbers

$$Re = \frac{nD_i^2\rho_L}{\mu_L}$$

are 9.27×10^6 for the large and 4.95×10^6 for the small tank, respectively, so the assumption of 0.6 power number is satisfactory.

From the definition of the power number,

$$\psi_n = \frac{P}{\rho_L n^3 D_i^5}$$

Therefore,

$$P = \psi_n \rho_L n^3 D_i^5\ (kg)(m^5)/(dm^3)(s^3)$$

For the two cases, the power consumption is calculated to be 91,417 N·m/s and 22,150 N·m/s for a large and small tank, respectively (1 watt = 1 N·m/s). The power consumption on an annual basis is 745,960 kW for the large tank and 361,490 kW for two small tanks. Therefore, the annual cost advantage is as follows:

System	Depreciation	Power	Total cost
1 tank	$12,026	$22,379	$34,405
2 tanks	10,321	10,845	21,166

The benefit for the two-tank system is $13,239 per year.

[1]Barona, N., *Hydrocarbon Proc. 59*(7), 1979.

5-9 Determination of Reaction-Rate Expressions from Plug-Flow-Reactor Data

A 25-cm-long by 1-cm-diameter plug-flow reactor was used to investigate the homogeneous kinetics of benzene dehydrogenation. The stoichiometric equations are as follows:

Reaction 1: $2C_6H_6 \rightleftarrows C_{12}H_{10} + H_2$ $[2\ Bz \rightleftarrows Bi + H_2]$

Reaction 2: $C_6H_6 + C_{12}H_{10} \rightleftarrows C_{18}H_{14} + H_2$ $[Bz + Bi \rightleftarrows Tri + H_2]$

At 760°C (1400°F) and 101.325 kPa (1 atm), the data in Table 5-2 were collected. Find rate equations for the production of biphenyl (Bi) and triphenyl (Tri) at 760°C by the differential method.

TABLE 5-2 Kinetics of Benzene Dehydrogenation (Example 5-9)

Residence time		Mole fraction in product			
$\frac{(ft^3)(h)}{lb \cdot mol}$	$\frac{(dM^3)(h)}{kg \cdot mol}$	Benzene	Biphenyl	Triphenyl	Hydrogen
0	0	1.0	0	0	0
0.01	0.129	0.941	0.0288	0.00051	0.0298
0.02	0.257	0.888	0.0534	0.00184	0.0571
0.06	0.772	0.724	0.1201	0.0119	0.1439
0.12	1.543	0.583	0.163	0.0302	0.224
0.22	2.829	0.477	0.179	0.0549	0.289
0.30	3.858	0.448	0.175	0.0673	0.310
∞	∞	0.413	0.157	0.091	0.339

Calculation Procedure:

1. Calculate equilibrium constants.

From the data, the equilibrium constants for the two reactions, respectively, may be determined as follows (see Sec. 4):

$$K_1 = \frac{a_{Bi}a_{H2}}{a_{Bz}^2} \qquad K_2 = \frac{a_{Tri}a_{H2}}{a_{Bz}a_{Bi}}$$

where a_i is the activity of species i. For the standard state being ideal gases at 101.325 kPa and 760°C, the activity for the Bi component, for example, is

$$a_{Bi} = \frac{\hat{f}_{Bi}}{f_{Bi}^{\circ}} = \frac{\hat{f}_{Bi}}{101.325} = \frac{\hat{\phi}_{Bi}y_{Bi}P}{101.325}$$

where f is fugacity, ϕ is fugacity coefficient, y is mole fraction, and P is pressure. Assuming that the gases are ideal, $\phi_{Bi} = 1$, so $a_{Bi} = y_{Bi}$, and in general, $a_i = y_i$.

The mole fractions at infinite residence time can be taken as the equilibrium mole fractions. Then, from the preceding data, $K_1 = a_{Bi}a_{H2}/a_{Bz}^2 = (0.157)(0.339)/(0.413)^2 = 0.312$, and $K_2 = a_{Tri}a_{H2}/a_{Bz}a_{Bi} = (0.091)(0.339)/(0.413)(0.157) = 0.476$.

2. *Develop the reactor mass balance.*

Let x = mol/h of benzene reacted by reaction 1 and y = mol/h of benzene reacted by reaction 2, and let $F_{i,0}$ and F_i be the moles per hour of species i in the inflow and outflow, respectively. Then, noting that each mole of benzene in reaction 1 involves one-half mole each of biphenyl and hydrogen,

$$F_{\mathrm{Bz}} = F_{\mathrm{Bz},0} - x - y$$

$$F_{\mathrm{Bi}} = F_{\mathrm{Bi},0} + x/2 - y$$

$$F_{\mathrm{H2}} = F_{\mathrm{H2},0} + x/2 + y$$

$$F_{\mathrm{Tri}} = F_{\mathrm{Tri},0} + y$$

Adding these four equations and letting subscript t stand for moles per hour,

$$F_t = F_{t,0} = F_{\mathrm{Bz},0} \qquad \text{and} \qquad F_{\mathrm{Bi},0} = F_{\mathrm{H2},0} = F_{\mathrm{Tri},0} = 0$$

Now, for a plug-flow reactor, the material balance is as follows for a differential volume at steady state:

(Rate of A input by flow) − (rate of A output by flow) + (rate of A generated) = 0

or

$$F_{A,0} - (F_{A,0} + dF_A) + r_A dV = 0$$

where r_A is the reaction rate for A and V is the reaction volume. Therefore, $r_A = dF_A/dV$. In this case, $F_A = y_A F_t = y_A F_{t,0}$; thus, $dF_A = F_{t,0} dy_A$, and the material balance becomes

$$r_A = F_{t,0}\frac{dy_A}{dV} = \frac{dy_A}{d(V/F_{t,0})}$$

where $V/F_{t,0}$ is residence time in the reactor. Thus the reaction rate is the slope of the concentration-versus-residence-time plot.

3. *Calculate reaction rates.*

For the homogeneous plug-flow reactor, the conversion is a function of residence time. By carrying out a series of experiments at various residence times ($V/F_{t,0}$) in a reactor of fixed volume for a constant feed composition, one obtains the same concentration-versus-residence-time plot as if an infinitely long reactor had been used and the composition had been sampled along the reactor length. Thus the data given can be plotted to give a continuous concentration-versus-residence-time plot that may be differentiated according to the mass-balance equation to give rates of reaction. The concentrations corresponding to those rates are obtained from the data plot at the time for which the rate is evaluated. The differentiation may be accomplished by drawing tangents on the graph at various times to the concentration curves.

The concentration data given can be used to determine net rates of reaction by the material-balance expressions. These rates must be analyzed in terms of the stoichiometry to get the individual rates of reaction. Thus, for the trimer,

$$r_{\text{Tri}} = r_{\text{Tri by reaction 2}}$$

$$r_{\text{Bi}} = r_{\text{Bi by reaction 1}} + r_{\text{Bi by reaction 2}}$$

$$r_{\text{Tri by reaction 2}} = -r_{\text{Bi by reaction 2}}$$

Therefore,

$$r_{\text{Bi by reaction 1}} = r_{\text{Bi}} + r_{\text{Tri by reaction 2}}$$

The net triphenyl and biphenyl rates r_{Tri} and r_{Bi} can be found by plotting the mole fraction of triphenyl and biphenyl versus residence time and taking tangents. Figure 5-4 shows such a plot and Table 5-3 presents the results.

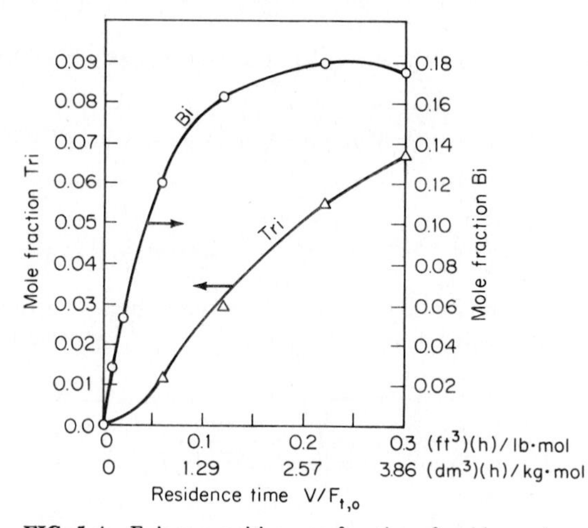

FIG. 5-4 Exit composition as a function of residence time (Example 5-9).

TABLE 5-3 Rates of Reaction for Triphenyl and Biphenyl (Example 5-9)

Residence time $V/F_{t,0}$, $(dm^3)(h)/(kg \cdot mol)$	Rate of reaction, $kg \cdot mol/(dm^3)(h)$	
	r_{Tri}	$r_{\text{Bi (by reaction 1)}}$
0.129	0.00715	0.2148
0.257	0.01166	0.1893
0.772	0.02916	0.1322
1.543	0.02138	0.0467
2.829	0.01361	0.0194

Note: To obtain reaction rate in pound-moles per cubic foot per hour, multiply the preceding rates by 12.86.

Assume the reactions are elementary, as a first guess. Then, letting P_t be total pressure and k_i be the specific reaction-rate constant for reaction 1, the two required equations are

$$r_{\text{Bi by reaction 1}} = k_1 P_t^2 (y_{\text{Bz}}^2 - y_{\text{H2}} y_{\text{Bi}}/K_1)$$

and

$$r_{\text{Tri by reaction 2}} = k_2 P_t^2 (y_{\text{Bz}} y_{\text{Bi}} - y_{\text{Tri}} y_{\text{H2}}/K_2)$$

In order to determine k_1 and k_2, and to check the suitability of the assumption, plot $r_{\text{Bi by reaction 1}}$ against $(y_{\text{Bz}}^2 - y_{\text{H2}} y_{\text{Bi}}/K_1)$ and plot r_{Tri} (which equals $r_{\text{Tri by reaction 2}}$) against $(y_{\text{Bz}} y_{\text{Bi}} - y_{\text{Tri}} y_{\text{H2}}/K_2)$. Check to see that the lines are indeed straight, and measure their slope (see Fig. 5-5). Define the slope as $k_i^* = k_i P_t^2$. Then $k_1 = k_1^* = 0.2496$ (by measurement of the slope), and $k_2 = k_2^* = 0.3079$.

With the numerical values for the reaction-rate constants and the equilibrium constants inserted, then,

$$r_{\text{Bi by reaction 1}} = 0.2496(y_{\text{Bz}}^2 - y_{\text{H2}} y_{\text{Bi}}/0.312)$$

and

$$r_{\text{Tri by reaction 2}} = 0.3079(y_{\text{Bz}} y_{\text{Bi}} - y_{\text{Tri}} y_{\text{H2}}/0.476)$$

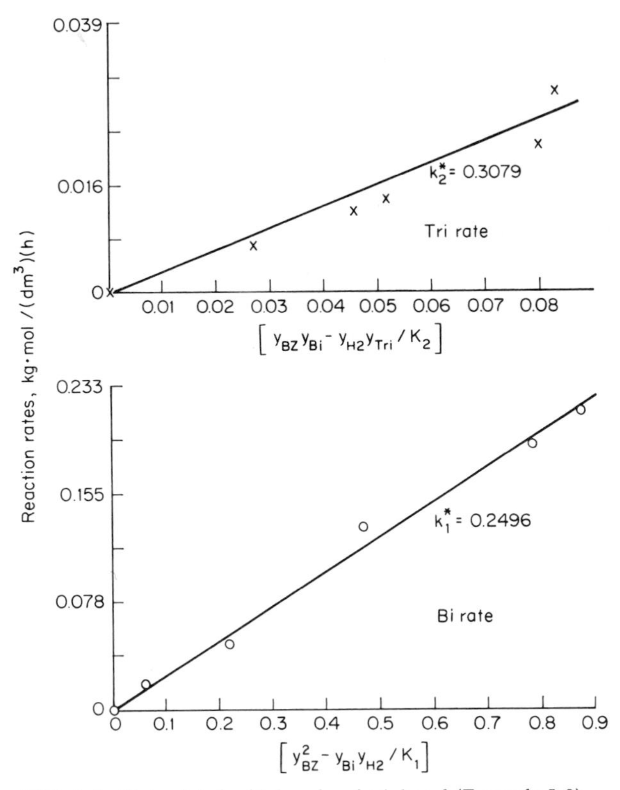

FIG. 5-5 Rate plots for biphenyl and triphenyl (Example 5-9).

Related Calculations: For a batch reactor, the material balance is Rate of accumulation of species A = rate of generation of species A, or $dN_A/dt = r_A$, where N is number of moles at time t and r is rate of reaction (which can be, for example, per unit of catalyst mass in the reactor, in which case it must be multiplied by the number of such units present). The rate at any given time can be found by plotting N_A against residence time and measuring the slope, but this technique can lead to large errors. A better approach is to use the Taylor-series interpolation formula (see mathematics handbooks for details).

6

*Flow of Fluids and Solids**

*The material in this section is taken from T. G. Hicks, *Standard Handbook of Engineering Calculations,* McGraw-Hill Book Co., Inc.

6-1 Bernoulli's Theorem, and Equation of Continuity

A piping system is conveying 10 ft^3/s (0.28 m^3/s) of ethanol. At a particular cross section of the system, section 1, the pipe diameter is 12 in (0.30 m), the pressure is 18 lb/in^2 (124 kPa), and the elevation is 140 ft (42.7 m). At another cross section further downstream, section 2, the pipe diameter is 8 in (0.20 m), and the elevation is 106 ft (32.3 m). If there is a head loss of 9 ft (2.74 m) between these sections due to pipe friction, what is the pressure at section 2? Assume that the specific gravity of the ethanol is 0.79.

Calculation Procedure:

1. Compute the velocity at each section.

Use the equation of continuity

$$Q = A_1V_1 = A_2V_2$$

where Q is volumetric rate of flow, A is cross-sectional area, V is velocity, and the subscripts refer to sections 1 and 2. Now, $A = (\pi/4)d^2$, where d is (inside) pipe diameter, so $A_1 = (\pi/4)(1\text{ ft})^2 = 0.785\text{ ft}^2$, and $A_2 = (\pi/4)(8/12\text{ ft})^2 = 0.349\text{ ft}^2$; and $Q = 10\text{ ft}^3/\text{s}$. Therefore, $V_1 = 10/0.785 = 12.7$ ft/s, and $V_2 = 10/0.349 = 28.7$ ft/s.

2. Compute the pressure at section 2.

Use Bernoulli's theorem, which in one form can be written as

$$\frac{V_1^2}{2g} + \frac{p_1}{\rho} + z_1 = \frac{V_2^2}{2g} + \frac{p_2}{\rho} + z_2 + h_L$$

where g is the acceleration due to gravity, 32.2 ft/s^2; p is pressure; ρ is density; z is elevation; and h_L is loss of head between two sections. In this case, $\rho = 0.79(62.4\text{ lb/ft}^3) = 49.3\text{ lb/ft}^3$. Upon rearranging the equation for Bernoulli's theorem,

$$\frac{p_2 - p_1}{\rho} = \frac{V_1^2 - V_2^2}{2g} + z_1 - z_2 - h_L$$

or $(p_2 - p_1)/49.3 = (12.7^2 - 28.7^2)/64.4 + 140 - 106 - 9$, so $(p^2 - p^1) = 725.2$ lb/ft^2, or $725.2/144 = 5.0$ lb/in^2. Therefore, $p_2 = 18 + 5 = 23$ lb/in^2 (159 kPa).

6-2 Specific Gravity and Viscosity of Liquids

An oil has a specific gravity of 0.8000 and a viscosity of 200 SSU (Saybolt Seconds Universal) at 60°F (289 K). Determine the API gravity and Bé gravity of this oil at 70°F (294 K) and its weight in pounds per gallon. What is the kinematic viscosity in centistokes? What is the absolute viscosity in centipoise?

Calculation Procedure:

1. Determine the API gravity of the liquid.

For any oil at 60°F, its specific gravity S, in relation to water at 60°F, is $S = 141.5/(131.5 + °\text{API})$; or $\text{API} = (141.5 - 131.5S)/S$. For this oil, $°\text{API} = [141.5 - 131.5(0.80)]/0.80 = 45.4°\text{API}$.

2. Determine the Bé gravity of the liquid.

For any liquid lighter than water, $S = 140/(130 + \text{Bé})$; or $\text{Bé} = (140 - 130S)/S$. For this oil, $\text{Bé} = [140 - 130(0.80)]/0.80 = 45$ Bé.

3. Compute the weight per gallon of liquid.

With a specific gravity of S, the weight of 1 ft^3 oil equals (S)(weight of 1 ft^3 fresh water at 60°F) $= (0.80)(62.4) = 49.92$ lb/ft^3. Since 1 gal liquid occupies 0.13368 ft^3, the weight of this oil per gal is $(49.92(0.13368)) = 6.66$ lb/gal (800 kg/m^3).

4. Compute the kinematic viscosity of the liquid.

For any liquid having a viscosity between 32 and 99 SSU, the kinematic viscosity $k = 0.226\ \text{SSU} - 195/\text{SSU}$ Cst. For this oil, $k = 0.226(200) - 195/200 = 44.225$ Cst.

5. Convert the kinematic viscosity to absolute viscosity.

For any liquid, the absolute viscosity, cP, equals (kinematic viscosity, Cst)(specific gravity). Thus, for this oil, the absolute viscosity $= (44.225)(0.80) = 35.38$ cP.

Related Calculations: For liquids *heavier* than water, $S = 145/(145 - \text{Bé})$. When the SSU viscosity is greater than 100 s, $k = 0.220\ \text{SSU} - 135/\text{SSU}$. Use these relations for any liquid—brine, gasoline, crude oil, kerosene, Bunker C, diesel oil, etc. Consult the *Pipe Friction Manual* and King and Crocker—*Piping Handbook* for tabulations of typical viscosities and specific gravities of various liquids.

6-3 Pressure Loss in Piping with Laminar Flow

Fuel oil at 300°F (422 K) and having a specific gravity of 0.850 is pumped through a 30,000-ft-long 24-in pipe at the rate of 500 gal/min (0.032 m^3/s). What is the pressure loss if the viscosity of the oil is 75 cP?

Calculation Procedure:

1. Determine the type of flow that exists.

Flow is laminar (also termed viscous) if the Reynolds number Re for the liquid in the pipe is less than about 2000. Turbulent flow exists if the Reynolds number is greater than about 4000. Between these values is a zone in which either condition may exist, depending on the roughness of the pipe wall, entrance conditions, and other factors. Avoid sizing a pipe for flow in this critical zone because excessive pressure drops result without a corresponding increase in the pipe discharge.

Compute the Reynolds number from $\text{Re} = 3.162G/kd$, where G = flow rate, gal/min; k = kinematic viscosity of liquid, Cst = viscosity z cP/specific gravity of the liquid S; d = inside diameter of pipe, in. From a table of pipe properties, $d = 22.626$ in. Also, $k = z/S = 75/0.85 = 88.2$ Cst. Then, $\text{Re} = 3162(500)/[88.2(22.626)] = 792$. Since Re < 2000, laminar flow exists in this pipe.

2. Compute the pressure loss using the Poiseuille formula.

The Poiseuille formula gives the pressure drop p_d lb/in² $= 2.73(10^{-4})luG/d^4$, where l = total length of pipe, including equivalent length of fitting, ft; u = absolute viscosity of liquid, cP; G = flow rate, gal/min; d = inside diameter of pipe, in. For this pipe, $p_d = 2.73\,(10^{-4})(30{,}000)(75)(500)/262{,}078 = 1.17$ lb/in² (8.07 kPa).

Related Calculations: Use this procedure for any pipe in which there is laminar flow of liquid. Table 6-1 gives a quick summary of various ways in which the Reynolds num-

TABLE 6-1 Reynolds Number

		Numerator			Denominator	
Reynolds number Re	Coefficient	First symbol	Second symbol	Third symbol	Fourth symbol	Fifth symbol
Dvp/μ	—	ft	ft/s	lb/ft³	lb mass/ft·s	—
$124dvp/z$	124	in	ft/s	lb/ft³	cP	—
$50.7Gp/dz$	50.7	gal/min	lb/ft³	—	in	cP
$6.32W/dz$	6.32	lb/h	—	—	in	cP
$35.5Bp/dz$	35.5	bbl/h	lb/ft³	—	in	cP
$7{,}742dv/k$	7,742	in	ft/s	—	—	cP
$3{,}162G/dk$	3,162	gal/min	—	—	in	cP
$2{,}214B/dk$	2,214	bbl/h	—	—	in	cP
$22{,}735qp/dz$	22,735	ft³/s	lb/ft³	—	in	cP
$378.9Qp/dz$	378.9	ft³/min	lb/ft³	—	in	cP

ber can be expressed. The symbols in Table 6-1, in the order of their appearance, are D = inside diameter of pipe, ft; v = liquid velocity, ft/s; ρ = liquid density, lb/ft^3; μ = absolute viscosity of liquid, lb mass/ft·s; d = inside diameter of pipe, in. From a table of pipe properties, d = 22.626 in. Also, k = z/S liquid flow rate, lb/h; B = liquid flow rate, bbl/h; k = kinematic viscosity of the liquid, Cst; q = liquid flow rate, ft^3/s; Q = liquid flow rate, ft^3/min. Use Table 6-1 to find the Reynolds number for any liquid flowing through a pipe.

6-4 Determining the Pressure Loss in Pipes

What is the pressure drop in a 5000-ft-long 6-in oil pipe conveying 500 bbl/h (0.022 m^3/s) kerosene having a specific gravity of 0.813 at 65°F, which is the temperature of the liquid in the pipe? The pipe is schedule 40 steel.

Calculation Procedure:

1. Determine the kinematic viscosity of the oil.

Use Fig. 6-1 and Table 6-2 or the Hydraulic Institute—*Pipe Friction Manual* kinematic viscosity and Reynolds number chart to determine the kinematic viscosity of the liquid. Enter Table 6-2 at kerosene and find the coordinates as X = 10.2, Y = 16.9. Using these coordinates, enter Fig. 6-1 and find the absolute viscosity of kerosene at 65°F as 2.4 cP. Using the method of Example 6-2, the kinematic viscosity, in cSt, equals absolute viscosity, cP/specific gravity of the liquid = 2.4/0.813 = 2.95 cSt. This value agrees closely with that given in the *Pipe Friction Manual.*

2. Determine the Reynolds number of the liquid.

The Reynolds number can be found from the *Pipe Friction Manual* chart mentioned in step 1 or computed from $\text{Re} = 2214\ B/dk = 2214(500)/[(6.065)(2.95)] = 61{,}900$.

To use the *Pipe Friction Manual* chart, compute the velocity of the liquid in the pipe by converting the flow rate to cubic feet per second. Since there are 42 gal/bbl and 1 gal = 0.13368 ft^3, 1 bbl = (42)(0.13368) = 5.6 ft^3. With a flow rate of 500 bbl/h, the equivalent flow in ft^3 = (500)(5.6) = 2800 ft^3/h, or 2800/3600 s/h = 0.778 ft^3/s. Since 6-in schedule 40 pipe has a cross-sectional area of 0.2006 ft^2 internally, the liquid velocity, in ft/s, equals 0.778/0.2006 = 3.88 ft/s. Then, the product (velocity, ft/s)(internal diameter, in) = (3.88)(6.065) = 23.75. In the *Pipe Friction Manual,* project horizontally from the kerosene specific-gravity curve to the vd product of 23.75 and read the Reynolds number as 61,900, as before. In general, the Reynolds number can be found faster by computing it using the appropriate relation given in Table 6-1, unless the flow velocity is already known.

3. Determine the friction factor of this pipe.

Enter Fig. 6-2 at the Reynolds number value of 61,900 and project to the curve 4 as indicated by Table 6-3. Read the friction factor as 0.0212 at the left. Alternatively, the *Pipe Friction Manual* friction-factor chart could be used, if desired.

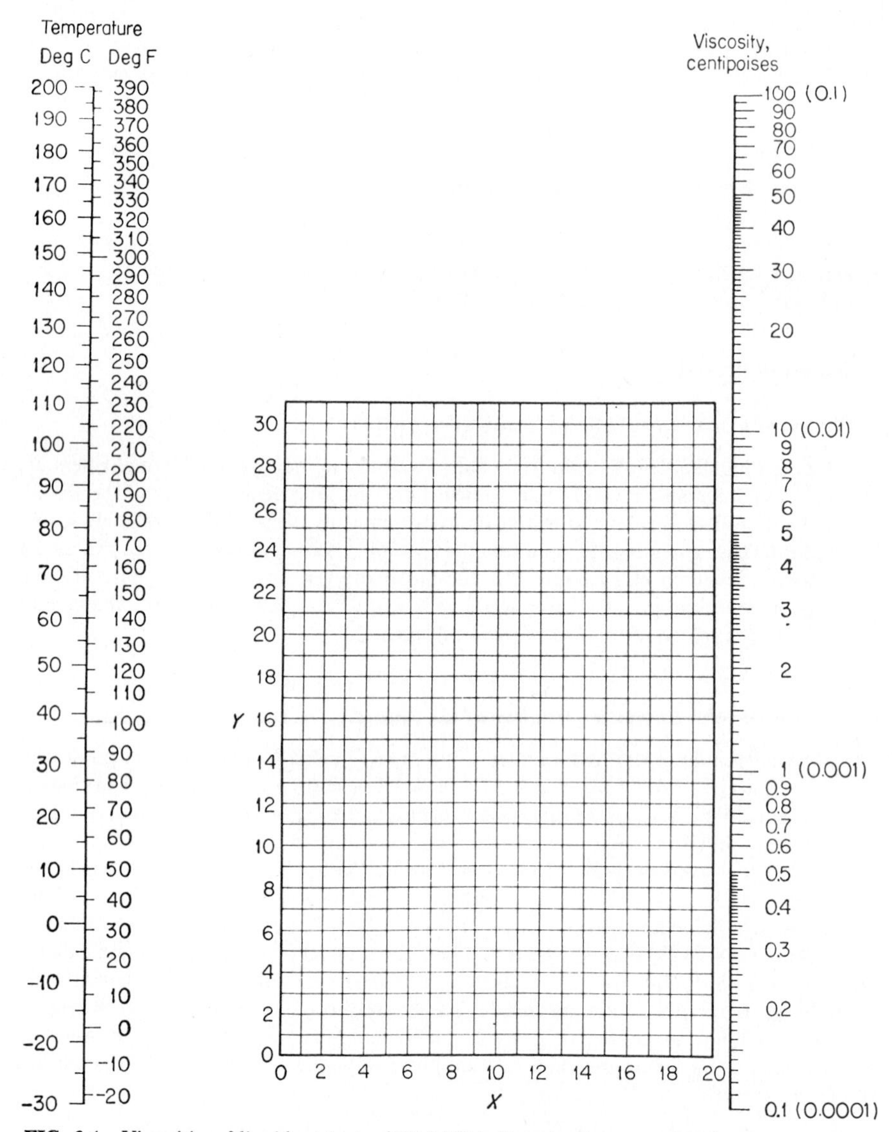

FIG. 6-1 Viscosities of liquids at 1 atm (101.3 kPa). For coordinates, see Table 6-2.

TABLE 6-2 Viscosities of Liquids (coordinates for use with Fig. 6-1)

No.	Liquid	X	Y
1	Acetaldehyde	15.2	4.8
	Acetic acid:		
2	100%	12.1	14.2
3	70%	9.5	17.0
4	Acetic anhydride	12.7	12.8
	Acetone:		
5	100%	14.5	7.2
6	35%	7.9	15.0
7	Allyl alcohol	10.2	14.3
	Ammonia:		
8	100%	12.6	2.0
9	26%	10.1	13.9
10	Amyl acetate	11.8	12.5
11	Amyl alcohol	7.5	18.4
12	Aniline	8.1	18.7
13	Anisole	12.3	13.5
14	Arsenic trichloride	13.9	14.5
15	Benzene	12.5	10.9
	Brine:		
16	$CaCl_2$, 25%	6.6	15.9
17	NaCl, 25%	10.2	16.6
18	Bromine	14.2	13.2
19	Bromotoluene	20.0	15.9
20	Butyl acetate	12.3	11.0
21	Butyl alcohol	8.6	17.2
22	Butyric acid	12.1	15.3
23	Carbon dioxide	11.6	0.3
24	Carbon disulfide	16.1	7.5
25	Carbon tetrachloride	12.7	13.1
26	Chlorobenzene	12.3	12.4
27	Chloroform	14.4	10.2
28	Chlorosulfonic acid	11.2	18.1
	Chlorotoluene:		
29	Ortho	13.0	13.3
30	Meta	13.3	12.5
31	Para	13.3	12.5
32	Cresol, meta	2.5	20.8
33	Cyclohexanol	2.9	24.3
34	Dibromoethane	12.7	15.8
35	Dichloroethane	13.2	12.2
36	Dichloromethane	14.6	8.9
37	Diethyl oxalate	11.0	16.4
38	Dimethyl oxalate	12.3	15.8
39	Diphenyl	12.0	18.3
40	Dipropyl oxalate	10.3	17.7
41	Ethyl acetate	13.7	9.1
	Ethyl alcohol:		
42	100%	10.5	13.8
43	95%	9.8	14.3
44	40%	6.5	16.6
45	Ethyl benzene	13.2	11.5
46	Ethyl bromide	14.5	8.1
47	Ethyl chloride	14.8	6.0
48	Ethyl ether	14.5	5.3
49	Ethyl formate		
50	Ethyl iodide	14.7	10.3
51	Ethylene glycol	6.0	23.6
52	Formic acid	10.7	15.8
53	Freon-11	14.4	9.0
54	Freon-12	16.8	5.6
55	Freon-21	15.7	7.5
56	Freon-22	17.2	4.7
57	Freon-13	12.5	11.4
	Glycerol:		
58	100%	2.0	30.0
59	50%	6.9	19.6
60	Heptene	14.1	8.4
61	Hexane	14.7	7.0
62	Hydrochloric acid, 31.5%	13.0	16.6
63	Isobutyl alcohol	7.1	18.0
64	Isobutyric acid	12.2	14.4
65	Isopropyl alcohol	8.2	16.0
66	Kerosene	10.2	16.9
67	Linseed oil, raw	7.5	27.2
68	Mercury	18.4	16.4
	Methanol		
69	100%	12.4	10.5
70	90%	12.3	11.8
71	40%	7.8	15.5
72	Methyl acetate	14.2	8.2
73	Methyl chloride	15.0	3.8
74	Methyl ethyl ketone	13.9	8.6
75	Naphthalene	7.9	18.1
	Nitric acid:		
76	95%	12.8	13.8
77	60%	10.8	17.0
78	Nitrobenzene	10.6	16.2
79	Nitrotoluene	11.0	17.0
80	Octane	13.7	10.0
81	Octyl alcohol	6.6	21.1
82	Pentachloroethane	10.9	17.3
83	Pentane	14.9	5.2
84	Phenol	6.9	20.8
85	Phosphorus tribromide	13.8	16.7
86	Phosphorus trichloride	16.2	10.9
87	Propionic acid	12.8	13.8
88	Propyl alcohol	9.1	16.5
89	Propyl bromide	14.5	9.6
90	Propyl chloride	14.4	7.5
91	Propyl iodide	14.1	11.6
92	Sodium	16.4	13.9
93	Sodium hydroxide, 50%	3.3	25.8
94	Stannic chloride	13.5	12.8
95	Sulfur dioxide	15.2	7.1
	Sulfuric acid:		
96	110%	7.2	27.4
97	98%	7.0	24.8
98	60%	10.2	21.3
99	Sulfuryl chloride	15.2	12.4
100	Tetrachloroethane	11.9	15.7
101	Tetrachloroethylene	14.2	12.7
102	Titanium tetrachloride	14.4	12.3
103	Toluene	13.7	10.4
104	Trichloroethylene	14.8	10.5
105	Turpentine	11.5	14.9
106	Vinyl acetate		
107	Water	10.2	13.0
	Xylene:		
108	Ortho	13.5	12.1
109	Meta	13.9	10.6
110	Para	13.9	10.9

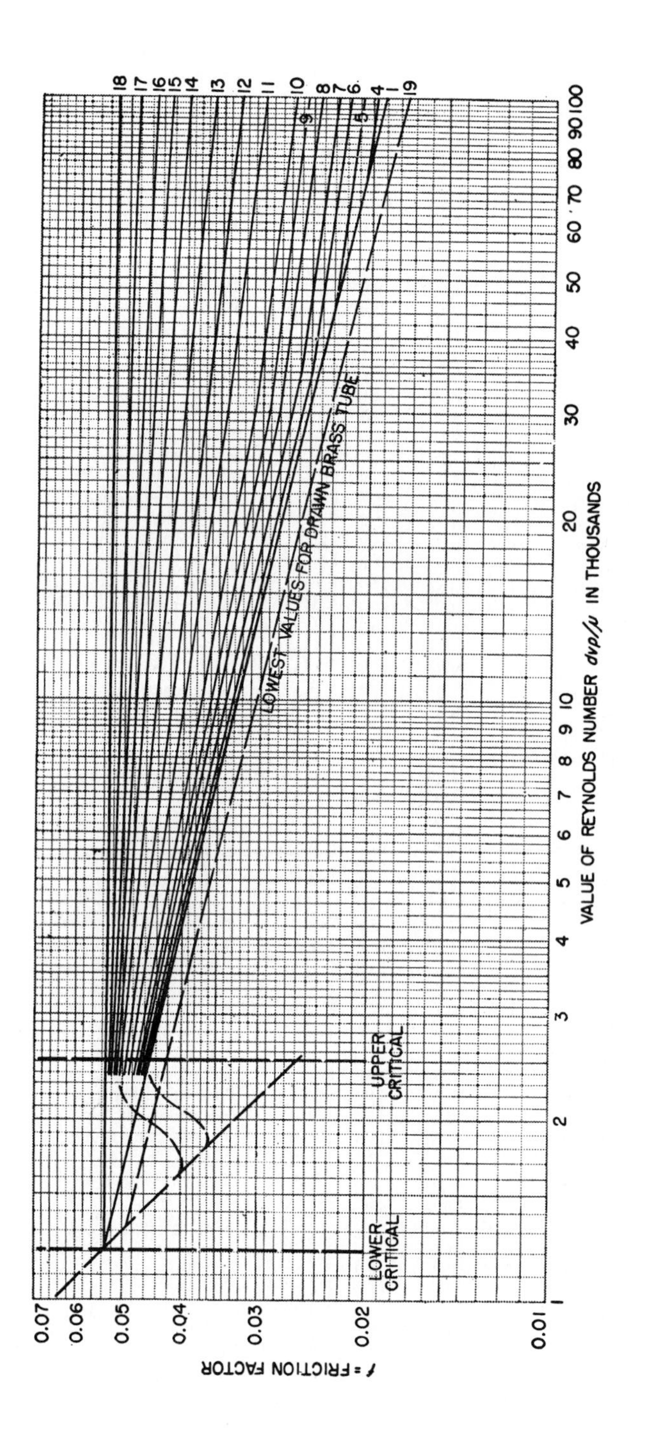

f = FRICTION FACTOR
0.07
0.06
0.05
0.04
0.03
0.02
0.01
VALUE OF REYNOLDS NUMBER dvρ/μ IN THOUSANDS
1
2
3
4
5
6
7
8
9
10
20
30
40
50
60
70
80
90
100
LOWER CRITICAL
UPPER CRITICAL
LOWEST VALUES FOR DRAWN BRASS TUBE
18
17
16
15
14
13
12
11
10
9
8
7
6
5
4
1
19

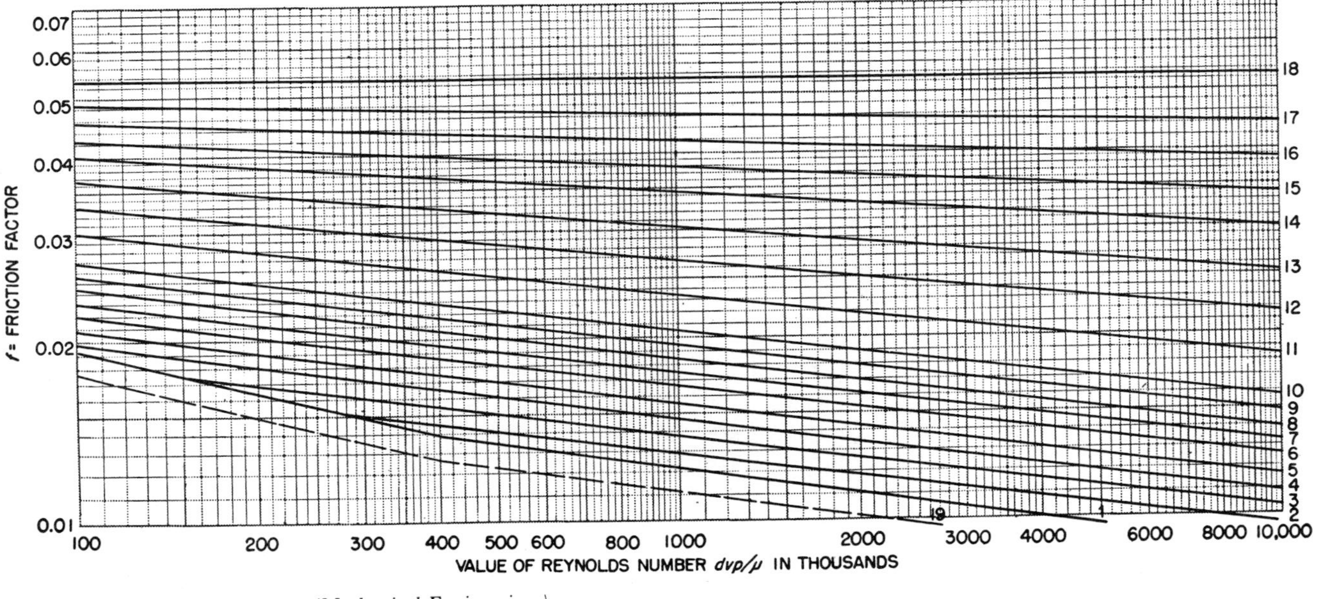

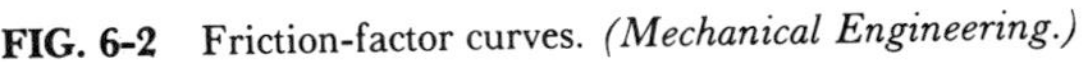
FIG. 6-2 Friction-factor curves. *(Mechanical Engineering.)*

TABLE 6-3 Data for Fig. 6-2

Percent roughness	For value of f see curve	Diameter (actual of drawn tubing, nominal of standard-weight pipe)											
		Drawn tubing, brass, tin, lead, glass		Clean steel, wrought iron		Clean, galvanized		Best cast iron		Average cast iron		Heavy riveted, spiral riveted	
		in	mm	in	mm	in	mm	in	mm	in	mm	in	mm
0.2	1	0.35 up	8.89 up	72	1829	—	—	—	—	—	—	—	—
1.35	4	—	—	6–12	152–305	10–24	254–610	20–48	508–1219	42–96	1067–2438	84–204	2134–5182
2.1	5	—	—	4–5	102–127	6–8	152–203	12–16	305–406	24–36	610–914	48–72	1219–1829
3.0	6	—	—	2–3	51–76	305	76–127	5–10	127–254	10–20	254–508	20–42	508–1067
3.8	7	—	—	1½	38	2½	64	3–4	76–102	6–8	152–203	16–18	406–457
4.8	8	—	—	1–1¼	25–32	1½–2	38–51	2–2½	51–64	4–5	102–127	10–14	254–356
6.0	9	—	—	¾	19	1¼	32	1½	38	3	76	8	203
7.2	10	—	—	½	13	1	25	1¼	32	—	—	5	127
10.5	11	—	—	⅜	9.5	¾	19	1	35	—	—	4	102
14.5	12	—	—	¼	6.4	½	13	—	—	—	—	3	76
24.0	14	0.125	3.18	—	—	⅜	9.5	—	—	—	—	—	—
31.5	16	—	—	—	—	¼	6.4	—	—	—	—	—	—
37.5	18	0.0625	1.588	—	—	⅛	3.2	—	—	—	—	—	—

4. Compute the pressure loss in the pipe.

Use the Fanning formula $p_d = 1.06(10^{-4})f\rho lB^2/d^5$. In this formula, ρ = density of the liquid, lb/ft³. For kerosene, p = (density of water, lb/ft³)(specific gravity of the kerosene) = (62.4)(0.813) = 50.6 lb/ft³. Then, $p_d = 1.06(10^{-4})(0.0212)(50.6)(5000)(500)^2/8206 = 17.3$ lb/in² (119 kPa).

Related Calculations: The Fanning formula is popular with oil-pipe designers and can be stated in various ways: (1) with velocity v, in ft/s, $p_d = 1.29(10^{-3})f\rho v^2l/d$; (2) with velocity V, in ft/min, $p_d = 3.6(10^{-7})f\rho V^2l/d$; (3) with flow rate G, in gal/min, $p_d = 2.15(10^{-4})f\rho lG^2/d^2$; (4) with the flow rate W, in lb/h, $p_d = 3.36(10^{-6})flW^2/d^5\rho$.

Use this procedure for any fluid—crude oil, kerosene, benzene, gasoline, naptha, fuel oil, Bunker C, diesel oil toluene, etc. The tables and charts presented here and in the *Pipe Friction Manual* save computation time.

6-5 Equivalent Length of a Complex-Series Pipeline

Figure 6-3 shows a complex-series pipeline made up of four lengths of different size pipe. Determine the equivalent length of this pipe if each size of pipe has the same friction factor.

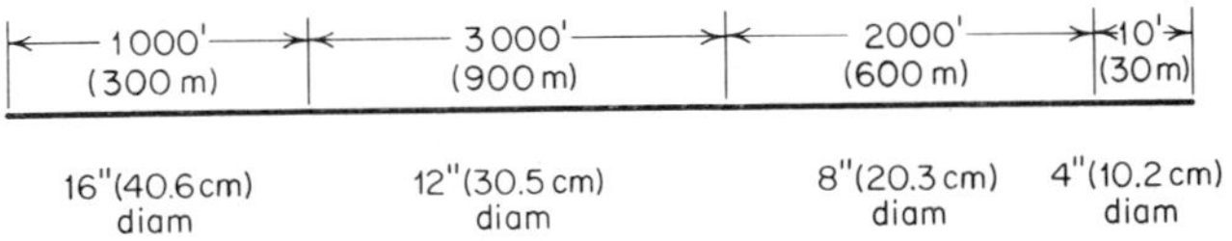

1000' (300 m) | 3000' (900 m) | 2000' (600 m) | 10' (30 m)

16"(40.6 cm) diam | 12"(30.5 cm) diam | 8"(20.3 cm) diam | 4"(10.2 cm) diam

FIG. 6-3 Complex-series pipeline.

Calculation Procedure:

1. Select the pipe size for expressing the equivalent length.

The usual procedure when analyzing complex pipelines is to express the equivalent length in terms of the smallest, or next-to-smallest, diameter pipe. Choose the 8-in size as being suitable for expressing the equivalent length.

2. Find the equivalent length of each pipe.

For any complex-series pipeline having equal friction factors in all the pipes, L_e = equivalent length, ft, of a section of constant diameter = (actual length of section, ft) (inside diameter, in, of pipe used to express the equivalent length/inside diameter, in, of section under consideration)^5.

For the 16-in pipe, $L_e = (1000)(7.981/15.000)^5 = 42.6$ ft. The 12-in pipe is next; for it, $L_e = (3000)(7.981/12.00)^5 = 390$ ft. For the 8-in pipe, the equivalent length = actual length = 2000 ft. For the 4-in pipe, $L_e = (10)(7.981/4.026)^5 = 306$ ft. Then, the total equivalent length of 8-in pipe = sum of the equivalent lengths = 42.6 + 390 + 2000 + 306 = 2738.6 ft, or rounding off, 2740 ft of 8-in pipe (835 m of 0.2-m pipe) will have a frictional resistance equal to the complex-series pipeline shown in Fig. 6-3. To compute the actual frictional resistance, use the methods given in previous Calculation Procedures.

Related Calculations: Use this general procedure for any complex-series pipeline conveying water, oil, gas, steam, etc. See King and Crocker—*Piping Handbook* for derivation of the flow equations. Use the tables in King and Crocker to simplify finding the fifth power of the inside diameter of a pipe.

Choosing a flow rate of 1000 gal/min and using the tables in the Hydraulic Institute *Pipe Friction Manual* gives an equivalent length of 2770 ft for the 8-in pipe. This compares favorably with the 2740 ft computed above. The difference of 30 ft is negligible.

The equivalent length is found by summing the friction-head loss for 1000 gal/min flow for each length of the four pipes—16, 12, 8, and 4 in—and dividing this by the friction-head loss for 1000 gal/min flowing through an 8-in pipe. Be careful to observe the units in which the friction-head loss is stated, because errors are easy to make if the units are ignored.

6-6 Hydraulic Radius and Liquid Velocity in Pipes

What is the velocity of 1000 gal/min (0.064 m^3/s) of water flowing through a 10-in inside-diameter cast-iron water-main pipe? What is the hydraulic radius of this pipe when it is full of water? When the water depth is 8 in (0.203 m)?

Calculation Procedure:

1. Compute the water velocity in the pipe.

For any pipe conveying liquid, the liquid velocity, in ft/s, is v = (gal/min)/($2.448d^2$), where d = internal pipe diameter, in. For this pipe, v = 1000/[2.448(100)] = 4.08 ft/s, or (60)(4.08) = 244.8 ft/min.

2. Compute the hydraulic radius for a full pipe.

For any pipe, the hydraulic radius is the ratio of the cross-sectional area of the pipe to the wetted perimeter, or $d/4$. For this pipe, when full of liquid, the hydraulic radius = 10/4 = 2.5.

3. Compute the hydraulic radius for a partially full pipe.

Use the hydraulic radius tables in King and Brater—*Handbook of Hydraulics* or compute the wetted perimeter using the geometric properties of the pipe, as in step 2. Using the King and Brater table, the hydraulic radius = Fd, where F = table factor for the ratio of the depth of liquid, in/diameter of channel, in = 8/10 = 0.8. For this ratio, F = 0.304. Then, hydraulic radius = (0.304)(10) = 3.04 in.

6-7 Friction-Head Loss in Water Piping of Various Materials

Determine the friction-head loss in 2500 ft of clean 10-in new tar-dipped cast-iron pipe when 2000 gal/min (0.126 m^3/s) of cold water is flowing. What is the friction-head loss 20 years later? Use the Hazen-Williams and Manning formulas and compare the results.

Calculation Procedure:

1. Compute the friction-head loss using the Hazen-Williams formula.

The Hazen-Williams formula is $h_f = (v/1.318CR_h^{0.63})^{1.85}$, where h_f = friction-head loss per foot of pipe, in feet of water; v = water velocity, in ft/s; C = a constant depending on the condition and kind of pipe; and R_h = hydraulic radius of pipe, in ft.

For a water pipe, $v = (\text{gal/min})/(2.44d^2)$; for this pipe, $v = 2000/[2.448(10)^2] = 8.18$ ft/s. From Table 6-4 or King and Crocker—*Piping Handbook,* C for new pipe = 120; for 20-year-old pipe, $C = 90$; $R_h = d/4$ for a full-flow pipe = 10/4 = 2.5 in, or 2.5/12 = 0.208 ft. Then, $h_f = (8.18/1.318 \times 120 \times 0.208^{0.63})^{1.85} = 0.0263$ ft of water per foot of pipe. For 2500 ft of pipe, the total friction-head loss = 2500(0.0263) = 65.9 ft (20.1 m) of water for the new pipe.

TABLE 6-4 Values of C in Hazen-Williams Formula

Type of pipe	C*	Type of pipe	C*
Cement-asbestos	140	Cast iron or wrought iron	100
Asphalt-lined iron or steel	140	Welded or seamless steel	100
Copper or brass	130	Concrete	100
Lead, tin, or glass	130	Corrugated steel	60
Wood stave	110		

*Values of C commonly used for design. The value of C for pipes made of corrosive materials decreases as the age of the pipe increases; the values given are those which apply at an age of 15 to 20 years. For example, the value of C for cast-iron pipes 30 in in diameter or greater at various ages is approximately as follows: new, 130; 5 years old, 120; 10 years old, 115; 20 years old, 100; 30 years old, 90; 40 years old, 80; and 50 years old, 75. The value of C for smaller-size pipes decreases at a more rapid rate.

For 20-year-old pipe using the same formula, except with $C = 90$, $h_f = 0.0451$ ft of water per foot of pipe. For 2500 ft of pipe, the total friction-head loss = 2500(0.0451) = 112.9 ft (34.4 m) of water. Thus the friction-head loss nearly doubles (from 65.9 to 112.9 ft) in 20 years. This shows that it is wise to design for future friction losses; otherwise, pumping equipment may become overloaded.

2. Compute the friction-head loss using the Manning formula.

The Manning formula is $h_f = n^2v^2/(2.208R_h^{4/3})$, where n = a constant depending on the condition and kind of pipe; other symbols as before.

Using $n = 0.011$ for new coated cast-iron pipe from Table 6-5 or King and Crocker—*Piping Handbook,* $h_f = (0.011)^2(8.18)^2/[2.208(0.208)^{4/3}] = 0.0295$ ft of water per foot of pipe. For 2500 ft of pipe, the total friction-head loss = 2500(0.0295) = 73.8 ft (22.5 m) of water, as compared with 65.9 ft of water computed with the Hazen-Williams formula.

For coated cast-iron pipe in fair condition, $n = 0.013$, and $h_f = 0.0411$ ft of water. For 2500 ft of pipe, the total friction-head loss = 2500(0.0411) = 102.8 ft (31.4 m) of water, as compared with 112.9 ft of water computed with the Hazen-Williams formula. Thus the Manning formula gives results higher than the Hazen-Williams in one case and lower in another. However, the differences in each case are not excessive; (73.8 − 65.9)/65.9 = 0.12, or 12 percent higher, and (112.9 − 102.8)/102.8 = 0.0983, or 9.83

TABLE 6-5 Roughness Coefficients (Manning's n) for Closed Conduits

Type of conduit			Manning's n: Good construction*	Manning's n: Fair construction*
Concrete pipe			0.013	0.015
Corrugated metal pipe or pipe arch, $2\frac{2}{3} \times \frac{1}{2}$ in corrugation, riveted:				
Plain			0.024	—
Paved invert:				
Percent of circumference paved	25	50		
Depth of flow:				
Full	0.021	0.018		
$0.8D$	0.021	0.016		
$0.6D$	0.019	0.013		
Vitrified clay pipe			0.012	0.014
Cast-iron pipe, uncoated			0.013	—
Steel pipe			0.011	—
Brick			0.014	0.017
Monolithic concrete:				
Wood forms, rough			0.015	0.017
Wood forms, smooth			0.012	0.014
Steel forms			0.012	0.013
Cemented-rubble masonry walls:				
Concrete floor and top			0.017	0.022
Natural floor			0.019	0.025
Laminated treated wood			0.015	0.017
Vitrified-clay liner plates			0.015	—

*For poor-quality construction, use larger values of n.

percent lower. Both these differences are within the normal range of accuracy expected in pipe friction-head calculations.

Related Calculations: The Hazen-Williams and Manning formulas are popular with many piping designers for computing pressure losses in cold-water piping. To simplify calculations, most designers use the precomputed tabulated solutions available in King and Crocker—*Piping Handbook,* King and Brater—*Handbook of Hydraulics,* and similar publications. In the rush of daily work these precomputed solutions are also preferred over the more complex Darcy-Weisbach equation used in conjunction with the friction factor f, the Reynolds number Re, and the roughness-diameter ratio.

Use the method given here for sewer lines, water-supply pipes for commercial, industrial, or process plants, and all similar applications where cold water at temperatures of 33 to 90°F flows through a pipe made of cast iron, riveted steel, welded steel, galvanized iron, brass, glass, wood-stove, concrete, vitrified, common clay, corrugated metal, unlined rock, or enameled steel. Thus either of these formulas, used in conjunction with a suitable constant, gives the friction-head loss for a variety of piping materials. Suitable constants are given in Tables 6-4 and 6-5 and in the preceding references. For the Hazen-Williams formula, the constant C varies from about 70 to 140, while n in the Manning formula varies from about 0.017 for $C = 70$ to $n = 0.010$ for $C = 140$. Values obtained with these formulas have been used for years with satisfactory results. At present, the Manning formula appears the more popular of the two.

6-8 Relative Carrying Capacity of Pipes

What is the equivalent steam-carrying capacity of a 24-in-inside-diameter pipe in terms of a 10-in-inside-diameter pipe? What is the equivalent water-carrying capacity of a 23-in-inside-diameter pipe in terms of a 13.25-in-inside-diameter pipe?

Calculation Procedure:

1. Compute the relative carrying capacity of the steam pipes.

For steam, air, or gas pipes, the number N of small pipes of inside diameter d_2 in equal to one pipe of larger inside diameter d_1 in is $N = (d_1^3\sqrt{d_2} + 3.6)/(d_2^3 + \sqrt{d_1} + 3.6)$. For this piping system, $N = (24^3 + \sqrt{10} + 3.6)/(10^3 + \sqrt{24} + 3.6) = 9.69$, say 9.7. Thus a 24-in-inside-diameter steam pipe has a carrying capacity equivalent to 9.7 pipes having a 10-in inside diameter.

2. Compute the relative carrying capacity of the water pipes.

For water, $N = (d_2/d_1)^{2.5} = (23/13.25)^{2.5} = 3.97$. Thus one 23-in-inside-diameter pipe can carry as much water as 3.97 pipes of 13.25 in inside diameter.

Related Calculations: King and Crocker—*Piping Handbook* and certain piping catalogs (Crane, Walworth, National Valve and Manufacturing Company) contain tabulations of relative carrying capacities of pipes of various sizes. Most piping designers use these tables. However, the equations given here are useful for ranges not covered by the tables and when the tables are unavailable.

6-9 Flow Rate and Pressure Loss in Compressed-Air and Gas Piping

Dry air at 80°F (300 K) and 150 psia (1034 kPa) flows at the rate of 500 ft^3/min (0.24 m^3/s) through a 4-in schedule 40 pipe from the discharge of an air compressor. What is the flow rate in pounds per hour and the air velocity in feet per second? Using the Fanning formula, determine the pressure loss if the total equivalent length of the pipe is 500 ft (153 m).

Calculation Procedure:

1. Determine the density of the air or gas in the pipe.

For air or a gas, $pV = MRT$, where p = absolute pressure of the gas, in lb/ft^2; V = volume of M lb of gas, in ft^3; M = weight of gas, in lb; R = gas constant, in ft·lb/(lb)(°F); T = absolute temperature of the gas, in R. For this installation, using 1 ft^3 of air, $M = pV/RT$, $M = (150)(144)/[(53.33)(80 + 459.7)] = 0.754$ lb/ft^3. The value of R in this equation was obtained from Table 6-6.

TABLE 6-6 Gas Constants

Gas	R, ft·lb/(lb)(°F)	R, J/(kg)(K)	C for critical-velocity equation
Air	53.33	286.9	2870
Ammonia	89.42	481.1	2080
Carbon dioxide	34.87	187.6	3330
Carbon monoxide	55.14	296.7	2820
Ethane	50.82	273.4	
Ethylene	54.70	294.3	2480
Hydrogen	767.04	4126.9	750
Hydrogen sulfide	44.79	240.9	
Isobutane	25.79	138.8	
Methane	96.18	517.5	2030
Natural gas	—	—	2070–2670
Nitrogen	55.13	296.6	2800
n-butane	25.57	137.6	
Oxygen	48.24	259.5	2990
Propane	34.13	183.6	
Propylene	36.01	193.7	
Sulfur dioxide	23.53	126.6	3870

2. *Compute the flow rate of the air or gas.*

For air or a gas, the flow rate W_h, in lb/h, = (60)(density, lb/ft^3)(flow rate, ft^3/min); or $W_h = (60)(0.754)(500) = 22{,}620$ lb/h.

3. *Compute the velocity of the air or gas in the pipe.*

For any air or gas pipe, velocity of the moving fluid v, in ft/s, $= 183.4W_h/(3600\ d^2\rho)$, where d = internal diameter of pipe, in; ρ = density of fluid, lb/ft^3. For this system, $v = (183.4)(22{,}620)/[(3600)(4.026)^2(0.754)] = 95.7$ ft/s.

4. *Compute the Reynolds number of the air or gas.*

The viscosity of air at 80°F is 0.0186 cP, obtained from King and Crocker—*Piping Handbook,* Perry et al.—*Chemical Engineers' Handbook,* or a similar reference. Then, using the Reynolds number relation given in Table 6-1, $\text{Re} = 6.32W/dz = 6.32(22{,}620)/[4.026(0.0186)] = 3{,}560{,}000$.

5. *Compute the pressure loss in the pipe.*

Using Fig. 6-2 or the Hydraulic Institute *Pipe Friction Manual,* $f = 0.0142$ to 0.0162 for a 4-in schedule 40 pipe when the Reynolds number = 3,560,000. Using the Fanning formula from Example 6-4 and the higher value of f, $p_d = 3.36\ (10^{-6})flW^2/(d^5\rho)$, or $p_d = 3.36(10^{-6})(0.0162)(500)(22{,}620)^2/[(4.026)^5(0.754)] = 17.52$ lb/in^2 (121 kPa).

Related Calculations: Use this procedure to compute the pressure loss, velocity, and flow rate in compressed-air and gas lines of any length. Gases for which this procedure can be used include ammonia, carbon dioxide, carbon monoxide, ethane, ethylene, hydrogen, hydrogen sulfide, isobutane, methane, nitrogen, *n*-butane, oxygen, propane, propylene, and sulfur dioxide.

Alternate relations for computing the velocity of air or gas in a pipe are $v = 144W_s/(a\rho)$; $v = 183.4W_s/(d^2\rho)$; $v = 0.0509W_sv_g/d^2$, where W_s = flow rate, in lb/s; a = cross-sectional area of pipe, in in^2; v_g = specific volume of the air or gas at the operating pressure and temperature, in ft^3/lb.

6-10 Flow Rate and Pressure Loss in Gas Pipelines

Using the Weymouth formula, determine the flow rate in a 10-mile-long 4-in schedule 40 gas pipeline when the inlet pressure is 200 psig (1480 kPa), the outlet pressure is 20 psig (239 kPa), the gas has a specific gravity of 0.80, a temperature of 60°F (289 K), and the atmospheric pressure is 14.7 psia (101.4 kPa).

Calculation Procedure:

1. *Compute the flow rate using the Weymouth formula.*

The Weymouth formula for flow rate Q, in lb/h, is $Q = 28.05[(p_i^2 - p_o^2)d^{5.33}/(sL)]^{0.5}$, where p_i = inlet pressure, in psia; p_o = outlet pressure, in psia; d = inside diameter of pipe, in in; s = specific gravity of gas; L = length of pipeline, in miles. For this pipe, $Q = 28.05 \times [(214.7^2 - 34.7^2)4.026^{5.33}/(0.8 \times 10)]^{0.5} = 86{,}500$ lb/h (10.9 kg/s).

2. *Determine if the acoustic velocity limits flow.*

If the outlet pressure of a pipe is less than the critical pressure p_c, in psia, the flow rate in the pipe cannot exceed that obtained with a velocity equal to the critical acoustic velocity, i.e., the velocity of sound in the gas. For any gas, $p_c = Q(T_i)^{0.5}/(d^2C)$, where T_i = inlet temperature, in R; C = a constant for the gas being considered.

Using $C = 2070$ from Table 6-6 or King and Crocker—*Piping Handbook*, $p_c = (86{,}500)(60 + 460)^{0.5}/[(4.026)^2(2070)] = 58.8$ psia. Since the outlet pressure $p_o = 34.7$ psia, the critical or acoustic velocity limits the flow in this pipe because $p_c > p_o$. When $p_c < p_o$, critical velocity does not limit the flow.

Related Calculations: Where a number of gas pipeline calculations must be made, use the tabulations in King and Crocker—*Piping Handbook* and Bell—*Petroleum Transportation Handbook*. These tabulations will save much time. Other useful formulas for gas flow include the Panhandle, Unwin, Fritzche, and rational. Results obtained with these formulas agree within satisfactory limits for normal engineering practice.

Where the outlet pressure is unknown, assume a value for it and compute the flow rate that will be obtained. If the computed flow is less than desired, check to see if the outlet pressure is less than the critical. If it is, increase the diameter of the pipe. Use this procedure for natural gas from any gas field, manufactured gas, or any other similar gas.

To find the volume of gas, in ft^3, that can be stored per mile of pipe, solve $V_m = 1.955p_md^2K$, where p_m = mean pressure in pipe, in psia, $\approx (p_i + p_o)/2$; $K = (1/Z)^{0.5}$, where Z = supercompressibility factor of the gas, as given in Baumeister—*Standard Handbook for Mechanical Engineers* and Perry—*Chemical Engineer's Handbook*. For exact computation of p_m, use $p_m = \frac{2}{3}[p_i + p_o - p_ip_o/(p_i + p_o)]$.

6-11 Friction Loss in Pipes Handling Solids in Suspension

What is the friction loss in 800 ft of 6-in schedule 40 pipe when 400 gal/min (0.025 m^3/s) of sulfate paper stock is flowing? The consistency of the sulfate stock is 6 percent.

Calculation Procedure:

1. Determine the friction loss in the pipe.

There are few general equations for friction loss in pipes conveying liquids having solids in suspension. Therefore, most practicing engineers use plots of friction loss available in engineering handbooks, *Cameron Hydraulic Data, Standards of the Hydraulic Institute,* and from pump engineering data. Figure 6-4 shows one set of typical friction-loss curves based on work done at the University of Maine on the data of Brecht and Heller of the Technical College, Darmstadt, Germany, and published by Goulds Pumps, Inc. There is a similar series of curves for commonly used pipe sizes from 2 through 36 in.

Enter Fig. 6-4 at the pipe flow rate, 400 gal/min, and project vertically upward to the 6 percent consistency curve. From the intersection, project horizontally to the left to read the friction loss as 60 ft of liquid per 100 ft of pipe. Since this pipe is 800 ft long, the total friction-head loss in the pipe is (800/100)(60) = 480 ft (146 m) of liquid flowing.

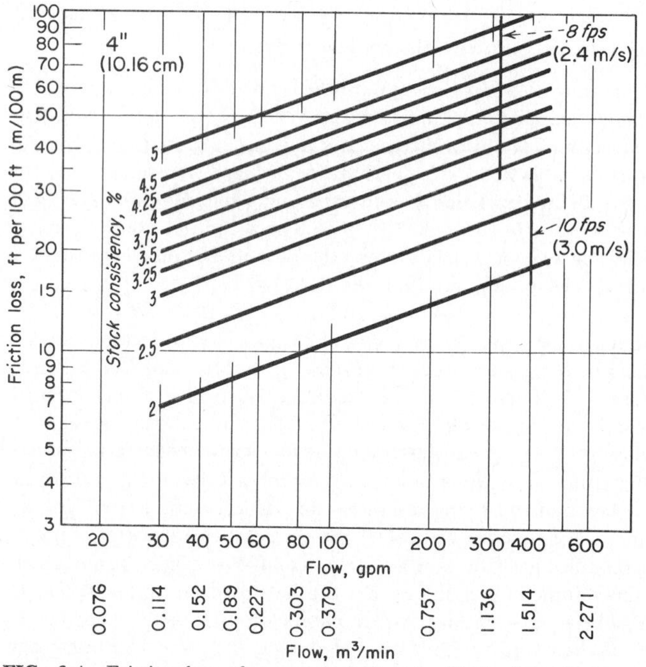

FIG. 6-4 Friction loss of paper stock in 4-in (0.10-m) steel pipe. *(Goulds Pumps, Inc.)*

2. Correct the friction loss for the liquid consistency.

Friction-loss factors are usually plotted for one type of liquid, and correction factors are applied to determine the loss for similar, but different, liquids. Thus, with the Goulds charts, a factor of 0.9 is used for soda, sulfate, bleached sulfite, and reclaimed paper stocks. For ground wood, the factor is 1.40.

When the stock consistency is less than 1.5 percent, water-friction values are used. Below a consistency of 3 percent, the velocity of flow should not exceed 10 ft/s. For suspensions of 3 percent and above, limit the maximum velocity in the pipe to 8 ft/s.

Since the liquid flowing in this pipe is sulfate stock, use the 0.9 correction factor, or the actual total friction head = (0.9)(480) = 432 ft (132 m) of sulfate liquid. Note that Fig. 6-4 shows that the liquid velocity is less than 8 ft/s.

Related Calculations: Use this procedure for soda, sulfate, bleached sulfite, and reclaimed and ground-wood paper stock. The values obtained are valid for both suction and discharge piping. The same general procedure can be used for sand mixtures, sewage, trash, sludge, foods in suspension in a liquid, and other slurries.

6-12 Determining the Pressure Loss in Steam Piping

Use a suitable pressure-loss chart to determine the pressure loss in 510 ft of 4-in flanged steel pipe containing two 90° elbows and four 45° bends. The schedule 40 piping conveys 13,000 lb/h (1.64 kg/s) of 40-psig 350°F superheated steam. List other methods of determining the pressure loss in steam piping.

Calculation Procedure:

1. Determine the equivalent length of the piping.

The equivalent length of a pipe L_e, in ft, equals length of straight pipe, ft + equivalent length of fittings, ft. Using data from the Hydraulic Institute, King and Crocker—*Piping Handbook,* or Fig. 6-5, find the equivalent of a 90° 4-in elbow as 10 ft of straight pipe. Likewise, the equivalent length of a 45° bend is 5 ft of straight pipe. Substituting in the preceding relation and using the straight lengths and the number of fittings of each type, $L_e = 510 + (2)(10) + 4(5) = 550$ ft of straight pipe.

2. Compute the pressure loss using a suitable chart.

Figure 6-6 presents a typical pressure-loss chart for steam piping. Enter the chart at the top left at the superheated steam temperature of 350°F and project vertically downward until the 40-psig superheated steam pressure curve is intersected. From here, project horizontally to the right until the outer border of the chart is intersected. Next, project through the steam flow rate, 13,000 lb/h on scale B of Fig. 6-6 to the pivot scale C. From this point, project through 4-in (101.6-mm) schedule 40 pipe on scale D of Fig. 6-6. Extend this line to intersect the pressure-drop scale and read the pressure loss as 7.25 lb/in^2 (5. kPa) per 100 ft (30.4 m) of pipe.

Since the equivalent length of this pipe is 550 ft (167.6 m), the total pressure loss in the pipe is $(550/100)(7.25) = 39.875$ lb/m^2 (274.9 kPa), say 40 lb/in^2 (275.8 kPa).

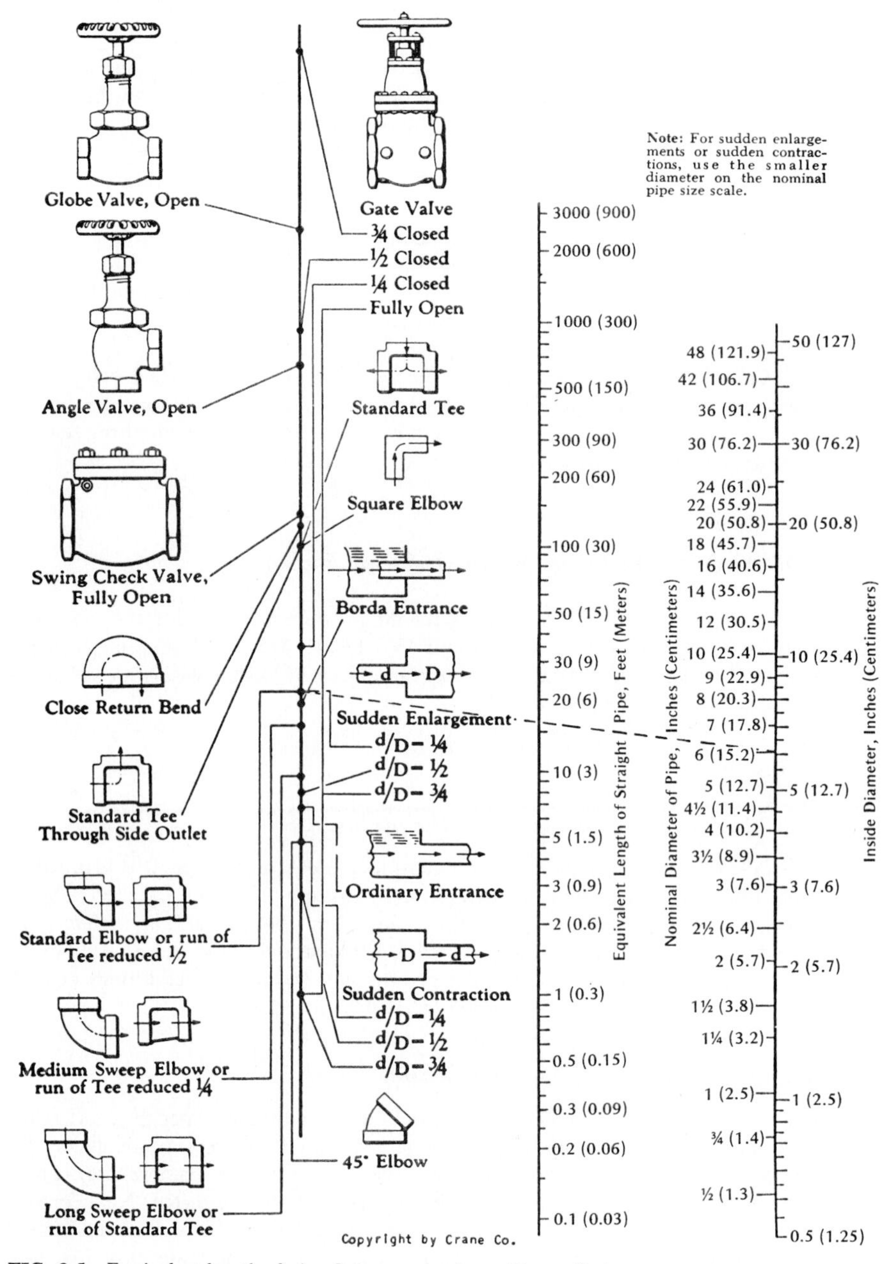

FIG. 6-5 Equivalent length of pipe fittings and valves. *(Crane Co.)*

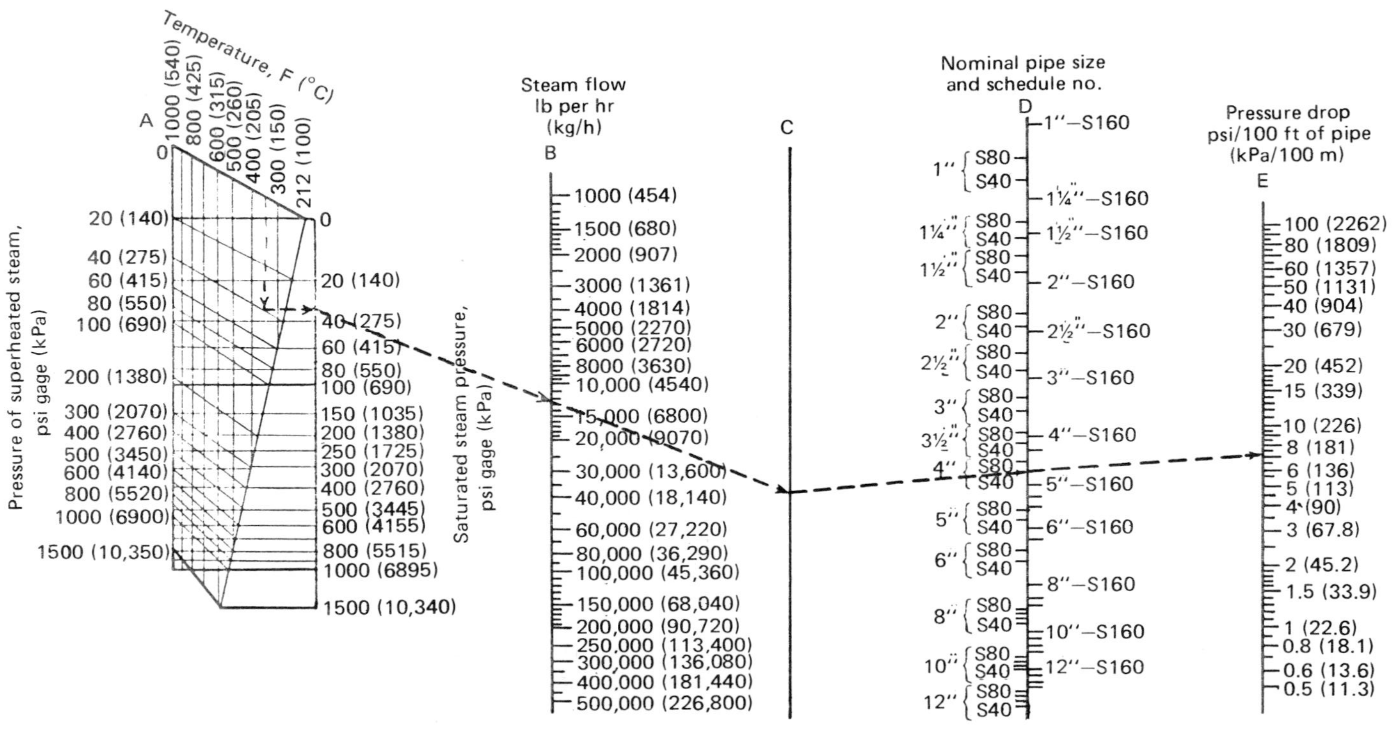

FIG. 6-6 Pressure loss in steam pipes based on the Fritzche formula. *(Power.)*

3. List the other methods of computing pressure loss.

Numerous pressure-loss equations have been developed to compute the pressure drop in steam piping. Among the better-known equations are those of Unwin, Fritzche, Spitzglass, Babcock, Gutermuth, and others. These equations are discussed in some detail in King and Crocker—*Piping Handbook* and in the engineering data published by valve and piping manufacurers.

Most piping designers use a chart to determine the pressure loss in steam piping because a chart saves time and reduces the effort involved. Further, the accuracy obtained is sufficient for all usual design practice.

Figure 6-7 is a popular flowchart for determining steam flow rate, pipe size, steam pressure, or steam velocity in a given pipe. Using this chart, the designer can determine any one of the four variables listed above when the other three are known. In solving a problem on the chart in Fig. 6-7, use the steam-quantity lines to intersect pipe sizes and the steam-pressure lines to intersect steam velocities. Here are two typical applications of this chart.

Example: What size schedule 40 pipe is needed to deliver 8000 lb/h (3600 kg/h) of 120-psig (827.3-kPa) steam at a velocity of 5000 ft/min (1524 m/min)?

Solution: Enter Fig. 6-7 at the upper left at a velocity of 5000 ft/min and project along this velocity line until the 120-psig pressure line is intersected. From this intersection, project horizontally until the 8000 lb/h (3600 kg/h) vertical line is intersected. Read the *nearest* pipe size as 4 in (101.6 mm) on the *nearest* pipe-diameter curve.

Example: What is the steam velocity in a 6-in (152.4-mm) pipe delivering 20,000 lb/h (9000 kg/h) of steam at 85 psig (586 kPa)?

Solution: Enter the bottom of Fig. 6-7 at the flow rate, 20,000 lb/h, and project vertically upward until the 6-in pipe (152.4-mm) curve is intersected. From this point, project horizontally to the 85-psig (586-kPa) curve. At the intersection, read the velocity as 7350 ft/min (2240.3 m/min).

Table 6-7 shows typical steam velocities for various industrial and commercial applications. Use the given values as guides when sizing steam piping.

6-13 Steam-Trap Selection for Process Applications

Select steam traps for the following five types of equipment: (1) where the steam directly heats solid materials, as in autoclaves, retorts, and sterilizers; (2) where the steam indirectly heats a liquid through a metallic surface, as in heat exchangers and kettles where the quantity of liquid heated is known and unknown; (3) where the steam indirectly heats a solid through a metallic surface, as in dryers using cylinders or chambers and platen presses; and (4) where the steam indirectly heats air through metallic surfaces, as in unit heaters, pipe coils, and radiators.

Calculation Procedure:

1. Determine the condensate load.

The first step in selecting a steam trap for any type of equipment is determination of the condensate load. Use the following general procedure.

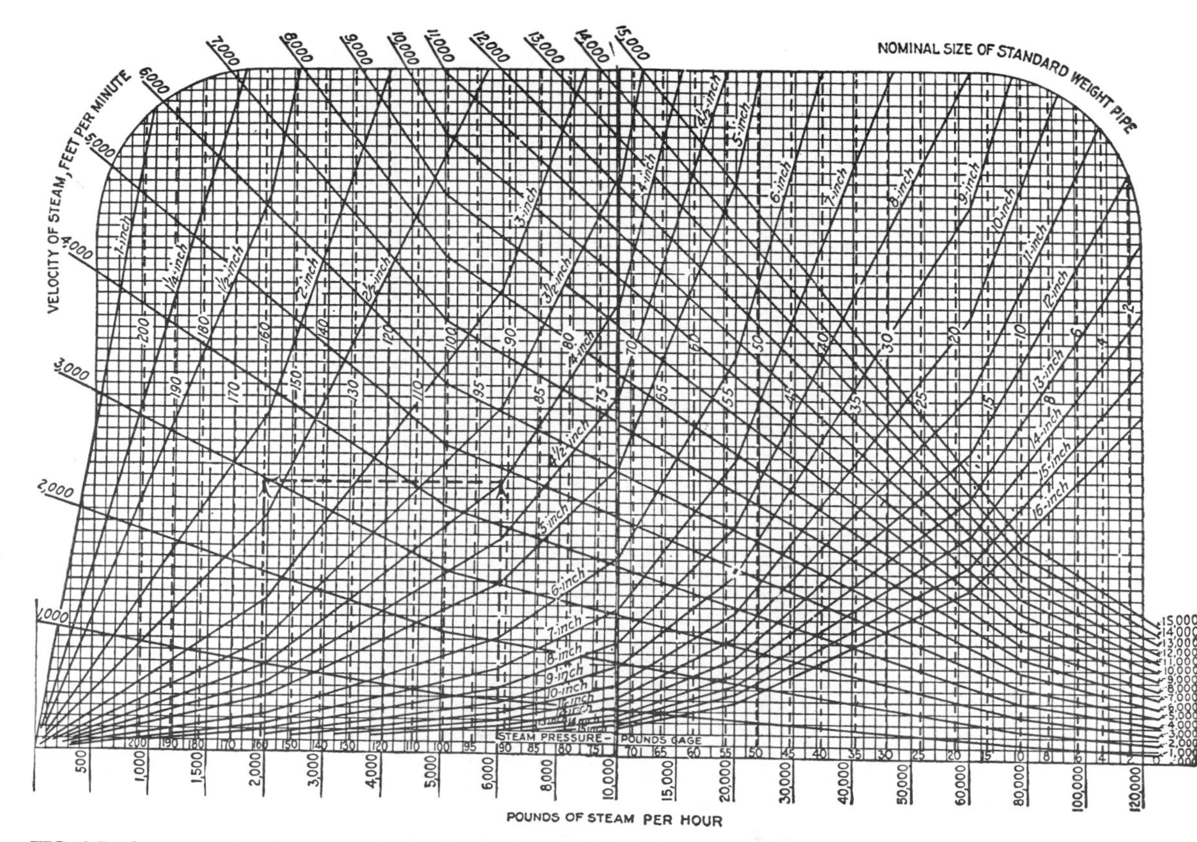

FIG. 6-7 Spitzglass chart for saturated steam flowing in schedule 40 pipe.

TABLE 6-7 Steam Velocities Used in Pipe Design

Steam condition	Steam pressure		Steam use	Steam velocity	
	lb/in²	kPa		ft/min	m/min
Saturated	0–15	0–103.4	Heating	4000–6000	1219.2–1828.8
Saturated	50–150	344.7–1034.1	Process	6000–10,000	1828.8–3048.0
Superheated	200 and higher	1378.8 and higher	Boiler leads	10,000–15,000	3048.0–4572.0

a. *Solid materials in autoclaves, retorts, and sterilizers.* How much condensate is formed when 2000 lb of solid material with a specific heat of 1.0 is processed in 15 min at 240°F by 25-psig steam from an initial temperature of 60°F in an insulated steel retort?

For this type of equipment, use $C = WsP$, where C = condensate formed, in lb/h; W = weight of material heated, in lb; s = specific heat, in Btu/(lb)(°F); P = factor from Table 6-8. Thus, for this application, $C = (2000)(1.0)(0.193) = 386$ lb of condensate. Note that P is based on a temperature rise of $240 - 60 = 180$°F and a steam pressure of 25 psig. For the retort, using the specific heat of steel from Table 6-9, $C = (4000)(0.12)(0.193) = 92.6$ lb of condensate, say 93 lb (41.9 kg). The total weight of condensate formed in 15 min $= 386 + 93 = 479$ lb (215.6 kg). In 1 h, $479(60/15) =$ 1916 lb (862.2 kg) of condensate is formed.

TABLE 6-8 Factors, $P = (T - t)/L$, to Find Condensate Load

Pressure		Temperature		
psia	kPa	160°F (71.1°C)	180°F (82.2°C)	200°F (93.3°C)
20	137.8	0.170	0.192	0.213
25	172.4	0.172	0.193	0.214
30	206.8	0.172	0.194	0.215

TABLE 6-9 Use These Specific Heats When Calculating Condensate Load

Solids	Btu/(lb)(°F)	kJ/(kg)(°C)	Liquids	Btu/(lb)(°F)	kJ/(kg)(°C)
Aluminum	0.23	0.96	Alcohol	0.65	2.7
Brass	0.10	0.42	Carbon tetrachloride	0.20	0.84
Copper	0.10	0.42	Gasoline	0.53	2.22
Glass	0.20	0.84	Glycerin	0.58	2.43
Iron	0.13	0.54	Kerosene	0.47	1.97
Steel	0.12	0.50	Oils	0.40–0.50	1.67–2.09

A safety factor must be applied to compensate for radiation and other losses. Typical safety factors used in selecting steam traps are

Steam mains and headers	2–3
Steam heating pipes	2–6
Purifiers and separators	2–3
Retorts for process	2–4
Unit heaters	3
Submerged pipe coils	2–4
Cylinder dryers	4–10

Using a safety factor of 4 for this process retort, the trap capacity = (4)(1916) = 7664 lb/h (3449 kg/h), say 7700 lb/h (3465 kg/h).

b(1). *Submerged heating surface and a known quantity of liquid.* How much condensate forms in the jacket of a kettle when 500 gal (1892.5 L) of water is heated in 30 min from 72 to 212°F (22.2 to 100°C) with 50-psig (344.7-kPa) steam?

For this type of equipment, $C = GwsP$, where G = gallons of liquid heated; w = weight of liquid, in lb/gal. Substitute the appropriate values as follows: $C = (500)(8.33)(1.0) \times (0.154) = 641$ lb, or (641)(60/30) = 1282 lb/h. Using a safety factor of 3, the trap capacity = (3)(1282) = 3846 lb/h; say 3900 lb/h.

b(2). *Submerged heating surface and an unknown quantity of liquid.* How much condensate is formed in a coil submerged in oil when the oil is heated as quickly as possible from 50 to 250°F by 25-psig steam if the coil has an area of 50 ft^2 and the oil is free to circulate around the coil?

For this condition, $C = UAP$, where U = overall coefficient of heat transfer, in Btu/(h)(ft^2)(°F), from Table 6-10; A = area of heating surface, in ft^2. With free convection and a condensing-vapor-to-liquid type of heat exchanger, U = 10 to 30. Using an average value of $U = 20$, $C = (20)(50)(0.214) = 214$ lb/h of condensate. Choosing a safety factor 3, the trap capacity = (3)(214) = 642 lb/h; say 650 lb/h.

b(3). *Submerged surfaces having more area then needed to heat a specified quantity of liquid in a given time with condensate withdrawn as rapidly as formed.* Use Table 6-11 instead of steps *b*(1) or *b*(2). Find the condensation rate by multiplying the submerged area by the appropriate factor from Table 6-11. Use this method for heating water, chemical solutions, oils, and other liquids. Thus, with steam at 100 psig and a temperature of 338°F and heating oil from 50 to 226°F with a submerged surface having an area of 500 ft^2, the mean temperature difference equals steam temperature minus the average liquid temperature = $Mtd = 338 - (50 + 226/2) = 200$°F. The factor from Table 6-11 for 100-psig steam and a 200°F Mtd is 56.75. Thus the condensation rate = (56.75) × (500) = 28,375 lb/h. With a safety factor of 2, the trap capacity = (2)(28,375) = 56,750 lb/h.

c. *Solids indirectly heated through a metallic surface.* How much condensate is formed in a chamber dryer when 1000 lb of cereal is dried to 750 lb by 10-psig steam? The initial temperature of the cereal is 60°F and the final temperature equals that of the steam.

For this condition, $C = 970(W - D)/h_{fg} + WP$, where D = dry weight of the material, in lb; h_{fg} = enthalpy of vaporization of the steam at the trap pressure, in Btu/lb. Using the steam tables and Table 6-8, $C = 970(1000 - 750)/952 + (1000)(0.189) = 443.5$ lb/h of condensate. With a safety factor of 4, the trap capacity = (4)(443.5) = 1774 lb/h.

d. *Indirect heating of air through a metallic surface.* How much condensate is formed in a unit heater using 10-psig steam if the entering-air temperature is 30°F and the leaving-air temperature is 130°F? Airflow is 10,000 ft^3/min.

Use Table 6-12, entering at a temperature difference of 100°F and projecting to a steam pressure of 10 psig. Read the condensate formed as 122 lb/h per 1000 ft^3/min. Since 10,000 ft^3/min of air is being heated, the condensation rate = (10,000/1000)(122) = 1220 lb/h. With a safety factor of 3, the trap capacity = (3)(1220) = 3660 lb/h, say 3700 lb/h.

Table 6-13 shows the condensate formed by radiation from bare iron and steel pipes in still air and with forced-air circulation. Thus, with a steam pressure of 100 psig and an initial air temperature of 75°F, 1.05 lb/h of condensate will be formed per square foot

TABLE 6-10 Ordinary Ranges of Overall Coefficients of Heat Transfer

Type of heat exchanger	State of controlling resistance		Typical fluid	Typical apparatus
	Free convection, U	Forced convection, U		
Liquid to liquid	25–60 [141.9–340.7]	150–300 [851.7–1703.4]	Water	Liquid-to-liquid heat exchangers
Liquid to liquid	5–10 [28.4–56.8]	20–50 [113.6–283.9]	Oil	
Liquid to gas*	1–3 [5.7–17.0]	2–10 [11.4–56.8]	—	Hot-water radiators
Liquid to boiling liquid	20–60 [113.6–340.7]	50–150 [283.9–851.7]	Water	Brine coolers
Liquid to boiling liquid	5–20 [28.4–113.6]	25–60 [141.9–340.7]	Oil	
Gas* to liquid	1–3 [5.7–17.0]	2–10 [11.4–56.8]	—	Air coolers, economizers
Gas* to gas	0.6–2 [3.4–11.4]	2–6 [11.4–34.1]	—	Steam superheaters
Gas* to boiling liquid	1–3 [5.7–17.0]	2–10 [11.4–56.8]	—	Steam boilers
Condensing vapor to liquid	50–200 [283.9–1136]	150–800 [851.7–4542.4]	Steam to water	Liquid heaters and condensers
Condensing vapor to liquid	10–30 [56.8–170.3]	20–60 [113.6–340.7	Steam to oil	
Condensing vapor to liquid	40–80 [227.1–454.2]	60–150 [340.7–851.7]	Organic vapor to water	
Condensing vapor to liquid	— —	15–300 [85.2–1703.4]	Steam-gas mixture	
Condensing vapor to gas*	1–2 [5.7–11.4]	2–10 [11.4–56.8]	—	Steam pipes in air, air heaters
Condensing vapor to boiling liquid	40–100 [227.1–567.8]	—	—	Scale-forming evaporators
Condensing vapor to boiling liquid	300–800 [1703.4–4542.4]	—	Steam to water	
Condensing vapor to boiling liquid	50–150 [283.9–851.7]	—	Steam to oil	

*At atmospheric pressure.

Note: U = Btu/(h)(ft²)(°F) [W/(m²)(°C)]. Under many conditions, either higher or lower values may be realized.

TABLE 6-11 Condensate Formed in Submerged Steel* Heating Elements, lb/(ft²)(h) [kg/(m²)(min)]

Mtd†		Steam pressure				
°F	°C	75 psia (517.1 kPa)	100 psia (689.4 kPa)	150 psia (1034.1 kPa)	Btu/(ft²)(h)	kW/m²
175	97.2	44.3 (3.6)	45.4 (3.7)	46.7 (3.8)	40,000	126.2
200	111.1	54.8 (4.5)	56.8 (4.6)	58.3 (4.7)	50,000	157.7
250	138.9	90.0 (7.3)	93.1 (7.6)	95.7 (7.8)	82,000	258.6

*For copper, multiply table data by 2.0. For brass, multiply table data by 1.6.

†Mean temperature difference, °F or °C, equals temperature of steam minus average liquid temperature. Heat-transfer data for calculating this table obtained from and used by permission of the American Radiator & Standard Sanitary Corp.

TABLE 6-12 Steam Condensed by Air, lb/h at 1000 ft³/min (kg/h at 28.3 m³/min)*

Temperature difference		Pressure		
°F	°C	5 psig (34.5 kPa)	10 psig (68.9 kPa)	50 psig (344.7 kPa)
50	27.8	61 (27.5)	61 (27.5)	63 (28.4)
100	55.6	120 (54.0)	122 (54.9)	126 (56.7)
150	83.3	180 (81.0)	183 (82.4)	189 (85.1)

*Based on 0.0192 Btu (0.02 kJ) absorbed per cubic foot (0.028 m³) of saturated air per °F (0.556°C) at 32°F (0°C). For 0°F (−17.8°C), multiply by 1.1.

TABLE 6-13 Condensate Formed by Radiation from Bare Iron and Steel*, lb/(ft²)(h) [kg/(m²)(h)]

Air temperature		Steam pressure			
°F	°C	50 psig (344.7 kPa)	75 psig (517.1 kPa)	100 psig (689.5 kPa)	150 psig (1034 kPa)
65	18.3	0.82 (3.97)	1.00 (5.84)	1.08 (5.23)	1.32 (6.39)
70	21.2	0.80 (3.87)	0.98 (4.74)	1.06 (5.13)	1.21 (5.86)
75	23.9	0.77 (3.73)	0.88 (4.26)	1.05 (5.08)	1.19 (5.76)

*Based on still air; for forced-air circulation, multiply by 5.

of heating surface in still air. With forced air circulation, the condensate rate is (5)(1.05) = 5.25 lb/h per square foot of heating surface.

Unit heaters have a *standard rating* based on 2-psig steam with entering air at 60°F. If the steam pressure or air temperature is different from these standard conditions, multiply the heater Btu/h capacity rating by the appropriate correction factor from Table 6-14. Thus a heater rated at 10,000 Btu/h with 2-psig steam and 60°F air would have an output of (1.290)(10,000) = 12,900 Btu/h with 40°F inlet air and 10-psig steam. Trap manufacturers usually list heater Btu ratings and recommend trap model numbers and sizes in their trap engineering data. This allows easier selection of the correct trap.

TABLE 6-14 Unit-Heater Correction Factors

Steam pressure		Temperature of entering air		
psig	kPa	20°F (−6.7°C)	40°F (4.4°C)	60°F (15.6 °C)
5	34.5	1.370	1.206	1.050
10	68.9	1.460	1.290	1.131
15	103.4	1.525	1.335	1.194

Source: Yarway Corporation; SI values added by *Handbook* editor.

2. *Select the trap size based on the load and steam pressure.*

Obtain a chart or tabulation of trap capacities published by the manufacturer whose trap will be used. Figure 6-8 is a capacity chart for one type of bucket trap manufactured by Armstrong Machine Works. Table 6-15 shows typical capacities of impulse traps manufactured by the Yarway Company.

To select a trap from Fig. 6-8, when the condensation rate is uniform and the pressure across the trap is constant, enter at the left at the condensation rate—say 8000 lb/h (3600 kg/h) (as obtained from step 1)—and project horizontally to the right to the vertical ordinate representing the pressure across the trap (= Δp = steam-line pressure, in psig − return-line pressure with trap valve closed, in psig). Assume Δp = 20 psig (138 kPa) for this trap. The intersection of the horizontal 8000 lb/h (3600 kg/h) projection and the vertical 20-psig (137.9-kPa) projection is on the sawtooth capacity curve for a trap having a 9/16-in (14.3-mm) orifice. If these projections intersected beneath this curve, a 9/16-in (14.3-mm) orifice would still be used if the point was between the verticals for this size orifice.

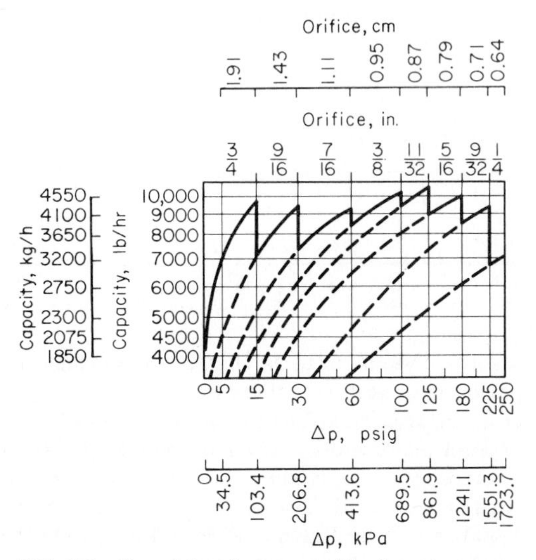

FIG. 6-8 Capacities of one type of bucket steam trap. *(Armstrong Machine Works.)*

TABLE 6-15 Capacities of Impulse Traps, lb/h (kg/h) [maximum continuous discharge of condensate, based on condensate at 30°F (16.7°C) below steam temperature]

Pressure at trap inlet		Trap nominal size	
psig	kPa	1.25 in (38.1 mm)	2.0 in (50.8 mm)
125	861.8	6165 (2774)	8530 (3839)
150	1034.1	6630 (2984)	9075 (4084)
200	1378.8	7410 (3335)	9950 (4478)

Source: Yarway Corporation.

The dashed lines extending downward from the sawtooth curves show the capacity of a trap at reduced Δp. Thus the capacity of a trap with a ⅜-in (9.53-mm) orifice at Δp = 30 psig (207 kPa) is 6200 lb/h (2790 kg/h), read at the intersection of the 30-psig (207-kPa) ordinate and the dashed curve extended from the ⅜-in (9.53-mm) solid curve.

To select an impulse trap from Table 6-15, enter the table at the trap inlet pressure—say 125 psig (862 kPa)—and project to the desired capacity—say 8000 lb/h (3600 kg/h), determined from step 1. Table 6-15 shows that a 2-in (50.8-mm) trap having an 8530 lb/h (3839 kg/h) capacity must be used because the next smallest size has a capacity of 5165 lb/h (2324 kg/h). This capacity is less than that required.

Some trap manufacturers publish capacity tables relating various trap models to specific types of equipment. Such tables simplify trap selection, but the condensation rate must still be computed as given here.

Related Calculations: Use the procedure given here to determine the trap capacity required for any industrial, commercial, or domestic application.

When using a trap-capacity diagram or table, be sure to determine the basis on which it was prepared. Apply any necessary correction factors. Thus *cold-water capacity ratings* must be corrected for traps operating at higher condensate temperatures. Correction factors are published in trap engineering data. The capacity of a trap is greater at condensate temperatures less than 212°F (100°C), because at or above this temperature condensate forms flash steam when it flows into a pipe or vessel at atmospheric [14.7 psia (101.3 kPa)] pressure. At altitudes above sea level, condensate flashes into steam at a lower temperature, depending on the altitude.

The method presented here is the work of L. C. Campbell, Yarway Corporation, as reported in *Chemical Engineering*.

6-14 Orifice-Meter Selection for a Steam Pipe

Steam is metered with an orifice meter in a 10-in boiler lead having an internal diameter of d_p = 9.760 in. Determine the maximum rate of steam flow that can be measured with a steel orifice plate having a diameter of d_o = 5.855 in at 70°F (294 K). The upstream pressure tap is 1D ahead of the orifice, and the downstream tap is 0.5D past the orifice. Steam pressure at the orifice inlet p_p = 250 psig (1825 kPa); temperature is 640°F (611 K). A differential gage fitted across the orifice has a maximum range of 120 in of water. What is the steam flow rate when the observed differential pressure is 40 in of water?

Use the ASME Research Committee on Fluid Meters method in analyzing the meter. Atmospheric pressure is 14.696 psia.

Calculation Procedure:

1. Determine the diameter ratio and steam density.

For any orifice meter, diameter ratio $= \beta =$ meter orifice diameter, in/pipe internal diameter, in $= 5.855/9.760 = 0.5999$.

Determine the density of the steam by entering the superheated steam table at $250 + 14.696 = 264.696$ psia and 640°F and reading the specific volume as 2.387 ft^3/lb. For steam, the density $=$ 1/specific volume $= d_s = 1/2.387 = 0.4193$ lb/ft^3.

2. Determine the steam viscosity and meter flow coefficient.

From the ASME publication *Fluid Meters—Their Theory and Application,* the steam viscosity gu_1 for a steam system operating at 640°F is $gu_1 = 0.0000141$ in·lb/(°F)(s)(ft^2).

Find the flow coefficient K from the same ASME source by entering the 10-in nominal pipe diameter table at $\beta = 0.5999$ and projecting to the appropriate Reynolds number column. Assume that the Reynolds number $= 10^7$, approximately, for the flow conditions in this pipe. Then, $K = 0.6486$. Since the Reynolds number for steam pressures above 100 lb/in^2 ranges from 10^6 to 10^7, this assumption is safe because the value of K does not vary appreciably in this Reynolds number range. Also, the Reynolds number cannot be computed yet because the flow rate is unknown. Therefore, assumption of the Reynolds number is necessary. The assumption will be checked later.

3. Determine the expansion factor and the meter area factor.

Since steam is a compressible fluid, the expansion factor Y_1 must be determined. For superheated steam, the ratio of the specific heat at constant pressure c_p to the specific heat at constant volume c_v is $k = c_p/c_v = 1.3$. Also, the ratio of the differential maximum pressure reading h_w, in in of water, to the maximum pressure in the pipe, in psia, equals $120/246.7 = 0.454$. Using the expansion-factor curve in the ASME *Fluid Meters,* $Y_1 = 0.994$ for $\beta = 0.5999$, and the pressure ratio $= 0.454$. And, from the same reference, the meter area factor $F_a = 1.0084$ for a steel meter operating at 640°F.

4. Compute the rate of steam flow.

For square-edged orifices, the flow rate, in lb/s, is $w = 0.0997\ F_aKd^2Y_1(h_wd_s)^{0.5} = (0.0997)(1.0084)(0.6486)(5.855)^2(0.994)(120 \times 0.4188)^{0.5} = 15.75$ lb/s.

5. Compute the Reynolds number for the actual flow rate.

For any steam pipe, the Reynolds number $\text{Re} = 48w/d_p gu_1 = 48(15.75)/[3.1416(0.760)(0.0000141)] = 1{,}750{,}000$.

6. Adjust the flow coefficient for the actual Reynolds number.

In step 2, $\text{Re} = 10^7$ was assumed and $K = 0.6486$. For $\text{Re} = 1{,}750{,}000$, $K = 0.6489$, from ASME *Fluid Meters,* by interpolation. Then, the actual flow rate $w_h =$ (computed flow rate)(ratio of flow coefficients based on assumed and actual Reynolds numbers) = (15.75)(0.6489/0.6486)(3600) = 56,700 lb/h, closely, where the value 3600 is a conversion factor for changing lb/s to lb/h.

7. Compute the flow rate for a specific differential gage deflection.

For a 40-in H_2O deflection, F_a is unchanged and equals 1.0084. The expansion factor changes because h_w/p_p = 40/264.7 = 0.151. Using the ASME *Fluid Meters,* Y_1 = 0.998. Assuming again that Re = 10^7, K = 0.6486, as before; then w = $0.0997(1.0084)(0.6486)(5.855)^2(0.998)(40 \times 0.4188)^{0.5}$ = 9.132 lb/s. Computing the Reynolds number as before, Re = 40(0.132)/[3.1416(0.76)(0.0000141)] = 1,014,000. The value of K corresponding to this value, as before, is from ASME *Fluid Meters;* K = 0.6497. Therefore, the flow rate for a 40-in H_2O reading, in lb/h, is w_h = 0.132(0.6497/0.6486)(3600) = 32,940 lb/h (4.15 kg/s).

Related Calculations: Use these steps and the ASME *Fluid Meters* or comprehensive meter engineering tables giving similar data to select or check an orifice meter used in any type of steam pipe—main, auxiliary, process, industrial, marine, heating, or commercial—conveying wet, saturated, or superheated steam.

6-15 Selection of a Pressure-Regulating Valve for Steam Service

Select a single-seat spring-loaded diaphragm-actuated pressure-reducing valve to deliver 350 lb/h (0.044 kg/s) of steam at 50 psig when the initial pressure is 225 psig. Also select an integral pilot-controlled piston-operated single-seat pressure-regulating valve to deliver 30,000 lb/h (3.78 kg/s) of steam at 40 psig with an initial pressure of 225 psig saturated. What size pipe must be used on the downstream side of the valve to produce a velocity of 10,000 ft/min (50.8 m/s)? How large should the pressure-regulating valve be if the steam entering the valve is at 225 psig and 600°F (589 K)?

Calculation Procedure:

1. Compute the maximum flow for the diaphragm-actuated valve.

For best results in service, pressure-reducing valves are selected so that they operate 60 to 70 percent open at normal load. To obtain a valve sized for this opening, divide the desired delivery, in lb/h, by 0.7 to obtain the maximum flow expected. For this valve, then, the maximum flow is 350/0.7 = 500 lb/h.

2. Select the diaphragm-actuated valve size.

Using a manufacturer's engineering data for an acceptable valve, enter the appropriate valve-capacity table at the valve inlet steam pressure, 225 psig, and project to a capacity of 500 lb/h, as in Table 6-16. Read the valve size as ¾ in at the top of the capacity column.

3. Select the size of the pilot-controlled pressure-regulating valve.

Enter the capacity table in the engineering data of an acceptable pilot-controlled pressure-regulating valve, similar to Table 6-17, at the required capacity, 30,000 lb/h, and project

TABLE 6-16 Pressure-Reducing-Valve Capacity, lb/h (kg/h)

Inlet pressure		Valve size		
psig	kPa	½ in (12.7 mm)	¾ in (19.1 mm)	1 in (25.4 mm)
200	1379	420 (189)	460 (207)	560 (252)
225	1551	450 (203)	500 (225)	600 (270)
250	1724	485 (218)	560 (252)	650 (293)

Source: Clark-Reliance Corporation.

TABLE 6-17 Pressure-Regulating-Valve Capacity

Steam capacity		Initital steam pressure, saturated			
lb/h	kg/h	40 psig (276 kPa)	175 psig (1206 kPa)	225 psig (1551 kPa)	300 psig (2068 kPa)
20,000	9,000	6* (152.4)	4 (101.6)	4 (101.6)	3 (76.2)
30,000	13,500	8 (203.2)	5 (127.0)	4 (101.6)	4 (101.6)
40,000	18,000	— —	5 (127.0)	5 (127.0)	4 (101.6)

*Value diameter measured in inches (millimeters).
Source: Clark-Reliance Corporation.

across until the correct inlet steam pressure column, 225 psig, is intercepted, and read the required valve size as 4 in.

Note that it is not necessary to compute the maximum capacity before entering the table, as in step 1, for the pressure-reducing valve. Also note that a capacity table such as Table 6-17 can be used only for valves conveying saturated steam, unless the table notes state that the values listed are valid for other steam conditions.

4. Determine the size of the downstream pipe.

Enter Table 6-17 at the required capacity, 30,000 lb/h (13,500 kg/h), and project across to the valve *outlet pressure,* 40 psig (275.8 kPa), and read the required pipe size as 8 in (203.2 mm) for a velocity of 10,000 ft/min (3048 m/min). Thus the pipe immediately downstream from the valve must be enlarged from the valve size, 4 in (101.6 mm), to the required pipe size, 8 in (203.2 mm), to obtain the desired steam velocity.

5. Determine the size of the valve handling superheated steam.

To determine the correct size of a pilot-controlled pressure regulating valve handling superheated steam, a correction must be applied. Either a factor may be used or a tabulation of corrected pressures, such as Table 6-18. To use Table 6-18, enter at the valve inlet pressure, 225 psig (1551.2 kPa), and project across to the total temperature, 600°F (316°C), to read the corrected presssure, 165 psig (1137.5 kPa). Enter Table 6-17 at the *next highest* saturated steam pressure, 175 psig, and project down to the required capacity, 30,000 lb/h (13,500 kg/h), and read the required valve size as 5 in (127 mm).

Related Calculations: To simplify pressure-reducing and pressure-regulating valve selection, become familiar with two or three acceptable valve manufacturers' engineering data. Use the procedures given in the engineering data or those given here to select valves

TABLE 6-18 Equivalent Saturated Steam Values for Superheated Steam at Various Pressures and Temperatures

Steam pressure		Steam temp		Total temperature					
				500°F	600°F	700°F	260.0°C	315.6°C	371.1°C
psig	kPa	°F	°C	Steam values, psig			Steam values, kPa		
205	1413.3	389	198	171	149	133	1178.9	1027.2	916.9
225	1551.2	397	203	190	165	147	1309.9	1137.5	1013.4
265	1826.9	411	211	227	200	177	1564.9	1378.8	1220.2

Source: Clark-Reliance Corporation.

for industrial, marine, utility, heating, process, laundry, kitchen, or hospital service with a saturated or superheated steam supply.

Do not oversize reducing or regulating valves. Oversizing causes chatter and excessive wear.

When an anticipated load on the downstream side will not develop for several months after installation of a valve, fit to the valve a reduced-area disk sized to handle the present load. When the load increases, install a full-size disk. Size the valve for the ultimate load, not the reduced load.

Where there is a wide variation in demand for steam at the reduced pressure, consider installing two regulators piped in parallel. Size the smaller regulator to handle light loads and the larger regulator to handle the difference between 60 percent of the light load and the maximum heavy load. Set the larger regulator to open when the minimum allowable reduced pressure is reached. Then both regulators will be open to handle the heavy load. Be certain to use the actual regulator inlet pressure and not the boiler pressure when sizing the valve if this is different from the inlet pressure. Data in this calculation procedure are based on valves built by the Clark-Reliance Corporation, Cleveland, Ohio.

Some valve manufacturers use the valve-flow coefficient C_v for valve sizing. This coefficient is defined as the flow rate, in lb/h, through a valve of given size when the pressure loss across the valve is 1 lb/in^2. Tabulations such as Tables 6-16 and 6-17 incorporate this flow coefficient and are somewhat easier to use. These tables make the necessary allowances for downstream pressures less than the critical pressure (= 0.55 × absolute upstream pressure, in lb/in^2, for superheated steam and hydrocarbon vapors, and 0.58 × absolute upstream pressure, in lb/in^2, for saturated steam). The accuracy of these tabulations equals that of valve size determined by using the flow coefficient.

6-16 Pressure-Reducing-Valve Selection for Water Piping

What size pressure-reducing valve should be used to deliver 1200 gal/h (1.26 L/s) of water at 40 lb/in^2 (275.8 kPa) if the inlet pressure is 140 lb/in^2 (965.2 kPa)?

Calculation Procedure:

1. Determine the valve capacity required.

Pressure-reducing valves in water systems operate best when the nominal load is 60 to 70 percent of the maximum load. Using 60 percent, the maximum load for this valve = 1200/0.6 = 2000 gal/h (2.1 L/s).

2. *Determine the valve size required.*

Enter a valve-capacity table in suitable valve engineering data at the valve inlet pressure and project to the exact, or next higher, valve capacity. Thus, enter Table 6-19 at 140 lb/in^2 (965.2 kPa) and project to the next higher capacity, 2200 gal/h (2.3 L/s), since a capacity of 2000 gal/h (2.1 L/s) is not tabulated. Read at the top of the column the required valve size as 1 in (25.4 mm).

TABLE 6-19 Maximum Capacities of Water Pressure-Reducing Valves, gal/h (L/h)

Inlet pressure		Valve size		
psig	kPa	¾ in (19.1 mm)	1 in (25.4 mm)	1¼ in (31.8 mm)
120	827.3	1550 (5867)	2000 (7570)	4500 (17,033)
140	965.2	1700 (6435)	2200 (8327)	5000 (18,925)
160	1103.0	1850 (7002)	2400 (9084)	5500 (20,818)

Source: Clark-Reliance Corporation.

Some valve manufacturers present the capacity of their valves in graphic instead of tabular form. One popular chart, Fig. 6-9, is entered at the difference between the inlet and outlet pressures on the abscissa, or 140 − 40 = 100 lb/in^2 (689.4 kPa). Project vertically to the flow rate of 2000/60 = 33.3 gal/min (2.1 L/s). Read the valve size on the intersecting valve-capacity curve, or on the next curve if there is no intersection with the curve. Figure 6-9 shows that a 1-in valve should be used. This agrees with the tabulated capacity.

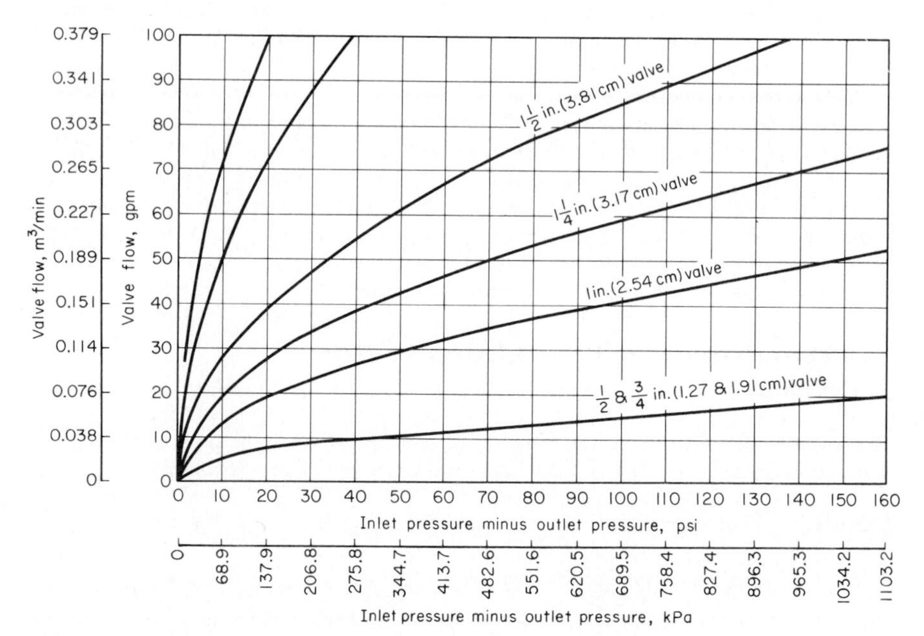

FIG. 6-9 Pressure-reducing valve flow capacity. *(Foster Engineering Co.)*

Related Calculations: Use this method for pressure-reducing valves in any type of water piping—process, domestic, commercial—where the water temperature is 100°F (37.8 °C) or less. Table 6-19 is from data prepared by the Clark-Reliance Corporation; Figure 6-9 is from Foster Engineering Company data.

Some valve manufacturers use the valve-flow coefficient C_v for valve sizing. This coefficient is defined as the flow rate, in gal/min, through a valve of given size when the pressure loss across the valve is 1 lb/in^2. Tabulations such as Table 6-19 and flowcharts such as Fig. 6-9 incorporate this flow coefficient and are somewhat easier to use. Their accuracy equals that of the flow-coefficient method.

6-17 Similarity or Affinity Laws for Centrifugal Pumps

A centrifugal pump designed for an 1800-r/min operation and a head of 200 ft (61 m) has a capacity of 3000 gal/min (0.19 m^3/s) with a power input of 175 hp. What effect will a speed reduction to 1200 r/min have on the head, capacity, and power input of the pump? What will be the change in these variables if the impeller diameter is reduced from 12 to 10 in while the speed is held constant at 1800 r/min?

Calculation Procedure:

1. Compute the effect of a change in pump speed.

For any centrifugal pump in which the effects of fluid viscosity are negligible or are neglected, the similarity or affinity laws can be used to determine the effect of a speed, power, or head change. For a *constant impeller diameter,* these laws are $Q_1/Q_2 = N_1/N_2$; $H_1/H_2 = (N_1/N_2)^2$; $P_1/P_2 = (N_1/N_2)^3$. For a *constant speed,* $Q_1/Q_2 = D_1/D_2$; $H_1/H_2 = (D_1/D_2)^2$; $P_1/P_2 = (D_1/D_2)^3$. In both sets of laws, Q = capacity, in gal/min; N = impeller r/min; D = impeller diameter, in in; H = total head, in ft of liquid; P = bhp input. The subscripts 1 and 2 refer to the initial and changed conditions, respectively.

For this pump, with a constant impeller diameter, $Q_1/Q_2 = N_1/N_2$; $3000/Q_2 = 1800/1200$; $Q_2 = 2000$ gal/min (0.13 m^3/s). And, $H_1/H_2 = (N_1/N_2)^2 = 200/H_2 = (1800/1200)^2$; $H_2 = 88.9$ ft (27.1 m). Also, $P_1/P_2 = (N_1/N_2)^3 = 175/P_2 = (1800/1200)^3$; $P_2 = 51.8$ bhp.

2. Compute the effect of a change in impeller diameter.

With the speed constant, use the second set of laws. Or for this pump, $Q_1/Q_2 = D_1/D_2$; $3000/Q_2 = 12/10$; $Q_2 = 2500$ gal/min (0.016 m^3/s). And, $H_1/H_2 = (D_1/D_2)^2$; $200/H_2 = (12/10)^2$; $H^2 = 138.8$ ft (42.3 m). Also, $P_1/P_2 = (D_1/D_2)^3$; $175/P_2 = (12/10)^3$; $P_2 = 101.2$ bhp.

Related Calculations: Use the similarity laws to extend or change the data obtained from centrifugal-pump characteristic curves. These laws are also useful in field calculations when the pump head, capacity, speed, or impeller diameter is changed.

The similarity laws are most accurate when the efficiency of the pump remains nearly constant. Results obtained when the laws are applied to a pump having a constant impeller diameter are somewhat more accurate than for a pump at constant speed with a changed impeller diameter. The latter laws are more accurate when applied to pumps having a low specific speed.

If the similarity laws are applied to a pump whose impeller diameter is increased, be certain to consider the effect of the higher velocity in the pump suction line. Use the similarity laws for any liquid whose viscosity remains constant during passage through the pump. However, the accuracy of the similarity laws decreases as the liquid viscosity increases.

6-18 Similarity or Affinity Laws in Centrifugal-Pump Selection

A test-model pump delivers, at its best efficiency point, 500 gal/min (0.03 m^3/s) at a 350-ft (107-m) head with a required net positive suction head (NPSH) of 10 ft (3.05 m) and a power input of 55 hp (41 kW) at 3500 r/min, when using a 10.5-in-diameter impeller. Determine the performance of the model at 1750 r/min. What is the performance of a full-scale prototype pump with a 20-in impeller operating at 1170 r/min? What are the specific speeds and the suction specific speeds of the test-model and prototype pumps?

Calculation Procedure:

1. Compute the pump performance at the new speed.

The similarity or affinity laws can be stated in general terms, with subscripts p and m for prototype and model, respectively, as $Q_p = K_d^3 K_n Q_m$; $H_p = K_d^2 K_n^2 H_m$; $\text{NPSH}_p = K_2^2 K_n^2 \text{NPSH}_m$; $P_p = K_d^5 K_n^5 P_m$, where K_d = size factor = prototype dimension/model dimension. The usual dimension used for the size factor is the impeller diameter. Both dimensions should be in the same units of measure. Also, K_n = prototype speed, r/min/model speed, r/min. Other symbols are the same as in the previous example.

When the model speed is reduced from 3500 to 1750 r/min, the pump dimensions remain the same and $K_d = 1.0$; $K_n = 1750/3500 = 0.5$. Then $Q = (1.0)(0.5)(500) =$ 250 r/min; $H = (1.0)^2(0.5)^2(350) = 87.5$ ft (26.7 m); $\text{NPSH} = (1.0)^2(0.5)^2(10) = 2.5$ ft (0.76 m); $P = (1.0)^5(0.5)^3(55) = 6.9$ hp. In this computation, the subscripts were omitted from the equation because the same pump, the test model, was being considered.

2. Compute performance of the prototype pump.

First, K_d and K_n must be found. $K_d = 20/10.5 = 1.905$; $K_n = 1170/3500 = 0.335$. Then, $Q_p = (1.905)^3(0.335)(500) = 1158$ gal/min (0.073 m^3/s); $H_p = (1.905)^2(0.335)^2(350) = 142.5$ ft (43.4 m); $\text{NPSH}_p = (1.905)^2(0.335)^2(10) = 4.06$ ft; $P_p = (1.905)^5(0.335)^3(55) = 51.8$ hp.

3. Compute the specific speed and suction specific speed.

The specific speed or, as Horwitz[1] says, "more correctly, discharge specific speed," $N_S = N(Q)^{0.5}/(H)^{0.75}$, while the suction specific speed $S = N(Q)^{0.5}/\text{NPSH}^{0.75}$, where all values are taken at the best efficiency point of the pump.

[1]R. P. Horwitz, "Affinity Laws and Specific Speed Can Simplify Centrifugal Pump Selection," *Power,* November 1964.

For the model, $N_S = 3500(500)^{0.5}/350^{0.75} = 965$; $S = 3500(500)^{0.5}/10^{0.75} = 13{,}900$. For the prototype, $N_S = 1170(1158)^{0.5}/142.5^{0.75} = 965$; $S = 1170(1156)^{0.5}/4.06^{0.75} = 13{,}900$. The specific speed and suction specific speed of the model and prototype are equal because these units are geometrically similar or homologous pumps and both speeds are mathematically derived from the similarity laws.

Related Calculations: Use the procedure given here for any type of centrifugal pump where the similarity laws apply. When the term "model" is used, it can apply to a production test pump or to a standard unit ready for installation. The procedure presented here is the work of R. P. Horwitz, as reported in *Power* magazine.[1]

[1]R. P. Horwitz, "Affinity Laws and Specific Speed Can Simplify Centrifugal Pump Selection," *Power*, November 1964.

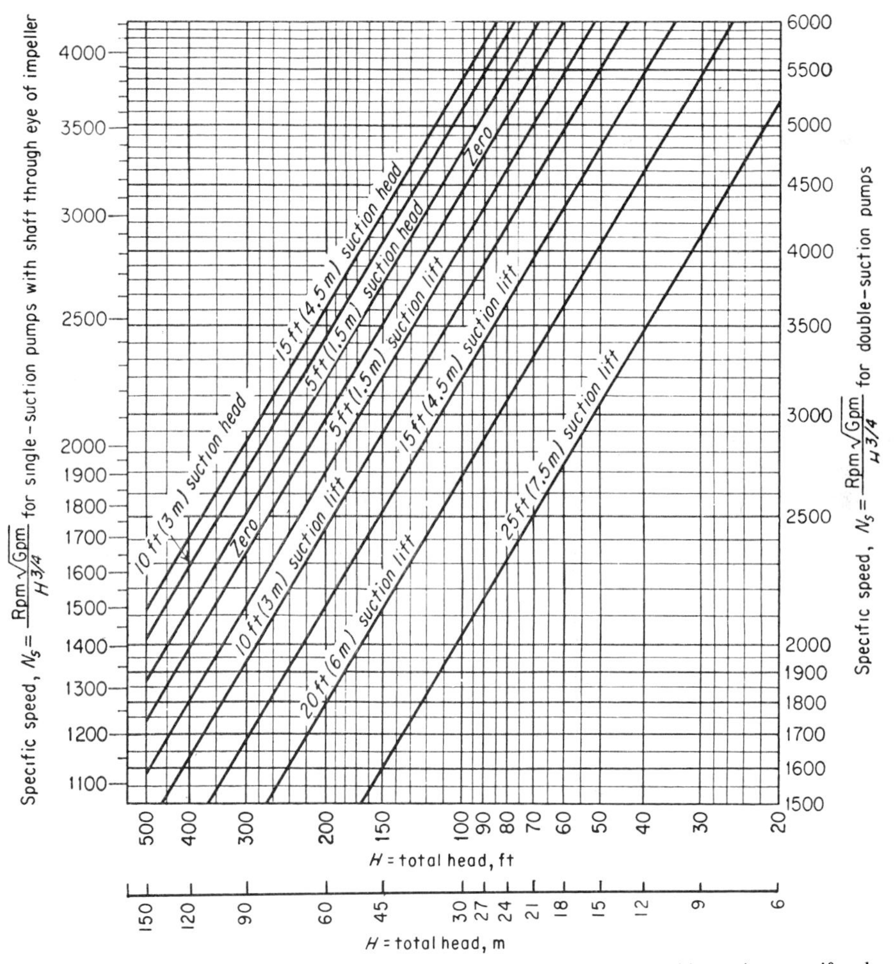

FIG. 6-10 Upper limits of specific speeds of single-stage single- and double-suction centrifugal pumps handling clear water at 85°F (29°C) at sea level. *(Hydraulic Institute.)*

6-19 Specific-Speed Considerations in Centrifugal-Pump Selection

What is the upper limit of specific speed and capacity of a 1750-r/min single-stage double-suction centrifugal pump having a shaft that passes through the impeller eye if it handles clear water at 85°F (302 K) at sea level at a total head of 280 ft with a 10-ft suction lift? What is the efficiency of the pump and its approximate impeller shape?

Calculation Procedure:

1. Determine the upper limit of specific speed.

Use the Hydraulic Institute upper-specific-speed curve, Fig. 6-10, for centrifugal pumps or a similar curve, Fig. 6-11, for mixed- and axial-flow pumps. Enter Fig. 6-10 at the bottom at 280-ft total head and project vertically upward until the 10-ft suction-lift curve is intersected. From here, project horizontally to the right to read the specific speed N_S = 2000. Figure 6-11 is used in a similar manner.

2. Compute the maximum pump capacity.

For any centrifugal, mixed- or axial-flow pump, $N_S = (\text{gal/min})^{0.5}(\text{r/min})/H_t^{0.75}$, where H_t = total head on the pump, in ft of liquid. Solving for the maximum capacity, gal/min $= [N_S H_t^{0.75}/(\text{r/min})]^2 = (2000 \times 280^{0.75}/1750)^2 = 6040$ gal/min.

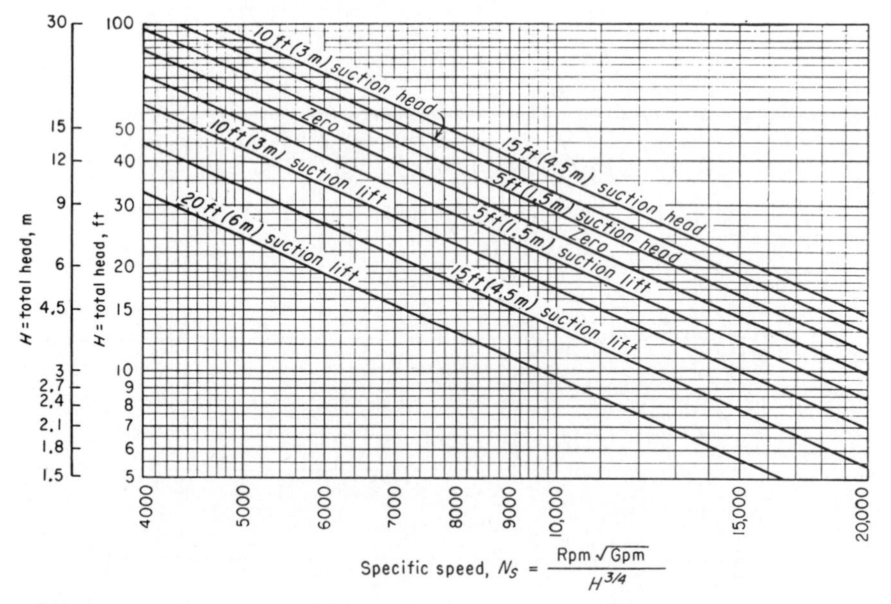

FIG. 6-11 Upper limits of specific speeds of single-suction mixed-flow and axial-flow pumps. *(Hydraulic Institute.)*

3. *Determine the pump efficiency and impeller shape.*

Figure 6-12 shows the general relation between impeller shape, specific speed, pump capacity, efficiency, and characteristic curves. At N_S = 2000, efficiency = 87 percent. The impeller, as shown in Fig. 6-12, is moderately short and has a relatively large discharge area. A cross section of the impeller appears directly under the N_S = 2000 ordinate.

Related Calculations: Use the method given here for any type of pump whose variables are included in the Hydraulic Institute curves (Figs. 6-10 and 6-11) and in similar curves available from the same source. *Operating specific speed,* computed as above, is sometimes plotted on the performance curve of a centrifugal pump so that the characteristics of the unit can be better understood. *Type specific speed* is the operating specific speed giving maximum efficiency for a given pump and is a number used to identify a pump. Specific speed is important in cavitation and suction-lift studies. The Hydraulic Institute curves (Figs. 6-10 and 6-11) give upper limits of speed, head, capacity, and suction lift for cav-

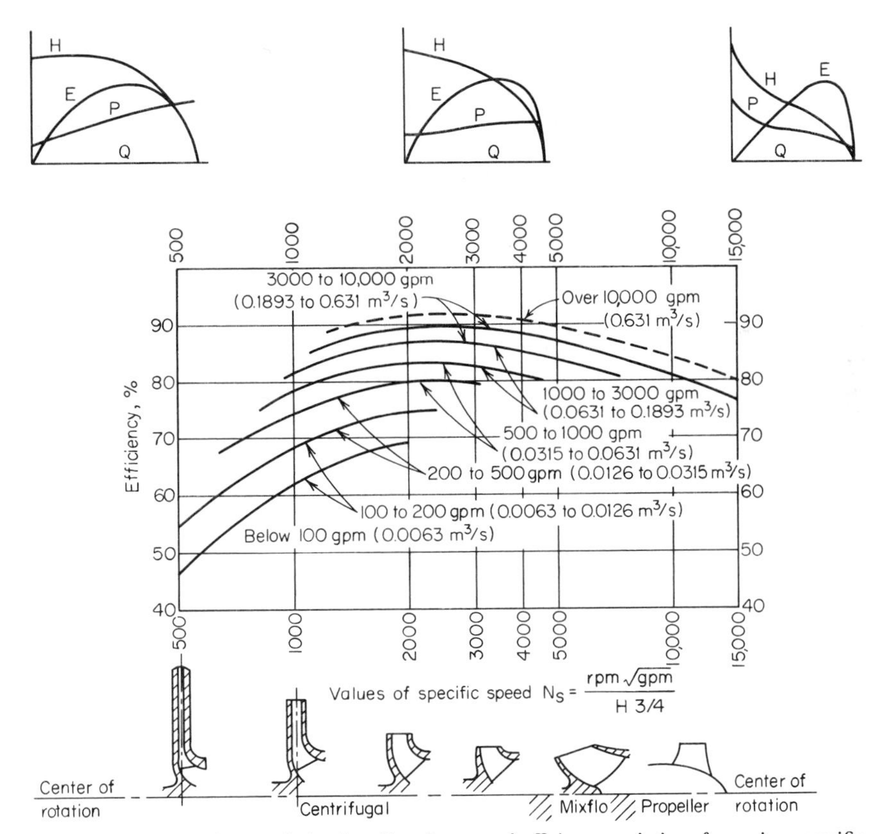

FIG. 6-12 Approximate relative impeller shapes and efficiency variations for various specific speeds of centrifugal pumps. *(Worthington Corp.)*

itation-free operation. When making actual pump analyses, be certain to use the curves (Figs. 6-10 and 6-11 herewith) in the latest edition of the *Standards of the Hydraulic Institute*.

6-20 Selecting the Best Operating Speed for a Centrifugal Pump

A single-suction centrifugal pump is driven by a 60-Hz ac motor. The pump delivers 10,000 gal/min (0.63 m^3/s) of water at a 100-ft (30.5-m) head. The available net positive suction head is 32 ft (9.75 m) of water. What is the best operating speed for this pump if the pump operates at its best efficiency point?

Calculation Procedure:

1. Determine the specific speed and suction specific speed.

Alternating-current motors can operate at a variety of speeds, depending on the number of poles. Assume that the motor driving this pump might operate at 870, 1160, 1750, or 3500 r/min. Compute the specific speed $N_S = N(Q)^{0.5}/H^{0.75} = N(10{,}000)^{0.5}/100^{0.75} = 3.14N$, and the suction specific speed $S = N(Q)^{0.5}/\text{NPSH}^{0.75} = N(10{,}000)^{0.5}/32^{0.75} = 7.43N$ for each of the assumed speeds, and tabulate the results as follows:

Operating speed, r/min	Required specific speed	Required suction specific speed
870	2,740	6,460
1,160	3,640	8,620
1,750	5,500	13,000
3,500	11,000	26,000

2. Choose the best speed for the pump.

Analyze the specific speed and suction specific speed at each of the various operating speeds using the data in Tables 6-20 and 6-21. These tables show that at 870 and 1160 r/min, the suction specific-speed rating is poor. At 1750 r/min, the suction specific-speed rating is excellent, and a turbine or mixed-flow type of pump will be suitable. Operation at 3500 r/min is unfeasible because a suction specific speed of 26,000 is beyond the range of conventional pumps.

TABLE 6-20 Pump Types Listed by Specific Speed

Specific speed range	Type of pump
Below 2,000	Volute, diffuser
2,000–5,000	Turbine
4,000–10,000	Mixed-flow
9,000–15,000	Axial-flow

Source: Peerless Pump Division, FMC Corporation.

TABLE 6-21 Suction Specific-Speed Ratings

Single-suction pump	Double-suction pump	Rating
Above 11,000	Above 14,000	Excellent
9,000–11,000	11,000–14,000	Good
7,000–9,000	9,000–11,000	Average
5,000–7,000	7,000–9,000	Poor
Below 5,000	Below 7,000	Very poor

Source: Peerless Pump Division, FMC Corporation.

Related Calculations: Use this procedure for any type of centrifugal pump handling water for plant services, cooling, process, fire protection, and similar requirements. This procedure is the work of R. P. Horwitz, Hydrodynamics Division, Peerless Pump, FMC Corporation, as reported in *Power* magazine.

6-21 Total Head on a Pump Handling Vapor-Free Liquid

Sketch three typical pump piping arrangements with static suction lift and submerged, free, and varying discharge head. Prepare similar sketches for the same pump with static suction head. Label the various heads. Compute the total head on each pump if the elevations are as shown in Fig. 6-13 and the pump discharges a maximum of 2000 gal/min (0.126 m^3/s) of water through 8-in schedule 40 pipe. What horsepower is required to drive the pump? A swing check valve is used on the pump suction line and a gate valve on the discharge line.

Calculation Procedure:

1. Sketch the possible piping arrangements.

Figure 6-13 shows the six possible piping arrangements for the stated conditions of the installation. Label the total static head—i.e., the *vertical* distance from the surface of the source of the liquid supply to the free surface of the liquid in the discharge receiver, or to the point of free discharge from the discharge pipe. When both the suction and discharge surfaces are open to the atmosphere, the total static head equals the vertical difference in elevation. Use the free-surface elevations that cause the maximum suction lift and discharge head—i.e., the *lowest* possible level in the supply tank and the *highest* possible level in the discharge tank or pipe. When the supply source is *below* the pump centerline, the vertical distance is called the "static suction lift." With the supply *above* the pump centerline, the vertical distance is called "static suction head." With variable static suction head, use the lowest liquid level in the supply tank when computing total static head. Label the diagrams as shown in Fig. 6-13.

2. Compute the total static head on the pump.

The total static head, in feet, is H_{ts} = static suction lift, in feet, h_{sl} + static discharge head, in feet, h_{sd}, where the pump has a suction lift, *s* in Fig. 6-13*a*, *b*, and *c*. In these

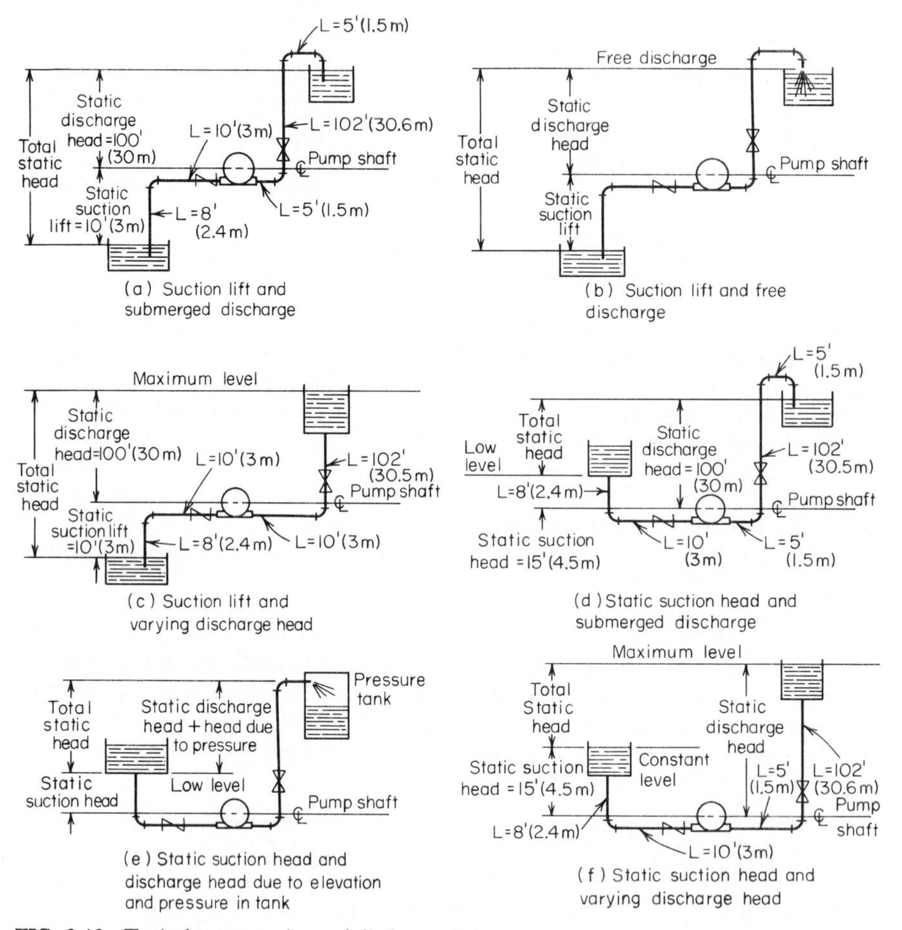

(a) Suction lift and submerged discharge
(b) Suction lift and free discharge
(c) Suction lift and varying discharge head
(d) Static suction head and submerged discharge
(e) Static suction head and discharge head due to elevation and pressure in tank
(f) Static suction head and varying discharge head

FIG. 6-13 Typical pump suction and discharge piping arrangements.

installations, H_{ts} = 10 + 100 = 110 ft. Note that the static discharge head is computed between the pump centerline and the water level with an underwater discharge (Fig. 6-13*a*), to the pipe outlet with a free discharge (Fig. 6-13*b*), and to the maximum water level in the discharge tank (Fig. 6-13*c*). When a pump is discharging into a closed compression tank, the total discharge head equals the static discharge head plus the head equivalent, in feet of liquid, of the internal pressure in the tank, or 2.31 × tank pressure, in lb/in^2.

Where the pump has a static suction head, as in Fig. 6-13*d*, *e*, and *f*, the total static head, in feet, is $H_{ts} = h_{sd}$ − static suction head, in feet, h_{sh}. In these installations, H_t = 100 − 15 = 85 ft.

The total static head, as computed above, refers to the head on the pump without liquid flow. To determine the total head on the pump, the friction losses in the piping system during liquid flow must also be determined.

3. *Compute the piping friction losses.*

Mark the length of each piece of straight pipe on the piping drawing. Thus, in Fig. 6-13*a*, the total length of straight pipe L_t, in feet, is $8 + 10 + 5 + 102 + 5 = 130$ ft, starting at the suction tank and adding each length until the discharge tank is reached. To the total length of straight pipe must be added the *equivalent* length of the pipe fittings. In Fig. 6-13*a* there are four long-radius elbows, one swing check valve, and one globe valve. In addition, there is a minor head loss at the pipe inlet and at the pipe outlet.

The equivalent length of one 8-in-long-radius elbow is 14 ft of pipe, from Table 6-22. Since the pipe contains four elbows, the total equivalent length is $4(14) = 56$ ft of straight pipe. The open gate valve has an equivalent resistance of 4.5 ft, and the open swing check valve has an equivalent resistance of 53 ft.

The entrance loss h_e, in feet, assuming a basket-type strainer is used at the suction-pipe inlet, is $Kv^2/(2g)$, where K = a constant from Fig. 6-14; v = liquid velocity, in ft/s; and $g = 32.2$ ft/s^2. The exit loss occurs when the liquid passes through a sudden enlargement, as from a pipe to a tank. Where the area of the tank is large, causing a final velocity that is zero, $h_{ex} = v^2/2g$.

The velocity v, in feet per second, in a pipe is (gal/min)/($2.448d^2$). For this pipe, $v = 2000/[2.448(7.98)^2] = 12.82$ ft/s. Then, $h_e = 0.74(12.82)^2/[2(32.2)] = 1.89$ ft, and $h_{ex} = (12.82)^2/[2(32.2)] = 2.56$ ft (0.78 m). Hence the total length of the piping system in Fig. 6-13*a* is $130 + 56 + 4.5 + 53 + 1.89 + 2.56 = 247.95$ ft (75.6 m), say 248 ft (75.6 m).

Use a suitable head-loss equation, or Table 6-23, to compute the head loss for the pipe and fittings. Enter Table 6-23 at an 8-in (203.2-mm) pipe size and project horizontally across to 2000 gal/min (126.2 L/s) and read the head loss as 5.86 ft of water per 100 ft (1.8 m/30.5 m) of pipe.

The total length of pipe and fittings computed above is 288 ft (87.8 m). Then, total friction-head loss with a 2000-gal/min (126.2 L/s) flow is $H_f = (5.86)(248/100) = 14.53$ ft (4.5 m).

4. *Compute the total head on the pump.*

The total head on the pump $H_t = H_{ts} + H_f$. For the pump in Fig. 6-13*a*, $H_t = 110 + 14.53 = 124.53$ ft (38.0 m), say 125 ft (38.0 m). The total head on the pump in Fig. 6-13*b* and *c* would be the same. Some engineers term the total head on a pump the "total dyamic head" to distinguish between static head (no-flow vertical head) and operating head (rated flow through the pump).

The total head on the pumps in Fig. 6-13*d*, *c*, and *f* is computed in the same way as described above, except that the total static head is less because the pump has a static suction head—that is, the elevation of the liquid on the suction side reduces the total distance through which the pump must discharge liquid; thus the total static head is less. The static suction head is *subtracted* from the static discharge head to determine the total static head on the pump.

5. *Compute the horsepower required to drive the pump.*

The brake horsepower input to a pump equals (gal/min)(H_t)(s)/3960e, where s = specific gravity of the liquid handled, and e = hydraulic efficiency of the pump, expressed as a decimal. The usual hydraulic efficiency of a centrifugal pump is 60 to 80 percent;

TABLE 6-22 Resistance of Fittings and Valves (length of straight pipe giving equivalent resistance)

Pipe size		Standard ell		Medium-radius ell		Long-radius ell		45° Ell		Tee		Gate valve, open		Globe valve, open		Swing check, open	
in	mm	ft	m	ft	m	ft	m	ft	m	ft	m	ft	m	ft	m	ft	m
6	152.4	16	4.9	14	4.3	11	3.4	7.7	2.3	33	10.1	3.5	1.1	160	48.8	40	12.2
8	203.2	21	6.4	18	5.5	14	4.3	10	3.0	43	13.1	4.5	1.4	220	67.0	53	16.2
10	254.0	26	7.9	22	6.7	17	5.2	13	3.9	56	17.1	5.7	1.7	290	88.4	67	20.4
12	304.8	32	9.8	26	7.9	20	6.1	15	4.6	66	20.1	6.7	2.0	340	103.6	80	24.4

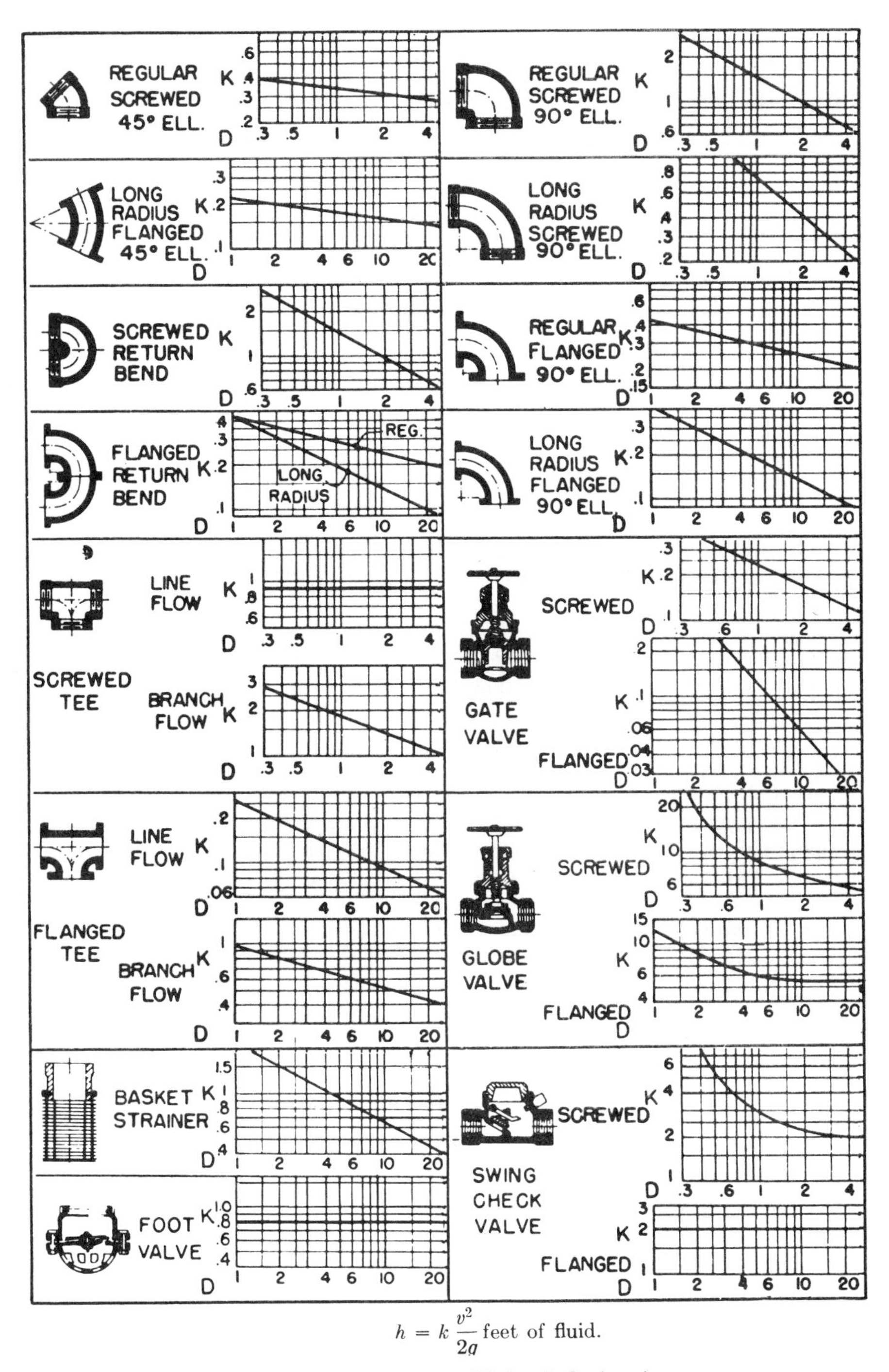

$$h = k\frac{v^2}{2g} \text{ feet of fluid.}$$

FIG. 6-14 Resistance coefficients of pipe fittings. *(Hydraulic Institute.)*

TABLE 6-23 Pipe Friction Loss for Water (wrought-iron or steel schedule 40 pipe in good condition)

Diameter		Flow		Velocity		Velocity head		Friction loss/100 ft (30.5 m) pipe	
in	mm	gal/min	L/s	ft/s	m/s	ft water	m water	ft water	m water
6	152.4	1000	63.1	11.1	3.4	1.92	0.59	6.17	1.88
6	152.4	2000	126.2	22.2	6.8	7.67	2.3	23.8	7.25
6	152.4	4000	252.4	44.4	13.5	30.7	9.4	93.1	28.4
8	203.2	1000	63.1	6.41	1.9	0.639	0.195	1.56	0.475
8	203.2	2000	126.2	12.8	3.9	2.56	0.78	5.86	1.786
8	203.2	4000	252.4	25.7	7.8	10.2	3.1	22.6	6.888
10	254.0	1000	63.1	3.93	1.2	0.240	0.07	0.497	0.151
10	254.0	3000	189.3	11.8	3.6	2.16	0.658	4.00	1.219
10	254.0	5000	315.5	19.6	5.9	5.99	1.82	10.8	3.292

reciprocating pumps, 55 to 90 percent; rotary pumps, 50 to 90 percent. For each class of pump, the hydraulic efficiency decreases as the liquid viscosity increases.

Assume that the hydraulic efficiency of the pump in this system is 70 percent and the specific gravity of the liquid handled is 1.0. Then, input brake horsepower equals (2000)(125)(1.0)/[3960(0.70)] = 90.2 hp (67.4 kW).

The theoretical or *hydraulic horsepower* equals (gal/min)(H_t)(s)/3960 = (2000)(125)(1.0)/3900 = 64.1 hp (47.8 kW).

Related Calculations: Use this procedure for any liquid–water, oil, chemical, sludge, etc.—whose specific gravity is known. When liquids other than water are being pumped, the specific gravity and viscosity of the liquid must be taken into consideration. The procedure given here can be used for any class of pump—centrifugal, rotary, or reciprocating.

Note that Fig. 6-14 can be used to determine the equivalent length of a variety of pipe fittings. To use Fig. 6-14, simply substitute the appropriate K value in the relation $h = Kv^2/2g$, where h = equivalent length of straight pipe; other symbols as before.

6-22 Pump Selection for Any Pumping System

Give a step-by-step procedure for choosing the class, type, capacity, drive, and materials for a pump that will be used in an industrial pumping system.

Calculation Procedure:

1. Sketch the proposed piping layout.

Use a single-line diagram (Fig. 6-15) of the piping system. Base the sketch on the actual job conditions. Show all the piping, fittings, valves, equipment, and other units in the system. Mark the *actual* and *equivalent* pipe length (see the previous example) on the sketch. Be certain to include all vertical lifts, sharp bends, sudden enlargements, storage tanks, and similar equipment in the proposed system.

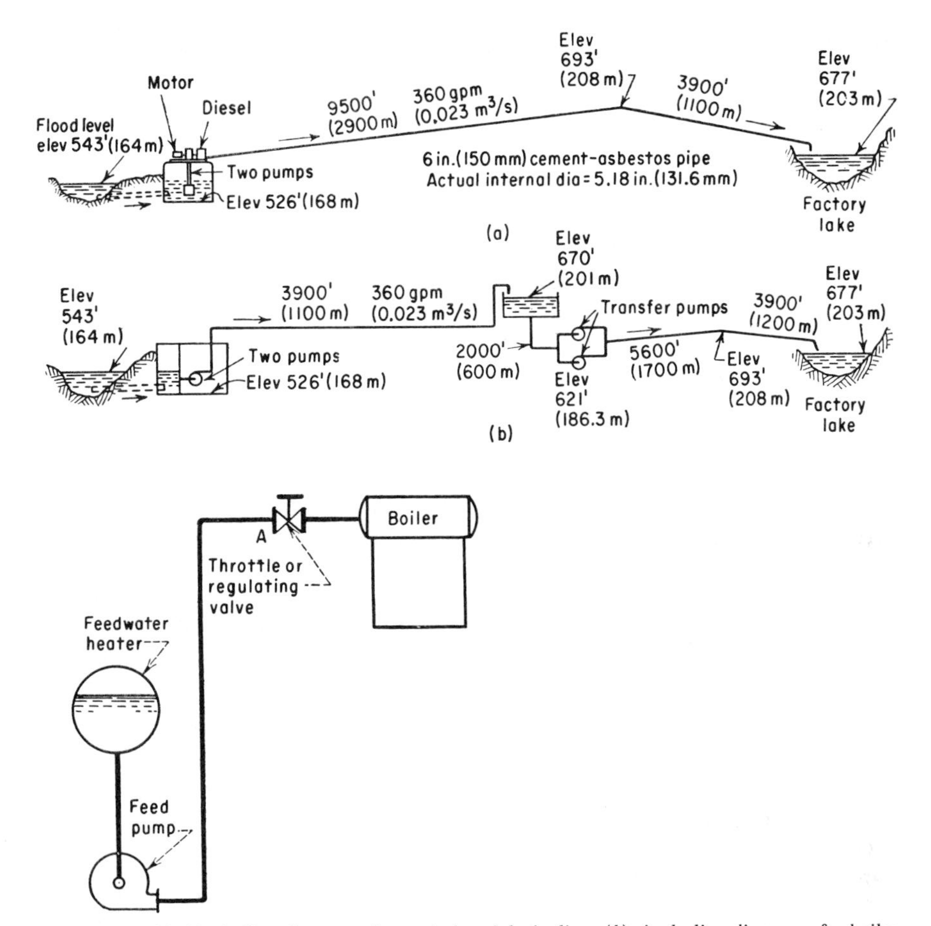

FIG. 6-15 (*a*) Single-line diagrams for an industrial pipeline; (*b*) single-line diagram of a boiler feed system. *(Worthington Corp.)*

2. *Determine the required capacity of the pump.*

The required capacity is the flow rate that must be handled in gal/min, million gal/day, ft^3/s, gal/h, bbl/day, lb/h, acre-ft/day, mil/h, or some similar measure. Obtain the required flow rate from the process conditions—for example, boiler feed rate, cooling-water flow rate, chemical feed rate, etc. The required flow rate for any process unit is usually given by the manufacturer.

Once the required flow rate is determined, apply a suitable factor of safety. The value of this factor of safety can vary from a low of 5 percent of the required flow to a high of 50 percent or more, depending on the application. Typical safety factors are in the 10 percent range. With flow rates up to 1000 gal/min, and in the selection of process pumps, it is common practice to round off a computed required flow rate to the next highest round-number capacity. Thus, with a required flow rate of 450 gal/min and a 10 percent safety factor, the flow of 450 + 0.10(450) = 495 gal/min would be rounded off to 500 gal/min *before* selecting the pump. A pump of 500 gal/min, or larger, capacity would be selected.

3. *Compute the total head on the pump.*

Use the steps given in the previous example to compute the total head on the pump. Express the result in feet of water—this is the most common way of expressing the head

Summary of Essential Data Required in Selection of Centrifugal Pumps

1. Number of Units Required

2. Nature of the Liquid to Be Pumped
   Is the liquid:
   a. Fresh or salt water, acid or alkali, oil, gasoline, slurry, or paper stock?
   b. Cold or hot and if hot, at what temperature? What is the vapor pressure of the liquid at the pumping temperature?
   c. What is its specific gravity?
   d. Is it viscous or nonviscous?
   e. Clear and free from suspended foreign matter or dirty and gritty? If the latter, what is the size and nature of the solids, and are they abrasive? If the liquid is of a pulpy nature, what is the consistency expressed either in percentage or in lb per cu ft of liquid? What is the suspended material?
   f. What is the chemical analysis, pH value, etc.? What are the expected variations of this analysis? If corrosive, what has been the past experience, both with successful materials and with unsatisfactory materials?

3. Capacity
   What is the required capacity as well as the minimum and maximum amount of liquid the pump will ever be called upon to deliver?

4. Suction Conditions
   Is there:
   a. A suction lift?
   b. Or a suction head?
   c. What are the length and diameter of the suction pipe?

5. Discharge Conditions
   a. What is the static head? Is it constant or variable?
   b. What is the friction head?
   c. What is the maximum discharge pressure against which the pump must deliver the liquid?

6. Total Head
   Variations in items 4 and 5 will cause variations in the total head.

7. Is the service continuous or intermittent?

8. Is the pump to be installed in a horizontal or vertical position? If the latter,
   a. In a wet pit?
   b. In a dry pit?

9. What type of power is available to drive the pump and what are the characteristics of this power?

10. What space, weight, or transportation limitations are involved?

11. Location of installation
    a. Geographical location
    b. Elevation above sea level
    c. Indoor or outdoor installation
    d. Range of ambient temperatures

12. Are there any special requirements or marked preferences with respect to the design, construction, or performance of the pump?

FIG. 6-16 Typical selection chart for centrifugal pumps. *(Worthington Corp.)*

on a pump. Be certain to use the exact specific gravity of the liquid handled when expressing the head in feet of water. A specific gravity less than 1.00 *reduces* the total head when expressed in feet of water, whereas a specific gravity greater than 1.00 *increases* the total head when expressed in feet of water. Note that variations in the suction and discharge conditions can affect the total head on the pump.

4. Analyze the liquid conditions.

Obtain complete data on the liquid pumped. These data should include the name and chemical formula of the liquid, maximum and minimum pumping temperature, corresponding vapor pressure at these temperatures, specific gravity, viscosity at the pumping temperature, pH, flash point, ignition temperature, unusual characteristics (such as tendency to foam, curd, crystallize, become gelatinous or tacky), solids content, type of solids and their size, and variation in the chemical analysis of the liquid.

Enter the liquid conditions on a pump-selection form such as that in Fig. 6-16. Such forms are available from many pump manufacturers or can be prepared to meet special job conditions.

5. Select the class and type of pump.

Three *classes* of pumps are used today—centrifugal, rotary, and reciprocating (Fig. 6-17). Note that these terms apply only to the mechanics of moving the liquid—not to the service for which the pump was designed. Each class of pump is further subdivided into a number of *types* (Fig. 6-17).

Use Table 6-24 as a general guide to the class and type of pump to be used. For example, when a large capacity at moderate pressure is required. Table 6-24 shows that a centrifugal pump would probably be best. Table 6-24 also shows the typical characteristics of various classes and types of pumps used in industrial process work.

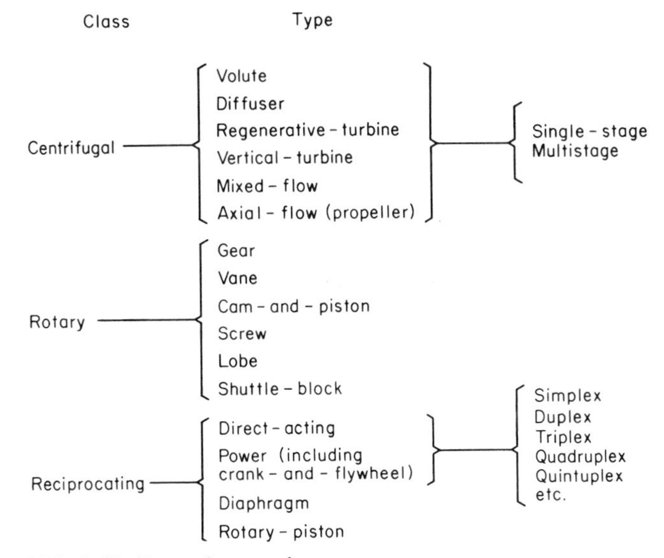

FIG. 6-17 Pump classes and types.

TABLE 6-24 Characteristics of Modern Pumps

	Centrifugal		Rotary	Reciprocating		
	Volute and diffuser	Axial flow	Screw and gear	Direct acting steam	Double acting power	Triplex
Discharge flow	Steady	Steady	Steady	Pulsating	Pulsating	Pulsating
Usual maximum suction lift, ft (m)	15 (4.6)	15 (4.6)	22 (6.7)	22 (6.7)	22 (6.7)	22 (6.7)
Liquids handled	Clean, clear; dirty, abrasive; liquids with high solids content		Viscous, nonabrasive	Clean and clear		
Discharge pressure range	Low to high		Medium	Low to highest produced		
Usual capacity range	Small to largest available		Small to medium	Relatively small		
How increased head affects:						
Capacity	Decrease		None	Decrease	None	None
Power input	Depends on specific speed		Increase	Increase	Increase	Increase
How decreased head affects:						
Capacity	Increase		None	Small increase	None	None
Power input	Depends on specific speed		Decrease	Decrease	Decrease	Decrease

Consider the liquid properties when choosing the class and type of pump, because exceptionally severe conditions may rule out one or another class of pump at the start. Thus, screw- and gear-type rotary pumps are suitable for handling viscous, nonabrasive liquid (Table 6-24). When an abrasive liquid must be handled, either another class of pump or another type of rotary pump must be used.

Also consider all the operating factors related to the particular pump. These factors include the type of service (continuous or intermittent), operating-speed preferences, future load expected and its effect on pump head and capacity, maintenance facilities available, possibility of parallel or series hookup, and other conditions peculiar to a given job.

Once the class and type of pump are selected, consult a rating table (Table 6-25) or rating chart (Fig. 6-18) to determine if a suitable pump is available from the manufacturer whose unit will be used. When the hydraulic requirements fall between two standard pump models, it is usual practice to choose the next larger size of pump, unless there is some reason why an exact head and capacity are required for the unit. When one manufacturer does not have the desired unit, refer to the engineering data of other man-

TABLE 6-25 Typical Centrifugal-Pump Rating Table

Size		Total head			
gal/min	L/s	20 ft, r/min—hp	6.1 m, r/min—kW	25 ft, r/min—hp	7.6 m, r/min—kW
3 CL:					
200	12.6	910—1.3	910—0.97	1010—1.6	1010—1.19
300	18.9	1000—1.9	1000—1.41	1100—2.4	1100—1.79
400	25.2	1200—3.1	1200—2.31	1230—3.7	1230—2.76
500	31.5	—	—	—	—
4 C:					
400	25.2	940—2.4	940—1.79	1040—3	1040—2.24
600	37.9	1080—4	1080—2.98	1170—4.6	1170—3.43
800	50.5	—	—	—	—

Example: 1080—4 indicates pump speed is 1080 r/min; actual input required to operate the pump is 4 hp (2.98 kW).

Source: Condensed from data of Goulds Pumps, Inc.; SI values added by *Handbook* editor.

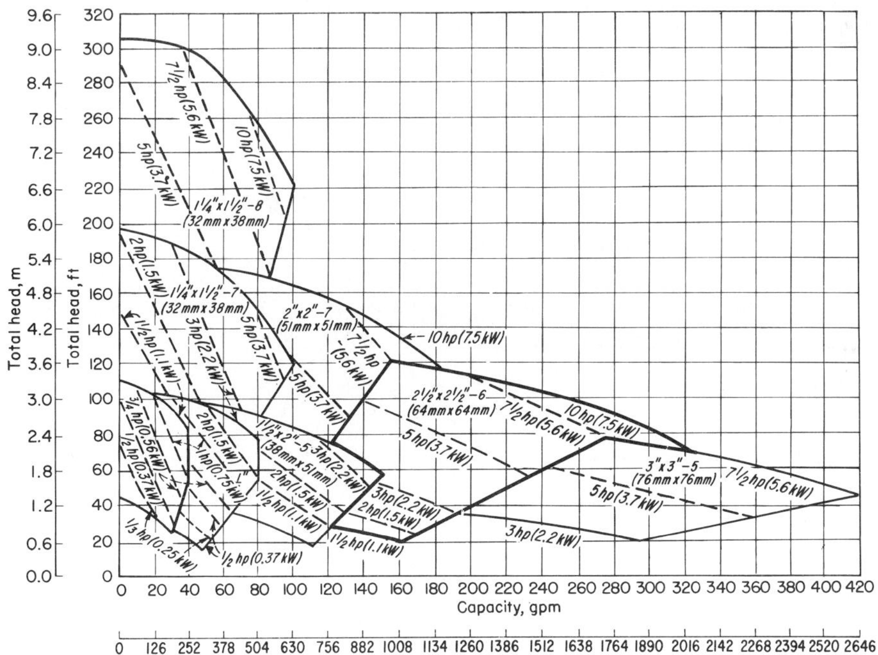

FIG. 6-18 Composite rating chart for a typical centrifugal pump. *(Goulds Pumps, Inc.)*

ufacturers. Also keep in mind that some pumps are custom-built for a given job when precise head and capacity requirements must be met.

Other pump data included in manufacturer's engineering information include characteristic curves for various diameter impellers in the same casing (Fig. 6-19) and variable-speed head-capacity curves for an impeller of given diameter (Fig. 6-20). Note that

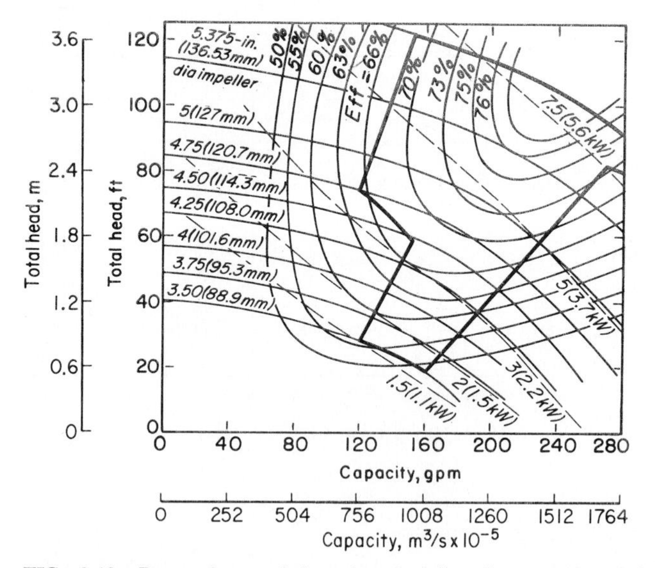

FIG. 6-19 Pump characteristics when impeller diameter is varied within the same casing.

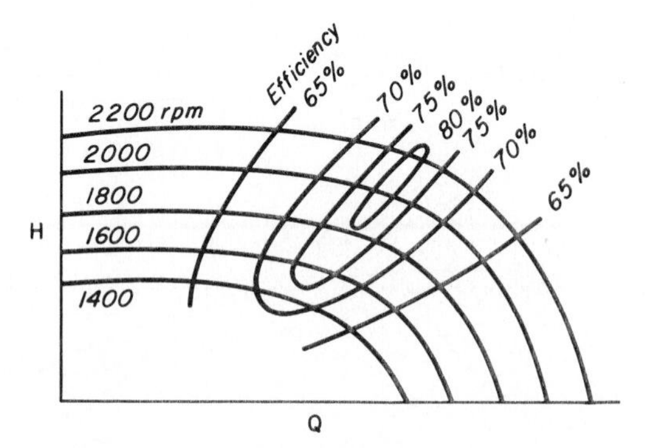

FIG. 6-20 Variable-speed head-capacity curves for a centrifugal pump.

the required power input is given in Figs. 6-18 and 6-19 and may also be given in Fig. 6-20. Use of Table 6-25 is explained in the table.

Performance data for rotary pumps is given in several forms. Figure 6-21 shows a typical plot of the head and capacity ranges of different types of rotary pumps. Reciprocating-pump capacity data are often tabulated, as in Table 6-26.

6. Evaluate the pump chosen for the installation.

Check the specific speed of a centrifugal pump using the method given in Example 6-20. Once the specific speed is known, the impeller type and approximate operating efficiency can be found from Fig. 6-12.

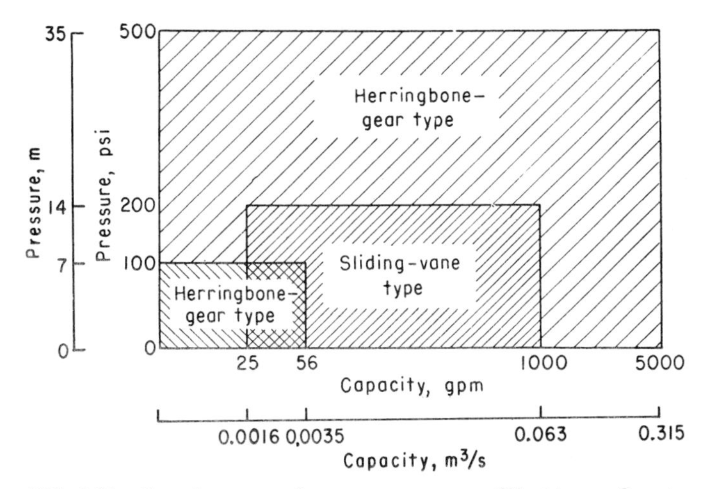

FIG. 6-21 Capacity ranges of some rotary pumps. *(Worthington Corp.)*

TABLE 6-26 Capacities of Typical Horizontal Duplex Plunger Pumps

Size		Cold-water pressure service			
				Piston speed	
in	cm	gal/min	L/s	ft/min	m/min
6 × 3½ × 6	15.2 × 8.9 × 15.2	60	3.8	60	18.3
7½ × 4½ × 10	19.1 × 11.4 × 25.4	124	7.8	75	22.9
9 × 5 × 10	22.9 × 12.7 × 25.4	153	9.7	75	22.9
10 × 6 × 12	25.4 × 15.2 × 30.5	235	14.8	80	24.4
12 × 7 × 12	30.5 × 17.8 × 30.5	320	20.2	80	24.4

Size		Boiler-feed service					
				Boiler		Piston speed	
in	cm	gal/min	L/s	hp	kW	ft/min	m/min
6 × 3½ × 6	15.2 × 8.9 × 15.2	36	2.3	475	354.4	36	10.9
7½ × 4½ × 10	19.1 × 11.4 × 25.4	74	4.7	975	727.4	45	13.7
9 × 5 × 10	22.9 × 12.7 × 25.4	92	5.8	1210	902.7	45	13.7
10 × 6 × 12	25.4 × 15.2 × 30.5	141	8.9	1860	1387.6	48	14.6
12 × 7 × 12	30.5 × 17.8 × 30.5	192	12.1	2530	1887.4	48	14.6

Source: Courtesy of Worthington Corporation.

Check the piping system, using the method of Example 6-20, to see if the available net positive suction head equals, or is greater than, the required net positive suction head of the pump.

Determine whether a vertical or horizontal pump is more desirable. From the standpoint of floor space occupied, required NPSH, priming, and flexibility in changing the pump use, vertical pumps may be preferable to horizontal designs in some installations. But where headroom, corrosion, abrasion, and ease of maintenance are important factors, horizontal pumps may be preferable.

As a general guide, single-suction centrifugal pumps handle up to 50 gal/min (0.0032 m^3/s) at total heads up to 50 ft (15 m); either single- or double-suction pumps are used for the flow rates to 1000 gal/min (0.063 m^3/s) and total heads to 300 ft (91 m); beyond these capacities and heads, double-suction or multistage pumps are generally used.

Mechanical seals have fully established themselves for all types of centrifugal pumps in a variety of services. Though more costly than packing, the mechanical seal reduces pump maintenance costs.

Related Calculations: Use the procedure given here to select any class of pump—centrifugal, rotary, or reciprocating—for any type of service—power plant, atomic energy, petroleum processing, chemical manufacture, paper mills, textile mills, rubber factories, food processing, water supply, sewage and sump service, air conditioning and heating, irrigation and flood control, mining and construction, marine services, industrial hydraulics, iron and steel manufacture, etc.

6-23 Analysis of Pump and System Characteristic Curves

Analyze a set of pump and system characteristic curves for the following conditions: friction losses without static head, friction losses with static head, pump without lift, system with little friction and much static head, system with gravity head, system with different pipe sizes, system with two discharge heads, system with diverted flow, and effect of pump wear on characteristic curve.

Calculation Procedure:

1. Plot the system-friction curve.

Without static head, the system friction curve passes through the origin (0,0) (Fig. 6-22), because when no head is developed by the pump, flow through the piping is zero. For most piping systems, the friction-head loss varies as the square of the liquid flow rate in the system. Hence, a system-friction curve, also called a "friction-head curve," is parabolic—the friction head increasing as the flow rate or capacity of the system increases. Draw the curve as shown in Fig. 6-22.

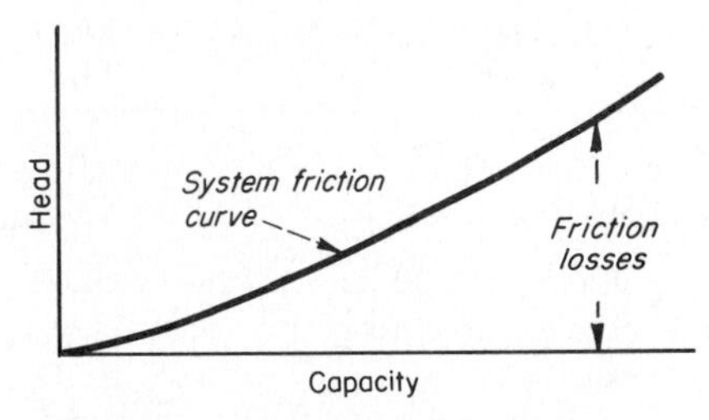

FIG. 6-22 Typical system-friction curve.

2. Plot the piping system and system-head curve.

Figure 6-23*a* shows a typical piping system with a pump operating against a static discharge head. Indicate the total static

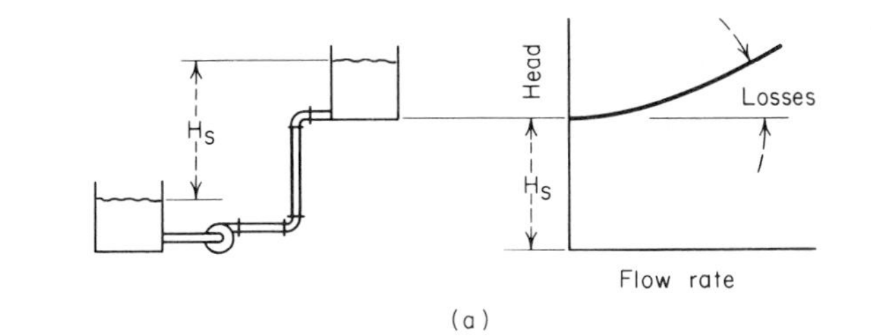

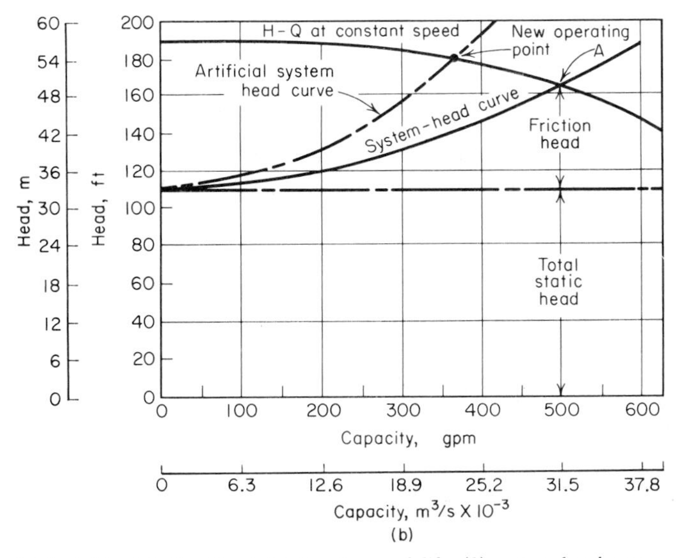

FIG. 6-23 (*a*) Significant friction loss and lift; (*b*) system-head curve superimposed on pump head-capacity curve. *(Peerless Pumps.)*

head (Fig. 6-23*b*) by a dashed line—in this installation $H_{ts} = 110$ ft. Since static head is a physical dimension, it does not vary with flow rate and is a constant for all flow rates. Draw the dashed line parallel to the abscissa (Fig. 6-23*b*).

From the point of no flow—zero capacity—plot the friction-head loss at various flow rates—100, 200, 300 gal/min, etc. Determine the friction-head loss by computing it as shown in Example 6-7. Draw a curve through the points obtained. This is called the "system-head curve."

Plot the pump head-capacity (*H-Q*) curve of the pump on Fig. 6-23*b*. The *H-Q* curve can be obtained from the pump manufacturer or from a tabulation of *H* and *Q* values for the pump being considered. The point of intersection, *A*, between the *H-Q* and system-head curves is the operating point of the pump.

Changing the resistance of a given piping system by partially closing a valve or making some other change in the friction alters the position of the system-head curve and pump operating point. Compute the frictional resistance as before and plot the artificial system-head curve as shown. Where this curve intersects the *H-Q* curve is the new operating point of the pump. System-head curves are valuable for analyzing the suitability of a given pump for a particular application.

3. Plot the no-lift system-head curve and compute the losses.

With no static head or lift, the system-head curve passes through the origin (0,0) (Fig. 6-24). For a flow of 900 gal/min (56.8 L/s), in this system, compute the friction loss as

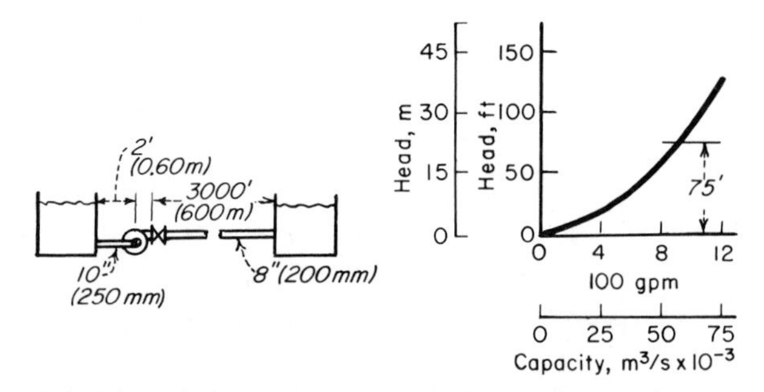

FIG. 6-24 No lift; all friction head. *(Peerless Pumps.)*

follows using the Hydraulic Institute—*Pipe Friction Manual* tables or the method of Example 6-7:

	ft	m
Entrance loss from tank into 10-in (254-mm) suction pipe, $0.5v^2/2g$	0.10	0.03
Friction loss in 2 ft (0.61 m) of suction pipe	0.02	0.01
Loss in 10-in (254-mm) 90° elbow at pump	0.20	0.06
Friction loss in 3000 ft (914.4 m) of 8-in (203.2-mm) discharge pipe	74.50	22.71
Loss in fully open 8-in (203.2-mm) gate valve	0.12	0.04
Exit loss from 8-in (203.2-mm) pipe into tank, $v^2/2g$	0.52	0.16
Total friction loss	75.46	23.01

Compute the friction loss at other flow rates in a similar manner and plot the system-head curve (Fig. 6-24). Note that if all losses in this system except the friction in the discharge pipe are ignored, the total head would not change appreciably. However, for the purposes of accuracy, all losses should always be computed.

4. Plot the low-friction, high-head system-head curve.

The system-head curve for the vertical pump installation in Fig. 6-25 starts at the total static head, 15 ft (4.6 m), and zero flow. Compute the friction head for 15,000 gal/min (946.4 L/s) as follows:

	ft	m
Friction in 20 ft (6.1 m) of 24-in pipe	0.40	0.12
Exit loss from 24-in pipe into tank, $v^2/2g$	1.60	0.49
Total friction loss	2.00	0.61

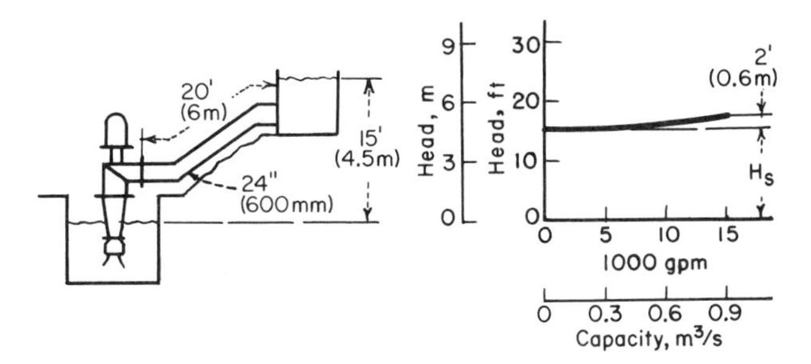

FIG. 6-25 Mostl', lift; little friction head. *(Peerless Pumps.)*

Hence, almost 90 percent of the total head of $15 + 2 = 17$ ft at 15,000-gal/min (946.4-L/s) flow is static head. But neglect of the pipe friction and exit losses could cause appreciable error during selection of a pump for the job.

5. *Plot the gravity-head system-head curve.*

In a system with gravity head (also called "negative lift"), fluid flow will continue until the system friction loss equals the available gravity head. In Fig. 6-26 the available gravity head is 50 ft (15.2 m). Flows up to 7200 gal/min (454.3 L/s) are obtained by gravity

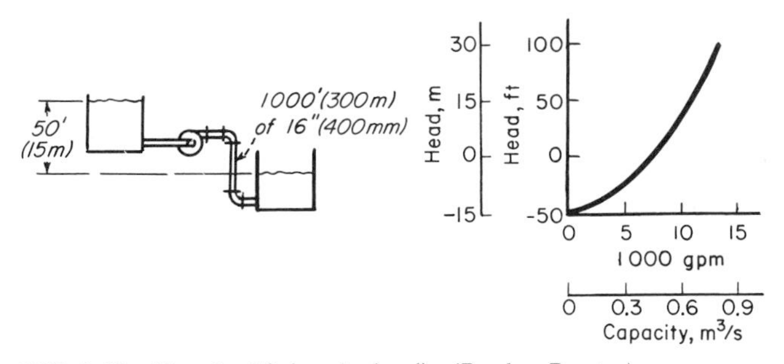

FIG. 6-26 Negative lift (gravity head). *(Peerless Pumps.)*

head alone. To obtain larger flow rates, a pump is needed to overcome the friction in the piping between the tanks. Compute the friction loss for several flow rates as follows:

	ft	m
At 5000 gal/min (315.5 L/s) friction loss in 1000 ft (305 m) of 16-in pipe	25	7.6
At 7200 gal/min (454.3 L/s) friction loss = available gravity head	50	15.2
At 13,000 gal/min (820.2 L/s) friction loss	150	45.7

Using these three flow rates, plot the system-head curve (Fig. 6-26).

6. Plot the system-head curves for different pipe sizes.

When different diameter pipes are used, the friction-loss-vs.-flow rate is plotted independently for the two pipe sizes. At a given flow rate, the total friction loss for the system is the sum of the loss for the two pipes. Thus the combined system-head curve represents the sum of the static head and the friction losses for all portions of the pipe.

Figure 6-27 shows a system with two different pipe sizes. Compute the friction losses as follows:

	ft	m
At 150 gal/min (9.5 L/s), friction loss in 200 ft (60.9 m) of 4-in (102-mm) pipe	5	1.52
At 150 gal/min (9.5 L/s), friction loss in 200 ft (60.9 m) of 3-in (76.2-mm) pipe	19	5.79
Total static head for 3- (76.2-mm) and 4-in (102-mm) pipes	10	3.05
Total head at 150-gal/min (9.5 L/s) flow	34	10.36

Compute the total head at other flow rates and plot the system-head curve as shown in Fig. 6-27.

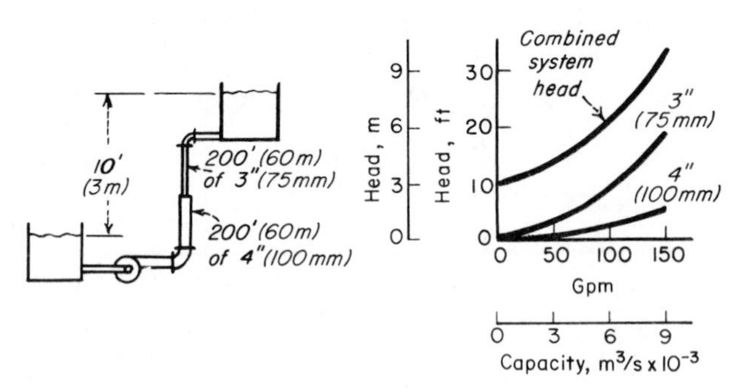

FIG. 6-27 System with two different pipe sizes. *(Peerless Pumps.)*

7. Plot the system-head curve for two discharge heads.

Figure 6-28 shows a typical pumping system having two different discharge heads. Plot separate system-head curves when the discharge heads are different. Add the flow rates for the two pipes at the same head to find points on the combined system-head curve (Fig. 6-28). Thus,

	ft	m
At 550 gal/min (34.7 L/s), friction loss in 1000 ft (305 m) of 8-in pipe	= 10	3.05
At 1150 gal/min (72.6 L/s), friction	= 38	11.6
At 1150 gal/min (72.6 L/s), friction + lift in pipe 1 = 38 + 50	= 88	26.8
At 550 gal/min (34.7 L/s), friction + lift in pipe 2 = 10 + 78	= 88	26.8

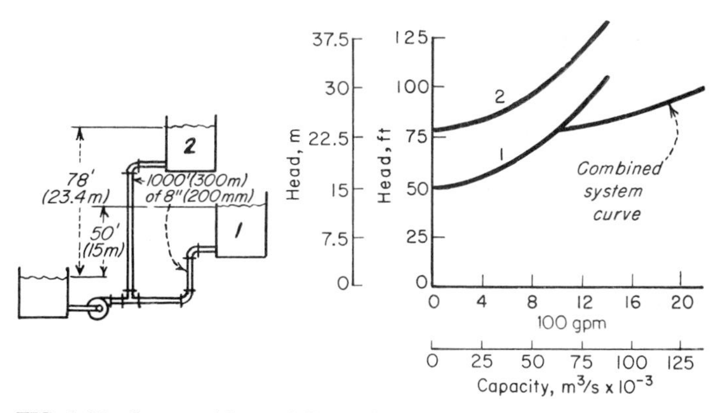

FIG. 6-28 System with two different discharge heads. *(Peerless Pumps.)*

The flow rate for the combined system at a head of 88 ft is 1150 + 550 = 1700 gal/min (107.3 L/s). To produce a flow of 1700 gal/min (107.3 L/s) through this system, a pump capable of developing an 88-ft (26.8-m) head is required.

8. Plot the system-head curve for diverted flow.

To analyze a system with diverted flow, assume that a constant quantity of liquid is tapped off at the intermediate point. Plot the friction-loss-vs.-flow rate in the normal manner for pipe 1 (Fig. 6-29). Move the curve for pipe 3 to the right at zero head by an

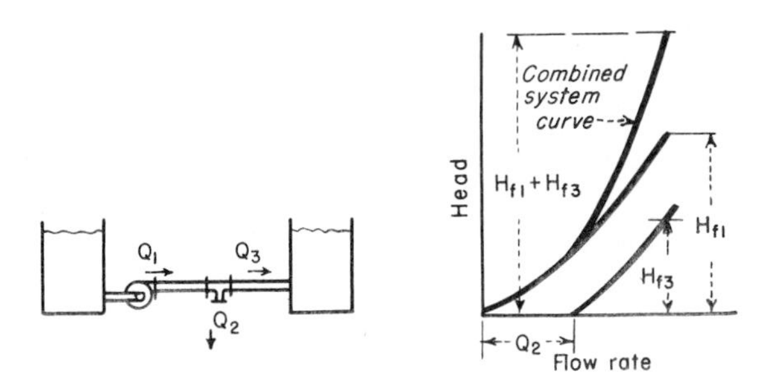

FIG. 6-29 Part of the fluid flow diverted from the main pipe. *(Peerless Pumps.)*

amount equal to Q_2, since this represents the quantity passing through pipes 1 and 2 but not through pipe 3. Plot the combined system-head curve by adding, at a given flow rate, the head losses for pipes 1 and 3. With Q = 300 gal/min (18.9 L/s), pipe 1 = 500 ft (152.4 m) of 10-in (254-mm) pipe, and pipe 3 = 50 ft (15.2 m) of 6-in (152.4-mm) pipe:

	ft	m
At 1500 gal/min (94.6 L/s) through pipe 1, friction loss	= 11	3.35
Friction loss for pipe 3 (1500 − 300 = 1200 gal/min) (75.7 L/s)	= 8	2.44
Total friction loss at 1500-gal/min (94.6-L/s) delivery	= 19	5.79

9. Plot the effect of pump wear.

When a pump wears, there is a loss in capacity and efficiency. The amount of loss depends, however, on the shape of the system-head curve. For a centrifugal pump (Fig. 6-30), the capacity loss is greater for a given amount of wear if the system-head curve is flat, as compared with a steep system-head curve.

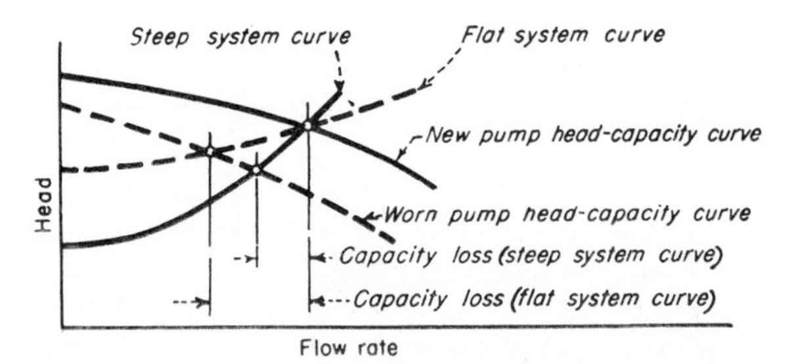

FIG. 6-30 Effect of pump wear on pump capacity. *(Peerless Pumps.)*

Determine the capacity loss for a worn pump by plotting its *H-Q* curve. Find this curve by testing the pump at different capacities and plotting the corresponding head. On the same chart, plot the *H-Q* curve for a new pump of the same size (Fig. 6-30). Plot the system-head curve and determine the capacity loss as shown in Fig. 6-30.

Related Calculations: Use the techniques given here for any type of pump—centrifugal, reciprocating, or rotary—handling any type of liquid—oil, water, chemicals, etc. The methods given here are the work of Melvin Mann, as reported in *Chemical Engineering,* and Peerless Pump Div. of FMC Corp.

6-24 Net Positive Suction Head for Hot-Liquid Pumps

What is the maximum capacity of a double-suction condensate pump operating at 1750 r/min if it handles 100°F (311 K) water from a hot well in a condenser having an absolute pressure of 2.0 in Hg (6.8 kPa) if the pump centerline is 10 ft (3.05 m) below the hot-well liquid level and the friction-head loss in the suction piping and fitting is 5 ft (1.5 m) of water?

Calculation Procedure:

1. Compute the net positive suction head on the pump.

The net positive suction head h_n on a pump when the liquid supply is *above* the pump inlet equals pressure on liquid surface + static suction head − friction-head loss in suction piping and pump inlet − vapor pressure of the liquid, all expressed in feet absolute of liquid handled. When the liquid supply is *below* the pump centerline—i.e., there is a static suction lift—the vertical distance of the lift is *subtracted* from the pressure on the liquid surface instead of added as in the preceding relation.

The density of 100°F water is 62.0 lb/ft^3. The pressure on the liquid surface, in absolute feet of liquid, is (2.0 inHg)(1.133)(62.4/62.0) = 2.24 ft. In this calculation, 1.133 = ft of 39.2°F water = 1 inHg; 62.4 = lb/ft^3 of 39.2°F water. The temperature of 39.2°F is used because at this temperature water has its maximum density. Thus, to convert inches of mercury to feet of absolute of water, find the product of (inHg)(1.133)(water density at 39.2°F)/(water density at operating temperature). Express both density values in the same unit, usually lb/ft^3.

The static suction head is a physical dimension that is measured in feet of liquid at the operating temperature. In this installation, h_{sh} = 10 ft absolute.

The friction-head loss is 5 ft of water at maximum density. To convert to feet absolute, multiply by the ratio of water densities at 39.2°F and the operating temperature or (5)(62.4/62.0) = 5.03 ft.

The vapor pressure of water at 100°F is 0.949 psia, from the steam tables. Convert any vapor pressure to feet absolute by finding the result of (vapor pressure, psia)(144 in^2/ft^2)/liquid density at operating temperature, or (0.949)(144)/62.0 = 2.204 ft absolute.

With all the heads known, the net positive suction head is h_n = 2.24 + 10 − 5.03 − 2.204 = 5.01 ft (1.53 m) absolute.

2. Determine the capacity of the condensate pump.

Use the Hydraulic Institute curve (Fig. 6-31) to determine the maximum capacity of the pump. Enter at the left of Fig. 6-31 at a net positive suction head of 5.01 ft and project horizontally to the right until the 3500-r/min curve is intersected. At the top, read the capacity as 278 gal/min (0.0175 m^3/s).

Related Calculations: Use this procedure for any condensate or boiler-feed pump handling water at an elevated temperature. Consult the *Standards of the Hydraulic Institute* for capacity curves of pumps having different types of construction. In general, pump manufacturers who are members of the Hydraulic Institute rate their pumps in accordance with the *Standards,* and a pump chosen from a catalog capacity table or curve will deliver the stated capacity. A similar procedure is used for computing the capacity of pumps handling volatile petroleum liquids. When using this procedure, be certain to refer to the latest edition of the *Standards.*

6-25 Minimum Safe Flow for a Centrifugal Pump

A centrifugal pump handles 220°F (377 K) water and has a shutoff head (with closed discharge valve) of 3200 ft (975 m). At shutoff, the pump efficiency is 17 percent and the input brake horsepower is 210. What is the minimum safe flow through this pump to prevent overheating at shutoff? Determine the minimum safe flow if the NPSH is 18.8 ft (5.73 m) of water and the liquid specific gravity is 0.995. If the pump contains 500 lb (227 kg) of water, determine the rate of the temperature rise at shutoff.

Calculation Procedure:

1. Compute the temperature rise in the pump.

With the discharge valve closed, the power input to the pump is converted to heat in the casing and causes the liquid temperature to rise. The temperature rise $t = (1 - e) \times$

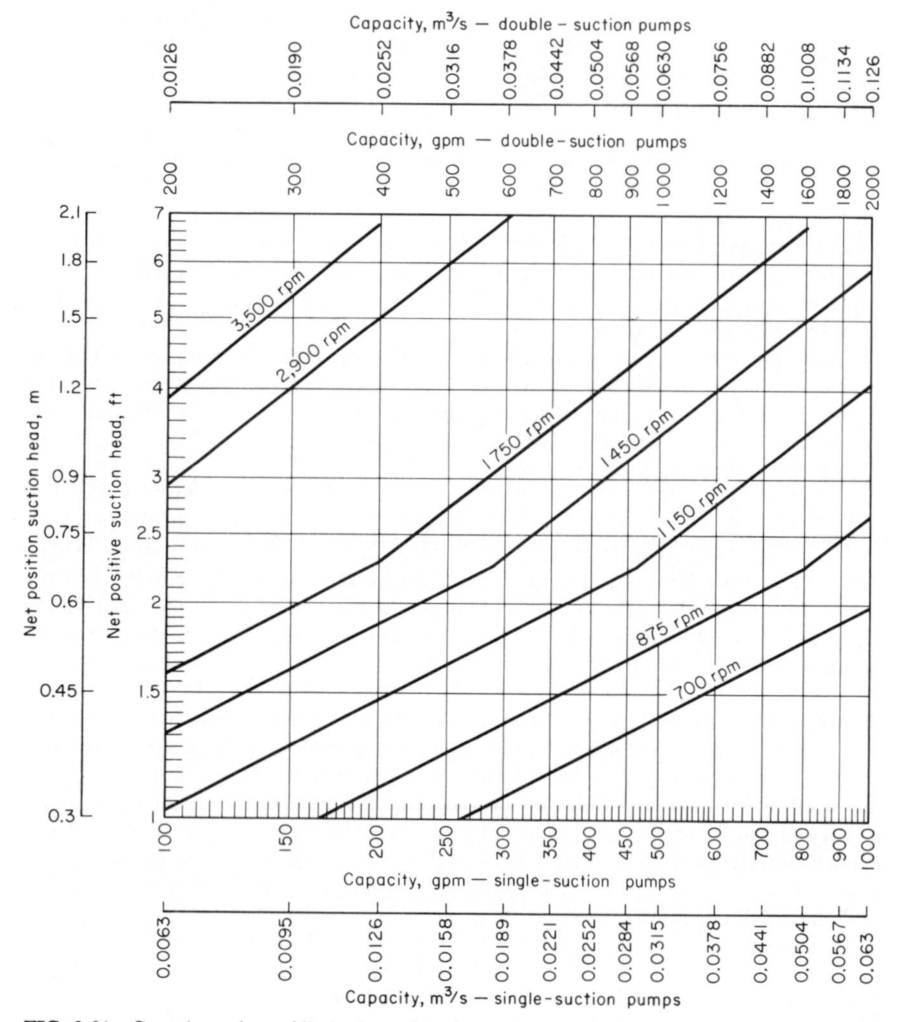

FIG. 6-31 Capacity and speed limitations of condensate pumps with the shaft through the impeller eye. *(Hydraulic Institute.)*

$H_s/778e$, where t is temperature rise during shutoff, °F; e is pump efficiency, expressed as a decimal; H_s is shutoff head, ft. For this pump, $t = (1 - 0.17)(3200)/[778(0.17)] = 20.4$°F (11.3°C).

2. *Compute the minimum safe liquid flow.*

For general-service pumps, the minimum safe flow M, in gal/min, is 6.0(bhp input at shutoff)/t. Or, $M = 6.0(210)/20.4 = 62.7$ gal/min (0.00396 m³/s). This equation includes a 20 percent safety factor.

Centrifugal boiler-feed pumps usually have a maximum allowable temperature rise of 15°F. The minimum allowable flow through the pump to prevent the water temperature from rising more than 15°F is 30 gal/min for each 100 bhp input at shutoff.

3. Compute the temperature rise for the operating NPSH.

An NPSH of 18.8 ft is equivalent to a pressure of 18.8(0.433)(0.995) = 7.78 psia at 220°F, where the factor 0.433 converts feet of water to pounds per square inch. At 220°F, the vapor pressure of the water is 17.19 psia, from the steam tables. Thus the total vapor pressure the water can develop before flashing occurs equals NPSH pressure + vapor pressure at operating temperature = 7.78 + 17.19 = 24.97 psia. Enter the steam tables at this pressure and read the corresponding temperature as 240°F. The allowable temperature rise of the water is then 240 − 220 = 20°F. Using the safe-flow relation of step 2, the minimum safe flow is 62.9 gal/min (0.00397 m^3/s).

4. Compute the rate of temperature rise.

In any centrifugal pump, the rate of temperature rise t_r, in °F per minute, is 42.4(bhp input at shutoff)/wc, where w is weight of liquid in the pump, lb; c is specific heat of the liquid in the pump, Btu/(lb)(°F). For this pump containing 500 lb of water with a specific heat c of 1.0, t_r = 42.4(210)/[500(1.0)] = 17.8°F/min (0.16 K/s). This is a very rapid temperature rise and could lead to overheating in a few minutes.

Related Calculations: Use this procedure for any centrifugal pump handling any liquid in any service—power, process, marine, industrial, or commercial. Pump manufacturers can supply a temperature-rise curve for a given model pump if it is requested. This curve is superimposed on the pump characteristic curve and shows the temperature rise accompanying a specific flow through the pump.

6-26 Selecting a Centrifugal Pump to Handle a Viscous Liquid

Select a centrifugal pump to deliver 750 gal/min (0.047 m^3/s) of 1000-SSU oil at a total head of 100 ft (30.5 m). The oil has a specific gravity of 0.90 at the pumping temperature. Show how to plot the characteristic curves when the pump is handling the viscous liquid.

Calculation Procedure:

1. Determine the required correction factors.

A centrifugal pump handling a viscous liquid usually must develop a greater capacity and head, and it requires a larger power input than the same pump handling water. With the water performance of the pump known—either from the pump characteristic curves or a tabulation of pump performance parameters—Fig. 6-32, prepared by the Hydraulic Institute, can be used to find suitable correction factors. Use this chart only within its scale limits; do not extrapolate. Do not use the chart for mixed-flow or axial-flow pumps or for pumps of special design. Use the chart only for pumps handling uniform liquids; slurries, gels, paper stock, etc. may cause incorrect results. In using the chart, the available net positive suction head is assumed adequate for the pump.

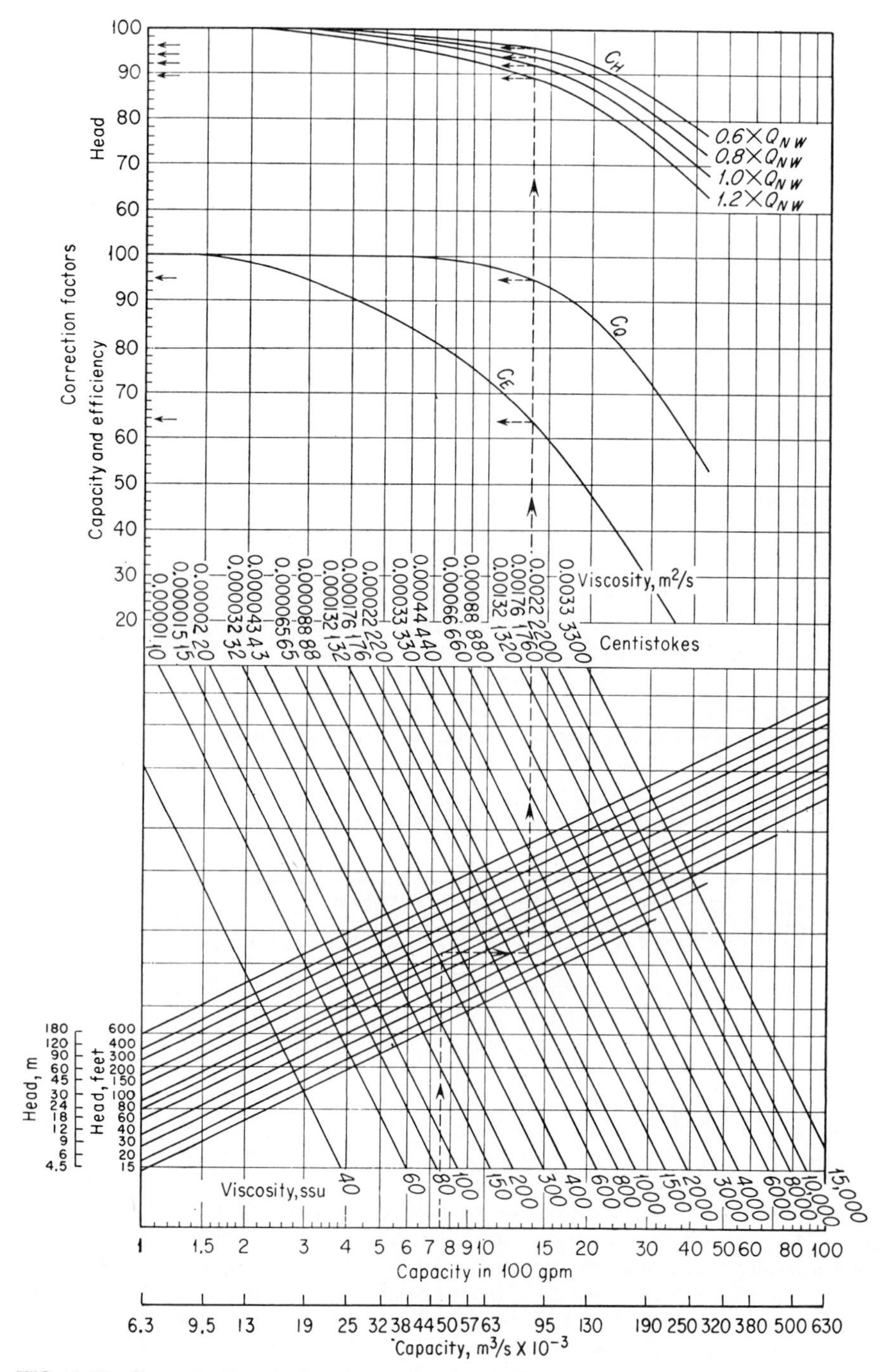

FIG. 6-32 Correction factors for viscous liquids handled by centrifugal pumps. *(Hydraulic Institute.)*

To use Fig. 6-32, enter at the bottom at the required capacity, 750 gal/min, and project vertically to intersect the 100-ft head curve, the required head. From here project horizontally to the 1000-SSU viscosity curve and then vertically upward to the correction-factor curves. Read $C_E = 0.635$; $C_Q = 0.95$; $C_H = 0.92$ for $1.0Q_{NW}$. The subscripts E, Q, and H refer to correction factors for efficiency, capacity, and head, respectively, and NW refers to the water capacity at a particular efficiency. At maximum efficiency, the water capacity is given as $1.0Q_{NW}$; other efficiencies, expressed by numbers equal to or less than unity, give different capacities.

2. Compute the water characteristics required.

The water capacity required for the pump $Q_w = Q_v/C_Q$, where Q_v is viscous capacity, gal/min. For this pump, $Q_w = 750/0.95 = 790$ gal/min. Likewise, water head $H_w = H_v/C_H$, where H_v is viscous head. Or, $H_w = 100/0.92 = 108.8$, say 109 ft water.

Choose a pump to deliver 790 gal/min of water at 109-ft head of water and the required viscous head and capacity will be obtained. Pick the pump so that it is operating at or near its maximum efficiency on water. If the water efficiency $E_w = 81$ percent at 790 gal/min for this pump, the efficiency when handling the viscous liquid $E_v = E_w C_E$. Or, $E_v = 0.81(0.635) = 0.515$, or 51.5 percent.

The power input to the pump when handling viscous liquids is given by $P_v = Q_v H_v s/(3960E_v)$, where s is specific gravity of the viscous liquid. For this pump, $P_v = (750)(100)(0.90)/[3960(0.515)] = 33.1$ hp (24.7 kW).

3. Plot the characteristic curves for viscous-liquid pumping.

Follow these eight steps to plot the complete characteristic curves of a centrifugal pump handling a viscous liquid when the water characteristics are known: (*a*) Secure a complete set of characteristic curves (H, Q, P, E) for the pump to be used. (*b*) Locate the point of maximum efficiency for the pump when handling water. (*c*) Read the pump capacity, Q gal/min, at this point. (*d*) Compute the values of $0.6Q$, $0.8Q$, and $1.2Q$ at the maximum efficiency. (*e*) Using Fig. 6-32, determine the correction factors at the capacities in steps *c* and *d*. Where a multistage pump is being considered, use the head per stage (= total pump head, ft/number of stages), when entering Fig. 6-32. (*f*) Correct the head, capacity, and efficiency for each of the flow rates in *c* and *d* using the correction factors from Fig. 6-32. (*g*) Plot the corrected head and efficiency against the corrected capacity, as in Fig. 6-33. (*h*) Compute the power input at each flow rate and plot. Draw smooth curves through the points obtained (Fig. 6-33).

Related Calculations: Use the method given here for any uniform viscous liquid—oil, gasoline, kerosene, mercury, etc.—handled by a centrifugal pump. Be careful to use Fig. 6-32 only within its scale limits; *do not extrapolate*. The method presented here is that developed by the Hydraulic Institute. For new developments in the method, be certain to consult the latest edition of the Hydraulic Institute—*Standards*.

6-27 Effect of Liquid Viscosity on Regenerative Pump Performance

A regenerative (turbine) pump has the water head-capacity and power-input characteristics shown in Fig. 6-34. Determine the head-capacity and power-input characteristics

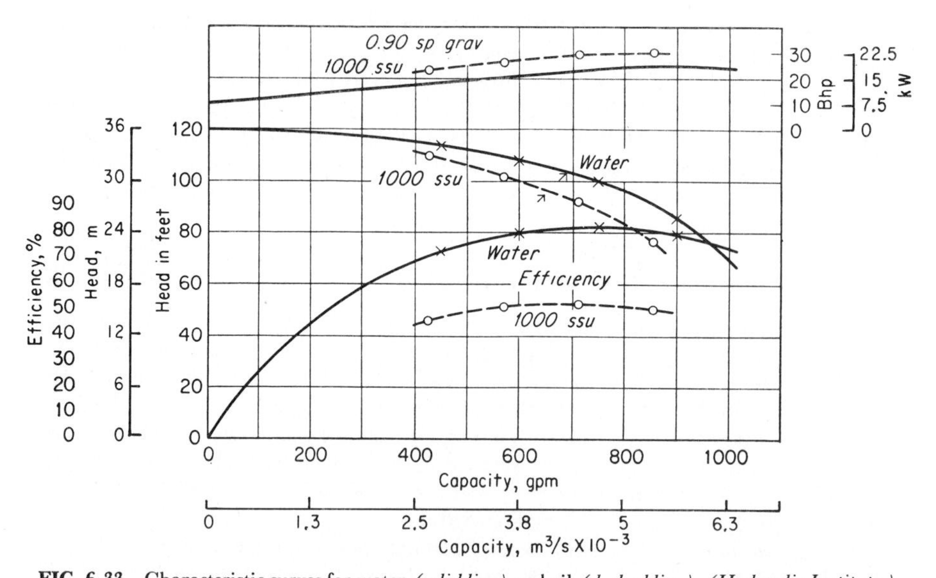

FIG. 6-33 Characteristic curves for water *(solid line)* and oil *(dashed line)*. *(Hydraulic Institute.)*

for four different viscosity oils to be handled by the pump—400, 600, 900, and 1000 SSU. What effect does increased viscosity have on the performance of the pump?

Calculation Procedure:

1. Plot the water characteristics of the pump.

Obtain a tabulation or plot of the water characteristics of the pump from the manufacturer or from the engineering data. With a tabulation of the characteristics, enter the various capacity and power points given and draw a smooth curve through them (Fig. 6-34).

2. Plot the viscous-liquid characteristics of the pump.

The viscous-liquid characteristics of regenerative-type pumps are obtained by test of the actual unit. Hence the only source of this information is the pump manufacturer. Obtain these characteristics from the pump manufacturer or the test data and plot them on Fig. 6-34, as shown, for each oil or other liquid handled.

3. Evaluate the effect of viscosity on pump performance.

Study Fig. 6-34 to determine the effect of increased liquid viscosity on the performance of the pump. Thus at a given head—say 100 ft—the capacity of the pump decreases as the liquid viscosity increases. At 100-ft head, this pump has a water capacity of 43.5 gal/min (Fig. 6-34). The pump capacity for the various oils at 100-ft head is 36 gal/min for 400 SSU, 32 gal/min for 600 SSU, 28 gal/min for 900 SSU, and 26 gal/min for 1000 SSU, respectively. There is a similar reduction in capacity of the pump at the other heads plotted in Fig. 6-34. Thus, as a general rule, it can be stated that the capacity of a regen-

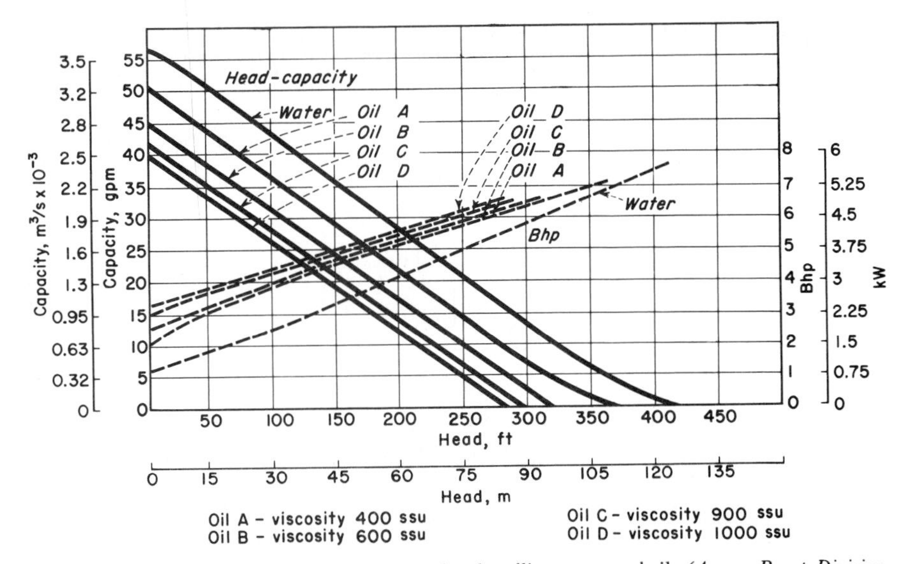

FIG. 6-34 Regenerative pump performance when handling water and oil. *(Aurora Pump Division, The New York Air Brake Co.)*

erative pump decreases with an increase in liquid viscosity at constant head. Or conversely, at constant capacity, the head developed decreases as the liquid viscosity increases.

Plots of the power input to this pump show that the input power increases as the liquid viscosity increases.

Related Calculations: Use this procedure for a regenerative-type pump handling any liquid—water, oil, kerosene, gasoline, etc. A decrease in the viscosity of a liquid—as compared with the viscosity of water—will produce the opposite effect from that of increased viscosity.

6-28 Effect of Liquid Viscosity on Reciprocating-Pump Performance

A direct-acting steam-driven reciprocating pump delivers 100 gal/min (0.0063 m^3/s) of 70°F (294 K) water when operating at 50 strokes/min. How much 2000-SSU crude oil will this pump deliver? How much 125°F (325 K) water will this pump deliver?

Calculation Procedure:

1. Determine the recommended change in the pump performance.

Reciprocating pumps of any type—direct-acting or power—having any number of liquid-handling cylinders—1 to 5, or more—are usually rated for maximum delivery when handling 250-SSU liquids or 70°F (21° C) water. At higher liquid viscosities or water temperatures, the speed—strokes or revolutions per minute—is reduced. Table 6-27 shows typical recommended speed-correction factors for reciprocating pumps for various liquid viscosities and water temperatures. This table shows that with a liquid viscosity of 2000

TABLE 6-27 Speed-Correction Factors

Liquid viscosity, SSU	Speed reduction, %	Water temperature °F	Water temperature °C	Speed reduction, %
250	0	70	21.1	0
500	4	80	26.2	9
1000	11	100	37.8	18
2000	20	125	51.7	25
3000	26	150	65.6	29
4000	30	200	97.3	34
5000	35	250	121.1	38

SSU, the pump speed should be reduced 20 percent. When handling 125°F (51.7°C) water, the pump speed should be reduced 25 percent, as shown in Table 6-27.

2. *Compute the delivery of the pump.*

The delivery capacity of any reciprocating pump is directly proportional to the number of strokes per minute it makes or to its revolutions per minute.

When handling 2000-SSU oil, the pump strokes/min must be reduced 20 percent, or (50)(0.20) = 10 strokes/min. Hence the pump speed will be 50 − 10 = 40 strokes/min. Since the delivery is directly proportional to speed, the delivery of 2000-SSU oil is (40/50)(100) = 80 gal/min (5.1 L/s).

When handling 125°F (51.7°C) water, the pump strokes per minute must be reduced 25 percent, or (50)(0.5) = 12.5 strokes/min. Hence the pump speed will be 50.0 − 12.5 = 37.5 strokes/min. Since the delivery is directly proportional to speed, the delivery of 125°F (51.7°C) water is (37.5/50)(100) = 75 gal/min (4.7 L/s).

Related Calculations: Use this procedure for any type of reciprocating pump handling liquids falling within the range of Table 6-27. Such liquids include oil, kerosene, gasoline, brine, water, etc.

6-29 Effect of Viscosity and Dissolved Gas on Rotary Pumps

A rotary pump handles 8000-SSU liquid containing 5% entrained gas and 10% dissolved gas at a 20-in(508-mm)Hg pump inlet vacuum. The pump is rated at 1000 gal/min (63.1 L/s) when handling gas-free liquids at viscosities less than 600 SSU. What is the output of this pump without slip? With 10 percent slip?

Calculation Procedure:

1. *Compute the required speed reduction of the pump.*

When the liquid viscosity exceeds 600 SSU, many pump manufacturers recommend that the speed of a rotary pump be reduced to permit operation without excessive noise or vibration. The speed reduction ususally recommended is shown in Table 6-28.

TABLE 6-28 Rotary-Pump Speed Reduction for Various Liquid Viscosities

Liquid viscosity, SSU	Speed reduction, percent of rated pump speed
600	2
800	6
1,000	10
1,500	12
2,000	14
4,000	20
6,000	30
8,000	40
10,000	50
20,000	55
30,000	57
40,000	60

With this pump handling 8000-SSU liquid, a speed reduction of 40 percent is necessary, as shown in Table 6-28. Since the capacity of a rotary pump varies directly with its speed, the output of this pump when handling 8000-SSU liquid is $(1000 \text{ gal/min}) \times (1.0 - 0.40) = 600 \text{ gal/min}$ (37.9 L/s).

2. *Compute the effect of gas on the pump output.*

Entrained or dissolved gas reduces the output of a rotary pump, as shown in Table 6-29. The gas in the liquid expands when the inlet pressure of the pump is below atmospheric and the gas occupies part of the pump chamber, reducing the liquid capacity.

With a 20-in(508-mm)Hg inlet vacuum, 5% entrained gas, and 10% dissolved gas, Table 6-29 shows that the liquid displacement is 74 percent of the rated displacement. Thus, the output of the pump when handling this viscous, gas-containing liquid will be $(600 \text{ gal/min})(0.74) = 444 \text{ gal/min}$ (28.0 L/s) without slip.

3. *Compute the effect of slip on the pump output.*

Slip reduces rotary-pump output in direct proportion to the slip. Thus, with 10 percent slip, the output of this pump is $(444 \text{ gal/min})(1.0 - 0.10) = 399.6 \text{ gal/min}$ (25 L/s).

Related Calculations: Use this procedure for any type of rotary pump—gear, lobe, screw, swinging-vane, sliding-vane, or shuttle-block—handling any clear, viscous liquid. Where the liquid is gas-free, apply only the viscosity correction. Where the liquid viscosity is less than 600 SSU but the liquid contains gas or air, apply the entrained or dissolved gas correction, or both corrections.

6-30 Selecting Forced- and Induced-Draft Fans

Combustion calculations show that an oil-fired watertube boiler requires 200,000 lb/h (25.2 kg/s) for air of combustion at maximum load. Select forced- and induced-draft fans for this boiler if the average temperature of the inlet air is 75°F (297 K) and the average

TABLE 6-29 Effect of Entrained or Dissolved Gas on the Liquid Displacement of Rotary Pumps (liquid diplacement: percent of displacement)

Vacuum at pump inlet, inHg (mmHg)	Gas entrainment					Gas solubility					Gas entrainment and gas solubility combined				
	1%	2%	3%	4%	5%	2%	4%	6%	8%	10%	1% 2%	2% 4%	3% 6%	4% 8%	5% 10%
5 (127)	99	97½	96½	95	93½	99½	99	98½	97	97½	98½	96½	96	92	91
10 (254)	98½	97¼	95½	94	92	99	97½	97	95	95	97½	95	90	90	88¼
15 (381)	98	96½	94½	92½	90½	97	96	94	92	90½	96	93	89½	86½	83¼
20 (508)	97½	94½	92	89	86½	96	92	89	86	83	94	88	83	78	74
25 (635)	94	89	84	79	75½	90	83	76½	71	66	85½	75½	68	61	55

For example: with 5 percent gas entrainment at 15 in Hg (381 mmHg) vacuum, the liquid displacement will be 90½ percent of the pump displacement neglecting slip or with 10 percent dissolved gas liquid displacement will be 90½ percent of pump displacement, and with 5 percent entrained gas combined with 10 percent dissolved gas, the liquid displacement will be 83¼ percent of pump replacement.

Source: Courtesy of Kinney Mfg. Div., The New York Air Brake Co.

temperature of the combustion gas leaving the air heater is 350°F (450 K) with an ambient barometric pressure of 29.9 inHg. Pressure losses on the air-inlet side are, in inH_2O: air heater, 1.5; air supply ducts, 0.75; boiler windbox, 1.75; burners, 1.25. Draft losses in the boiler and related equipment are, in inH_2O: furnace pressure, 0.20; boiler, 3.0; superheater, 1.0; economizer, 1.50; air heater, 2.00; uptake ducts and dampers, 1.25. Determine the fan discharge pressure and horsepower input. The boiler burns 18,000 lb/h (2.27 kg/s) of oil at full load.

Calculation Procedure:

1. Compute the quantity of air required for combustion.

The combustion calculations show that 200,000 lb/h of air is theoretically required for combustion in this boiler. To this theoretical requirement must be added allowances for excess air at the burner and leakage out of the air heater and furnace. Allow 25 percent excess air for this boiler. The exact allowance for a given installation depends on the type of fuel burned. However, a 25 percent excess-air allowance is an average used by power-plant designers for coal, oil, and gas firing. Using this allowance, the required excess air is $200{,}000(0.25) = 50{,}000$ lb/h.

Air-heater air leakage varies from about 1 to 2 percent of the theoretically required airflow. Using 2 percent, the air-heater leakage allowance is $200{,}000(0.02) = 4{,}000$ lb/h.

Furnace air leakage ranges from 5 to 10 percent of the theoretically required airflow. Using 7.5 percent, the furnace leakage allowance is $200{,}000(0.075) = 15{,}000$ lb/h.

The total airflow required is the sum of the theoretical requirement, excess air, and leakage, or $200{,}000 + 50{,}000 + 4000 + 15{,}000 = 269{,}000$ lb/h. The forced-draft fan must supply at least this quantity of air to the boiler. Usual practice is to allow a 10 to 20 percent safety factor for fan capacity to ensure an adequate air supply at all operating conditions. This factor of safety is applied to the total airflow required. Using a 10 percent factor of safety, fan capacity is $269{,}000 + 269{,}000(0.1) = 295{,}000$ lb/h. Round this off to 296,000 lb/h (37.3 kg/s) fan capacity.

2. Express the required airflow in cubic feet per minute.

Convert the required flow in pounds per hour to cubic feet per minute. To do this, apply a factor of safety to the ambient air temperature to ensure an adequate air supply during times of high ambient temperature. At such times, the density of the air is lower and the fan discharges less air to the boiler. The usual practice is to apply a factor of safety of 20 to 25 percent to the known ambient air temperature. Using 20 percent, the ambient temperature for fan selection is $75 + 75(0.20) = 90°F$. The density of air at 90°F is 0.0717 lb/ft^3, found in Baumeister—*Standard Handbook for Mechanical Engineers.* Converting, $ft^3/min = lb/h/60(lb/ft^3) = 296{,}000/60(0.0717) = 69{,}400\ ft^3/min$. This is the minimum capacity the forced-draft fan may have.

3. Determine the forced-draft discharge pressure.

The total resistance between the forced-draft fan outlet and furnace is the sum of the losses in the air heater, air-supply ducts, boiler windbox, and burners. For this boiler, the total resistance, in inH_2O, is $1.5 + 0.75 + 1.75 + 1.25 = 5.25\ inH_2O$. Apply a

15 to 30 percent factor of safety to the required discharge pressure to ensure adequate airflow at all times. Or fan discharge pressure, using a 20 percent factor of safety, is $5.25 + 5.25(0.20) = 6.30$ inH_2O. The fan must therefore deliver at least 69,400 ft^3/min (32.7 m^3/s) at 6.30 inH_2O.

4. Compute the power required to drive the forced-draft fan.

The air horsepower for any fan is $0.0001753H_fC$, where H_f is total head developed by fan, in inH_2O; C is airflow, in ft^3/min. For this fan, air hp $= 0.0001753(6.3)(69{,}400)$, $= 76.5$ hp. Assume or obtain the fan and fan-driver efficiencies at the rated capacity (69,400 ft^3/min) and pressure (6.30 inH_2O). With a fan efficiency of 75 percent and assuming the fan is driven by an electric motor having an efficiency of 90 percent, the overall efficiency of the fan-motor combination is $(0.75)(0.90) = 0.675$, or 67.5 percent. Then the motor horsepower required equals air horsepower/overall efficiency $= 76.5/0.675 = 113.2$ hp (84.4 kW). A 125-hp motor would be chosen because it is the nearest, next larger, unit readily available. Usual practice is to choose a *larger* driver capacity when the computed capacity is lower than a standard capacity. The next larger standard capacity is generally chosen, except for extremely large fans where a special motor may be ordered.

5. Compute the quantity of flue gas handled.

The quantity of gas reaching the induced draft fan is the sum of the actual air required for combustion from step 1, air leakage in the boiler and furnace, and the weight of fuel burned. With an air leakage of 10 percent in the boiler and furnace (this is a typical leakage factor applied in practice), the gas flow is as follows:

	lb/h	kg/s
Actual airflow required	296,000	37.3
Air leakage in boiler and furnace	29,600	3.7
Weight of oil burned	18,000	2.3
Total	343,600	43.3

Determine from combustion calculations for the boiler the density of the flue gas. Assume that the combustion calculations for this boiler show that the flue-gas density is 0.045 lb/ft^3 (0.72 kg/m^3) at the exit-gas temperature. To determine the exit-gas temperature, apply a 10 percent factor of safety to the given exit temperature, 350°F (176.6°C). Hence exit-gas temperature is $350 + 350(0.10) = 385$°F (196.1°C). Then, flue-gas flow, in ft^3/min, is (flue-gas flow, lb/h)/(60)(flue-gas density, lb/ft^3) $= 343{,}600/[(60)(0.045)] = 127{,}000$ ft^3/min (59.9 m^3/s). Apply a 10 to 25 percent factor of safety to the flue-gas quantity to allow for increased gas flow. Using a 20 percent factor of safety, the actual flue-gas flow the fan must handle is $127{,}000 + 127{,}000(0.20) = 152{,}400$ ft^3/min (71.8 m^3/s), say 152,500 ft^3/min for fan-selection purposes.

6. Compute the induced-draft fan discharge pressure.

Find the sum of the draft losses from the burner outlet to the induced-draft inlet. These losses are, for this boiler:

	inH_2O	kPa
Furnace draft loss	0.20	0.05
Boiler draft loss	3.00	0.75
Superheater draft loss	1.00	0.25
Economizer draft loss	1.50	0.37
Air heater draft loss	2.00	0.50
Uptake ducts and damper draft loss	1.25	0.31
Total draft loss	8.95	2.23

Allow a 10 to 25 percent factor of safety to ensure adequate pressure during all boiler loads and furnace conditions. Using a 20 percent factor of safety for this fan, the total actual pressure loss is $8.95 + 8.95(0.20) = 10.74$ inH_2O (2.7 kPa). Round this off to 11.0 inH_2O (2.7 kPa) for fan-selection purposes.

7. *Compute the power required to drive the induced-draft fan.*

As with the forced-draft fan, air horsepower is $0.0001753\ H_fC = 0.0001753(11.0)(127{,}000) = 245$ hp (182.7 kW). If the combined efficiency of the fan and its driver, assumed to be an electric motor, is 68 percent, the motor horsepower required is $245/0.68 = 360.5$ hp (268.8 kW). A 375-hp (279.6-kW) motor would be chosen for the fan driver.

8. *Choose the fans from a manufacturer's engineering data.*

Use Example 6-31 to select the fans from the engineering data of an acceptable manufacturer. For larger boiler units, the forced-draft fan is usually a backward-curved blade centrifugal-type unit. Where two fans are chosen to operate in parallel, the pressure curve of each fan should decrease at the same rate near shutoff so that the fans divide the load equally. Be certain that forced-draft fans are heavy-duty units designed for continuous operation with well-balanced rotors. Choose high-efficiency units with self-limiting power characteristics to prevent overloading the driving motor. Airflow is usually controlled by dampers on the fan discharge.

Induced-draft fans handle hot, dusty combustion products. For this reason, extreme care must be used to choose units specifically designed for induced-draft service. The usual choice for large boilers is a centrifugal-type unit with forward- or backward-curved, or flat blades, depending on the type of gas handled. Flat blades are popular when the flue gas contains large quantities of dust. Fan bearings are generally water-cooled.

Related Calculations: Use this procedure for selecting draft fans for all types of boilers—fire-tube, packaged, portable, marine, and stationary. Obtain draft losses from the boiler manufacturer. Compute duct pressure losses using the methods given in later Procedures in this Handbook.

6-31 Power-Plant Fan Selection from Capacity Tables

Choose a forced-draft fan to handle 69,400 ft^3/min (32.7 m^3/s) of 90°F (305 K) air at 6.30 inH_2O static pressure and an induced-draft fan to handle 152,500 ft^3/min (71.9 $m^3/$

s) of 385°F (469 K) gas at 11.0 inH_2O static pressure. The boiler that these fans serve is installed at an elevation of 5000 ft (1524 m) above sea level. Use commercially available capacity tables for making the fan choice. The flue-gas density is 0.045 lb/ft^3 (0.72 kg/m^3) at 385°F (469 K).

Calculation Procedure:

1. Compute the correction factors for the forced-draft fan.

Commercial fan-capacity tables are based on fans handling standard air at 70°F at a barometric pressure of 29.92 inHg and having a density of 0.075 lb/ft^3. Where different conditions exist, the fan flow rate must be corrected for temperature and altitude.

Obtain the engineering data for commercially available forced-draft fans and turn to the temperature and altitude correction-factor tables. Pick the appropriate correction factors from these tables for the prevailing temperature and altitude of the installation. Thus, in Table 6-30, select the correction factors for 90°F air and 5000-ft altitude. These correction factors are $C_T = 1.018$ for 90°F air and $C_A = 1.095$ for 5000-ft altitude.

TABLE 6-30 Fan Correction Factors

Temperature		Correction factor	Altitude		Correction factor
°F	°C		ft	m	
80	26.7	1.009	4500	1371.6	1.086
90	32.2	1.018	5000	1524.0	1.095
100	37.8	1.028	5500	1676.4	1.106
375	190.6	1.255			
400	204.4	1.273			
450	232.2	1.310			

Find the composite correction factor CCF by taking the product of the temperature and altitude correction factors, or $\text{CCF} = C_T C_A = 1.018(1.095) = 1.1147$. Now divide the given ft^3/min by the composite correction factor to find the capacity-table ft^3/min. Or, capacity-table ft^3/min is $69{,}400/1.1147 = 62{,}250$ ft^3/min.

2. Choose the fan size from the capacity table.

Turn to the fan-capacity table in the engineering data and look for a fan delivering 62,250 ft^3/min at 6.3 inH_2O static pressure. Inspection of the table shows that the capacities are tabulated for pressures of 6.0 and 6.5 inH_2O static pressure. There is no tabulation for 6.3 inH_2O. The fan must therefore be selected for 6.5 inH_2O static pressure.

Enter the table at the nearest capacity to that required, 62,250 ft^3/min, as shown in Table 6-31. This table, excerpted with permission from the American Standard Inc. engineering data, shows that the nearest capacity of this particular type of fan is 62,595 ft^3/min. The difference, or $62{,}595 - 62{,}250 = 345$ ft^3/min, is only $345/62{,}250 = 0.0055$, or 0.55 percent. This is a negligible difference, and the 62,595-ft^3/min fan is well suited for its intended use. The extra static pressure, $6.5 - 6.3 = 0.2$ inH_2O, is desirabe in a forced-draft fan because furnace or duct resistance may increase during the life of the boiler. Also, the extra static pressure is so small that it will not markedly increase the fan power consumption.

TABLE 6-31 Typical Fan Capacities

Capacity		Outlet velocity		Outlet velocity pressure		Ratings at 6.5 inH_2O (1.6 kPa) static pressure		
ft^3/min	m^3/s	ft/min	m/s	inH_2O	kPa	r/min	bhp	kW
61,204	28.9	4400	22.4	1.210	0.3011	1083	95.45	71.2
62,595	29.5	4500	22.9	1.266	0.3150	1096	99.08	73.9
63,975	30.2	4600	23.4	1.323	0.3212	1109	103.0	76.8

3. *Compute the fan speed and power input.*

Multiply the capacity-table r/min and bhp by the composite correction factor to determine the actual r/min and bhp. Thus, using data from Table 6-31, the actual r/min is (1096)(1.1147) = 1221.7 r/min. Actual bhp is (99.08)(1.1147) = 110.5 hp. This is the horsepower input required to drive the fan and is close to the 113.2 hp computed in the previous example. The actual motor horsepower would be the same in each case because a standard-size motor would be chosen. The difference of 113.2 − 110.5 = 2.7 hp results from the assumed efficiencies that depart from the actual values. Also, a sea-level altitude was assumed in the previous example. However, the two methods used show how accurately fan capacity and horsepower input can be estimated by judicious evaluation of variables.

4. *Compute the correction factors for the induced-draft fan.*

The flue-gas density is 0.045 lb/ft^3 at 385°F. Interpolate in the temperature correction-factor table because a value of 385°F is not tabulated. Find the correction factor for 385°F thus: (Actual temperature − lower temperature)/(higher temperature − lower temperature) × (higher temperature-correction factor − lower temperature-correction factor) + lower-temperature-correction factor. Or, [(385 − 375)/(400 − 375)](1.273 − 1.255) + 1.255 = 1.262.

The altitude-correction factor is 1.095 for an elevation of 5000 ft, as shown in Table 6-30. As for the forced-draft fan, $CCF = C_T C_A = (1.262)(1.095) = 1.3819$. Use the CCF to find the capacity-table ft^3/min in the same manner as for the forced-draft fan. Or, capacity-table ft^3/min is (given ft^3/min)/CCF = 152,500/1.3819 = 110,355 ft^3/min.

5. *Choose the fan size from the capacity table.*

Check the capacity table to be sure that it lists fans suitable for induced-draft (elevated temperature) service. Turn to the 11-in static-pressure-capacity table and find a capacity equal to 110,355 ft^3/min. In the engineering data used for this fan, the nearest capacity at 11-in static pressure is 110,467 ft^3/min, with an outlet velocity of 4400 ft^3/min, an outlet velocity pressure of 1.210 inH_2O, a speed of 1222 r/min, and an input horsepower of 255.5 bhp. The tabulation of these quantities is of the same form as that given for the forced-draft fan (step 2). The selected capacity of 110,467 ft^3/min is entirely satisfactory because it is only 110,467 − 110,355/110,355 = 0.00101, or 0.1 percent, higher than the desired capacity.

6. Compute the fan speed and power input.

Multiply the capacity-table r/min and brake horsepower by the CCF to determine the actual r/min and brake horsepower. Thus, the actual r/min is (1222)(1.3819) = 1690 r/min. Actual brake horsepower is (255.5)(1.3819) = 353.5 bhp (263.7 kW). This is the horsepower input required to drive the fan and is close to the 360.5 hp computed in the previous example. The actual motor horsepower would be the same in each case because a standard-size motor would be chosen. The difference in horsepower of 360.5 − 353.5 = 7.0 hp results from the same factors discussed in step 3.

Note: The static pressure is normally used in most fan-selection procedures because this is the pressure value used in computing pressure and draft losses in boilers, economizers, air heaters, and ducts. In any fan system, the total air pressure equals static pressure + velocity pressure. However, the velocity pressure at the fan discharge is not considered in draft calculations unless there are factors requiring its evaluation. These requirements are generally related to pressure losses in the fan-control devices.

6-32 Determination of the Most Economical Fan Control

Determine the most economical fan control for a forced- or induced-draft fan designed to deliver 140,000 ft^3/min (66.03 m^3/s) at 14 inH_2O (3.5 kPa) at full load. Plot the power-consumption curve for each type of control device considered.

Calculation Procedure:

1. Determine the types of controls to consider.

There are five types of controls used for forced- and induced-draft fans: (*a*) a damper in the duct with constant-speed fan drive, (*b*) two-speed fan driver, (*c*) inlet vanes or inlet louvres with a constant-speed fan drive, (*d*) multiple-step variable-speed fan drive, and (*e*) hydraulic or electric coupling with constant-speed driver giving wide control over fan speed.

2. Evaluate each type of fan control.

Tabulate the selection factors influencing the control decision as follows, using the control letters in step 1:

Control type	Control cost	Required power input	Advantages (A), and disadvantages (D)
a	Low	High	(A) Simplicity; (D) High power input
b	Moderate	Moderate	(A) Lower input power; (D) Higher cost
c	Low	Moderate	(A) Simplicity; (D) ID fan erosion
d	Moderate	Moderate	(D) Complex; also needs dampers
e	High	Low	(A) Simple; no dampers needed

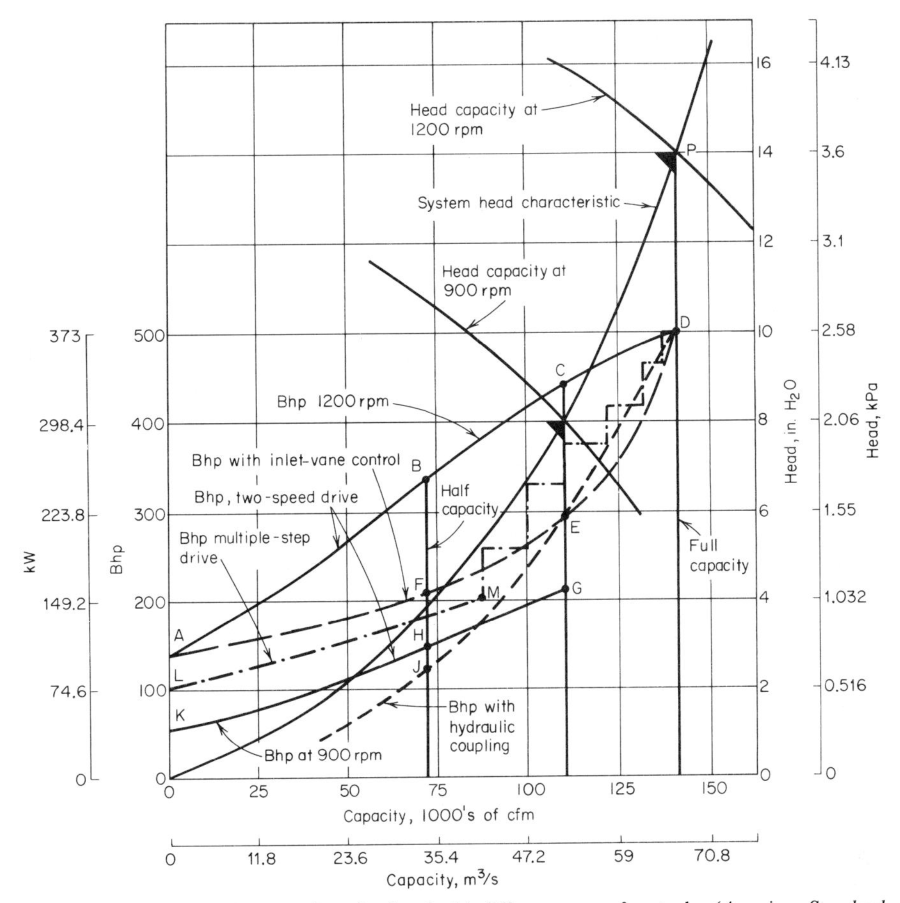

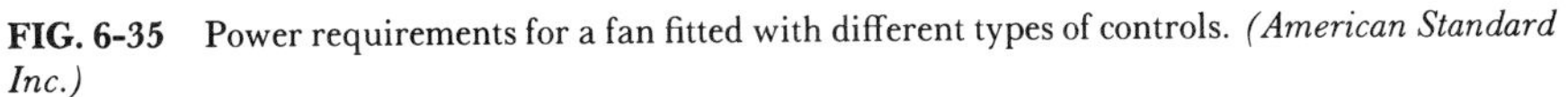

FIG. 6-35 Power requirements for a fan fitted with different types of controls. *(American Standard Inc.)*

3. Plot the control characteristics for the fans.

Draw the fan head-capacity curve for the airflow or gasflow range considered (Fig. 6-35). This plot shows the maximum capacity of 140,000 ft^3/min and required static head of 14 inH_2O, point *P*.

Plot the power-input curve *ABCD* for a constant-speed motor or turbine drive with damper control—type *a,* listed above—after obtaining from the fan manufacturer or damper builder the input power required at various static pressures and capacities. Plotting these values gives curve *ABCD.* Fan speed is 1200 r/min.

Plot the power-input curve *GHK* for a two-speed drive, type *b.* This drive might be a motor with an additional winding, or it might be a second motor for use at reduced boiler capacities. With either arrangement, the fan speed at lower boiler capacities is 900 r/min.

Plot the power-input curve *AFED* for inlet-vane control on the forced-draft fan or inlet-louvre control on induced-draft fans. The data for plotting this curve can be obtained from the fan manufacturer.

Multiple-step variable-speed fan control, type *d*, is best applied with steam-turbine drives. In a plant with ac auxiliary motor drives, slip-ring motors with damper integration must be used between steps, making the installation expensive. Although dc motor drives would be less costly, few power plants other than marine propulsion plants have direct current available. And since marine units normally operate at full load 90 percent or more of the time, part-load operating economics are unimportant. If steam-turbine drive will be used for the fans, plot the power-input curve *LMD*, using data from the fan manufacturer.

A hydraulic coupling or electric magnetic coupling, type *e*, with a constant-speed motor drive would have the power-input curve *DEJ*.

Study of the power-input curves shows that the hydraulic and electric couplings have the smallest power input. Their first cost, however, is usually greater than any other types of power-saving devices. To determine the return on any extra investment in power-saving devices, an economic study, including a load-duration analysis of the boiler load, must be made.

4. Compare the return on the extra investment.

Compute and tabulate the total cost of each type of control system. Then determine the extra investment for each of the more costly systems by subtracting the cost of type *a* from the cost of each of the other types. With the extra investment known, compute the lifetime savings in power input for each of the more efficient control methods. With the extra investment and savings resulting from it known, compute the percentage return on the extra investment. Tabulate the findings as in Table 6-32.

In Table 6-32, considering control type *c*, the extra cost of type *c* over type *b* is $75,000 − 50,000 = $25,000. The total power saving of $6500 is computed on the basis of the cost of energy in the plant for the life of the control. The return on the extra investment then is $6500/$25,000 = 0.26, or 26 percent. Type *e* control provides the highest percentage return on the extra investment. It would probably be chosen if the only measure of investment desirability is the return on the extra investment. However, if other criteria are used—such as a minimum rate of return on the extra investment—one of the other control types might be chosen. This is easily determined by studying the tabulation in conjunction with the investment requirement.

TABLE 6-32 Fan Control Comparison

	Type of control used				
	a	*b*	*c*	*d*	*e*
Total cost, $	30,000	50,000	75,000	89,500	98,000
Extra cost, $	—	20,000	25,000	14,500	8,500
Total power saving, $	—	8,000	6,500	3,000	6,300
Return on extra investment, %	—	40	26	20.7	74.2

Related Calculations: The procedure used here can be applied to heating, power, marine, and portable boilers of all types. Follow the same steps given above, changing the values to suit the existing conditions. Work closely with the fan and drive manufacturer when analyzing drive power input and costs.

6-33 Vacuum-Pump Selection for High-Vacuum Systems

Choose a mechanical vacuum pump for use in a laboratory fitted with a vacuum system having a total volume, including the piping, of 12,000 ft^3 (340 m^3). The operating pressure of the system is 0.10 torr, and the optimum pump-down time is 150 min. (*Note:* 1 torr = 1 mmHg.)

Calculation Procedure:

1. Make a tentative choice of pump type.

Mechanical vacuum pumps of the reciprocating type are well suited for system pressures in the 0.0001- to 760-torr range. Hence, this type of pump will be considered first to see if it meets the desired pump-down time.

2. Obtain the pump characteristic curves.

Many manufacturers publish pump-down factor curves such as those in Fig. 6-36*a* and *b*. These curves are usually published as part of the engineering data for a given line of pumps. Obtain the curves from the manufacturers whose pumps are being considered.

3. Compute the pump-down time for the pumps being considered.

Three reciprocating pumps can serve this system: (*a*) a single-stage pump, (*b*) a compound or two-stage pump, or (*c*) a combination of a mechanical booster and a single-stage backing or roughing-down pump. Figure 6-36 gives the pump-down factor for each type of pump.

To use the pump-down factor, apply this relation: $t = VF/d$, where t is pump-down time, min; V is system volume, ft^3; F is pump-down factor for the pump; d is pump displacement, ft^3/min.

Thus, for a single-stage pump, Fig. 6-36*a* shows that $F = 10.8$ for a pressure of 0.10 torr. Assuming a pump displacement of 1000 ft^3/min, $t = 12{,}000(10.8)/1000 = 129.6$ min; say 130 min.

For a compound pump, $F = 9.5$ from Fig. 6-36*a*. Hence, a compound pump having the same displacement, or 1000 ft^3/min, will require $t = 12{,}000(9.5)/1000 = 114.0$ min.

With a combination arrangement, the backing or roughing pump, a 130-ft^3/min unit, reduces the system pressure from atmospheric, 760 torr, to the economical transition pressure, 15 torr (Fig. 6-36*b*). Then the single-stage mechanical booster pump, a 1200-ft^3/min unit, takes over and in combination with the backing pump reduces the pressure to the desired level, or 0.10 torr. During this part of the cycle, the unit operates as a two-stage pump. Hence the total pump-down time consists of the sum of the backing-pump and booster-pump times. The pump-down factors are, respectively, 4.2 for the backing pump at 15 torr and 6.9 for the booster pump at 0.10 torr. Hence the respective pump-down times are $t_1 = 12{,}000(4.2)/130 = 388$ min; $t_2 = 12{,}000(6.9)/1200 = 69$ min. The total time is thus $388 + 69 = 457$ min.

The pump-down time with the combination arrangement is greater than the optimum

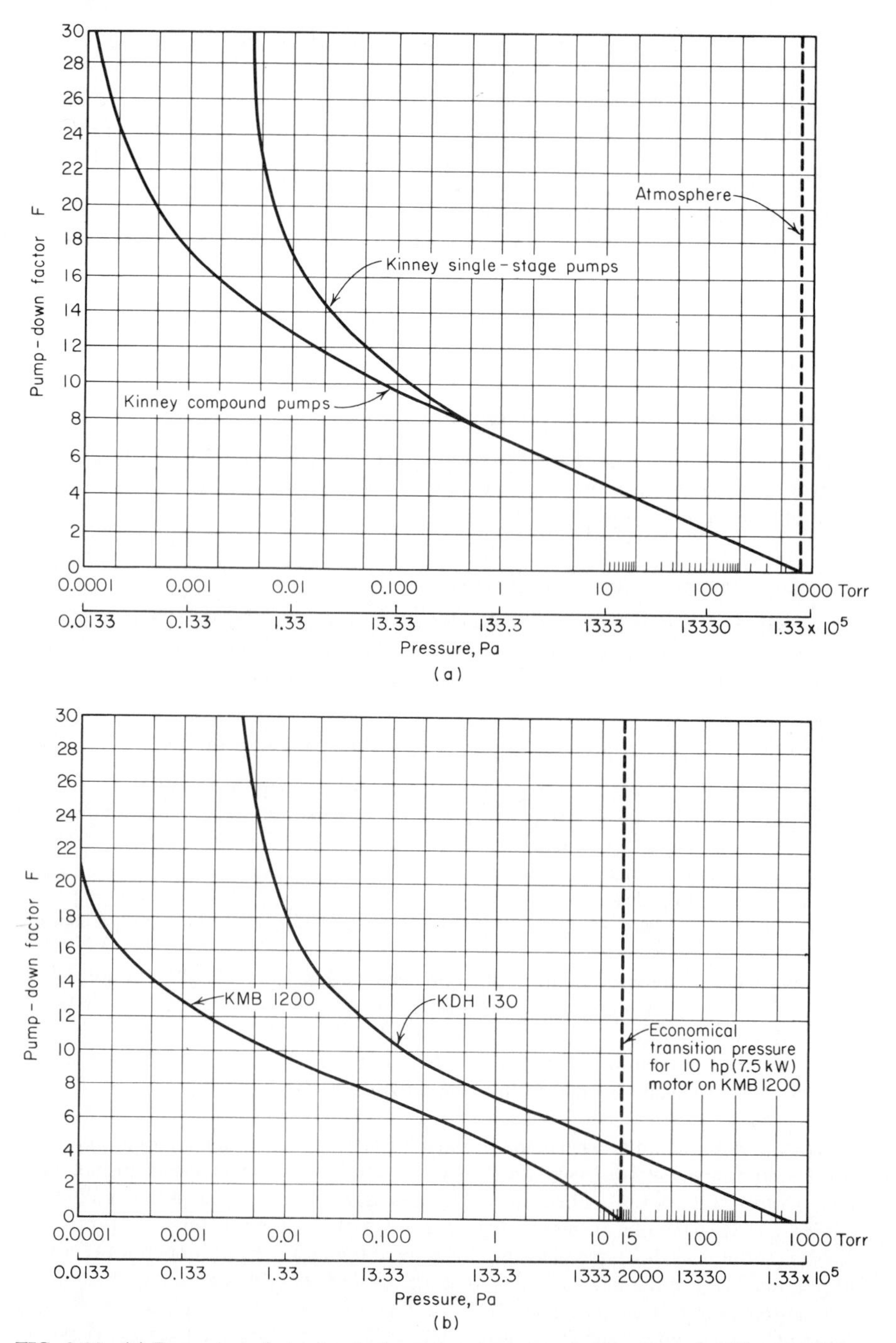

FIG. 6-36 (*a*) Pump-down factor for single-stage and compound vacuum pumps; (*b*) pump-down factor for mechanical booster and backing pump. *(After Kinney Vacuum Division, The New York Air Brake Co., and Van Atta.)*

150 min. Where a future lower operating pressure is anticipated, making the combination arrangement desirable, an additional large-capacity single-stage roughing pump can be used to assist the 130-ft^3/min unit. This large-capacity unit is operated until the transition pressure is reached and roughing down is finished. The pump is then shut off and the balance of the pumping down is carried on by the combination unit. This keeps the power consumption at a minimum.

Thus, if a 1200-ft^3/min single-stage roughing pump were used to reduce the pressure to 15 torr, its pump-down time would be $t = 12{,}000(4.0)/1200 = 40$ min. The total pump-down time for the combination would then be $40 + 69 = 109$ min, using the time computed above for the two pumps in combination.

4. Apply the respective system factors.

Studies and experience show that the calculated pump-down time for a vacuum system must be corrected by an appropriate system factor. This factor makes allowance for the normal outgassing of surfaces exposed to atmospheric air. It also provides a basis for judging whether a system is pumping down normally or whether some problem exists that must be corrected. Table 6-33 lists typical system factors that have proven reliable in many tests. To use the system factor for any pump, apply it this way: $t_a = tS$, where t_a is actual pump-down time, in min; t is computed pump-down time from step 3, in min; S is system factor for the type of pump being considered.

TABLE 6-33 Recommended System Factors

Pressure range		System factors		
torr	Pa	Single-stage mechanical pump	Compound mechanical pump	Mechanical booster pump*
760–20	115.6 kPa–3000	1.0	1.0	—
20–1	3000–150	1.1	1.1	1.15
1–0.5	150–76	1.25	1.25	1.15
0.5–0.1	76–15	1.5	1.25	1.35
0.1–0.02	15–3	—	1.25	1.35
0.02–0.001	3–0.15	—	—	2.0

*Based on bypass operation until the booster pump is put into operation. Larger system factors apply if rough pumping flow must pass through the idling mechanical booster. Any time needed for operating valves and getting the mechanical booster pump up to speed must also be added.

Source: From Van Atta—*Vacuum Science and Engineering,* McGraw-Hill.

Thus, using the appropriate system factor for each pump, the actual pump-down time for the single-stage mechanical pump is $t_a = 130(1.5) = 195$ min. For the compound mechanical pump, $t_a = 114(1.25) = 142.5$ min. For the combination mechanical booster pump, $t_a = 109(1.35) = 147$ min.

5. Choose the pump to use.

Based on the actual pump-down time, either the compound mechanical pump or the combination mechanical booster pump can be used. The final choice of the pump should take other factors into consideration—first cost, operating cost, maintenance cost, reliability, and probable future pressure requirements in the system. Where future lower pressure requirements are not expected, the compound mechanical pump would be a good choice. However, if lower operating pressures are anticipated in the future, the combination mechanical booster pump would probably be a better choice.

Van Atta[1] gives the following typical examples of pumps chosen for vacuum systems:

Pressure range, torr	Typical pump choice
Down to 50 (7.6 kPa)	Single-stage oil-sealed rotary; large water or vapor load may require use of refrigerated traps
0.05 to 0.01 (7.6 to 1.5 Pa)	Single-stage or compound oil-sealed pump plus refrigerated traps, particularly at the lower pressure limit
0.01 to 0.005 (1.5 to 0.76 Pa)	Compound oil-sealed plus refrigerated traps, or single-stage pumps backing diffusion pumps if a continuous large evolution of gas is expected
1 to 0.0001 (152.1 to 0.015 Pa)	Mechanical booster and backing pump combination with interstage refrigerated condenser and cooled vapor trap at the high-vacuum inlet for extreme freedom from vapor contamination
0.0005 and lower (0.076 Pa and lower)	Single-stage pumps backing diffusion pumps, with refrigerated traps on the high-vacuum side of the diffusion pumps and possibly between the single-stage and diffusion pumps if evolution of condensable vapor is expected

6-34 Vacuum-System Pumping Speed and Pipe Size

A laboratory vacuum system has a volume of 500 ft^3 (14 m^3). Leakage into the system is expected at the rate of 0.00035 ft^3/min. What backing pump speed, i.e., displacement, should an oil-sealed vacuum pump serving this system have if the pump blocking pressure is 0.150 mmHg and the desired operating pressure is 0.0002 mmHg? What should the speed of the diffusion pump be? What pump size is needed for the connecting pipe of the backing pump if it has a displacement or pumping speed of 388 ft^3/min (0.18 m^3/s) at 0.150 mmHg and a length of 15 ft (4.6 m)?

Calculation Procedure:

1. Compute the required backing pump speed.

Use the relation $d_b = G/P_b$, where d_b is backing pump speed or pump displacement, in ft^3/min; G is gas leakage or flow rate, in mm/(ft^3/min). To convert the gas or leakage flow rate to mm/(ft^3/min), multiply the ft^3/min by 760 mm, the standard atmospheric pressure, in mmHg. Thus, $d_b = 760(0.00035)/0.150 = 1.775$ ft^3/min.

[1]C. M. Van Atta—*Vacuum Science and Engineering,* McGraw-Hill, New York, 1965.

2. Select the actual backing pump speed.

For practical purposes, since gas leakage and outgassing are impossible to calculate accurately, a backing pump speed or displacement of at least twice the computed value, or $2(1.775) = 3.550$ ft^3/min—say 4 ft^3/min (0.002 m^3/s)—would probably be used.

If this backing pump is to be used for pumping down the system, compute the pump-down time as shown in the previous example. Should the pump-down time be excessive, increase the pump displacement until a suitable pump-down time is obtained.

3. Compute the diffusion pump speed.

The diffusion pump reduces the system pressure from the blocking point, 0.150 mmHg, to the system operating pressure of 0.0002 mmHg. (*Note:* 1 torr = 1 mmHg.) Compute the diffusion pump speed from $d_d = G/P_d$, where d is diffusion pump speed, in ft^3/min; P_d is diffusion-pump operating pressure, mmHg. Or $d_d = 760(0.00035)/0.0002 = 1330$ ft^3/min (0.627 m^3/s). To allow for excessive leaks, outgassing, and manifold pressure loss, a 3000- or 4000-ft^3/min diffusion pump would be chosen. To ensure reliability of service, two diffusion pumps would be chosen so that one could operate while the other was being overhauled.

4. Compute the size of the connection pipe.

In usual vacuum-pump practice, the pressure drop in pipes serving mechanical pumps is not allowed to exceed 20 percent of the inlet pressure prevailing under steady operating conditions. A correctly designed vacuum system, where this pressure loss is not exceeded, will have a pump-down time which closely approximates that obtained under ideal conditions.

Compute the pressure drop in the high-pressure region of vacuum pumps from $p_d = 1.9d_bL/d^4$, where p_d is pipe pressure drop, in μ; d_b is backing pump displacement or speed, in ft^3/min; L is pipe length, in ft; d is inside diameter of pipe, in in. Since the pressure drop should not exceed 20 percent of the inlet or system operating pressure, the drop for a backing pump is based on its blocking pressure, or 0.150 mmHg, or 150 μ. Hence $p_d = 0.20(150) = 30\ \mu$. Then, $30 = 1.9(380)(15)/d^4$, and $d = 4.35$ in (0.110 m). Use a 5-in-diameter pipe.

In the low-pressure region, the diameter of the converting pipe should equal, or be larger than, the pump inlet connection. Whenever the size of a pump is increased, the diameter of the pipe should also be increased to conform with the above guide.

Related Calculations: Use the general procedures given here for laboratory- and production-type high-vacuum systems.

6-35 Bulk-Material Elevator and Conveyor Selection

Choose a bucket elevator to handle 150 tons/h (136.1 tonnes/h) of abrasive material weighing 50 lb/ft^3 (800.5 kg/m^3) through a vertical distance of 75 ft (22.9 m) at a speed of 100 ft/min (30.5 m/min). What horsepower input is required to drive the elevator? The bucket elevator discharges onto a horizontal conveyor which must transport the material 1400 ft (426.7 m). Choose the type of conveyor to use and determine the required power input needed to drive it.

Calculation Procedure:

1. Select the type of elevator to use.

Table 6-34 summarizes the various characteristics of bucket elevators used to transport bulk materials vertically. This table shows that a continuous bucket elevator would be a good choice, because it is a recommended type for abrasive materials. The second choice would be a pivoted bucket elevator. However, the continuous bucket type is popular and will be chosen for this application.

TABLE 6-34 Bucket Elevators

	Centrifugal discharge	Perfect discharge	Continuous bucket	Gravity discharge	Pivoted bucket
Carrying paths	Vertical	Vertical to inclination 15° from vertical	Vertical to inclination 15° from vertical	Vertical and horizontal	Vertical and horizontal
Capacity range, tons/h (tonnes/h), material weighing 50 lb/ft^3 (800.5 kg/m^3)	78 (70.8)	34 (30.8)	345 (312.9)	191 (173.3)	255 (231.3)
Speed range, ft/min (m/min)	306 (93.3)	120 (36.6)	100 (30.5)	100 (30.5)	80 (24.4)
Location of loading point	Boot	Boot	Boot	On lower horizontal run	On lower horizontal run
Location of discharge point	Over head wheel	Over head wheel	Over head wheel	On horizontal run	On horizontal run
Handling abrasive materials	Not preferred	Not preferred	Recommended	Not recommended	Recommended

Source: Link-Belt Div. of FMC Corp.

2. Compute the elevator height.

To allow for satisfactory loading of the bulk material, the elevator length is usually increased by about 5 ft (1.5 m) more than the vertical lift. Hence the elevator height is 75 + 5 = 80 ft (24.4 m).

3. Compute the required power input to the elevator.

Use the relation $hp = 2CH/1000$, where C is elevator capacity, in tons/h; H is elevator height, in ft. Thus, for this elevator, $hp = 2(150)(80)/1000 = 24.9$ hp (17.9 kW).

The power input relation given above is valid for continuous bucket, centrifugal-discharge, perfect-discharge, and supercapacity elevators. A 25-hp (18.7-kW) motor would probably be chosen for this elevator.

4. Select the type of conveyor to use.

Since the elevator discharges onto the conveyor, the capacity of the conveyor should be the same, per unit time, as the elevator. Table 6-35 lists the characteristics of various

TABLE 6-35 Conveyor Characteristics

	Belt conveyor	Apron conveyor	Flight conveyor	Drag chain	En masse conveyor	Screw conveyor	Vibratory conveyor
Carrying paths	Horizontal to 18°	Horizontal to 25°	Horizontal to 45°	Horizontal or slight incline, 10°	Horizontal to 90°	Horizontal to 15°; may be used up to 90° but capacity falls off rapidly	Horizontal or slight incline, 5° above or below horizontal
Capacity range, tons/h (tonnes/h) material weighing 50 lb/ft^3	2160 (1959.5)	100 (90.7)	360 (326.6)	20 (18.1)	100 (90.7)	150 (136.1)	100 (90.7)
Speed range, ft/min	600 (182.9 m/min)	100 (30.5 m/min)	150 (45.7 m/min)	20 (6.1 m/min)	80 (24.4 m/min)	100 (30.5 m/min)	40 (12.2 m/min)
Location of loading point	Any point	Any point	Any point	Any point	On horizontal runs	Any point	Any point
Location of discharge point	Over end wheel and intermediate points by tripper or plow	Over end wheel	At end of trough and intermediate points by gates	At end of trough	Any point on horizontal runs by gate	At end of trough and intermediate points by gates	At end of trough
Handling abrasive materials	Recommended	Recommended	Not recommended	Recommended with special steels	Not recommended	Not preferred	Recommended

Source: Link-Belt Div. of FMC Corp.

types of conveyors. Study of the tabulation shows that a belt conveyor would probably be best for this application, based on the speed, capacity, and type of material it can handle, hence, it will be chosen for this installation.

5. Compute the required power input to the conveyor.

The power input to a conveyor is composed of two portions: (*a*) the power required to move the empty belt conveyor, and (*b*) the power required to move the load horizontally.

Determine from Fig. 6-37 the power required to move the empty belt conveyor, after choosing the required belt width. Determine the belt width from Table 6-36.

Thus, for this conveyor, Table 6-36 shows that a belt width of 42 in (106.7 cm) is required to transport up to 150 tons/h (136.1 tonnes/h) at a belt speed of 100 ft/min

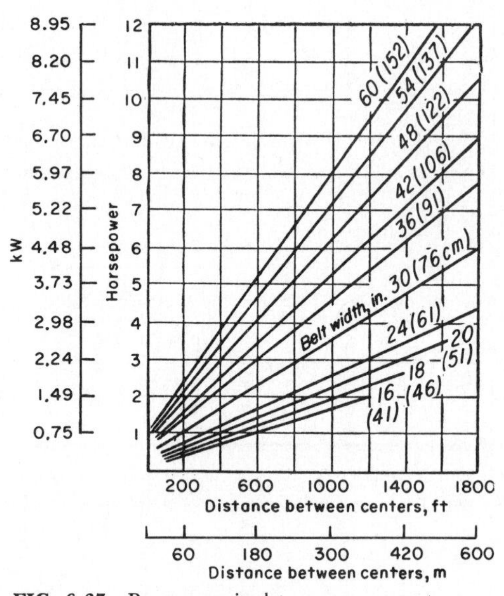

FIG. 6-37 Power required to move an empty conveyor belt at 100 ft/min (0.508 m/s).

TABLE 6-36 Capacities of Troughed Rest [tons/h (tonnes/h) with belt speed of 100 ft/min (30.5 m/min)]

Belt width, in (cm)	Weight of material, lb/ft³ (kg/m³)			
	30 (480.3)	50 (800.5)	100 (1601)	150 (2402)
30 (9.1)	47 (42.6)	79 (71.7)	158 (143.3)	237 (214.9)
36 (10.9)	69 (62.6)	114 (103.4)	228 (206.8)	342 (310.2)
42 (12.8)	97 (87.9)	162 (146.9)	324 (293.9)	486 (440.9)
48 (14.6)	130 (117.9)	215 (195.0)	430 (390.1)	645 (585.1)
60 (18.3)	207 (187.8)	345 (312.9)	690 (625.9)	1035 (938.9)

Source: United States Rubber Co.

(30.5 m/min). (Note that the next *larger* capacity, 162 tons/h (146.9 tonnes/h), is used when the exact capacity required is not tabulated.) Find the horsepower required to drive the empty belt by entering Fig. 6-37 at the belt distance between centers, 1400 ft (426.7 m), and projecting vertically upward to the belt width, 42 in (106.7 cm). At the left, read the required power input as 7.2 hp (5.4 kW).

Compute the power required to move the load horizontally from $hp = (C/100)(0.4 + 0.00345L)$, where L is distance between conveyor centers, in ft; other symbols as before. For this conveyor, $hp = (150/100)(0.4 + 0.00325 \times 1400) = 6.83$ hp (5.1 kW). Hence the total horsepower to drive this horizontal conveyor is $7.2 + 6.83 = 14.03$ hp (10.5 kW).

The total horsepower input to this conveyor installation is the sum of the elevator and conveyor belt horsepowers, or $14.03 + 24.0 = 38.03$ hp (28.4 kW).

Related Calculations: This procedure is valid for conveyors using rubber belts reinforced with cotton duck, open-mesh fabric, cords, or steel wires. It is also valid for stitched-canvas belts, balata belts, and flat-steel belts. The required horsepower input includes any power absorbed by idler pulleys.

Table 6-37 shows the minimum recommended belt widths for lumpy materials of various sizes. Maximum recommended belt speeds for various materials are shown in Table 6-38.

TABLE 6-37 Minimum Belt Width for Lumps

Belt width, in (mm)	24 (609.6)	36 (914.4)	42 (1066.8)	48 (1219.2)
Sized materials, in (mm)	4½ (114.3)	8 (203.2)	10 (254)	12 (304.9)
Unsized material, in (mm)	8 (203.2)	14 (355.6)	20 (508)	35 (889)

TABLE 6-38 Maximum Belt Speeds for Various Materials

Width of belt		Light or free-flowing materials, grains dry sand, etc.		Moderately free-flowing sand, gravel, fine stone, etc.		Lump coal, coarse stone, crushed ore		Heavy sharp lumpy materials, heavy ores, lump coke	
in	mm	ft/min	m/min	ft/min	m/min	ft/min	m/min	ft/min	m/min
12–14	305–356	400	122	250	76	—	—	—	—
16–18	406–457	500	152	300	91	250	76	—	—
20–24	508–610	600	183	400	122	350	107	250	76
30–26	762–914	750	229	500	152	400	122	300	91

When a conveyor belt is equipped with a tripper, the belt must rise about 5 ft (1.5 m) above its horizontal plane of travel.

This rise must be included in the vertical lift power input computation. When the tripper is driven by the belt, allow 1 hp (0.75 kW) for a 16-in (406.4-mm) belt, 3 hp (2.2 kW) for a 36-in (914.4-mm) belt, and 7 hp (5.2 kW) for a 60-in (1524-mm) belt. Where a rotary cleaning brush is driven by the conveyor shaft, allow about the same power input to the brush for belts of various widths.

6-36 Screw-Conveyor Power Input and Capacity

What is the required input for a 100-ft (30.5-m) long screw conveyor handling dry coal ashes having a maximum density of 40 lb/ft^3 if the conveyor capacity is 30 tons/h (27.2 tonnes/h)?

Calculation Procedure:

1. Select the conveyor diameter and speed.

Refer to a manufacturer's engineering data or Table 6-39 for a listing of recommended screw-conveyor diameters and speeds for various types of materials. Dry coal ashes are commonly rated as group 3 materials (Table 6-40)—i.e., materials with small mixed lumps with fines.

TABLE 6-39 Screw-Conveyor Capacities and Speeds

Material group	Max material density, lb/ft^3 (kg/m^3)	Max r/min for diameters of	
		6 in (152 mm)	20 in (508 mm)
1	50 (801)	170	110
2	50 (801)	120	75
3	75 (1201)	90	60
4	100 (1601)	70	50
5	125 (2001)	30	25

To determine a suitable screw diameter, assume two typical values and obtain the recommended r/min from the sources listed above or Table 6-39. Thus the maximum r/min recommended for a 6-in (152.4-mm) screw when handling group 3 material is 90, as shown in Table 6-39; for a 20-in (508.0-mm) screw, 60 r/min. Assume a 6-in (152.4-mm) screw as a trial diameter.

2. Determine the material factor for the conveyor.

A material factor is used in the screw conveyor power input computation to allow for the character of the substance handled. Table 6-40 lists the material factor for dry ashes as $F = 4.0$. Standard references show that the average weight of dry coal ashes is 35 to 40 lb/ft^3 (640.4 kg/m^3).

3. Determine the conveyor size factor.

A size factor that is a function of the conveyor diameter is also used in the power input computation. Table 6-41 shows that for a 6-in diameter conveyor the size factor $A = 54$.

4. Compute the required power input to the conveyor.

Use the relation $hp = 10^{-6}(ALN + CWLF)$, where hp is hp input to the screw conveyor head shaft; A is size factor from step 3; L is conveyor length, in ft; N is conveyor r/min; C is quantity of material handled, in ft^3/h; W is density of material, in lb/ft^3; F

TABLE 6-40 Material Factors for Screw Conveyors

Material group	Material type	Material factor
1	Lightweight:	
	Barley, beans, flour, oats, pulverized coal, etc.	0.5
2	Fines and granular:	
	Coal—slack or fines	0.9
	Sawdust, soda ash	0.7
	Flyash	0.4
3	Small lumps and fines:	
	Ashes, dry alum	4.0
	Salt	1.4
4	Semiabrasives; small lumps:	
	Phosphate, cement	1.4
	Clay, limestone;	2.0
	Sugar, white lead	1.0
5	Abrasive lumps:	
	Wet ashes	5.0
	Sewage sludge	6.0
	Flue dust	4.0

TABLE 6-41 Screw Conveyor Size Factors

Conveyor diameter, in (mm)	Size factor	Conveyor diameter, in (mm)	Size factor
6 (152.4)	54	16 (406.4)	336
9 (228.6)	96	18 (457.2)	414
10 (254)	114	20 (508)	510
12 (304.8)	171	24 (609.6)	690

is material factor from step 2. For this conveyor, using the data listed above $hp = 10^{-6}(54 \times 100 \times 60 + 1500 \times 40 \times 100 \times 4.0) = 24.3$ hp (18.1 kW). With a 90 percent motor efficiency, the required motor rating would be $24.3/0.90 = 27$ hp (20.1 kW). A 30-hp (22.4-kW) motor would be chosen to drive this conveyor. Since this is not an excessive power input, the 6-in (152.4-mm) conveyor is suitable for this application.

If the calculation indicates that an excessively large power input—say 50 hp (37.3 kW) or more—is required, the larger-diameter conveyor should be analyzed. In general, a higher initial investment in conveyor size that reduces the power input will be more than recovered by the savings in power costs.

Related Calculations: Use this procedure for screw or spiral conveyors and feeders handling any material that will flow. The usual screw or spiral conveyor is suitable for conveying materials for distances up to about 200 ft (60 m), although special designs can be built for greater distances. Conveyors of this type can be sloped upward to angles of 35° with the horizontal. However, the capacity of the conveyor decreases as the angle of inclination is increased. Thus the reduction in capacity at a 10° inclination is 10 percent over the horizontal capacity; at 35° the reduction is 78 percent.

The capacities of screw and spiral conveyors are generally stated in ft^3/h of various classes of materials at the maximum recommended shaft r/min. As the size of the lumps in the material conveyed increases, the recommended shaft r/min decreases. The capacity of a screw or spiral conveyor at a lower speed is found from (capacity at given speed, in ft^3/h)(lower speed, r/min/higher speed, r/min). Table 6-39 shows typical screw conveyor capacities at usual operating speeds.

Various types of screws are used for modern conveyors. These include short-pitch, variable-pitch, cut flights, ribbon, and paddle screws. The Procedure given above also applies to these screws.

7

Heat Transfer

Paul E. Minton
Principal Engineer
Union Carbide Corporation
South Charleston, WV

Edward S. S. Morrison
Senior Staff Engineer
Union Carbide Corporation
Houston, TX

REFERENCES: [1] McAdams—*Heat Transmission,* 3d ed., McGraw-Hill; [2] Krieth—*Principles of Heat Transfer,* 3d ed., Intext Educational Publishers; [3] Kern—*Process Heat Transfer,* McGraw-Hill; [4] Holman—*Heat Transfer,* 2d ed., McGraw-Hill; [5] Collier—*Convective Boiling and Condensation,* McGraw-Hill; [6] Rohsenow and Hartnett (eds.)—*Handbook of Heat Transfer,* McGraw-Hill; [7] Chapman—*Heat Transfer,* 3d ed., McMillan; [8] Oppenheim—"Radiation Analysis by the Network Method," in Hactwett et al.—*Recent Advances in Heat and Mass Transfer,* McGraw-Hill; [9] Siegel and Howell—*Thermal Radiation Heat Transfer,* NASA SP-164 vols. I, II, and III, Lewis Research Center, 1968, 1969, and 1970. Office of Technology Utilization; [10] Butterworth—*Introduction to Heat Transfer,* Oxford University Press; [11] *Standards of Tubular Exchangers Manufacturers Association,* 6th ed., Tubular Exchangers Manufacturers Association; [12] Blevins—*Flow-Induced Vibration,* Van Nostrand-Reinhold; [13] Daniels—*Terrestrial Environment (Climatic) Criteria Guidelines for Use in Aerospace Vehicle Development,* 1973 Revision. NASA TMS-64757, NASA/MSFC; [14] *The American Ephemeris and Nautical Almanac,* U.S.

Government Printing Office; [15] Bell—*University of Delaware Experimental Station Bulletin 5,* also *Petro/Chem Engineer,* October 1960, p. C-26; [16] Grant and Chisholm—ASME paper 77-WA/HT-22; [17] Butterworth—ASME paper 77-WA/HT-24; [18] Dukler—*Chemical Engineering Progress Symposium Series No. 30,* vol. 56, 1, 1960; [19] Akers and Rosson—*Chemical Engineering Progress Symposium Series "Heat Transfer—Storrs,"* vol. 56, 3, 1960; [20] Caglayan and Buthod—*Oil Gas J.,* Sept. 6, 1976, p. 91; [21] Bergles—*Sixth International Heat Transfer Conference, Toronto, 1978,* vol. 6, p. 89; [22] Briggs and Young—*Chemical Engineering Progress Symposium Series,* vol. 59, no. 41, pp. 1–10; [23] Fair—*Petroleum Refiner,* February 1960, p. 105; [24] Gilmour—*Chem. Engineer.,* October 1952, p. 144; March 1953, p. 226; April 1953, p. 214; October 1953, p. 203; February 1954, p. 190; March 1954, p. 209; August 1954, p. 199; [25] Lord, Minton, and Slusser—*Chem. Engineer.,* Jan. 26, 1970, p. 96; [26] Minton—*Chem. Engineer.,* May 4, 1970, p. 103; [27] Minton—*Chem. Engineer.,* May 18, 1970, p. 145; [28] Lord, Minton, and Slusser—*Chem. Engineer.,* March 23, 1970, p. 127; [29] Lord, Minton, and Slusser—*Chem. Engineer.,* June 1, 1970, p. 153; [30] Michiyoshi—*Sixth International Heat Transfer Conference, Toronto, 1978,* vol. 6, p. 219; [31] Wallis—*One-Dimensional Two-Phase Flow,* McGraw-Hill; [32] Gottzmann, O'Neill, and Minton—*Chem. Engineer.,* July 1973, p. 69.

7-1 Specifying Radiation Shielding

A furnace is to be located next to a dense complex of cryogenic propane piping. To protect the cold equipment from excessive heat loads, reflective aluminum radiation shielding sheets are to be placed between the piping and the 400°F (477 K) furnace wall. The space between the furnace wall and the cold surface is 2 ft (0.61 m). The facing surfaces of the furnace and the piping array are each 25 × 40 ft (7.6 × 12.2 m). With the ice-covered surface of the cold equipment at an average temperature of 35°F (275 K), how many 25 × 40 ft aluminum sheets must be installed between the two faces to keep the last sheet at or below 90°F (305 K)? Emittances of the furnace wall and cryogenic equipment are 0.90 and 0.65, respectively; that of the aluminum shields is 0.1.

Calculation Procedure:

1. Analyze the arrangement to assess the type(s) of heat transfer involved.

The distance separating the hot and cold surfaces is small compared with the size of the surfaces. The approximation can thus be made that the furnace wall, the dense network of cryogenic piping, and the radiation shields are all infinitely extended parallel planes. This is a conservative assumption, since the effect of proximity to an edge is to introduce a source of moderate temperature, thus allowing the hot wall to cool off. Convection is omitted with the same justification. So the problem can be treated as pure radiation.

Radiant heat transfer between two parallel, infinite plates is given by

$$\frac{q_{1\text{-}2}}{A} = \frac{\sigma(T_1^4 - T_2^4)}{(1/\epsilon_1) + (1/\epsilon_2) - 1}$$

where q_{1-2}/A is the heat flux between hotter surface 1 and colder surface 2, σ is the Stefan-Boltzmann constant, 0.1713×10^{-8} Btu/(h)(ft^2)(°R^4), T_1 and T_2 are absolute temperatures of the hotter and colder surfaces, and ϵ_1 and ϵ_2 are the emittances of the two surfaces. Emittance is the ratio of radiant energy emitted by a given real surface to the

TABLE 7-1 Emittances of Some Real Surfaces

Material	Theoretical emittance ϵ_{th}
Aluminum foil, bright, foil, at 700°F (644 K)	0.04
Aluminum alloy 6061T6 H_2CrO_4 anodized, 300°F (422 K)	0.17
Aluminum alloy 6061T6, H_2SO_4 anodized, 300°F (422 K)	0.80
Cast iron, polished, at 392°F (473 K)	0.21
Cast iron, oxidized, at 390°F (472 K)	0.64
Black lacquer on iron at 76°F (298 K)	0.88
White enamel on iron at 66°F (292 K)	0.90
Lampblack on iron at 68°F (293 K)	0.97
Firebrick at 1832°F (1273 K)	0.75
Roofing paper at 69°F (294 K)	0.91

Source: Chapman [7].

radiant energy that would be emitted by a theoretical, perfectly radiating black surface. Emittances of several materials are given in Table 7-1. The variations indicated between values of ϵ for apparently similar surfaces are not unusual, and they indicate the advisability of using measured data whenever available.

Note that the numerator expresses a potential and the denominator a resistance. If a series of n radiation shields with emittance ϵ_s is interspersed between the two infinite parallel plates, the equivalent resistance can be shown to be

$$\left[\frac{1}{\epsilon_1} + \frac{1}{\epsilon_2} - 1\right] + n\left[\frac{2}{\epsilon_s} - 1\right]$$

2. *Set up an equation for heat flux from the furnace to the cold equipment, expressed in terms of n.*

Substituting into the equation in step 1,

$$\frac{q}{A} = \frac{(0.1713 \times 10^{-8})[(400 + 460)^4 - (35 + 460)^4]}{[(1/0.9) + (1/0.65) - 1] + n[(2/0.1) - 1]} = 834.19/(1.65 + 19n)$$

3. *Set up an equation for heat flux from the furnace to the last shield, expressed in terms of n.*

In this case, the heat flows through $n - 1$ shields en route to the last shield, whose temperature is to be kept at or below 90°F and whose emittance (like that of the other shields) is 0.1. Substituting again into the equation in step 1,

$$\frac{q}{A} = \frac{(0.1713 \times 10^{-8})[(400 + 460)^4 - (90 + 460)^4]}{[(1/0.9) + (1/0.1) - 1] + (n - 1)[(2/0.1) - 1]}$$

$$= \frac{780.27}{10.11 + (n - 1)(19)}$$

$$= \frac{780.27}{(-8.89 + 19n)}$$

4. Solve for n.

Since all the heat must pass through all shields, the expressions of steps 2 and 3 are equal. Setting them equal to each other and solving for n, it is found to be 8.499. Rounding off, we specify 9 shields.

5. Check the answer by back-calculating the temperature of the last shield.

Substituting 9 for n in the equation in step 2, the heat transferred is

$$\frac{q}{A} = \frac{(0.1713 \times 10^{-8})(860^4 - 495^4)}{[(1/0.9) + 1/(0.65) - 1] + 9[(2/0.1) - 1]}$$
$$= 4.83 \text{ Btu/(h)(ft}^2\text{)}$$

Next, let T_2 in the equation in step 3, that is, (90 + 460), become an unknown, while substituting 9 for n and 4.83 for q/A. Thus, $4.83 = (0.1713 \times 10^8)(860^4 - T^4)/\{[(1/0.9) + (1/0.1) - 1] + [9 - 1][(2/0.1) - 1]\}$. Solving for T_2, we find it to be 547.6°R (or 87.6°F), thus satisfying the requirement of the problem.

7-2 Radiant Interchange inside a Black-Surfaced Enclosure

A furnace firebox is 20 ft (6.1 m) long, 10 ft (3.05 m) wide, and 5 ft (1.5 m) high. Because of a rich fuel-air mixture, all surfaces have become coated with lampblack, so that they all act as black-body surfaces with virtually complete absorption and emittance of radiant energy; emittance ϵ is 0.97. The furnace is overfired; i.e., its cold surface is the floor, composed of closely spaced tubes flowing water at 250°F (394 K). When the furnace is operating, its roof is at 1150°F (894 K), the sidewalls are at 920°F (766 K), and the end walls at around 800°F (700 K). A plant emergency suddenly shuts the furnace down. Determine the initial rate of heat transfer from each interior surface if the water in the tubes remains at 250°F. Assume that the tube surface is at the water temperature.

Calculation Procedure:

1. Analyze the situation to determine the type(s) of heat transfer involved.

The temperatures are relatively high and the problem does not involve material flow; it is safe to treat it as a radiation problem. In view of the high emittance, consider the entire enclosure black in a radiation sense. Since all surfaces are black, they absorb all radiation incident upon them, with no reflection. Therefore, the radiant-interchange factors depend only on the geometric relationships among the surfaces. A radiant-interchange factor is the proportionality constant that indicates the fraction of radiant-energy-transfer *potential* between two surfaces, as measured by Boltzmann's constant times the difference between the fourth powers of their absolute temperatures, that shows up as actual heat flux between them. Thus, $\mathcal{F}_{1-2}$ is the radiant interchange factor in the equation $q/A = \sigma\mathcal{F}_{1-2}(T_1^4 - T_2^4)$, where q/A is heat flux, σ is Boltzmann's constant, and T_i is the absolute temperature of surface i. In Example 7-1, $\mathcal{F}_{1-2}$ is given by $1/\{[(1/\epsilon_1) + (1/\epsilon_2) - 1] + n[(2/\epsilon_s) - 1]\}$.

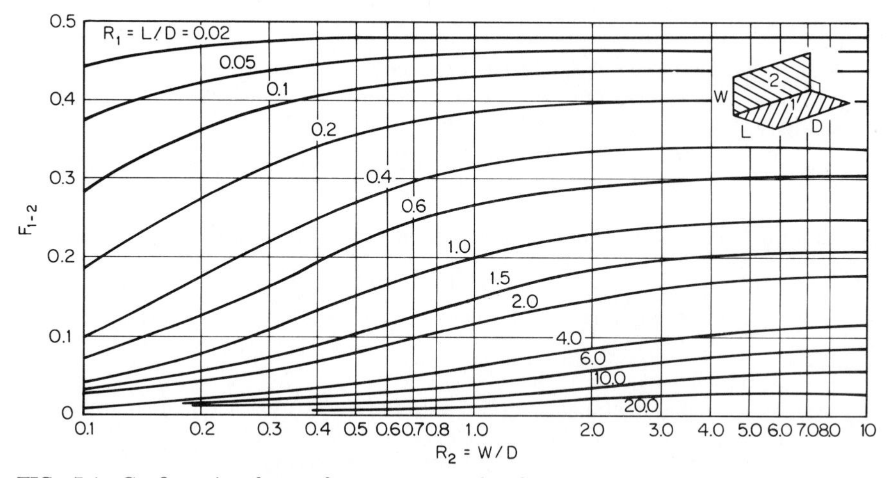

FIG. 7-1 Configuration factors for two rectangular figures with a common edge, at right angles [7].

In particular, for any two Lambertian surfaces (surfaces that emit or reflect with an intensity independent of angle, a condition approximately satisfied by most nonmetallic, tarnished, oxidized, or rough surfaces), the fraction F_{1-2} of total energy from one of them, designated 1, that is intercepted by the other, designated 2, is given by

$$F_{1\text{-}2} = \frac{1}{A_1} \int_{A1} \int_{A2} \left(\frac{\cos \theta_1 \cos \theta_2}{\pi r_{1,2}^2} \right) dA_2 \, dA_1$$

where A_1 and A_2 are the areas of the emitting and receiving surfaces, $r_{1,2}$ is the distance between the surfaces, and θ_1 and θ_2 are the angles between line $r_{1,2}$ and the normals to the two surfaces.

For black surfaces, the purely geometrically derived factor F_{1-2} is an acceptable substitute for the radiant interchange factor, evaluation of which requires one to consider the emittances of all surfaces and is beyond the scope of this book. For instance, in Example 7-1 the expression for $\mathcal{F}_{1-2}$ reduces to unity if $\epsilon_1 = \epsilon_2 = \epsilon_s = 1.0$, which is equal to the solution for F_{1-2} for closely spaced parallel rectangles.

F_{1-2} is called the "configuration factor," "the geometric factor," or "the shape factor." This has been integrated for many common configurations, two of which are plotted in Figs. 7-1 and 7-2. Other figures are presented in the literature, particularly Krieth [2], Chapman [7], and Siegel and Howell [9].

For configurations not presented specifically, geometric relationships exist whereby existing data can be combined to yield the required data. These relationships are called "shape-factor algebra" and are summarized as follows:

1. *Reciprocity:* For two areas 1 and 2,

$$A_1 F_{1\text{-}2} = A_2 F_{2\text{-}1}$$

2. *Addition:* If a surface i is subdivided into any number n of subsurfaces, $i1$, $i2$, . . . , in,

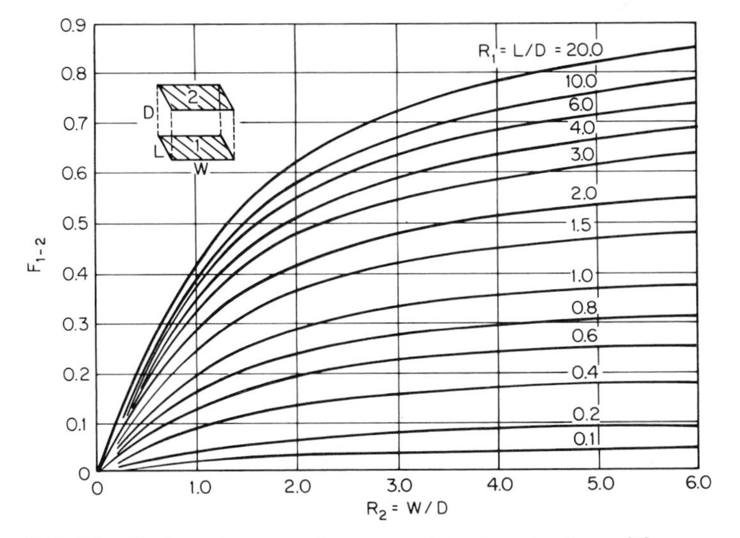

FIG. 7-2 Configuration factors for two parallel rectangular figures [7].

$$A_i F_{i-j} = \sum_n A_{in} F_{in} - j$$

3. *Enclosure:* If a surface i is completely enclosed by n other surfaces, the sum of all configuration factors from i to the other surfaces is 1:

$$\sum_{j=1}^{n} F_{i-j} = 1.0$$

The enclosure property also serves as a check to determine whether all the separately determined configuration factors are correct.

2. *Evaluate the configuration factors.*

The furnace can be sketched as in Fig. 7-3. Each of its six faces has a roman numeral and (for convenience later in the problem) the temperature of that face is shown. Since heat radiated from each face will impinge on every other face, there are 6(6 − 1) = 30

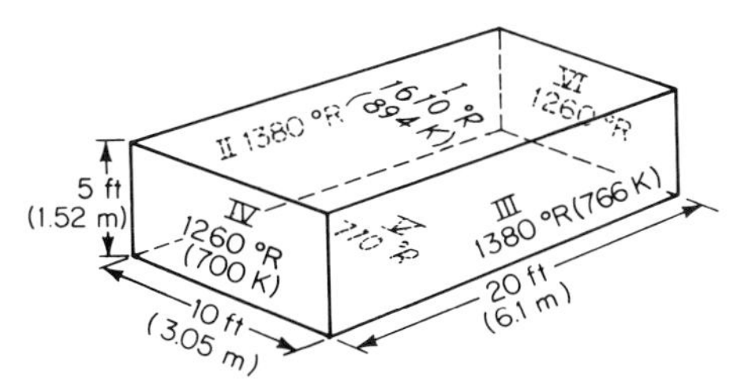

FIG. 7-3 Sketch of furnace for Example 7-2.

configuration factors to be determined. However, because of the reciprocity property and symmetry, fewer calculation steps will be needed.

1. $A_I F_{I-VI}$: Refer to Fig. 7-1. Here, $D = 10$, $L = 20$, and $W = 5$. Then $R_1 = L/D = 2.0$ and $R_2 = W/D = 0.5$. From the graph, $F_{I-VI} = 0.08$. Now, $A_I = 10 \times 20 = 200$ ft^2, so $A_I F_{I-VI} = 16$ ft^2. By reciprocity, $A_{VI} F_{VI-I} = 16$ ft^2 also. And by symmetry, $F_{I-VI} = F_{I-IV}$, so $A_I F_{I-IV} = A_{IV} F_{IV-I} = 16$ ft^2 as well.
2. $A_{II} F_{II-III}$: See Fig. 7-2. Here, $D = 10$, $L = 20$, and $W = 5$. Then $R_1 = 2.0$ and $R_2 = 0.5$. From the graph, $F_{II-III} = F_{III-II} = 0.167$. Now, $A_{II} = 100$ ft^2, so $A_{II} F_{II-III} = 16.7$ ft^2 $= A_{III} F_{III-II}$
3. $A_{IV} F_{IV-V}$: $D = 10$, $L = 5$, and $W = 20$. $R_1 = 0.5$, $R_2 = 0.25$, and $F_{IV-V} = 0.31$. $A_{IV} = 50$ ft^2, so $A_{IV} F_{IV-V} = 15.5$ ft^2 $= A_V F_{V-IV} = A_{VI} F_{VI-V} = A_V F_{V-VI}$.
4. $A_I F_{I-III}$: $D = 20$, $L = 10$, and $W = 5$. $R_1 = L/D = 0.50$, $R_2 = W/D = 0.25$, and $F_{I-III} = 0.168$. $A_I = 200$ ft^2, so $A_I F_{I-III} = 33.60$ ft^2 $= A_I F_{I-II} = A_{III} F_{III-I} = A_{II} F_{II-I}$.
5. $A_I F_{I-V}$: $D = 5$, $L = 20$, and $W = 10$. $R_1 = 4.0$, $R_2 = 2.0$, and $F_{I-V} = 0.508$. $A_I = 200$ ft^2, so $A_I F_{I-V} = 116$ ft^2 $= A_V F_{V-I}$.
6. $A_{II} F_{II-IV}$: $D = 5$, $L = 20$, $W = 10$. $R_1 = 4.0$, $R_2 = 2.0$, and $F_{II-IV} = 0.085$. $A_{II} = 100$ ft^2, so $A_{II} F_{II-IV} = 8.5$ ft^2 $= A_{IV} F_{IV-II} = A_{II} F_{II-VI} = A_{VI} F_{VI-II}$. And by symmetry, $A_{III} F_{III-IV} = A_{IV} F_{IV-III} = A_{III} F_{III-VI} = A_{VI} F_{VI-III} = 8.5$ ft^2.
7. $A_{II} F_{II-V}$: $D = 20$, $L = 5$, and $W = 10$. $R_1 = 0.25$, $R_2 = 0.5$, and $F_{II-V} = 0.33$. $A_{II} = 100$ ft^2, so $A_{II} F_{II-V} = 33$ ft^2 $= A_V F_{V-II} = A_{III} F_{III-V} = A_V F_{V-III}$.
8. $A_{IV} F_{IV-VI}$: $D = 20$, $L = 10$, and $W = 5$. $R_1 = 0.5$, $R_2 = 0.25$, $F_{IV-VI} = 0.038$. $A_{IV} = 50$ ft^2, so $A_{IV} F_{IV-VI} = 1.90$ ft^2 $= A_{VI} F_{VI-IV}$.

   Throughout, $F_{i,j}$ is found by dividing $A_i F_{i,j}$ by A_i, if $F_{i,j}$ is not already explicit.

3. Check the accuracy of all configuration factors.

Use the enclosure principle described in step 1. For surface I, $F_{I-II} + F_{I-III} + F_{I-IV} + F_{I-V} + F_{I-VI} = 0.168 + 0.168 + 0.08 + 0.508 + 0.08 = 1.004$, which is close enough to 1.0.

For surface II, the sum of the configuration factors is $(A_{II} F_{II-I}/A_{II}) + F_{II-III} + F_{II-IV} + F_{II-V} + F_{II-VI} = 33.6/100 + 0.167 + 0.085 + 0.33 + 0.085 = 1.003$. Close enough. The sums of the configuration factors for the other four surfaces are calculated similarly.

4. Calculate heat transfer from each surface.

Let q_i equal the heat emitted by surface i, and $q_{i,j}$ the heat emitted by surface i that impinges on surface j. Then, $q_i = \sum_j q_{i,j}$. For surface I,

$$q_I = q_{I\text{-}II} + q_{I\text{-}III} + q_{I\text{-}IV} + q_{I\text{-}V} + q_{I\text{-}VI} = \sigma A_I[F_{I\text{-}II}(T_I^4 - T_{II}^4) + F_{I\text{-}III}(T_I^4 - T_{III}^4) + F_{I\text{-}IV}(T_I^4 - T_{IV}^4) + F_{I\text{-}V}(T_I^4 - T_V^4) + F_{I\text{-}VI}(T_I^4 - T_{VI}^4)]$$

This can be simplified by considering symmetry; thus,

$$q_I = \sigma A_I[2F_{I\text{-}II}(T_I^4 - T_{II}^4) + 2F_{I\text{-}IV}(T_I^4 - T_{IV}^4) + F_{I\text{-}V}(T_I^4 - T_V^4)]$$

where σ is the Boltzmann constant and T is absolute temperature. Thus,

$$q_I = (0.1713 \times 10^{-8})(200)[2(0.168)(1610^4 - 1380^4) + 2(0.08)(1610^4 - 1260^4) + 0.508(1610^4 - 710^4)] = 171.1 \times 10^4 \text{ Btu/h (496 kW)}$$

Similarly, for the other five surfaces: $q_{II} = 4.489 \times 10^4$ Btu/h (13 kW) $= q_{III}$; $q_{IV} = -8.711 \times 10^4$ Btu/h (-25.3 kW) $= q_{VI}$; and $q_V = -162.7 \times 10^4$ Btu/h (-472 kW).

5. Check the results.

Since the system as a whole neither gains nor loses heat, the sum of the heat transferred should algebraically be zero. Now, $171.1 + 2(4.489) + 2(-8.711) + (-162.7) = 0$, so the results do check.

Related Calculations: If the six surfaces are not black but gray (in the radiation sense), it is nominally necessary to set up and solve six simultaneous equations in six unknowns. In practice, however, the network can be simplified by combining two or more surfaces (the two smaller end walls, for instance) into one node. Once this is done and the configuration factors are calculated, the next step is to construct a radiosity network (since each surface is assumed diffuse, all energy leaving it is equally distributed directionally and can therefore be taken as the radiosity of the surface rather than its emissive power). Then, using standard mathematical network-solution techniques, create and solve an equivalent network with direct connections between nodes representing the surfaces. For details, see Oppenheim [8].

7-3 Effect of Solar Heat on a Storage Tank

A flat-topped, nitrogen-blanketed atmospheric-pressure tank in a plant at Texas City, Texas, has a diameter of 30 ft and a height of 20 ft (9.1 m diameter and 6.1 m high) and is half full of ethanol at 85°F (302 K). As a first step in calculating nitrogen flow rates into and out of the tank during operations, calculate the solar heating of the tank and the tank skin temperature in the ullage space at a maximum-temperature condition. The tank has a coating of white zinc oxide paint, whose solar absorptance is 0.18. The latitude of Texas City is about N29°20′. For the maximum-temperature condition, select noon on June 20, the summer solstice, when the solar declination is 23.5°. Assume that the solar constant (the solar flux on a surface perpendicular to the solar vector) is 343 Btu/(h)(ft^2) (1080 W/m^2), the air temperature is 90°F (305 K), and the effective sky temperature is 5°F (258 K). Also assume that surrounding structural and other elements (such as hot pipes) are at 105°F (314 K) and have a radiant interchange factor of 0.2 with the tank and that the effective film coefficients for convection heat transfer between (1) the air and the outside of the tank and (2) the inside of the tank and the contained material are 0.72 and 0.75 Btu/(h)(ft^2)(°F) [4.08 and 4.25 W/(m^2)(K)], respectively.

Calculation Procedure:

1. Calculate the solar-heat input to the tank.

Since the sun, although an extremely powerful emitter, subtends a very small solid angle, it has an only minute radiant interchange factor with objects on earth. The earth's orbital distance from the sun is nearly constant throughout the year. Therefore, it is a valid simplification to consider solar radiation simply as a heat source independent of the radiation environment and governed solely by the solar absorptance of each surface and the angular relationship of the surface to the solar vector.

The magnitude of the solar heating is indicated by the so-called solar constant. In space, at the radius of the earth's orbit, the solar constant is about 443 Btu/(h)(ft^2) (1396 W/m^2). However, solar radiation is attenuated by passage through the atmosphere; it is also reflected diffusely by the atmosphere, which itself varies greatly in composition. Table 7-2 provides representative values of the solar constant for use at ground level, as well as of the apparent daytime temperature of the sky for radiation purposes.

Since the solar constant G_n is a measure of total solar radiation *perpendicular to* the solar vector, it is necessary to also factor in the actual angle which the solar vector makes with the surface(s) being heated. This takes into account the geographic location, the time of year, the time of day, and the geometry of the surface and gives a corrected solar constant G_i.

For a horizontal surface, such as the tank roof, $G_i = G_n \cos Z$, where $Z = \cos^{-1} (\sin \phi \sin \delta_s + \cos \phi \cos \delta_s \cos h)$, ϕ is the latitude, δ_s is the solar declination, and h the hour angle, measured from 0° at high noon. In the present case, $Z = \cos^{-1} (\sin 29°20' \sin 23.5° + \cos 29°20' \cos 23.5° \cos 0°) = \cos^{-1} 0.995 = 5°50'$, and $G_i = 343(0.995) = 341.2$ Btu/(h)(ft^2).

For a surface that is tilted ψ° from horizontal and whose surface normal has an azimuth of α° from due south (westward being positive), $G_i = G_n(\cos |Z - \psi| - \sin Z \sin \psi + \sin Z \sin \psi \cos |A - \alpha|)$, where $A = \sin^{-1}\{\cos \delta_s \sin h/[\cos (90 - Z)]\}$. In the present case, $A = 0°$ because $\sin h = 0$.

Because the solar effect is distributed around the vertical surface of the tank to a varying degree (the effect being strongest from the south, since the sun is in the south), select wall segments 30° apart and calculate each separately. In the G_i equation, α will thus

TABLE 7-2 Representative Values of Solar Constant and Sky Temperature

Conditions for total normal incident solar radiation	Solar constant		Effective sky temp., °F	Effective sky temp., K
	Btu/(h)(ft^2)	kJ/(s)(m^2)		
Southwestern United States, June, 6:00 A.M., extreme	252.3	0.797	−30 to −22	239 to 243
Southwestern United States, June, 12:00 M., extreme	385.1	1.216	−30 to −22	239 to 243
Southwestern United States, December, 9:00 A.M., extreme	307.7	0.972	−30 to −22	239 to 243
Southwestern United States, December, 12:00 M., extreme	396.2	1.251	−30 to −22	239 to 243
NASA recommended high design value, 12:00 M.	363.0	1.146	—	—
NASA recommended low design value, 12:00 M.	75.0	0.237	—	—
Southern United States, maximum for extremely bad weather	111.0	0.350	—	—
Southern United States desert, maximum for extremely bad weather	177.0	0.559	—	—

Source: Daniels [13].

assume values ranging from $-90°$ (facing due east) to $+90°$ (due west); the northern half of the tank will be in shadow. Because the wall is vertical, $\psi = 90°$.

It is also necessary to take sky radiation into account, that is, sunlight scattered by the atmosphere and reflected diffusely and which reaches all surfaces of the tank, including those not hit by direct sunlight because they are in shadow. This diffuse radiation G_s varies greatly but is generally small, between about 2.2 Btu/(h)(ft^2) (6.93 W/m^2) on a clear day and 44.2 Btu/(h)(ft^2) (139 W/m^2) on a cloudy day. For the day as described, assume that G_s is 25 Btu/(h)(ft^2). This value must be added to all surfaces, including those in shadow.

Finally, take the solar absorptance of the paint α_s into account. Thus, the solar heat absorbed q_s equals $\alpha_s(G_i + G_s)$. The calculations can be summarized as follows:

	Roof	1	2	3	4	5	6	7
Segment (azimuth)		−90	−60	−30	0	30	60	90
G_i	341.2	0	17.4	30.2	34.9	30.2	17.4	0
$G_i + G_s$	366.2	25	42.4	55.2	59.9	55.2	42.4	25
q_s/A, Btu/(h)(ft^2)	65.92	4.50	7.63	9.94	10.8	9.94	7.63	4.50
q_s/A, W/m^2	208	14.2	24.0	31.3	34.0	31.3	24.0	14.2

Since this calculation procedure treats the tank as if it had 12 flat sides, the calculated G_i for segments 1 and 7 is zero. Of course, G_i is also zero for the shaded half of the tank [and $q_s/A = 4.5$ Btu/(h)(ft^2)].

2. *Calculate the equilibrium temperature of each of the tank surfaces.*

It can be shown that conduction between the segments is negligible. Then, at equilibrium, each segment must satisfy the heat-balance equation, that is, solar-heat absorption + net heat input by radiation + heat transferred in by outside convection + heat transferred in by inside convection = 0, or

$$q_s/A + \sigma\mathcal{F}_o(T_o^4 - T_w^4) + \sigma\mathcal{F}_R(T_R^4 - T_w^4) + h_a(T_a - T_w) + h_i(T_i - T_w) = 0$$

where T_w is the tank-wall temperature; subscript o refers to surrounding structural and other elements having a radiant interchange factor $\mathcal{F}_o$ with the segments; subscript R refers to the atmosphere, having an equivalent radiation temperature T_R and a radiant interchange factor $\mathcal{F}_R$; subscript a refers to the air surrounding the tank, and subscript i refers to the gas inside the tank. The heat balance for the roof is solved as follows (similar calculations can be made for each segment of the tank wall):

Now, $q_s/A = 65.92$ Btu/(h)(ft^2), $T_o = 105°\text{F} = 565°\text{R}$ (due to hot pipes and other equipment in the vicinity), $\mathcal{F}_o = 0.2$, $T_R = 5°\text{F} = 465°\text{R}$, a good assumption for $\mathcal{F}_R$ is 0.75, $h_a = 0.72$, and $h_i = 0.75$. Therefore,

$$65.92 + (0.1713 \times 10^{-8})(0.2)(565^4 - T_w^4) + (0.1713 \times 10^{-8})(0.75)(465^4 - T_w^4) + 0.72(90 + 460 - T_w) + 0.75(85 + 460 - T_w) = 0$$

$$65.92 + 34.91 - (3.426 \times 10^{-10} + 12.848 \times 10^{-10})T_w^4 + 60.07 + 396 - (0.72 + 0.75)T_w + 408 = 0$$

$$1.6274 \times 10^{-9}T_w^4 + 1.47T_w = 965$$

This is solved by trial and error, to yield $T_w = 553°R = 93°F$ (307 K). Note that if the paint had been black, α_s might have been 0.97 instead of 0.18. In that case, the temperature would have been about 200°F (366 K).

The same procedure is then applied to each of the other tank segments.

7-4 Heat Loss from an Uninsulated Surface to Air

A steam line with a diameter of 3.5 in (0.089 m) and a length of 50 ft (15.2 m) transports steam at 320°F (433 K). The carbon steel pipe [thermal conductivity of 25 Btu/(h)(ft)(°F) or 142 W/(m²)(K)] is not insulated. Its emissivity is 0.8. Calculate the heat loss for calm air and also for a wind velocity of 15 mi/h (24 km/h), if the air temperature is 68°F (293 K).

Calculation Procedure:

1. Calculate the heat loss due to radiation.

Because the coefficient for heat transfer from the outside of the pipe as a result of radiation and convection is much less than all other heat-transfer coefficients involved in this example, the surface temperature of the pipe can be assumed to be that of the steam. To calculate the heat loss, use the straightforward radiation formula

$$\frac{Q}{A} = 0.1713\epsilon\left[\left(\frac{T_s}{100}\right)^4 - \left(\frac{T_a}{100}\right)^4\right]$$

where Q is heat loss in British thermal units per hour, A is heat-transfer area in square feet, T_s is absolute temperature of the surface in degrees Rankine, T_a is absolute temperature of the air, and ϵ is the emissivity of the pipe. (Note that in this version of the formula, the 10^{-8} portion of the Stefan-Boltzmann constant is built into the temperature terms.)

Thus,

$$\frac{Q}{A} = 0.1713(0.8)\left[\left(\frac{460 + 320}{100}\right)^4 - \left(\frac{460 + 68}{100}\right)^4\right] = 401 \text{ Btu/(h)(ft}^2\text{) (1264 W/m}^2\text{)}$$

2. Calculate the heat loss as a result of natural convection in calm air.

Use the formula

$$\frac{Q}{A} = \frac{0.27\,\Delta T^{1.25}}{D^{0.25}}$$

where $\Delta T = T_s - T_a$ in degrees Fahrenheit and D is pipe diameter in feet. Thus,

$$\frac{Q}{A} = \frac{0.27(320 - 68)^{1.25}}{(3.5/12)^{0.25}} = 369 \text{ Btu/(h)(ft}^2\text{) (1164 W/m}^2\text{)}$$

3. Calculate the total heat loss for the pipe in calm air.

Now, $Q = (Q/A)A$, and $A = \pi(3.5/12)50 = 45.81$ ft^2 (4.26 m^2), so $Q = (401 + 369)(45.81) = 35{,}270$ Btu/h (10,330 W).

4. Calculate the heat loss by convection for a wind velocity of 15 mi/h.

First, determine the mass velocity G of the air: $G = \rho v$, where ρ is density and v is linear velocity. For air, $\rho = 0.075$ lb/ft^3 (1.20 kg/m^3). In this problem, $v = 15$ mi/h (5280 ft/mi) = 79,200 ft/h (24,140 m/h), so $G = 0.075(79{,}200) = 5940$ lb/(h)(ft^2) [29,000 kg/(h)(m^2)].

Next, determine the heat-transfer coefficient, using the formula $h = 0.11cG^{0.6}/D^{0.4}$, where h is heat-transfer coefficient in British thermal units per hour per square foot per degree Fahrenheit, c is specific heat in British thermal units per pound per degree Fahrenheit (0.24 for air), G is mass velocity in pounds per hour per square foot, and D is diameter in feet. Thus, $h = 0.11(0.24)(5940)^{0.6}/(3.5/12)^{0.4} = 7.94$ Btu/(h)(ft^2)(°F) [45.05 W/(m^2)(K)].

Finally, use this coefficient to determine the heat loss due to convection via the formula $Q/A = h(T_s - T_a)$. Thus, $Q/A = 7.94(320 - 68) = 2000.9$ Btu/(h)(ft^2) (6307 W/m^2).

5. Calculate the total heat loss for the pipe when the wind velocity is 15 mi/h.

As in step 3, $Q = (401 + 2000.9)(45.81) = 110{,}030$ Btu/h (32,240 W).

7-5 Heat Loss from an Insulated Surface to Air

Calculate the heat loss from the steam line in Example 7-4 if it has insulation 2 in (0.050 m) thick having a thermal conductivity of 0.05 Btu/(h)(ft)(°F) [0.086 W/(m)(K)]. The inside diameter of the pipe is 3 in (0.076 m), and the heat-transfer coefficient from the condensing steam to the pipe wall is 1500 Btu/(h)(ft^2)(°F) [8500 W/(m^2)(K)]. Assume that the wind velocity is 15 mi/h (24 km/h). The pipe is illustrated in Fig. 7-4.

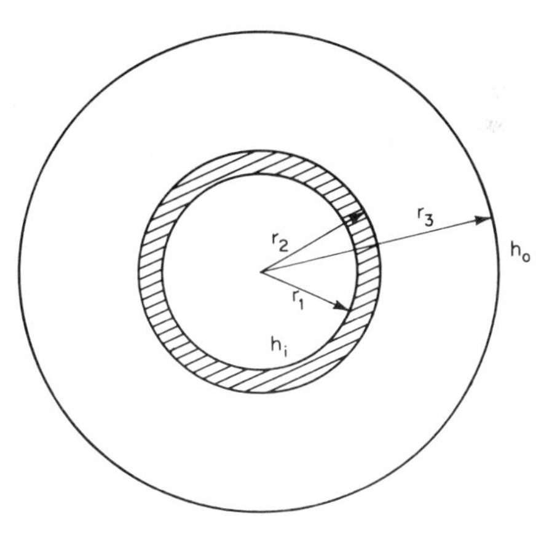

r_1 = 3.0/2 in = 1.5 in = 0.125 ft (0.0381 m)
r_2 = 3.5/2 in = 1.75 in = 0.1458 ft (0.0445 m)
r_3 = 7.5/2 in = 3.75 in = 0.3125 ft (0.0953 m)

FIG. 7-4 Cross section of insulated pipe in Example 7-5.

Calculation Procedure:

1. Set out the appropriate overall heat-transfer equation.

The heat loss can be calculated from the equation $Q = UA_o(T_s - T_a)$, where Q is heat loss, U is overall heat-transfer coefficient, A_o is area of the outside surface of the insulation ($= 2\pi Lr_3$; see Fig. 7-4), T_s

is steam temperature, and T_a is air temperature. The overall heat-transfer coefficient can be calculated with the equation

$$U = \frac{1}{\dfrac{r_3}{r_1 h_i} + \dfrac{r_3 \ln (r_2/r_1)}{k_1} + \dfrac{r_3 \ln (r_3/r_2)}{k_2} + \dfrac{1}{h_o}}$$

where the r_i are as described in Fig. 7-4, k_1 is the thermal conductivity of the pipe, k_2 is that of the insulation, and h_o and h_i are the outside and inside heat-transfer coefficients, respectively.

2. *Calculate h_o.*

The outside heat-transfer coefficient h_o is the sum of the heat-transfer coefficient for convection to the wind h_c and the transfer of heat as a result of radiation h_r. The latter is approximately 1.0 Btu/(h)(ft²)(°F) [5.68 W/(m²)(K)], and we will use that value here. From step 4 of Example 7-4, $h_c = 7.94$. Thus, $h_o = h_c + h_r = 7.94 + 1.0 = 8.94$ Btu/(h)(ft²)(°F) [50.78 W/(m²)(K)].

3. *Calculate U.*

Substituting into the equation in step 1,

$$U = \frac{1}{\dfrac{0.3125}{(0.125)(1500)} + \dfrac{0.3125 \ln (1.75/1.5)}{25} + \dfrac{0.3125 \ln (3.75/1.75)}{0.05} + \dfrac{1}{8.94}}$$

$$= 0.205 \text{ Btu/(h)(ft}^2\text{)(°F) [1.16 W/(m}^2\text{)(K)]}$$

4. *Calculate the heat loss.*

From step 1, $Q = 0.205(2\pi)(50)(0.3125)(320 - 68) = 5070$ Btu/h (1486 W).

7-6 Heat Loss from a Buried Line

A steam line with a diameter of 12.75 in (0.3239 m) is buried with its center 6 ft (1.829 m) below the surface in soil having an average thermal conductivity of 0.3 Btu/(h)(ft)(°F) [0.52 W/(m)(K)]. Calculate the heat loss if the pipe is 200 ft (60.96 m) long, the steam is saturated at a temperature of 320°F (433 K), and the surface of the soil is at a temperature of 40°F (277 K).

Calculation Procedure:

1. *Select the appropriate heat-transfer equation.*

The heat loss can be calculated from the equation

$$Q = Sk(T_1 - T_2)$$

where Q is heat loss, k is thermal conductivity of the soil, T_1 is surface temperature of the pipe, T_2 is surface temperature of the soil, and S is a shape factor.

2. Determine the shape factor.

Use the equation $S = 2\pi L/\cosh^{-1}(2z/D)$ (when z is much less than L), where L is length, z is distance from ground surface to the center of the pipe, and D is diameter. Now,

$$\cosh^{-1}(2z/D) = \ln\{(2z/D) + [(2z/D)^2 - 1]^{1/2}\}$$

For this example, $2z/D = 2(6)/(12.75/12) = 11.294$, and $\cosh^{-1}(2z/D) = \ln[11.294 + (11.294^2 - 1)^{1/2}] = 3.116$. Thus, $S = 2\pi 200/3.116 = 403.29$ ft (122.92 m).

3. Calculate the heat loss.

Because the heat-transfer resistance through the soil is much greater than all other resistances to heat transfer, the surface of the pipe can be assumed to be at the temperature of the steam. Thus, $Q = 403.29(0.3)(320 - 40) = 33{,}876$ Btu/h (9926 W).

Related Calculations: This approach can be used for any geometry for which shape factors can be evaluated. Shape factors for many other geometries are presented in several of the references, including Krieth [2] and Holman [4].

7-7 Conduction of Heat in the Unsteady State: Cooling Time

A steel sphere with a radius of 0.5 in (0.0127 m) at a temperature of 500°F (533 K) is suddenly immersed in a water bath. The water temperature is 68°F (293 K) and the heat-transfer coefficient from the steel surface to the water is 200 Btu/(h)(ft^2)(°F) [1130 W/(m^2)(K)]. Calculate the time required for the center of the sphere to reach a temperature of 250°F (394 K). The thermal diffusivity α of the steel is 0.45 ft^2/h (0.0418 m^2/h), and the thermal conductivity is 25 Btu/(h)(ft)(°F) [43 W/(m)(K)].

Calculation Procedure:

1. Referring to the curves in Fig. 7-5, calculate $k/(hr_o)$.

$$\frac{k}{hr_o} = \frac{25}{200(0.5/12)} = 3.0$$

2. Calculate $\alpha\tau/r_o^2$.

$$\frac{\theta_o}{\theta_i} = \frac{250 - 68}{500 - 68} = 0.421$$

3. Determine $\alpha\tau/r_o^2$.

From Fig. 7-5, with $k/(hr_o) = 3.0$ and $\theta_o/\theta_i = 0.421$, $\alpha\tau/r_o^2 = 1.1$.

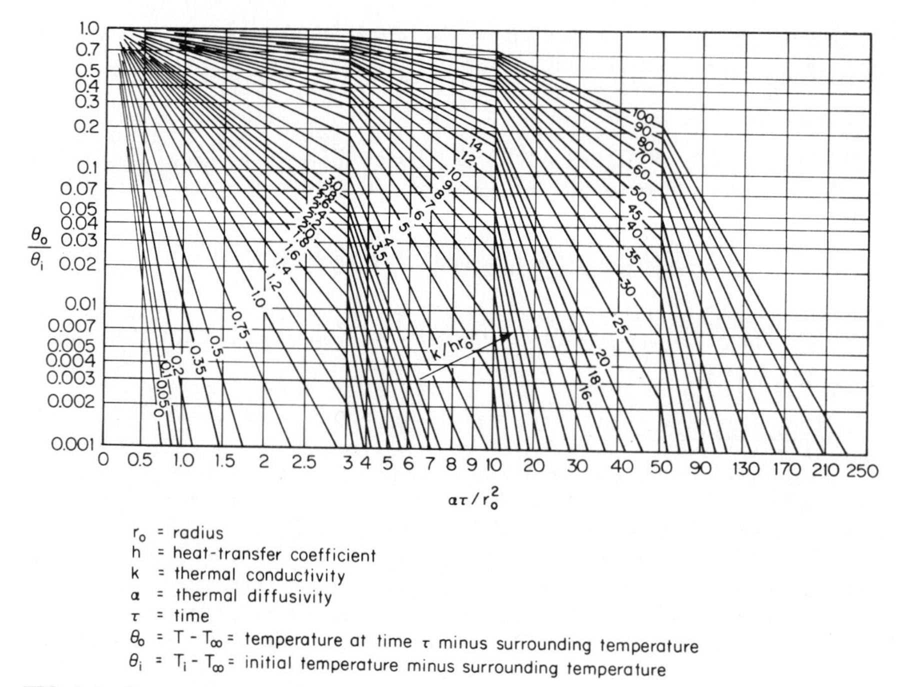

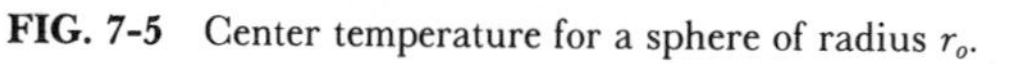
FIG. 7-5 Center temperature for a sphere of radius r_o.

4. Calculate the time required.

Now, $\alpha\tau/r_o^2 = 1.1$, so

$$\tau = \frac{(1.1r_o^2)}{\alpha}$$

$$= \frac{1.1(0.5/12)^2}{0.45}$$

$$= 0.0042 \text{ h}$$

$$= 15.3 \text{ s}$$

7-8 Conduction of Heat in the Unsteady State: Temperature

For the sphere described in the preceding example, calculate the temperature at the center of the sphere after a period of 1 min.

Calculation Procedure:

1. Referring to the curves in Fig. 7-5, calculate $\alpha\tau/r_o^2$.

Here, $\alpha\tau/r_o^2 = 0.45(1/60)/(0.5/12)^2 = 4.32$.

2. Determine θ_o/θ_i.

From Fig. 7-5, with $\alpha\tau/r_o^2 = 4.32$ and $k/(hr_o) = 3.0$, $\theta_o/\theta_i = 0.02$.

3. Calculate the temperature after a period of 1 min.

Now, $\theta_o/\theta_i = (T - T_\infty)/(T_i - T_\infty) = 0.02$, so

$$T = T_\infty + 0.02(T_i - T_\infty)$$
$$= 68 + 0.02(500 - 68)$$
$$= 76.6°\text{F (298 K)}$$

Related Calculations: The procedure outlined in these two problems can also be used for geometries other than spheres. Figure 7-6 can be used for cylinders, and Fig. 7-7 can be used for slabs.

7-9 Conduction of Heat in the Unsteady State: Temperature Distribution

Calculate the temperature distribution in a steel cylinder 1 min after the surrounding temperature is suddenly changed. Conditions are as follows:

D_o = outside diameter = 2 in (0.05 m)
k = thermal conductivity = 25 Btu/(h)(ft)(°F) [43 W/(m)(K)]
α = thermal diffusivity = 0.45 ft^2/h (0.0418 m^2/h)
h = heat-transfer coefficient = 300 Btu/(h)(ft^2)(°F) [1700 W/(m^2)(K)]
T_1 = initial cylinder temperature = 68°F (293 K)
T_∞ = surrounding temperature = 1000°F (811 K)

Calculation Procedure:

1. With regard to the curves in Figs. 7-6 and 7-9, calculate $k/(hr_o)$.

Now, $k/(hr_o) = 25/300(1/12) = 1.0$.

2. Calculate $\alpha\tau/r_o^2$.

Now, $\alpha\tau/r_o^2 = 0.45(1/60)/(1/12)^2 = 1.08$.

3. Determine θ_o/θ_i.

From Fig. 7-6, for $k/(hr_o) = 1.0$ and $\alpha\tau/r_o^2 = 1.08$, $\theta_o/\theta_i = 0.20$.

4. Determine θ_o.

Since $\theta_o/\theta_i = 0.2 = (T - T_\infty)/(T_i - T_\infty)$, $\theta_o = 0.2(68 - 1000) = -186.4°\text{F}$.

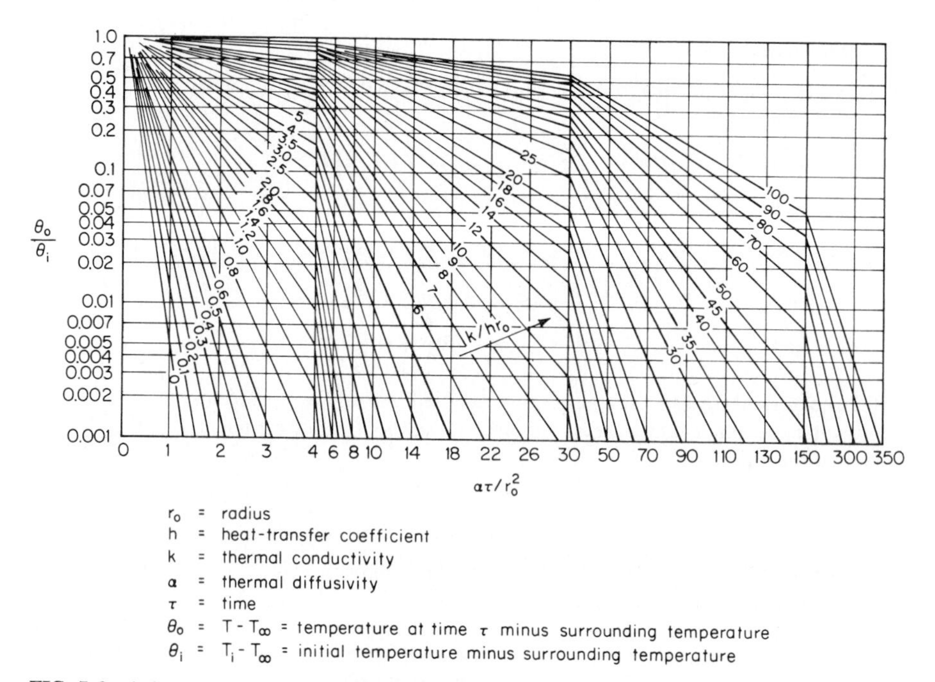

FIG. 7-6 Axis temperature for a cylinder of radius r_o.

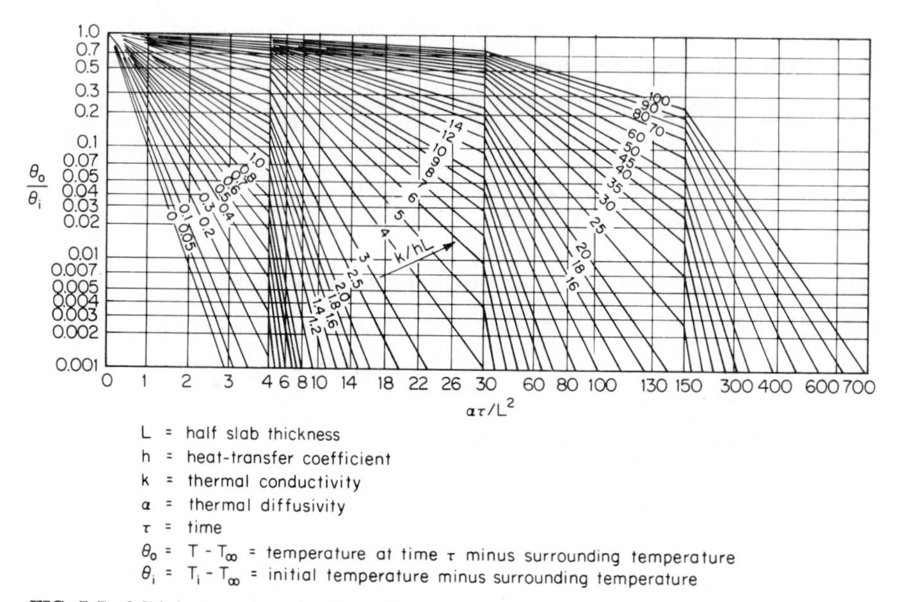

FIG. 7-7 Midplane temperature for a plate of thickness $2L$.

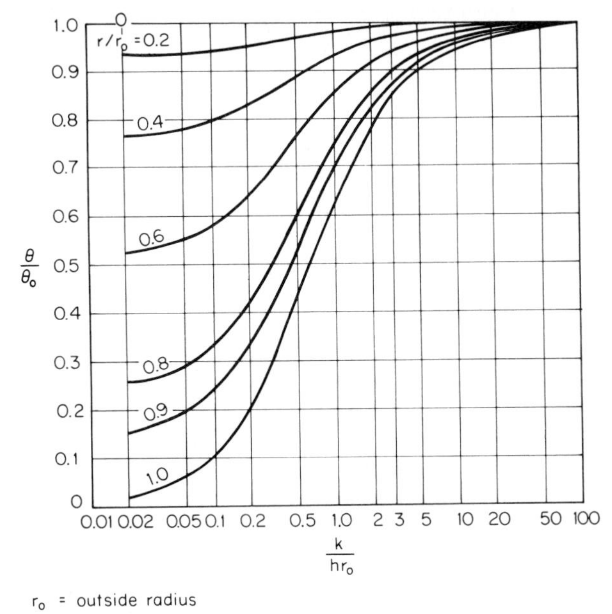

FIG. 7-8 Temperature as a function of center temperature for a sphere.

5. Determine θ_o/θ_i as a function of radius.

From Fig. 7-9, for $k/(hr_o) = 1.0$, the following parameters are obtained and the following values of θ calculated:

r/r_o	θ/θ_o	$\theta = (\theta/\theta_o)(-186.4)$
0	1.0	−186.4°F
0.2	0.98	−182.7
0.4	0.93	−173.4
0.6	0.86	−160.3
0.8	0.75	−139.8
1.0	0.64	−119.3

6. Determine temperature as a function of radius.

Since $T = T_\infty + \theta = 1000 + \theta$, the following temperatures are found:

r/r_o	T, °F	T, K
0	813.6	707.2
0.2	817.3	709.3
0.4	826.6	714.4
0.6	839.7	721.7
0.8	860.2	733.1
1.0	880.7	744.5

Related Calculations: The procedure outlined here can also be used for solids with other geometries. Figures 7-5 and 7-8 can be used for spheres, and Figs. 7-7 and 7-10 with slabs.

7-10 Conduction of Heat from a Belt Cooler

A stainless steel belt cooler with a width of 3 ft (0.914 m), a thickness of ⅛ in (3.175 mm), and a length of 100 ft (30.48 m) is to be used to cool 10,000 lb/h (4535.9 kg/h) of material with the physical properties listed below. The bottom of the belt is sprayed with water at 86°F (303 K), and the heat-transfer coefficient h from the belt to the water is 500 Btu/

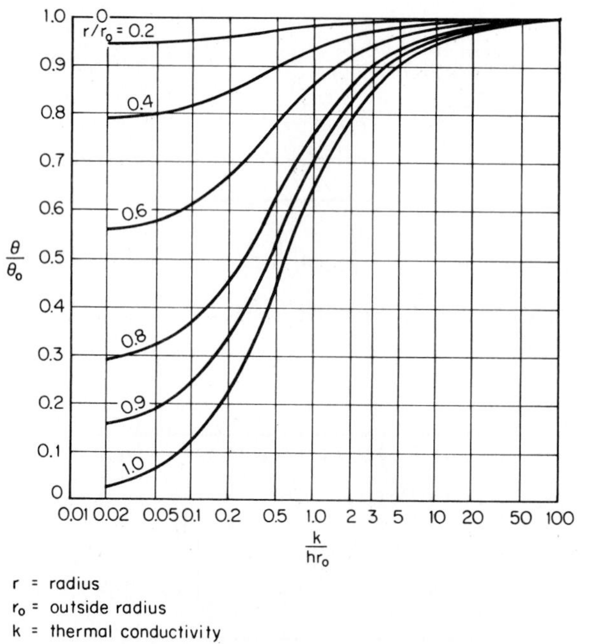

FIG. 7-9 Temperature as a function of axis temperature in a cylinder.

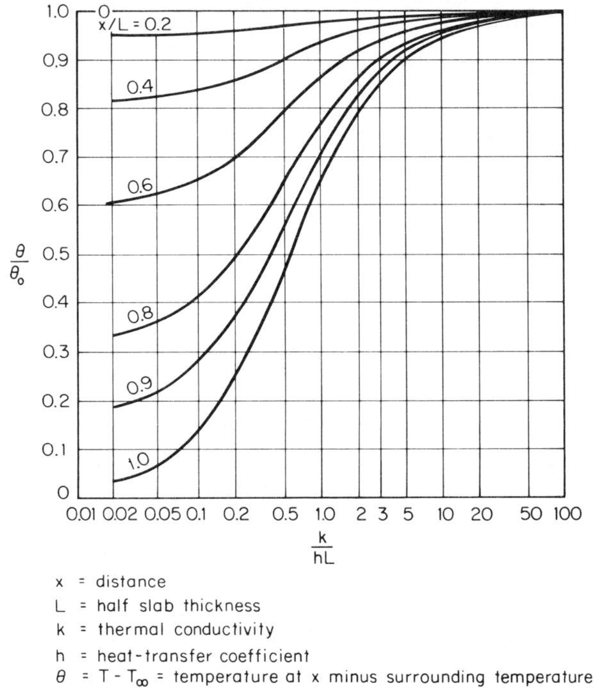

x = distance
L = half slab thickness
k = thermal conductivity
h = heat-transfer coefficient
θ = $T - T_\infty$ = temperature at x minus surrounding temperature
θ_o = $T_i - T_\infty$ = center temperature minus surrounding temperature

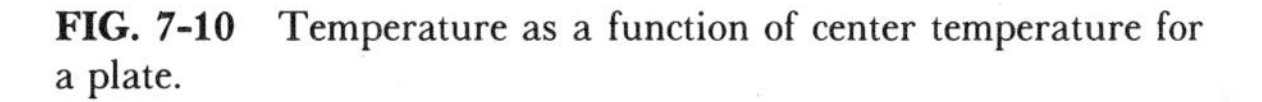

FIG. 7-10 Temperature as a function of center temperature for a plate.

(h)(ft^2)(°F) [2835 W/(m^2)(K)]. Calculate the surface temperature of the material as it leaves the belt if the material is placed on the belt at a temperature of 400°F (477 K) and the belt moves at a speed of 150 ft/min (45.72 m/min).

Physical Properties of the Material

ρ = density = 60 lb/ft^3 (961.1 kg/m^3)
k = thermal conductivity = 0.10 Btu/(h)(ft)(°F) [0.17 W/(m)(K)]
α = thermal diffusivity = 0.0042 ft^2/h (3.871 cm^3/h)
c = specific heat = 0.4 Btu/(lb)(°F) [1.7 kJ/(kg)(K)]

Physical Properties of the Belt

ρ = density = 500 lb/ft^3 (8009.2 kg/m^3)
k = thermal conductivity = 10 Btu/(h)(ft)(°F) [17 W/(m)(K)]
c = specific heat = 0.11 Btu/(lb)(°F) [0.46 kJ/(kg)(K)]
α = thermal diffusivity = 0.182 ft^2/h (0.169 m^2/h)

Calculation Procedure:

1. Calculate the thickness of the material on the belt.

By dimensional analysis, (lb/h)(ft^3/lb)(min/ft belt)(h/min)(1/ft belt) = ft, so (10,000)(1/60)(1/150)(1/60)(1/3) = 0.00617 ft (1.881 mm).

2. Calculate the time the material is in contact with the cooled belt.

Here, (ft)(min/ft)(h/min) = h, so 100(1/150)(1/60) = 0.0111 h.

3. Calculate $\alpha\tau/L^2$ for the belt.

Use the curves in Fig. 7-7. Assume that the water flow rate is high enough to neglect the temperature rise as the water removes heat from the material. Then, $\alpha\tau/L^2 = 0.182(0.0111)/(0.125/12)^2 = 18.62$.

4. Calculate $k/(hL)$ for the belt.

Now, $k/(hL) = 10/[500(0.125/12)] = 1.92$.

5. Determine θ_o/θ_i for the belt.

From Fig. 7-7, for $\alpha\tau/L^2 = 18.62$ and $k/(hL) = 1.92$, θ_o/θ_i is less than 0.001. Therefore, pending the outcome of steps 6 through 8, the conduction of heat through the belt will presumably be negligible compared with the conduction of heat through the material to be cooled.

6. Calculate $\alpha\tau/L^2$ for the material.

Now, $\alpha\tau/L^2 = 0.0042(0.0111)/0.00617^2 = 1.225$.

7. Calculate $k/(hL)$ for the material.

Now, $k/(hL) = 0.10/[500(0.00617)] = 0.0324$.

8. Determine θ_o/θ_i.

From Fig. 7-7, for $\alpha\tau/L^2 = 1.225$ and $k/(hL) = 0.0324$, $\theta_o/\theta_i = 0.075$.

9. Determine the final surface temperature of the material.

Now, $\theta_o/\theta_i = (T - T_\infty)/(T_i - T_\infty)$, where T is the temperature at the end of the belt, T_i is the initial material temperature, and T_∞ is the water temperature. So, $\theta_o/\theta_i = (T - 86)/(400 - 86) = 0.075$, and $T = 109.6°F$ (316.3 K).

7-11 Sizing a Belt Cooler

For the conditions in the preceding example, calculate the length of belt required if the belt speed is reduced to 100 ft/min (30.48 m/min) and an outlet temperature of 125°F (324.8 K) is acceptable.

Calculation Procedure:

The curves in Fig. 7-7 will be used.

1. Calculate the thickness of the material on the belt.

By dimensional analysis, (lb/h)(ft^3/lb)(min/ft belt)(h/min)(1/ft belt) = ft, so (10,000)(1/60)(1/100)(1/60)(1/3) = 0.0093 ft (2.82 mm).

2. Calculate $k/(hL)$ for the material.

Now, $k/(hL) = 0.10/[500(0.0093)] = 0.0215$.

3. Calculate θ_o/θ_i.

Thus, $\theta_o/\theta_i = (125 - 86)/(400 - 86) = 0.1242$.

4. Determine $\alpha\tau/L^2$.

From Fig. 7-7, for $\theta_o/\theta_i = 0.1242$ and $k/(hL) = 0.0215$, $\alpha\tau/L^2 = 1.0$.

5. Determine the time required for contact with the water.

Since $\alpha\tau/L^2 = 1.0 = 0.0042\tau/0.0093^2$, $\tau = 0.0206$ h.

6. Calculate the length of belt required.

Now, τ = h = ft(min/ft)(h/min) = 0.0206, which is feet of belt(1/100)(1/60), so the length of belt is (0.0206)(100)(60) = 123.6 ft (37.67 m).

7-12 Batch Heating: Internal Coil, Isothermal Heating Medium

A tank containing 50,000 lb (22,679.5 kg) of material with a specific heat of 0.5 Btu/(lb)(°F) [2.1 kJ/(kg)(K)] is to be heated from 68°F (293 K) to 257°F (398 K). The tank contains a heating coil with a heat-transfer surface of 100 ft^2 (9.29 m^2), and the overall heat-transfer coefficient from the coil to the tank contents is 150 Btu/(h)(ft^2)(°F) [850 W/(m^2)(K)]. Calculate the time required to heat the tank contents with steam condensing at 320°F (433 K).

Calculation Procedure:

1. Select and apply the appropriate heat-transfer formula.

When heating a batch with an internal coil with an isothermal heating medium, the following equation applies:

$$\ln\left(\frac{T_1 - t_1}{T_1 - t_2}\right) = \left(\frac{UA}{Mc}\right)\theta$$

where T_1 = heating-medium temperature
t_1 = initial batch temperature
t_2 = final batch temperature
U = overall heat-transfer coefficient
A = heat-transfer surface
M = weight of batch
c = specific heat of batch
θ = time

For this problem, $\ln[(320 - 68)/(320 - 257)] = \{150(100)/[50{,}000(0.5)]\}\theta$, so θ = 2.31 h.

Related Calculations: This procedure can also be used for batch cooling with internal coils and isothermal cooling media. The equation in such cases is

$$\ln\left(\frac{T_1 - t_1}{T_2 - t_1}\right) = \left(\frac{UA}{Mc}\right)\theta$$

where T_1 = initial batch temperature
T_2 = final batch temperature
t_1 = cooling-medium temperature

7-13 Batch Cooling: Internal Coil, Nonisothermal Cooling Medium

For the tank described in the preceding example, calculate the time required to cool the batch from 257°F (398 K) to 104°F (313 K) if cooling water is available at a temperature of 86°F (303 K) and a flow rate of 10,000 lb/h (4535.9 kg/h).

Calculation Procedure:

1. Select and apply the appropriate heat-transfer formula.

When cooling a batch with an internal coil and a nonisothermal cooling medium, the following equation applies:

$$\ln\left(\frac{T_1 - t_1}{T_2 - t_1}\right) = \frac{w_c c_c}{Mc}(K_2 - 1)\left(\frac{1}{K_2}\right)\theta$$

where $K_2 = e^{UA/w_c c_c}$
T_1 = initial batch temperature
T_2 = final batch temperature
t_1 = initial coolant temperature
w_c = coolant flow rate
c_c = coolant specific heat
U = overall heat-transfer coefficient
A = heat-transfer surface
M = weight of batch
c = specific heat of batch
θ = time

For this problem, $K_2 = \exp 150(100)/[10{,}000(1.0)] = 4.4817$, so

$$\ln \frac{257 - 86}{104 - 86} = \frac{(10{,}000)1.0}{(50{,}000)0.5} \frac{4.4817 - 1}{4.4817} \theta$$

Therefore, $\theta = 7.245$ h.

Related Calculations: This procedure can also be used for batch heating with internal coils and isothermal heating media. The equations in such cases are

$$\ln \frac{T_1 - t_1}{T_1 - t_2} = \frac{W_h c_h}{Mc} (K_3 - 1) \frac{1}{K_3} \theta \qquad \text{and} \qquad K_3 = \exp \frac{UA}{W_h c_h}$$

where T_1 = heating-medium temperature
t_1 = initial batch temperature
t_2 = final batch temperature
W_h = heating-medium flow rate
c_h = heating-medium specific heat

7-14 Batch Cooling: External Heat Exchanger (Counterflow), Nonisothermal Cooling Medium

Calculate the time required to cool the batch described in the preceding example if an external heat exchanger with a heat-transfer surface of 200 ft^2 (18.58 m^2) is available. The batch material is circulated through the exchanger at the rate of 25,000 lb/h (11,339.8 kg/h). The overall heat-transfer coefficient in the heat exchanger is 200 Btu/(h)(ft^2)(°F) [1134 W/(m^2)(K)].

Calculation Procedure:

1. Select and apply the appropriate heat-transfer formula.

When cooling a batch with an external heat exchanger and a nonisothermal cooling medium, the following equations apply:

$$\ln \frac{T_1 - t_1}{T_2 - t_1} = \frac{K_4 - 1}{M} \frac{W_b w_c c_c}{K_4 w_c c_c - W_b c} \theta \qquad \text{and} \qquad K_4 = \exp UA \left(\frac{1}{W_b c} - \frac{1}{w_c c_c} \right)$$

where T_1 = initial batch temperature
T_2 = final batch temperature
t_1 = initial coolant temperature
w_c = coolant flow rate
c_c = coolant specific heat
W_b = batch flow rate
c = batch specific heat
M = weight of batch
U = overall heat-transfer coefficient
A = heat transfer surface
θ = time

For this problem,

$$K_4 = \exp\,(200)(200)\left[\frac{1}{(25{,}000)0.5} - \frac{1}{(10{,}000)1.0}\right]$$
$$= 0.4493$$

So,

$$\ln\frac{257 - 86}{104 - 86} = \frac{0.4493 - 1}{50{,}000}\,\frac{25{,}000(10{,}000)(1.0)}{[0.4493(10{,}000)(1.0) - 25{,}000(0.5)]}\,\theta$$

Therefore, $\theta = 6.547$ h.

Related Calculations: This procedure can also be used for batch heating with external heat exchangers and nonisothermal heating media. The equations in such cases are

$$\ln\frac{T_1 - t_1}{T_1 - t_2} = \frac{K_5 - 1}{M}\,\frac{W_b W_h c_h}{K_5 W_h c_h - W_b c}\,\theta \qquad \text{and} \qquad K_5 = \exp UA\left(\frac{1}{W_b c} - \frac{1}{W_h c_h}\right)$$

where T_1 = heating-medium initial temperature
t_1 = initial batch temperature
t_2 = final batch temperature
W_h = heating-medium flow rate
c_h = specific heat of heating medium

7-15 Batch Cooling: External Heat Exchanger (1–2 Multipass), Nonisothermal Cooling Medium

Calculate the time required to cool the batch described in the preceding example if the external heat exchanger is a 1–2 multipass unit rather than counterflow.

Calculation Procedure:

1. Select and apply the appropriate heat-transfer formula.

When cooling a batch with an external 1–2 multipass heat exchanger and a nonisothermal cooling medium, the following equations apply:

$$\ln\frac{T_1 - t_1}{T_2 - t_1} = S\,\frac{w_c c_c}{Mc}\,\theta$$

and

$$S = \frac{2(K_7 - 1)}{K_7[R + 1 + (R^2 + 1)^{1/2}] - [R + 1 - (R^2 + 1)^{1/2}]}$$

where $K_7 = \exp\dfrac{UA}{w_c c_c}(R^2 + 1)^{1/2}$

$R = \dfrac{w_c c_c}{W_b c}$

T_1 = initial batch temperature
T_2 = final batch temperature
t_1 = initial coolant temperature
w_c = coolant flow rate
c_c = coolant specific heat
M = weight of batch
W_b = batch flow rate
c = specific heat of batch
θ = time required to cool the batch
U = overall heat-transfer coefficient
A = heat-transfer surface

For the problem here, $R = 10{,}000(1.0)/[25{,}000(0.5)] = 0.80$, so

$$K_7 = \exp \frac{200(200)}{10{,}000(1.0)} (0.80^2 + 1)^{1/2}$$
$$= 167.75$$

and

$$S = \frac{2(167.75 - 1)}{167.75[0.80 + 1 + (0.8^2 + 1)^{1/2}] - [0.8 + 1 - (0.8^2 + 1)^{1/2}]}$$
$$= 0.646;$$

Therefore,

$$\ln \frac{257 - 86}{104 - 86} = \frac{0.646(10{,}000)(1.0)}{50{,}000(0.5)} \theta$$

and $\theta = 8.713$ h

Related Calculations: This procedure can also be used for batch heating with external 1–2 multipass heat exchangers and nonisothermal heating media. The equation in such a case is

$$\ln \frac{T_1 - t_1}{T_1 - t_2} = \frac{SW_h}{M} \theta$$

where S is defined by the preceding equation, and

$$R = \frac{W_b c}{W_h c_h}$$

$$K_7 = \exp \frac{UA}{W_b c} (R^2 + 1)^{1/2}$$

T_1 = initial temperature of heating medium
t_1 = initial temperature of batch
t_2 = final batch temperature
W_h = heating-medium flow rate
c_h = specific heat of heating medium

7-16 Heat Transfer in Agitated Vessels

Calculate the heat-transfer coefficient from a coil immersed in an agitated vessel with a diameter of 8 ft (2.44 m). The agitator is a turbine 3 ft (0.91 m) in diameter and turns at 150 r/min. The fluid has these properties:

ρ = density = 45 lb/ft^3 (720.8 kg/m^3)

μ = viscosity = 10 lb/(ft)(h) (4.13 cP)

c = specific heat = 0.7 Btu/(lb)(°F) [2.9 kJ/(kg)(K)]

k = thermal conductivity = 0.10 Btu/(h)(ft)(°F) [0.17 W/(m)(K)]

The viscosity may be assumed to be constant with temperature.

Calculation Procedure:

1. Select and apply the appropriate heat-transfer formula.

The following equation can be used to predict heat-transfer coefficients from coils or tank walls in agitated tanks:

$$\frac{hD_j}{k} = a\left(\frac{L^2N\rho}{\mu}\right)^{2/3}\left(\frac{c\mu}{k}\right)^{1/3}\left(\frac{\mu_b}{\mu_w}\right)^{0.14}$$

The term a has these values:

Agitator	Surface	a
Turbine	Jacket	0.62
Turbine	Coil	1.50
Paddle	Jacket	0.36
Paddle	Coil	0.87
Anchor	Jacket	0.46
Propeller	Jacket	0.54
Propeller	Coil	0.83

The other variables in the equation are

h = heat-transfer coefficient

D_j = diameter of vessel

k = thermal conductivity

L = diameter of agitator

N = speed of agitator in revolutions per hour

ρ = density

μ = viscosity

c = specific heat

μ_b = viscosity at bulk fluid temperature

μ_w = viscosity at surface temperature

Therefore, $h = 1.50(k/D_j)(L^2N\rho/\mu)^{2/3}(c\mu/k)^{1/3}(\mu_b/\mu_w)^{0.14}$. As noted above, assume that $(\mu_b/\mu_w)^{0.14} = 1.0$. Then,

$$h = 1.50(0.10/8)\left[\frac{3^2(150)(60)(45)}{10}\right]^{2/3}\left[\frac{0.7(10)}{0.1}\right]^{1/3}(1.0)$$

$$= 394 \text{ Btu/(h)(ft}^2\text{)(°F) [2238 W/(m}^2\text{)(K)]}$$

7-17 Natural-Convection Heat Transfer

Calculate the heat-transfer coefficient from a coil immersed in water with the physical properties listed below. The coil has a diameter of 1 in (0.025 m), and the temperature difference between the surface of the coil and the fluid is 10°F (5.56 K). The properties of the water are

c = specific heat = 1.0 Btu/(lb)(°F) [4.19 kJ/(kg)(K)]

ρ = liquid density = 60 lb/ft^3 (961.1 kg/m^3)

k = thermal conductivity = 0.395 Btu/(h)(ft)(°F)[0.683 W/(m)(K)]

μ = viscosity = 0.72 lb/(ft)(h) (0.298 cP)

β = coefficient of expansion = 0.0004°F^{-1}(0.00022 K^{-1})

Calculation Procedure:

1. Consider the natural-convection equations available.

Heat-transfer coefficients for natural convection may be calculated using the equations presented below. These equations are also valid for horizontal plates or discs. For horizontal plates facing upward which are heated or for plates facing downward which are cooled, the equations are applicable directly. For heated plates facing downward or cooled plates facing upward, the heat-transfer coefficients obtained should be multiplied by 0.5.

	Vertical surfaces	Horizontal cylinders
For Reynolds numbers greater than 10,000:		
$[h/(cG)](c\mu/k)^{2/3}$	$= 0.13/(LG/\mu)^{1/3}$	$= 0.13/(DG/\mu)^{1/3}$
For Reynolds numbers from 100 to 10,000:		
$[h/(cG)](c\mu/k)^{3/4}$	$= 0.59/(LG/\mu)^{1/2}$	$= 0.53/(DG/\mu)^{1/2}$
For Reynolds numbers less than 100:		
$[h/(cG)](c\mu/k)^{5/6}$	$= 1.36/(LG/\mu)^{2/3}$	$= 1.09/(DG/\mu)^{2/3}$

In these equations, G is mass velocity, that is, $(g\beta\,\Delta T\rho^2L)^{1/2}$ for vertical surfaces or $(g\beta\Delta T\rho^2D)^{1/2}$ for horizontal cylinders, and,

h = heat-transfer coefficient
c = specific heat
L = length
D = diameter
μ = viscosity
g = acceleration of gravity (4.18×10^8 ft/h^2)
β = coefficient of expansion
ΔT = temperature difference between surface and fluid
ρ = density
k = thermal conductivity

2. *Calculate mass velocity.*

Thus, $G = (g\beta\Delta T\rho^2 D)^{1/2} = [(4.18 \times 10^8)(0.0004)(10)60^2(1/12)]^{1/2} = 22{,}396.4$ lb/(ft^2)(h).

3. *Calculate the Reynolds number DG/μ.*

Thus, $DG/\mu = (1/12)(22{,}396.4)/0.72 = 2592.2$.

4. *Calculate the heat-transfer coefficient h.*

Thus, $h = 0.53cG/[(c\mu/k)^{3/4}(DG/\mu)^{1/2}] = 0.53(1.0)(22{,}396.4)/\{[1.0(0.72)/0.395]^{3/4}(2592.2)^{1/2}\} = 148.6$ Btu/(h)(ft^2)(°F) [844 W/(m^2)(K)].

Related Calculations: The equations presented above can be simplified for natural-convection heat transfer in air. The usual cases yield the following equations: For vertical surfaces,

$$h = 0.29(\Delta T/L)^{1/4}$$

for horizontal cylinders,

$$h = 0.27(\Delta T/D)^{1/4}$$

In these, h is in Btu/(h)(ft^2)(°F), ΔT is in °F, and L and D are in feet.

7-18 Heat-Transfer Coefficients for Fluids Flowing inside Tubes: Forced Convection, Sensible Heat

Calculate the heat-transfer coefficient for a fluid with the properties listed below flowing through a tube 20 ft (6.1 m) long and of 0.62-in (0.016-m) inside diameter. The bulk fluid temperature is 212°F (373 K), and the tube surface temperature is 122°F (323 K). Calculate the heat-transfer coefficient if the fluid is flowing at a rate of 2000 lb/h (907.2 kg/h). Also calculate the heat-transfer coefficient if the flow rate is reduced to 100 lb/h (45.36 kg/h).

Physical Properties of the Fluid

c = specific heat = 0.65 Btu/(lb)(°F) [2.72 kJ/(kg)(K)]

k = thermal conductivity = 0.085 Btu/(h)(ft)(°F) [0.147 W/(m)(K)]

μ_w = viscosity at 122°F = 4.0 lb/(ft)(h) (1.65 cP)

μ_b = viscosity at 212°F = 1.95 lb/(ft)(h) (0.806 cP)

Calculation Procedure:

1. Select the appropriate heat-transfer coefficient equation.

Heat-transfer coefficients for fluids flowing inside tubes or ducts can be calculated using these equations:

a. For Reynolds numbers (DG/μ) greater than 8000,

$$\frac{h}{cG} = \frac{0.023}{(c\mu/k)^{2/3}(D_iG/\mu)^{0.2}(\mu_w/\mu_b)^{0.14}}$$

b. For Reynolds numbers (DG/μ) less than 2100,

$$\frac{h}{cG} = \frac{1.86}{(c\mu/k)^{2/3}(D_iG/\mu)^{2/3}(L/D_i)^{1/3}(\mu_w/\mu_b)^{0.14}}$$

In these equations,

h = heat-transfer coefficient

c = specific heat

G = mass velocity (mass flow rate divided by cross-sectional area)

μ = viscosity

μ_w = viscosity at the surface temperature

μ_b = viscosity at the bulk fluid temperature

k = thermal conductivity

D_i = inside diameter

L = length

2. Calculate D_iG/μ for a 2000 lb/h flow rate.

$$\frac{D_iG}{\mu} = \frac{0.62}{12}\,\frac{2000}{(0.62/12)^2(\pi/4)}\,\frac{1}{1.95}$$

$$= 25{,}275$$

3. Calculate h for the 2000 lb/h flow rate.

Because DG/μ is greater than 8000,

$$\frac{h}{cG} = \frac{0.023}{(c\mu/k)^{2/3}(D_iG/\mu)^{0.2}(\mu_w/\mu_b)^{0.14}}$$

$$= \frac{0.023(0.65)[2000/(0.62/12)^2(\pi/4)]}{[0.65(1.95)/0.085]^{2/3}25{,}275^{0.2}(4.0/1.95)^{0.14}}$$

$$= 280.3 \text{ Btu/(h)(ft}^2\text{)(K) [1592 W/(m}^2\text{)(K)]}$$

4. Calculate DG/μ for a 100 lb/h flow rate.

$$D_iG/\mu = 25{,}275(100/2000) = 1263.8$$

5. Calculate h for the 100 lb/h flow rate.

Because DG/μ is less than 2100,

$$\frac{h}{cG} = \frac{1.86}{(c\mu/k)^{2/3}(DG/\mu)^{2/3}(L/D)^{1/3}(\mu_w/\mu_b)^{0.14}}$$

$$= \frac{1.86(0.65)\{100/[(0.62/12)^2(\pi/4)]\}}{\left[\frac{0.65(1.95)}{0.085}\right]^{2/3} 1263.8^{2/3} \left[\frac{20}{(0.62/12)}\right]^{1/3} \left(\frac{4.0}{1.95}\right)^{0.14}}$$

$$= 10.1 \text{ Btu/(h)(ft}^2\text{)(°F) [57.4 W/(m}^2\text{)(K)].}$$

Related Calculations: Heat transfer for fluids with Reynolds numbers between 2100 and 8000 is not stable, and the heat-transfer coefficients in this region cannot be predicted with certainty. Equations have been presented in many of the references. The heat-transfer coefficients in this region can be bracketed by calculating the values using both the preceding equations for the Reynolds number in question.

The equations presented here can also be used to predict heat-transfer coefficients for the shell side of shell-and-tube heat exchangers in which the baffles have been designed to produce flow parallel to the axis of the tube. For such cases, the diameter that should be used is the equivalent diameter

$$D_e = \frac{4a}{P}$$

where a = flow area
P = wetted perimeter

Here, $a = (D_s^2 - nD_o^2)(\pi/4)$, where D_s is the shell inside diameter, D_o is the tube outside diameter, and n is the number of tubes; and $P = \pi(D_s + nD_o)$.

For shells with triple or double segmental baffles, the heat-transfer coefficient calculated for turbulent flow (DG/μ greater than 8000) should be multiplied by a value of 1.3.

For gases, the equation for heat transfer in the turbulent region (DG/μ greater than 8000) can be simplified because the Prandtl number ($c\mu/k$) and the viscosity for most gases are approximately constant. Assigning the values $c\mu/k = 0.78$ and $\mu = 0.0426$ lb/(h)(ft) (0.0176 cP) results in the following equation for gases:

$$h = 0.0144 \frac{cG^{0.8}}{D_i^{0.2}}$$

with the variables defined in English units.

7-19 Heat-Transfer Coefficients for Fluids Flowing inside Helical Coils

Calculate the heat-transfer coefficient for a fluid with a flow rate of 100 lb/h (45.36 kg/h) and the physical properties outlined in Example 7-18. The inside diameter of the tube is 0.62 in (0.016 m), and the tube is fabricated into a helical coil with a helix diameter of 24 in (0.61 m).

Calculation Procedure:

1. Select the appropriate heat-transfer coefficient equation.

Heat-transfer coefficients for fluids flowing inside helical coils can be calculated with modifications of the equations for straight tubes. The equations presented in Example 7-18 should be multiplied by the factor $1 + 3.5D_i/D_c$, where D_i is the inside diameter and D_c is the diameter of the helix or coil. In addition, for laminar flow, the term $(D_c/D_i)^{1/6}$ should be substituted for the term $(L/D)^{1/3}$. The Reynolds number required for turbulent flow is $2100[1 + 12(D_i/D_c)^{1/2}]$.

2. Calculate the minimum Reynolds number for turbulent flow.

Now,

$$\begin{aligned} (DG/\mu)_{\min} &= 2100[1 + 12(D_i/D_c)^{1/2}] \\ &= 2100[1 + 12(0.62/24)^{1/2}] \\ &= 6150 \end{aligned}$$

3. Calculate h.

From the preceding calculations, $DG/\mu = 1263.8$ at a flow rate of of 100 lb/h. Therefore,

$$\begin{aligned} \text{h} &= \frac{1.86cG(1 + 3.5D_i/D_c)}{(c\mu/k)^{2/3}(DG/\mu)^{2/3}(D_c/D_i)^{1/6}(\mu_w/\mu_b)^{0.14}} \\ &= \frac{1.86(0.65)\{100/[(0.62/12)^2(\pi/4)]\}[1 + 3.5(0.62/24)]}{[0.65(1.95)/0.085]^{2/3}1263.8^{2/3}(24/0.62)^{1/6}(4.0/1.95)^{0.14}} \\ &= 43.7 \text{ Btu/(h)(ft}^2\text{)(°F) [248 W/(m}^2\text{)(K)]} \end{aligned}$$

7-20 Heat-Transfer Coefficients: Fluids Flowing across Banks of Tubes; Forced Convection, Sensible Heat

Calculate the heat-transfer coefficient for a fluid with the properties listed in Example 7-18 if the fluid is flowing across a tube bundle with the following geometry. The fluid flows at a rate of 50,000 lb/h (22,679.5 kg/h). Calculate the heat-transfer coefficient for both clean and fouled conditions.

Tube Bundle Geometry

D_s = shell diameter = 25 in (2.08 ft or 0.635 m)

B = baffle spacing = 9.5 in (0.79 ft or 0.241 m)

D_o = outside tube diameter = 0.75 in (0.019 m)

s = tube center-to-center spacing = 0.9375 in (0.0238 m)

The tubes are spaced on a triangular pattern.

Calculation Procedure:

1. Select the appropriate heat-transfer coefficient equation.

Heat-transfer coefficients for fluids flowing across ideal-tube banks may be calculated using the equation

$$\frac{h}{cG} = \frac{a}{(c\mu/k)^{2/3}(D_oG/\mu)^m(\mu_w/\mu_b)^{0.14}}$$

The values of a and m are as follows:

Reynolds number	Tube pattern	m	a
Greater than 200,000	Staggered	0.300	0.166
Greater than 200,000	In-line	0.300	0.124
300 to 200,000	Staggered	0.365	0.273
300 to 200,000	In-line	0.349	0.211
Less than 300	Staggered	0.640	1.309
Less than 300	In-line	0.569	0.742

In these equations,

h = heat-transfer coefficient

c = specific heat

G = mass velocity = W/a_c

a_c = flow area

W = flow rate
k = thermal conductivity
D_o = outside tube diameter
μ = viscosity
μ_w = viscosity at wall temperature
μ_b = viscosity at bulk fluid temperature

For triangular and square tube patterns,

$$a_c = \frac{BD_s(s - D_o)}{s}$$

For rotated square tube patterns,

$$a_c = \frac{1.5BD_s(s - D_o)}{s}$$

where B = baffle spacing
D_s = shell diameter
s = tube center-to-center spacing
D_o = tube outside diameter

The values of a in the preceding table are based on heat exchangers that have no bypassing of the bundle by the fluid. For heat exchangers built to the standards of the Tubular Exchangers Manufacturers Association [11] and with an adequate number of sealing devices, the heat-transfer coefficient calculated with the preceding equation must be corrected as below to reflect the bypassing of the fluid:

$$h_o = hF_1F_r$$

where $F_1 = 0.8(B/D_s)^{1/6}$ for bundles with typical fouling
$F_1 = 0.8(B/D_s)^{1/4}$ for bundles with no fouling
$F_r = 1.0$ for D_oG/μ greater than 100
$F_r = 0.2(D_oG/\mu)^{1/3}$ for D_oG/μ less than 100

2. *Calculate D_oG/μ.*

$$a_c = \frac{BD_s(s - D_o)}{s}$$

$$= \frac{(9.5/12)(25/12)(0.9375 - 0.75)}{0.9375}$$

$$= 0.3299 \text{ ft}^2 \ (0.0306 \text{ m}^2)$$

$$G = \frac{W}{a_c} = \frac{50{,}000}{0.3299} = 151{,}578.9 \text{ lb/(h)(ft}^2\text{)}$$

$$\frac{D_oG}{\mu} = \frac{(0.75/12)151{,}578.9}{1.95} = 4858.3$$

3. Calculate h.

Now,

$$h = \frac{0.273cG}{(c\mu/k)^{2/3}(D_oG/\mu)^{0.365}(\mu_w/\mu_b)^{0.14}}$$

Thus,

$$h = \frac{0.273(0.65)(151{,}578.9)}{[0.65(1.95)/0.085]^{2/3}(4858.3)^{0.365}(4.0/1.95)^{0.14}}$$

$$= 181.2 \text{ Btu/(h)(ft}^2\text{)(°F) [1029 W/(m}^2\text{)(K)]}$$

4. Calculate h_o for the fouled condition.

Now, $h_o = hF_1F_r$, where $F_r = 1.0$ and $F_1 = 0.8(B/D_s)^{1/6} = 0.8(9.5/25)^{1/6} = 0.6809$, so $h_o = 181.5(0.6809)(1.0) = 123.6$ Btu/(h)(ft^2)(°F) [702 W/(m^2)(K)].

5. Calculate h_o for the clean condition.

Here, $F_1 = 0.8(B/D_s)^{1/4} = 0.8(9.5/25)^{1/4} = 0.6281$, so $h_o = 181.5(0.6281)(1.0) =$ 114.0 Btu/(h)(ft^2)(°F) [647.5 W/(m^2)(K)].

Related Calculations: The preceding equations for F_1 and F_r are based on heat exchangers that have been fabricated to minimize bypassing of the bundle by the fluid. The assumption has also been made that the baffle cut for segmental baffles is 20 percent of the shell diameter and that the layout includes tubes in the baffle window areas. For conditions removed from these assumptions, the effects of fluid bypassing should be evaluated. Several methods have been presented; the most widely used is that of Bell [15].

For gases, the preceding equations for the turbulent regime can be simplified because the Prandtl number ($c\mu/k$) and viscosity for most gases are approximately constant. Assigning the values $c\mu/k = 0.78$ and $\mu = 0.0426$ lb/(ft)(h) (0.0176 cP) results in the equation

$$h = \frac{bcG^{1-m}}{D_o^m}$$

with the variables defined in English units and with b having these values:

Reynolds number	Tube pattern	b
300 to 200,000	Staggered	0.102
300 to 200,000	In-line	0.083
Above 200,000	Staggered	0.076
Above 200,000	In-line	0.057

7-21 Mean Temperature Difference for Heat Exchangers

A warm stream enters a heat exchanger at 200°C (392°F), T_h, and is cooled to 100°C (212°F), T_c. The cooling stream enters the exchanger at 20°C (68°F), t_c, and is heated to 95°C (203°F), t_h. Calculate the mean temperature difference for the following cases:

1. Countercurrent flow
2. Cocurrent flow (also called "parallel flow")
3. 1–2 Multipass (one pass on the shell side, two passes on the tube side)
4. 2–4 Multipass (two passes on the shell side, four passes on the tube side)
5. 1–1 Cross flow (one cross-flow pass on the shell side, one pass on the tube side)
6. 1–2 Cross flow (one cross-flow pass on the shell side, two passes on the tube side)

These flow arrangements are illustrated in Fig. 7-11.

Calculation Procedure:

1. Use equation for countercurrent flow.

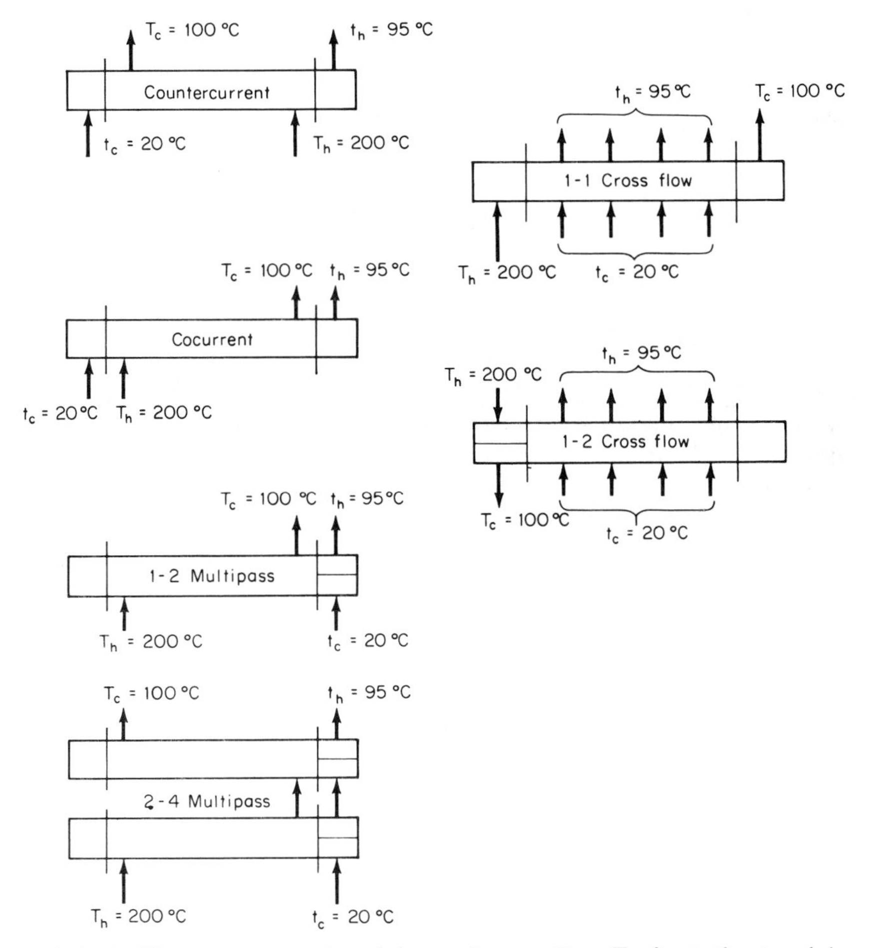

FIG. 7-11 Flow arrangements through heat exchangers. Note: T refers to the steam being cooled; t refers to the steam being heated.

The mean temperature difference for countercurrent flow is the log mean temperature difference as calculated from the equation

$$\Delta T_{LM} = \frac{(T_h - t_h) - (T_c - t_c)}{\ln [(T_h - t_h)/(T_c - t_c)]}$$

Thus,

$$\Delta T_{LM} = \frac{(200 - 95) - (100 - 20)}{\ln [(200 - 95)/(100 - 20)]}$$
$$= 91.93°\text{C } (165.47°\text{F})$$

2. *Use the equation for cocurrent flow.*

The mean temperature difference for cocurrent flow is the log mean temperature difference as calculated from the equation

$$\Delta T_{LM} = \frac{(T_h - t_c) - (T_c - t_h)}{\ln [(T_h - t_c)/(T_c - t_h)]}$$

Thus,

$$\Delta T_{LM} = \frac{(200 - 20) - (100 - 95)}{\ln [(200 - 20)/(100 - 95)]}$$
$$= 48.83°\text{C } (87.89°\text{F})$$

3. *Use equation for 1–2 multipass.*

The mean temperature difference for a 1–2 multipass heat exchanger can be calculated from the equation

$$\Delta T_m = \frac{M}{\ln [(P + M)/(P - M)]}$$

where $P = (T_h - t_h) + (T_c - t_c)$ and $M = [(T_h - T_c)^2 + (t_h - t_c)^2]^{1/2}$. Thus, $P = (200 - 95) + (100 - 20) = 185$, $M = [(200 - 100)^2 + (95 - 20)^2]^{1/2} = 125$, and

$$\Delta T_m = \frac{125}{\ln [(185 + 125)/(185 - 125)]}$$
$$= 76.12°\text{C } (137.02°\text{F})$$

4. *Use equation for 2–4 multipass.*

The mean temperature difference for a 2–4 multipass heat exchanger can be calculated from the equation

$$\Delta T_m = \frac{M/2}{\ln [(Q + M)/(Q - M)]}$$

where M is defined above and

$$Q = [(T_h - t_h)^{1/2} + (T_c - t_c)^{1/2}]^2$$

Thus, $Q = [(200 - 95)^{1/2} + (100 - 20)^{1/2}]^2 = 368.30$, and

$$\Delta T_m = \frac{125/2}{\ln [(368.30 + 125)/(368.30 - 125)]}$$

$$= 88.42°\text{C } (159.16°\text{F})$$

5. Use the correction factor for a 1–1 cross flow.

The mean temperature difference for a 1–1 cross-flow heat exchanger can be calculated by using the correction factor determined from Fig. 7-12. The mean temperature difference will be the product of this factor and the log mean temperature difference for countercurrent flow. To obtain the correction factor F, calculate the value of two parameters P and R:

$$P = \frac{t_h - t_c}{T_h - t_c}$$

$$= \frac{95 - 20}{200 - 20} = 0.42$$

$$R = \frac{T_h - T_c}{t_h - t_c}$$

$$= \frac{200 - 100}{95 - 20} = 1.33$$

From Fig. 7-12; with these values of P and R, F is found to be 0.91. Then,

$$\Delta T_m = F\Delta T_{LM} = 0.91(91.93) = 83.66°\text{C } (150.58°\text{F})$$

6. Use the correction factor for a 1–2 cross flow.

The mean temperature difference for a 1–2 cross-flow heat exchanger can be calculated using the same procedure as used for a 1–1 cross-flow heat exchanger, except that the value of F is determined from Fig. 7-13.

For $P = 0.42$ and $R = 1.33$, $F = 0.98$ (from Fig. 7-13), so

$$\Delta T_m = F\Delta T_{LM} = 0.98(91.93) = 90.09°\text{C } (162.16°\text{F})$$

Related Calculations: Mean temperature differences for multipass heat exchangers may also be calculated by using appropriate correction factors for the log mean temperature difference for countercurrent flow

$$\Delta T_m = F\Delta T_{LM}$$

Curves for determining values of F are presented in many of the references [1–4,6,10,11].

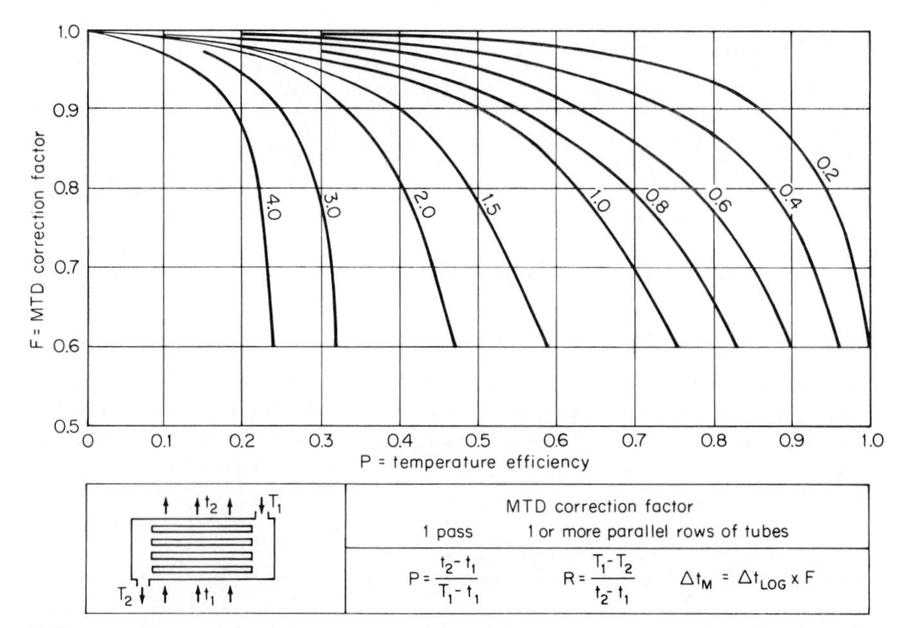

FIG. 7-12 Correction factor for one-pass crossflow. *(From "Engineering Data Book, 1966," Natural Gas Processors Suppliers Association.)*

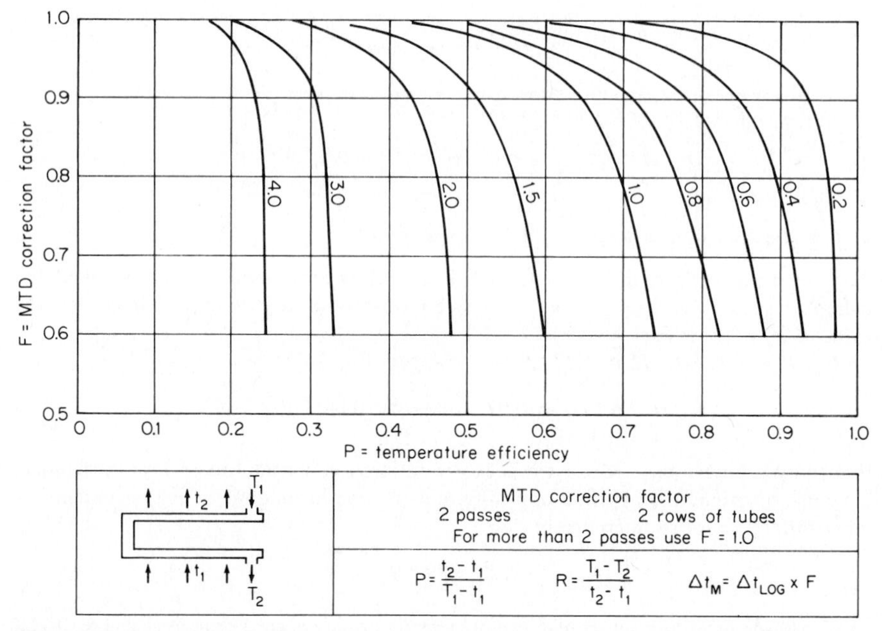

FIG. 7-13 Correction factor for two-pass crossflow. *(From "Engineering Data Book, 1966," Natural Gas Processors Suppliers Association.)*

For shell-and-tube heat exchangers with cross-flow baffles, the preceding methods assume that an adequate number of baffles has been provided. If the shell-side fluid makes less than eight passes across the tube bundle, the mean temperature difference may need to be corrected for this cross-flow condition. Appropriate curves are presented in Caglayan and Buthod [20]. The curves in this reference may also be used to determine correction factors for cross-flow exchangers with one shell pass and more than two tube passes.

The methods presented above are applicable only for conditions in which the heat transferred is a straight-line function of temperature. For systems that do not meet this condition, the total heat-release curve can be treated in sections, each section of which closely approximates the straight-line requirement. A log mean temperature difference can then be calculated for each section. Common examples in which this approach is encountered include (1) total condensers in which the condensate is subcooled after condensation, and (2) vaporizers in which the fluid enters as a subcooled liquid, the liquid is heated to the saturation temperature, the fluid is vaporized, and the vapor is heated and leaves in a superheated state.

Other types of heat-exchanger configurations not illustrated here are sometimes used. Curves for determining the value of F for these are presented in Ref. 11. These configurations include divided-flow and split-flow exchangers.

7-22 Overall Heat-Transfer Coefficient for Shell-and-Tube Heat Exchanger

A shell-and-tube exchanger with the following geometry is available:

D_s = shell diameter = 25 in (0.635 m)
n = number of tubes = 532
D_o = tube outside diameter = 0.75 in (0.019 m)
D_i = tube inside diameter = 0.62 in (0.016 m)
L = tube length = 16 ft (4.88 m)
s = tube spacing = 0.9375 in, triangular (0.024 m)
B = baffle spacing = 9.5 in (0.241 m)

There is one tube-side pass and one shell-side pass, and the tube material is stainless steel with a thermal conductivity of 10 Btu/(h)(ft)(°F) [17 W/(m)(K)].

Calculate the overall heat-transfer coefficient for this heat exchanger under the following service conditions:

Tube Side

Liquid undergoing sensible-heat transfer:

W_t = flow rate = 500,000 lb/h (226,795 kg/h)
c = specific heat = 0.5 Btu/(lb)(°F) [2.1 kJ/(kg)(K)]
μ = viscosity = 1.21 lb/(ft)(h) (0.5 cP)
specific gravity = 0.8
k = thermal conductivity = 0.075 Btu/(h)(ft)(°F) [0.13 W/(m)(K)]

Shell Side

Liquid undergoing sensible-heat transfer:

W_s = flow rate = 200,000 lb/h (90,718 kg/h)
c = specific heat = 1.0 Btu/(lb)(°F) [4.19 kJ/(kg)(K)]
μ = viscosity = 2.0 lb/(ft)(h) (0.83 cP)
specific gravity = 1.0
k = thermal conductivity = 0.36 Btu/(h)(ft)(°F) [0.62 W/(m)(K)]

In addition, the fouling heat-transfer coefficient is 1000 Btu/(h)(ft^2)(°F) [5670 W/(m^2)(K)]. Assume that the change in viscosity with temperature is negligible, that is, $\mu_w/\mu_b = 1.0$.

Calculation Procedure:

1. Calculate the Reynolds number inside the tubes.

The Reynolds number is DG/μ, where G is the mass flow rate.

$$G = \frac{500{,}000}{532 \text{ tubes } (0.62/12)^2(\pi/4)} = 448{,}278 \text{ lb/(h)(ft}^2\text{)}$$

$$DG/\mu = \frac{(0.62/12)448{,}278}{1.21} = 19{,}141$$

2. Calculate h_i, the heat-transfer coefficient inside the tubes.

For Reynolds numbers greater than 8000,

$$h_i = \frac{0.023cG}{(c\mu/k)^{2/3}(DG/\mu)^{0.2}}$$

Thus,

$$h_i = \frac{0.023(0.5)(448{,}278)}{[0.5(1.21)/0.075]^{2/3}19{,}141^{0.2}}$$

$$= 178.6 \text{ Btu/(h)(ft}^2\text{)(°F) [1013 W/(m}^2\text{)(K)]}$$

3. Calculate the Reynolds number for the shell side.

For the shell side, $G = W_s/a_c$, where the flow area $a_c = BD_s(s - D_o)/s$. Then,

$$a_c = \frac{(9.5/12)(25/12)(0.9375 - 0.75)}{0.9375}$$

$$= 0.33 \text{ ft}^2 \text{ (0.0307 m}^2\text{)}$$

So,

$$D_oG/\mu = \frac{(0.75/12)(200{,}000/0.33)}{2.0}$$

$$= 18{,}940$$

4. Calculate h_o, the heat-transfer coefficient outside the tubes.

Now,

$$h = \frac{0.273cG}{(c\mu/k)^{2/3}(D_oG/\mu)^{0.365}}$$

$$= \frac{0.273(1.0)(200{,}000/0.33)}{[1.0(2.0)/0.36]^{2/3}18{,}940^{0.365}}$$

$$= 1450.7 \text{ Btu/(h)(ft}^2\text{)(°F)}$$

Then, correcting for flow bypassing (see Example 7-20),

$$h_o = hF_1F_r = h(0.8)(B/D_s)^{1/6}(1.0)$$

So,

$$h_o = 1450.7(0.8)(9.5/25)^{1/6}(1.0)$$

$$= 987.7 \text{ Btu/(h)(ft}^2\text{)(°F) [5608 W/(m}^2\text{)(K)]}$$

5. Calculate h_w, the heat-transfer coefficient across the tube wall.

The heat-transfer coefficient across the tube wall h_w can be calculated from $h_w = 2k/(D_o - D_i)$. Thus,

$$h_w = \frac{2(10)}{(0.75/12) - (0.62/12)}$$

$$= 1846 \text{ Btu/(h)(ft}^2\text{)(°F) [10,480 W/(m}^2\text{)(K)]}$$

6. Calculate the overall heat-transfer coefficient.

The formula is

$$\frac{1}{U} = \frac{1}{h_o} + \frac{1}{h_i(D_i/D_o)} + \frac{1}{h_w} + \frac{1}{h_s}$$

where U = overall heat-transfer coefficient
h_o = outside heat-transfer coefficient
h_i = inside heat-transfer coefficient
h_w = heat transfer across tube wall
h_s = fouling heat-transfer coefficient
D_i = inside diameter
D_o = outside diameter

Then,

$$\frac{1}{U} = \frac{1}{987.7} + \frac{1}{178.6(0.62/0.75)} + \frac{1}{1846} + \frac{1}{1000}$$

So, $U = 107.2$ Btu/(h)(ft²)(°F) [608.8 W/(m²)(K)].

Related Calculations: The method described for calculating the overall heat-transfer coefficient is also used to calculate the overall resistance to conduction of heat through a composite wall containing materials in series that have different thicknesses and thermal conductivities. For this case, each individual heat-transfer coefficient is equal to the thermal conductivity of a particular material divided by its thickness. The amount of heat transferred by conduction can then be determined from the formula

$$Q = (t_1 - t_2)\frac{1}{R}A$$

where Q = heat transferred
t_1 = temperature of the hot surface
t_2 = temperature of the cold surface
A = area of wall
R = overall resistance, which equals

$$\frac{x_1}{k_1} + \frac{x_2}{k_2} + \frac{x_3}{k_3} + \cdots$$

where x = thickness
k = thermal conductivity

It is seen from inspection that $U = 1/R$.

7-23 Outlet Temperatures for Countercurrent Heat Exchanger

For the heat exchanger described in Example 7-22, calculate the outlet temperatures and the amount of heat transferred if the tube-side fluid enters at 68°F (293 K) and the shell-side fluid enters at 500°F (533 K).

Calculation Procedure:

1. Determine the heat-transfer surface.

The surface $A = \pi n D_o L = \pi(532)(0.75/12)(16)$, so $A = 1670$ ft² (155.2 m²).

2. Determine the thermal effectiveness.

The thermal effectiveness of a countercurrent heat exchanger can be determined from Fig. 7-14. Now,

$$\frac{UA}{wc} = \frac{107.2(1670)}{500{,}000(0.5)}$$
$$= 0.716$$

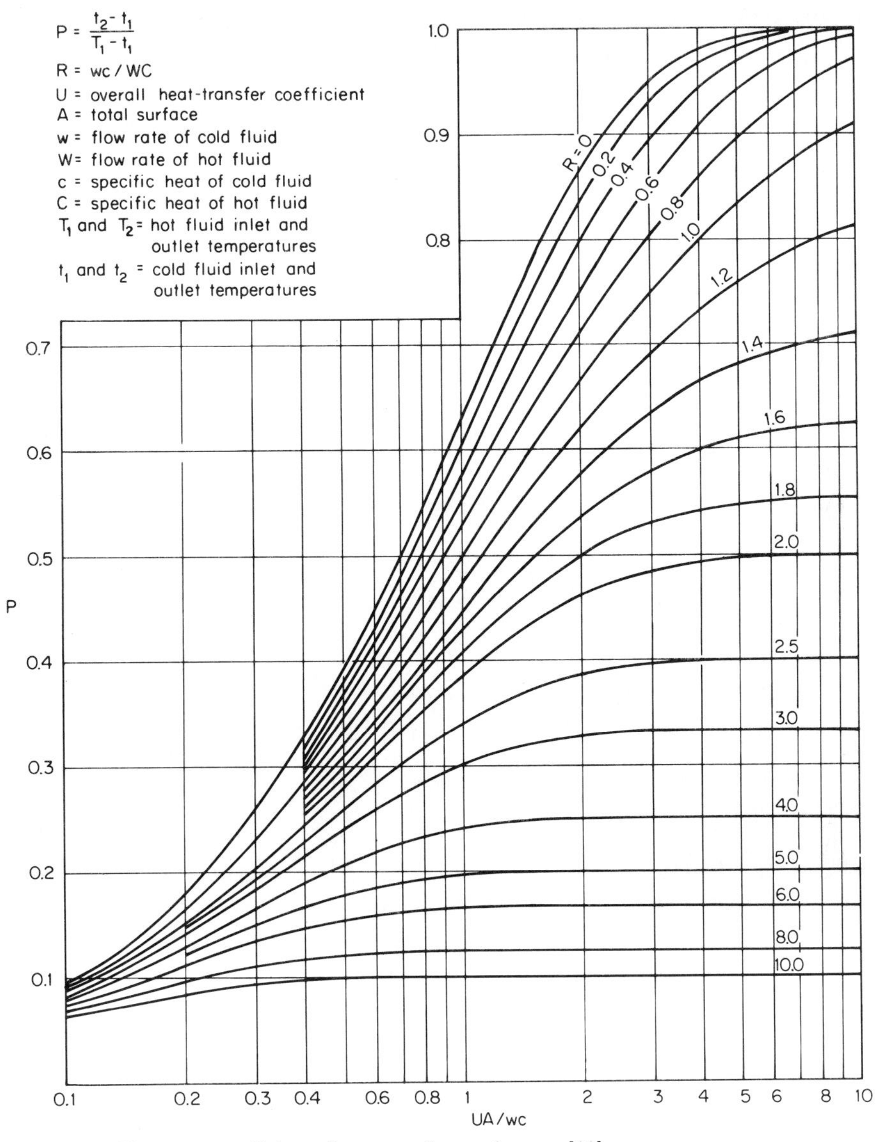

FIG. 7-14 Temperature efficiency for counterflow exchangers [11].

and,

$$R = \frac{500{,}000(0.5)}{200{,}000(1.0)}$$

$$= 1.25$$

So, from Fig. 7-14, the thermal effectiveness $P = 0.39$.

3. Determine outlet temperatures.

Because $P = (t_2 - t_1)/(T_1 - t_1) = 0.39$,

$$t_2 = t_1 + P(T_1 - t_1) = 68 + (0.39)(500 - 68)$$
$$= 236.5°\text{F}$$

Now, by definition, $wc(t_2 - t_1) = WC(T_1 - T_2)$, and $R = wc/(WC)$, so $T_2 = T_1 - R(t_2 - t_1)$. Thus,

$$T_2 = 500 - 1.25(236.5 - 68)$$
$$= 289.4°\text{F (416 K)}$$

4. Determine the amount of heat transferred.

Use the formula,

$$Q = wc(t_2 - t_1)$$

where Q = heat rate
w = flow rate for fluid being heated
c = specific heat for fluid being heated

Then,

$$Q = 500{,}000(0.5)(236.5 - 68)$$
$$= 42{,}120{,}000 \text{ Btu/h } (12{,}343{,}000 \text{ W})$$

(The calculation can instead be based on the temperatures, flow rate, and specific heat for the fluid being cooled.)

Related Calculations: Figure 7-14 can also be used to determine the thermal effectiveness for exchangers in which one fluid is isothermal. For this case, $R = 0$.

7-24 Outlet Temperatures for 1–2 Multipass Heat Exchanger

Calculate the outlet temperatures and the amount of heat transferred for the exchanger described in Example 7-23 if the tube side is converted from single-pass to two-pass.

Calculation Procedure:

1. Determine the overall heat-transfer coefficient.

The heat-transfer coefficient on the tube side is proportional to $G^{0.8}$. The mass velocity G will be doubled because the exchanger is to be converted to two passes on the tube side. Therefore, $h_i = 178.6(2)^{0.8} = 310.9$ Btu/(h)(ft²)(°F) [1764 W/(m²)(K)]. The new

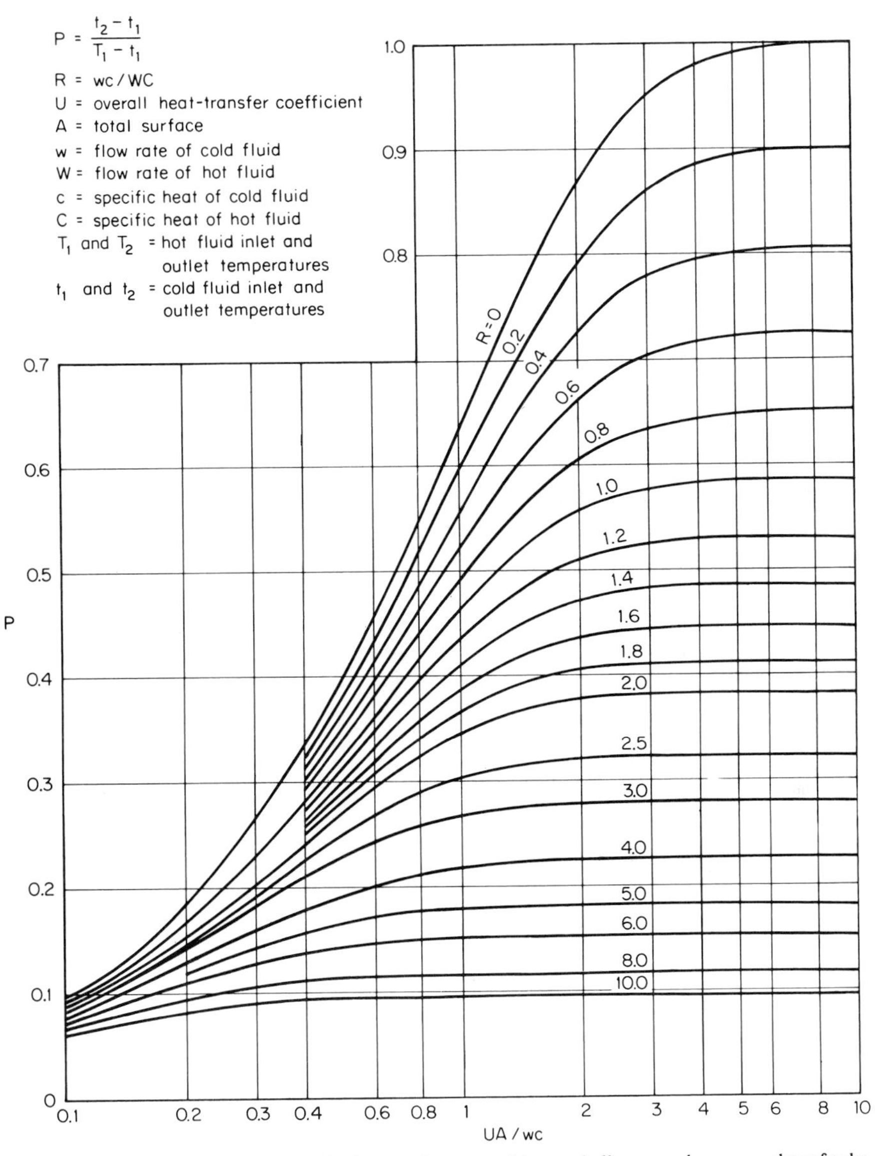

FIG. 7-15 Temperature efficiency for heat exchangers with one shell pass and even number of tube passes [11].

overall heat-transfer coefficient can now be calculated:

$$\frac{1}{U} = \frac{1}{987.7} + \frac{1}{(310.9)(0.62/0.75)} + \frac{1}{1846} + \frac{1}{1000}$$

$$= 155.2 \text{ Btu/(h)(ft}^2\text{)(}^\circ\text{F) [880.4 W/(m}^2\text{)(K)]}$$

2. Determine the thermal effectiveness.

The thermal effectiveness of a 1–2 multipass exchanger can be determined from Fig. 7-15. Now,

$$\frac{UA}{wc} = \frac{155.2(1670)}{500{,}000(0.5)}$$

$$= 1.037$$

and $R = 1.25$, so from Fig. 7-15, the thermal effectiveness $P = 0.43$.

3. Determine the outlet temperatures.

See Example 7-23, step 3. Now, $t_2 - t_1 = 0.43(500 - 68) = 185.8°F$, so $t_2 = 185.8 + 68 = 253.8°F$; and $T_2 - T_1 = R(t_2 - t_1) = 1.25(185.8) = 232.2°F$, so $T_2 = 500 - 232.2 = 267.8°F$ (404 K).

4. Calculate the amount of heat transferred.

See Example 7-23, step 4. Now, $Q = 500{,}000(0.5)(185.8) = 46{,}450{,}000$ Btu/h (13,610,000 W).

Related Calculations: The thermal effectiveness of a 2–4 multipass heat exchanger can be determined from Fig. 7-16. The value of R for exchangers that have one isothermal fluid is zero, for both 1–2 and 2–4 multipass heat exchangers.

7-25 Condensation for Vertical Tubes

Calculate the condensing coefficient for a vertical tube with an inside diameter of 0.62 in (0.016 m) if steam with these properties is condensing on the inside of the tube at a rate of 50 lb/h (22.68 kg/h):

ρ_L = liquid density = 60 lb/ft^3 (961.1 kg/m^3)

ρ_v = vapor density = 0.0372 lb/ft^3 (0.60 kg/m^3)

μ_L = liquid viscosity = 0.72 lb/(ft)(h) (0.298 cP)

μ_v = vapor viscosity = 0.0313 lb/(ft)(h) (0.0129 cP)

c = liquid specific heat = 1.0 Btu/(lb)(°F) [4.19 kJ/(kg)(K)]

k = liquid thermal conductivity, = 0.395 Btu/(h)(ft)(°F) [0.683 W/(m)(K)]

Calculation Procedure:

1. Select the calculation method to be used.

Use the Dukler theory [18], which assumes that three fixed factors must be known to establish the value of the average heat-transfer coefficient for condensing inside vertical tubes. These are the terminal Reynolds number ($4\Gamma/\mu$), the Prandtl number ($c\mu/k$) of the condensed phase, and a dimensionless group designated A_d and defined as follows:

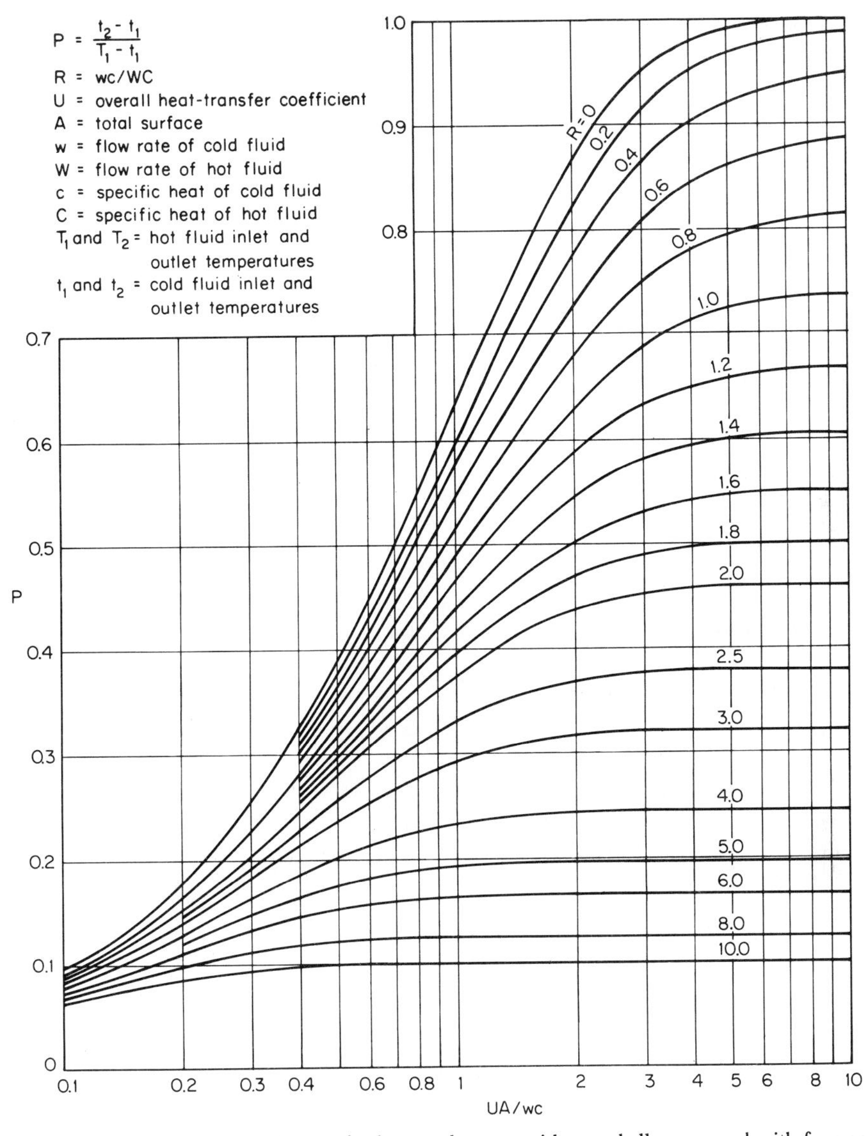

FIG. 7-16 Temperature efficiency for heat exchangers with two shell passes and with four or a multiple of four tube passes [11].

$$A_d = \frac{0.250\mu_L^{1.173}\mu_v^{0.16}}{g^{2/3}D_i^2\rho_L^{0.553}\rho_v^{0.78}}$$

In these equations,

$$\Gamma = W/(n\pi D_i)$$

where W = mass flow rate
n = number of tubes
D_i = inside tube diameter
ρ_L = liquid density
ρ_v = vapor density
g = gravitational constant
μ_L = liquid viscosity
μ_v = vapor viscosity

The Reynolds and Prandtl numbers are related to the condensing coefficient in Fig. 7-17. However, that figure is based on $A_d = 0$, which assumes no interfacial shear. The following factors can be used to evaluate the effects of any interfacial shear (h = coefficient with interfacial shear; h_0 = coefficient with no interfacial shear):

Terminal Reynolds number $4\Gamma/\mu$	h/h_0			
	$A_d = 0$	$A_d = 10^{-5}$	$A_d = 10^{-4}$	$A_d = 2 \times 10^{-4}$
200	1.0	1.0	1.03	1.05
500	1.0	1.03	1.15	1.28
1000	1.0	1.07	1.40	1.68
3000	1.0	1.25	2.25	2.80
10,000	1.0	1.90	4.35	6.00
30,000	1.0	3.30	8.75	13.00

2. *Calculate the terminal Reynolds number $4\Gamma/\mu$.*

Since $\Gamma = W/(n\pi D_i) = 50/[1.0\pi(0.62/12)] = 308$ lb/(h)(ft), $4\Gamma/\mu = 4(308)/0.72 = 1711.3$.

3. *Calculate the Prandtl number.*

Thus, $c\mu/k = 1.0(0.72)/0.395 = 1.82$.

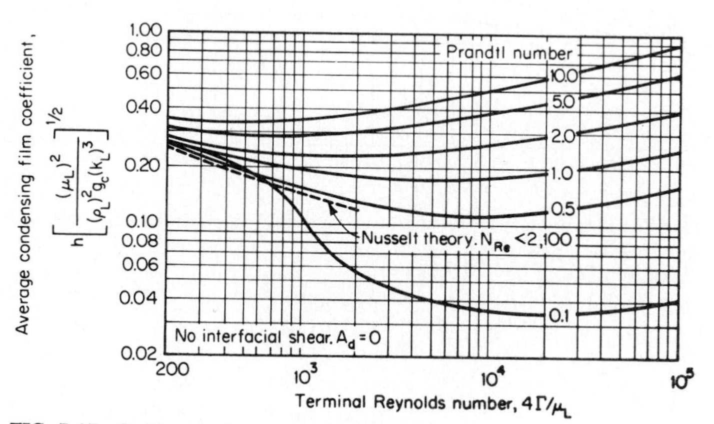

FIG. 7-17 Dukler plot for condensing-film coefficients.

4. Calculate h_0.

Refer to Fig. 7-17. For $4\Gamma/\mu = 1711.3$ and $c\mu/k = 1.82$, $(h\mu_L^2/\rho_L^2 g k^3)^{1/3} = 0.21$. Therefore,

$$h = 0.21\left[\frac{60^2(4.18 \times 10^8)0.395^3}{0.72^2}\right]^{1/3}$$

$$= 1183 \text{ Btu/(h)(ft}^2\text{)(°F) [6720 W/(m}^2\text{)(K)]}$$

5. Calculate A_d.

Thus,

$$A_d = \frac{0.250\mu_L^{1.173}\mu_v^{0.16}}{g^{2/3}D_i^2\rho_L^{0.553}\rho_v^{0.78}}$$

$$= \frac{(0.250)(0.72)^{1.173}(0.0313)^{0.16}}{(4.18 \times 10^8)^{2/3}(0.62/12)^2(60)^{0.553}(0.0372)^{0.78}}$$

$$= 8.876 \times 10^{-5}$$

6. Calculate h.

For $A_d = 8.876 \times 10^{-5}$ and $4\Gamma/\mu = 1711.3$, h/h_0 is approximately 1.65 from step 1. Then, $h = 1.65h_0 = 1.65(1183) = 1952$ Btu/(h)(ft^2)(°F) [11,088 W/(m^2)(K)].

Related Calculations: For low values of Reynolds number $(4\Gamma/\mu)$, the Nusselt equation can be used to predict condensing heat-transfer coefficients for vertical tubes:

$$h = 0.925k\left(\frac{\rho^2 g}{\mu\Gamma}\right)^{1/3}$$

where h = heat-transfer coefficient
k = liquid thermal conductivity
ρ = liquid density
g = gravitational constant
μ = liquid viscosity
Γ = condensate rate per unit periphery

It can be seen from Fig. 7-17 that the condensing heat-transfer coefficient for a fluid with a Prandtl number of approximately 2 (for instance, steam) is not strongly dependent on flow rate or Reynolds number. For this reason, heat-transfer coefficients for steam condensing on vertical tubes are frequently not calculated, but are assigned a value of 1500 to 2000 Btu/(h)(ft^2)(°F) [8500 to 11,340 W/(m^2)(K)].

This approach can be used for condensing on the outside of vertical tubes. The equivalent diameter should be used in evaluating the value of A_d, and the outside tube diameter should be used in calculating the terminal Reynolds number.

Heat-transfer coefficients for falling films can also be predicted from Fig. 7-17. The coefficient obtained should be multiplied by 0.75 to obtain heat-transfer coefficients for falling-film mechanisms.

A minimum flow rate is required to produce a falling film on vertical tubes when not condensing. The minimum flow rate can be predicted with the equation

$$\Gamma_{min} = 19.5(\mu s \sigma^3)^{1/5}$$

where Γ_{min} = tube loading, in lb/(h)(ft)
μ = liquid viscosity, in centipoise
s = liquid specific gravity
σ = surface tension, in dyn/cm

If this minimum is not achieved, the perimeter of the tube will not be uniformly wetted. Once a falling film has been induced, however, a lower terminal flow rate can be realized without destroying the film. This minimum terminal rate can be predicted with the equation

$$\Gamma_T = 2.4(\mu s \sigma^3)^{1/5}$$

where Γ_T = terminal tube loading, in lb/(h)(ft)

This criterion establishes the maximum amount of material that can be vaporized with a falling-film vaporizer. Approximately 85 percent of the entering material can be vaporized in a single pass without destroying the film. If tube loadings below the terminal loading are attempted, the film will break and form rivulets. Part of the tube surface will not be wetted, and the result will be reduced heat transfer with possible increased fouling of the heat-transfer surface.

For vertical condensers, condensate can be readily subcooled if required. The subcooling occurs as falling-film heat transfer, so the procedure discussed for falling-film heat exchangers can be used to calculate heat-transfer coefficients.

Condensation of mixed vapors of immiscible liquids is not well understood. The conservative approach is to assume that two condensate films are present and all the heat must be transferred through both films in series. Another approach is to use a mass fraction average thermal conductivity and calculate the heat-transfer coefficient using the viscosity of the film-forming component (the organic component for water-organic mixtures).

The recommended approach is to use a shared-surface model and calculate the effective heat-transfer coefficient as

$$h_L = V_A h_A + (1 - V_A) h_B$$

where h_L = the effective heat-transfer coefficient
h_A = the heat-transfer coefficient for liquid A assuming it only is present
h_B = the heat-transfer coefficient for liquid B assuming it only is present
V_A = the volume fraction of liquid A in the condensate

7-26 Condensation inside Horizontal Tubes

Calculate the effective condensing coefficient for a horizontal tube with an inside diameter of 0.62 in (0.016 m) and a length of 9 ft (2.74 m) for a fluid with the following properties condensing at a rate of 126 lb/h (57.15 kg/h):

ρ_L = liquid density = 50 lb/ft^3 (800.9 kg/m^3)
ρ_v = vapor density = 1.0 lb/ft^3 (16.02 kg/m^3)
μ = liquid viscosity = 0.25 lb/(ft)(h) (0.104 cP)
c = liquid specific heat = 0.55 Btu/(lb)(°F) [2.30 kJ/(kg)(K)]
k = liquid thermal conductivity = 0.08 Btu/(h)(ft)(°F) [0.14 W/(m)(K)]

Calculation Procedure:

1. Select the calculation method to be used.

Condensation inside horizontal tubes can be predicted assuming two mechanisms. The first assumes stratified flow, with laminar film condensation. The second assumes annular flow and is approximated with single-phase heat transfer using an equivalent mass velocity to reflect the two-phase flow. For the stratified-flow assumption, the further assumption is made that the rate of condensation on the stratified layer of liquid running along the bottom of the tube is negligible. Consequently, this layer of liquid must not exceed values assumed without being appropriately accounted for.

The following equations can be used to predict heat-transfer coefficients for condensation inside horizontal tubes: For stratified flow,

$$h_c = 0.767k\left(\frac{\rho_L^2 gL}{n\mu W}\right)^{1/3}$$

For annular flows,

a. $h_a D_i/k = 0.0265(c\mu/k)^{1/3}(DG_E/\mu)^{0.8}$
when Re_L is greater than 5000 and Re_v greater then 20,000, or

b. $h_a D_i/k = 5.03(c\mu/k)^{1/3}(DG_E/\mu)^{1/3}$
when Re_L is less than 5000 or Re_v less than 20,000.

In these equations, $Re_L = DG_L/\mu$, $Re_v = (DG_v/\mu)(\rho_L/\rho_v)^{1/2}$, and $G_E = G_L + G_v(\rho_L/\rho_v)^{1/2}$,

where h = heat-transfer coefficient
ρ_L = liquid density
ρ_v = vapor density
g = gravitational constant
k = liquid thermal conductivity
μ = liquid viscosity
L = tube length
n = number of tubes
W = condensate flow rate
G_E = equivalent mass velocity
G_L = liquid mass velocity assuming only liquid is flowing
G_v = vapor mass velocity assuming only vapor is flowing
D_i = inside tube diameter
c = liquid specific heat
Re_L = liquid Reynolds number
Re_v = vapor Reynolds number, defined above

Calculate the heat-transfer coefficient using both mechanisms and select the higher value calculated as the effective heat-transfer coefficient h_L. The annular-flow assumption results in heat-transfer coefficients that vary along the tube length. The condenser should be broken into increments, with the average vapor and liquid flow rates for each increment used to calculate heat-transfer coefficients. The total is the integrated value of all the increments.

For this problem, we calculate the heat-transfer coefficient for nine increments, each 1 ft (0.305 m) long and each condensing the same amount. The actual calculations are shown for only three increments—the first, the middle, and the last. All nine increments, however, are shown in the summary.

2. *Calculate h_c, the condensing coefficient if stratified flow is assumed.*

Because an equal amount condenses in each increment, h_c will be the same in each increment. Thus, $W_{c,av} = 126/9 = 14$ lb/(h)(ft). Therefore,

$$h_c = 0.767k\left(\frac{\rho_L^2 gL}{n\mu W_{c,av}}\right)^{1/3}$$

$$= 0.767(0.08)\left[\frac{50^2(4.18 \times 10^8)(1.0)}{1(0.25)(14)}\right]^{1/3}$$

$$= 410.1 \text{ Btu/(h)(ft}^2\text{)(°F) [2330 W/(m}^2\text{)(K)]}$$

3. *Evaluate the effect of condensate loading on h_c.*

The preceding equation for h_c assumes a certain condensate level on the bottom of the tube. This should be evaluated, which can be done by comparing the value of $W/(n\rho D_i^{2.56})$ with the values shown in Table 7-3, W being the condensate flow rate at the end of the tube, in pounds per hour, ρ is the density, in pounds per cubic foot, and D_i is the inside diameter, in feet.

Now, $W/(n\rho D_i^{2.56}) = 126/[1(50)(0.62/12)^{2.56}] = 4961$. From Table 7-3, $h/h_c = 0.92$, so $h = 0.92(410.1) = 377.3$ Btu/(h)(ft^2)(°F) [2144 W/(m^2)(K)].

4. *Calculate h_a, the condensing coefficient if annular flow is assumed.*

As noted earlier, this coefficient must be calculated separately for each increment of tube length. Here are the calculations for three of the nine increments:

TABLE 7-3 Effect of Condensate Loading on Condensing Coefficient

$W/(n\rho D_i^{2.56})$	h/h_c	$W/(n\rho D_i^{2.56})$	h/h_c
0	1.30	7000	0.85
50	1.25	9000	0.80
200	1.20	10,000	0.75
500	1.15	12,000	0.70
750	1.10	17,000	0.60
1500	1.05	20,000	0.50
2650	1.00	25,000	0.40
4000	0.95	30,000	0.25
6000	0.90	35,000	0.00

First Increment

a. *Determine average liquid and vapor flow rates.* Of the total vapor input, 126 lb/h, there is 14 lb/h condensed in this increment. Therefore, $W_{L1} = (0 + 14)/2 = 7$ lb/h of liquid, and $W_{v1} = 126 - 7 = 119$ lb/h of vapor.

b. *Calculate G_E and D_iG_E/μ.* Thus,

$$G_L = \frac{W_{L1}}{a} = \frac{7}{(0.62/12)^2(\pi/4)} = 3338.8 \text{ lb/(h)(ft}^2\text{)}$$

$$G_v = \frac{W_{v1}}{a} = \frac{119}{(0.62/12)^2(\pi/4)} = 56{,}759.2$$

$$G_E = G_L + G_v(\rho_L/\rho_v)^{1/2} = 3338.8 + (56{,}759.2)(50/1.0)^{1/2}$$

$$= 3338.8 + 401{,}348.1 = 404{,}686.9 \text{ lb/(h)(ft}^2\text{)}$$

$$\frac{D_iG_E}{\mu} = \frac{(0.62/12)404{,}686.9}{0.25} = 83{,}635.3$$

c. *Determine Re_v and Re_L.* Thus,

$$Re_v = \frac{D_iG_v(\rho_L/\rho_v)^{1/2}}{\mu} = \frac{(0.62/12)401{,}348.1}{0.25}$$

$$= 82{,}945.3$$

$$Re_L = \frac{D_iG_L}{\mu} = \frac{(0.62/12)3338.8}{0.25}$$

$$= 690.0$$

d. *Calculate h_a.* Since Re_v is greater than 20,000 but Re_L less than 5000, $h_a = 5.03(k/D_i)(c\mu/k)^{1/3}(D_iG_E/\mu)^{1/3}$. Now, $(c\mu/k)^{1/3} = [0.55(0.25)/0.08^{1/3}] = 1.1979$, so

$$h_a = 5.03 \frac{0.08}{(0.62/12)} 1.1979(83{,}635.3)^{1/3}$$

$$= 408.0 \text{ Btu/(h)(ft}^2\text{)(°F) [2318 W/(m}^2\text{)(K)]}$$

Middle (Fifth) Increment

a. *Determine average liquid and vapor flow rates.* Since the four upstream increments each condensed 14 lb/h, $W_{L5} = 14(4 + 5)/2 = 63$ lb/h and $W_{v5} = 126 - 63 =$ lb/h.

b. *Calculate G_E and D_iG_E/μ.* Now, $G_L = G_v = W_{L5}/a = 63/[(0.62/12)^2(\pi/4)] =$ 30,049.0 lb/(h)(ft²); $G_E = G_L + G_v(\rho_L/\rho_v)^{1/2} = 30{,}049.0 + 30{,}049.0(50/1.0)^{1/2} =$ $30{,}049.0 + 212{,}478.4 = 242{,}527.4$ lb/(h)(ft²); and $D_iG_E/\mu = (0.62/12)242{,}527.4/0.25$ $= 50{,}122.3$.

c. *Determine Re_v and Re_L.* Now,

$$Re_v = \frac{D_iG_v(\rho_L/\rho_v)^{1/2}}{\mu} = \frac{(0.62/12)212{,}478.4}{0.25}$$

$$= 43{,}912.2$$

$$Re_L = \frac{D_i G_L}{\mu} = \frac{(0.62/12)30{,}049.0}{0.25}$$

$$= 6210.1$$

d. *Calculate h_a.* Since Re_v is greater than 20,000 and Re_L is greater than 5000,

$$h_a = 0.0265(k/D_i)(c\mu/k)^{1/3}(D_i G_E/\mu)^{0.8}$$

$$= 0.0265[0.08/(0.62/12)]1.1979(50{,}122.3)^{0.8}$$

$$= 282.8 \text{ Btu/(h)(ft}^2\text{)(°F) [1607 W/(m}^2\text{)(K)]}$$

Last Increment

a. *Determine average liquid and vapor flow rates.* Now, $W_{L9} = 14(8 + 9)/2 = 119$ lb/h and $W_{v9} = 126 - 119 = 7$ lb/h.

b. *Calculate G_E and $D_i G_E/\mu$.* Now, $G_L = 119/[(0.62/12)^2(\pi/4)] = 56{,}759.2$ lb/(h)(ft²); $G_v = 7/[(0.62/12)^2(\pi/4)] = 3338.8$ lb/(h)(ft²); $G_E = G_L + G_v(\rho_L/\rho_v)^{1/2} = 56{,}759.2 + 3338.8(50/1.0)^{1/2} = 56{,}759.2 + 23{,}608.7 = 80{,}367.9$ lb/(h)(ft²); and $D_i G_E/\mu = (0.62/12)80{,}367.9/0.25 = 16{,}609.4$.

c. *Determine Re_v and Re_L.* Thus, $Re_L = D_i G_L/\mu = (0.62/12)56{,}759.2/0.25 = 11{,}730.2$; $Re_v = D_i G_v(\rho_L/\rho_v)^{1/2}/\mu = (0.62/12)23{,}608.7/0.25 = 4879.1$.

d. *Calculate h_a.* Since Re_v is less than 20,000 and Re_L greater than 5000,

$$h_a = 5.03(k/D_i)(c\mu/k)^{1/3}(D_i G_E/\mu)^{1/3}$$

$$= 5.03[0.08/(0.62/12)]1.1979(16{,}609.4)^{1/3}$$

$$= 238.0 \text{ Btu/(h)(ft}^2\text{)(°F) [1353 W/(m}^2\text{)(K)]}$$

5. Compare heat-transfer coefficients.

The effective heat-transfer coefficient h_L is the larger value of h_c and h_a, as indicated here:

Increment	h_c	h_a	h_L
1	377.3	408.0	408.0
2	377.3	392.7	392.7
3	377.3	378.7	378.7
4	377.3	362.2	377.3
5	377.3	282.8	377.3
6	377.3	244.3	377.3
7	377.3	204.3	377.3
8	377.3	272.8	377.3
9	377.3	238.0	377.3

The average value of h_L is 382.6 Btu/(h)(ft²)(°F) [2174 W/(m²)(K)].

Related Calculations: Condensation of mixed vapors of immiscible liquids can be treated in the same manner as outlined in Example 7-25.

No good methods are available for calculating heat-transfer coefficients when appreciable subcooling of the condensate is required. A conservative approach is to calculate a superficial mass velocity assuming the condensate fills the entire tube and use the equations presented above for single-phase heat transfer inside tubes. This method is less conservative for higher condensate loads.

7-27 Condensation outside Horizontal Tubes

For the following conditions, calculate the heat-transfer coefficient when condensing at a rate of 54,000 lb/h (24,493.9 kg/h) on the outside of a tube bundle with a diameter of 25 in (0.635 m) with nine baffle sections each 12 in (0.305 m) long. The bundle contains 532 tubes with an outside diameter of 0.75 in (0.019 m). The tubes are on a triangular pitch and are spaced 0.9375 in (0.02381 m) center to center. Assume equal amounts condense in each baffle section.

Physical Properties

ρ_L = liquid density = 87.5 lb/ft^3 (1401.6 kg/m^3)
ρ_v = vapor density = 1.03 lb/ft^3 (16.5 kg/m^3)
μ = liquid viscosity = 0.7 lb/(ft)(h) (0.29 cP)
c = liquid specific heat = 0.22 Btu/(lb)(°F) [0.92 kJ/(kg)(K)]
k = liquid thermal conductivity = 0.05 Btu/(h)(ft)(°F) [0.086 W/(m)(K)]

Calculation Procedure:

1. *Select the calculation method to be used.*

Condensation on the outside of banks of horizontal tubes can be predicted assuming two mechanisms. The first assumes laminar condensate flow; the second assumes that vapor shear dominates the heat transfer. The following equations can be used to predict heat-transfer coefficients for condensation on banks of horizontal tubes: For laminar-film condensation,

$$h_c = ak\left(\frac{\rho_L^2 gnL}{\mu W}\right)^{1/3}\left(\frac{1}{N_r}\right)^{1/6}$$

where a = 0.951 for triangular tube patterns, 0.904 for rotated square tube patterns, or 0.856 for square tube patterns.

For vapor-shear-dominated condensation,

$$\frac{h_s D_o}{k} = b\left(\frac{D_o \rho_L v_G}{\mu}\right)^{1/2}\left(\frac{1}{N_r}\right)^{1/6}$$

where b = 0.42 for triangular tube patterns, 0.39 for square tube patterns, or 0.43 for rotated square tube patterns. In these equations, h_c is the laminar-film heat-transfer coefficient, h_s is the vapor-shear-dominated heat-transfer coefficient, k is liquid thermal conductivity, ρ_L is liquid density, g is the gravitational constant, μ is liquid viscosity, D_o is outside tube diameter, L is tube length, W is condensate flow rate, n is the number of tubes, v_G is maximum vapor velocity (defined below), and N_r is the number of vertical tube rows (defined below). Specifically,

$$N_r = \frac{mD_s}{s}$$

where D_s = shell diameter
s = tube center-to-center spacing
m = 1.0 for square tube patterns, 1.155 for triangular tube patterns, or 0.707 for rotated square tube patterns

And,

$$v_G = \frac{W_v}{\rho_v a_c}$$

where W_v = vapor flow rate
ρ_v = vapor density
a_c = minimum flow area

In this equation, $a_c = BD_s(s - D_o)/s$ for square and triangular tube patterns or $a_c = 1.5BD_s(s - D_o)/s$ for rotated square tube patterns, B being the baffle spacing.

Calculate the heat-transfer coefficient using both mechanisms, and select the higher value calculated as the effective heat-transfer coefficient h_L. The vapor-shear effects vary for each typical baffle section. The condenser should be calculated in increments, with the average vapor velocity for each increment used to calculate vapor-shear heat-transfer coefficients.

When the heat-transfer coefficients for laminar flow and for vapor shear are nearly equal, the effective heat-transfer coefficient is increased above the higher of the two values. The following table permits the increase to be approximated:

h_s/h_c	h_L/h_c
0.5	1.05
0.75	1.125
1.0	1.20
1.25	1.125
1.5	1.05

For this problem, we calculate the heat-transfer coefficient for the first increment, the middle increment, and the last increment. The summary will show the results of all increments. The increment chosen for illustration is one baffle section.

2. *Calculate h_c, the condensing coefficient if laminar-film condensation is assumed.*

Because an equal amount condenses in each increment, h_c will be the same in each increment. The average condensate flow rate in each increment will be $W_i = W_c/9 = 54{,}000/9 = 6000$ lb/h. Now,

$$h_c = 0.951k\left(\frac{\rho_L^2 gnL}{\mu W_i}\right)^{1/3}\left(\frac{1}{N_r}\right)^{1/6}$$

where $N_r = 1.155D_s/s = 1.155(25)/0.9375 = 30.8$. So,

$$h_c = 0.951(0.05)\left[\frac{87.5^2(4.18 \times 10^8)(532)1.0}{0.7(6000)}\right]^{1/3}\left(\frac{1}{30.8}\right)^{1/6}$$

$$= 198.7 \text{ Btu/(h)(ft}^2\text{)(°F) [1129 W/(m}^2\text{)(K)]}$$

3. Calculate h_s, the coefficient if vapor-shear domination is assumed.

As noted earlier, this must be calculated separately for each increment. Here are the calculations for three of the nine increments:

First Increment

a. *Calculate average vapor velocity.* Now, $W_{v1} = (54{,}000 + 54{,}000 - 6000)/2 = 51{,}000$ lb/h. And,

$$a_c = \frac{BD_s(s - D_o)}{s}$$

$$= \frac{(12/12)(25/12)(0.9375 - 0.75)}{0.9375}$$

$$= 0.4167 \text{ ft}^2 (0.0387 \text{ m}^2)$$

Therefore,

$$v_G = \frac{W_{v1}}{\rho a_c}$$

$$= \frac{51{,}000}{1.03(0.4167)}$$

$$= 118{,}825.4 \text{ ft/h}$$

b. *Calculate h_s.* Now,

$$h_s = 0.42\left(\frac{k}{D_o}\right)\left(\frac{D_o \rho_L v_G}{\mu}\right)^{1/2}\left(\frac{1}{N_r}\right)^{1/6}$$

$$= 0.42\frac{0.05}{(0.75/12)}\left[\frac{(0.75/12)(87.5)118{,}825.4}{0.7}\right]^{1/2}\left(\frac{1}{30.8}\right)^{1/6}$$

$$= 182.9 \text{ Btu/(h)(ft}^2\text{(°F) [1039 W/(m}^2\text{)(K)]}$$

c. *Calculate h_L.* Now, $h_s/h_c = 182.9/198.7 = 0.921$. From the preceding table, $h_L/h_c = 1.18$. So, $h_L = 1.18(198.7) = 234.5$ Btu/(h)(ft^2)(°F) [1332 W/(m^2)(K)].

Middle (Fifth) Increment

a. *Determine average vapor flow rate.* Thus, $W_{v5} = 54{,}000 - 4(6000) - 0.5(6000) = 27{,}000$ lb/h.

b. *Calculate h_s.* Refer to the calculation for the first increment. Then, $h_s = 182.9(27{,}000/51{,}000)^{1/2} = 133.1$ Btu/(h)(ft^2)(°F).

c. *Calculate h_L.* Thus, $h_s/h_c = 133.1/198.7 = 0.67$. From the preceding table, $h_L/h_c = 1.105$. So, $h_L = 1.105(198.7) = 219.6$ Btu/(h)(ft^2)(°F) [1248 W/(m^2)(K)].

Last Increment

a. *Determine average vapor flow rate.* Thus, $W_{v9} = 54{,}000 - 8(6000) - 0.5(6000) = 3000$ lb/h.

b. *Calculate h_s.* Refer to the calculation for the first increment. Then, $h_s = 182.9(3000/51{,}000)^{1/2} = 44.4$ Btu/(h)(ft^2)(°F).

c. *Calculate h_L.* Now, $h_s/h_c = 44.4/198.7 = 0.223$. For this low value, assume that $h_L = h_c = 198.7$ Btu/(h)(ft^2)(°F) [1129 W/(m^2)(K)].

4. Summarize the results.

The following table summarizes the calculations for all nine increments:

Increment	h_c	h_s	h_L
1	198.7	182.7	234.5
2	198.7	171.8	231.5
3	198.7	159.9	227.5
4	198.7	147.1	223.5
5	198.7	133.1	219.6
6	198.7	117.4	213.6
7	198.7	99.2	198.7
8	198.7	76.8	198.7
9	198.7	44.4	198.7

The average value of h_L is 216.3 Btu/(h)(ft^2)(°F) [1229 W/(m^2)(K)].

Related Calculations: For most cases, vapor-shear condensation is not important, and the condensing heat-transfer coefficient can be calculated simply as the laminar-film coefficient.

It is important that the shell side of horizontal shell-side condensers be designed to avoid excessive condensate holdup caused by the baffle or nozzle types selected by the designer.

For bundles with slight slopes, the following correction should be applied:

$$h_t = h_c(\cos\alpha)^{1/3} \qquad \text{for } L/D_o > 1.8 \tan\alpha$$

where h_t = heat-transfer coefficient for sloped tube
h_c = heat-transfer coefficient for horizontal tube
α = angle from horizontal, in degrees
L = tube length
D_o = outside tube diameter

Condensation of mixed vapors of immiscible fluids can be treated in the same manner as outlined in Example 7-25.

Condensate subcooling when condensing on banks of horizontal tubes can be accomplished in two ways. The first method requires holding a condensate level on the shell side; heat transfer can then be calculated using the appropriate single-phase correlation. The second method requires that the vapor make a single pass across the bundle in a vertical downflow direction. Subcooling heat transfer can then be calculated using falling-film correlations.

For low-fin tubes, the laminar condensing coefficient can be calculated by applying an appropriate correction factor F to the value calculated using the preceding equation for laminar-film condensation. The factor F is defined thus:

$$F = \left(\eta \frac{A_t}{A_i}\frac{D_i}{D_r}\right)^{1/4}$$

where η = weighted fin efficiency
A_t = total outside surface
A_i = inside surface
D_i = inside tube diameter
D_r = diameter at root of fins

7-28 Condensation in the Presence of Noncondensables

A mixture of vapor and noncondensable gases flows through a vertical tube bundle containing 150 tubes with an inside diameter of 0.62 in (0.016 m) and an outside diameter of 0.75 in (0.019 m) at a rate of 25,000 lb/h (11,339.8 kg/h). The mixture enters at a temperature of 212°F (373 K) and is cooled to a temperature of 140°F (333 K). As cooling occurs, 15,000 lb/h (6803.9 kg/h) of the mixture condenses. The condensation may be assumed to be straight-line condensation. Calculate the tube length required if the outside heat-transfer coefficient is 300 Btu/(h)(ft^2)(°F) [1700 W/(m^2)(K)] and the temperature is isothermal at 104°F (313 K). Assume the fouling heat-transfer coefficient is 1000 Btu/(h)(ft^2)(°F) [5680 W/(m^2)(K)]. The tube-wall thermal conductivity is 10 Btu/(h)(ft)(°F) [17.28 W/(m)(K)]. The physical properties of the system are as follows:

Vapor specific heat = 0.35 Btu/(lb)(°F) [1.47 kJ/(kg)(K)]
Liquid specific heat = 0.7 Btu/(lb)(°F) [2.93 kJ/(kg)(K)]
Vapor viscosity = 0.048 lb/(ft)(h) (0.020 cP)
Liquid viscosity = 0.24 lb/(ft)(h) (0.10 cP)
Vapor thermal conductivity = 0.021 Btu/(h)(ft)(°F) [0.036 W/(m)(K)]
Liquid thermal conductivity = 0.075 Btu/(h)(ft)(°F) [0.13 W/(m)(K)]
Heat of vaporization = 200 Btu/lb (465 kJ/kg)
Liquid density = 30 lb/ft^3 (481 kg/m^3)
Vapor density = 0.9 lb/ft^3 (14.4 kg/m^3)

Calculation Procedure:

1. Select the calculation method to be used.

A heat-transfer coefficient that predicts heat transfer when both sensible heat and latent heat are being transferred can be calculated using the equation

$$h_{c,g} = \frac{1}{\frac{Q_g}{Q_T}\frac{1}{h_g} + \frac{1}{h_c}}$$

where $h_{c,g}$ = combined cooling-condensing heat-transfer coefficient
h_g = heat-transfer coefficient for gas cooling only
h_c = condensing heat-transfer coefficient
Q_g = heat-transfer rate for cooling the gas only
Q_T = total heat-transfer rate (includes sensible heat for gas and condensate cooling and latent heat for condensing)

2. *Calculate the sensible and latent heat loads.*

Since straight-line condensing occurs, the mean condensing temperature may be taken as the arithmetic mean (176°F; 353 K). The heat loads are calculated assuming that all gas and vapor are cooled to the mean condensing temperature, that all condensing then occurs at this temperature, and that the uncondensed mixture plus the condensate are further cooled to the outlet temperature.

Inlet Vapor Cooling

$$Q_i = 25{,}000(0.35)(212 - 176) = 315{,}000 \text{ Btu/h}$$

Latent Heat of Condensation

$$Q_c = 15{,}000(200) = 3{,}000{,}000 \text{ Btu/h}$$

Outlet Vapor Cooling

$$Q_o = 10{,}000(0.35)(176 - 140) = 126{,}000 \text{ Btu/h}$$

Condensate Cooling

$$Q_s = 15{,}000(0.7)(176 - 140) = 378{,}000 \text{ Btu/h}$$

Therefore, the total heat load Q_T equals $Q_i + Q_c + Q_o + Q_s = 3{,}819{,}000$ Btu/h. And $Q_g = Q_i + Q_o = 441{,}000$ Btu/h, so $Q_T/Q_g = 3{,}819{,}000/441{,}000 = 8.66$.

3. *Calculate the gas cooling heat-transfer coefficient.*

Use the equation $h = 0.023cG/[(c\mu/k)^{2/3}(DG/\mu)^{0.2}]$. Base the mass velocity on the average vapor flow rate, that is, $(25{,}000 + 10{,}000)/2 = 17{,}500$ lb/h. Then $G = 17{,}500/[150(0.62/12)^2(\pi/4)] = 55{,}646.3$ lb/(h)(ft^2), and

$$h_g = \frac{0.023(0.35)(55{,}646.3)}{[0.35(0.048)/0.021]^{2/3}[(0.62/12)55{,}646.3/0.048]^{0.2}}$$

$$= 57.58 \text{ Btu/(h)(ft}^2\text{)(°F) [327 W/(m}^2\text{)(K)]}$$

4. Calculate the condensing coefficient.

See Example 7-25. Now, $4\Gamma/\mu = 4[15{,}000/150\pi(0.62/12)]/0.24 = 10{,}268$, and $c\mu/k = 0.7(0.24)/0.075 = 2.24$. From Fig. 7-17, for the parameters calculated above, $h_c[\mu^2/(\rho^2 g k^3)]^{1/3} = 0.28$. Therefore,

$$h_c = 0.28[30^2(4.18 \times 10^8)0.075^3/0.24^2]^{1/3}$$
$$= 392.5 \text{ Btu/(h)(ft}^2\text{)(°F) [2229 W/(m}^2\text{)(K)]}$$

5. Calculate $h_{c,g}$.

Now,

$$h_{c,g} = 1/[(Q_g/Q_T)(1/h_g) + (1/h_c)]$$
$$= 1/[(1/8.66)(1/57.58) + (1/392.5)]$$
$$= 219.6 \text{ Btu/(h)(ft}^2\text{)(°F) [1247 W/(m}^2\text{)(K)]}$$

6. Calculate the overall heat-transfer coefficient.

Use the equation

$$\frac{1}{U} = \frac{1}{h_o} + \frac{1}{h_s} + \frac{1}{h_w} + \frac{1}{h_{c,g}\,(d_i/d_o)}$$

where U = overall heat-transfer coefficient
h_o = outside heat-transfer coefficient
h_s = fouling heat-transfer coefficient
h_w = heat-transfer coefficient through the wall (thermal conductivity of the wall divided by its thickness)
$h_{c,g}$ = inside heat-transfer coefficient
d_i = inside tube diameter
d_o = outside tube diameter

Then,

$$\frac{1}{U} = \frac{1}{300} + \frac{1}{1000} + \frac{1}{10/[(0.75 - 0.62)/12(2)]} + \frac{1}{219.6(0.62/0.75)}$$
$$= 96.3 \text{ Btu/(h)(ft}^2\text{)(°F) [547 W/(m}^2\text{)(K)]}$$

7. Calculate the log mean temperature difference.

See Example 7-21 and use the equation for countercurrent flow. Thus,

$$\Delta T_{LM} = \frac{(212 - 104) - (140 - 104)}{\ln[(212 - 104)/(140 - 104)]}$$
$$= 65.5\text{°F (36.4 K)}$$

8. Calculate the heat-transfer surface required.

The equation is

$$A = \frac{Q_T}{U\Delta T_{LM}} = \frac{3{,}819{,}000}{96.3(65.5)}$$

$$= 605.4 \text{ ft}^2 \ (56.24 \text{ m}^2)$$

9. Calculate the tube length required.

Since $A = n\pi D_o L$, $L = 605.4/[150\pi(0.75/12)] = 20.6$ ft (6.27 m).

Related Calculations: For condensers that do not exhibit straight-line condensing, this procedure can be used by treating the problem as a series of condensing zones that approximate straight-line segments. The vapor and condensate flow rates, the heat load, the overall heat-transfer coefficient, and the log mean temperature difference will vary with each zone. The total answer is obtained by integrating all the segments. It is usually not necessary to use a large number of zones. The accuracy lost by using less than 10 zones is not significant in most cases.

For horizontal shell-side condensers, the condensate falls to the bottom of the shell, and vapor and liquid do not coexist, as assumed by the preceding method. The effect this has on the heat transfer must be considered. It is recommended that shell-side condensers with noncondensable gases present be somewhat overdesigned; perhaps 20 percent excess surface should be provided.

7-29 Maximum Vapor Velocity for Condensers with Upflow Vapor

Calculate the maximum velocity to avoid flooding for the vapor conditions of Example 7-28. Flooding occurs in upflow condensers when the vapor velocity is too high to permit the condensate to drain. Unstable conditions exist when flooding occurs.

Calculation Procedure:

1. Select the appropriate equation.

The following equation can be used to establish the condition for flooding of vertical-upflow vapor condensers:

$$v_v^{1/2}\rho_v^{1/4} + v_L^{1/2}\rho_L^{1/4} \leq 0.6[gD_i(\rho_L - \rho_v)]^{1/4}$$

where v_v = vapor velocity assuming only vapor is flowing in the tube
v_L = liquid velocity of the condensate assuming only condensate is flowing in the tube
ρ_v = vapor density
ρ_L = liquid density
D_i = inside tube diameter
g = acceleration of gravity

2. Calculate the maximum allowable vapor mass-flow rate G_v.

Now, $v_L = G_L/\rho_L$, and $v_v = G_v/\rho_v$, where G is mass flow rate and the subscripts L and v refer to liquid and vapor. For the problem at hand, $G_L = (15{,}000/25{,}000)G_v = 0.6G_v$; therefore, $v_L = 0.6G_v/\rho_L$. Then, substituting into the expression in step 1, $(G_v/\rho_v)^{1/2}\rho_v^{1/4} + (0.6G_v/\rho_L)^{1/2}\rho_L^{1/4} = 0.6[gD_i(\rho_L - \rho_v)]^{1/4}$, or $G_v^{1/2}/\rho_v^{1/4} + 0.6^{1/2}G_v^{1/2}/\rho_L^{1/4} = 0.6[gD_i(\rho_L - \rho_v]^{1/4}$. Then, $G_v^{1/2}/0.9^{1/4} + 0.6^{1/2}G_v^{1/2}/30^{1/4} = 0.6[32.2(0.62/12)(30 - 0.9)]^{1/4}$, and $G_v = 1.3589$ lb/(ft^2)(s).

3. Calculate the maximum allowable velocity.

Now, $G_v = \rho_v v_v$, so $v_v = G_v/\rho_v = 1.3589/0.9 = 1.51$ ft/s (0.46 m/s).

Related Calculations: The velocity calculated above is the threshold for flooding of the condenser. Flooding causes unstable operation. A safety factor of 0.85 is commonly used when designing upflow condensers. The maximum velocity at which flooding occurs can be increased by cutting the bottom of the tube at an angle. Improvements that can be achieved are as follows:

Angle of cut (measured from horizontal)	Increase in maximum flooding velocity
30°	5%
60°	25%
75°	55%

7-30 Nucleate-Boiling Heat Transfer

A steel tube 1 in (0.025 m) in diameter and 12 ft (3.66 m) long is cooled by employing it to boil water at 1 atm (101.35 kPa). Calculate the nucleate-boiling heat-transfer coefficient if 300 lb/h (136.1 kg/h) of vapor is generated.

The physical properties are as follows:

c = liquid specific heat = 1.0 Btu/(lb)(°F) [4.2 kJ/(kg)(K)]

ρ_L = liquid density = 60 lb/ft^3 (961.1 kg/m^3)

ρ_v = vapor density = 0.0372 lb/ft^3 (0.596 kg/m^3)

k = liquid thermal conductivity = 0.396 Btu/(h)(ft)(°F) [0.684 W/(m)(K)]

μ = liquid viscosity = 0.72 lb/(ft)(h) (0.298 cP)

σ = surface tension = 0.0034 lb/ft (50 dyn/cm)

(Note that 1 atm = 2116.8 lb/ft^2.)

Calculation Procedure:

1. Select the appropriate equation.

Nucleate-boiling heat-transfer coefficients can be calculated from the equation

$$\frac{h}{cG} = \frac{\phi}{(c\mu/k)^{0.6}(DG/\mu)^{0.3}(\rho_L\sigma/P^2)^{0.425}}$$

where $G = W\rho_L/(A\rho_v)$
h = heat-transfer coefficient
G = mass velocity
P = absolute pressure
A = heat-transfer surface (*not* cross-sectional flow area)
D = diameter
W = vapor flow rate
ϕ = a constant that depends on the surface condition (number of nucleation sites that can support boiling), e.g., 0.001 for steel and copper, 0.0006 for stainless steel, and 0.0004 for polished surfaces

2. *Calculate G.*

Thus, $G = W\rho_L/(A\rho_v) = 300(60)/[(1/12)\pi 12(0.0372)] = 154{,}020.9$.

3. *Calculate h.*

For a steel tube,

$$h = \frac{0.001cG}{(c\mu/k)^{0.6}(DG/\mu)^{0.3}(\rho_L\sigma/P^2)^{0.425}}$$

$$= \frac{0.001(1.0)(154{,}020.9)}{\left[\dfrac{1.0(0.72)}{0.395}\right]^{0.6}\left[\dfrac{(1/12)154{,}020.9}{0.72}\right]^{0.3}\left[\dfrac{60(0.0034)}{2116.8^2}\right]^{0.425}}$$

$$= 7517.2 \text{ Btu/(h)(ft}^2\text{)(°F) [42,698 W/(m}^2\text{)(K)]}$$

Related Calculations: The mechanism of nucleate boiling has not been clearly established, but several expressions are available from which reasonable values of heat-transfer coefficients may be obtained. These do not yield exactly the same numerical results even though based on much of the same data. There is thus neither a prominent nor a unique equation for nucleate-boiling heat transfer. Either convenience or familiarity usually governs the user's selection.

Surface conditions have a profound effect on boiling heat transfer. The values of ϕ presented above have been obtained from plots of data. Extreme accuracy cannot be claimed for these values because of the variable condition of the surfaces in these tests.

There are upper and lower limits of applicability of the equation above. The lower limit results because natural-convection heat transfer governs at low temperature differences between the surface and the fluid. The upper limit results because a transition to film boiling occurs at high temperature differences. In film boiling, a layer of vapor blankets the heat-transfer surface and no liquid reaches the surface. Heat transfer occurs as a result of conduction across the vapor film as well as by radiation. Film-boiling heat-transfer coefficients are much less than those for nucleate boiling. For further discussion of boiling heat transfer, see Refs. 5 and 6.

7-31 Minimum Temperature Difference to Achieve Nucleate Boiling

For water boiling on a 1-in (0.025-m) tube under the conditions of Example 7-30, determine the minimum difference between the temperature of the surface of the tube and the bulk fluid temperature required in order for nucleate boiling to occur. The coefficient of expansion for water is 0.0004 per degree Fahrenheit, and the heat of vaporization is 970 Btu/lb (2256 kJ/kg).

Calculation Procedure:

1. Consider the criteria required for nucleate boiling.

Nucleate boiling occurs when the difference between the temperature of the hot surface and the bulk fluid temperature is above a certain value. At temperature differences less than this value, heat transfer occurs as a result of natural convection. Nucleate-boiling heat-transfer coefficients for a steel tube may be calculated using the equation

$$\frac{h}{cG} = \frac{0.001}{(c\mu/k)^{0.6}(DG/\mu)^{0.3}(\rho_L\sigma/P^2)^{0.425}}$$

where the nomenclature is the same as listed for Example 7-30.

Natural-convection heat-transfer coefficients may be calculated using the equations

$$\frac{h}{cG} = \frac{0.53}{(c\mu/k)^{3/4}(DG/\mu)^{1/2}} \qquad \text{and} \qquad G = (g\beta\rho_L^2 D\,\Delta T)^{1/2}$$

where g is the acceleration of gravity (4.18×10^8 ft/h^2) and β is the coefficient of expansion for water (for further discussion, see Example 7-17).

The equation for nucleate-boiling heat transfer can be rearranged to become a function of ΔT, the temperature difference between the surface and the fluid. The minimum temperature difference required to effect nucleate boiling will occur when the heat-transfer coefficients for nucleate boiling and natural convection are equal. This will permit solution for the temperature difference ΔT.

2. Obtain an equation for nucleate-boiling heat-transfer coefficient as a function of ΔT.

The heat load $Q = hA\,\Delta T = W\lambda$, where λ is the heat of vaporization. Therefore, $W = hA\,\Delta T/\lambda$. Now, $G = W\rho_L/(A\rho_v) = hA\,\Delta T\rho_L/(A\rho_v\lambda) = h\,\Delta T\rho_L/(\rho_v\lambda)$. Furthermore,

$$h = \frac{0.001cG}{(c\mu/k)^{0.6}(DG/\mu)^{0.3}(\rho_L\sigma/P^2)^{0.425}}$$

$$= \frac{0.001cG^{0.7}}{(c\mu/k)^{0.6}(D/\mu)^{0.3}(\rho_L\sigma/P^2)^{0.425}}$$

From above, $G^{0.7} = [h\ \Delta T\ \rho_L/(\rho_v\lambda)]^{0.7}$. Therefore, $h = 0.001c[h\ \Delta T\ \rho_L/(\rho_v\lambda)]^{0.7}/[(c\mu/k)^{0.6}(D/\mu)^{0.3}(\rho_L\sigma/P^2)^{0.425}]$, or $h^{0.3} = 0.001c[\rho_L/(\rho_v\lambda)]^{0.7}(\Delta T)^{0.7}/[c\mu/k)^{0.6}(D/\mu)^{0.3}(\rho_L\sigma/P^2)^{0.425}]$, or $h = \{0.001c[\rho_L/(\rho_v\lambda]^{0.7}/[(c\mu/k)^{0.6}(D/\mu)^{0.3}(\rho_L\sigma/P^2)^{0.425}]\}^{1/0.3}\ \Delta T^{0.7/0.3}$. Thus,

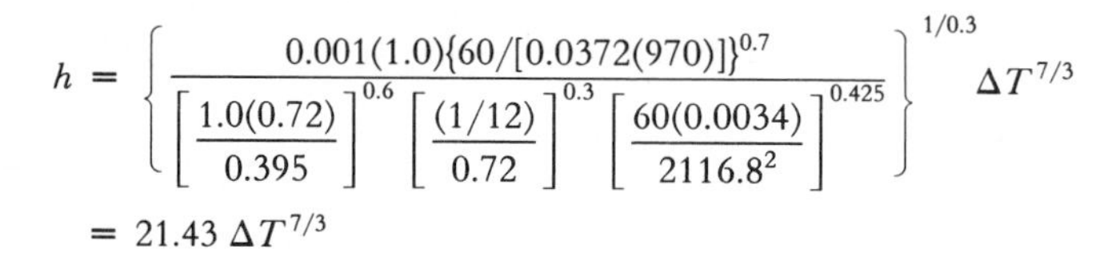

$$h = \left\{ \frac{0.001(1.0)\{60/[0.0372(970)]\}^{0.7}}{\left[\dfrac{1.0(0.72)}{0.395}\right]^{0.6}\left[\dfrac{(1/12)}{0.72}\right]^{0.3}\left[\dfrac{60(0.0034)}{2116.8^2}\right]^{0.425}} \right\}^{1/0.3} \Delta T^{7/3}$$

$$= 21.43\ \Delta T^{7/3}$$

where h is in Btus per hour per square foot per degree Fahrenheit, and the temperature difference is in degrees Fahrenheit.

3. *Obtain an equation for natural-convection heat transfer as a function of ΔT.*

From step 1, $h = 0.53cG/[(c\mu/k)^{3/4}(DG/\mu)^{1/2}]$, and $G = (g\beta\rho_L^2 D\ \Delta T)^{1/2}$. By algebra, $h = 0.53cG^{1/2}/[(c\mu/k)^{3/4}(D/\mu)^{1/2}]$, and by substitution,

$$h = 0.53c[(g\beta\rho_L^2 D\ \Delta T)^{1/2}]^{1/2}/[(c\mu/k)^{3/4}(D/\mu)^{1/2}]$$

$$= 0.53c(g\beta\rho_L^2 D)^{1/4}\ \Delta T^{1/4}/[(c\mu/k)^{3/4}(D/\mu)^{1/2}]$$

Thus,

$$h = \frac{0.53(1.0)[(4.18 \times 10^8)(0.0004)(60)^2(1/12)]^{1/4}}{[1.0(0.72)/0.395]^{3/4}[(1/12)/0.72]^{1/2}} \Delta T^{1/4}$$

$$= 83.57\ \Delta T^{1/4}$$

where h is in Btus per hour per square foot per degree Fahrenheit, and the temperature difference is in degrees Fahrenheit.

4. *Calculate the minimum required temperature difference required for nucleate boiling.*

The minimum temperature difference at which nucleate boiling will occur can be calculated by equating the heat-transfer coefficients for natural convection and nucleate boiling and solving for the temperature difference. Thus, $h = 21.43\ \Delta T^{7/3} = 83.57\ \Delta T^{1/4}$. Therefore, $\Delta T^{(7/3-1/4)} = \Delta T^{25/12} = 83.57/21.43$, or $\Delta T = 1.92°\text{F}$ (1.07 K). Thus the minimum temperature difference between the tube surface and the water at which nucleate boiling will occur is 1.92°F.

Related Calculations: The minimum temperature difference required for nucleate boiling to occur can also be determined by plotting the equations for nucleate-boiling and natural-convection heat-transfer coefficients. The intersection of these two lines represents the required temperature difference ΔT.

The procedure illustrated here can be used to obtain the minimum temperature difference required for nucleate boiling for any correlation for nucleate-boiling heat-transfer coefficients preferred by the user.

Low-fin tubes, often used for horizontal-pool boiling, reduce the minimum temperature difference required to achieve nucleate boiling. In addition, the boiling coefficients for low-fin tubes are higher than those for bare tubes at a given temperature difference.

Special boiling surfaces are available commercially that permit nucleate-boiling heat transfer at extremely low temperature differences. These surfaces also achieve much higher heat-transfer coefficients than conventional tubes; see Gottzmann, O'Neill, and Minton [32].

7-32 Maximum Heat Flux for Kettle-Type Reboiler

A kettle-type reboiler with a shell diameter of 30 in (0.76 m) contains a tube bundle with a diameter of 15 in (0.38 m). The bundle contains 80 tubes, each with a diameter of 1 in (0.025 m) and a length of 12 ft (3.66 m). Determine the maximum heat flux for a fluid with the following physical properties: heat of vaporization of 895 Btu/lb (2082 kJ/kg), surface tension of 0.00308 lb/ft (0.045 N/m), liquid density of 56.5 lb/ft^3 (905 kg/m^3), and vapor density of 0.2 lb/ft^3 (3.204 kg/m^3).

Calculation Procedure:

1. Determine the maximum superficial vapor velocity.

The maximum superficial vapor velocity to avoid film boiling can be calculated using the equation

$$v_c = \frac{3(\rho_L - \rho_v)^{1/4}\sigma^{1/4}}{\rho_v^{1/2}}$$

where v_c = superficial vapor velocity, in ft/s
ρ_L = liquid density, in lb/ft^3
ρ_v = vapor density, in lb/ft^3
σ = surface tension, in lb/ft

For this problem,

$$v_c = \frac{3(56.5 - 0.2)^{1/4}0.00308^{1/4}}{0.2^{1/2}}$$
$$= 4.33 \text{ ft/s } (1.329 \text{ m/s})$$

2. Determine the maximum heat rate.

The superficial vapor velocity is based on the projected area of the tube bundle

$$a = D_b L$$

where a = projected area of the bundle
D_b = bundle diameter
L = bundle length

In this case, $a = (15/12)12 = 15$ ft^2 (1.39 m^2).

The maximum heat transferred is determined from the equation

$$Q = 3600 a v_c \rho_v \lambda$$

where Q = heat transferred, in Btu/h
a = projected area, in ft^2
v_c = superficial vapor velocity, in ft/s
ρ_v = vapor density, in lb/ft^3
λ = heat of vaporization, in Btu/lb

In this case, $Q = 3600(15)(4.33)(0.2)(895) = 41{,}854{,}000$ Btu/h (12,266 kW).

3. Calculate the maximum heat flux.

Maximum heat flux = Q/A, where Q is maximum heat transferred and A is total heat-transfer surface for all n tubes. In this case, $A = n\pi DL = 80\pi(1/12)12 = 251$ ft^2 (23.35 m^2), so maximum heat flux = 41,854,000/251 = 166,750 Btu/(h)(ft^2) (525,600 W/m^2).

Related Calculations: The preceding equation for maximum superficial vapor velocity is applicable only to kettle-type reboilers having a shell diameter 1.3 to 2.0 times greater than the diameter of the tube bundle. (For small-diameter bundles, the ratio required is greater than that for large bundles.) This ratio is generally sufficient to permit liquid circulation adequate to obtain the superficial vapor velocity predicted by the equation presented here. For single tubes or for tube bundles with geometries that do not permit adequate liquid circulation, the superficial vapor velocity calculated using this equation should be multiplied by 0.3.

For vertical tubes, the superficial vapor velocity (based on the total heat-transfer surface) can be obtained by multiplying the value calculated from the preceding equation by 0.22. This assumes that there is adequate liquid circulating past the surface to satisfy the mass balance. For thermosiphon reboilers, a detailed analysis must be made to establish circulation rate, boiling pressure, sensible heat-transfer zone, boiling heat-transfer zone, and mean temperature difference. If liquid circulation rates are not adequate, all liquid will be vaporized and superheating of the vapor will occur with a resultant decrease in heat-transfer rates.

The procedure for design of thermosiphon reboilers presented by Fair [23] has been widely used. Special surfaces are available commercially that permit much higher superficial vapor velocities than calculated by the method presented here; see Gottzmann, O'Neill, and Minton [32].

7-33 Double-Pipe Heat Exchangers with Bare Tubes

Calculate the outlet temperature for air entering the annulus of a double-pipe exchanger at 68°F (293K) at a flow rate of 500 lb/h (226.8 kg/h) if steam is condensing inside the tube at 320°F (433 K). The shell of the double-pipe exchanger is 3.068 in (0.0779 m) in diameter. The steel tube is 1.9 in (0.048 m) in outside diameter and 1.61 in (0.0409 m) in inside diameter and has a thermal conductivity of 25 Btu/(h)(ft)(°F) [43 W/(m)(K)]. The tube length is 20 ft (6.1 m). Air has a viscosity of 0.0426 lb/(ft)(h) and a specific heat of 0.24 Btu/(lb)(°F) [1.01 kJ/(kg)(K)]. The fouling coefficient is 1000 Btu/(h)(ft^2)(°F) [5680 W/(m^2)(K)].

The physical properties for the steam condensate are:

μ = liquid viscosity = 0.5 lb/(ft)(h) (0.207 cP)

ρ = liquid density = 55.5 lb/ft^3 (889 kg/m^3)

λ = heat of vaporization = 895 Btu/lb (2,081,770 J/kg)

k = liquid thermal conductivity = 0.395 Btu/(h)(ft)(°F) [0.683 W/(m)(K)]

Calculation Procedure:

1. *Calculate condensing coefficient h_i.*

The steam-condensing coefficient will be much larger than the air-side coefficient; this permits us to approximate the condensing coefficient by assuming the maximum steam condensate loading. The maximum heat transferred would occur if the air were heated to the steam temperature. Thus, $Q_{max} = W_{air}c(T_{steam} - t_{air})$, where W is mass flow rate and c is specific heat. Or, $Q_{max} = 500(0.24)(320 - 68) = 30{,}240$ Btu/h. Then, maximum condensate flow $W_{c,max} = Q_{max}/\lambda = 30{,}240/895 = 33.8$ lb/h.

Now, with reference to Example 7-26, $W_c/n\rho D_i^{2.56} = 33.8/[1(55.5)(1.61/12)^{2.56}] = 104$. This is a relatively low value, so stratified flow can be assumed. Then,

$$h_c = 0.767k\left(\frac{\rho^2 gL}{n\mu W}\right)^{1/3}$$

$$= 0.767(0.39)\left(\frac{55.5^2(4.18 \times 10^8)(20)}{1(0.5)(33.8)}\right)^{1/3}$$

$$= 3486 \text{ Btu/(h)(ft}^2\text{)(°F)}$$

Now, for a $W_c/(n\rho D_i^{2.56})$ of 104, $h_i = 1.2h_c$, as indicated in Example 7-26. Thus, $h_i = 1.2(3486) = 4183$ Btu/(h)(ft^2)(°F), and $h_i(D_i/D_o) = 4183(1.61/1.9) = 3545$ Btu/(h)(ft^2)(°F) [20,136 W/(m^2)(K)].

2. *Calculate h_w, the heat-transfer coefficient through the tube wall.*

By definition, $h_w = 2k/(D_o - D_i) = 2(25)/(1.9 - 1.61)(1/12) = 2068$ Btu/(h)(ft^2)(°F) [11,750 W/(m^2)(K)].

3. *Calculate h_o, the heat-transfer coefficient for the outside of the tube.*

Because the fluid flowing is air, we can use simplified equations for heat-transfer coefficients. Cross-sectional area of the annulus $a_c = (\pi/4)(D_s^2 - D_o^2) = (\pi/4)(3.068^2 - 1.9^2)(1/12)^2 = 0.0316$ ft^2. Then, mass flow rate through the annulus $G = W_{air}/a_c = 500/0.0316 = 15{,}800$ lb/(h)(ft^2), and equivalent diameter $D_e = 4a_c/P$, where $P = \pi(D_s + D_o)$. Thus, by algebraic simplification, $D_e = D_s - D_o = (3.068/12) - (1.9/12) = 0.0973$ ft (0.0297 m).

Since the Reynolds number, that is, $D_eG/\mu = 0.0973(15{,}800)/0.0426 = 36{,}088$, is greater than 8000 (see Related Calculations in Example 7-18),

$$h_o = 0.0144cG^{0.8}/D_e^{0.2}$$
$$= 0.0144(0.24)(15{,}800)^{0.8}/0.0973^{0.2}$$
$$= 12.58 \text{ Btu/(h)(ft}^2\text{)(°F) [71.45 W/(m}^2\text{)(K)]}$$

4. Calculate U, the overall heat-transfer coefficient.

Now,

$$\frac{1}{U} = \frac{1}{h_o} + \frac{1}{h_i(D_i/D_o)} + \frac{1}{h_w} + \frac{1}{h_s}$$
$$= \frac{1}{12.58} + \frac{1}{3545} + \frac{1}{2068} + \frac{1}{1000}$$

So, $U = 12.32$ Btu/(h)(ft^2)(°F) [69.97 W/(m^2)(K)].

5. Calculate outlet air temperature.

Use Fig. 7-14. Now, $A = \pi D_oL = \pi(1.9/12)(20) = 9.95$ ft^2, so $UA/(wc) = 12.32(9.95)/[500(0.24)] = 1.02$. In addition, $R = 0$ (isothermal on steam side).

For these values of $UA/(wc)$ and R, $P = 0.64$ from Fig. 7-14. Then $t_2 = t_1 + P(T_s - t_1) = 68 + 0.64(320 - 68) = 229.2$°F (382.8 K).

7-34 Double-Pipe Heat Exchangers with Longitudinally Finned Tubes

Calculate the outlet air temperature for the double-pipe heat exchanger under the conditions of Example 7-33 if the tube has 24 steel fins 0.5 in (0.013 m) high and 0.03125 in (0.794 mm) thick.

Calculation Procedure:

1. Calculate the relevant fin areas.

The relevant areas are cross-sectional flow area a_c, fin surface A_f, outside bare surface A_o, and inside surface A_i. Now,

$$a_c = (\pi/4)(D_s^2 - D_o^2) - nlb$$

where n = number of fins
l = fin height
b = fin thickness

Thus, $a_c = 0.0316 - 24(0.5/12)(0.03125/12) = 0.029$ ft^2 (0.0028 m^2).

Further, A_f is the heat-transfer surface on both sides and the tip of the fins, and it equals $2nl + nb = 2(24)(0.5/12) + 24(0.03125/12) = 2.0625$ ft^2 per foot of tube length. And, A_o is the outside bare surface exclusive of the area beneath the fins, and it equals $\pi D_o - nb = \pi(1.9/12) - 24(0.03125/12) = 0.4349$ ft^2 per foot of tube length. Finally, $A_i = \pi D_i = \pi(1.61/12) = 0.4215$ ft^2 per foot of tube length.

2. *Calculate D_e for the fin tube.*

Use the formula $D_e = 4a_c/P$, where $P = \pi(D_s + D_o) + 2nl$. Now, $P = \pi(3.068 + 1.9)(1/12) + 2(24)(0.5/12) = 3.3006$ ft. Then, $D_e = 4(0.029)/3.3006 = 0.0351$ ft (0.0107 m).

3. *Calculate h_o.*

Mass flow rate through the annulus $G = W_{air}/a_c = 500/0.029 = 17{,}241.4$ lb/(h)(ft^2). Since the Reynolds number, that is, $D_eG/\mu = 0.0351(17{,}241.4)/0.0426 = 14{,}206$, is greater than 8000,

$$\begin{aligned} h_o &= 0.0144cG^{0.8}/D_e^{0.2} \\ &= 0.0144(0.24)(17{,}241.4)^{0.8}/0.0351^{0.2} \\ &= 16.55 \text{ Btu/(h)(ft}^2\text{)(°F) [93.99 W/(m}^2\text{)(K)]} \end{aligned}$$

4. *Calculate fin efficiency.*

Because the heat must be transferred through the fin by conduction, the fin is not as effective as a bare tube with the same heat-transfer surface and heat-transfer coefficient. The fin efficiency is a measure of the actual heat transferred compared with the amount that could be transferred if the fin were uniformly at the temperature of the base of the fin.

For a fin with rectangular cross section,

$$\eta = \frac{\tanh ml}{ml}$$

where η = fin efficiency
l = fin height
$m = (2h_o/kb)^{1/2}$, where k is fin thermal conductivity, and b is fin thickness

For this problem, $m = \{2(16.55)/[25(0.03125/12)]\}^{1/2} = 22.55$, so $ml = 22.55(0.5/12) = 0.9395$ and $\eta = (\tanh 0.9395)/0.9395$. Now, $\tanh x = (e^x - e^{-x})/(e^x + e^{-x})$, so

$$\begin{aligned} \tanh 0.9395 &= \frac{e^{0.9395} - e^{-0.9395}}{e^{0.9395} + e^{-0.9395}} \\ &= 0.735 \end{aligned}$$

Therefore, $\eta = 0.735/0.9395 = 0.782$.

5. *Calculate $h_{f,i}$, pertaining to the outside of the fin tube.*

It is convenient to base the overall heat-transfer coefficient on the inside area of a fin tube. Then the relevant outside coefficient is

$$h_{f,i} = (\eta A_f + A_o)(h_o/A_i)$$

$$= [0.782(2.0625) + 0.4349]\frac{16.55}{0.4215}$$

$$= 80.40 \text{ Btu/(h)(ft}^2\text{)(°F) [456.6 W/(m}^2\text{)(K)]}$$

6. *Calculate U_i.*

The formula is

$$\frac{1}{U_i} = \frac{1}{h_{f,i}} + \frac{1}{h_i} + \frac{1}{h_w} + \frac{1}{h_s}$$

Thus, $1/U_i = 1/80.4 + 1/4183 + 1/2068 + 1/1000$, so $U_i = 70.6$ Btu/(h)(ft²)(°F) [401 W/(m²)(K)].

7. *Calculate outlet air temperature.*

From step 1, total area $A_i = 0.4215(20) = 8.43$ ft² (0.783 m²). Then, $U_iA_i/(wc) = 70.6(8.43)/[500(0.24)] = 4.96$. From Fig. 7-14, for $R = 0$, $P = 0.99$. Then $t_2 = t_1 + P(T_s - t_1) = 68 + 0.99(320 - 68) = 317.5$°F (431.8 K).

Related Calculations: This procedure can also be used for fins with cross sections other than rectangular. Fin-efficiency curves for some of these shapes are presented in Refs. 2 through 4.

7-35 Heat Transfer for Low-Fin Tubes

An existing heat exchanger with the following geometry must be retubed. Bare copper-alloy tubes [$k = 65$ Btu/(h)(ft)(°F) or 112 W/(m)(K)] cost $1 per foot; low-fin copper-alloy tubes cost $1.75 per foot. Heat is to be exchanged between two process streams operating under the following conditions. The cool stream must be further heated downstream with steam that has a heat of vaporization of 895 Btu/lb (2082 kJ/kg). The warm stream must be further cooled downstream with cooling water that can accept a maximum temperature rise of 30°F (16.67 K). Pressure drop for the tube side is not a penalty because the tube-side fluid must be throttled downstream. Is it economical to retube the exchanger with low-fin tubes if the evaluated cost of the steam is $50 per pound per hour and the evaluated cost of the cooling water is $25 per gallon per minute?

Tube-Side Fluid

Condition = liquid, sensible-heat transfer

Flow rate = 50,000 lb/h (22,679.5 kg/h)

c = specific heat, 1.0 Btu/(lb)(°F) [4.2 kJ/(kg)(K)]

μ = viscosity, 1.21 lb/(ft)(h) (0.5 cP)

k = thermal conductivity, 0.38 Btu/(h)(ft)(°F) [0.66 W/(m)(K)]

Inlet temperature = 320°F (433 K)

Assume that $(\mu_w/\mu_b)^{0.14} = 1.0$

Shell-Side Fluid

Condition = liquid, sensible-heat transfer

Flow rate = 30,000 lb/h (13,607.7 kg/h)

c = specific heat, 0.7 Btu/(lb)(°F) [2.9 kJ/(kg)(K)]

μ = viscosity, 10 lb/(h)(ft) (4.13 cP)

k = thermal conductivity, 0.12 Btu/(h)(ft)(°F) [0.21 W/(m)(K)]

Inlet temperature = 41°F (278 K)

Assume that $(\mu_w/\mu_b)^{0.14} = 0.9$

Heat-Exchanger Geometry

D_s = shell diameter, 8.071 in (0.205 m)

n = number of tubes, 36

D_o = tube outside diameter, 0.75 in (0.019 m)

D_i = tube inside diameter, 0.62 in (0.016 m) for bare tube and 0.495 in (0.0126 m) for low-fin tube

L = tube length, 16 ft (4.88 m)

B = baffle spacing, 4 in (0.102 m)

s = tube spacing, 0.9375 in (0.02381 m) with a triangular pitch

D_r = root diameter of fin tube, 0.625 in (0.0159 m)

$A_o/A_i = 3.84$

There is one tube pass, one shell pass. Assume an overall fouling coefficient of 1000 Btu/(h)(ft²)(°F) [5680 W/(m²)(K)].

Calculation Procedure:

1. Calculate h_i, the inside film coefficient, for bare tubes.

See Example 7-22. Here, $G = W/a_c = 50{,}000/[36(0.62/12)^2(\pi/4)] = 662{,}455.5$ lb/(h)(ft²). Then $D_iG/\mu = (0.62/12)662{,}455.5/1.21 = 28{,}286.7$, and

$$h_i = \frac{0.023cG}{(c\mu/k)^{2/3}(D_iG/\mu)^{0.2}}$$

$$= \frac{0.023(1.0)(662{,}455.5)}{[1.0(1.21)/0.38]^{2/3}28{,}286.7^{0.2}}$$

$$= 906.9 \text{ Btu/(h)(ft}^2\text{)(°F) [5151 W/(m}^2\text{)(K)]}$$

2. *Calculate h_i for low-fin tubes.*

Use the relationship $h_{i,\text{fin}} = h_{i,\text{bare}}(D_{i,\text{bare}}/D_{i,\text{fin}})^{1.8}$. Thus, $h_{i,\text{fin}} = 906.9(0.62/0.495)^{1.8} = 1360.1$ Btu/(h)(ft²)(°F) [7725 W/(m²)(K)].

3. *Calculate h_o, the outside film coefficient, for bare tubes.*

See Example 7-22. Here, $a_c = D_s B(s - D_o)/s = (8.071/12)(4/12)(0.9375 - 0.75)/0.9375 = 0.0448$ ft². Then, $G = W/a_c = 30{,}000/0.0448 = 669{,}062$ lb/(h)(ft²), and the Reynolds number $D_o G/\mu = (0.75/12)669{,}062/10 = 4181.6$. Therefore, $h = 0.273cG/[(c\mu/k)^{2/3}(D_o G/\mu)^{0.365}(\mu_w/\mu_o)^{0.14}]$. Now, $(c\mu/k)^{2/3} = [0.7(10)/0.12]^{2/3} = 15.04$, and $h = 0.273(0.7)(669{,}062)/[15.04(4181.6)^{0.365}0.9] = 450.2$ Btu/(h)(ft²)(°F). Finally, $h_o = hF_1F_r$, $F_r = 1.0$, and $F_1 = 0.8(B/D_s)^{1/6}$. Thus, $h_o = 450.2(0.8)(4/8.071)^{1/6}(1.0) = 320.4$ Btu/(h)(ft²)(°F) [1819 W/(m²)(K)].

4. *Calculate h_o for low-fin tubes.*

For low-fin tubes, the shell-side mass velocity is reduced because of the space between the fins. This reduction can be closely approximated with the expression $(s - D_o)/(s - D_o + 0.09)$, each term being expressed in inches.

For this problem, the expression has the value $(0.9375 - 0.75)/(0.9375 - 0.75 + 0.09) = 0.676$. The diameter that should be used for calculating the Reynolds number is the root diameter of the fin tube. By applying the diameter ratio and the velocity reduction to the Reynolds number from step 3, the result is a Reynolds number for this case of $D_r G/\mu = 4181.6(0.625/0.75)(0.676) = 2355.6$. Then,

$$h = \frac{0.273cG}{(c\mu/k)^{2/3}(D_r G/\mu)^{0.365}(\mu_w/\mu_b)^{0.14}}$$

$$= \frac{0.273(0.7)(669{,}062)(0.676)}{15.04(2355.6)^{0.365}0.9}$$

$$= 375.3 \text{ Btu/(h)(ft}^2\text{)(°F)}$$

Then, as in step 3, $h_o = 375.3(0.8)(4/8.071)^{1/6}1.0 = 267.1$ Btu/(h)(ft²)(°F) [1517 W/(m²)(K)].

5. *Determine weighted fin efficiency.*

Weighted fin efficiencies for low-fin tubes are functions of the outside heat-transfer coefficient. Weighted fin efficiencies η can be determined from curves provided by various manufacturers. Table 7-4 permits approximation of weighted fin efficiencies. For this problem, the weighted efficiency η is 0.94.

6. *Calculate $h_{f,i}$, pertaining to the outside of the fin tubes.*

For fin tubes, it is convenient to base the overall heat-transfer coefficient on the inside surface of the tube. Then the relevant outside coefficient is $h_{f,i} = \eta(A_o/A_i)h_o = 0.94(3.84)(267.1) = 964.1$ Btu/(h)(ft²)(°F) [5476.2 W/(m²)(K)].

TABLE 7-4 Weighted Fin Efficiency for Low-Fin Tubes

	Weighted fin efficiency η				
h_o	k = 10 [17.3]	k = 25 [43.3]	k = 65 [112]	k = 125 [216]	k = 225 [389]
10 [56.7]	0.97	0.98	0.99	1.00	1.00
50 [284]	0.94	0.97	0.98	0.99	1.00
100 [567]	0.89	0.94	0.97	0.98	0.99
200 [1134]	0.81	0.90	0.95	0.97	0.98
500 [2840]	0.68	0.80	0.90	0.93	0.96

Note: h_o, the outside film coefficient, is in Btu/(h)(ft^2)(°F) [W/(m^2)(K)]; k, the thermal conductivity, is in Btu/(h)(ft)(°F) [W/(m)(K)].

7. Calculate h_w, the coefficient for heat transfer through the tube wall.

The formulas are $h_w = 2k/(D_o - D_i)$ for bare tubes, and $h_w = 2k/(D_r - D_i)$ for low-fin tubes. For this problem, $D_o - D_i = D_r - D_i = 0.13/12$, so $h_w = 2(65)/(0.13/12) = 12{,}000$ Btu/(h)(ft^2)(°F) [68,160 W/(m^2)(K)].

8. Calculate U_o for bare tubes.

The formula is

$$\frac{1}{U_o} = \frac{1}{h_i(D_i/D_o)} + \frac{1}{h_o} + \frac{1}{h_w} + \frac{1}{h_s}$$

Thus, $1/U_o = 1/[906.9(0.62/0.75)] + 1/320.4 + 1/12{,}000 + 1/1000$, so $U_o =$ 180.6 Btu/(h)(ft^2)(°F) [1026 W/(m^2)(K)].

9. Calculate U_i for low-fin tubes.

Again, $1/U_i = 1/h_i + 1/h_{f,i} + 1/h_w + 1/h_s$. Thus, $1/U_i = 1/1360.1 + 1/964.1 + 1/12{,}000 + 1/1000$, so $U_i = 350.2$ Btu/(h)(ft^2)(°F) [1989 W/(m^2)(K)].

10. Determine outlet temperatures for bare tubes.

Use Fig. 7-14. Now, $A_o = n\pi D_o L = 36\pi(0.75/12)16 = 113.1$ ft^2, so $UA/(wc) = 180.6(113.1)/[30{,}000(0.7)] = 0.973$, and $R = wc/(WC) = 30{,}000(0.7)/[50{,}000(1.0)] = 0.420$. For these values, $P = 0.57$. Therefore, $t_2 = t_1 + P(T_1 - t_1) = 41 + 0.57(320 - 41) = 200.0$°F (366.5 K).

11. Determine outlet temperatures for low-fin tubes.

Use Fig. 7-14 again. In this case, the calculation is based on the inside diameter. So, $A_i = n\pi D_i L = 36\pi(0.495/12)16 = 74.6$ ft^2. And $UA/(wc) = 350.2(74.6)/[30{,}000(0.7)] = 1.244$. For this value and $R = 0.42$, $P = 0.65$. Therefore, $t_2 = t_1 + P(T_1 - t_1) = 41 + 0.65(320 - 41) = 222.4$°F (378.9 K).

12. Calculate water savings using low-fin tubes.

The tube-side fluid must be further cooled; the water savings is represented by the difference in heat recovery between bare and low-fin tubes. Thus,

$$Q_{\text{saved}} = wc\,\Delta t = 30{,}000(0.7)(222.4 - 200.0)$$
$$= 470{,}400 \text{ Btu/h}$$

Since the cooling water can accept a temperature rise of 30°C, the water rate $= Q_{\text{saved}}/(c\,\Delta T) = 470{,}400/[1(30)] = 15{,}680$ lb/h. Then, dollars saved (lb water/h)[(gal/min)/(lb/h)][$/(gal/min)] = 15,680(1/500)(25) = $784.

13. Calculate steam savings using low-fin tubes.

The shell-side fluid must be further heated; the steam savings is represented by the difference in heat recovery between bare and low-fin tubes:

$$\text{Steam rate} = Q_{\text{saved}}/\lambda = 470{,}400/895 = 525.6 \text{ lb/h}$$

Then dollars saved = (lb steam/h)[$/(lb/h)] = equivalent savings = 525.6(50) = $26,280.

14. Compare energy savings with additional tubing cost.

Additional cost for retubing is $nL(\$1.75 - \$1.00) = 36(16)(1.75 - 1.00) = \$432$. The equivalent energy savings is $784 + $26,280 = $27,064.

Related Calculations: Low-fin tubes are tubes with extended surfaces that have the same outside diameter as bare tubes. They can therefore be used interchangeably with bare tubes in tubular exchangers. Various geometries and materials of construction are available from several manufacturers.

Low-fin tubes find wide application when the heat-transfer coefficient on the inside of the tube is much greater than the coefficient on the outside of the tube. A guideline suggests that low-fin tubes should be considered when the outside coefficient is less than one-third that on the inside. Low-fin tubes also find application in some fouling services because the fin tubes are more easily cleaned by hydroblasting than are bare tubes. In addition, low-fin tubes are used when boiling at low temperature differences, because the minimum temperature difference required for nucleate boiling is reduced with the use of low-fin tubes. Low-fin tubes are also used for condensing, primarily for materials with low surface tension.

The procedure outlined in this example can also be used when designing equipment using low-fin tubes. The same approach is used when condensing or boiling on the outside of low-fin tubes.

7-36 Heat Transfer for Banks of Finned Tubes

Calculate the outlet temperature from a duct cooler if hydrogen is flowing at a rate of 1000 lb/h (453.6 kg/h) with a duct velocity of 500 ft/min (152.4 m/min). The duct cooler contains a bank of finned tubes described below. Hydrogen enters the cooler at

200°F (366.5 K) and has a specific heat of 3.4 Btu/(lb)(°F) [14.2 kJ/(kg)(K)], a viscosity of 0.225 lb/(ft)(h), (0.0093 cP), a thermal conductivity of 0.11 Btu/(h)(ft)(°F) [0.19 W/(m)(K)], and a density of 0.0049 lb/ft^3(0.0785 kg/m^3). The coolant is water entering the tubes at 86°F (303 K) and a rate of 10,000 lb/h (4535.0 kg/h). The water-side heat-transfer coefficient h_i is 1200 Btu/(h)(ft^2)(°F) [6800 W/(m^2)(K)], the heat-transfer coefficient through the tube wall h_w is 77,140 Btu/(h)(ft^2)(°F) [437,380 W/(m^2)(K)], and the overall fouling coefficient is 1000 Btu/(h)(ft^2)(°F) [5680 W/(m^2)(K)].

Duct Cooler (Finned-Tube Bank)

D_r = root diameter (bare-tube outside diameter) = 0.625 in (0.0159 m)
l = fin height = 0.5 in (0.013 m)
b = fin thickness = 0.012 in (0.305 mm)
n = number of fins per inch = 12
s = tube spacing = 1.75 in (0.0445 m)
x = tube-wall thickness = 0.035 in (0.889 mm)
A_o = bare heat-transfer surface = 50 ft^2 (4.64 m^2)

The tube pattern is triangular, the fin material is aluminum, and the tube material is copper.

Calculation Procedure:

1. Select the appropriate equation for the outside heat-transfer coefficient.

The heat-transfer coefficient for finned tubes can be calculated using the equation

$$\frac{hD_r}{k} = a\left(\frac{c\mu}{k}\right)^{1/3}\left(\frac{D_rG}{\mu}\right)^{0.681}\left(\frac{y^3}{l^2b}\right)^{0.1}$$

where a = 0.134 for triangular tube patterns or $0.128(y/l)^{0.15}/[1 + (s - D_f)/(s - D_r)]$ for inline tube patterns
h = outside heat-transfer coefficient
D_r = root diameter (diameter at base of fins)
D_f = diameter of fins (equal to $D_r + 2l$)
k = thermal conductivity
c = specific heat
μ = viscosity
G = mass velocity, based on minimum flow area
l = fin height
b = fin thickness
y = distance between fins = $(1/n) - b$
n = number of fins per unit length
s = tube spacing perpendicular to flow
p = tube spacing parallel to flow

The equation is valid for triangular spacings with s/p ranging from 0.7 to 1.1.

2. Calculate the relevant fin areas.

These are heat-transfer surface of fins A_f, outside heat-transfer surface of bare tube A_o, total outside heat-transfer surface A_t (equal to $A_f + A_o$), and ratio of maximum flow area to minimum flow area a_r. For this problem

$$A_f = (D_f^2 - D_r^2)(\pi/4)2n$$
$$= (1.625^2 - 0.625^2)(1/12)^2(\pi/4)(2)(12)(12)$$
$$= 3.5343 \text{ ft}^2 \text{ per foot of tube length}$$

And $A_o = \pi D_o = \pi(0.625/12) = 0.1636$ ft² per foot of tube length. Then, $A_t = A_f + A_o = 3.5343 + 0.1636 = 3.6979$ ft² per foot of tube length, and $A_t/A_o = 3.6979/0.1636 = 22.6$. Finally, $a_r = s/(s - D_o - 2nlb) = 1.75/[1.75 - 0.625 - 2(12)(0.5)(0.012)] = 1.7839$.

3. Calculate h_o, the outside film coefficient.

From step 1, $h_o = 0.134(k/D_r)(c\mu/k)^{1/3}(D_rG/\mu)^{0.681}(y^3/l^2b)^{0.1}$. Now, $G = 60\rho FVa_r$, where ρ is the density in pounds per cubic foot and FV is the face velocity (duct velocity) in feet per minute. Then, $G = 60(0.0049)(500)(1.7839) = 262.2$ lb/(h)(ft²), and $D_rG/\mu = (0.625/12)262.2/0.0225 = 607.0$. Now, $y = (1/n) - b = (1/12) - 0.012 = 0.0713$ in (1.812 mm).

The Prandtl number for hydrogen $(c\mu/k) = 0.7$, so

$$h_o = 0.134\left(\frac{0.11}{0.625/12}\right)0.7^{1/3}607.0^{0.681}\left[\frac{0.0713^3}{0.5^2(0.012)}\right]^{0.1}$$
$$= 15.99 \text{ Btu/(h)(ft}^2\text{)(°F) [90.82 W/(m}^2\text{)(K)]}$$

4. Calculate $h_{f,o}$, pertaining to the outside of the finned tubes.

For transverse fins fabricated by finning a bare tube, it is convenient to base the overall heat-transfer coefficient on the outside surface of the bare tube. Thus

$$h_{f,o} = h_o(A_t/A_o)\eta$$

where η is the weighted fin efficiency.

The weighted fin efficiency may be approximated from Table 7-5. For the present problem, use a value of 0.80. Then, $h_{f,o} = 15.99(22.6)(0.80) = 289.1$ Btu/(h)(ft²)(°F) [1642 W/(m²)(K)].

5. Calculate U_o.

The formula is

$$\frac{1}{U_o} = \frac{1}{h_{f,o}} + \frac{1}{h_i(D_i/D_o)} + \frac{1}{h_w} + \frac{1}{h_s}$$
$$= \frac{1}{289.1} + \frac{1}{1200(0.555/0.625)} + \frac{1}{77{,}140} + \frac{1}{1000}$$
$$= 185 \text{ Btu/(h)(ft}^2\text{)(°F) [1050 W/(m}^2\text{)(K)]}$$

TABLE 7-5 Weighted Fin Efficiency for Transverse Fins

	Weighted fin efficiency η			
h_o	Copper, k = 220 [381]	Aluminum, k = 125 [216]	Steel, k = 25 [43.3]	Stainless steel, k = 10 [17.3]
5 [28.4]	0.95	0.90	0.70	0.50
10 [56.7]	0.90	0.85	0.60	0.35
25 [142]	0.85	0.70	0.40	0.20
50 [284]	0.70	0.55	0.25	0.15
100 [567]	0.55	0.40	0.15	0.10

Note; h_o, the outside film coefficient, is in Btu/(h)(ft^2)(°F) [W/(m^2)(K)]; k, the thermal conductivity, is in Btu/(h)(ft)(°F) [W/(m)(K)]. This table is valid for fin heights in the vicinity of 0.5 in and fin thicknesses around 0.012 in.

6. Calculate the outlet temperature.

Assume that there are several tube-side passes. The outlet temperature can then be calculated using Fig. 7-14. This figure can be employed to directly calculate the outlet temperature of the hot fluid (which is of interest in this example); in such a case, the abscissa becomes $UA/(WC)$, $R = WC/(wc)$, and $P = (T_2 - T_1)/(t_1 - T_1)$. Then, for the present example, $UA/(WC) = 185(50)/[1000(3.4)] = 2.72$, and $R = 1000(3.4)/[10{,}000(1.0)] = 0.34$. For these parameters, Fig. 7-14 shows P to be 0.88. Therefore, $T_2 = T_1 + P(t_1 - T_1) = 200 + 0.88(86 - 200) = 99.7$°F (310.6 K).

7-37 Air-Cooled Heat Exchangers

Design an air-cooled heat exchanger to cool water under the following conditions. The design ambient air temperature is 35°C (95°F). The tubes to be used are steel tubes [thermal conductivity = 25 Btu/(h)(ft^2)(°F), or 43 W/(m)(K)] with aluminum fins. The steel tube is 1 in (0.0254 m) outside diameter and 0.834 in (0.0212 m) inside diameter. The inside heat-transfer coefficient h_i and the fouling coefficient h_s are each 1000 Btu/(h)(ft^2)(°F) [5680 W/(m^2)(K)]. Heat-exchanger purchase cost is $22 per square foot for four-tube-row units, $20 per square foot for five-tube-row units, and $18 per square foot for six-tube-row units.

Water Conditions

Flow rate = 500,000 lb/h (226,795 kg/h), multipass

Inlet temperature = 150°C (302°F)

Outlet temperature = 50°C (122°F)

Calculation Procedure:

1. Decide on the design approach.

A four-tube-row unit, a five-tube-row unit, and a six-tube-row unit will be designed and the cost compared. Standard tube geometries for air-cooled exchangers are 1 in (0.0254 m) outside diameter with ⅝ in (0.0159 m) high aluminum fins spaced 10 fins per inch. Also available are 8 fins per inch. Standard spacings for tubes are listed in Table 7-6.

TABLE 7-6 Design Face Velocities for Air-Cooled Exchangers

	Face velocity, ft/min (m/s)		
Number of tube rows	8 fins/in (315 fins/m), 2.375-in (0.0603-m) pitch	10 fins/in (394 fins/m), 2.375-in (0.0603-m) pitch	10 fins/in (394 fins/m), 2.5-in (0.0635-m) pitch
3	650 (3.30)	625 (3.18)	700 (3.56)
4	615 (3.12)	600 (3.05)	660 (3.35)
5	585 (2.97)	575 (2.92)	625 (3.18)
6	560 (2.84)	550 (2.79)	600 (3.05)

Typical face velocities (*FV*s) used for design are also tabulated in Table 7-6. These values result in air-cooled heat exchangers that approach an optimum cost, taking into account the purchase cost, the cost for installation, and the cost of power to drive the fans. Each designer may wish to establish his or her own values of typical design face velocities; these should not vary greatly from those tabulated.

For air-cooled equipment, the tube spacing is normally determined by the relative values of the inside heat-transfer coefficient and the air-side heat-transfer coefficient. For inside coefficients much greater than the air-side coefficient, tubes spaced on 2.5-in (0.064-m) centers are normally justified. For low values of the inside coefficient, tubes spaced on 2.375-in (0.060-m) centers are normally justified. In the present case, because water is being cooled, the inside coefficient is indeed likely to be much greater, so specify the use of tubes on 2.5-in centers.

2. *Determine h_a, the air-side heat-transfer coefficient.*

The air-side coefficient is frequently calculated on the basis of the outside surface of a bare tube. The equations for air can be simplified as follows:

$$h_a = 8(FV)^{1/2} \qquad \text{for 10 fins per inch}$$

$$h_a = 6.75(FV)^{1/2} \qquad \text{for 8 fins per inch}$$

where h_a is in Btus per hour per square foot per degree Fahrenheit and FV is in feet per minute.

For this problem, FV for 5 rows = 625 ft/min, so $h_a = 8(625)^{1/2} = 200$ Btu/(h)(ft^2)(°F) [1136 W/(m^2)(K)]. FV for 4 rows = 660 ft/min, so $h_a = 8(660)^{1/2} =$ 205 Btu/(h)(ft^2)(°F) [1164 W/(m^2)(K)]. FV for 6 rows = 600 ft/min,so $h_a =$ $8(600)^{1/2} = 196$ Btu/(h)(ft^2)(°F) [1113 W/(m^2)(K)].

3. *Calculate h_w, the coefficient of heat transfer through the tube wall.*

The formula is $h_w = 2k/(D_o - D_i) = 2(25)/[(1 - 0.834)(1/12)] = 3614$ Btu/(h)(ft^2)(°F) [20,520 W/(m^2)(K)].

4. *Calculate U, the overall heat-transfer coefficient.*

The formula is

$$\frac{1}{U} = \frac{1}{h_a} + \frac{1}{h_i(D_i/D_o)} + \frac{1}{h_w} + \frac{1}{h_s}$$

For five-tube rows, $1/U = 1/200 + 1/[1000(0.834/1)] + 1/3614 + 1/1000$, so $U = 134$ Btu/(h)(ft^2)(°F). For four-tube rows, $1/U = 1/205 + 1/[1000(0.834/1)] + 1/3614 + 1/1000$, so $U = 136$ Btu/(h)(ft^2)(°F). For six-tube rows, $1/U = 1/196 + 1/[1000(0.834/1)] + 1/3614 + 1/1000$, so $U = 132$ Btu/(h)(ft^2)(°F). Since the values are so close, use the same value of U for four, five, and six rows, namely, $U = 135$ Btu/(h)(ft^2)(°F) [767 W/(m^2)(K)].

5. *Design a five-row unit.*

Air-cooled equipment is fabricated in standard modules. The standards begin with fin-tube bundles 48 in (1.22m) wide and increase in 6-in (0.152-m) increments up to a 144-in (3.66-m) maximum. These modules can then be placed in parallel to obtain any size exchanger needed. The maximum tube length is 48 ft (14.63 m). In general, long tubes result in economical heat exchangers. In the present case, assume that the plant layout allows a maximum tube length of 40 ft (12.2 m).

The design of heat-transfer equipment is a trial-and-error procedure because various design standards are followed in order to reduce equipment cost. To design an air-cooled exchanger, an outlet air temperature and a tube length are assumed, which establishes the amount of air to be pumped by the fan. The amount of air to be pumped establishes a face area, because we have assumed a face velocity. The face area fixes the heat-transfer area for a given tube length and number of tube rows. Table 7-7 permits an estimate of the outlet air temperature, based on 90 to 95°F (305 to 308 K) design ambient air temperature.

For this problem, $U = 135$ and inlet process temperature $= 150°C$, so assume an outlet air temperature of 83°C. Then, Q = heat load = $wc(t_2 - t_1)$ = $500{,}000(1.0)(150 - 50)(1.8°F/°C) = 90{,}000{,}000$ Btu/h. The face area FA can be estimated from the equation

$$FA = \frac{Q}{FV(T_2 - T_1)(1.95)}$$

TABLE 7-7 Estimated Outlet Air Temperatures for Air-Cooled Exchangers

Process inlet temperature, °C	Outlet air temperature, °C		
	$U = 50$	$U = 100$	$U = 150$
175	90	95	100
150	75	80	85
125	70	75	80
100	60	65	70
90	55	60	65
80	50	55	60
70	48	50	55
60	45	48	50
50	40	41	42

Note: U is the overall heat-transfer coefficient in Btu/(h)(ft^2)(°F) [1 Btu/(h)(ft^2)(°F) = 5.67 W/(m^2)(K)].

where Q is the heat load in Btu/h, FA is the face area in square feet, FV is the face velocity in feet per minute, and T_1 and T_2 are inlet and outlet air temperatures in degrees Celsius. Thus, $FA = 90{,}000{,}000/[625(83 - 35)(1.95)] = 1540 \text{ ft}^2$.

The exchanger width can now be determined:

$$Y = \text{width} = \text{face area/tube length} = FA/L = 1540/40 = 38.5 \text{ ft}$$

However, standard widths have 6-in increments. Therefore, assume four 9.5-ft-wide bundles. Then, $FA = 4(9.5) = 38$ ft.

For this face area, calculate the outlet air temperature. First, air temperature rise (in °C) is

$$\Delta T_a = \frac{Q}{Y(FV)(L)1.95} = \frac{90{,}000}{38(625)(40)1.95} = 48.5°\text{C}$$

Then, $T_2 = T_1 + \Delta T_a = 35 + 48.5 = 83.5°\text{C}$. Next, calculate mean temperature difference. The formula is

$$\Delta T_m = \frac{(t_1 - T_2) - (t_2 - T_1)}{\ln[(t_1 - T_2)/(t_2 - T_1)]}$$

Thus, $\Delta T_m = [(150 - 83.5) - (50 - 35)]/\ln[(150 - 83.5)/(50 - 35)] = 34.6°\text{C}$ (62.3°F).

Now, calculate the area required if this is the available temperature difference. Thus, $A = Q/(U\,\Delta T_m) = 90{,}000{,}000/[135(34.6)(1.8)] = 10{,}705 \text{ ft}^2$ (994.4 m²). Next, calculate the area actually available. The number of tubes per row N_t can be approximated by dividing the bundle width by the tube spacing, that is, $N_t = Y/s = 38/(2.5/12) = 182$ per row. Then, letting N_r equal the number of rows, $A = N_r N_t \pi D_o L = 5(182)\pi(1/12)(40) = 9530 \text{ ft}^2$. Therefore, the area available is less than that required. Accordingly, increase the bundle width from 9.5 to 10 ft and calculate the new ΔT_a:

$$\Delta T_a = 48.5(9.5/10) = 46.1°\text{C}$$

Therefore, $T_2 = 35 + 46.1 = 81.1°\text{C}$.

Next, calculate the new ΔT_m:

$$\Delta T_m = [(150 - 81.1) - (50 - 35)]/\ln[(150 - 81.1)/(50 - 35)] = 35.4°\text{C}\ (63.7°\text{F})$$

And calculate the new area required

$$A = 90{,}000{,}000/[135(35.4)(1.8)] = 10{,}460 \text{ ft}^2$$

Finally, calculate the new area available: $N_t = 40/(2.5/12) = 192$, so $A = 5(192)\pi(1/12)(40) = 10{,}050 \text{ ft}^2$. Or more precisely, the actual tube count provided by manufacturers' standards for a bundle 10 ft wide with five tube rows is 243. Letting N_b equal the number of bundles, $A = N_b N \pi D_o L = 4(243)\pi(1/12)(40) = 10{,}180$ ft^2 (945.7 m^2). This area would normally be accepted because the relatively high design air temperature occurs for relatively short periods and the fouling coefficient is usually arbitrarily assigned. If the design air temperature must be met at all times, then a wider bundle or greater face velocity would be required.

With the dimensions of the cooler established, the next step is to calculate the air-side pressure drop. For air, the following relations can be used:

$$\Delta P_a = 0.0047 N_r (FV/100)^{1.8} \qquad \text{for 10 fins per inch, 2.375-in spacing}$$

$$\Delta P_a = 0.0044 N_r (FV/100)^{1.8} \qquad \text{for 8 fins per inch, 2.375-in spacing}$$

$$\Delta P_a = 0.0037 N_r (FV/100)^{1.8} \qquad \text{for 10 fins per inch, 2.5-in spacing}$$

In these equations, ΔP_a is air-side pressure drop, in inches of water, N_r is the number of tube rows, and FV is face velocity, in feet per minute. For this problem, $\Delta P_a = 0.0037(5)(625/100)^{1.8} = 0.501$ in water.

Now, calculate the power required to pump the air. Use the formula

$$bhp = (FV)(FA)(T_2 + 273)(\Delta P_a + 0.1)/(1.15 \times 10^6)$$

where bhp = brake horsepower
FV = face velocity, in ft/min
FA = face area, in ft^2 (equals $N_b YL$)
T_2 = outlet air temperature, in °C
ΔP_a = air-side pressure drop, in inches of water

In this case, $FA = N_b YL = 4(10)(40) = 1600$ ft^2, so $bhp = 625(1600)(81.1 + 273)(0.501 + 0.1)/(1.15 \times 10^6) = 185$ hp (138 kW).

6. *Compare costs.*

The same procedure can be followed to design a six-row unit and a four-row unit. The following comparison can then be made:

	Number of tube rows		
	4	5	6
Number of bundles	4	4	4
Bundle width, ft	12	10	9.5
Heat transfer area, ft^2	9,800	10,180	11,510
Equipment cost, dollars	215,600	203,600	207,180
Power required, bhp	195	185	189

From this comparison, a five-row unit would be the most economical.

7-38 Pressure Drop for Flow inside Tubes: Single-Phase Fluids

Calculate the pressure drop for the water flowing through the air-cooled heat exchanger designed in Example 7-37 if the number of tube-side passes is 10. The density of the water is 60 lb/ft^3 (961.1 kg/m^3), and the viscosity is 0.74 lb/(ft)(h) (0.31 cP). Assume that the velocity in the nozzles is 10 ft/s (3.05 m/s) and that the viscosity change with temperature is negligible.

Calculation Procedure:

1. Consider the causes of the pressure drop and the equations to find each.

The total pressure drop for fluids flowing through tubes results from frictional pressure drop as the fluid flows along the tube, from pressure drop as the fluid enters and leaves the tube-side heads or channels, and from pressure drop as the fluid enters and leaves the tubes from the heads or channels.

The frictional pressure drop can be calculated from the equation

$$\Delta P_f = \frac{4fG^2LN_p}{2(144)g\rho D}$$

where ΔP_f = pressure drop, in lb/in^2
f = friction factor
G = mass velocity, in $lb/(h)(ft^2)$
L = tube length, in ft
N_p = number of tube-side passes
g = the gravitational constant
ρ = density, in lb/ft^3
D = tube inside diameter, in ft

For fluids with temperature-dependent viscosities, the pressure drop must be corrected by the ratio of

$$(\mu_w/\mu_b)^n$$

where μ_w = viscosity at the surface temperature
μ_b = viscosity at the bulk fluid temperature
n = 0.14 for turbulent flow or 0.25 for laminar flow

The friction factor f can be calculated from the equations

$$f = 16/(DG/\mu) \qquad \text{for } DG/\mu \text{ less than 2100}$$

$$f = 0.054/(DG/\mu)^{0.2} \qquad \text{for } DG/\mu \text{ greater than 2100}$$

where μ is viscosity, in lb/(ft)(h).

The pressure drop as the fluid enters and leaves a radial nozzle at the heads or channels can be calculated from

$$\Delta P_n = k\rho v_n^2/9266$$

where ΔP_n = pressure drop, in lb/in^2
ρ = density, in lb/ft^3
v_n = velocity in the nozzle, in ft/s
k = 0 for inlet nozzles or 1.25 for outlet nozzles

The pressure drop associated with inlet and outlet nozzles can be reduced by selection of other channel types, but the expense is not warranted except for situations in which pressure drop is critical or costly.

The pressure drop as a result of entry and exit from the tubes can be calculated from

$$\Delta P_e = kN_p \rho v_t^2/9266$$

where ΔP_e = pressure drop, in lb/in^2
N_p = number of tube-side passes
v_t = velocity in tube, in ft/s
ρ = density, in lb/ft^3
k = 1.8

The total pressure drop is the sum of all these components:

$$\Delta P_t = \Delta P_f + \Delta P_n + \Delta P_e$$

2. *Calculate frictional pressure drop.*

The total number of tubes from the previous design is $4(243) = 972$. The number of tubes per pass is $972/10 = 97.2$. Letting a_c be flow area,

$$G = \frac{W}{a_c} = \frac{500{,}000}{97.2(0.834/12)^2(\pi/4)}$$

$$= 1{,}355{,}952 \text{ lb/(h)(ft}^2\text{) [6,619,758 kg/(h)(m}^2\text{)]}$$

Then, $DG/\mu = (0.834/12)1{,}355{,}952/0.74 = 127{,}350$. Therefore, $f = 0.054/(DG/\mu)^{0.2} = 0.054/127{,}350^{0.2} = 0.0051$, and

$$\Delta P_f = \frac{4fG^2LN_p}{2(144)g\rho D}$$

$$= \frac{4(0.0051)(1{,}355{,}952)^2(40)10}{2(144)(4.18 \times 10^8)(60)(0.834/12)}$$

$$= 29.89 \text{ lb/in}^2$$

3. *Calculate nozzle pressure drop.*

From step 1, $\Delta P_n = 1.25\rho v_n^2/9266 = 1.25(60)(10)^2/9266 = 0.81$ lb/in^2

4. *Calculate tube entry/exit pressure drop.*

From step 1, $\Delta P_e = 1.8N_p \rho v_t^2/9266$. Now, by definition, $v_t = G/3600\rho = 1{,}355{,}952/[3600(60)] = 6.28$ ft/s (1.91 m/s). So, $\Delta P_e = 1.8(10)(60)(6.28)^2/9266 = 4.59$ lb/in^2.

5. *Calculate total pressure drop.*

Thus, $\Delta P_t = \Delta P_f + \Delta P_n + \Delta P_e = 29.89 + 0.81 + 4.59 = 35.29$ lb/in^2 (243.32 kPa).

Related Calculations: *Helical Coils.* The same procedure can be used to calculate the pressure drop in helical coils. For turbulent flow, a friction factor for curved flow is substituted for the friction factor for straight tubes. For laminar flow, the friction loss for a curved tube is expressed as an equivalent length of straight tube and the friction factor for straight tubes is used. The Reynolds number required for turbulent flow is $2100[1 + 12(D_i/D_c)^{1/2}]$, where D_i is the inside diameter of the tube and D_c is the coil diameter.

The friction factor for turbulent flow is calculated from the equation $f_c(D_c/D_i)^{1/2} = 0.0073 + 0.076[(DG/\mu)(D_i/D_c)^2]^{-1/4}$, for $(DG/\mu)(D_i/D_c)^2$ between 0.034 and 300, where f_c is the friction factor for curved flow. For values of $(DG/\mu)(D_i/D_c)^2$ below 0.034, the friction factor for curved flow is practically the same as that for straight pipes.

For laminar flow, the equivalent length L_e can be predicted as follows: For $(DG/\mu)(D_i/D_c)^{1/2}$ between 150 and 2000,

$$L_e/L = 0.23[(DG/\mu)(D_i/D_c)^{1/2}]^{0.4}$$

For $(DG/\mu)(D_i/D_c)^{1/2}$ between 10 and 150,

$$L_e/L = 0.63[(DG/\mu)(D_i/D_c)^{1/2}]^{0.2}$$

For $(DG/\mu)(D_i/D_c)^{1/2}$ less than 10,

$$L_e/L = 1$$

In these equations, L is straight length, and L_e is equivalent length of a curved tube.

Longitudinal fin tubes. The same procedure can be used for longitudinally finned tubes. The equivalent diameter D_e is substituted for D_i. The friction factor can be determined from these equations: For Reynolds numbers below 2100,

$$f_{lf} = \frac{16}{(D_e G/\mu)}$$

For Reynolds numbers greater than 2100,

$$f_{lf} = \frac{0.103}{(D_e G/\mu)^{0.25}}$$

In these equations, f_{lf} is the friction factor for longitudinally finned tubes, and D_e is equivalent diameter ($= 4a_c/P$, where a_c is the cross-sectional flow area and P is the wetted perimeter).

7-39 Pressure Drop for Flow inside Tubes: Two-Phase Fluids

Calculate the frictional pressure drop for a two-phase fluid flowing through a tube 0.62 in (0.0158 m) in inside diameter D and 20 ft (6.1 m) in length L at a rate of 100 lb/h (45.4 kg/h). The mixture is 50 percent gas by weight and 50 percent liquid by weight, having the following properties:

Liquid Properties

ρ = density = 50 lb/ft³ (800 kg/m³)

μ = viscosity = 2.0 lb/(ft)(h) (0.84 cP)

Gas Properties

ρ = density = 0.1 lb/ft^3 (1.6 kg/m^3)

μ = viscosity = 0.045 lb/(ft)(h) (0.019 cP)

Calculation Procedure:

1. Select the method to be used.

For two-phase flow, the friction pressure drop inside a tube can be calculated using the equation [31]

$$\Delta P_{tp} = [\Delta P_L^{1/n} + \Delta P_G^{1/n}]^n$$

where ΔP_{tp} = two-phase pressure drop
ΔP_L = pressure drop for the liquid phase, assuming only the liquid phase is present
ΔP_G = pressure drop for the gas phase, assuming only the gas phase is present
n = 4.0 when both phases are in turbulent flow or 3.5 when one or both phases are in laminar flow

2. Calculate the liquid-phase pressure drop.

Let W_L be the mass flow rate of the liquid, equal to 50 percent of 100 lb/h, or 50 lb/h, and let a_c be the cross-sectional area of the tube. Then G_L = mass velocity of liquid phase = $W_L/a_c = 50/[(0.62/12)^2(\pi/4)] = 23{,}848.4$ lb/(h)(ft^2). Then the Reynolds number $DG_L/\mu = (0.62/12)23{,}848.4/2.0 = 616.1$. Then, as in step 1 of Example 7-38, f = friction factor = $16/(DG_L/\mu) = 16/616.1 = 0.026$, and

$$\Delta P_L = \frac{4fG_L^2LN_p}{2(144)g\rho D}$$

$$= \frac{4(0.026)(23{,}848.4)^2(20)(1)}{2(144)(4.18 \times 10^8)(50)(0.62/12)}$$

$$= 0.0038 \text{ lb/in}^2$$

3. Calculate the gas-phase pressure drop.

The mixture is 50 percent each phase, so G_G = mass velocity of gas phase = G_L from step 2. Then $DG_G/\mu = (0.62/12)23{,}848.4/0.045 = 27{,}381.5$. Again referring to step 1 of Example 7-38, f = friction factor = $0.054/(DG_G/\mu)^{0.2} = 0.054/27{,}381.5^{0.2} =$ 0.0070, and

$$\Delta P_G = \frac{4fG_G^2LN_p}{2(144)g\rho D}$$

$$= \frac{4(0.0070)(23{,}848.4)^2(20)(1)}{2(144)(4.18 \times 10^8)(0.1)(0.62/12)}$$

$$= 0.5120 \text{ lb/in}^2$$

4. Calculate the two-phase pressure drop.

The liquid phase is in laminar flow; the gas phase is in turbulent flow. Therefore,

$$\begin{aligned}\Delta P_{tp} &= [\Delta P_L^{1/3.5} + \Delta P_G^{1/3.5}]^{3.5} \\ &= (0.0038^{1/3.5} + 0.5120^{1/3.5})^{3.5} \\ &= 1.107\ \text{lb/in}^3\ (7.633\ \text{kPa})\end{aligned}$$

Related Calculations: *Homogeneous flow method.* Two-phase flow pressure drop can also be calculated using a homogeneous-flow model that assumes that gas and liquid flow at the same velocity (no slip) and that the physical properties of the fluids can be suitably averaged. The correct averages are

$$\rho_{ns} = \rho_L \lambda_L + \rho_G(1 - \lambda_L) \qquad \text{and} \qquad \mu_{ns} = \mu_L \lambda_L + \mu_G(1 - \lambda_L)$$

where $\lambda_L = Q_L/(Q_L + Q_G)$
ρ_{ns} = no-slip density
μ_{ns} = no-slip viscosity
ρ_L = liquid density
ρ_G = gas density
μ_L = liquid viscosity
μ_G = gas viscosity
Q_L = liquid volumetric flow rate
Q_G = gas volumetric flow rate

Once the average density and viscosity are calculated, an average no-slip Reynolds number can be calculated:

$$\text{No-slip Reynolds number} = DG_T/\mu_{ns}$$

where G_T = total mass velocity = $G_L + G_G$
D = tube diameter

The frictional pressure drop is then

$$\Delta P_f = \frac{4fG_T^2 LN_p}{2(144)g\rho_{ns}D}$$

where f is the friction factor calculated using the method presented for single-phase fluids but based on the no-slip Reynolds number.

In addition to the frictional pressure drop, there will be a pressure loss associated with the expansion of the gas, termed an "acceleration loss" and calculated from

$$\Delta P_a = \frac{G_T^2}{2(144)g}\left(\frac{1}{\rho_{G2}} - \frac{1}{\rho_{G1}}\right)$$

where ΔP_a = acceleration pressure drop, in lb/in²
G_T = total mass velocity = $G_L + G_G$, in lb/(h)(ft²)
ρ_{G2} = gas density at outlet, in lb/ft³
ρ_{G1} = gas density at inlet, in lb/ft³
g = gravitational constant, in ft/h²

The total pressure drop ΔP_{tp} is, then,

$$\Delta P_{tp} = \Delta P_f + \Delta P_a$$

For flashing flow, the pressure drop is

$$\Delta P_{tp} = \frac{G_T^2}{2(144)g}\left[2\left(\frac{x_2}{\rho_{G2}} - \frac{x_1}{\rho_{G1}}\right) + \frac{4fLN_p}{D\rho_{ns,\text{avg}}}\right]$$

where x_1 = mass fraction vapor at inlet
x_2 = mass fraction vapor at outlet
$\rho_{ns,\text{avg}}$ = average no-slip density

Pressure drop for condensers. For condensers, the frictional pressure drop can be estimated using the relation

$$\Delta P_c = \tfrac{1}{2}\,\frac{1 + v_2}{v_1}\,\Delta P_1$$

where ΔP_c = condensing pressure drop
ΔP_1 = pressure drop based on the inlet conditions of flow rate, density, and viscosity
v_2 = vapor velocity at the outlet
v_1 = vapor velocity at the inlet

For a total condenser, this becomes

$$\Delta P_c = \frac{\Delta P_1}{2}$$

7-40 Pressure Drop for Flow across Tube Banks: Single-Phase Fluids

Calculate the pressure drop for the conditions of Example 7-20. Assume that the nozzle velocities are 5 ft/s (1.52 m/s), that the fluid density is 55 lb/ft^3 (881 kg/m^3), and that there are 24 baffles. Assume that there is also an impingement plate at the inlet nozzle. Calculate the pressure drop for both fouled and clean conditions.

Calculation Procedure:

1. Consider the causes of the pressure drop, and select equations to calculate each.

The pressure drop for fluids flowing across tube banks may be determined by calculating the following components:

1. Inlet-nozzle pressure drop
2. Outlet-nozzle pressure drop
3. Frictional pressure drop for inlet and outlet baffle sections

4. Frictional pressure drop for intermediate baffle sections

5. Pressure drop for flow through the baffle windows

Heat exchangers with well-constructed shell sides will have a certain amount of bypassing that will reduce the pressure drop experienced with an ideal-tube bundle (one with no fluid bypassing or leakage). The amount of bypassing for a clean heat exchanger is more than that for a fouled heat exchanger. The following leakage factors are based on data from operating heat exchangers and include the typical effects of fouling on pressure drop.

Pressure drop for nozzles may be calculated from

$$\Delta P_n = \frac{k\rho v_n^2}{9266}$$

where k = 0 for inlet nozzles with no impingement plate, 1.0 for inlet nozzles with impingement plates, or 1.25 for outlet nozzles

ΔP_n = pressure drop, in lb/in^2

ρ = density, in lb/ft^3

v_n = velocity in the nozzle, in ft/s

Frictional pressure drop for tube banks may be calculated as follows: For intermediate baffle sections,

$$\Delta P_f = \frac{4fG^2N_r(N_b - 1)R_1R_b\phi}{2(144)g\rho}$$

and for inlet and outlet baffle sections combined,

$$\Delta P_{fi} = \frac{4(2.66)fG^2N_rR_b\phi}{2(144)g\rho}$$

where $R_1 = 0.6(B/D_s)^{1/2}$ for clean bundles or $0.75(B/D_s)^{1/3}$ for bundles with assumed fouling

$R_b = 0.80(D_s)^{0.08}$ for clean bundles or $0.85(D_s)^{0.08}$ for bundles with assumed fouling

$\phi = (\mu_w/\mu_b)^n$, where n = 0.14 for DG/μ greater than 300 or 0.25 for DG/μ less than 300

$N_r = bD_s/s$, where b = 0.7 for triangular tube spacing, 0.6 for square tube spacing, or 0.85 for rotated square tube spacing

The friction factor f can be calculated. For D_gG/μ greater than 100,

$$f = \frac{z}{(D_gG/\mu)^{0.25}}$$

where z = 1.0 for square and triangular tube patterns or 0.75 for rotated square tube patterns. For D_gG/μ less than 100,

$$f = \frac{r}{(D_gG/\mu)^{0.725}}$$

where r = 10 for triangular tube patterns or 5.7 for square and rotated square tube patterns. In these equations, D_g is defined as the gap between the tubes, that is, $D_g = s - D_o$.

Pressure drop for baffle windows may be calculated as follows: For D_oG/μ greater than 100,

$$\Delta P_w = \frac{G^2}{2(144)g\rho}\frac{a_c}{a_w}(2 + 0.6N_w)N_bR_1$$

in which the factor $(2 + 0.6N_w)$ can be approximated with the term $2 + 0.6N_w = m(D_s)^{5/8}$, where m = 3.5 for triangular tube patterns, 3.2 for square tube patterns, or 3.9 for rotated square tube patterns. For D_oG/μ less than 100,

$$\Delta P_w = \frac{26\mu G}{g\rho}\left(\frac{a_c}{a_w}\right)^{1/2}\left(\frac{N_w}{s - D_o} + \frac{B}{D_e^2}\right) + \frac{2G^2}{2g\rho}\frac{a_c}{a_w}$$

Baffles are normally cut on the centerline of a row of tubes. Baffles should be cut on the centerline of a row whose location is nearest the value of 20% of the shell diameter. For baffle cuts meeting this criterion, the following values can be used:

Tube geometry	$N_w/(s - D_o)$	D_e, ft
D_o = ⅝ in (0.016 m); s = ¹³⁄₁₆ in (0.020 m); triangular pitch	$173D_s$	0.059
D_o = ⅝ in (0.016 m); s = ⅞ in (0.022 m); square pitch	$105D_s$	0.090
D_o = ¾ in (0.019 m); s = ¹⁵⁄₁₆ in (0.024 m); triangular pitch	$151D_s$	0.063
D_o = ¾ in (0.019 m); s = 1 in (0.025 m); square pitch	$92D_s$	0.097
D_o = 1 in (0.025 m); s = 1¼ in (0.032 m); triangular pitch	$85D_s$	0.083
D_o = 1 in (0.025 m); s = 1¼ in (0.032 m); square pitch	$73.5D_s$	0.104

In these equations, $a_c = BD_s(s - D_o)/s$ for triangular and square tube patterns or 1.5 $BD_s(s - D_o)/s$ for rotated square tube patterns, and $a_w = 0.055D_s^2$ for triangular tube patterns or $0.066D_s^2$ for square and rotated square tube patterns. Moreover, the units for the various terms are as follows:

ΔP_f = total pressure drop for intermediate baffle sections, in lb/in²
ΔP_{fi} = total pressure drop for inlet and outlet baffle sections, in lb/in²
ΔP_w = total pressure drop for baffle windows, in lb/in²
f = friction factor (dimensionless)
G = mass velocity, in lb/(h)(ft²) (= W/a_c, where W = flow rate)
g = gravitational constant, in ft/h²
ρ = density, in lb/ft³
N_r = number of tube rows crossed
N_b = number of baffles
a_c = minimum flow area for cross flow, in ft²
a_w = flow area in baffle window, in ft²

D_s = shell diameter, in ft
s = tube center-to-center spacing, in ft
B = baffle spacing, in ft
D, D_o = outside tube diameter, in ft
D_g = gap between tubes, in ft ($= s - D_o$)
μ = viscosity, in lb/(ft)(h)
μ_w = viscosity at wall temperature, in lb/(ft)(h)
μ_b = viscosity at bulk fluid temperature, in lb/(ft)(h)
R_1 = correction factor for baffle leakage
R_b = correction factor for bundle bypass
D_e = equivalent diameter of baffle window, in ft

The total pressure drop is the sum of all these components:

$$\Delta P = \Delta P_n + \Delta P_f + \Delta P_{fi} + \Delta P_w$$

2. *Calculate the total pressure drop assuming fouled conditions.*

a. *Calculate nozzle pressure drops.* The equation is

$$\Delta P_n = \frac{k\rho v_n^2}{9266} = \frac{2.25\rho v_n^2}{9266}$$

Thus, with k equal to 1.0 plus 1.25,

$$\Delta P_n = \frac{2.25(55)(5)^2}{9266} = 0.334 \text{ lb/in}^2 \text{ (2.303 kPa)}$$

b. *Calculate frictional pressure for intermediate baffle sections.* Now, $a_c = BD_s(s - D_o)/s = (9.5/12)(25/12)(0.9375 - 0.75)/0.9375 = 0.3299$ ft^2 (0.0306 m^2). Therefore, $G = W/a_c = 50{,}000/0.3299 = 151{,}561.1$ lb/(h)(ft^2), and $D_gG/\mu = (0.9375 - 0.75)(1/12)151{,}561.1/1.95 = 1214.4$. Accordingly, $f = 1/(D_gG/\mu)^{0.25} = 1/1214.4^{0.25} = 0.1694$. And in the equation $\Delta P_f = 4fG^2N_r(N_b - 1)R_1R_b\phi/[2(144)g\rho]$, $R_1 = 0.75(B/D_s)^{1/3} = 0.75(9.5/25)^{1/3} = 0.5432$, $R_b = 0.85(D_s)^{0.08} = 0.85(25/12)^{0.08} = 0.9014$, $\phi = (4.0/1.95)^{0.14} = 1.1058$, and $N_r = 0.7D_s/s = 0.7(25)/0.9375 = 18.7$. Therefore,

$$\Delta P_f = \frac{4(0.1694)(151{,}561.1)^2(18.7)(24 - 1)(0.5432)(0.9014)1.1058}{2(144)(4.18 \times 10^8)55} = 0.547 \text{ lb/in}^2 \text{ (3.774 kPa)}$$

c. *Calculate frictional pressure drop for inlet and outlet baffle sections.* The equation is

$$\Delta P_{fi} = \frac{4(2.66)fG^2N_rR_b\phi}{2(144)g\rho}$$

Thus,

$$\Delta P_{fi} = \frac{4(2.66)(0.1694)(151{,}561.1)^2(18.7)(0.9014)1.1058}{2(144)(4.18 \times 10^8)55}$$

$$= 0.117 \text{ lb/in}^2 \text{ (0.803 kPa)}$$

d. *Calculate pressure drop for window sections.* The equation is

$$\Delta P_w = \frac{3.5G^2}{2(144)g\rho}\frac{a_c}{a_w}D_s^{5/8}N_bR_1$$

Now, $a_w = 0.055D_s^2 = 0.055(25/12)^2 = 0.2387$ ft². Therefore,

$$\Delta P_w = \frac{3.5(151{,}561.1)^2(0.3299/0.2387)(25/12)^{5/8}(24)0.5432}{2(144)(4.18 \times 10^8)55}$$

$$= 0.346 \text{ lb/in}^2 \text{ (2.38 kPa)}$$

e. *Calculate total pressure drop.* The equation is

$$\Delta P = \Delta P_n + \Delta P_f + \Delta P_{fi} + \Delta P_w$$

Thus, $\Delta P = 0.334 + 0.547 + 0.117 + 0.346 = 1.344$ lb/in² (9.267 kPa).

3. *Calculate the total pressure drop assuming clean conditions.*

a. *Calculate R_1 and R_b for clean conditions.* Here, $R_1 = 0.6(B/D_s)^{1/2} = 0.6(9.5/25)^{1/2} = 0.3699$, and $R_b = 0.80(D_s)^{0.08} = 0.80(25/12)^{0.08} = 0.848$.

b. *Calculate total pressure drop.* The clean-pressure-drop components will be the fouled-pressure-drop components multiplied by the appropriate ratios of R_1 and R_b for the clean and fouled conditions. Thus,

$$\Delta P = \Delta P_n + \Delta P_f \frac{R_1}{R_1}\frac{R_b}{R_b} + \Delta P_{fi}\frac{R_b}{R_b} + \Delta P_w \frac{R_1}{R_1}$$

$$= 0.334 + 0.547\frac{0.3699}{0.5432}\frac{0.848}{0.9014} + 0.117\frac{0.848}{0.9014} + 0.346\frac{0.3699}{0.5432}$$

$$= 0.334 + 0.350 + 0.110 + 0.236$$

$$= 1.03 \text{ lb/in}^2 \text{ (7.102 kPa)}$$

Related Calculations: The preceding equations for pressure drop assume a well-constructed tube bundle with baffle cuts amounting to 20 percent of the shell diameter and tubes included in the baffle windows. The corrections for fluid bypassing (R_1 and R_b) are based on the standards of the Tubular Exchangers Manufacturers Association [11] and assume an adequate number of sealing devices. The values for fouling are based on plant data for typical services. Methods that evaluate the effects of the various bypass streams have been presented; the most widely used is that of Bell [15]. For poorly constructed tube bundles or for conditions greatly different from those assumed, this reference should be used for evaluating the effects on the calculated pressure drop.

Bundles with tubes omitted from baffle windows. Frequently, tubes are omitted from the baffle-window areas. For this configuration, maldistribution of the fluid as it flows across the bank of tubes may occur as a result of the momentum of the fluid as it flows through the baffle window. For this reason, baffle cuts less than 20 percent of the shell diameter should only be used with caution. Maldistribution will normally be minimized if the fluid velocity in the baffle window is equal to or less than the fluid velocity in cross-flow across the bundle. This frequently requires baffle cuts greater than 20 percent of the shell diameter. For such cases, the number of tubes in cross-flow will be less than that assumed in the preceding methods, so a correction is required. In addition, the pressure drop for the first and last baffle sections assumes tubes in the baffle windows; the factor 2.66 should be reduced to 2.0 for the case of 20 percent baffle cuts.

For baffles with no tubes in the baffle windows, the pressure drop for the window section can be calculated from

$$\Delta P_w = \frac{1.8\rho v_w^2 N_b}{9266}$$

where ΔP_w = pressure drop, in lb/in^2
ρ = density, in lb/ft^3
v_w = velocity in the baffle window, in ft/s
N_b = number of baffles

The flow area may be calculated from $a_w = 0.11D_s^2$ for 20 percent baffle cuts or $0.15D_s^2$ for 25 percent baffle cuts, where D_s is the shell diameter.

Low-fin tubes. The method presented above can be used to predict pressure drop for banks of low-fin tubes. For low-fin tubes, the pressure drop is calculated assuming that the tubes are bare. The mass velocity and tube diameter used for calculation are those for a bare tube with the same diameter as the fins of the low-fin tube.

Banks of finned tubes. For fin tubes other than low-fin tubes, the pressure drop for flowing across banks of transverse fin tubes can be calculated from

$$\Delta P_f = \frac{4f_r G_m^2 N_r \phi}{2(144)g\rho}$$

In this equation, for values of $D_r G_m/\mu$ from 2000 to 50,000,

$$f_r = \frac{a}{(D_r G_m/\mu)^{0.316}(s/D_r)^{0.927}}$$

where $a = 10.5(a/p)^{1/2}$ for triangular tube patterns or $6.5(D_f/s)^{1/2}$ for square tube patterns. The units for the various terms are as follows:

ΔP_f = frictional pressure drop, in lb/in^2
f_r = friction factor (dimensionless)

G_m = mass velocity based on minimum flow area ($= W/a_c$) in lb/(h)(ft^2)
N_r = number of tube rows crossed by the fluid
ρ = density, in lb/ft^3
g = gravitational constant, in ft/h^2
D_r = root diameter of fin, in ft
μ = viscosity, in lb/(ft)(h)
s = tube spacing perpendicular to flow, in ft
p = tube spacing parallel to flow, in ft
D_f = diameter of fins, in ft
a_c = minimum flow area, in ft^2

And,

$$\phi = (\mu_w/\mu_b)^{0.14}$$

where μ_w is the viscosity at the wall temperature and μ_b is the viscosity at the bulk fluid temperature, and

$$a_c = s - D_r - 2nlb$$

where n = number of fins per foot
l = fin height, in ft
b = fin thickness, in ft

7-41 Pressure Drop for Flow across Tube Bundles: Two-Phase Flow

Calculate the pressure drop for the conditions of Example 7-40 if 10,000 lb/h (4535.9 kg/h) of the total fluid is a gas with a density of 0.5 lb/ft^3 (8.01 kg/m^3) and a viscosity of 0.05 lb/(ft)(h) (0.0207 cP). Assume the nozzle velocities are 50 ft/s (15.24 m/s). The flow is vertical up-and-down flow.

Calculation Procedure:

1. Select the appropriate equation.

Two-phase-flow pressure drop for flow across tube bundles may be calculated using the equation

$$\frac{\Delta P_{tp}}{\Delta P_{LO}} = 1 + (K^2 - 1)[Bx^{(2-n)/n}(1 - x)^{(2-n)/n} + x^{2-n}]$$

where ΔP_{tp} = the two-phase pressure drop
ΔP_{LO} = the pressure drop for the total mass flowing as liquid
x = the mass fraction vapor

$K = (\Delta P_{GO}/\Delta P_{LO})^{1/2}$, where ΔP_{GO} is the pressure drop for the total mass flowing as vapor

B = 1.0 for vertical up-and-down flow, 0.75 for horizontal flow other than stratified flow, or 0.25 for horizontal stratified flow

The value of n can be calculated from the relation

$$K = (\Delta P_{GO}/\Delta P_{LO})^{1/2} = (\rho_L/\rho_G)^{1/2}(\mu_G/\mu_L)^{n/2}$$

where ρ_G = gas density
ρ_L = liquid density
μ_G = gas viscosity
μ_L = liquid viscosity

For the baffle windows, $n = 0$, and the preceding equation becomes

$$\frac{\Delta P_{tp}}{\Delta P_{LO}} = 1 + (K^2 - 1)[Bx(1 - x) + x^2]$$

where $B = (\rho_{ns}/\rho_L)^{1/4}$ for vertical up-and-down flow or $2/(K + 1)$ for horizontal flow, and where ρ_L is liquid density and ρ_{ns} is no-slip density (homogeneous density).

2. *Determine ΔP_{LO} for cross flow.*

The pressure drop for the total mass flowing as liquid was calculated in the preceding example. The total friction pressure drop across the bundle is $\Delta P_f + \Delta P_{fi}$. Thus, $\Delta P_f + \Delta P_{fi} = 0.547 + 0.117 = 0.664$ lb/in^2 (4.578 kPa).

3. *Calculate ρ_{ns}.*

The formula is

$$\rho_{ns} = \rho_L\lambda_L + \rho_G(1 - \lambda_L)$$

where $\lambda_L = Q_L/(Q_L + Q_G)$
Q_L = liquid volumetric flow rate = 40,000/55 = 727.3 ft^3/h
Q_G = gas volumetric flow rate = 10,000/0.5 = 20,000 ft^3/h

Thus $\lambda_L = 727.3/(727.3 + 20{,}000) = 0.0351$, so $\rho_{ns} = 55(0.0351) + 0.5(1 - 0.0351) = 2.412$ lb/ft^3 (38.64 kg/m^3).

4. *Calculate ΔP_{GO} for cross flow.*

Analogously to step 2, $\Delta P_{GO} = \Delta P_f + \Delta P_{fi}$. From the preceding example, G = 151,561.1 lb/(h)(ft^2). Then, $D_gG/\mu = (0.9375 - 0.75)(1/12)151{,}561.1/0.05 = 47{,}362.8$, and $f = 1/(D_gG/\mu)^{0.25} = 1/47{,}362.8^{0.25} = 0.0678$. Now,

$$\Delta P_f = \frac{4fG^2N_r(N_b - 1)R_1R_b\phi}{2(144)g\rho}$$

From the preceding example,

$N_r = 18.7$
$N_b - 1 = 23$
$R_1 = 0.5432$
$R_b = 0.9014$

For the gas, $\phi = 1.0$. Then,

$$\Delta P_f = \frac{4(0.0678)(151{,}561.1)^2(18.7)(23)(0.5432)(0.9014)1.0}{2(144)(4.18 \times 10^8)0.5}$$

$$= 21.796 \text{ lb/in}^2 \text{ (150.28 kPa)}$$

Also,

$$\Delta P_{fi} = \frac{4(2.66)fG^2N_rR_b\phi}{2(144)g\rho}$$

$$= \frac{4(2.66)(0.0678)(151{,}561.1)^2(18.7)(0.9014)1.0}{2(144)(4.18 \times 10^8)0.5}$$

$$= 4.641 \text{ lb/in}^2 \text{ (32.00 kPa)}$$

Thus, $\Delta P_{GO} = 21.796 + 4.641 = 26.437$ lb/in^2 (182.28 kPa).

5. *Calculate K for cross flow.*

As noted above, $K = (\Delta P_{GO}/\Delta P_{LO})^{1/2} = (26.437/0.664)^{1/2} = 6.3099$.

6. *Calculate n for cross flow.*

As noted above, $K = (\rho_L/\rho_G)^{1/2}(\mu_G/\mu_L)^{n/2}$. Then, $K = 6.3099 = (55/0.5)^{1/2}(0.05/1.95)^{n/2}$, so $n = 0.2774$.

7. *Calculate x, the mass fraction of vapor.*

Straightforwardly, $x = 10{,}000/50{,}000 = 0.2$.

8. *Calculate P_{tp} for cross flow.*

As noted above, $\Delta P_{tp}/\Delta P_{LO} = 1 + (K^2 - 1)[Bx^{(2-n)/2}(1 - x)^{(2-n)/2} + x^{2-n}]$. Thus,

$$\frac{\Delta P_{tp}}{\Delta P_{LO}} = 1 + (6.3099^2 - 1)[1.0(0.2)^{(2-0.2774)/2}(1 - 0.2)^{(2-0.2774)/2} + (0.2)^{(2-0.2774)}]$$

$$= 11.434$$

Therefore, $\Delta P_{tp} = 11.434\Delta P_{LO} = 11.434(0.664) = 7.592$ lb/in^2 (52.35 kPa).

9. Calculate ΔP_w, the pressure drop for the baffle windows, for the total mass flowing as gas.

From the preceding example, ΔP_w for the total mass flowing as a liquid is 0.346 lb/in². Now, ΔP_w for the gas = ΔP_w for the liquid (ρ_L/ρ_G). So, $\Delta P_w = 0.346(55/0.5) = 38.06$ lb/in² (262.42 kPa).

10. Calculate K for the flow through the baffle windows.

Here, $K = (\Delta P_{GO}/\Delta P_{LO})^{1/2} = (38.06/0.346)^{1/2} = 10.4881$.

11. Calculate ΔP_{tp} for the baffle windows.

As noted in step 1, $\Delta P_{tp}/\Delta P_{LO} = 1 + (K^2 - 1)[Bx(1 - x) + x^2]$. Now, $B = (\rho_{ns}/\rho_L)^{1/4} = (2.412/55)^{1/4} = 0.4576$. So, $\Delta P_{tp}/\Delta P_{LO} = 1 + (10.4881^2 - 1)[0.4576(0.2)(1 - 0.2) + 0.2^2] = 13.34$. Therefore, $\Delta P_{tp} = 13.34\Delta P_{LO} = 13.34(0.346) = 4.616$ lb/in² (31.82 kPa).

12. Calculate ΔP_n, the pressure drop through the nozzles.

Use the same approach as in step 2a of the previous example, employing the no-slip density. Thus, $\Delta P_n = 2.25\rho_{ns}v_n^2/9266 = 2.25(2.412)(50)^2/9266 = 1.464$ lb/in² (10.09 kPa).

13. Calculate total two-phase pressure drop.

The total is the sum of the pressure drops for cross flow, window flow, and nozzle flow. Therefore $\Delta P_{tp} = 7.592 + 4.616 + 1.464 = 13.672$ lb/in² (94.27 kPa).

Related Calculations: For situations where $\Delta P_{tp}/\Delta P_{GO}$ is much less than $1/K^2$ and $1/K^2$ is much less than 1, the equation for two-phase pressure drop can be written as

$$\frac{\Delta P_{tp}}{\Delta P_{GO}} = Bx^{(2-n)/2}(1 - x)^{(2-n)/2} + x^{(2-n)}$$

This form may be more convenient for condensers.

The method presented here is taken from Grant and Chisholm [16]. This reference should be consulted to determine the flow regime for a given two-phase system.

Acceleration pressure drop. When the gas density or the vapor mass fraction changes, there is an acceleration pressure drop calculated from

$$\Delta P_a = \frac{G_T^2}{144g}\left(\frac{x_2}{\rho_{G2}} - \frac{x_1}{\rho_{G1}}\right)$$

where ΔP_a = acceleration pressure drop, in lb/in²
G_T = total mass velocity, in lb/(h)(ft²)
x_2 = mass fraction gas at outlet
x_1 = mass fraction gas at inlet
ρ_{G2} = gas density at outlet, in lb/ft³

ρ_{G1} = gas density at inlet, in lb/ft^3
g = gravitational constant, in ft/(h)^2

Shell-side condensation. The frictional pressure drop for shell-side condensation can be calculated from the equation

$$\Delta P_c = \tfrac{1}{2}\left(1 + \frac{v_2}{v_1}\right)\Delta P_1$$

where ΔP_c = condensing pressure drop
ΔP_1 = pressure drop based on the inlet conditions of flow rate, density, and viscosity
v_1 = vapor velocity at the inlet
v_2 = vapor velocity at the outlet

For a total condenser, this becomes

$$\Delta P_c = \frac{\Delta P_1}{2}$$

7-42 Condenser-Subcooler Test Data

Test data were obtained on a vertical condenser-subcooler for a pure component. Compare the observed heat-transfer coefficient to that expected, and calculate the apparent fouling coefficient. Data were obtained in late autumn and the water flow was cocurrent with the process flow in order to minimize the amount of subcooling. Design inlet water temperature was 30°C (86°F).

The *condenser data* are as follows:

D_s = shell diameter = 36 in (0.914 m)
D_o = tube outside diameter = 0.625 in (0.0159 m)
D_i = tube inside diameter = 0.495 in (0.0126 m)
L = effective tube length = 21 ft (6.4 m)
B = baffle spacing = 36 in (0.914 m)
n = number of tubes = 1477
s = tube spacing (center-to-center) = 0.8125 in (0.02064 m)
k_w = tube-wall thermal conductivity = 220 Btu/(h)(ft)(°F) [380 W/(m)(K)].

Tubes have been omitted from the baffle window areas. The *performance data* are as follows:

W_v = vapor flow rate = 170,860 lb/h (77,500.4 kg/h)
ρ_L = liquid density = 54.3 lb/ft^3 (869.6 kg/m^3)
ρ_v = vapor density = 0.49 lb/ft^3 (7.85 kg/m^3)
μ_L = liquid viscosity = 0.484 lb/(ft)(h) (0.2 cP)

μ_v = vapor viscosity = 0.036 lb/(ft)(h) (0.015 cP)
c = liquid specific heat = 0.44 Btu/(lb)(°F) [1.84 kJ/(kg)(K)]
k = liquid thermal conductivity = 0.089 Btu/(h)(ft)(°F) [0.15 W/(m)(K)]
λ = heat of vaporization = 250 Btu/lb (582 kJ/kg)
t_1 = inlet vapor temperature = 54°C (129.2°F)
t_2 = outlet condensate temperature = 40.5°C (104.9°F)
T_1 = inlet water temperature = 16.1°C (61°F)
T_2 = outlet water temperature = 40°C (104°F)

The *water properties* are as follows:

c = specific heat = 1.0 Btu/(lb)(°F)
μ = viscosity = 2.1 lb/(ft)(h)
k = thermal conductivity = 0.35 Btu/(h)(ft)(°F) [0.60 W/(m)(K)]

Calculation Procedure

1. *Set out the overall approach.*

This exchanger must be evaluated in two sections, the condensing section and the subcooling section. The observed and calculated heat-transfer coefficients will be compared. An apparent fouling coefficient can then be calculated by trial and error.

2. *Calculate water flow rate.*

Total heat load Q_T is the sum of condensing heat load Q_c and subcooling heat load Q_s. Now,

$$Q_c = W_v \lambda \qquad \text{and} \qquad Q_s = W_v c(t_2 - t_1)$$

Thus, $Q_c = 170{,}860(250) = 42{,}715{,}000$ Btu/h, and $Q_s = 170{,}860(0.44)(54 - 40.5)(1.8) = 1{,}827{,}000$ Btu/h. Then $Q_T = Q_c + Q_s = 42{,}715{,}000 + 1{,}827{,}000 = 44{,}542{,}000$ Btu/h (13,051,000 W). Then, letting W_w be the mass flow rate for the water

$$W_w = \frac{Q_T}{c(T_2 - T_1)} = \frac{44{,}542{,}000}{1.0(40 - 16.1)(1.8)} = 1{,}035{,}380 \text{ lb/h } (469{,}638 \text{ kg/h})$$

3. *Calculate intermediate water temperature.*

The water-temperature rise for the condensing zone can be calculated as follows, letting T_m be the intermediate water temperature:

$$T_m - T_1 = \frac{Q_c}{Q_T}(T_2 - T_1)$$

Then $T_m = 16.1 + (42{,}715{,}000/44{,}542{,}000)(40 - 16.1) = 39.0$°C (102.2°F).

4. *Calculate mean temperature differences.*

See Example 7-21. In the present case, on the process side, $t_1 = 54°C$, $t_m = 54°C$, and $t_2 = 40.5°C$. On the water side, $T_1 = 16.1°C$, $T_m = 39°C$, and $T_2 = 40°C$. For the condensing zone, then

$$\Delta T_{mc} = \frac{(t_1 - T_1) - (t_m - T_m)}{\ln [(t_1 - T_1)/(t_m - T_m)]}$$

$$= \frac{(54 - 16.1) - (54 - 39)}{\ln [(54 - 16.1)/(54 - 39)]}$$

$$= 24.71°C\ (44.47°F)$$

And for the subcooling zone

$$\Delta T_{ms} = \frac{(t_m - T_m) - (t_2 - T_2)}{\ln [(t_m - T_m)/(t_2 - T_2)]}$$

$$= \frac{(54 - 39) - (40.5 - 40)}{\ln [(54 - 39)/(40.5 - 40)]}$$

$$= 4.26°C\ (7.67°F)$$

5. *Calculate h_o.*

Use the procedure outlined in Example 7-20. In the present case,

$$a_c = BD_s(s - D_o)/s$$

$$= (36/12)(36/12)(0.8125 - 0.625)/0.8125$$

$$= 2.0769\ \text{ft}^2\ (0.1929\ \text{m}^2)$$

Therefore, $G = W_w/a_c = 1{,}035{,}380/2.0769 = 498{,}520$ lb/(h)(ft²), and $D_oG/\mu = (0.625/12)498{,}520/2.1 = 12{,}364$. Therefore,

$$h = \frac{0.273cG}{(c\mu/k)^{2/3}(D_oG/\mu)^{0.365}}$$

$$= \frac{0.273(1.0)498{,}520}{[1.0(2.1)/0.35]^{2/3}12{,}364^{0.365}}$$

$$= 1322.6\ \text{Btu/(h)(ft}^2\text{)(°F)}$$

Correcting for fluid bypassing, $h_o = hF_1F_r$; $F_1 = 0.8(B/D_s)^{1/6}$; $F_r = 1.0$. Thus, $h_o = 1322.6(0.8)(36/36)^{1/6}1.0 = 1058.1$ Btu/(h)(ft²)(°F) [6009.9 W/(m²)(K)].

6. *Calculate condensing coefficient.*

Use the procedure outlined in Example 7-25. In this case, $c\mu/k = 0.44(0.484)/0.089 = 2.393$, and $4\Gamma/\mu = 4\{170{,}860/[1477\pi(0.495/12)]\}/0.484 = 7377$. From Fig. 7-17, for $c\mu/k = 2.393$ and $4\Gamma/\mu = 7377$, $h_o(\mu^2/\rho^2gk^3)^{1/3} = 0.26$, so $h_o = 0.26(\rho^2gk^3/$

$\mu^2)^{1/3} = 0.26[54.3^2(4.18 \times 10^8)0.089^3/0.484^2]^{1/3} = 402.5$ Btu/(h)(ft²)(°F) [2285 W/(m²)(K)].

7. *Calculate the dimensionless group A_d and the condensing coefficient corrected for shear.*

From Example 7-25,

$$A_d = \frac{0.250\mu_L^{1.173}\mu_v^{0.16}}{g^{2/3}D_i^2\rho_L^{0.553}\rho_v^{0.78}}$$

$$= \frac{0.250(0.484)^{1.173}0.036^{0.16}}{(4.18 \times 10^8)^{2/3}(0.495/12)^2 54.3^{0.553}0.49^{0.78}}$$

$$= 1.25 \times 10^{-5}$$

For $A_d = 1.25 \times 10^{-5}$, $h/h_o = 1.7$, from Example 7-25. Therefore, $h = 1.7h_o = 1.7(402.5) = 684.3$ Btu/(h)(ft²)(°F) [3887 W/(m²)(K)].

8. *Calculate subcooling coefficient.*

No interfacial shear will occur in the subcooling zone. Therefore, taking into account the falling-film behavior (see Example 7-25, under Related Calculations), $h_{sc} = 0.75\ h_o = 0.75(402.5) = 301.9$ Btu/(h)(ft²)(°F) [1714 W/(m²)(K)].

9. *Calculate h_w, the coefficient of heat transfer through the tube wall.*

The formula is

$$h_w = \frac{2k}{D_o - D_i} = \frac{2(220)}{(0.625 - 0.495)(1/12)}$$

$$= 40{,}615 \text{ Btu/(h)(ft}^2\text{)(°F) [230,700 W/(m}^2\text{)(K)]}$$

10. *Calculate overall coefficients (clean condition).*

For the condensing zone,

$$\frac{1}{U_c} = \frac{1}{h_c(D_i/D_o)} + \frac{1}{h_o} + \frac{1}{h_w}$$

$$= \frac{1}{684.3(0.495/0.625)} + \frac{1}{1058.1} + \frac{1}{40{,}615}$$

Therefore, $U_c = 355.3$ Btu/(h)(ft²)(°F) [2018 W/(m²)(K)]. For the subcooling zone,

$$\frac{1}{U_{sc}} = \frac{1}{h_{sc}(D_i/D_o)} + \frac{1}{h_o} + \frac{1}{h_w}$$

$$= \frac{1}{301.9(0.495/0.625)} + \frac{1}{1058.1} + \frac{1}{40{,}615}$$

Therefore, $U_{sc} = 194.1$ Btu/(h)(ft^2)(°F) [1102 W/(m^2)(K)].

11. Calculate required areas.

For the condensing zone,

$$A_c = \frac{Q_c}{U_c \Delta T_{mc}} = \frac{42{,}715{,}000}{355.3(24.71)(1.8)}$$

$$= 2702 \text{ ft}^2 \ (251.1 \text{ m}^2)$$

For the subcooling zone,

$$A_s = \frac{Q_s}{U_{sc} \Delta T_{ms}} = \frac{1{,}827{,}000}{194.1(4.26)(1.8)}$$

$$= 1228 \text{ ft}^2 \ (114.0 \text{ m}^2)$$

Then, A_t = total area = $A_c + A_s = 2702 + 1228 = 3930$ ft^2 (365.1 m^2).

12. Calculate available area.

The formula is $A = n\pi D_o L = 1477\pi(0.625/12)(21) = 5075$ ft^2 (471.5 m^2).

13. Calculate apparent fouling coefficients h_s.

Assume that fouling is uniform along the tube length. The calculation for fouling coefficient is then a trial-and-error solution. As a first approximation, assume the observed coefficient U_o will be the clean coefficient multiplied by the ratio of calculated to available areas. Thus, $U_{oc} = U_c(A_t/A) = 355.3\ (3930/5075) = 275.1$ Btu/(h)(ft^2)(°F) and $U_{os} = U_s(A_t/A) = 194.1\ (3930/5075) = 150.3$ Btu/(h)(ft^2)(°F). Now, $1/U_{oc} = 1/U_c + 1/h_s = 1/355.3 + 1/h_s = 1/275.1$. Therefore, $h_s = 1219$ Btu/(h)(ft^2)(°F), and $1/U_{os} = 1/U_s + 1/h_s = 1/194.1 + 1/h_s = 1/150.3$, so $h_s = 666$ Btu/(h)(ft^2)(°F).

However, the value of h_s should be constant. Assume a new value of h_s, calculate U_{oc} and U_{os}, and compare A to A_t. For the first trial, assume $h_s = 1000$ Btu/(h)(ft^2)(°F). Then $1/U_{oc} = 1/355.3 + 1/1000$ and $U_{oc} = 262.2$, and $1/U_{os} = 1/194.1 + 1/1000$ and $U_{os} = 162.5$. Therefore, $A_c = 42{,}715{,}000/262.2(24.71)1.8 = 3663$ ft^2, $A_s = 1{,}827{,}000/162.5(4.26)1.8 = 1466$ ft^2, and therefore, $A_t = 3663 + 1466 = 5129$ ft^2.

The value of A_t is now greater than A. Therefore, h_s must be greater than 1000. Repeating the procedure above leads to a value of h_s that satisfies the parameters: $h_s = 1050$ Btu/(h)(ft^2)(°F). The observed apparent fouling coefficient is thus 1050 Btu/(h)(ft^2)(°F) [5960 W/(m^2)(K)].

8

Distillation

Otto Frank

Supervisor, Process Engineering
Allied Corporation
Morristown, NJ

8-1 Calculation of Equilibrium Stages

Design a distillation column to separate benzene, toluene, and xylene, using (1) the McCabe-Thiele *xy* diagram and (2) the Fenske-Underwood-Gilliland (FUG) method. Compare the results with each other. Assume that the system is ideal.

Feed consists of 60 mol/h benzene, 30 mol/h toluene, and 10 mol/h xylene. There must be no xylene in the overhead, and the concentration of toluene must be no greater than 0.2 percent. Benzene concentration in the bottoms should be minimized; however, recognizing that only one end of a column can be closely controlled (in this case, the overhead is being so controlled), a 2% concentration is specified. Feed temperature is 40°C (104°F); the reflux is not subcooled. The column will be operated at atmospheric conditions, with an average internal back pressure of 850 mmHg (113.343 kPa or 1.13 bar).

Calculation Procedure:

1. Assess the problem to make sure that a hand-calculation approach is adequate.

The separation of multicomponent and nonideal mixtures is difficult to calculate by hand and is evaluated largely by the application of proper computer programs. However, under certain conditions, hand calculation can be justified, and it will be necessary for the designer to have access to suitable shortcut procedures. This may be the case when (1) making preliminary cost estimates, (2) performing parametric evaluations of operating variables, (3) dealing with situations that call for separations having only coarse purity requirements, or (4) dealing with a system that is thermodynamically ideal or nearly so. For guidelines as to ideality, see Table 8-1.

In the present case, hand calculation and shortcut procedures are adequate because the benzene-toluene-xylene system is close to ideal.

2. Establish the equilibrium relationship among the constituents.

The equilibrium data are based on vapor pressures. Therefore, this step consists of plotting the vapor pressure–temperature curves for benzene, toluene, and xylene. The vapor pressures can be determined by methods such as those discussed in Sec. 1 or can be found in the literature. In any case, the results are shown in Fig. 8-1.

3. Convert the given ternary system to a pseudobinary system.

This step greatly reduces the complexity of calculation, although at some expense in accuracy. (Based on limited data, a rough estimate as to the loss in accuracy is 10 percent, i.e., a 10 percent error). Benzene is the lightest-boiling component in the bottoms stream, so it is designated the "light key" and taken as one component of the binary system. Toluene is the higher-boiling of the two constituents in the overhead stream, so it is designated the "heavy key" and taken as the other component of the binary system.

TABLE 8-1 Rules of Thumb on Equilibrium Properties of Vapor-Liquid Mixtures

Declining ideality ↓	
	Mixtures of isomers usually form ideal solutions.
	Mixtures of close-boiling aliphatic hydrocarbons are nearly ideal below a pressure of 10 atm.
	Mixtures of compounds close in molecular weight and structure frequently do not deviate greatly from ideality (e.g., ring compounds, unsaturated compounds, naphthenes, etc.).
	Mixtures of simple aliphatics with aromatic compounds deviate modestly from ideality.
	"Inerts," such as CO_2, H_2S, H_2, N_2, etc., that are present in mixtures of heavier components tend to behave nonideally with respect to the other compounds.
	Mixtures of polar and nonpolar compounds are always strongly nonideal. (Look for polarity in molecules containing oxygen, chlorine, fluorine, or nitrogen, in which electrons in bonds between these atoms and hydrogen are not equally shared).
	Azeotropes and phase separation represent the ultimate in nonideality, and their occurrence should always be confirmed before detailed distillation studies are undertaken.

Source: Otto Frank, "Shortcuts for Distillation Design," *Chemical Engineering,* March 14, 1977.

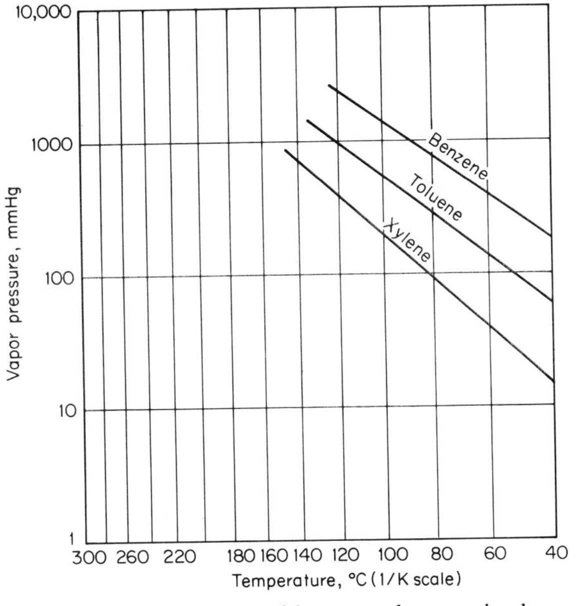

FIG. 8-1 Vapor pressures of benzene, toluene, and xylene (Example 8-1). Note: 1 mmHg = 0.133 kPa.

Since there are 60 mol/h benzene and 30 mol/h toluene in the feed stream, the percentage composition of the pseudobinary feed is 100[60/(30 + 60)], i.e., 66.7% benzene and 100[30/(30 + 60)] = 33.3% toluene.

The overhead composition is stipulated to be 99.8% benzene and 0.2% toluene. By material-balance calculation, the bottoms is found to be 2.6% benzene and 97.4% toluene.

4. *Apply the McCabe-Thiele method of column design, based on an xy diagram.*

a. *Develop the equilibrium line or xy curve.* This can be done using the procedure described in Example 3-12. The result appears as the upper curve in Fig. 8-2 (including both the full figure and the portion that depicts in expanded form the enriching section). This curve shows the composition of vapor y that is in equilibrium with any given liquid of composition x in the benzene-toluene binary system.

Another way of plotting the curve is to employ the equation

$$y = \frac{\alpha x}{1 + (\alpha - 1)x}$$

where α is the relative volatility of benzene with respect to toluene (see Sec. 3). Relative volatility may vary with temperature from point to point along the column; if the values at the top and bottom of the column are within 15 percent of each other, an average can be used to establish the equilibrium line. The present example meets this criterion, and its average volatility is 2.58.

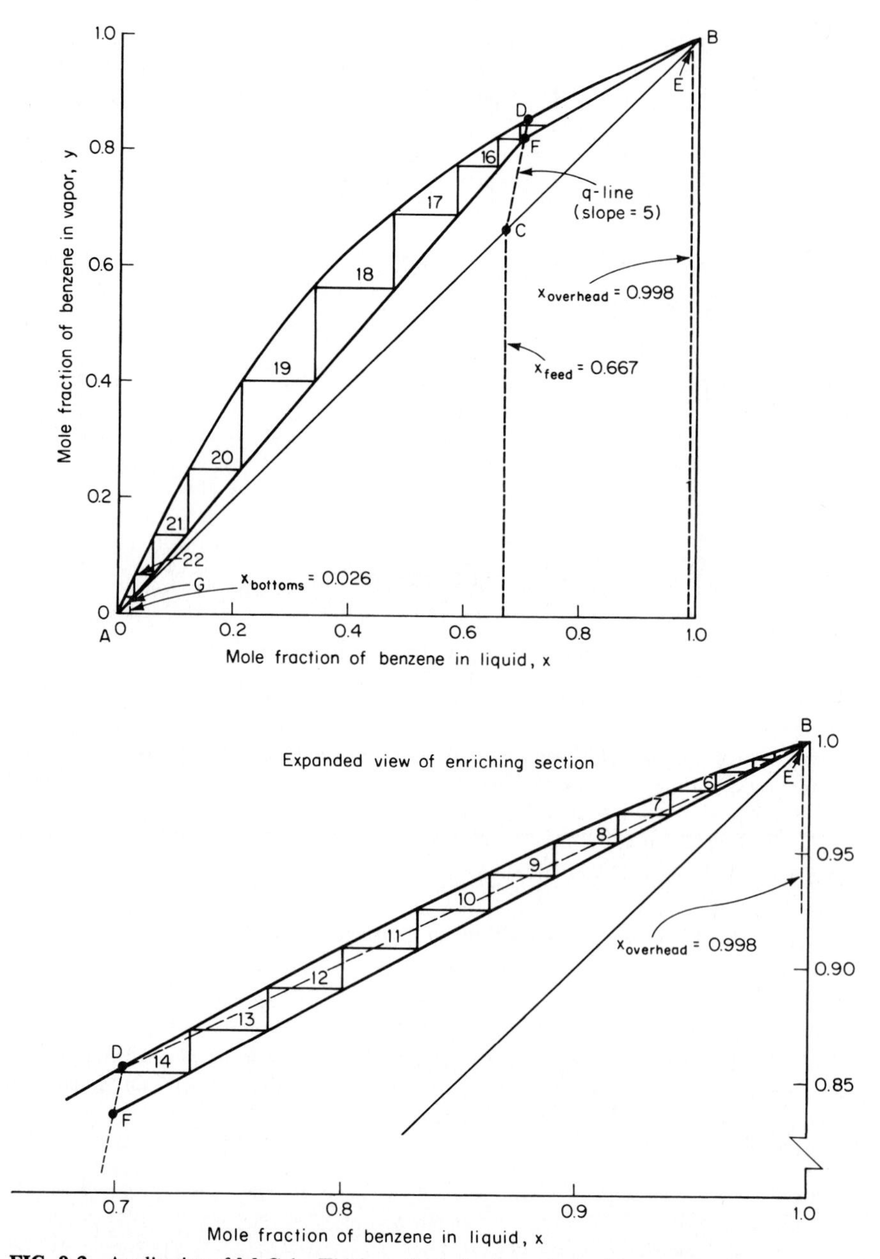

FIG. 8-2 Application of McCabe-Thiele method of column design (Example 8-1).

b. *Calculate the bubble point of the feed.* This is done via procedures outlined in Sec. 3. In the present case, where both the vapor and liquid phases can be considered ideal, the vapor-liquid equilibrium ratio K_i equals vapor pressure of ith component divided by system pressure. The bubble point is found to be 94°C. At 40°C, then, the feed is subcooled 54°C.

c. *Calculate the q-line slope, to compensate for the feed not entering at its bubble point (40°C versus 94°C).* The thermal condition of the feed is taken into account by the slope of a q line, where

$$q = \frac{\text{heat needed to convert 1 mol feed to saturated vapor}}{\text{molar heat of vaporization}}$$

The enthalpy to raise the temperature of the benzene to the bubble point and vaporize it is Sensible heat required + latent heat of vaporization, or (94 − 40°C)[32.8 cal/(mol)(°C)] + 7566 cal/mol = 9337 cal/mol. The enthalpy to raise the temperature of the toluene to the bubble point and vaporize it is (94 − 40°C)[40.5 cal/(mol)(°C)] + 7912 cal/mol = 10,100 cal/mol.

Then

$$q = \frac{(0.667 \text{ mol benzene})9337 + (0.333 \text{ mol toluene})10{,}100}{(0.667 \text{ mol benzene})7566 + (0.333 \text{ mol toluene})7912}$$

$$= 1.25$$

And by definition of the q line, its slope is $q/(q - 1)$, or $1.25/0.25 = 5$.

d. *Plot the q line on the xy diagram.* This line, also known as the "feed line," has the slope calculated in the previous step, and it passes through the point on the xy diagram diagonal that has its abscissa (and ordinate) corresponding to the feed composition x_F. In Fig. 8-2, the diagonal is AB and the q line is CD. The operating lines, calculated in the next step, intersect with each other along the q line.

e. *Construct the operating lines.* There are two operating lines, one for the enriching (upper) section of the column and the other for the stripping (lower) section, the feed plate marking the point of separation between the two. The upper line intersects the diagonal at the abscissa corresponding to the overhead-product composition (in Fig. 8-2, at point E); the lower line intersects it at the abscissa corresponding to the bottoms composition. The slope of each is the ratio of (descending) liquid flow to (ascending) vapor flow, or L/V, in that particular section of the column. The two lines intersect along the q line, as noted earlier.

The ratio L/V depends on the amount of reflux in the column. At minimum reflux, the operating lines intersect each other and the q line at the point where the q line intersects the xy curve (in Fig. 8-2, at point D).

In this example, the approach used for constructing the operating lines is based on the relationship that typically holds, in practice, between the minimum reflux and the amount of reflux actually employed. First, draw the operating line that would pertain for minimum reflux in the enriching section. As noted earlier, this line is defined by points D and E. Next, measure its slope $(L/V)_{\text{min}}$ on the diagram. This is found to be 0.5.

For water- or air-cooled columns, the actual reflux ratio R_{actual} is normally 1.1 to 1.3 times the minimum reflux ratio R_{min}. The optimal relationship between R_{actual} and R_{min} can be established by an economic analysis that compares the cost of energy (which rises with rising reflux) with number of column trays (which declines with rising reflux). For the present example, assume that $R_{actual}/R_{min} = 1.2$.

In the enriching section of the column, the descending liquid consists of the reflux. So, $R = L/(V - L)$. By algebra, $1/R = (V - L)/L = V/L - 1$. Since $(L/V)_{min}$ in the present example is 0.5, $1/R_{min} = 1/0.5 - 1 = 1$. Therefore, $R_{min} = 1$, and $R_{actual} = 1.2R_{min} = 1.2(1) = 1.2$. Therefore, $1/1.2 = V/L - 1$, so $V/L = 1.83$, and L/V, the slope of the actual operating line for the enriching section, is 0.55. This line, then, is constructed by passing a line through point E having a slope of 0.55. This intersects the q line at F in Fig. 8-2. Finally, the operating line for the stripping section is constructed by drawing a line connecting point F with the point on the diagonal that corresponds to the bottoms composition, namely, point G. The two operating lines for the actual column, then, are EF and FG.

f. Step off the actual number of theoretical stages. Draw a horizontal line through point E, intersecting the xy curve. Through that intersection, draw a vertical line intersecting the operating line EF. Through the latter intersection, draw a horizontal line intersecting the xy curve. Continue the process until a horizontal line extends to the left of point G (as for the steps after the one in which a horizontal line crosses the q line, the vertical lines, of course, intersect operating line FG rather than EF). Count the number of horizontal lines; this number is found to be 22.2 (taking into account the shortness of the last line). That is the number of theoretical stages required for the column. The number of lines through or above the q line, in this case 14, is the number of stages in the enriching section, i.e., above the feed plate.

g. Consider the advantages and disadvantages of the McCabe-Thiele method. This example illustrates both the advantages and the disadvantages. On the positive side, it is easy to alter the slope of the operating line and thus the reflux ratio, changes introduced by varying the feed quality and feed location can be judged visually, and the effect of the top and bottom concentrations on number of equilibrium stages can be readily established by moving the operating lines to new points of origin. On the negative side, the example indicates the limit to which the xy diagram can be reasonably applied: when the number of stages exceeds 25, the accuracy of stage construction drops off drastically. This is usually the case with a small relative volatility or high purity requirements or when a very low reflux ratio is called for.

5. *As an alternative to the McCabe-Thiele method, design the column by using the Fenske-Underwood-Gilliland correlations.*

a. *Calculate the minimum number of stages needed (implying total reflux).* Apply the Fenske correlation:

$$N_{min} = \frac{\log\,[(x_{LK}/x_{HK})_D(x_{HK}/x_{LK})_B]}{\log \alpha}$$

where N_{min} is the minimum number of stages, x is mole fraction, mole percent, or actual number of moles, α is relative volatility of the light key (benzene) with respect to the heavy key (toluene), and the subscripts LK, HK, D, and B refer to light key, heavy key, overhead (or distillate) product, and bottoms product, respectively. In this example, α is 2.58, so

$$N_{min} = \frac{\log\,(99.8/0.2)(97.4/2.6)}{\log 2.58} = 10.4 \text{ stages}$$

b. Estimate the feed tray. Again, apply the Fenske correlation, but this time replace the bottoms-related ratio in the numerator with one that relates to the feed conditions, namely, $(x_{HK}/x_{LK})_F$. Thus,

$$N_{\min} = \frac{\log (99.8/0.2)(33.3/66.7)}{\log 2.58} = 5.8 \text{ stages}$$

Therefore, the ratio of feed stages to total stages is $5.8/10.4 = 0.56$, so 56% of all trays should be located above the feed point.

c. Calculate the minimum reflux required for this specific separation. Apply the Underwood correlation: $(L/D)_{\min} + 1 = \Sigma(\alpha x_D/(\alpha - \theta)]$ and $1 - q = \Sigma[\alpha x_F/(\alpha - \theta)]$. The two summations are over each component in the distillate and feed, respectively (thus, the system is not treated as pseudobinary). Each relative volatility α is with respect to the heavy key (toluene), and $(L/D)_{\min}$ is minimum reflux, x is mole fraction, q is the heat needed to convert one mole of feed to saturated vapor divided by the molar heat of vaporization, and θ is called the "Underwood constant." The value of q is 1.25, as found in step 4*c*. The relative volatilities, handled as discussed in step 4*a*, are $\alpha_{\text{benzene-toluene}} = 2.58$, $\alpha_{\text{toluene-toluene}} = 1.00$, and $\alpha_{\text{xylene-toluene}} = 0.36$.

The calculation procedure is as follows: Estimate the value of θ by trial and error using the second of the two preceding equations; then employ this value in the first equation to calculate $(L/D)_{\min}$.

Thus, it is first necessary to assume values of θ in the equation

$$1 - q = \frac{\alpha_{B-T} x_B}{\alpha_{B-T} - \theta} + \frac{\alpha_{T-T} x_T}{\alpha_{T-T} - \theta} + \frac{\alpha_{X-T} x_X}{\alpha_{X-T} - \theta}$$

where the mole fractions pertain to the feed and where B, T, and X refer to benzene, toluene, and xylene, respectively. Trial and error shows the value of θ to be 1.22; in other words, $2.58(0.60)/(2.58 - 1.22) + 1(0.30)/(1 - 1.22) + 0.36(0.1)/(0.36 - 1.22) = -0.26$, which is close enough to $(1 - 1.25)$.

Now, substitute $\theta = 1.22$ into

$$(L/D)_{\min} + 1 = \frac{\alpha_{B-T} x_B}{\alpha_{B-T} - \theta} + \frac{\alpha_{T-T} x_T}{\alpha_{T-T} - \theta}$$

where the mole fractions pertain to the overhead product (there is no xylene-related term because there is no xylene in the overhead). Thus,

$$(L/D)_{\min} + 1 = \frac{2.58(0.998)}{2.58 - 1.22} + \frac{1(0.002)}{1 - 1.22}$$
$$= 0.88$$

(Note that because $(L/D)_{\min} = R_{\min}$ and $1/R = V/L - 1$, this $(L/D)_{\min}$ of 0.88 corresponds to an $(L/V)_{\min}$ of 0.47, versus the value of 0.5 that was found in the McCabe-Thiele method, step 4*e* above.)

A direct solution for θ can be obtained by using a graph developed by Van Winkle and Todd (Fig. 8-3) in which the abscissa is based on the system being treated as pseudobinary. Strictly speaking, this graph applies only to a liquid at its bubble point, but in the present case, with $(x_{LK}/x_{HK})_F = (x_B/x_T)_F = 66.7/33.3 = 2$, the graph shows for an α of 2.58 that $\theta = 1.26$. Then, from the Underwood equation, $(L/D)_{\min} = 0.94$ and $(L/V)_{\min} = 0.48$, so for typical feed conditions (only a limited amount of subcooling or flashing), the simplification introduced by the graph is acceptable.

d. *Estimate the actual number of theoretical stages using the Gilliland correlation.* This step employs Fig. 8-4. From step 5*c*, $(L/D)_{\min}$ or $R_{\min}$ is 0.88. Assume (as was done in the McCabe-Thiele procedure, step 4*e*) that the actual reflux ratio R is 1.2 times $R_{\min}$. Then $R = 1.2(0.88) = 1.06$, and the abscissa of Fig. 8-4 $(R - R_{\min})/(R + 1) = (1.06 - 0.88)/(1.06 + 1) = 0.087$. From the graph, $(N - N_{\min})/(N + 1) = 0.57$, where N is the actual number of stages and $N_{\min}$ is the minimum number. From step 5*a*, $N_{\min} = 10.4$. Then $(N - 10.4)/(N + 1) = 0.57$, so $N = 25.5$ theoretical stages. (Note that if the $R_{\min}$ of 0.94, based on the graphically derived Underwood constant, is employed, the resulting number of theoretical stages is nearly the same, at 24.9.)

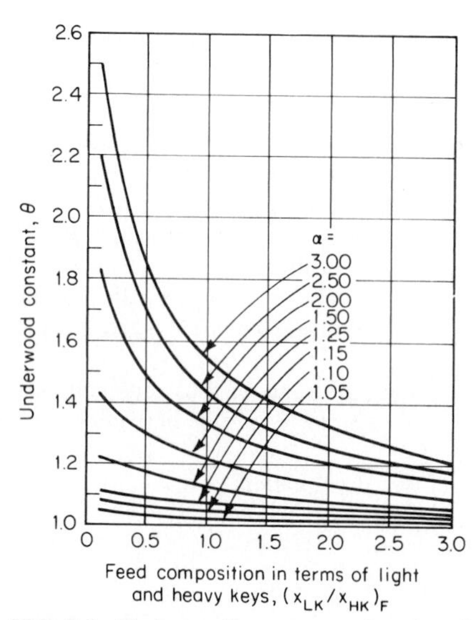

FIG. 8-3 Underwood constant as a function of feed composition and relative volatility.

e. *Comment on the usage of the Fenske-Underwood-Gilliland method.* A reasonably good estimate of N can be obtained by simply doubling the $N_{\min}$ that emerges from the Fenske correlation. The Underwood correlation can handle multicomponent systems, whereas the McCabe-Thiele *xy* diagram is confined to binary (or pseudobinary) systems. However, if there is a component that has a vapor pressure between that of the light and heavy keys, then the Underwood-calculation procedure becomes more complicated than the one outlined above.

6. *Compare the results of the McCabe-Thiele and the Fenske-Underwood-Gilliland methods.*

Summarizing the results found in this example, the comparison is as follows:

Method	Results		
	Reflux ratio	Number of theoretical stages	Optimal feed stage
McCabe-Thiele	1.2	22.2	14
Fenske-Underwood-Gilliland:			
Using calculated θ	1.06	25.5	14
Using θ from graph	1.13	24.9	14
Using nomograph	—	22	—
Using approximation, $2N_{\min}$	—	20.8	12

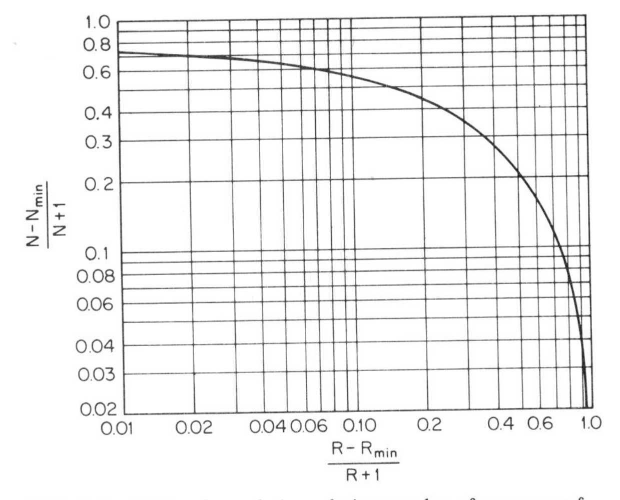

FIG. 8-4 Gilliland correlation relating number of stages to reflux ratio. *(From Chemical Engineering, McGraw-Hill, 1977.)*

Related Calculations: Special McCabe-Thiele graph paper that expands the top and bottom of the *xy* diagram is available. This makes stepping off stages easier and more precise. However, it does not materially improve the accuracy of the procedure.

Variations of the basic McCabe-Thiele method, such as incorporating tray efficiencies, nonconstant molal overflow, side streams, or a partial condenser, are often outlined in standard texts. These modifications usually are not justified; modern computer programs can calculate complex column arrangements and nonideal systems considerably faster and more accurately than use of any hand-drawn diagram can.

A nomograph for the overall Fenske-Underwood-Gilliland method has been derived that considerably reduces the required calculation effort without undue loss of accuracy. It is based on Fig. 8-5, where the subscripts D and B in the abscissa refer to overhead and bottoms product streams, respectively, and the other symbols are as used in step 5. Find the relevant abscissa; in this case, $(99.8/0.2)(97.4/2.6) = 18{,}700$. Erect a vertical line and let it intersect the positively sloped line that represents the assumed ratio of actual to minimum reflux ratios, in this case, 1.2. Through this intersection, draw a horizontal line that intersects the right-hand vertical border of the diagram. Finally, draw a line through the latter intersection and the relevant α_{LK-HK} point (in this case, 2.58) on the negatively sloped line that represents relative volatilities, and extend this last-constructed line until it intersects the left-hand border of the diagram. The point of intersection (in this case, it turns out to be 22) represents the number of theoretical stages.

8-2 Distillation-Tray Selection and Design

Design the trays for a distillation column separating dichlorobenzene (DCB) from a high-boiling reaction product. Include designs for sieve trays and valve trays, and discuss the applications of each. The product is temperature-sensitive, so sump pressure should be held at about 100 mmHg (3.9 inHg or 0.13 bar). The separation requires 20 actual trays.

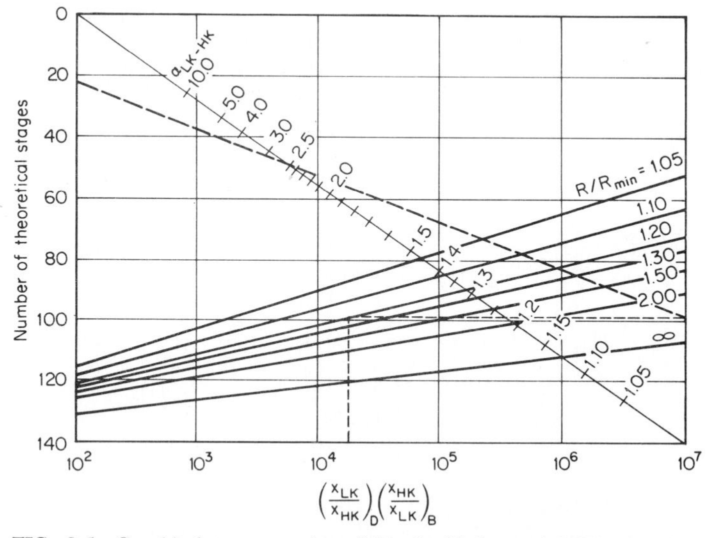

FIG. 8-5 Graphical representation of Fenske-Underwood-Gilliland procedure. *(From Chemical Engineering, McGraw-Hill, 1977.)*

Column conditions in the top section are as follows:

Vapor flow rate = 69,000 lb/h (31,300 kg/h)
Liquid flow rate = 20,100 lb/h (9120 kg/h)
Liquid viscosity = 0.35 cP (3.5×10^{-4} Pa·s)
Liquid density = 76.7 lb/ft^3 (1230 kg/m^3)
Surface tension of DCB liquid = 16.5 dyn/cm (1.65×10^{-4} N/cm)
Temperature (average for top section) = 200°F (366 K)
Pressure (average for top section) = 50 mmHg (6.67 kPa)
Molecular weight of DCB = 147

Note: Pressure drop for the vapor stream is assumed to be 3 mmHg (0.4 kPa) per tray; therefore, the pressure at the top of the column is set at 40 mmHg (5.3 kPa), given that the sump pressure is to be no greater than 100 mmHg (13.3 kPa).

Calculation Procedure:

1. Set tray spacing.

Tray spacing is selected to minimize entrainment. A large distance between trays is needed in vacuum columns, where vapor velocities are high and excessive liquid carry-over can drastically reduce tray efficiencies. A tradeoff between column diameter (affecting vapor velocity) and tray spacing (affecting the disengaging height) is often possible.

Trays are normally 12 to 30 in (0.305 to 0.762 m) apart. A close spacing of 12 to 15 in (0.305 to 0.381 m) is usually quite suitable for moderate- or high-pressure columns.

Frequently, 18 in (0.457 m) is selected for atmospheric columns. High-vacuum systems may require a tray spacing of 24 to 30 in (0.61 to 0.762 m). A spacing as low as 9 in (0.229 m) is sometimes found in high-pressure systems. Spacings greater than 30 in are seldom if ever justified. As a starting point for the column in question, select a 24-in (0.61-m) tray spacing.

2. *Estimate column diameter.*

In high-pressure columns, liquid flow is the dominant design consideration for calculating column diameter. However, at moderate and low pressures (generally below 150 psig or 10.34 bar), as is the case for the column in this example, vapor flow governs the diameter.

In vacuum columns, the calculations furthermore are based on conditions at the column top, where vapor density is lowest and vapor velocity highest. The *F*-factor method used here is quite satisfactory for single- and two-pass trays; it has an accuracy well within ±15 percent. The *F* factor F_c is *defined* as follows:

$$F_c = v\rho_v^{1/2}$$

where v is the superficial vapor velocity in the tower (in ft/s), and ρ_v is the vapor density (in lb/ft³). In the calculation procedure, first the appropriate value of F_c is determined from Fig. 8-6, which gives F_c as a function of column pressure (in this case, 0.97 psia) and tray spacing (in this case, 24 in, from step 1). The figure shows F_c to be 1.6. (Note that if the straight-line portion of the curve for 24-in tray spacing were continued at pressures below about 15 psia, the value of F_c would instead be about 1.87. The lower

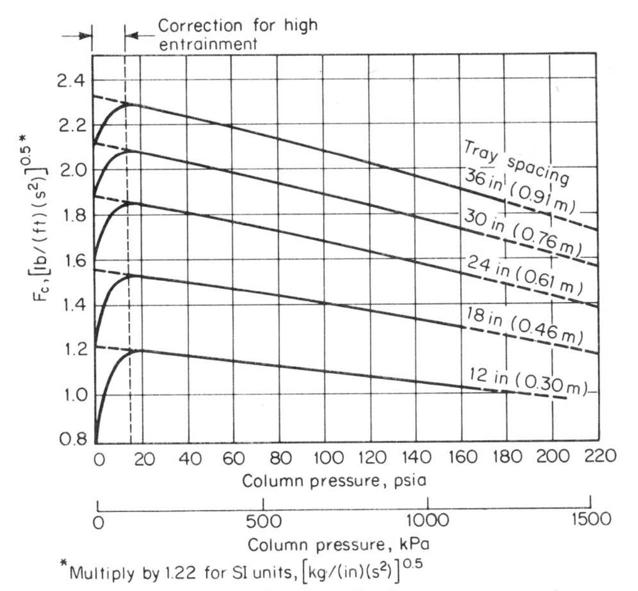

FIG. 8-6 *F* factor as a function of column pressure and tray spacing.

value, 1.6, takes into account the efficiency loss at low pressures that is due to entrainment.)

Next, the F_c value is used in calculating required free area for vapor flow, via the relation

$$A_F = \frac{W}{F_c \rho_v^{1/2}}$$

where A_F is free area (in ft^2), and W is vapor mass flow rate (in lb/s). (Note that this is merely a rearrangement of the preceding definition for F_c.) In this case, $W = (69{,}000 \text{ lb/h})/(3600 \text{ s/h}) = 19.17$ lb/s, and (from the ideal-gas law, which is acceptable at up to moderate pressures) $\rho = (0.97 \text{ psia})(147 \text{ lb/lb·mol})/(10.73 \text{ psia·ft}^3/R)[200 + 460)R] = 0.02 \text{ lb/ft}^3$. Therefore, $A_F = 19.17/1.6(0.02)^{1/2} = 84.72 \text{ ft}^2$.

Actual column cross-sectional area A_T is the sum of the free area A_F plus the downcomer area A_D. Downcomer cross-sectional area must be adequate to permit proper separation of vapor and liquid. Although downcomer size is a direct function of liquid flow, total downcomer area usually ranges between 3 and 20 percent of total column area, and for vacuum columns it is usually 3 to 5 percent. In this example, assume for now that it is 5 percent; then $A_T = A_F/0.95 = 84.72/0.95 = 89.18 \text{ ft}^2$. Then, by geometry, column diameter $= [A_T/(\pi/4)]^{1/2} = (89.18/0.785)^{1/2} = 10.66$ ft; say, 11 ft (3.4 m).

3. *Calculate the flooding velocity.*

Use the Fair calculation. Flooding velocity $U_f = C(\sigma/20)^{0.2}[(\rho_L - \rho_V)/\rho_V]^{0.5}$, where σ is surface tension (in dyn/cm), ρ is density, the subscripts L and V refer to liquid and vapor, respectively, and C is a parameter that is related to tray spacing, liquid flow rate L, vapor flow rate G, and liquid and vapor densities by the graph in Fig. 8-7. In this

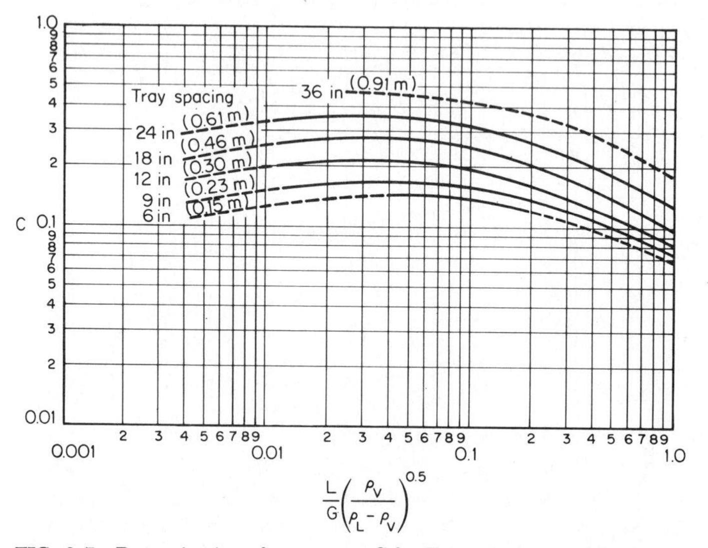

FIG. 8-7 Determination of parameter C for Fair calculation of flooding velocity.

case, the abscissa is $(20{,}100/69{,}000)[0.02/(76.7 - 0.02)]^{0.5} = 0.005$, and tray spacing is 24 in from step 1, so C is found from the graph to be 0.3. Then,

$$U_f = 0.3\left(\frac{16.5}{20}\right)^{0.2}\left(\frac{76.7 - .02}{0.02}\right)^{0.5}$$

$$= 17.9 \text{ ft/s}$$

The following adjustments, however, must be incorporated into the calculated flooding velocity to assure a reasonable design safety factor:

Open area (cap slots or holes)	Multiplying factor
10 percent of active tray area	1.0
8 percent of active tray area	0.95
6 percent of active tray area	0.90

System properties	Multiplying factor
Known nonfoaming system at atmospheric or moderate pressures	0.9
Nonfoaming systems; no prior experience	0.85
Systems thought to foam	0.75
Severely foaming systems	0.70
Vacuum systems (based on entrainment curves; see Fig. 8-17)	0.60 to 0.80
For downcomerless trays if the open area is less than 20 percent of active tray area (Note: it should never be less than 15 percent)	0.85

In this example, no adjustment is necessary for percent holes per active tray area, because in low-pressure systems, the hole area is usually greater than 10 percent. However, because we are dealing with a vacuum system, the flooding velocity must be downgraded. Referring to Fig. 8-17, take as the maximum acceptable fractional entrainment a ψ of 0.15. Then for an abscissa of $(L/G)(\rho_V/\rho_L)^{0.5} = (20{,}100/69{,}000)(0.02/76.7)^{0.5} = 0.005$, establish at what percent of flooding the column can be operated. The graph shows this to be 60 percent. Therefore, the column should be designed for a vapor velocity of $0.6(17.9) = 10.74$ ft/s (3.27 m/s).

4. Reestimate the required free cross-sectional area, this time to accommodate the maximum allowable vapor velocity.

Since mass flow rate equals density times cross-sectional area times velocity, free column area $A_F = (69{,}000 \text{ lb/h})/[(0.02 \text{ lb/ft}^3)(3600 \text{ s/h})(10.74 \text{ ft/s})] = 89.23 \text{ ft}^2$. Note that this turns out to be slightly larger than the area calculated in step 2.

5. Set the downcomer configuration.

A downcomer area can be selected from Fig. 8-8. In this case, the abscissa is (76.7 − 0.02) and the tray spacing is 24 in, so Design gal/(min)(ft^2) is read to be 175 (by extrapolating the horizontal lines on the graph to the right). Since $76.7 \text{ lb/ft}^3 = 10.25$ lb/gal, the gal/min rate is $(20{,}100 \text{ lb/h})/[(60 \text{ min/h})(10.25 \text{ lb/gal})] = 32.68$ gal/min. Therefore, the minimum required downcomer area is $32.68/175 = 0.187 \text{ ft}^2$.

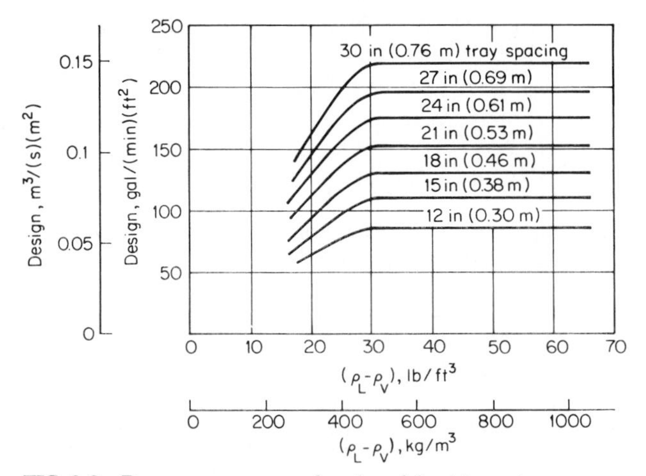

FIG. 8-8 Downcomer area as a function of densities and tray spacing. Note: Multiply clear liquid rate (ordinate) by 0.7 for foaming systems.

However, flow distribution across the tray must also be taken into account. To optimize this distribution, it is usually recommended that weir length be no less than 50 percent of the column diameter. Thus, from Fig. 8-9, the downcomer area subtended by a weir having half the length of the column diameter should be at least 3 percent of the column area (5 percent was assumed in step 2 during the initial estimate of column diameter).

Allowing 3% for downcomer area, then, total cross-sectional area for the column must be 89.23/0.97 = 91.99 ft^2 (8.55 m^2), and the downcomer area required for good flow distribution is (91.99 − 89.23) = 2.76 ft^2 (0.26 m^2). (Note that this is considerably larger than the minimum 0.187 ft^2 required for flow in the downcomer itself.) Finally, the required column diameter is $(91.99/0.785)^{1/2}$ = 10.83 ft; say, 11 ft (3.4 m). (This turns out the same as the diameter estimated in step 2.)

Because of the low liquid rates in vacuum systems, downcomers will usually be oversized, and specific flow rates across the weir will be low. However, liquid rates in high-pressure columns may exceed values recommended for optimum tray performance across a single weir. The maximum specific flow is 70 gal/(min·ft)[53 m^3/(h·m)] for a straight segmental weir and 80 gal/(min·ft)[60 m^3/(h·m)] for a weir with relief wings. Above 80 gal/(min·ft), a multiple downcomer arrangement should be considered.

When dealing with foaming or high-pressure systems, a frequent recommendation is the installation of sloped downcomers. This provides for adequate liquid-vapor disengaging volume at the top, while leaving a maximum active area on the tray below.

6. *Design suitable sieve trays.*

***a.** Determine the required hole area.* A workable hole-area calculation can be based on the empirically developed relationship shown in Fig. 8-10. In this *F* factor (the ordinate), v_h is the velocity of vapor passing through the holes. The graph shows that columns operating at low pressures have less flexibility (lower turndown ratios) than those operating at higher pressures.

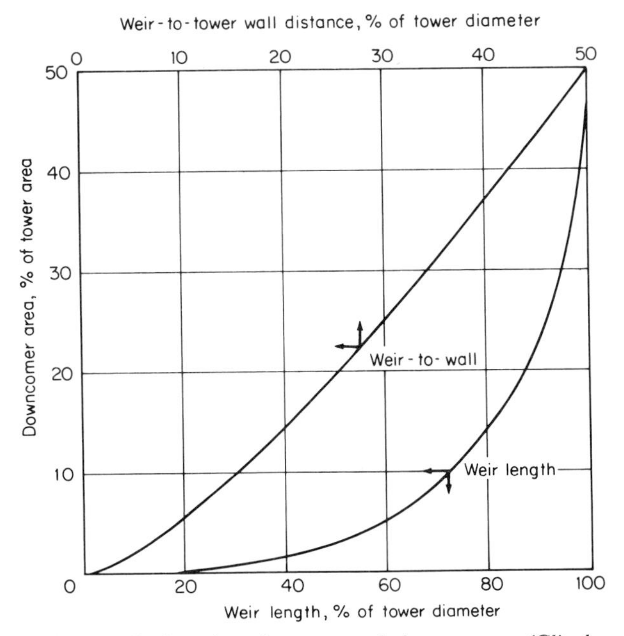

FIG. 8-9 Design chart for segmental downcomers. *(Glitsch, Inc.)*

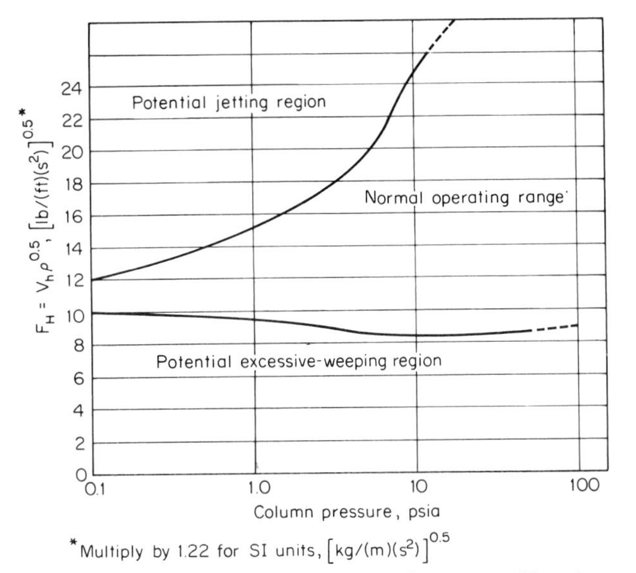

FIG. 8-10 Approximate operating range of sieve trays. Note: 1 psia = 6.895 kPa.

Assume that F_H is 12. Then $v_h = F_H/\rho_v^{0.5} = 12./0.02^{0.5} = 85$ ft/s (26 m/s). Therefore, total hole area required is (69,000 lb/h)/[(0.02 lb/ft³)(3600 s/h)(85 ft/s)] = 11.27 ft². Note that since each tray is fed by one downcomer and drained by another and each downcomer occupies 3 percent of the total tray area according to step 5, the hole area as a percent of active area is

$$\frac{(11.27 \text{ ft}^2)(100)}{[1.00 - (2)(0.03)](11 \text{ ft})^2(\pi/4)} = 12.62 \text{ percent}$$

b. *Specify hole size, weir height, and downcomer.* Guidelines for this step are as follows:

Hole pitch-to-diameter ratio	2 to 4.5
Hole size	Use ½ in (0.0127 m) for normal service, ⅜ or ¼ in (0.0095 or 0.0064 m) for clean vacuum systems, or ¾ in (0.0191 m) for fouling service.
Open area/active area	Use 4 to 16 percent, depending on system pressure and vapor velocity.
Weir height	Use 1 to 4 in (0.025 to 0.102 m) (but no higher than 15 percent of tray spacing) as follows: 1 to 2 in (0.025 to 0.051 m) for vacuum and atmospheric columns or 1½ to 3 in (0.038 to 0.076 m) for moderate to high pressures.
Downcomer seal	Use one-half of weir height or ¾ in (0.0191 m), whichever is greater. Clearance between downcomer and tray deck should never be less than ½ in (0.0127 m), and the velocity through the clearance should be under 1 ft/s (0.30 m/s).

Select ¼-in (6.4-mm) holes and a 1-in (2.54-mm) high weir, with a ¾-in downcomer clearance.

c. *Calculate pressure drop to confirm suitability of design.* The total pressure drop h_t is the sum of the pressure drop across the holes h_{dry} and the drop through the aerated material above the holes h_{liq}. Now,

$$h_{dry} = 0.186 \frac{\rho_V}{\rho_L}\left(\frac{v_h}{C}\right)^2$$

where C is a discharge coefficient obtainable from Fig. 8-11. From step 6*b*, hole area divided by active area is 0.1262, and for a 14-gauge tray (0.078 in or 2 mm, a reasonable thickness for this service), the tray thickness divided by hole diameter ratio is 0.078/0.25 = 0.31. Therefore, C is read as 0.74. Then, $h_{dry} = 0.186(0.02/76.7)(85/0.74)^2 = 0.64$ in of liquid. However, this must be adjusted for entrainment:

$$h_{dry}(\text{corrected}) = h_{dry}\left(1 + \frac{\psi}{1 - \psi}\right)$$

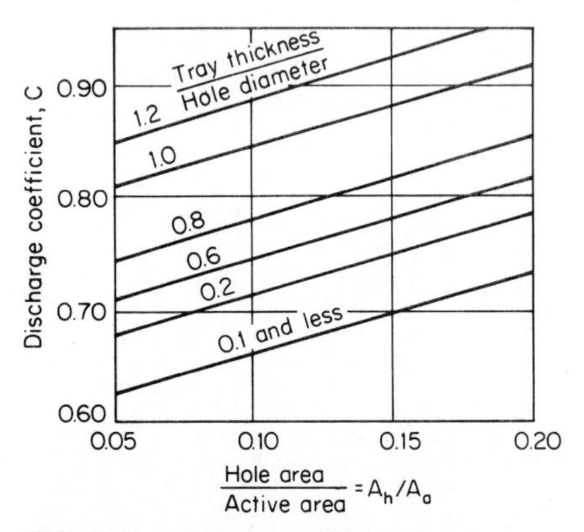

FIG. 8-11 Discharge coefficient for sieve-tray performance. Note: Most perforated trays are fabricated from 14-gauge stainless steel or 12-gauge carbon-steel plates.

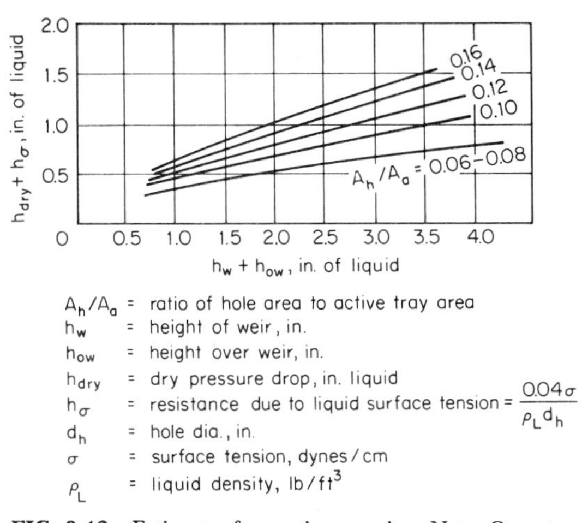

FIG. 8-12 Estimate of excessive weeping. Note: Operating points above the respective lines represent "safe" designs. A point below the line may indicate uncertainty, but not necessarily a dumping station. *(From Chemical Engineering, McGraw-Hill, 1977.)*

where ψ is the entrainment function. Since this is 0.15 (see step 3),

$$h_{\text{dry}}(\text{corrected}) = 0.64\left(1 + \frac{0.15}{0.85}\right) = 0.75 \text{ in liquid}$$

Dry pressure drop for sieve trays should fall between 0.75 and 3 in of liquid (19 and 76 mm) to prevent excessive weeping on the one hand or jetting on the other. Since, in this case, the dry pressure drop is barely within the recommended range, it is desirable that the tray be checked for weeping using Fig. 8-12.

Now, $A_h/A_A = 0.1262$, from step 6*a*. And

$$h_\sigma = \frac{0.04\sigma}{\rho_L\ (\text{hole diameter})} = \frac{0.04(16.5)}{76.7(0.25)} = 0.034 \text{ in liquid}$$

Therefore, the ordinate in Fig. 8-12, $(h_{\text{dry}} + h_\sigma)$, is $0.75 + 0.03 = 0.78$. As for the abscissa, h_w (weir height) was chosen to be 1 in, and $h_{ow} = 0.5[(\text{liquid flow, gal/min})/(\text{weir length, in})]^{0.67} = 0.5\{32.68/[½(11\text{ ft})(12\text{ in/ft})]\}^{0.67} = 0.31$ in liquid, so $h_w + h_{ow} = 1.00 + 0.31 = 1.31$. In Fig. 8-12, the point (1.31, 0.78) falls above the line $A_H/A_A = 0.12$; therefore, weeping will not be excessive.

As for the pressure drop through the aerated material,

$$h_{\text{liq}} = \beta(h_w + h_{ow})$$

where β is an aeration factor obtainable from Fig. 8-13. In the abscissa, the vapor-velocity term refers to velocity through the column (10.74 ft/s from step 3) rather than to velocity through the holes, so $F_c = 10.74(0.02)^{1/2} = 1.52$. Thus β is read to be 0.6. So, $h_{\text{liq}} = 0.6(1.31) = 0.79$ in liquid. Finally, the total pressure drop $h_t = 0.75 + 0.79 = 1.54$

in liquid. Liquid density is 76.7 lb/ft^3, so the equivalent pressure drop is 1.54(76.7/62.4) = 1.89 in water or 3.5 mmHg (0.47 kPa) per tray.

A calculated pressure drop of 3.5 mmHg, although somewhat higher than the assumed 3.0 mmHg, falls within the range of accuracy for the outlined procedure. However, if there is an overriding concern to avoid exceeding the specified 100 mmHg pressure in the column sump, the following options could be considered: (1) redesign the trays using an F_H factor (in step 6*a*) of 11.0 or 11.5, or (2) lower the design top pressure to 20 to 30 mmHg and redesign the trays for a pressure drop of 4 or 3.5 mmHg, respectively.

Note that the 3 mmHg pressure drop for this column can in fact be attained if turndown requirements are not excessive. A turndown to 80 percent of the design vapor load is probably the lower limit at which this column can be operated without loss of efficiency. At higher pressures, when pressure drop is not critical, the base design should be for a pressure drop greater than 4.5 mmHg; this will usually permit operation at turndown rates as high as 50 percent.

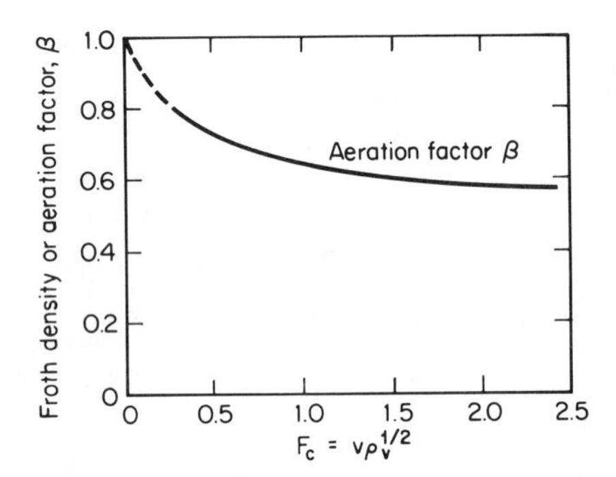

FIG. 8-13 Aeration factor for pressure-drop calculations.

d. Discuss the applicability of sieve trays. The sieve tray is probably the most versatile contacting device. It should be considered first for the design of a tray column. It has the lowest installed cost of any equilibrium-stage-type device, its fouling tendencies are low, and it offers good efficiency when properly designed.

However, sieve trays are not recommended for the following conditions: (1) when very low pressure drop (less than 2.5 mmHg, or 0.39 kPa) is required; (2) when high turndown ratios are required at low pressure drop; or (3) when very low liquid rates are required: below either 0.25 gal/(min)(ft^2) [0.6 m^3/(h)(m^2)] of active tray area, or 1 gal/(min)(ft) [0.75 m^3/(h)(m)] of average flow-path width.

7. *Design suitable valve trays (as an alternative to sieve trays).*

a. *Select the valve layout.* Although valve trays come in several configurations, all have the same basic operational principle: Vapor passing through orifices in the tray lifts small metal disks or strips, thereby producing a variable opening that is proportional to the flow rate.

Because of their proprietary nature, valve trays are usually designed by their respective vendors based on process specifications supplied by the customer. However, most fabricators publish technical manuals that make it possible to estimate some of the design parameters. The procedure for calculating valve-tray pressure drop outlined here has been adapted from the *Koch Design Manual.* As for the other column specifications required, they can be obtained via the same calculation procedures outlined above for the sieve-tray design.

The number of valve caps that can be fitted on a tray is at best an estimate unless a detailed tray layout is prepared. However, a standard has evolved for low- and moderate-pressure operations: a 3 × 2½ in pattern that is the tightest arrangement available, accommodating about 14 caps/ft^2 (150 caps/m^2). The active area does not take into

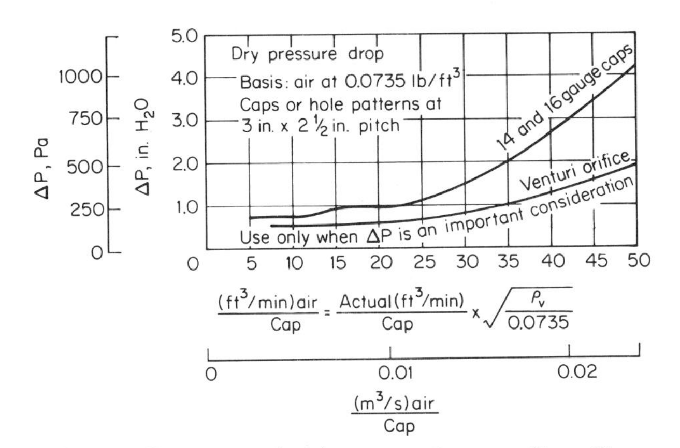

FIG. 8-14 Dry pressure drop h_{dry} across valve trays. *(From Chemical Engineering, McGraw-Hill, 1977.)*

account liquid-distribution areas at the inlet and outlet, nor edge losses due to support rings, nor unavailable space over tray-support beams. In smaller columns, it is possible that as much as 25 percent of the active tray area may not be available for functioning valves. For this column, which operates at low pressure, select the standard 3 × 2½ in pitch.

b. *Calculate the pressure drop per tray.* The vapor velocity at the top of the column is $(69{,}000 \text{ lb/h})/[(60 \text{ min/h})(0.02 \text{ lb/ft}^3)] = 57{,}500 \text{ ft}^3/\text{ min}$. Assume that 15 percent of the tray is not available for functioning valves. Then, since the active tray area (subtracting both the descending and ascending downcomers; see step 6*a*) is $0.94(11)^2(\pi/4) = 89.3 \text{ ft}^2$, the number of valves per tray is $0.85(89.3 \text{ ft}^2)(14 \text{ caps/ft}^2) = 1063$ valves. Then, in entering Fig. 8-14 so as to find the pressure drop h_{dry} through the valves, the abscissa (ft^3/min) air/cap is $[(57{,}500 \text{ ft}^3/\text{min})/1063 \text{ valves}](0.02/0.0735)^{0.5} = 28.2$. From Fig. 8-14, using the curve for the Venturi orifice valve, $\Delta P = 0.8$ in water, or 0.65 in liquid, or 1.5 mmHg, which when corrected for entrainment (same as for sieve trays; see step 6*c*) becomes $0.65(1 + 0.15/0.85) = 0.76$ in liquid.

The pressure drop h_{liq} through the aerated liquid can be obtained directly from Fig. 8-15. For a liquid rate of 32.68 gal/min (see step 5) flowing over a 5.5-ft-long, 1-in weir (see steps 5 and 6*b*), or 6 gal/(min)(ft), h_{liq} is found to be 0.5 in liquid. Therefore, the total pressure drop across a valve tray is $0.76 + 0.5 = 1.26$ in liquid, or 1.55 in water, or 2.9 mmHg (0.39 kPa).

c. *Discuss the applicability of valve trays.* The relatively low pressure drop that can be maintained without undue loss of turndown in vacuum columns is probably the valve tray's greatest attribute. Although the accuracy of either the sieve or valve-tray calculation procedure is probably no better than ± 20 percent, a lower pressure drop is likely to be achieved with Venturi orifice valves than with sieve trays if a reasonable turndown ratio (say 60 percent) is required. This is of little concern at column pressures above 400 mmHg (53 kPa), when pressure drop becomes a minor consideration.

A word of caution: When the valves are exposed to a corrosive environment, it is likely that their constant movement will induce fatigue stresses, which frequently lead to the rapid deterioration of the retaining lugs and valve caps. It is not unusual to find valves missing in that part of the column where corrosive constituents are concentrated.

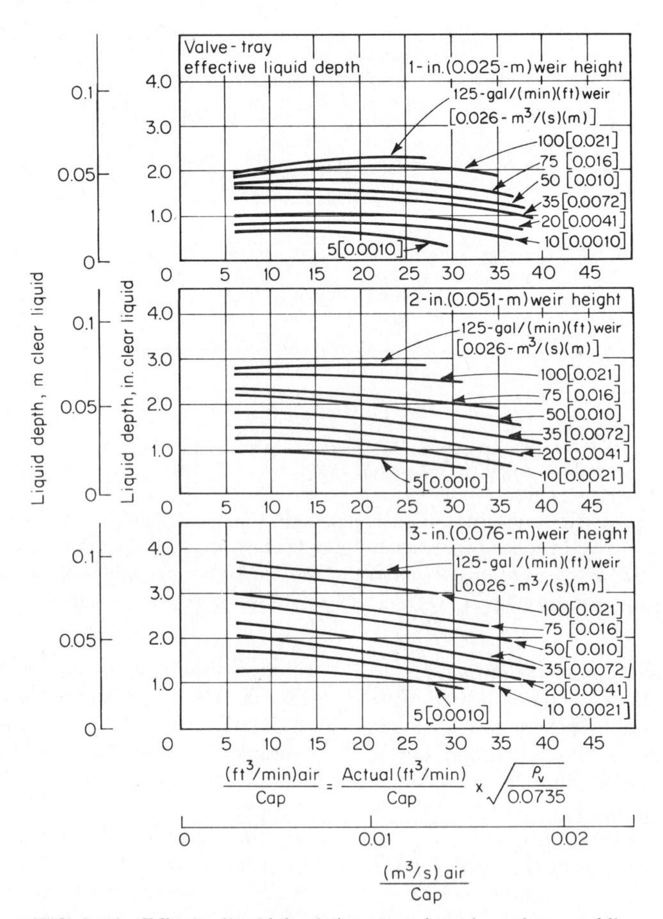

FIG. 8-15 Effective liquid depth (pressure drop through aerated liquid, h_{liq}) on valve trays. *(From Chemical Engineering, McGraw-Hill, 1977.)*

8-3 Column Efficiency

Determine the efficiency in the upper portion of the DCB distillation column designed in the preceding example. Relative volatility for the system is 3.6.

Calculation Procedure:

1. Determine the effects of the physical properties of the system on column efficiency.

Tray efficiency is a function of (1) physical properties of the system, such as viscosity, surface tension, relative volatility, and diffusivity; (2) tray hydraulics, such as liquid height, hole size, fraction of tray area open, length of liquid flow path, and weir config-

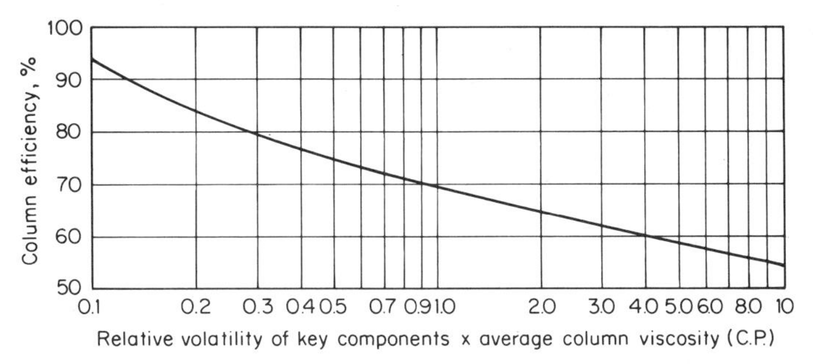

FIG. 8-16 Column efficiency as a function of average column viscosity and relative volatility.

uration; and (3) degree of separation of the liquid and vapor streams leaving the tray. Overall column efficiency is based on the same factors, but will ordinarily be less than individual-tray efficiency.

The effect of physical properties on column efficiency can be roughly estimated from Fig. 8-16. For this system, viscosity is 0.35 cP (see statement of previous example) and relative volatility is 3.6, so the abscissa is 1.26. The ordinate, or column efficiency, is read to be 68 percent.

2. *Determine the effects of tray hydraulics on the efficiency.*

Tray hydraulics will affect efficiency adversely only if submergence, hole size, open tray area, and weir configuration are outside the recommended limits outlined in the previous example. Since that is not the case, no adverse effects need be expected.

3. *Determine the effects of inadequate separation of liquid and vapor (e.g., entrainment) on the efficiency.*

The effects of entrainment on efficiency can be quite drastic, especially in vacuum columns, as the vapor rate in the column approaches flooding velocities. The corrected-for-entrainment efficiency can be calculated as follows:

$$E_c = \frac{E_i}{1 + E_i[\psi/(1 - \psi)]}$$

where E_c is the corrected efficiency, E_i is the efficiency that would prevail if entrainment were no problem (i.e., 68 percent, from step 1), and ψ is the fractional entrainment, whose relationship to other column parameters is defined in Fig. 8-17. In step 3 of the previous example, ψ was assumed to be 0.15. Therefore, $E_c = 0.68/\{1 + 0.68[0.15/(1 - 0.15)]\} = 0.607$; say, 60 percent.

Accordingly, an actual column efficiency of 60 percent for the top section of the column would be reasonable.

Related Calculations: Since efficiencies are likely to vary from column top to bottom, it is usually well to estimate them at various points (at least two) along the column.

"Normal" column efficiencies run between 60 and 85 percent. They will tend toward

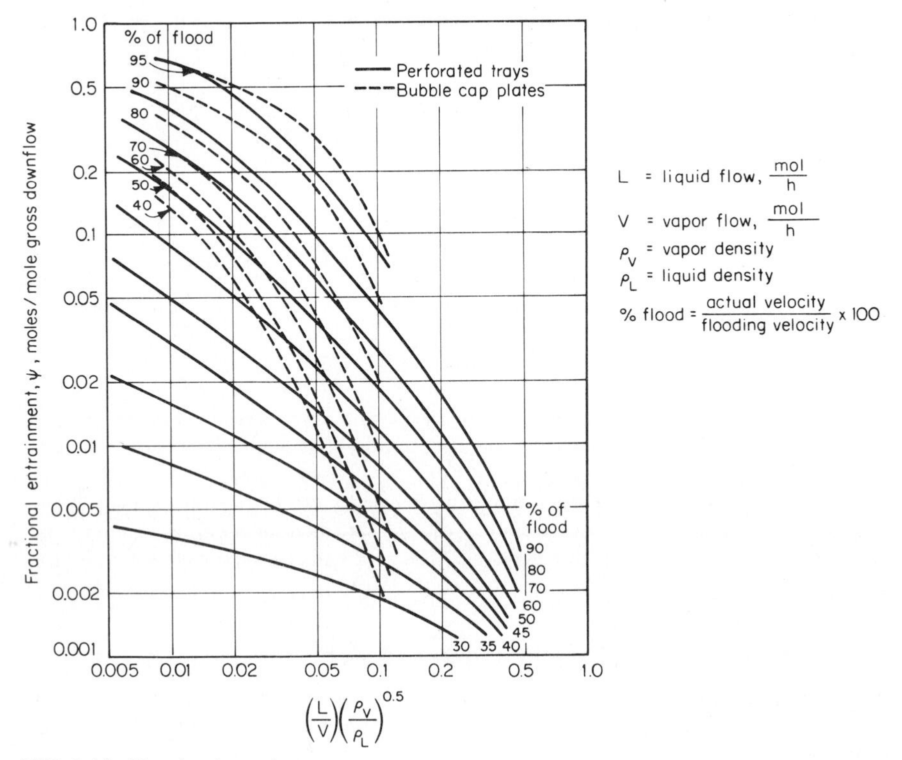

FIG. 8-17 Fractional entrainment.

the lower part of this range in vacuum columns, where entrainment can be a major factor, and in systems where high purities are demanded (<100 ppm of a contaminant). Moderate-pressure systems frequently show higher efficiencies. For instance, the benzene-toluene-xylene system outlined in Example 8-1 has a viscosity of 0.3 cP and a 2.58 relative volatility. From Fig. 8-16, this calls for an efficiency of 71 percent. Little if any correction for entrainment will be necessary, and a final assumed efficiency of 70 percent would not be unreasonable.

Apparent efficiencies for high-pressure systems have frequently been reported in the 90 to 100 percent range.

Weeping is considered excessive and will adversely affect efficiency when the major fraction of the liquid drops through the holes rather than flows over the weir. Figure 8-12 provides a good guide for selecting safe operating conditions of trays; no derating of the basic efficiency is necessary if the operating point falls above the appropriate area-ratio curve.

A correction for dynamic column instability must be made in order to adjust for the continuous shifting of the concentration profile that results from the interaction of controllers with changes in feed flow, cooling water rates, and the ambient temperature. To ensure that product quality will always stay within specifications, it is recommended that 10 percent, but not less than three trays, be added to the calculated number of trays. For instance, recall that the benzene-toluene-xylene system in Example 8-1 requires 25 theoretical stages. With a 70 percent column efficiency, this is now raised to 36 actual trays. Another 4 trays should be added to account for dynamic instability, making a total of 40 installed trays. (Although texts frequently suggest that reboilers and condensers be counted as a theoretical stage, this is, strictly speaking, true only of kettle-type reboilers and partial condensers. Generally, the safest approach in a column design is to ignore both the reboiler and condenser when counting equilibrium stages.)

There may be a drop in efficiency for very large diameter columns (>12 ft) due to their size, unless special jets or baffles are provided to ensure an even flow pattern.

8-4 Packed-Column Design

Specify a packing and the column dimensions for a distillation column separating ethyl benzene and styrene at 1200 mmHg (23.21 psia). The separation requires 30 theoretical stages. Vapor flow is 12,000 lb/h (5455 kg/h), average vapor density is 0.3 lb/ft^3 (4.8 kg/m^3), liquid flow is 10,000 lb/h (4545 kg/h), and average liquid density is 52 lb/ft^3 (833 kg/m^3). Liquid kinematic viscosity is 0.48 cSt (4.8×10^{-7} m^2/s).

Calculation Procedure:

1. Select a type, arrangement, and size of packing.

For this particular system, if one is dealing with a new column, the use of random (dumped) packing is probably a better economic choice than a systematically packed column. Although there is no clear line of demarcation between the two, the latter type is generally favored for very low pressure operations and for expanding the capacity of an existing column.

Somewhat arbitrarily, let us choose a metal slotted ring, say Hy-Pak, as the packing type to be used for this service. Hy-Pak is the Norton Company (Akron, Ohio) version of a slotted ring. There are other packing devices, such as metal saddles and half or pyramidal Pall rings, that would be equally suitable for this service.

A packing size should be selected so that the column-diameter-to-packing-size ratio is greater than 30 for Raschig rings, 15 for ceramic saddles, and 10 for slotted rings or plastic saddles. When dealing with distillation columns larger than 24 in (0.6 m) in diameter, a 2-in, or no. 2, packing should probably be given the first consideration. Assume (essentially based on trial and error) that this column will be larger than 24 in. Therefore, select a 2-in size for the packing.

2. Determine the column diameter.

The generally accepted design procedures for sizing randomly packed columns are modifications of the Sherwood correlation. The most widely applied version is that developed

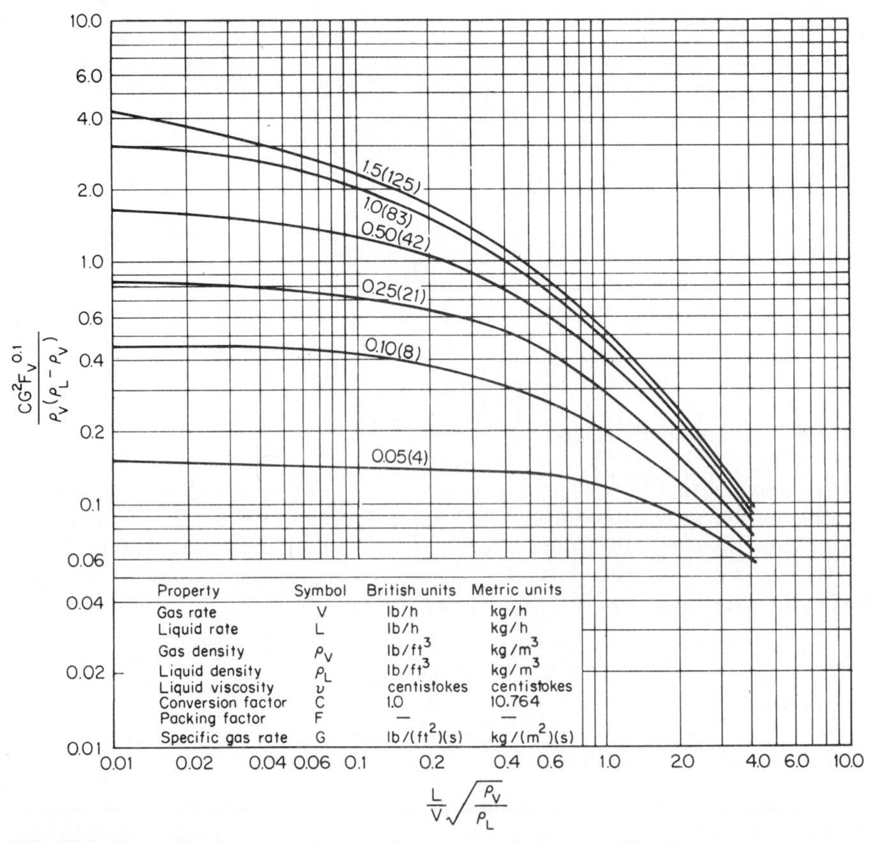

FIG. 8-18 Generalized pressure-drop correlation for packed towers. Note: Parameter of curves is pressure drop in inches of water per foot. Numbers in parentheses are millimeters of water per meter of packed height.

by the Norton Company. It has been adapted slightly for this text to permit its application to low-pressure systems.

Tower diameter is primarily a function of throughput rate and packing configuration. A specific design gas rate G can be determined from Fig. 8-18 if liquid and vapor flows are known and a suitable packing and the proper pressure drop have been selected. Recommended design pressure drops are as follows: 0.4 to 0.75 in water per foot of packing (32 to 63 mm water per meter) for moderate- and high-pressure distillation; 0.1 to 0.2 in water per foot of packing (8 to 16 mm water per meter) for vacuum distillation; and 0.2 to 0.6 in water per foot of packing (16 to 48 mm per meter) for absorbers and strippers.

First, evaluate the abscissa for Fig. 8-18. Thus, $(L/V)(\rho_V/\rho_L)^{0.5} = (10{,}000/12{,}000)(0.3/52)^{0.5} = 0.063$. (Note: ρ_V, especially for vacuum columns, should be determined for the top of the bed, because this is where the density is lowest and the vapor velocity highest.)

Next, select a pressure drop of 0.15 in/ft of packing, and on the figure read 0.55 as the ordinate. Thus, $0.55 = CG^2F\nu^{0.1}/[\rho_V(\rho_L - \rho_V)]$. From Table 8-2, F is 18. Solving for G, $G = [0.55(0.3)(52 - 0.3)/1(18)(0.48)^{0.1}]^{0.5} = 0.71$ lb/(ft^2)(s).

TABLE 8-2 Packing Factors for Column Packings (1 in = 0.0254 m)

Packing type	Material	Nominal packing size, in										
		¼	⅜	½	⅝	¾	1 or no. 1	1¼	1½	2 or no. 2	3	3½ or no. 3
Hy-Pak	Metal						43			18		15
Super Intalox saddles	Ceramic						60			30		
Super Intalox saddles	Plastic						33			21		16
Pall rings	Plastic				97		52		40	24		16
Pall rings	Metal				70		48		33	20		16
Intalox saddles	Ceramic	725	330	200		145	92		52	40	22	
Raschig rings	Ceramic	1600	1000	580	380	255	155	125	95	65	37	
Raschig rings	Metal, 1/32 in	700	390	300	170	155	115					
Raschig rings	Metal, 1/16 in			410	290	220	137	110	83	57	32	
Berl saddles	Ceramic	900		240		170	110		65	45		
Tellerettes	Plastic						38			19		
Mas Pac	Plastic									32		20
Quartz rock										160		
Cross partition	Ceramic										80	
Flexipac	Metal						33			22		16
Interlox	Metal						41			27		18
Chempak	Metal						29					

Note: Many of values are those listed in vendors' literature and are frequently based solely on pilot tests. It may be prudent on occasion to assume slightly larger packing factors in order to represent more precisely the pressure drop of newly marketed packing in commercial columns.

Accordingly, the required column cross-sectional area is (12,000 lb/h)/[3600 s/h)0.71 lb/(ft^2)(s)] = 4.7 ft^2. Finally, column diameter is $[(4.7 \text{ ft}^2)/(\pi/4)]^{0.5} = 2.45$ ft; say, 2.5 ft (0.75 m). (Thus the initial assumption in step 1 that column diameter would be greater than 24 in is valid.)

3. *Determine the column height.*

To translate the 30 required theoretical stages into an actual column height, use the height equivalent to a theoretical plate (HETP) concept. HETP values are remarkably constant for a large number of organic and inorganic systems.

As long as dumped packing in *commercial* columns is properly wetted [more than 1000 lb/(h)(ft^2) or 5000 kg/(h)(m^2)], the following HETP values will result in a workable column:

Nominal packing size, for slotted rings or Intalox saddles	HETP
1 in (or no. 1)	1.5 ft (0.46 m)
1½ in	2.2 ft (0.67 m)
2 in (or no. 2)	3.0 ft (0.91 m)

Because the irrigation rate in vacuum columns often falls below 1000 lb/(h)(ft^2), it may be wise to add another 6 in to the listed HETP values as a safety factor. For the ethylbenzene/styrene system in this example, the specific irrigation rate is (10,000 lb/h)/4.7 ft^2 = 2128 lb/(h)(ft^2), which is well above the minimum required for good wetting. Total packing height is 30(3.0) = 90 ft.

Since liquid maldistribution may become a problem unless the flow is periodically redistributed in a tall column, the total packing height must be broken up into a number of individual beds:

Packing	Maximum bed height
Raschig rings	2.5 to 3.0 bed diameters
Ceramic saddles	5 to 8 bed diameters
Slotted rings and plastic saddles	5 to 10 bed diameters

With 2-in slotted ring packing in a 2.5-ft (0.75-m) column, each bed should be no higher than 25 ft (7.6 m). Therefore, for the given system, four 23-ft (7-m) beds would be appropriate.

Related Calculations: With ceramic packing, the height of a single bed is, for structural reasons, frequently restricted to no more than 20 ft ($\simeq$6 m).

In step 1, the column-diameter-to-packing-size ratio for Raschig rings can be less than 30 for scrubbing applications in which the liquid-irrigation rate is high and the column is operated at above 70 percent of flooding.

If it is desirable to have systematically packed column internals, then the choice is from among various types of mesh pads, open-grid configurations, springs, spirals, and corrugated elements. Since no generalized design correlations can be applied to all configurations, it will be necessary to have the respective manufacturers develop the final packing design.

8-5 Batch Distillation

Establish the separation capability of a single-stage (differential) batch still processing a mixture of two compounds having a relative volatility of 4.0. At the start of the batch separation, there are 600 mol of the more-volatile compound A and 400 mol of compound B in the kettle. When the remaining charge in the kettle is 80 percent B, how much total material has been boiled off, and what is the composition of the accumulated distillate?

Calculation Procedure:

1. Calculate the amount of material boiled off.

Use the integrated form of the Rayleigh equation

$$\ln \frac{L_1}{L_2} = \frac{1}{\alpha - 1}\left(\ln \frac{x_1}{x_2} + \alpha \ln \frac{1 - x_2}{1 - x_1}\right)$$

where L_1 is the amount of initial liquid in the kettle (in moles), L_2 is the amount of final residual liquid (in moles), α is the relative volatility (4.0), x_1 is the initial mole fraction of more-volatile compound in the kettle (0.6), and x_2 is the final mole fraction of more-volatile component in the kettle (0.2). Then,

$$\ln \frac{L_1}{L_2} = \frac{1}{4 - 1}\left(\ln \frac{0.6}{0.2} + 4 \ln \frac{1 - 0.2}{1 - 0.6}\right) = 1.29$$

So, $L_1/L_2 = 3.63$, or $L_2 = 1000/3.63 = 275$ mol. Therefore, the amount of material distilled over is $1000 - 275 = 725$ mol.

2. Calculate the distillate composition.

Compound A in the initial charge consisted of 600 mol. Compound A in the residue amounts to $275(0.2) = 55$ mol. Therefore, compound A in the distillate amounts to $600 - 55 = 545$ mol. Therefore, mole fraction A in the distillate is $545/725 = 0.75$, and mole fraction B is $1 - 0.75 = 0.25$.

Related Calculations: Since a simple batch kettle provides only a single theoretical stage, it is impossible to achieve any reasonable separation unless the magnitude of the relative volatility approaches infinity. This is the case with the removal of very light components of a mixture, particularly of heavy residues. In this example, even with a comfortable relative volatility of 4, it was only possible to increase the concentration of A from 60 to 75% and to strip the residue to 20%.

Obviously, the more distillate that is boiled over, the lower will be the separation efficiency. Conversely, higher overhead concentrations can be obtained at the expense of a larger loss of compound A in the residue. For instance, if distillation in this example is stopped after 50 percent of the charge has been boiled off, the final concentration of A in the kettle x_2 can be obtained as follows:

$$\ln \frac{1000}{500} = \frac{1}{4-1}\left(\ln \frac{0.6}{x_2} + 4 \ln \frac{1-x_2}{1-0.6}\right)$$

from which $x_2 = 0.395$ (by trial-and-error calculation). The amount of A remaining in the residue is 198 mol, making a 40% concentration, and the amount of A in the distillate is 402 mol, making an 80% concentration.

8-6 Batch-Column Design

Estimate the required size of a batch still, with vapor rectification, to recover a dye intermediate from its coproduct and some low- and high-boiling impurities. It has been specified that 13,000 lb (5900 kg), consisting of fresh reactor product and recycled "slop" cuts, must be processed per batch.

The initial composition in the kettle is as follows:

Low-boiling impurities: 500 lb (227 kg)

Dye intermediate: 5500 lb (2500 kg)

Coproduct: 5000 lb (2273 kg)

High-boiling impurities: 2000 lb (909 kg)

Tests in laboratory columns have indicated that to ensure adequate removal of the low-boiling impurities, 500 lb (227 kg) of the dye intermediate is lost in the low boiler's cut. Similarly, high boilers remaining in the kettle at the end of the distillation will retain 500 lb (227 kg) of the coproduct. This leaves 9500 lb (4320 kg) of the two recoverable products. The specification for the dye intermediate requires a concentration of less than 0.5 mol % of the coproduct. Of the two, the coproduct has the higher boiling point.

The conditions for the separation of the dye intermediate from the coproduct are as follows:

Relative volatility $\alpha = 2$

Molecular weight of dye intermediate = 80

Molecular weight of coproduct = 100

Average liquid density = 62.0 lb/ft^3 (993 kg/m^3)

Column pressure (top) = 350 mmHg (46.6 kPa)

Column temperature (top) = 185°F (358 K)

Calculation Procedure:

1. Assess the applicability of a design not based on computer analyses.

The precise design of a batch still is extremely complex because of the transient behavior of the column. Not only do compositions change continuously during the rectification of a charge, but successive batches may start with varying compositions as "slop" cuts and heels are recycled. Only sophisticated computer programs can optimize the size, collection time, and reflux ratios for each cut. In addition to the recycle streams, these programs must also take into account nonideal equilibrium (where applicable) and the effect of holdup on trays. The following is not a detailed design for batch rectification, but instead an outline of how to estimate a "workable" facility with reasonable assurance that it will do the desired job.

Before continuing with a step-by-step procedure, consider these rules of thumb for batch stills:

1. Too low a reflux ratio cannot produce the required product specification no matter how many trays are installed. Conversely, even infinite reflux will not be sufficient if an inadequate number of equilibrium stages has been provided.
2. For optimum separation efficiency, reflux holdup should be minimized by eliminating surge drums and using flow splitters that retain little or no liquid.
3. Too little or too much holdup in the column is detrimental to separation efficiency. A reasonable amount provides a flywheel effect that dampens the effects of equilibrium-condition fluctuations; too much, especially at higher reflux ratios, makes it difficult to achieve good purity levels. A holdup equivalent to 10 or 15 percent of the initial batch charge is recommended.
4. Since the column consists solely of a rectifying section, there is a limit to how many trays can be profitably installed. The system will "pinch" regardless of stages once the low-boiler concentration in the reboiler approaches the intersection of the operating line with the equilibrium curve.
5. Once a workable column has been installed, capacity to produce at a given rate and product specification is only minimally affected by changes in reflux ratio or length of a cut.
6. As the more-volatile component is being removed from the reboiler, separation becomes progressively more difficult.
7. It is impossible to recover in a single operation, at high purity, a low-boiling component that represents only a small fraction of the initial charge.

2. *Set up an estimated batch-processing time schedule.*

A reasonable time schedule is as follows:

Charge new batch into kettle; heat up charge: 3 h

Run column at total reflux to stabilize concentration; distill off the low-boilers cut: 6 h

Recover (i.e., distill off) the dye intermediate while increasing the reflux ratio one or two times: 6 h

Distill off center (slop) cut while further increasing reflux ratio one or two times: 6 h

Distill off coproduct from its mixture with the high boilers while again increasing reflux ratio once (if necessary): 4 h

Drain the high-boiling residue and get ready for next batch; recycle the center cut and prepare to charge fresh feed: 3 h

Total elapsed time: 28 h

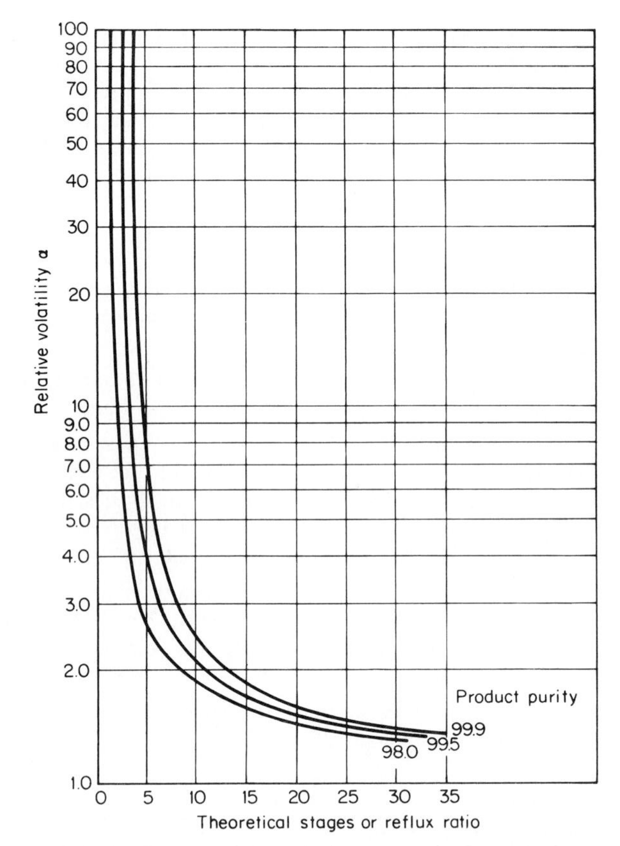

FIG. 8-19 Estimate of theoretical stages, or of reflux ratio, for batch distillation.

3. Estimate the number of theoretical trays needed to recover the dye intermediate.

Enter Fig. 8-19 along its ordinate at a relative volatility of 2. For product purity of 99.5 percent, the graph shows that 11 stages are needed.

4. Draw the relevant xy diagram (equilibrium curve).

The relevant diagram is one that pertains to a relative volatility α of 2. As indicated in Example 8-1, it can be plotted from the equation $y = \alpha x/[1 + (\alpha - 1)x]$. This is the uppermost curve in Fig. 8-20 (including both the full portion at the top of the diagram and the expanded upper-column section at the bottom).

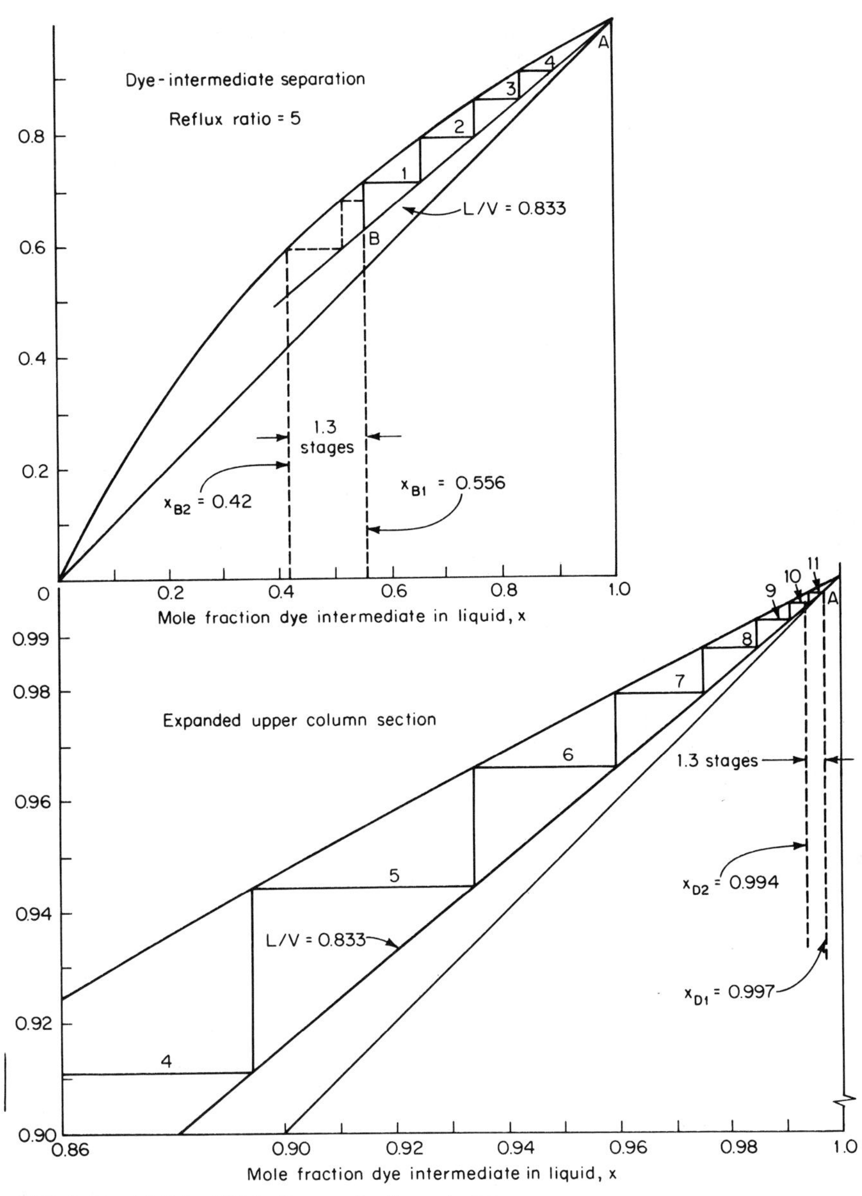

FIG. 8-20 McCabe-Thiele diagram for Example 8-6.

5. ***Applying the McCabe-Thiele principles, position an operating line on the xy diagram that will accommodate 11 stages between the initial kettle composition and the distillate composition that pertain to the time-schedule step in which the dye intermediate is recovered.***

The McCabe-Thiele principles applied here are outlined in more detail in Example 8-1. Using the dye intermediate and the coproduct as the two key components, the initial kettle composition after removal of the low boilers cut (based on the key components only) is $(5000/80)/[(5000/80) + (5000/100)] = 0.556$ mole fraction, or 55.6 mol %, dye intermediate and $(100 - 55.6) = 44.4$ mol % coproduct. The abscissa corresponding to this composition x_{B1} marks one end (the "feed" end) of the 11-stage separation; the other end (the overhead-product end) is marked by the abscissa x_{D1} that corresponds to the required dye-intermediate purity, 99.5%.

By trial-and-error positioning, it is found that an operating line having a slope of 0.833 will accommodate 11 stages (the overhead product corresponding to this line is very slightly purer than 99.5%, namely, 99.7%). In Fig. 8-20, the operating line is line *AB*. By definition (see Example 8-1), its 0.833 slope establishes (and is equal to) the ratio L/V of descending liquid to rising vapor. This in turn establishes the reflux ratio L/D, or $L/(V - L)$, where D is the amount of overhead product taken. Since $L/V = 0.833$, then $L/D = 0.833V/(V - 0.833V) = 5$. Thus, during this period, the column must be operated at a reflux ratio of 5.

6. ***Determine the effect of elapsed time on overhead-product concentration.***

It is reasonable to assume that about 35 to 50 percent of the dye intermediate can be removed efficiently at the initial, relatively low reflux ratio of 5. In this case, assume that 2100 lb of the 5000 lb is thus removed. It is also reasonable to assume that this takes place during half of the 6 h allotted in step 2 for recovering this intermediate. (Note that the shorter the allotted time, the bigger the required column diameter.)

Therefore, $D = 2100/3 = 700$ lb/h. Because $L/D = 5$, $L = 3500$ lb/h. And since $D = V - L$, $V = 4200$ lb/h. After 3 h (assuming negligible column holdup), the kettle will contain $(5000 - 2100)/80 = 36.25$ mol dye intermediate. Since the overhead stream contains (initially, at least) only 0.3% coproduct, the amount of coproduct in the kettle after 3 h is about $[5000 - 0.005(2100)]/100 = 49.9$ mol. The mole percentages thus are 42% intermediate and 58% coproduct. If the abscissa corresponding to this new kettle composition x_{B2} is extended upward to the operating line and equilibrium curve, it can be seen that this "feed" composition has shifted downward by 1.3 stages. This will likewise lower the overhead (distillate) composition by 1.3 stages, which means that the concentration of dye intermediate drops to 99.4% after 2100 lb of distillate has boiled over and been collected. Average concentration during this period of operating at a reflux ratio of 5 is thus $(99.7 + 99.4)/2 = 99.55$ percent, just slightly above the specified purity.

7. ***Adjust the reflux ratio so as to maintain the required overhead-product composition.***

It is necessary to raise the reflux ratio in order to keep the concentration of dye intermediate in the overhead product high enough. In this case, assume that an additional 1400 lb intermediate can be recovered in the remaining 3 h while maintaining the original vapor rate of 4200 lb/h. Then D for this latter 3 h is 1400/3 = 467 lb/h; L is (4200 − 467) = 3733 lb/h; and the reflux ratio L/D for this portion of operation must be 3733/467 = 8.

The slope L/V of the new operating line is 3733/4200 = 0.89. Drawing this line on the replotted xy diagram in Fig. 8-21 and stepping off 11 stages, we find that the initial overhead concentration x_{D3} while operating at the new reflux ratio is 99.7% dye intermediate. (Note that x_{B3} in Fig. 8-21 is a deliberate repetition of x_{B2} in Fig. 8-20, whereas x_{D3} is not a deliberate repetition but instead a coincidental result that emerges from the stepping off of the stages.)

8. *Again determine the effect of elapsed time on overhead-product composition and readjust the reflux ratio of step 7 if necessary.*

It follows from the preceding two steps that after the second 3 h, the kettle will contain (2900 − 1400)/80 = 18.75 mol dye intermediate and about [5000 − 0.005(3500)]/100 = 49.8 mol coproduct. Thus the mole percentages will be 27% intermediate and 73% coproduct. The abscissa x_{B4} corresponding to this in Fig. 8-21 indicates that the composition will have shifted downward about 1.6 stages. This in turn means that the distillate composition will have dropped 1.6 stages during the 3 h, to a composition x_{D4} of 99.2% dye intermediate. Average concentration while operating at a reflux ratio of 8 is therefore (99.7 + 99.2)/2 = 99.45 percent.

Since this average concentration is slightly below the required purity of 99.5%, it would be wise to recalculate step 7 at a higher reflux ratio with a correspondingly longer time for removing the same 1400 lb. A 4-h removal time, corresponding to a reflux ratio of 11, should be more than adequate.

9. *Determine the column diameter.*

The column diameter can be determined by taking the boil-up rate established for the separation of components having the smallest relative volatility (which, in this case, is given to be the dye-intermediate/coproduct separation) and using the design procedures outlined in Examples 8-2 and 8-4.

For a 4200 lb/h boil-up rate (from step 6) of dye intermediate having a density ρ_V of (80 lb/mol)(6.77 lb/in²)/[10.73(460 + 185R)] = 0.078 lb/ft³ (where 10.73 is the gas constant), the total vapor flow rate is 4200/[0.078(3600)] = 14.9 ft³/s. A tray spacing of 18 in seems appropriate. The column diameter required for an 18-in tray spacing can be obtained with the aid of Fig. 8-6, where F_c is found to be about 1.4 $[\text{lb}/(\text{ft}\cdot\text{s}^2)]^{0.5}$. Then free tray area $A_F = W/(F_c\rho_V^{0.5})$ = (4200 lb/h)/[3600 s/h)(1.4)(0.078 lb/ft³)$^{0.5}$] = 2.98 ft². Allowing 5 percent of the tray area for segmental downcomers, total tray area is 2.98/0.95 = 3.14 ft². Finally, column diameter is $[3.14/(\pi/4)]^{0.5}$ = 2 ft (0.41 m).

10. *Determine the column height.*

Efficiencies of batch columns vary greatly, since the concentration profile in the column shifts over a wide range. For appreciable intervals, separation may of necessity take place under pinched conditions, conducive to low efficiencies. So an overall column efficiency of 50 percent is not unreasonable. Thus the total number of trays provided is (11 stages)/

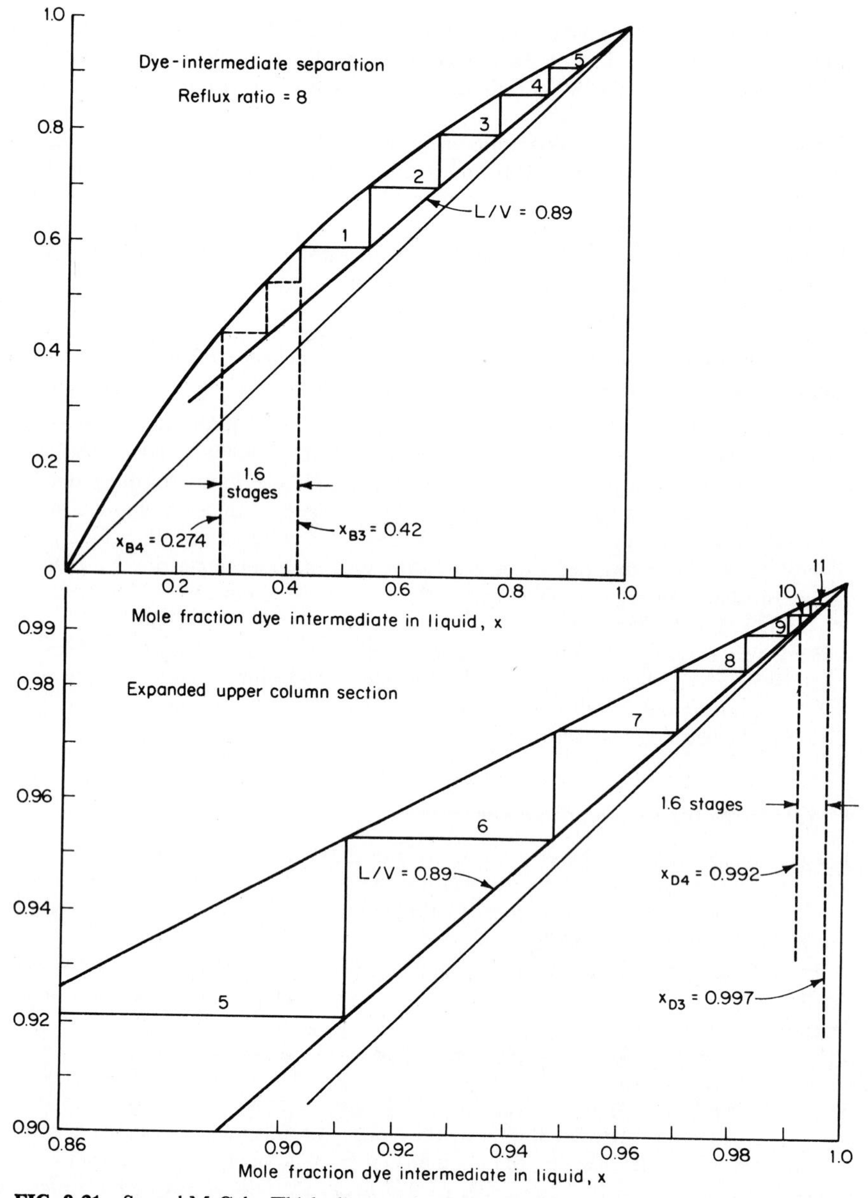

FIG. 8-21 Second McCabe-Thiele diagram for Example 8-6.

0.5 = 22 trays. At a spacing of 18 in, the required column height is 33 ft (10 m). The trays should be cartridge-type trays, because normal trays cannot be readily installed in a 2-ft column.

Related Calculations: Alternatively, a packed column can be considered for this separation. Its design should follow the procedure outlined in Example 8-4.

If the coproduct must also be recovered at a reasonably high purity, then steps 5 through 8 should also be repeated for distilling off the slop cut (essentially a mixture of dye intermediate and coproduct) and for the coproduct cut (a mixture of coproduct and high-boiling residue).

In the final analysis, processing time is the main criterion of batch-still design. To achieve optimal cost-effective performance requires a large number of trial calculations, such that the best combination of equilibrium stages, reflux ratio, batch size, and batch-processing time can be established. It is extremely difficult to successfully carry out such a procedure by hand calculation.

8-7 Overall Column Selection and Design

Select and specify an efficient distillation column to separate ethylbenzene (EB) and ethyl cyclohexane (ECH), and develop the appropriate heat and material balances. Feed rate is 10 lb·mol/h (4.54 kg·mol/h): 75 mol % ECH and 25 mol % EB. Concentration of EB in the overhead product must be less than 0.1%; concentration of ECH in the bottoms stream must be less than 5%. Feed is at ambient temperature (25°C). Specific heat of the distillate and bottoms streams can be taken as 0.39 and 0.45 Btu/(lb)(°F) [1.63 and 1.89 kJ/(kg)(K)], respectively. The normal boiling points for EB and ECH are 136.19°C and 131.78°C, respectively; their latent heats of vaporization can be taken as 153 and 147 Btu/lb (356 and 342 kJ/kg), respectively. It can be assumed that EB and ECH form an ideal mixture.

Calculation Procedure:

1. Make a rough estimate of the number of theoretical stages required.

This step employs the procedures developed in Example 8-1. Use the Fenske-Underwood-Gilliland approach rather than the McCabe-Thiele, because the small boiling-point difference indicates that a large number of stages will be needed.

The appropriate vapor pressures can be obtained from the Antoine equation:

$$\log P = A - \frac{B}{t + C}$$

where P is vapor pressure (in mmHg), t is temperature (in °C), and the other letters are the Antoine constants, which in this case are

	A	B	C
Ethylbenzene (EB)	6.96	1424	213
Ethyl cyclohexane (ECH)	6.87	1384	215

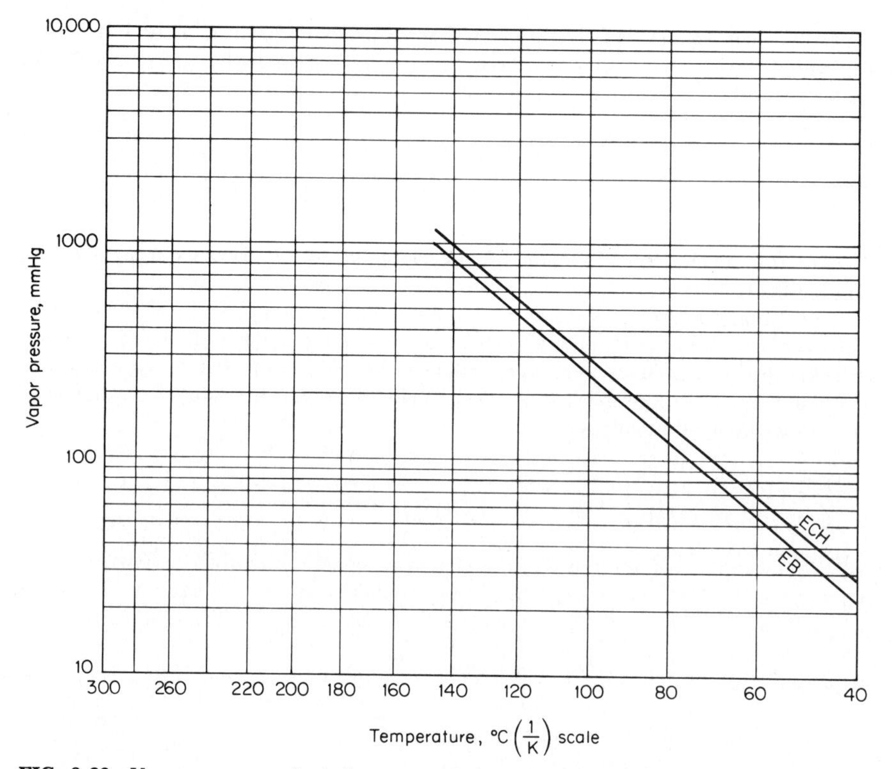

FIG. 8-22 Vapor pressures of ethylbenzene and ethyl cyclohexane (Example 8-7). Note: 1 mmHg = 0.133 kPa.

The resulting vapor-pressure plots are shown in Fig. 8-22.

Since the system is ideal, relative volatility α can be determined from the vapor-pressure ratio P_{ECH}/P_{EB}. If this is determined at, say, 50, 100, and 135°C, the effect of temperature (and thus, of system pressure) on relative volatility can be gaged as follows:

Temperature, °C (°F)	P_{EB}, mmHg	P_{ECH}, mmHg	α
50 (122)	35.1	44.4	1.265
100 (212)	257	300	1.167
135 (275)	738	824	1.117

There is a noticeable increase in the relative volatility as the system temperature and pressure are lowered. In the case of two closely boiling liquids, as is the present case, this can mean a substantial lessening of required equilibrium stages. This can be seen by employing the Fenske-Underwood-Gilliland correlations as expressed graphically in Fig. 8-5. The abscissa is (0.999/0.001)/(0.95/0.05) = 19,000, and assuming an R/R_{min} of 1.3, an α of 1.265 calls for 86 theoretical stages (as read along the ordinate), whereas an α of 1.117 calls for 176 stages (by extrapolating the ordinate downward). However, this comparison assumes an isobaric column. In actual practice, the bottoms pressures will be considerably higher, so the difference in stages, although substantial, will not be quite so dramatic. Depending on column pressures, it seems reasonable to assume that roughly 100 stages will be needed.

2. *Evaluate the effects of choosing trays, random packing, or systematic packing.*

It is recommended that the top column pressure be set for the lowest reasonable overhead pressure consistent with the use of a water- or air-cooled condenser. In this case, pressure would be 50 mmHg, corresponding to a temperature of 53°C (127°F).

The bottoms pressure is usually selected to permit use of a readily available heating medium (steam or hot oil), as well as to stay below a temperature that could cause product degradation. In the ECH-EB system, degradation is not considered a problem, and column bottoms pressure is solely a function of the pressure drop across the tower internals. Because, as seen in step 1, relative volatility can vary appreciably with pressure, it is advantageous in this case to install low-pressure-drop, high-efficiency tower internals.

***a.** Evaluation of cross-flow trays.* Assume as a first try that 110 equilibrium stages are needed. Assume further that a tray column in this service will operate at 70 percent efficiency. Then the actual number of trays needed is 110/0.7 = 157.

It is reasonable to allow a pressure drop of 3 mmHg per tray. Then the reboiler pressure will be 50 + 3(157) = 521 mmHg. At this pressure, the relative volatility (from Fig. 8-22) is 1.14, and the average relative volatility in the column is then (see step 1) (1.265 + 1.14)/2 = 1.2. From Fig. 8-5, the estimated number of equilibrium stages is 107, which confirms that the initially selected 110 trays was a reasonable assumption.

If the upper 25 percent of the trays (where the vapor velocity and therefore entrainment are highest) are spaced 24 in apart and the distance between the remaining trays is 18 in, then the total equipment height can be estimated as follows:

	ft (m)
Total height of trays: 40(2 ft) + 117(1.5 ft) =	256
Overhead disengaging area	4
Manways: 8(1.5 ft) =	12
Reboiler	6
Total column height	278
Skirt (minimum to ensure adequate pump suction head)	12
Total equipment height	290 (90)

***b.** Evaluation of random packing.* This step draws on information brought out in Example 8-4. Again, recognizing that relative volatility decreases with pressure, estimate the actual number of equilibrium stages and check the assumption from Fig. 8-5. If no. 2 slotted rings are selected, the pressure drop should be set at 0.15 in water per foot (12 mm water per meter), and the height equivalent to a theoretical-plate (or stage) HETP may be as large as 3.5 ft (*ca.* 1 m). For dumped packing, assume that the specified separation requires 100 stages. Then the total height of packing needed is 100(3.5 ft) = 350 ft, requiring at least 10 separate beds.

The pressure in the column reboiler can now be developed:

	mmHg
Pressure drop through packing (no. 2 slotted rings):	
(350 ft)(0.15 in water per foot) = 52.5 in water =	98
Pressure drop through 10 support plates	10
Pressure drop through 10 distributors	10
Column top pressure	50
Total reboiler pressure	168

Again using Fig. 8-22, the relative volatility at the bottom is found to be 1.16, resulting in an average of 1.21. Again from Fig. 8-5 we find that the assumption of 100 stages is reasonable.

Now calculate the column height:

	ft (m)
Total height of packing	350
Spacing between beds: (9)(4 ft) =	36
Disengaging and distribution	6
Reboiler	6
Total column height	398
Skirt	12
Total equipment height	410 (125)

c. Evaluation of a systematically packed column. One example of a high-efficiency packing (large number of stages per unit of pressure drop) is corrugated-wire-gauze elements, sections of which are assembled inside the tower. This type of packing has an HETP of 10 to 12 in (25 to 30 cm) and a pressure drop of 0.3 to 0.5 mmHg (53 Pa or 0.22 in water) per equilibrium stage. Assume that 95 stages are needed, each entailing 12 in of packing. Then the pressure at the bottom of 95 ft of packing can be determined as follows:

	mmHg
Pressure drop through wire-gauze packing: (95)(0.4 mmHg/stage) =	38
Pressure drop through four distributors and four packing supports	8
Column top pressure	50
Total reboiler pressure	96

The average relative volatility for the column is 1.22, and the assumption of 95 plates is reasonable.

The next step is to calculate the column height:

	ft (m)
Total height of packing	95
Spacing between beds: (3)(4 ft) =	12
Disengaging and distribution	6
Reboiler	6
Total column height	119
Skirt	12
Total equipment height	131 (40)

3. *Choose between the tray column, the randomly packed column, and the systematically packed column.*

It is unlikely that a single 300-ft tray column or a 400-ft randomly packed column would be installed for this system. In addition, multiple-column operation, by its nature, is expensive. So it is reasonable to opt for the 131-ft column containing systematically packed corrugated-wire-gauze packing. However, this type of packing is quite expensive. So a detailed economic analysis should be performed before a final decision is made.

4. *Estimate the column diameter.*

This step draws on information brought out in Example 8-2. If corrugated-wire-gauze packing is the final choice, the column size can be estimated by applying an appropriate F_c factor. The usual range for this packing is 1.6 to 1.9 $[\text{lb}/(\text{ft})(\text{s}^2)]^{0.5}$, with 1.7 being a good design point.

In order to establish the column vapor rate, it is necessary first to determine the required reflux ratio and then to set up a column material balance.

Using the Underwood equation (see Example 8-1) and noting from Fig. 8-3 that θ = 1.08,

$$(L/D)_{\min} = 1.22(0.999)/(1.22 - 1.08) + 1(0.001)/(1 - 1.08) - 1$$
$$= 7.69$$

With an arbitrarily selected $R/R_{\min}$ ratio of 1.3, then, the reflux ratio $L/D = 7.69(1.3) = 10$.

Material-balance calculations (see Sec. 2) indicate that the withdrawal rate for overhead product from the system is 7.38 mol/h; therefore, 73.8 mol/h must be refluxed, and the vapor flow from the top of the column must be 73.8 + 7.38 = 81.18 mol/h. The vapor consists of almost pure ECH, whose molecular weight is 112, so mass flow from the top is 81.18(112) = 9092 lb/h (4133 kg/h). The bottoms stream is 2.62 mol/h.

Take the average pressure in the upper part of the column to be 55 mmHg (1.06 psia). Assuming that this consists essentially of ECH and referring to Fig. 8-22, the corresponding temperature is 55°C (131°F). Then, by the ideal-gas law, vapor density $\rho_V = PM/RT = 1.06(112)/[10.73(460 + 131)] = 0.0187\ \text{lb/ft}^3$, where P is pressure, M is molecular weight, R is the gas-law constant, and T is absolute temperature. (Note: For high-pressure systems of greater than 5 atm, correlations for vapor-phase nonideality should be used instead; see Sec. 3.) Then, as discussed in Example 8-2, design vapor velocity V

$= F_c/\rho^{0.5} = 1.7/0.0187^{0.5} = 12.43$ ft/s. The volumetric flow rate is (9092 lb/h)/[(3600 s/h)(0.0187 lb/ft^3)] = 135 ft^3/s. Column cross-sectional area A, then, must be 135/12.43 = 10.9 ft^2, and column diameter must be $[A/(\pi/4)]^{0.5} = (10.9/0.785)^{0.5} =$ 3.73 ft (1.1 m).

5. *Calculate the heat duty of the reboiler and of the condenser.*

Assembly of a heat balance can be simplified by assuming constant molal overflow (not unreasonable in this case) and no subcooling in the condenser (usually, less than 5°C of cooling takes place in a well-designed heat exchanger). For constant molal overflow, the latent heats may be averaged, yielding 150 Btu/lb.

The reboiler duty is the sum of three parts:

1. Enthalpy to heat distillate (consisting essentially of ECH) from 25 to 53°C:

$$(7.38 \text{ lb}\cdot\text{mol/h})(112 \text{ lb/mol}) \times (53 - 25°\text{C})(1.8°\text{F}/°\text{C})[0.39 \text{ Btu/(lb)(°F)}] = 16{,}240 \text{ Btu/h}$$

2. Enthalpy to heat bottoms (essentially EB) from 25 to 72°C:

$$(2.62 \text{ lb}\cdot\text{mol/h})(106 \text{ lb/mol}) \times (72 - 25°\text{C})(1.8°\text{F}/°\text{C})[0.45 \text{ Btu/(lb)(°F)}] = 10{,}570 \text{ Btu/h}$$

3. Enthalpy to boil up the vapor:

$$(81.18 \text{ lb}\cdot\text{mol/h})(106 \text{ lb/mol})(153 \text{ Btu/lb}) = 1{,}316{,}580 \text{ Btu/h}$$

Therefore, total reboiler duty is 16,240 + 10,570 + 1,316,580 = 1,343,390 Btu/h (393,310 W). As for the condenser duty, it is (81.18 lb·mol/h)(112 lb/mol)(147 Btu/lb) = 1,336,550 Btu/h (391,350 W).

9

Extraction and Leaching

Frank H. Verhoff, Ph.D.

Director of Chemical Engineering Research
Miles Laboratories, Inc.
Elkhart, IN

9-1 Multistage Countercurrent Liquid-Liquid Extraction

Alcohol is to be extracted from an aqueous solution by pure ether in an extraction column. The alcohol solution, containing 30% alcohol by weight, enters the top of the column at a rate of 370 kg/h. The ether is to be fed to the column bottom at 350 kg/h. About 90 percent of the alcohol is to be extracted; that is, alcohol concentration in the exiting aqueous stream should be about 3%. Experimental data on the compositions of pairs of water-rich and ether-rich phases in equilibrium are given in Table 9-1. Calculate the flow rates and compositions of the exiting raffinate (i.e., alcohol-depleted, aqueous) phase and the extract (i.e., alcohol-enriched, ether) phase. Also calculate the number of extraction stages needed.

Calculation Procedure:

1. Plot the equilibrium-composition data on a right-triangular diagram.

The plot, shown in Fig. 9-1, is prepared as follows: Let the vertex labeled E represent 100% ether, let the one labeled A represent 100% alcohol, and let the one labeled W

TABLE 9-1 Equilibrium Data for Alcohol-Water-Ether System (Example 9-1)

Weight fraction in phase					
Water phase			Ether phase		
Alcohol	Ether	Water	Alcohol	Water	Ether
0.0	0.075	0.925	0.0	0.225	0.775
0.1	0.077	0.823	0.090	0.170	0.740
0.2	0.090	0.710	0.175	0.120	0.705
0.31	0.095	0.595	0.250	0.080	0.670
0.44	0.118	0.442	0.290	0.05	0.660
0.530	0.150	0.320	0.31	0.035	0.655
0.645	0.195	0.160	0.33	0.019	0.651
0.75	0.25	0.0	0.35	0.0	0.65

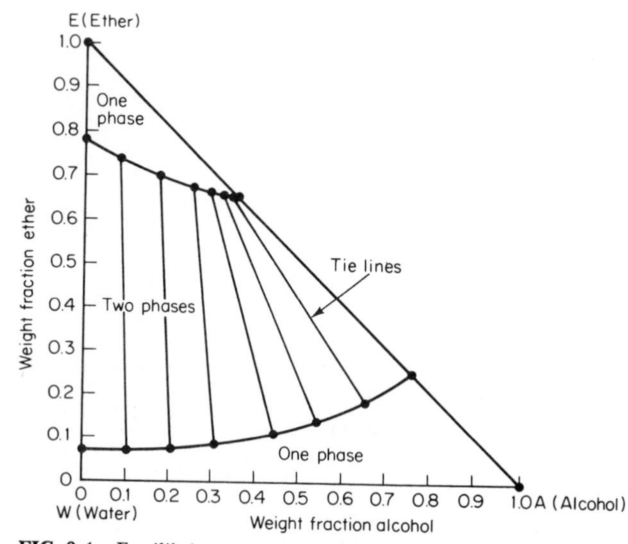

FIG. 9-1 Equilibrium phase diagram for alcohol, water, ether system (Example 9-1).

represent 100% water. Then the scale along the abscissa represents the weight fraction alcohol, and the scale along the ordinate represents the weight fraction ether. Take each pair of points in Table 9-1 and plot them, joining any given pair by a straight line, called a "tie line." (Note that for the first pair, the tie line coincides with part of the ordinate, since the alcohol concentration is 0.0 in each of the two phases.) Draw a curve through points at the lower ends of the tie lines and another curve through the points at the upper ends.

These two curves divide the diagram into three regions. Any composition falling within the uppermost region will consist solely of an ether-rich phase; any composition within the lowermost region will consist solely of an aqueous phase; any composition within the middle region will constitute a combination of those two liquid phases.

2. *Calculate the mean concentrations of ether, alcohol, and water within the system.*

This step is easier to visualize if a stage diagram is first drawn (see Fig. 9-2). Each box represents an extraction stage, with stage 1 being the stage at the top of the column. Stage N is at the bottom of the column. The streams labeled L are the aqueous-phase streams; those labeled S are the ether-phase (i.e., solvent-phase) streams. Let the given letter stand for the actual flow rates.

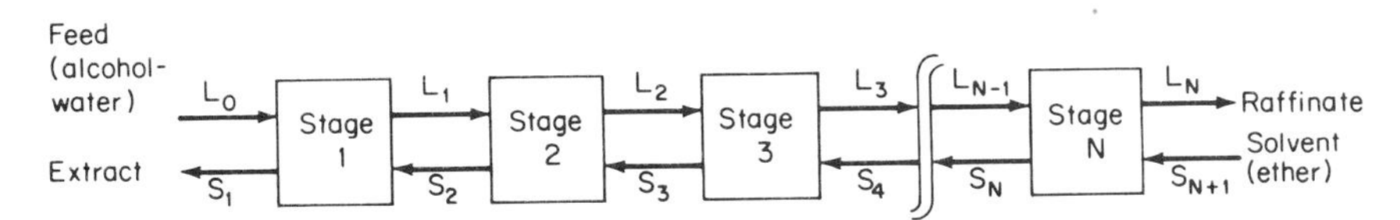

FIG. 9-2 Stage diagram for countercurrent multistage liquid-liquid extraction (Example 9-1).

Thus the alcohol-water feed mixture L_0 enters the top of the column, where the ether-solvent-phase stream S_1 leaves. At the bottom of the column, ether-solvent stream S_{N+1} enters, while the stripped water stream L_N leaves.

Now, the mean concentration of ether W_{EM} and alcohol W_{AM} in the system can be calculated from the equations

$$W_{EM} = \frac{L_0 W_{E,0} + S_{N+1} e_{E,N+1}}{L_0 + S_{N+1}}$$

$$W_{AM} = \frac{L_0 W_{A,0} + S_{N+1} e_{A,N+1}}{L_0 + S_{N+1}}$$

where L_0 is the feed rate of the water phase, S_{N+1} is the feed rate of the solvent, $W_{E,0}$ and $W_{A,0}$ are the concentrations of ether and alcohol, respectively, in the entering water phase, and $e_{E,N+1}$ and $e_{A,N+1}$ are the concentrations of ether and alcohol, respectively, in the entering ether (solvent) phase. Then

$$W_{EM} = \frac{370(0) + 350(1.0)}{370 + 350} = 0.49$$

$$W_{AM} = \frac{370(0.3) + 350(0)}{370 + 350} = 0.154$$

And by difference, the mean concentration of water is $(1.0 - 0.49 - 0.154) = 0.356$.

3. *Calculate the compositions of the exiting raffinate and extract streams.*

Replot (or trace) Fig. 9-1 without the tie lines (Fig. 9-3). On Fig. 9-3, plot the mean-concentration point M. Now from mass-balance considerations, the exit concentrations must lie on the two phase-boundary lines and on a straight line passing through the mean concentration point. We know we want the water phase to have an exit concentration of 3 wt % alcohol. Such a concentration corresponds to point L_N on the graph. At point L_N, the ether concentration is seen to be 7.6 wt % (this can be found more accurately in the present case by numerical extrapolation of the water-phase data in Table 9-1). Therefore,

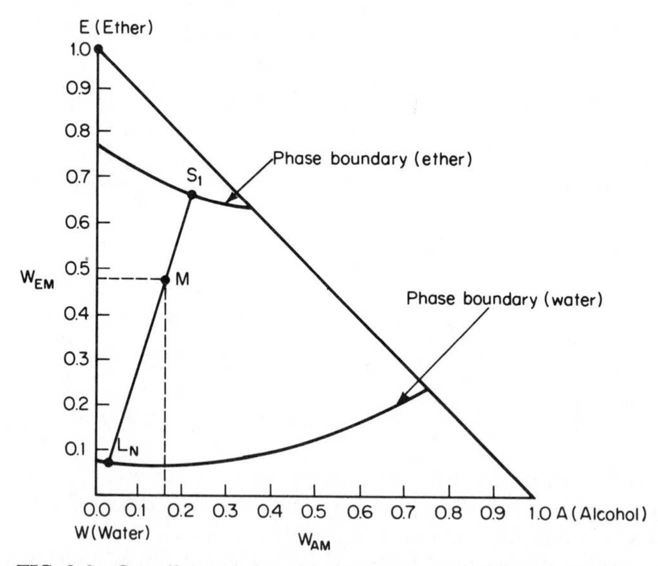

FIG. 9-3 Overall mass balance on extraction column (Example 9-1).

the composition of the raffinate stream is 3% alcohol, 7.6% ether, and (by difference) 89.4% water.

To determine the composition of the exiting extract stream, extend line $L_N M$ until it intersects the ether-phase boundary, at point S_1, and read the composition graphically. It is found to be 69% ether, 22% alcohol, and (by difference) 9% water.

4. *Calculate the flow rates of the exiting raffinate and extract streams.*

Set up an overall mass balance:

$$S_1 + L_N = L_0 + S_{N+1} = 370 + 350 = 720$$

where (see Fig. 9-2) L_0 and S_{N+1} are the entering streams (see step 2), and S_1 and L_N are the exiting extract and raffinate streams, respectively.

Now set up a mass balance for the ether, using the compositions found in step 3:

$$S_1(0.69) + L_N(0.076) = 370(0) + 350(1.0) = 350$$

Solving the overall and the ether balances simultaneously shows the exiting raffinate stream L_N to be 239 kg/h and the exiting extract stream S_1 to be 481 kg/h.

5. *Calculate the number of stages needed.*

The number of stages can be found graphically on Fig. 9-1 (repeated for convenience as Fig. 9-4), including the tie lines. Plot on the graph the points corresponding to L_0 (30% alcohol, 0% ether) and S_1 (69% ether, 22% alcohol), draw a straight line through them, and extend the line upward. Similarly, plot L_N (3% alcohol, 7.6% ether) and S_{N+1} (100% ether), and draw a line through them, extending it upward. Designate the intersection of the two lines as Δ.

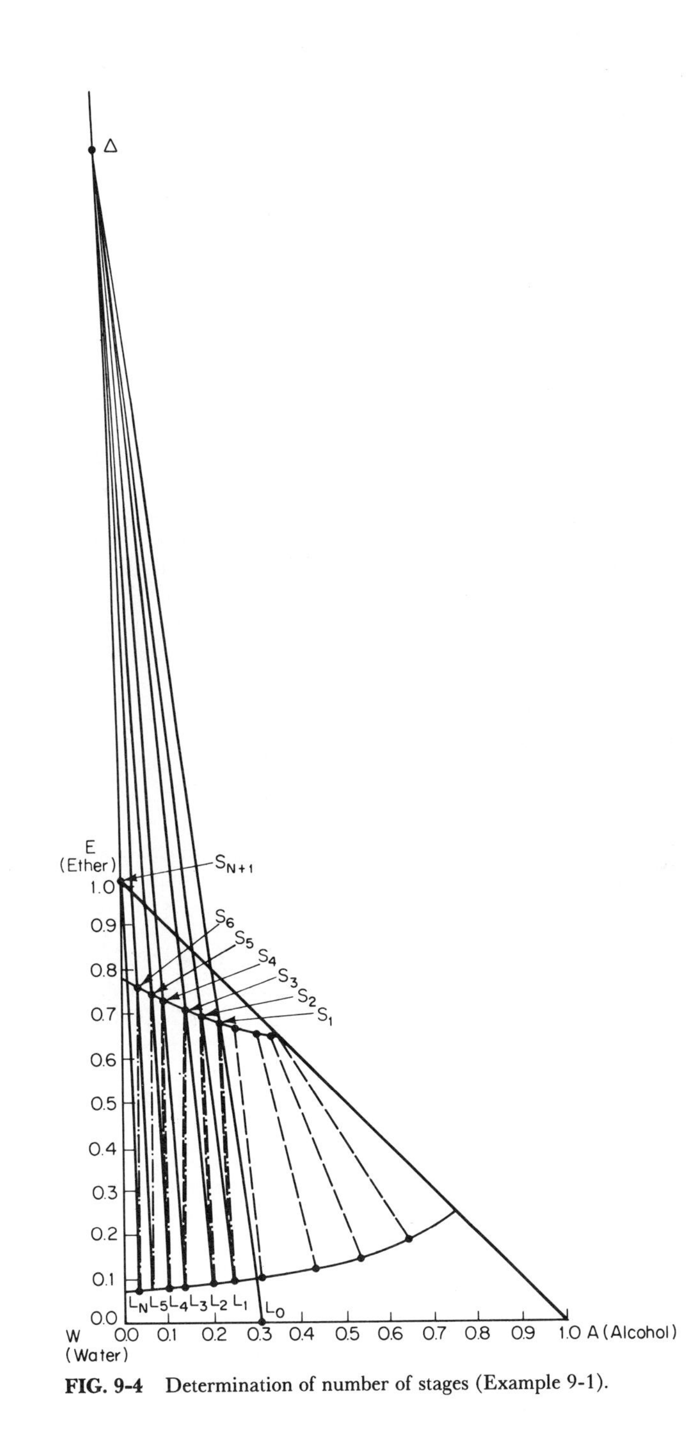

FIG. 9-4 Determination of number of stages (Example 9-1).

Now, begin to step off the stages by drawing an interpolated tie line from point S_1 down to the water-phase boundary. Designate their intersection as point L_1. Draw a line joining L_1 with Δ, and label the intersection of this line with the ether-phase boundary as S_2. Then repeat the procedure, drawing a tie line through S_2 and intersecting the water-phase boundary at L_2. This sequence is repeated until a point is reached on the water-phase boundary that has an alcohol content less than that of L_N. Count the number of steps involved. In this case, 6 steps are required. Thus the extraction operation requires 6 stages.

Related Calculations: This basic calculation procedure can be extended to the case of countercurrent multistage extraction with reflux. A schematic of the basic extractor is shown in Fig. 9-5. For this extractor there are N stages in the extracting section, $1E$ to NE, and there are M stages in the stripping section, $1S$ to MS.

The overall mass balance on the entire extractor as well as the mass balances on the solvent separator, feed stage, and solvent mixer are performed separately. The calculations on the extraction section and the stripping sections are then performed as described above.

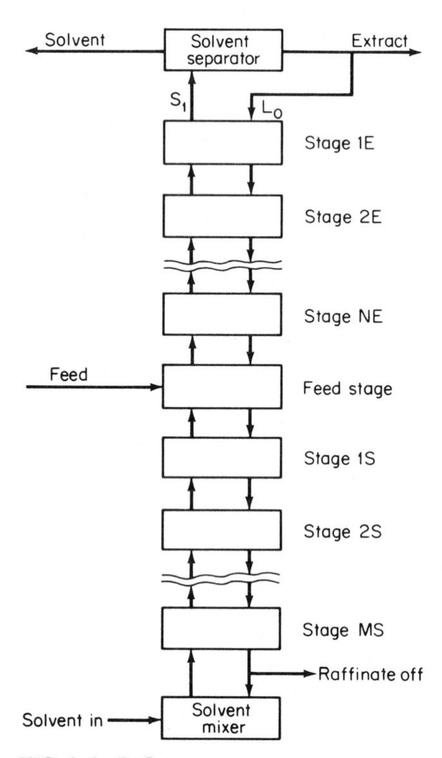

FIG. 9-5 Reflux extractor.

9-2 Multistage Countercurrent Leaching

Hot water is to be used to leach a protein out of seaweed in an isothermal multistage countercurrent system, as shown in Fig. 9-6. The seaweed slurry, consisting of 48.1% solids, 2.9% protein, and 49% water, enters at a rate of 400 kg/h. The hot water is fed at a rate of 500 kg/h. It is desired to have the outlet underflow (the spent seaweed) have a maximum residual concentration of 0.2% protein on a solids-free basis. Table 9-2 shows experimental data for the operation, taken by (1) contacting the seaweed with hot water for a period of time with mixing, (2) stopping the mixing and letting the seaweed settle, and (3) sampling the bottom slurry (underflow) phase and the upper extract (overflow) phase. Calculate the number of equilibrium leaching stages needed.

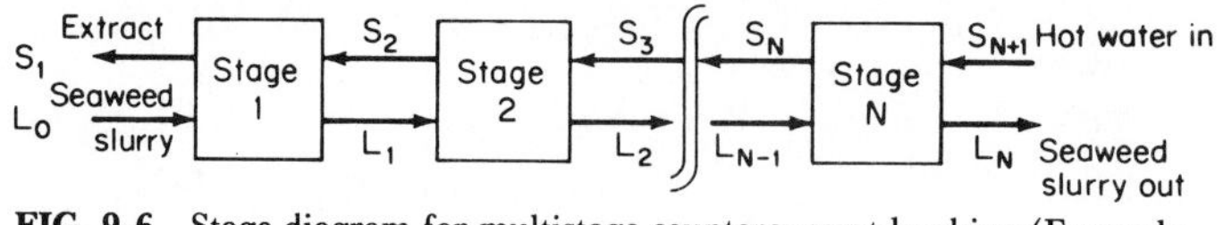

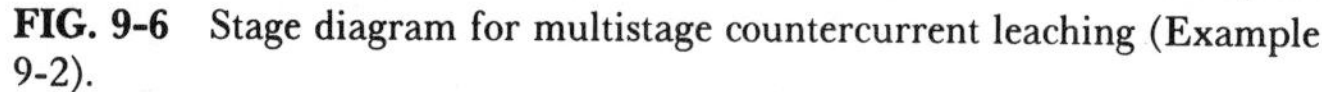

FIG. 9-6 Stage diagram for multistage countercurrent leaching (Example 9-2).

TABLE 9-2 Equilibrium Data for Seaweed-Water System (Example 9-2)

	Weight fraction in phase					
	Extract (overflow) phase			Slurry (underflow) phase		
Run number	Water	Solution protein	Solids	Water	Solution protein	Solids
1	0.952	0.046	0.002	0.542	0.026	0.432
2	0.967	0.032	0.001	0.564	0.019	0.417
3	0.979	0.021	0.00	0.586	0.013	0.401
4	0.989	0.011	0.0	0.5954	0.0066	0.398
5	0.994	0.006	0.0	0.5994	0.0036	0.397
6	0.998	0.002	0.0	0.6028	0.0012	0.396

Calculation Procedure:

1. Calculate the weight ratio of solids to total liquid and the weight concentration of protein on a solids-free basis for each of the two phases in each run of the experimental data.

Let N_o and N_u designate the weight ratio of solids to total liquid in the overflow and underflow phases, respectively. Similarly, let x_o and x_u represent the weight concentration of protein on a solids-free basis in the overflow and underflow phases, respectively. For run 1, for instance,

$$N_o = \frac{0.002}{0.952 + 0.046} = 0.002$$

$$N_u = \frac{0.432}{0.542 + 0.026} = 0.760$$

$$x_o = \frac{0.046}{0.952 + 0.046} = 0.046$$

$$x_u = \frac{0.026}{0.542 + 0.026} = 0.046$$

The calculated results for all runs are as follows (the sequence being rearranged for convenience in step 2):

Run number	x_u	N_u	x_o	N_o
1	0.046	0.760	0.046	0.002
2	0.032	0.715	0.032	0.001
3	0.022	0.669	0.021	0
4	0.011	0.661	0.011	0
5	0.006	0.658	0.006	0
6	0.002	0.656	0.002	0

2. Plot the equilibrium data.

See step 2 in Example 9-1. In the present case, however, a rectangular rather than a triangular diagram is employed. The abscissa is x, from above; similarly, the ordinate is

N. For each run, plot the points x_o,N_o and x_u,N_u and join the pair by a tie line. Then connect the x_o,N_o points, thereby generating the overflow equilibrium line, and similarly, connect the x_u,N_u points to generate the underflow equilibrium line. The result is shown as Fig. 9-7.

3. *Calculate the mean values for x and N.*

Refer to step 2 in Example 9-1. In the present case, the mean values can be found from the equations

$$x_M = \frac{L_0 W_{P,0} + S_{N+1} e_{P,N+1}}{L_0(W_{P,0} + W_{W,0}) + S_{N+1}(e_{P,N+1} + e_{W,N+1})}$$

and

$$N_M = \frac{L_0 W_{S,0} + S_{N+1} e_{S,N+1}}{L_0(W_{P,0} + W_{W,0}) + S_{N+1}(e_{P,N+1} + e_{W,N+1})}$$

where x_M is the mean weight concentration of protein on a solids-free basis, N_M is the mean weight ratio of solids to total liquid, L_0 is the feed rate of the seawater slurry, $W_{P,0}$, $W_{W,0}$, and $W_{S,0}$ are the weight fractions of protein, water, and solids, respectively, in the seawater slurry, S_{N+1} is the feed rate of the hot water, and $e_{P,N+1}$, $e_{W,N+1}$, and $e_{S,N+1}$ are the weight fractions of protein, water, and solids, respectively, in the entering hot water. Then,

$$x_M = \frac{400(0.029) + 500(0)}{400(0.029 + 0.49) + 500(0 + 1)}$$

$$= 0.0164$$

and

$$N_M = \frac{400(0.481) + 500(0)}{400(0.029 + 0.49) + 500(0 + 1)}$$

$$= 0.272$$

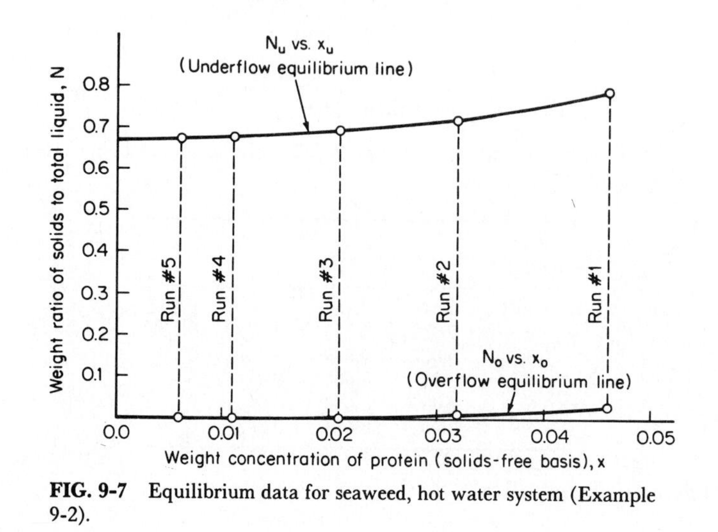

FIG. 9-7 Equilibrium data for seaweed, hot water system (Example 9-2).

4. Calculate the concentration of protein in the product extract (overflow) stream.

Replot (or trace) Fig. 9-7 without the tie lines (Fig. 9-8). On it, plot the point x_M,N_M. Locate along the underflow equilibrium line the point L_N that corresponds to the desired maximum residual concentration of protein, 0.2% or 0.002 weight fraction. Now, from mass-balance considerations, the exit concentrations must lie on the two equilibrium lines and on a straight line passing through x_M,N_M. So draw a line through L_N and x_M,N_M, and extend it to the overflow equilibrium line, labeling the intersection as S_1. The concentration of protein in the exiting overflow stream is read, then, as 0.027 weight fraction, or 2.7%. (Note also that there are virtually no solids in this stream; that is, the value for N at S_1 is virtually zero.)

5. Calculate the number of stages.

With reference to step 1, calculate the values of x and N for each of the two entering streams. Thus, for the seawater slurry,

$$N_{L,0} = \frac{0.481}{0.49 + 0.029} = 0.927$$

and

$$x_{L,0} = \frac{0.029}{0.49 + 0.029} = 0.0559$$

For the entering hot water, $N_{S,N+1}$ and $x_{S,N+1}$ both equal 0, because this stream contains neither solids nor protein.

Now, the number of stages can be found graphically on Fig. 9-7 (redrawn as Fig. 9-9), including the tie lines, in a manner analogous to that of step 5 in Example 9-1. Plot on Fig. 9-9 the points L_0 (that is, x = 0.0559, N = 0.927) and S_1 (from Fig. 9-8), draw a line through them, and extend it downward. Similarly, plot the points S_{N+1} (that is, x

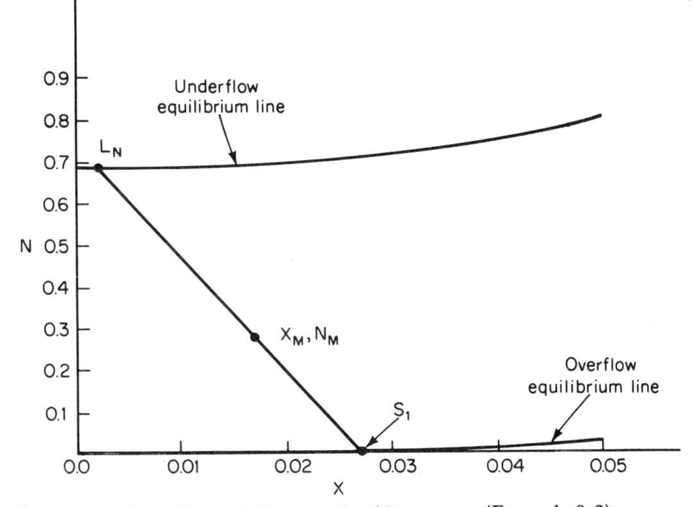

FIG. 9-8 Overall mass balance on leaching system (Example 9-2).

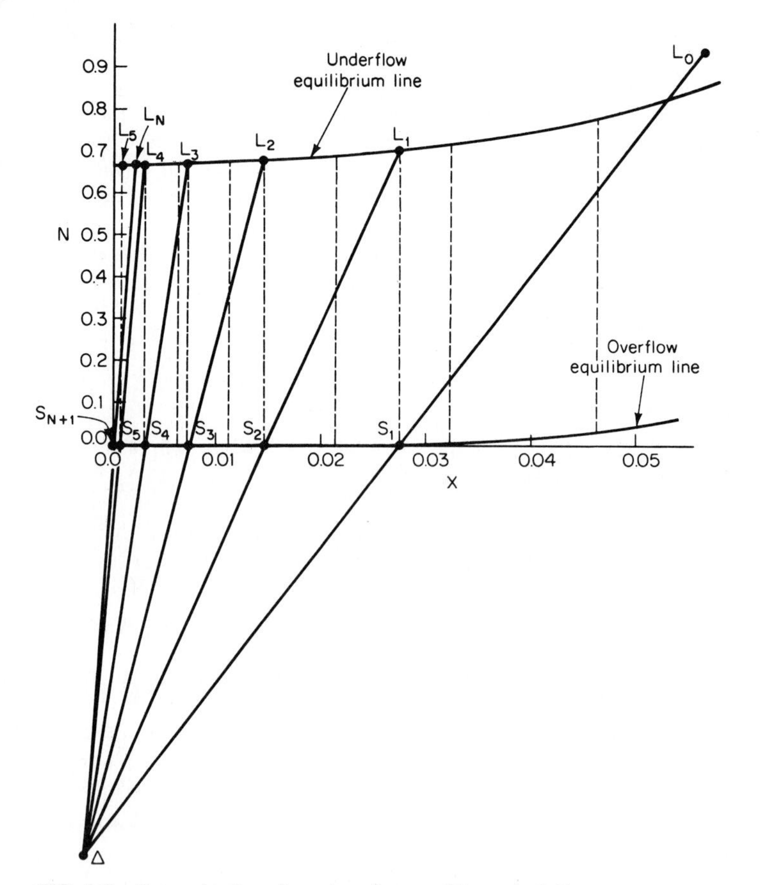

FIG. 9-9 Determination of number of stages (Example 9-2).

= 0, N = 0) and L_N, draw a line through them, and extend it downward. Designate the intersection of the two lines as Δ.

Now begin to step off the stages by drawing an interpolated tie line through point S_1 up to the underflow equilibrium line. Designate their intersection as point L_1. Draw a line joining L_1 with Δ, and label the intersection of this line with the overflow equilibrium line as S_2. Then repeat the procedure, drawing a tie line through S_2 and intersecting the underflow equilibrium line at L_2. This sequence is repeated until a point is reached on the underflow equilibrium line that has a protein content less than that of L_N. Count the number of steps involved. In this case, 5 steps are required. Thus the leaching operation requires 5 equilibrium stages.

Related Calculations: This example is a case of variable-underflow conditions. However, the same procedure can be applied to constant underflow.

10

Crystallization

James R. Beckman, Ph.D.

Associate Professor
Department of Chemical and Bio Engineering
Arizona State University
Tempe, AZ

REFERENCES: [1] Foust et al.—*Principles of Unit Operations,* Wiley; [2] Perry—*Chemical Engineers Handbook,* McGraw-Hill; [3] Randolph and Larson—*Theory of Particulate Processes,* Academic Press; [4] Mullin—*Crystallization,* CRC Press; [5] Bamforth—*Industrial Crystallization,* Macmillan; [6] Institute of Chemical Engineers—*Industrial Crystallization,* Hodgson; [7] Felder and Rousseau—*Elementary Principles of Chemical Processes,* Wiley.

10-1 Solid-Phase Generation of an Anhydrous Salt by Cooling

A 65.2 wt % aqueous solution of potassium nitrate originally at 100°C (212°F) is gradually cooled to 10°C (50°F). What is the yield of KNO_3 solids as a function of temperature? How many pounds of KNO_3 solids are produced at 10°C if the original solution weighed 50,000 lb (22,680 kg)?

Calculation Procedure:

1. Convert weight percent to mole percent.

In order to use Fig. 10-1 in the next step, the mole fraction of KNO_3 in the original solution must be determined. The calculations are as follows:

Compound	Pounds in original solution	÷	Molecular weight	=	Moles	Mole percent
KNO_3	0.652		101.1		0.00645	25.0
H_2O	0.348		18.0		0.01933	75.0
Total	1.000				0.02578	100.0%

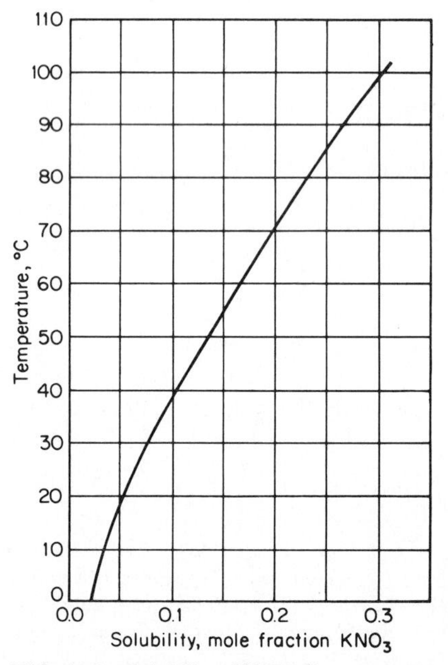

FIG. 10-1 Solubility of KNO_3 in water versus temperature.

2. Calculate yield of solids versus temperature.

Figure 10-1 shows the composition of saturated KNO_3 solution as a function of temperature. The solids formed during cooling will be 100 percent KNO_3, because KNO_3 is anhydrous. The yield of solids is the ratio of KNO_3 solidified to the KNO_3 originally dissolved. As can be seen from Fig. 10-1, no solids are formed from a 25 mol % solution until the solution is cooled to 85°C. As cooling proceeds, solid KNO_3 continues to form while the (saturated) solution concentration continues to decline. At 70°C, for instance, the solution will contain 20 mol % KNO_3 (80 mol % H_2O). If 100 mol of the original solution is assumed, then originally there were 25 mol KNO_3 and 75 mol H_2O. This amount of water present does not change during cooling and solids formation. At 70°C there are therefore $[(0.20\ KNO_3)/(0.80\ H_2O)]$ $(75\ mol\ H_2O) = 18.8\ mol\ KNO_3$ dissolved, or 25 − 18.8 = 6.2 mol KNO_3 solids formed. Therefore, the crystal yield at 70°C is (6.2/25)(100 percent) = 24.8 percent.

Similarly, at about 40°C, the solubility of KNO_3 is 10 percent, which leaves [(0.10 KNO_3)/(0.90 H_2O)](75 mol H_2O) = 8.3 mol KNO_3 in solution. Therefore, 25 − 8.3 = 16.7 mol KNO_3 will have precipitated by the time the solution has cooled to that temperature. Consequently, the yield of solids at 40°C is 16.7/25 = 66.8 percent. Finally, at 10°C, the KNO_3 solubility drops to 3 mol %, giving a yield of 91 percent. Figure 10-2 summarizes the yield of KNO_3 solids as a function of temperature.

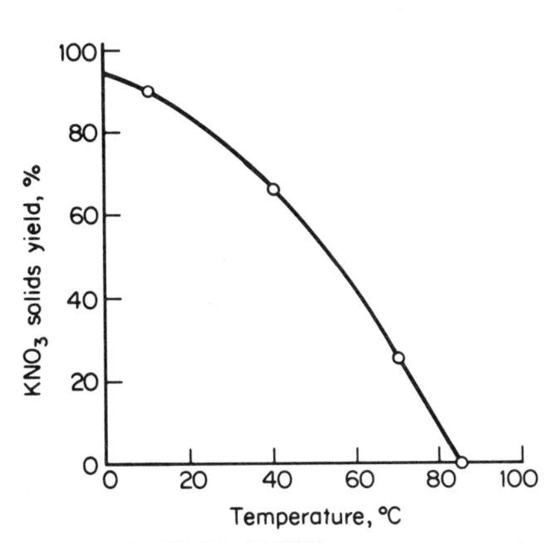

FIG. 10-2 Yield of KNO_3 versus temperature (Example 10-1).

3. Calculate the weight of solids at 10°C.

The weight of solids formed at 10°C is the solids yield (91 percent) multiplied by the weight of KNO_3 initially present in the 100°C mother liquor. The weight of KNO_3 initially in the mother liquor of a 50,000-lb solution is 50,000 lb × 0.652 = 32,600 lb. The weight of KNO_3 solids formed at 10°C is 32,600 lb × 0.91 = 29,670 lb (13,460 kg).

Related Calculations: This method can be used to calculate the yield of any anhydrous salt from batch or steady-state cooling crystallizers. For hydrated salts, see Example 10-3.

10-2 Solid-Phase Generation of an Anhydrous Salt by Boiling

A 70°C (158°F) aqueous solution initially containing 15 mol % KNO_3 is to be boiled so as to give a final yield of solid KNO_3 of 60 percent. How much of the initial water must be boiled off? What is the final liquid composition?

Calculation Procedure:

1. Find the final liquid composition.

Use Fig. 10-1 to determine the solubility of KNO_3 in saturated water solution at 70°C. From the figure, the KNO_3 solubility is 20 mol %.

2. Calculate the amount of water boiled off.

Take a basis of 100 mol of initial solution. Then 15 mol KNO_3 and 85 mol H_2O were initially present. To give a solids yield of 60 percent, then, 0.60 × 15 mol = 9 mol KNO_3 must be precipitated from the solution, leaving 15 − 9 = 6 mol KNO_3 in the solution. The solubility of KNO_3 at 70°C is 20 mol %, from step 1. Therefore, the amount of water still in solution is 6 mol × (0.80/0.20) = 24 mol H_2O, requiring that 85 − 24 = 61 mol had to be boiled off. The percent water boiled off is 61/85 = 72 percent.

Related Calculations: This method can be used to determine the amounts of water to be boiled from boiling crystallizers that yield anhydrous salts. For hydrated salts, see Example 10-4.

10-3 Solid-Phase Generation of a Hydrated Salt by Cooling

A 35 wt % aqueous $MgSO_4$ solution is originally present at 200°F (366 K). If the solution is cooled (with no evaporation) to 70°F (294 K), what solid-phase hydrate will form? If the crystallizer is operated at 10,000 lb/h (4540 kg/h) of feed, how many pounds of crystals will be produced per hour? What will be the solid-phase yield?

Calculation Procedure:

1. Determine the hydrate formation.

As the phase diagram (Fig. 10-3) shows, a solution originally containing 35 wt % $MgSO_4$ will, when cooled to 70°C, form a saturated aqueous solution containing 27 wt % $MgSO_4$ (corresponding to point A) in equilibrium with $MgSO_4 \cdot 7H_2O$ hydrated solids (point B). No other hydrate can exist at equilibrium under these conditions. Now since the molecular weights of $MgSO_4$ and $MgSO_4 \cdot 7H_2O$ are 120 and 246, respectively, the solid-phase hydrate is $(120/246)(100) = 48.8$ wt % $MgSO_4$; the rest of the solid phase is H_2O in the crystal lattice structure.

2. Calculate the crystal production rate and the solid-phase yield.

Let L be the weight of liquid phase formed and S the weight of solid phase formed. Then, for 10,000 lb/h of feed solution, $L + S = 10{,}000$, and (by making a material balance for the $MgSO_4$) $0.35(10{,}000) = 0.27L + 0.488S$. Solving these two equations gives $L = 6330$ lb/h of liquid phase and $S = 3670$ lb/h (1665 kg/h) of $MgSO_4 \cdot 7H_2O$.

Now the solid-phase yield is based on $MgSO_4$, not on $MgSO_4 \cdot 7H_2O$. The 3670 lb/h of solid phase is 48.8 wt % $MgSO_4$, from step 1, so it contains $3670(0.488) = 1791$ lb/h $MgSO_4$. Total $MgSO_4$ introduced into the system is $0.35(10{,}000) = 3500$ lb/h. Therefore, solid-phase yield is $1791/3500 = 51.2$ percent.

As a matter of interest, the amount of H_2O removed from the system by solid (hydrate) formation is $3670(1.0 - 0.488) = 1879$ lb/h.

Related Calculations: This method can be used to calculate the yield of any hydrated salt from a batch or a steady-state cooling crystallizer.

In step 2, L and S can instead be found by applying the inverse lever-arm rule to line segments $\overline{AB}$ and $\overline{AC}$ in Fig. 10-3. Thus, $S/(S + L) = S/10{,}000 = \overline{AB}/\overline{AC} = (0.35 - 0.27)/(0.488 - 0.27)$; therefore, $S = 3670$ lb/h.

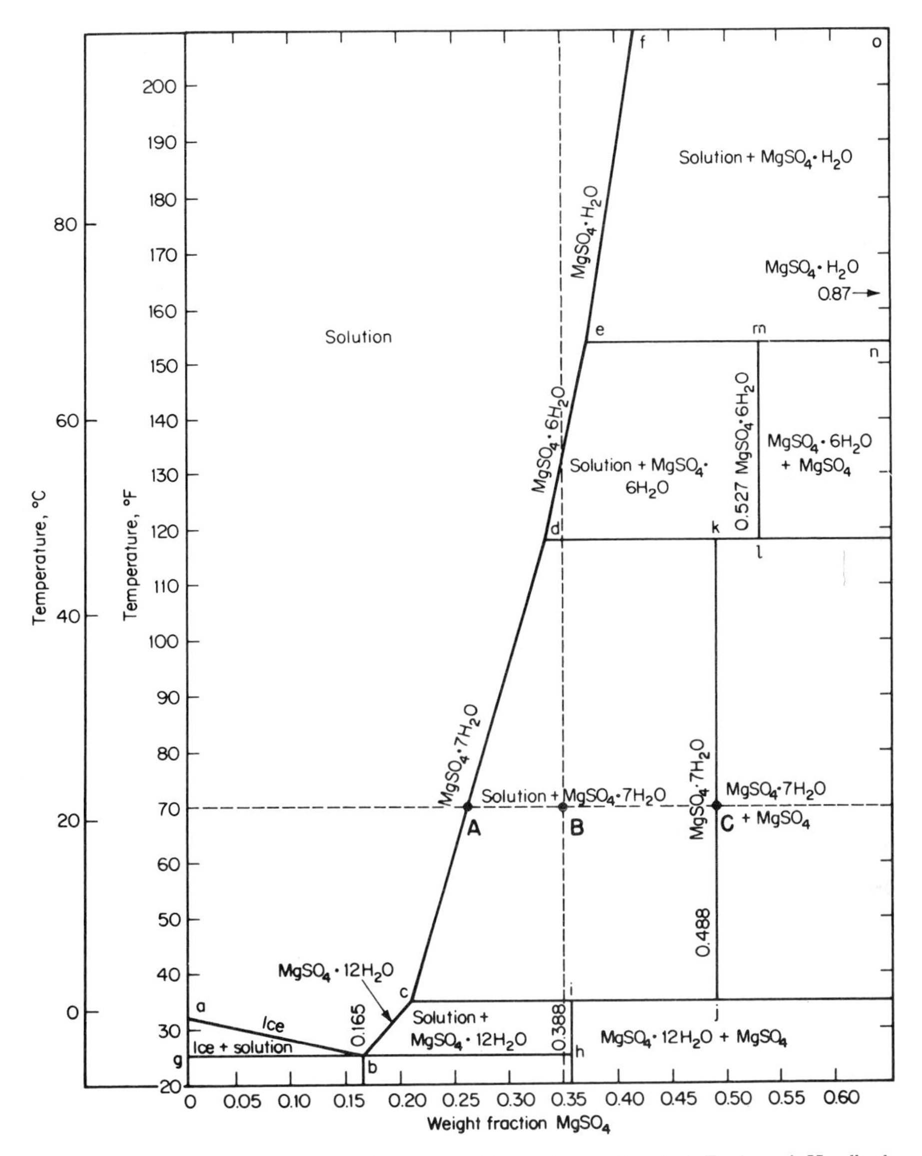

FIG. 10-3 Phase diagram for $MgSO_4 \cdot H_2O$. *(From Perry—Chemical Engineers' Handbook, McGraw-Hill, 1963.)*

10-4 Solid-Phase Generation of a Hydrated Salt by Boiling

Consider 40,000 lb/h (18,150 kg/h) of a 25 wt % $MgSO_4$ solution being fed at 200°F (366 K) to an evaporative crystallizer that boils off water at a rate of 15,000 lb/h (6800 kg/h). The crystallizer is operated at 130°F (327 K) under vacuum conditions. Determine the solid-phase composition, solid-phase production rate, and solid-phase yield. Also calculate the required energy addition rate for the process.

Calculation Procedure:

1. Determine the hydrate formation (solids composition).

Since 15,000 lb/h of water is removed, the product slurry will have an overall $MgSO_4$ composition of $0.25 \times 40{,}000 \text{ lb}/(40{,}000 - 15{,}000 \text{ lb}) = 40.0$ wt % $MgSO_4$. From Fig. 10-3, a system at 130°F and overall $MgSO_4$ composition of 40 wt % will yield $MgSO_4 \cdot 6H_2O$ solids in equilibrium with a 34.5 wt % $MgSO_4$ liquor. Since the molecular weights of $MgSO_4$ and $MgSO_4 \cdot 6H_2O$ are 120 and 228, respectively, the solid-phase hydrate is $(120/228)(100) = 52.7$ wt % $MgSO_4$, with 47.3 wt % water.

2. Calculate the solids production rate.

Let L be the weight of liquid phase formed and S the weight of solid phase formed. Then, for 40,000 lb/h of feed solution with 15,000 lb/h of water boil-off, $S + L = 40{,}000 - 15{,}000$, and (by making a material balance for the $MgSO_4$) $0.25(40{,}000) = 0.527S + 0.345L$. Solving these two equations gives $L = 17{,}450$ lb/h of liquid phase and $S = 7550$ lb/h (3425 kg/h) of $MgSO_4 \cdot 6H_2O$ solids.

3. Calculate the solid-phase yield.

The solid-phase yield is based on $MgSO_4$, not on $MgSO_4 \cdot 6H_2O$. From step 1, the 7550 lb/h of solid phase is 52.7 wt % $MgSO_4$, so it contains $7550(0.527) = 3979$ lb/h $MgSO_4$. Total $MgSO_4$ introduced into the system is $0.25(40{,}000) = 10{,}000$ lb/h. Therefore, solid-phase yield is $3979/10{,}000 = 39.8$ percent.

4. Calculate the energy addition rate.

Figure 10-4 shows the mass flow rates around the evaporative crystallizer, as well as an arrow symbolizing the energy addition. An energy balance around the crystallizer gives $Q = VH_V + LH_L + SH_S - FH_F$, where the H's are the stream enthalpies. From Fig. 10-5, $H_L = -32$ Btu/lb, $H_S = -110$ Btu/lb (extrapolated to 52.7 percent), and $H_F = 52$ Btu/lb. The value for the water vapor, H_V, takes a little more work to get. The enthalpy basis of water used in Fig. 10-5 is 32°F liquid; this can be deduced from the fact that the figure shows an enthalpy value of 0 for pure water (i.e., 0 wt % $MgSO_4$ solution) at 32°F. The basis of most steam tables is 32°F liquid water. From such a steam table an H_V value of about 1118 Btu/lb can be obtained for 130°F vapor water (the pressure correction is minor and can be neglected). Therefore, $Q = 15{,}000 \times 1118 + 17{,}450 \times (-32) + 7550 \times (-110) - 40{,}000 \times 52 = 13.3 \times 10^6$ Btu/h (3900 kW) energy addition to the crystallizer. Energy addition per pound of solids produced is $13.3 \times 10^6/7550 = 1760$ Btu (1860 kJ).

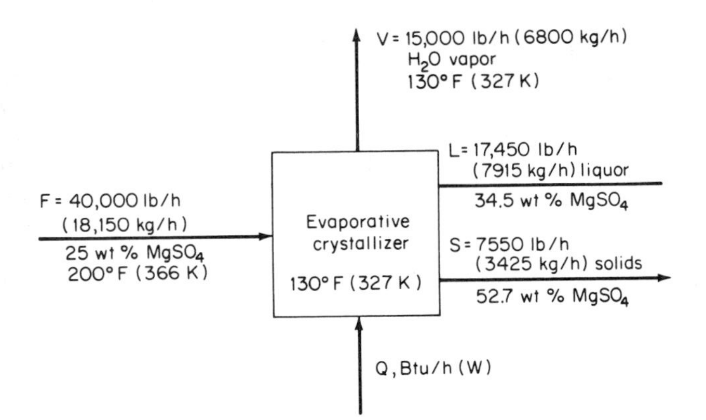

FIG. 10-4 Flow diagram for evaporative crystallizer (Example 10-4).

Related Calculations: This method can be used to calculate the yield, boiling (if any), and energy addition to an evaporative or cooling crystallizer that produces any hydrated or anhydrous crystal solid.

10-5 Separation of Benzene and Naphthalene by Crystallization

A 100,000 lb/h (4536 kg/h) 70°C (158°F) feed containing 80 wt % naphthalene is fed to a cooling crystallizer. At what temperature should the crystallizer operate for maximum naphthalene-only solids production? At this temperature, what is the solids yield of naphthalene? What is the total energy removed from the crystallizer? Naphthalene solids are removed from the mother liquor by centrifugation, leaving some of the solids liquor (10 wt % of the solids) adhering to the solids. After the solids are melted, what is the final purity of the naphthalene?

Calculation Procedure:

1. Determine the appropriate operating temperature for the crystallizer.

Figure 10-6 shows the mutual solubility of benzene and naphthalene. Most of the naphthalene can be crystallized by cooling to (i.e., operating the crystallizer at) the eutectic temperature of −3.5°C (25.7°F), where the solubility of naphthalene in the liquor is minimized to 18.9 wt %. (If one attempted to operate below this temperature, the whole system would become solid.)

2. Calculate the solids yield.

The solids yield is the ratio of the naphthalene solids produced (corresponding to point C in Fig. 10-6) to the naphthalene in the feed liquid (point B). Point A corresponds to the naphthalene remaining in the mother liquor. Then, using the inverse lever-arm rule, we

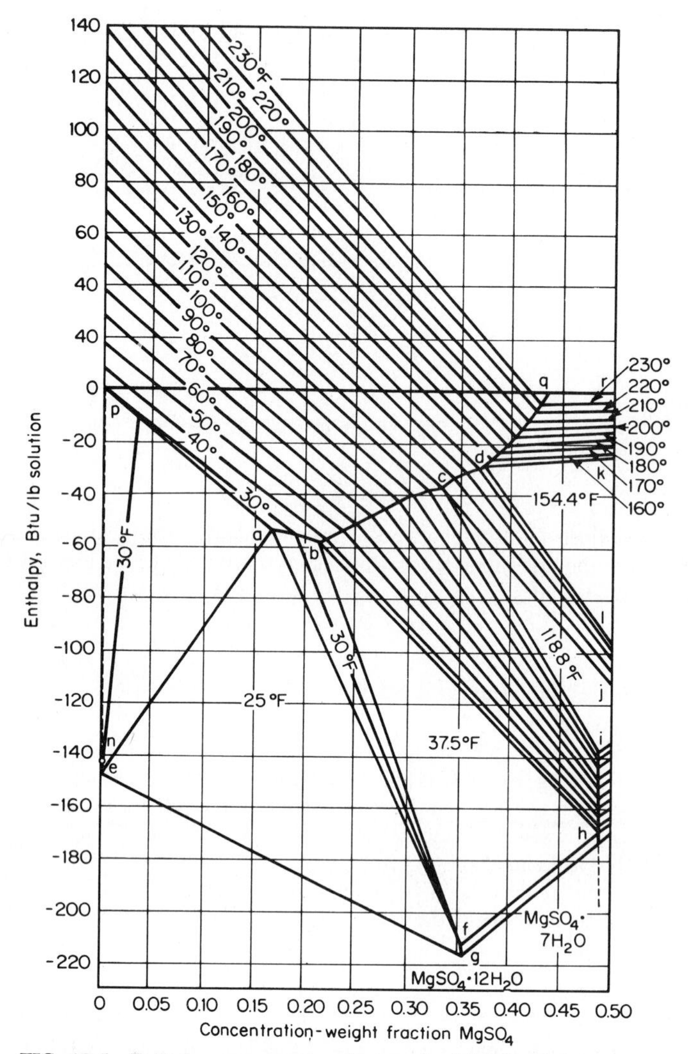

FIG. 10-5 Enthalpy-concentration diagram for $MgSO_4 \cdot H_2O$ system. (Note: 1 Btu/lb = 2.326 kJ/kg.) *(From Perry—Chemical Engineers' Handbook, McGraw-Hill, 1963.)*

find the naphthalene solids rate S as follows: $S = 100{,}000(\overline{AB}/\overline{AC}) = 100{,}000 \times (0.8 - 0.189)/(1.0 - 0.189) = 75{,}300$ lb/h. (This leaves $100{,}000 - 75{,}300 = 24{,}700$ lb/h in the mother liquor.) The solids yield is $75{,}300/(100{,}000 \times 0.8) = 94.1$ percent. The flows are shown in Fig. 10-7.

3. *Calculate the energy removal.*

An energy balance around the crystallizer (see Fig. 10-5) gives $Q = LH_L + SH_S - FH_F$, where Q is the heat added (or the heat removed, if the solved value proves to be negative); L, S, and F are the flow rates for mother liquor, solid product, and feed, respectively; and H_L, H_S, and H_F are the enthalpies of those streams relative to some base temperature. Select a base temperature T_R of 70°C, so that $H_F = 0$.

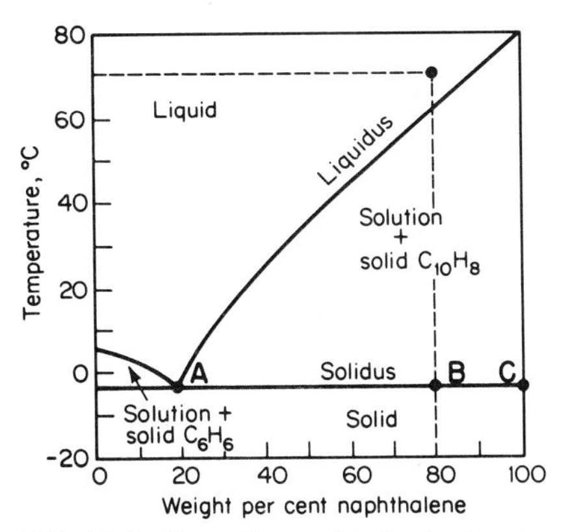

FIG. 10-6 Phase diagram for the simple eutectic system naphthalene-benzene.

For specifics of setting up an energy balance, see Example 2-7. From handbooks, the heat of fusion of naphthalene is found to be 64.1 Btu/lb, and over the temperature range considered here, the heat capacities of liquid benzene and naphthalene can be taken as 0.43 and 0.48 Btu/(lb)(°F), respectively.

Then, for the mother liquor (which consists of 18.9 wt % naphthalene and 81.1 wt % benzene), $H_L = (-3.5°C - 70°C)(1.8°F/°C)[0.48(0.189) + 0.43(0.811)] = -58.1$ Btu/lb. For the product naphthalene, which must cool from 70°C to −3.5°C and then solidify, $H_S = (-3.5°C - 70°C)(1.8°F/°C)(0.48) - 64.1 = -127.6$ Btu/lb.

Therefore, the heat added to the crystallizer is $Q = 24{,}700(-58.1) + 75{,}300(-127.6) - 100{,}000(0) = -11.0 \times 10^6$ Btu/h (3225 kW). Since the value for Q emerges negative, this is the energy *removed* from the crystallizer.

4. *Calculate the purity of the naphthalene obtained by melting the product crystals.*

The weight of mother liquor adhering to the solids is 10 percent of 75,300, or 7530 lb. The total amount of naphthalene present after melting is therefore 75,300 + 0.189(7530) = 76,720 lb. The weight of benzene present (owing to the benzene content of the mother liquor) is 7530(1.0 − 0.189) = 6100 lb. The product purity is therefore 76,720/(76,720 + 6100) = 92.6 percent naphthalene.

Related Calculations: This method can be used to separate organic mixtures having components of different freezing points, such as the xylenes. Organic separations by crystallization have found industrial importance in situations in which close boilers have

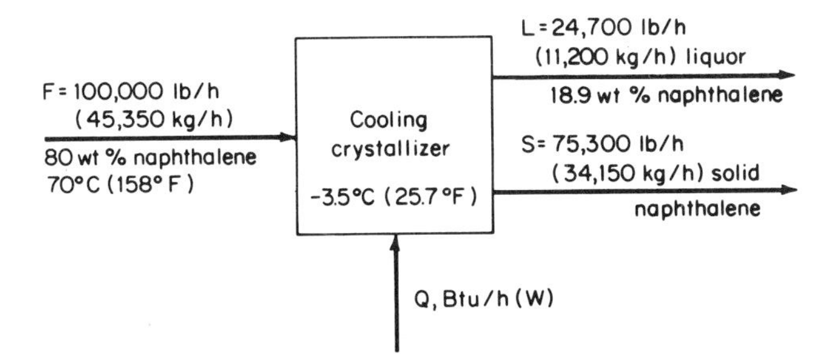

FIG. 10-7 Flow diagram for cooling crystallizer (Example 10-5).

widely separated freezing temperatures. Less energy is related to freezing as opposed to boiling processes because of the low ratio of heat of fusion to heat of vaporization.

10-6 Analysis of a Known Crystal Size Distribution (CSD)

A slurry contains crystals whose size-distribution function is known to be $n = 2 \times 10^5 L \exp(-L/10)$, where n is the number of particles of any size L (in μm) per cubic centimeter of slurry. The crystals are spherical, with a density of 2.5 g/cc. Determine the total number of crystals. Determine the total area, volume, and mass of the solids per volume of slurry. Determine the number-weighted average, the length-weighted average, and the area-weighted average particle size of the solids. What is the coefficient of variation of the particles? Generate a plot of the cumulative weight fraction of particles that are undersize in terms of particle size L.

Calculation Procedure:

1. Calculate the total number of particles per volume of slurry.

This step and the subsequent calculation steps require finding $\int_0^\infty nL^j dL$, that is (from the equation for n above), $2 \times 10^5 \int_0^\infty L^{j+1} \exp(-L/10) dL$, where L is as defined above and j varies according to the particular calculation step. From a table of integrals, the general integral is found to be $2 \times 10^5 [(j + 1)!/(1/10)^{j+2}]$.

For calculating the number of particles, $j = 0$, and the answer is the zeroth moment (designated M_0) of the distribution. Thus the total number N_T of particles is $2 \times 10^5[(0 + 1)!/(1/10)^{0+2}] = 2 \times 10^7$ particles per cubic centimeter of slurry.

2. Calculate the first moment of the distribution.

This quantity, M_1, which corresponds to the "total length" of the particles per cubic centimeter of slurry, is not of physical significance in itself, but it is used in calculating the averages in subsequent steps. It corresponds to the integral in step 1 when $j = 1$. Thus, $M_1 = 2 \times 10^5[(1 + 1)!/(1/10)^{1+2}] = 4 \times 10^8$ μm per cubic centimeter of slurry.

3. Calculate the total area of the particles per volume of slurry.

The total area $A_T = k_A M_2$, where k_A is a shape factor (see below) and M_2 is the second moment of the distribution, i.e., the value of the integral in step 1 when $j = 2$. Some shape factors are as follows:

Crystal shape	Value of k_A
Cube	6
Sphere	π
Octahedron	$2\sqrt{3}$

In the present case, then, $A_T = \pi(2 \times 10^5)[(2 + 1)!/(1/10)^{2+2}] = 3.77 \times 10^{10}$ μm^2 (377 cm^2) per cubic centimeter of slurry.

4. *Calculate the total volume of crystals per volume of slurry.*

The volume of solids per volume of slurry $V_T = k_V \int_0^\infty nL^3 dL = k_V M_3$, where k_V is a so-called volume shape factor (see below) and M_3 is the third moment of the distribution, i.e., the value of the integral in step 1 when $j = 3$. Some volume shape factors are as follows:

Crystal shape	Value of k_V
Cube	1
Sphere	$\pi/6$
Octahedron	$\sqrt{2}/3$

In the present case, then, $V_T = (\pi/6)(2 \times 10^5)[(3 + 1)!/(1/10)^{3+2}] = 2.51 \times 10^{11}$ μm^3 (0.251 cm^3) per cubic centimeter of slurry.

5. *Calculate the total mass of solids per volume of slurry.*

Total mass of solids $M_T = \rho_S V_T$, where ρ_S is the crystal density. Thus, $M_T = (2.5$ $g/cm^3)(0.251$ cm^3 per cubic centimeter of slurry) = 0.628 g per cubic centimeter of slurry.

6. *Calculate the average crystal size.*

The number-weighted average crystal size $\overline{L}_{1,0} = M_1/M_0$. Thus, $\overline{L}_{1,0} = (4 \times 10^8)/(2 \times 10^7) = 20$ μm. The length weighted average $\overline{L}_{2,1} = M_2/M_1$. Thus, $\overline{L}_{2,1} = 2 \times 10^5[(2 + 1)!/(1/10)^{2+2}]/(4 \times 10^8) = (12 \times 10^9)/(4 \times 10^8) = 30$ μm. And the area-weighted average $\overline{L}_{3,2} = M_3/M_2$. Thus, $\overline{L}_{3,2} = (2 \times 10^5)[(3 + 1)!/(1/10)^{3+2}]/(12 \times 10^9) = 40$ μm.

7. *Calculate the variance of the particle size distribution.*

The variance σ^2 of the particle size distribution equals $\int_0^\infty (\overline{L}_{1,0} - L)^2 n dL/M_0 = M_2/M_0 - (\overline{L}_{1,0})^2$. Thus, $\sigma^2 = (12 \times 10^9)/(2 \times 10^7) - 20^2 = 200$ μm^2.

8. *Calculate the coefficient of variation for the particle size distribution.*

The coefficient of variation c.v. equals $\sigma/\overline{L}_{1,0}$, where σ (the standard deviation) is the square root of the variance from step 7. Thus, c.v. $= 200^{1/2}/20 = 0.71$.

9. *Calculate and plot the cumulative weight fraction that is undersize.*

The weight fraction W undersize of a crystal size distribution is $W = \rho_s k_V \int_0^L nL^3 dL/$

$M_T = \int_0^L nL^3 dL/M_3 = 1 - [(L/10)^4/24 + (L/10)^3/6 + (L/10)^2/2 + L/10 + 1] \exp(-L/10)$. A plot of this function (Fig. 10-8) has the characteristic S-shaped curvature.

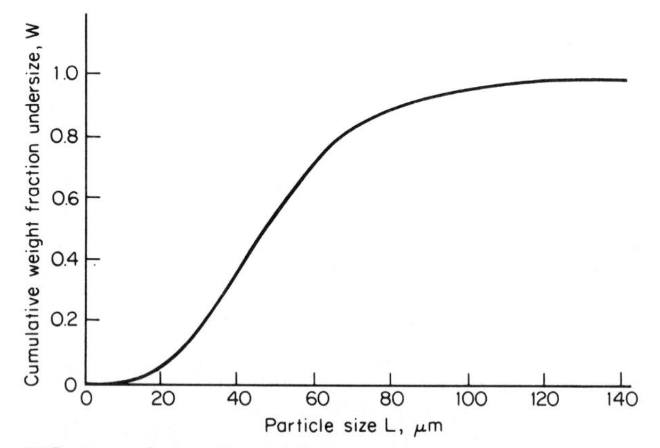

FIG. 10-8 Cumulative weight fraction undersize versus particle size (Example 10-6).

Related Calculations: This procedure can be used to calculate average sizes, moments, surface area, and mass of solids per volume of slurry for any known particle size distribution. The method can also be used for dry-solids distributions, say, from grinding operations. See Example 10-7 for an example of a situation in which the size distribution is based on an experimental sample rather than on a known size-distribution function.

10-7 Crystal Size Distribution of a Slurry Sample

The first three columns of Table 10-1 show a sieve-screen analysis of a 100-cc (0.0001-m^3 or 0.0035-ft^3) slurry sample. The crystals are cubic and have a solids density ρ_s of 1.77 g/cc (110.5 lb/ft^3). Calculate the crystal size distribution n of the solids, the average crystal size, and the coefficient of variation of the crystal size distribution.

Calculation Procedure:

1. Calculate the weight fraction retained on each screen.

The weight fraction ΔW_i retained on screen i equals the weight retained on that screen divided by the total solids weight, that is, 29.87 g. For instance, the weight fraction retained on screen 28 is $0.005/29.87 = 0.000167$. The weight fractions retained on all the screens are shown in the fourth column of Table 10-1.

2. Calculate the screen average sizes.

The screen average size L_i for a given screen reflects the average size of a crystal retained on that screen. Use the average of the size of the screen and the screen above it. For

TABLE 10-1 Crystal Size Distribution of a Slurry Sample (Example 10-7)

Sieve screen analysis			Summary of crystal size distribution analysis			
Tyler mesh	Opening, μm	Weight retained, g	Weight fraction retained ΔW_i	Average screen size L_i, μm	ΔL_i, μm	n_i, no./(cm^3)(μm)
(24) ficticious screen	(701)	—	—	—	—	—
28	589	0.005	0.000167	645	112	9.39×10^{-4}
32	495	0.016	0.00536	542	94	0.0605
35	417	0.096	0.00321	456	78	0.0733
42	351	0.315	0.0106	384	66	0.479
48	295	1.61	0.0539	323	56	4.827
60	248	3.42	0.1145	272	47	20.46
65	208	7.56	0.253	228	40	90.19
80	175	8.21	0.275	192	33	199.0
100	147	5.82	0.195	161	28	282.0
115	124	2.47	0.0827	136	23	241.5
150	104	0.32	0.0107	114	20	61.0
170	88	0.025	0.000837	96	16	9.99
200	74	0.0076	0.000254	81	14	5.77
	Total	29.87				

instance, the average crystal size of the solids on screen 28 is $(701 + 589)/2 = 645$ μm. The averages for all the screens appear in the fifth column of Table 10-1.

3. Calculate the size difference between screens.

The size difference between screens ΔL_i is the difference between the size of the screen in question and the size of the screen directly above it. For instance, the size difference ΔL_i for screen 60 is $(295\ \mu\text{m}) - (248\ \mu\text{m}) = 47\ \mu\text{m}$. Size differences for all the screens appear in the sixth column of Table 10-1.

4. Calculate the third moment of the crystal size distribution.

The third moment M_3 of the crystal size distribution equals $M_T/\rho_s k_V$, where M_T is the total weight of the crystals and k_V is the volume shape factor; see step 4 of Example 10-6. In this case, $M_3 = 29.87/[1.77(1)(100\ \text{cc})] = 0.169\ \text{cm}^3$ solids per cubic centimeter of slurry.

5. Calculate the crystal size distribution.

The crystal size distribution for the ith screen n_i equals $10^{12}M_3\Delta W_i/(L_i^3\Delta L_i)$, in number of crystals per cubic centimeter per micron. For instance, for screen 60, $n_i = 10^{12}(0.169)(0.1145)/[272^3(47)] = 20.46$ crystals per cubic centimeter per micron. The size distributions for all other screens appear in the seventh column of Table 10-1.

6. Calculate the zeroth, first, and second moments of the crystal size distribution.

The zeroth moment M_0 is calculated as follows: $M_0 = \sum_i n_i\Delta L_i = 2.64 \times 10^4$ crystals per cubic centimeter. The first moment M_1 is calculated by $M_1 = \sum_i n_i L_i \Delta L_i = 4.64 \times 10^6\ \mu\text{m/cm}^3$. The second moment M_2 is calculated by $M_2 = \sum_i n_i L_i^2 \Delta L_i = 8.62 \times 10^8\ \mu\text{m}^2/\text{cm}^3$. [The third moment M_3 can be calculated by $M_3 = \sum_i n_i L_i^3 \Delta L_i = 0.169 \times 10^{12}\ \mu\text{m}^3/\text{cm}^3$ ($0.169\ \text{cm}^3/\text{cm}^3$), which agrees with the calculation of the third moment from step 4.]

7. Calculate the average crystal size.

The number-weighted average crystal size is $\overline{L}_{1,0} = M_1/M_0 = (4.64 \times 10^6)/(2.64 \times 10^4) = 176\ \mu\text{m}$. The length-weighted average crystal size is $\overline{L}_{2,1} = M_2/M_1 = (8.62 \times 10^8)/(4.64 \times 10^6) = 186\ \mu\text{m}$. The area-weighted average crystal size is $\overline{L}_{3,2} = M_3/M_2 = (0.169 \times 10^{12})/(8.63 \times 18^8) = 196\ \mu\text{m}$.

8. Calculate the variance.

The variance of the crystal size distribution is $\sigma^2 = M_2/M_0 - (\overline{L}_{1,0})^2 = (8.62 \times 10^8)/(2.64 \times 10^4) - 176^2 = 1676\ \mu\text{m}^2$.

9. Calculate the coefficient of variation.

The coefficient of variation $\text{c.v.} = \sigma/\overline{L}_{1,0} = 1676^{1/2}/176 = 0.23$.

Related Calculations: This procedure can be used to analyze either wet or dry solids particle size distributions. Particle size distributions from grinding or combustion and particles from crystallizers are described by the same mathematics. See Example 10-6 for an example of a situation in which the size distribution is based on a known size-distribution function rather than on an experimental sample.

10-8 Product Crystal Size Distribution from a Seeded Crystallizer

A continuous crystallizer producing 25,000 lb/h (11,340 kg/h) of cubic solids is continuously seeded with 5000 lb/h (2270 kg/h) of crystals having a crystal size distribution as listed in Table 10-2. Predict the product crystal size distribution if nucleation is ignored. If the residence time of solids in the crystallizer is 2 h, calculate the average particle-diameter growth rate G.

TABLE 10-2 Size Distribution of Seed Crystals (Example 10-8)

Tyler mesh	Weight fraction retained ΔW_i	Average size L_i, μm (from Table 10-1)
(65)	—	—
80	0.117	192
100	0.262	161
115	0.314	136
150	0.274	114
170	0.032	96
200	0.001	81
Total	1.000	

Calculation Procedure:

1. *Calculate the crystal-mass-increase ratio.*

The crystal-mass-increase ratio is the ratio of crystallizer output to seed input; in this case, 25,000/5000 = 5.0.

2. *Calculate the increase ΔL in particle size.*

The increase in weight of a crystal is related to the increase in particle diameter. For any given screen size, that increase ΔL is related to the initial weight ΔM_s and initial size L_s of seed particles corresponding to that screen and to the product weight ΔM_p of particles corresponding to that screen, by McCabe's ΔL law:

$$\Delta M_p = \left(1 + \frac{\Delta L}{L_s}\right)^3 \Delta M_s$$

This equation can be solved for ΔL by trial and error. From step 1, and summing over all the screens, $\Sigma\Delta M_p/\Sigma\Delta M_s = 5.0$. The trial-and-error procedure consists of assuming a value for ΔL, calculating ΔM_p for each screen, summing the values of ΔM_p and ΔM_s, and repeating the procedure until the ratio of these sums is close to 5.0.

For a first guess, assume that $\Delta L = 100\ \mu m$. Assuming that the total seed weight is 1.0 (in any units), this leads to the results shown in Table 10-3. Since $\Sigma\Delta M_p$ is found to be 5.26, the ratio $\Sigma\Delta M_p/\Sigma\Delta M_s$ emerges as $5.26/1.0 = 5.26$, which is too high. A lower assumed value of ΔL is called for. At final convergence of the trial-and-error procedure, ΔL is found to be 96 μm, based on the results shown in the first five columns of Table 10-4.

This leads to the crystal size distribution shown in the last two columns of the table. The sixth column, weight fraction retained ΔW_i, is found (for each screen size) by dividing ΔM_p by $\Sigma\Delta M_p$. The screen size (seventh column) corresponding to each weight fraction consists of the original seed-crystal size L_s plus the increase ΔL.

TABLE 10-3 Results from (Incorrect) Guess that $\Delta L = 100\ \mu m$ (Example 10-8). Basis: Total seed mass = 1.0

Tyler mesh	Seed mass ΔM_s (from Table 10-2)	Seed size L_s (from Table 10-2)	$(1 + \Delta L/L_s)^3$	Calculated product mass Δm_p
80	0.117	192	3.52	0.412
100	0.262	161	4.26	1.12
115	0.314	136	5.23	1.64
150	0.274	114	6.61	1.81
170	0.032	96	8.51	0.272
200	0.001	81	11.16	0.011
	$\Sigma\Delta M_s = 1.000$			$\Sigma\Delta M_p = 5.26$

$$\frac{\Sigma\Delta M_p}{\Sigma\Delta M_s} = \frac{5.26}{1.00} = 5.26 \quad \text{which is too high}$$

TABLE 10-4 Results from (Correct) Guess that $\Delta L = 96\ \mu m$ and Resulting Crystal Size Distribution (Example 10-8). Basis: Total seed mass = 1.0

Tyler mesh	Seed mass ΔM_s	Seed size L_s	$(1 + \Delta L/L_s)^3$	Calculated product mass ΔM_p	Calculated weight fraction retained ΔW_i	Product screen size $(L_s + \Delta L)$
80	0.117	192	3.37	0.395	0.079	288
100	0.262	161	4.07	1.066	0.214	257
115	0.314	136	4.96	1.56	0.312	232
150	0.274	114	6.25	1.71	0.342	210
170	0.032	96	8.00	0.256	0.051	192
200	0.001	81	10.4	0.0104	0.002	177
	$\Sigma\Delta M_s = 1.000$			$\Sigma\Delta M_p = 4.997$		

$$\frac{\Sigma\Delta M_p}{\Sigma\Delta M_s} = \frac{4.997}{1.000} = 4.997 \quad \text{which is close enough to 5.0}$$

3. Calculate the growth rate.

The average particle-diameter growth rate G can be found thus:

$$G = \frac{\Delta L}{\text{(elapsed time)}} = \frac{96\ \mu\text{m}}{(2\ \text{h})(60\ \text{min/h})} = 0.8\ \mu\text{m/min}$$

Related Calculations: This method uses McCabe's ΔL law, which assumes total growth and no nucleation. For many industrial situations, these two assumptions seem reasonable. If significant nucleation is present, however, this method will overpredict product crystal size.

The presence of nucleation can be determined by product screening: If particles of size less than the seeds can be found, then nucleation is present. In such a case, prediction of product crystal size distribution requires a knowledge of nucleation kinetics; see Randolph and Larson [3] for the basic mathematics.

10-9 Analysis of Data from a Mixed Suspension–Mixed Product Removal Crystallizer (MSMPR)

The first three columns of Table 10-5 show sieve data for a 100-cc slurry sample containing 21.0 g of solids taken from a 20,000-gal (75-m^3) mixed suspension–mixed product removal crystallizer (MSMPR) producing cubic ammonium sulfate crystals. Solids density is 1.77 g/cm^3, and the density of the clear liquor leaving the crystallizer is 1.18 g/cm^3. The hot feed flows to the crystallizer at 374,000 lb/h (47 kg/s). Calculate the residence time τ, the crystal size distribution function n, the growth rate G, the nucleation density n^0, the nucleation birth rate B^0, and the area-weighted average crystal size $\bar{L}_{3,2}$ for the product crystals.

TABLE 10-5 Crystal Size Distribution from an MSMPR Crystallizer (Example 10-9)

			Summary of crystal size distribution analysis				
Screen number	Tyler mesh	Weight fraction retained ΔW_i	Screen size, μm	Average screen size L_i, μm	ΔL_i, μm	n_i	ln n_i
1	24	0.081	701	—	—	—	—
2	28	0.075	589	645	112	0.297	−1.21
3	32	0.120	495	542	94	0.954	−0.047
4	35	0.100	417	456	78	1.61	0.476
5	42	0.160	351	384	60	5.60	1.72
6	48	0.110	295	323	56	6.94	1.94
7	60	0.102	248	272	47	12.8	2.55
8	65	0.090	208	228	40	22.6	3.12
9	80	0.060	175	192	33	30.6	3.42
10	100	0.040	147	161	28	40.7	3.71
11	115	0.024	124	136	23	49.4	3.90
12	150	0.017	104	114	20	68.3	4.22
13	170	0.010	88	96	16	84.1	4.43
14	200	0.005	74	81	14	80.0	4.38
—	fines	0.006	—	—	—	—	—
	Total	1.000					

Calculation Procedure:

1. *Calculate the density of the crystallizer magma.*

The slurry density in the crystallizer is the same as the density of the product stream. Select as a basis 100 cm^3 slurry. The solids mass is 21.0 g; therefore, the solids volume is (21.0 g)/(1.77 g/cm^3) = 11.9 cm^3. The clear-liquor volume is 100 − 11.9 = 88.1 cm^3, and its mass is (88.1 cm^3)(1.18 g/cm^3) = 104 g. Therefore, the density of the slurry is (104 g + 21 g)/100 cm^3 = 1.25 g/cm^3 (78.0 lb/ft^3).

2. *Calculate the residence time in the crystallizer.*

The residence time τ in the crystallizer is based on the outlet conditions (which are the same as in the crystallizer). Thus, τ = (volume of crystallizer)/(outlet volumetric flow rate) = (20,000 gal)/[(374,000 lb/h)/(78.0 lb/ft^3)(0.1337 ft^3/gal)] = 0.557 h = 33.4 min.

3. *Calculate the third moment of the solids crystal size distribution.*

The third moment M_3 of the crystal size distribution equals $M_T/\rho_s k_V$, where M_T is the weight of crystals, ρ_s is the solids density, and k_V is the volume shape factor; see step 4 of Example 10-6. Thus, M_3 = 21.0 g/[(1.77 g/cm^3)(1)(100 cm^3)] = 0.119 cm^3 solids per cubic centimeter of slurry.

4. *Calculate the crystal size distribution function n.*

The crystal size distribution for the ith sieve tray is $n_i = 10^{12} M_3 \Delta W_i/(L_i^3 \Delta L_i)$, where ΔW_i is the weight fraction retained on the ith screen, L_i is the average screen size of material retained on the ith screen (see Example 10-7, step 2), and ΔL_i is the difference in particle sizes on the ith screen (see Example 10-7, step 3). For instance, for the Tyler mesh 100 screen, $n_{10} = 10^{12}(0.119)(0.040)/(161^3 \times 28) = 40.7$ crystals per cubic centimeter per micron. Table 10-5 shows the results for each sieve screen.

5. *Calculate the growth rate G.*

The growth rate for an MSMPR can be calculated from the slope of an ln n versus L diagram (Fig. 10-9). Here the slope equals $-[1/(G\tau)] = (\ln n_2 - \ln n_1)/(L_2 - L_1) = (-0.6 - 5.4)/(600 - 0) = -0.010\ \mu\text{m}^{-1}$ or $G\tau = 100\ \mu$m. Then the growth rate $G = 100\ \mu\text{m}/33.4\text{ min} = 3.0\ \mu$m/min.

6. *Calculate the nucleation density n^0.*

The nucleation density n^0 is the value of n at size $L = 0$. From Fig. 10-9, ln n^0 at size equal to zero is 5.4. So $n^0 = \exp(\ln n^0) = \exp 5.4 = 221$ particles per cubic centimeter per micron.

7. *Calculate the nucleation birth rate B^0.*

The nucleation birth rate is $B^0 = n^0 G = 221(3.0) = 663$ particles per cubic centimeter per minute.

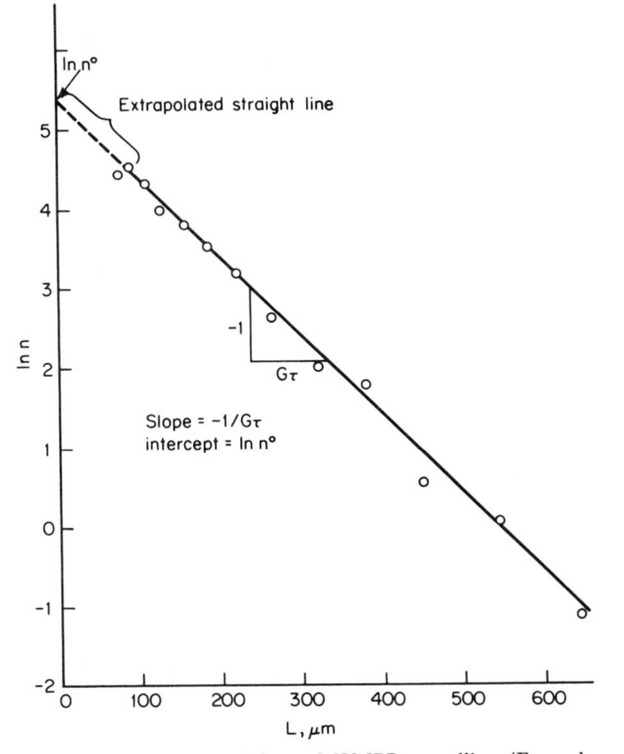

FIG. 10-9 Ln n versus L for an MSMPR crystallizer (Example 10-9).

8. Calculate area-weighted average size $\overline{L}_{3,2}$.

As shown in Examples 10-6 and 10-7, the area-weighted average size equals M_3/M_2. However, for an MSMPR, the area-weighted average particle size also happens to equal $3G\tau$. Thus, $\overline{L}_{3,2} = 3(3\ \mu\text{m/min})(33.4\ \text{min}) = 300\ \mu\text{m}$.

Related Calculations: Use this procedure to calculate the crystal size distribution from both class I and class II MSMPR crystallizers. This procedure cannot be used to calculate growth rates and nucleation with crystallizers having either fines destruction or product classification.

10-10 Product Screening Effectiveness

Figure 10-10 shows the sieve-screen analysis of a feed slurry, overflow slurry, and underflow slurry being separated by a 600-μm classifying screen. Calculate the overall effectiveness of the classifying screen.

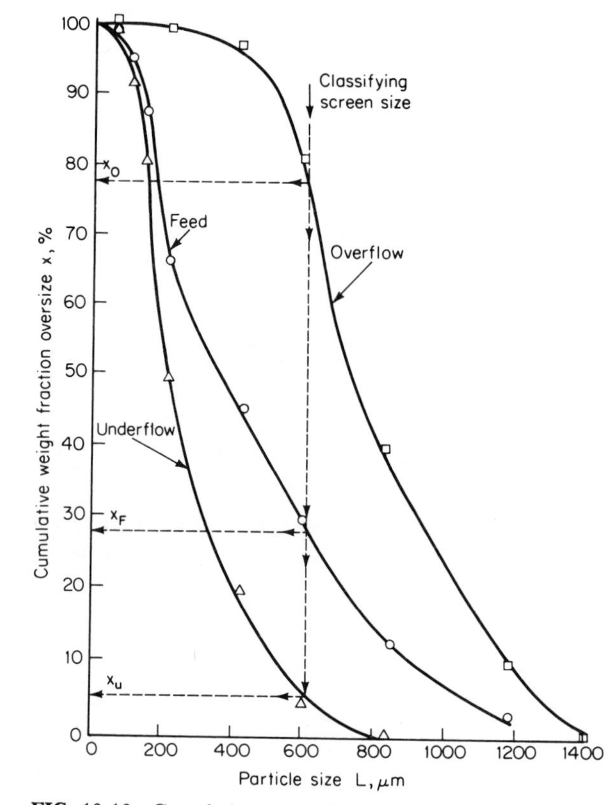

FIG. 10-10 Cumulative weight fraction oversize versus particle size (Example 10-10).

Calculation Procedure:

1. Calculate solids mass fractions in each stream.

On Fig. 10-10, draw a vertical line through the abscissa that corresponds to the screen size, that is, 600 μm. This line will intersect each of the three cumulative-weight-fraction curves. The ordinate corresponding to each intersection gives the mass fraction of total solids actually in that stream which would be in the overflow stream if the screen were instead perfectly effective. Thus the mass fraction in the feed x_F is found to be 0.28, the fraction in the overflow x_o is found to be 0.77, and the fraction in the underflow x_u is found to be 0.055.

2. Calculate solids-overflow to total-solids-feed ratio.

The ratio of total overflow-solids mass to total feed-solids mass q equals $(x_F - x_u)/(x_o - x_u)$; that is, $q = (0.28 - 0.055)/(0.77 - 0.055) = 0.315$.

3. Calculate the screen effectiveness based on oversize.

The screen effectiveness based on oversize material E_o equals $q(x_o/x_F)$. Thus, $E_o = 0.315(0.77/0.28) = 0.87$.

4. Calculate the screen effectiveness based on undersize.

The screen effectiveness based on undersize material E_u equals $(1 - q)(1 - x_u)/(1 - x_F)$. Thus, $E_u = (1 - 0.315)(1 - 0.055)/(1 - 0.28) = 0.90$.

5. Calculate the overall screen effectiveness.

Overall screen effectiveness E is the product of E_o and E_u. Thus, $E = 0.90(0.87) = 0.78$.

Related Calculations: This method can be used for determining separation effectiveness for classifying screens, elutriators, cyclones, or hydroclones in which a known feed of a known crystal size distribution is segregated into a fine and a coarse fraction. If a cut size cannot be predetermined, assume one at a time and complete the described effectiveness analysis. The assumed cut size that gives the largest effectiveness is the cut size that best describes the separation device.

11

Absorption and Stripping

K. J. McNulty, Sc.D.

Technical Director
Research and Development
Koch Engineering Co., Inc.
Wilmington, MA

11-1 Hydraulic Design of a Packed Tower

Gas and liquid are to be contacted countercurrently in a packed tower. The approach to flooding is not to exceed 80 percent as defined at constant liquid loading. What column diameter should be used for 2-in Koch Flexiring (FR) packing (i.e., 2-in slotted-ring random packing), and what diameter should be used for Koch Flexipac (FP) Type 2Y packing (i.e., structured packing having ½-in crimp height, with flow channels inclined at 45° to the axis of flow)? The packings are to be of stainless steel. Calculate and compare the pressure drops per foot of packing for these two packings. The operating conditions are

Maximum liquid rate: 150,000 lb/h

Maximum gas rate: 75,000 lb/h

Liquid density: 62.4 lb/ft³

Gas density: 0.25 lb/ft³

Calculation Procedure:

1. *Calculate the tower diameter.*

Various methods are available for the design and rating of packed towers. The method shown here is an extension of the CVCL model, which is more fundamentally sound than the generalized pressure drop correlation (GPDC). The basic flooding equation is

$$C_{VF}^{1/2} + sC_{LF}^{1/2} = c \tag{11-1}$$

where C_{VF} is the vapor or gas capacity factor at flooding, C_{LF} is the liquid capacity factor at flooding, and s and c are constants for a particular packing. The capacity factors for the vapor and liquid are defined as

$$C_V = U_g[\rho_g/(\rho_l - \rho_g)]^{1/2} \tag{11-2}$$

$$C_L = U_L[\rho_l/(\rho_l - \rho_g)] \tag{11-3}$$

where U_g is the superficial velocity of the gas based on the empty-tower cross-sectional area, U_L is the superficial liquid velocity, ρ_g is the gas density, and ρ_l is the liquid density. In U.S. engineering units, C_V and U_g are in feet per second, and C_L and U_L are in gallons per minute per square foot. The superficial velocities are related to the mass flow rates as follows:

$$U_g = w_g/\rho_g A \qquad \text{and} \qquad U_l = w_l/\rho_l A \tag{11-4}$$

where w_g and w_l are, respectively, the mass flow rates of gas and liquid, and A is the tower cross-sectional area. Here, U_l has units of feet per second; this can be converted to the more commonly used U_L (with an uppercase subscript) having units of gallons per minute per square foot, by multiplying by 7.48 gal/ft³ and 60 s/min.

For flooding that is defined at constant liquid loading L, the flooding capacity factor C_{LF} is the same as the design capacity factor C_L. The gas capacity factor at flood, C_{VF}, is equal to the design capacity factor divided by the fractional approach to flooding, f: $C_{VF} = C_V/f$. For this problem, f equals 0.80. Substituting the definitions and given values for this problem into Eq. (11-1), bearing in mind that for a circular cross section $A = \pi D^2/4$ and solving for tower diameter D gives

$$D = \frac{1}{c}\sqrt{\frac{4}{\pi}}\left\{\left[\frac{75000}{(3600)(0.8)(0.25)}\sqrt{\frac{0.25}{(62.4-0.25)}}\right]^{1/2} + s\left[\frac{(150000)(7.48)}{(62.4)(60)}\sqrt{\frac{62.4}{(62.4-0.25)}}\right]^{1/2}\right\} \tag{11-5}$$

where the quantity 3600 converts from hours to seconds, 60 converts from hours to minutes, and 7.48 converts from cubic feet to gallons.

It remains to determine values of s and c for the packings of interest. These values can

TABLE 11-1 Tower-packing Constants for Hydraulic Design and Rating of Packed Towers

Generic shape	Typical brand	Nominal size	Void fraction, ϵ	K_1, (in $wc \cdot s^2$)/lb	s, $[ft^3/(gal/min)(s)]^{1/2}$	c, $(ft/s)^{1/2}$
Random: slotted ring	Flexiring Packing	1 in	0.95	0.14	0.016	0.67
		1.5 in	0.96	0.11	0.015	0.71
		2 in	0.98	0.069	0.040	0.74
		3.5 in	0.98	0.044	0.034	0.78
Structured: corrugated sheet	Flexipac Packing	Type 1Y ¼ in, 45°	0.97	0.088	0.054	0.69
		Type 2Y ½ in, 45°	0.99	0.041	0.044	0.75
		Type 3Y 1 in, 45°	0.99	0.019	0.044	0.87
		Type 4Y 2 in, 45°	0.99	0.012	0.035	0.88

Note: Flexiring and Flexipac are trademarks of Koch Engineering Co., Inc.

be calculated from air-water pressure-drop data published by packing vendors. Table 11-1 provides values of s and c for several random and structured packings of stainless steel construction. With the appropriate constants from the table for 2-in FR and FP2Y substituted into Eq. (11-5), the diameter emerges as 4.96 ft for 2-in FR and 5.00 ft for FP2Y. Thus, for the conditions stated in this example, the capacities of these two packings are essentially identical. In subsequent calculations, a tower diameter of 5.0 ft will be used for both packings.

2. *Calculate the loadings, the capacity factors, and the superficial F factor at the design conditions.*

For the design conditions given and a tower of 5 ft diameter, the loadings and capacity factors are calculated as follows from Eqs. (11-2), (11-3), and (11-4):

$$U_L = \frac{w_l}{\rho_l A} = \frac{(150{,}000)(4)}{\pi(62.4)(5)^2}\,\frac{(7.48)}{(60)} = 15.26 \text{ gal/(min)(ft}^2\text{)}$$

$$C_L = 15.26\sqrt{\frac{62.4}{62.4 - 0.25}} = 15.29 \text{ gal/(min)(ft}^2\text{)}$$

$$U_g = \frac{w_g}{\rho_g A} = \frac{(75{,}000)(4)}{\pi(0.25)(5)^2}\,\frac{1}{3{,}600} = 4.24 \text{ ft/s}$$

$$C_V = U_g\sqrt{\frac{\rho_g}{\rho_l - \rho_g}} = 4.24\sqrt{\frac{0.25}{62.4 - 0.25}} = 0.269 \text{ gal/(min)(ft}^2\text{)}$$

The superficial F factor F_S is calculated as follows:

$$F_S = U_g\sqrt{\rho_g} = 4.24(0.25)^{1/2} = 2.12 \text{ (ft/s)(lbm/ft}^3\text{)}^{1/2}$$

3. *Construct the dry-pressure-drop line on log-log coordinates (optional).*

For turbulent flow, the gas-phase pressure drop for frictional loss, contraction and expansion loss, and directional change loss are all proportional to the square of the superficial F factor. For the dry packing the pressure drop can be calculated from the equation

$$\Delta P/\Delta Z = K_1 F_S^2 \tag{11-6}$$

where ΔZ represents a unit height of packing (e.g., 1 ft) and K_1 is a constant whose value depends on the packing size and geometry. Values of K_1 can be determined from pressure-drop data published by packing vendors and are included in Table 11-1. For the two packings of this problem, the values of K_1 are 0.069 for 2-in FR and 0.041 for FP2Y. Using Eq. (11-6) with the design F factor calculated in step 2 gives the following. For 2-in FR,

$$\Delta P/\Delta Z = (0.069)(2.12)^2 = 0.310 \text{ in } wc\text{/ft}$$

For FP2Y,

$$\Delta P/\Delta Z = (0.041)(2.12)^2 = 0.184 \text{ in } wc\text{/ft}$$

This calculation indicates that for the same nominal capacity, the random packing has a 68 percent higher pressure drop per foot than the structured packing.

To construct the dry-pressure-drop line, plot $\Delta P/\Delta Z$ vs. F_S on log-log coordinates. Select appropriate values of F_S (e.g., 1 and 3 for this example) and use Eq. (11-6) to calculate the corresponding pressure drop per foot (e.g., for 2-in FR, 0.069 in wc/ft at $F_S = 1$ and 0.621 in/ft at $F_S = 3$; and for FP2Y, 0.041 in/ft at $F_S = 1$ and 0.369 in/ft at $F_S = 3$). Plot the points on log-log coordinates and connect them with a straight line.

4. *Construct the wet-pressure-drop line, on log-log coordinates, for pressure drops below column loading.*

For most packings of commercial interest, the relationship between the wet and dry pressure drops can be given by

$$\frac{(\Delta P/\Delta Z)_W}{(\Delta P/\Delta Z)_D} = \frac{1}{\left(1 - \dfrac{0.02\, U_L^{0.8}}{\epsilon}\right)^{2.5}} \tag{11-7}$$

where the subscripts W and D refer to wet and dry, respectively, and ϵ is the void fraction of the dry packed bed. Values of ϵ for the various packings are given in Table 11-1. The liquid loading, U_L in Eq. (11-7), is in units of gallons per minute per square foot. For the random packing,

$$\frac{(\Delta P/\Delta Z)_W}{(\Delta P/\Delta Z)_D} = \frac{1}{\left(1 - \dfrac{0.02(15.26)^{0.8}}{0.98}\right)^{2.5}} = 1.64$$

For the structured packing, the numerical result is the same.

To construct the wet-pressure-drop line, select values of F_S that span the design value. For example, at $F_S = 1$, the pressure drop for random packing is (1.64)(0.069)(1) = 0.113 in *wc*/ft, and that for structured packing is (1.64)(0.041)(1) = 0.0672 in *wc*/ft. At $F_S = 3$, the pressure drop for random packing is (1.64)(0.069)(9) = 1.018 in *wc*/ft, and that for structured packing is (1.64)(0.041)(9) = 0.605 in *wc*/ft. Plot these points on log-log coordinates and connect the points for each packing with a straight line. The solid lines of Fig. 11-1 are the wet-pressure-drop lines for the two packings.

5. *Determine the pressure drop and superficial F factor at flooding.*

For the air-water system at ambient conditions, flooding occurs at a pressure drop of about 2 in *wc*/ft. The pressure drop at flooding for other systems can be calculated by

$$\Delta P_F = (2 \text{ in } wc/\text{ft})[(\rho_l - \rho_g)/(62.4 - 0.075)] \tag{11-8}$$

Employing Eq. (11-8) with the density values in this problem gives a pressure drop at flooding of 1.99 in *wc*/ft.

The superficial F factor at flooding F_{SF} is determined with the aid of the flooding equation, Eq. (11-1). Since the pressure drop curve is being constructed for a constant liquid loading, the value of C_{LF} in Eq. (11-1) is 15.29 gal/(min)(ft²) as determined in step 2. With appropriate values for s and c, solve Eq. (11-1) for C_{VF} and determine F_{SF} from

$$F_{SF} = C_{VF}(\rho_l - \rho_g)^{1/2}$$

Thus, for 2-in FR,

$$C_{VF} = [0.74 - (0.040)(15.29)^{1/2}]^2 = 0.340 \text{ ft/s}$$

and accordingly $F_{SF} = 0.340(62.4 - 0.25)^{1/2} = 2.68$ (ft/s)(lbm/ft³)$^{1/2}$.

For FP2Y,

$$C_{VF} = [0.75 - (0.044)(15.29)^{1/2}]^2 = 0.334 \text{ ft/s}$$

and $F_{SF} = 0.334(62.4 - 0.25)^{1/2} = 2.63$ (ft/s)(lbm/ft³)$^{1/2}$

Locate the flood points for the two packings on the pressure drop plot at the coordinates ΔP_F and F_{SF}. See Fig. 11-1.

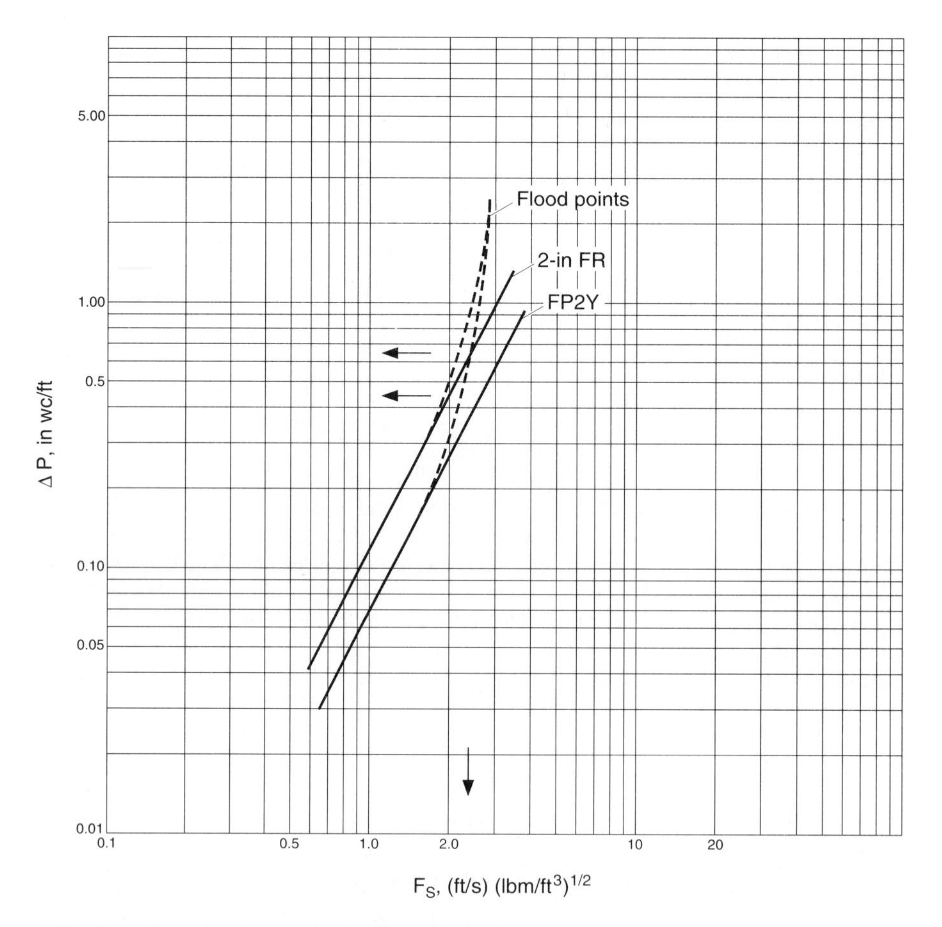

FIG. 11-1 Pressure-drop curves for Example 11-1.

6. Draw the wet-pressure-drop curve for the loading region.

The "loading region" covers a range of F factors from flooding downward to roughly half of the flooding F factor. This is the region in which the gas flow causes additional liquid holdup in the packing and produces a pressure drop higher than the one indicated by the straight lines of Fig. 11-1.

For each packing, sketch an empirical curve such that: (a) its upper end passes through the point F_{SF}, ΔP_F with a slope approaching infinity, and (b) its lower end becomes tangent to the straight, wet-pressure-drop line at an abscissa value of $F_{SF}/2$. These two pressure-drop curves for the loading regions are shown by the dashed lines of Fig. 11-1.

7. Determine the pressure drops from the curves.

The overall pressure-drop curve for each packing consists of the solid curve at low gas loadings and the dashed curve at loadings between the load point and the flood point. At the design F factor of 2.12 (ft/s)(lbm/ft^3)$^{1/2}$, the pressure drops determined from the curves are

For 2-in FR, 0.65 in wc/ft
For FP2Y, 0.44 in wc/ft

Thus, the pressure drop of the random packing is 48 percent higher than that of the structured packing at the design conditions.

Related Calculations: It is usual practice in distillation operations to keep the liquid-to-vapor ratio constant as the throughput is varied. When this is the case, the percent of flood is usually defined at constant L/V rather than at constant L. The procedures for solving for the tower diameter are similar to those in step 1, except that the liquid-capacity factor at flooding is instead given by $C_{LF} = C_L/f$. For the present case, this would introduce a factor of 0.8 in the denominator of the second term in the brackets of Eq. (11-5).

11-2 Hydraulic Rating of a Packed Tower

A 5.5-ft-diameter tower is to be used to countercurrently contact a vapor stream and a liquid stream. The mass flow rates are 150,000 lb/h for both. The liquid density is 50 lb/ft^3, and the gas density 1 lb/ft^3. Determine the approach to flooding at constant liquid loading L and at constant liquid-to-vapor ratio (L/V) for Flexipac type 2Y (FP2Y) structured packing.

Calculation Procedure:

1. Calculate the liquid and vapor capacity factors, C_L and C_V.

For information about these factors, see Problem 11-1. Use Eqs. (11-2), (11-3), and (11-4) from that problem to make the required calculation, bearing in mind that for a circular cross section, $A = \pi D^2/4$:

$$C_L = \frac{(150000)(4)}{(50)(\pi)(5.5)^2}\,\frac{7.48}{60}\sqrt{\frac{50}{50-1}} = 15.90 \text{ gal/(min)(ft}^2\text{)}$$

$$C_V = \frac{(150000)(4)}{(1)(\pi)(5.5)^2}\,\frac{1}{3600}\sqrt{\frac{1}{50-1}} = 0.250 \text{ ft/s}$$

2. Determine the flooding constants s and c.

From Table 11-1, the values of the flooding constants for FP2Y are

$$s = 0.044$$
$$c = 0.75$$

3. Determine the approach to flooding at constant L.

Use Eq. (11-1) with $C_{VF} = C_V/f$ and $C_{LF} = C_L$, where f is the fractional approach to flooding. Solve for f to give

$$f = \{(0.250)^{1/2}/[0.75 - (0.044)(15.90)^{1/2}]\}^2 = 0.757, \text{ or } 75.7 \text{ percent}$$

***4. Determine the approach to flooding at constant* L/V.**

Use Eq. (11-1) with $C_{VF} = C_V/f$ and $C_{LF} = C_L/f$ where f is the fractional approach to flooding. Solve for f to give

$$f = \{[(0.250)^{1/2} + (0.044)(15.90)^{1/2}]/0.75\}^2 = 0.811, \text{ or } 81.1 \text{ percent}$$

Related Calculation: To complete the hydraulic rating, the pressure drop at the design conditions is determined in the same way as for Problem 11-1.

11-3 Required Packing Height for Absorption with Straight Equilibrium and Operating Lines

An air stream containing 2% ammonia by volume (molecular weight = 28.96) is to be treated with water to remove the ammonia to a level of 53 ppm (the odor threshold concentration). The tower is to operate at 80 percent of flood defined at constant liquid to gas ratio (L/G). The liquid to vapor ratio is to be 25 percent greater than the minimum value. The absorption is to occur at 80°F and atmospheric pressure. What is the height of packing required for 2-in Flexiring random packing (2-in FR), and what height is required for Flexipac Type 2Y structured packing (FP2Y)? Determine the height using three methods: individual transfer units, overall transfer units, and theoretical stages.

Calculation Procedure:

1. Construct the equilibrium line.

The equilibrium line relates the mole fraction of ammonia in the gas phase to that in the liquid phase when the two phases are at equilibrium. Equilibrium is assumed to exist between the two phases only at the gas-liquid interface. For dilute systems, Henry's law will apply. It applies for liquid mole fractions less than 0.01 in systems in general, and, as can be seen in Example 11-5 , for the ammonia-water system it applies to liquid mole fractions as high as about 0.03. (Equilibrium data for this system are given in Perry's *Chemical Engineers' Handbook,* 4th ed., McGraw-Hill, New York, 1963, p. 14-4.)

From the data at low concentration, determine the Henry's law constant as a function of temperature for the temperature range of interest. To interpolate over a limited temperature range, fit the data to an equation of the form $\log_{10} \text{He} = a + (b/T)$. For the data cited,

$$\log_{10} \text{He} = 5.955 - (1778 \text{ K})/T \tag{11-9}$$

where T is absolute temperature in Kelvins and He is the Henry's law constant in atmospheres. Applying this equation at 80°F gives an He value of 1.06 atm.

Henry's law constant is defined as

$$\text{He} = Py_i/x_i \tag{11-10}$$

where P is the total pressure in atmospheres and y_i and x_i are, respectively, the gas-phase and liquid-phase mole fractions of ammonia in equilibrium with each other at the interface. The slope of the equilibrium line on the x-y operating diagram is

$$m = y_i/x_i = \text{He}/P \tag{11-11}$$

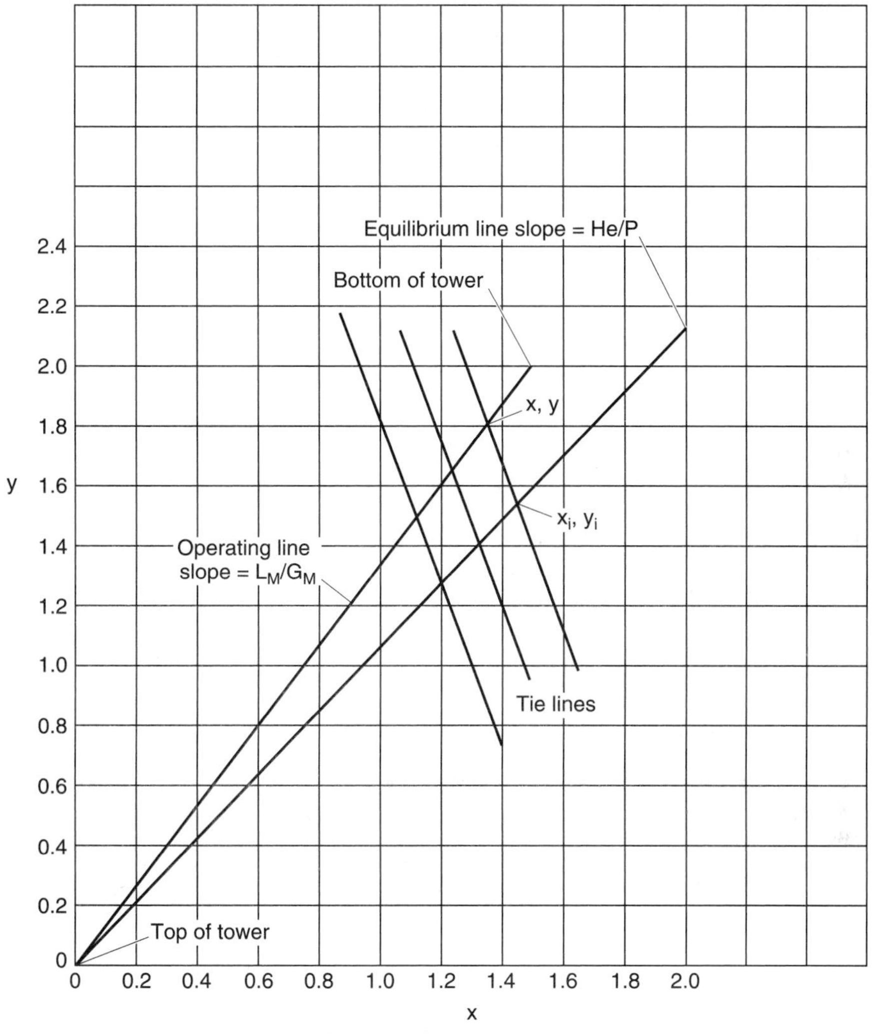

FIG. 11-2 *x-y* operating diagram for Example 11-3.

For operation at 1 atm, the equilibrium line is a straight line of slope 1.06 passing through the origin of the plot as shown in Fig. 11-2.

2. Locate the operating line of minimum slope.

The operating line gives the relationship between the bulk gas and liquid concentrations throughout the tower. A material balance around the tower is as follows:

$$L_M(x_1 - x_2) = G_M(y_1 - y_2) \tag{11-12}$$

or

$$L_M/G_M = (y_1 - y_2)/(x_1 - x_2)$$

where L_M and G_M are the liquid and gas molar fluxes (for example, in pound-moles per square foot second), respectively, and the subscripts 1 and 2 refer to the bottom and top of the tower, respectively. For dilute systems, L_M and G_M can be taken as constants over the tower. From Eq. (11-12), the slope of the operating line is L_M/G_M, which for this example may be considered constant. This gives a straight operating line that can be constructed on the x-y diagram from a knowledge of the mole fractions at the top and bottom of the tower.

The operating line of minimum slope is the operating line that just touches the equilibrium line at one end (in this case, the bottom) of the tower. At the top of the tower, $y_2 = 53/10^6 = 5.3 \times 10^{-5}$ and $x_2 = 0$. At the bottom of the tower, $y_1 = 0.020$ and, by Eq. (11-11), $x_1 = 0.020/1.06 = 0.01887$. The slope of the operating line of minimum slope is $(L_M/G_M)_{\min} = (0.020 - 5.3 \times 10^{-5})/(0.01887 - 0) = 1.057$. Because the conditions at the top of the tower put the end of the operating line so close to the origin of the x-y diagram, the operating line of minimum slope is essentially coincident with the equilibrium line. In Fig. 11-2, the operating line of minimum slope would lie just above the equilibrium line but would be indistinguishable from it unless the plot in the region of the origin were greatly expanded.

3. Construct the actual operating line.

The design specifies operating at an L/G value of 1.25 times the minimum value determined in step 2. Accordingly, $L_M/G_M = (1.25)(1.057) = 1.32$. The values of y_1, y_2, and x_2 are given and remain the same as for step 2. The value of x_1 is determined from Eq. (11-12): $x_1 = (0.020 - 5.3 \times 10^{-5})/1.32 = 1.51 \times 10^{-5}$. Construct a straight line through the end points x_1, y_1 and x_2, y_2 to give the operating line as shown in Fig. 11-2.

4. Determine the liquid and gas loadings.

The liquid and gas loadings are defined by the liquid-to-gas ratio determined in step 3 and by the stipulation that the tower is to operate at an approach to flooding of 80 percent at constant L/G. For flooding defined at constant L/G, Eq. (11-1) becomes

$$(C_V/f)^{1/2} + s(C_L/f)^{1/2} = c \qquad (11\text{-}13)$$

From the definitions of C_V and C_L, their ratio can be related to the ratio of molar fluxes by

$$C_L/C_V = (L_M/G_M)(M_{wl}/M_{wg})(\rho_g/\rho_l)^{1/2}\,(7.48)(60)\ (\text{gal}\cdot\text{s})/(\text{min})(\text{ft}^3) \qquad (11\text{-}14)$$

In this equation, ρ_g is the gas density, ρ_l is the liquid density (62.3 lb/ft³), M_{wl} is the molecular weight of the liquid (18.02), M_{wg} is the molecular weight of the gas, and the numerical values 7.48 and 60 are the number of gallons per cubic foot and the number of seconds per minute, respectively. Use of the ideal-gas law yields a gas density of 0.0735 lb/ft³. Accordingly, Eq. (11-14) becomes $C_L/C_V = (1.32)(18.02/28.96)(0.0735/62.3)^{1/2}(448.8) = 12.7$ (gal · s)/(min)(ft³). Therefore, C_L in Eq. (11-13) can be replaced by $12.7\,C_V$. Upon rearrangement of Eq. (11-13) and substitution of the appropriate values of c and s from Table 11-1, we have

For 2 in FR, $$C_V = \left(\frac{(0.74)(0.80)^{1/2}}{1 + 0.040\sqrt{12.7}}\right)^2 = 0.336 \text{ ft/s}$$

For FP2Y, $$C_V = \left(\frac{(0.75)(0.80)^{1/2}}{1 + 0.044\sqrt{12.7}}\right)^2 = 0.336 \text{ ft/s}$$

$$C_L = 12.7\,C_V = 4.27 \text{ gal/(min)(ft}^2)$$

Thus, for this case, the two packings give the same loadings and would require the same tower diameter to operate at 80 percent of flood.

From the definition of the capacity factors and the F factor (see Example 11-1), alternative expressions for the gas and liquid loadings are

$$F_S = (0.336)(62.3 - 0.0735)^{1/2} = 2.65\ (\text{ft/s})(\text{lb/ft}^3)^{1/2},$$

$$U_g = 2.65/(0.0735)^{1/2} = 9.77\ \text{ft/s},$$

and

$$U_L = (4.27)((62.3 - 0.0735)/62.3)^{1/2} = 4.27\ \text{gal/(min)(ft}^2)$$

5. Determine the HTUs for the packings at the design conditions.

For these packings, data available from the vendor give the values of H_g, the height of a transfer unit for the gas film, and H_l, the height of a transfer unit for the liquid film. At the design conditions, these values are

Packing	H_g, ft	H_l, ft
2-in FR	1.73	0.87
FP2Y	1.14	0.55

6. Determine the major resistance to mass transfer.

The major resistance to mass transfer can be in either the gas film or the liquid film. The film with the greater resistance will exhibit the greater driving force in consistent units of concentration. For dilute systems, the rate of transport from bulk gas to bulk liquid per unit tower volume $N_A\alpha$ is given, at a particular elevation in the tower, by

$$N_A\alpha = k_g\alpha(P/RT)(y - y_i) = k_l\alpha\rho_{Ml}(x_i - x) \tag{11-15}$$

where $k_g\alpha$ and $k_l\alpha$ are the gas-side and liquid-side mass transfer coefficients, respectively, both in units of s^{-1}, P/RT and ρ_{Ml} are the gas and liquid molar densities, respectively, (lb · mol/ft³), and $y - y_i$ and $x_i - x$ are the gas and liquid driving forces, respectively. The ratio of driving force R_{DF} in the gas to that in the liquid is

$$R_{DF} = (y - y_i)/(x_i - x)m = k_l\alpha\rho_{Ml}/k_g\alpha(P/RT)m \tag{11-16}$$

where m is the slope of the equilibrium line.

This equation can be put into a more useful form. The stripping factor λ is defined as the ratio of the slope of the equilibrium line to that of the operating line:

$$\lambda = m/(L_M/G_M) \tag{11-17}$$

For dilute systems, the height of a transfer unit for the gas resistance H_g and that for the liquid resistance H_l are related to their respective mass transfer coefficients by

$$H_g = U_g/k_g\alpha \quad \text{and} \quad H_l = U_l/k_l\alpha \tag{11-18}$$

Noting that $L_M = U_l\rho_{Ml}$ and $G_M = U_g(P/RT)$, we can use Eqs. (11-17) and (11-18) to convert Eq. (11-16) to

$$R_{DF} = H_g/\lambda H_l \tag{11-19}$$

If R_{DF} is greater than unity, the gas phase provides the major resistance to mass transfer; if R_{DF} is less than unity, the liquid phase provides the major resistance to mass transfer. From Eq. (11-17), $\lambda = 1.06/1.32 = 0.80$. For 2-in FR, $R_{DF} = 1.73/(0.80)(0.87) = 2.48$. For FP2Y, $R_{DF} = 1.14/(0.80)(0.55) = 2.59$. For both packings, the gas phase provides the major resistance to mass transfer. Therefore, greater precision will be obtained in the calculation of the packed height by using gas phase transfer units.

7. Construct tie lines on the x-y operating diagram.

Tie lines are straight lines that connect corresponding points on the operating and equilibrium lines. The intersection of the tie line with the operating line gives the bulk concentration at a particular point in the tower; the intersection of the tie line with the equilibrium line gives the interfacial concentration at the same point in the tower. The slope of the tie line *TLS* is obtained from Eq. (11-15):

$$TLS = (y - y_i)/(x - x_i) = -k_l a \rho_{Ml}/k_g a(P/RT) = (H_g/H_l)(L_M/G_M) \qquad (11\text{-}20)$$

Various tie lines can be drawn between the operating and equilibrium line as shown in Fig. 11-2. (The diagram ignores the slight difference between the slopes for 2-in FR and for FP2Y.) These lines are useful in understanding the relationships between the various concentrations in the operating diagram. In the general case, the tie lines are essential in relating the bulk and interfacial concentrations so that the mass-transfer equations can be integrated. For complete gas-phase control, the tie line will be vertical; for complete liquid-phase control, the tie line will be horizontal.

8. Determine the equations for the height of packing.

Since the gas phase provides the major resistance to mass transfer, use the rate equation for transport across the gas film in the material balance on the gas phase in a differential height of the packed bed. Upon integration, this leads to the general equation for the height of packing:

$$Z = \int_{y_2}^{y_1} \frac{U_g}{k_g a y_{BM}} \frac{y_{BM}\, dy}{(1-y)(y-y_i)} \qquad (11\text{-}21)$$

where Z is the height of packing and y_{BM} is the log-mean average concentration of the nondiffusing gas (air in this case) between the bulk and the interface at a particular elevation in the tower. This quantity is considered in more detail in Example 11-5. For dilute systems such as the current example, y_{BM} and $(1 - y)$ are both approximately unity throughout the tower. With this simplification, Eq. (11-21) becomes

$$Z = \int_{y_2}^{y_1} \frac{U_g}{k_g a} \frac{dy}{(y-y_i)} = H_g \int_{y_2}^{y_1} \frac{dy}{(y-y_i)} = H_g N_g \qquad (11\text{-}22)$$

In this equation, it is assumed that H_g is constant over the height of the packed bed and can therefore be removed from the integral. The remaining integral function of y is the number of transfer units based on the gas film resistance N_g.

For dilute systems in which the equilibrium line and operating lines are straight, various analytical solutions of Eq. (11-22) are available. These may be listed as follows.

Individual Gas-Side Transfer Units:

$$Z = H_g N_g$$

$$N_g = \frac{1 + (1/R_{DF})}{1-\lambda} \ln\left[(1-\lambda)\left(\frac{y_1 - mx_2}{y_2 - mx_2}\right) + \lambda\right] \qquad (11\text{-}23)$$

$$R_{DF} = \frac{H_g}{\lambda H_l}$$

$$\lambda = \frac{m}{L_M/G_M} = \frac{\text{He}/P}{L_M/G_M}$$

When the absorption is completely gas-phase limited, R_{DF} approaches infinity and the $1/R_{DF}$ term of the above equation for N_g drops out.

Overall Gas-Side Transfer Units: The overall gas-side transfer units are calculated by assuming that all of the resistance to mass transfer is in the gas phase. The effect of liquid-phase resistance is taken into account by adjusting the height of the transfer unit from H_g to H_{og}. The equations are

$$Z = H_{og} N_{og}$$

$$H_{og} = H_g + \lambda H_l$$

$$N_{og} = \frac{1}{1-\lambda} \ln \left[(1-\lambda) \frac{y_1 - mx_2}{y_2 - mx_2} + \lambda \right] \tag{11-24}$$

$$\lambda = \frac{m}{L_M/G_M} = \frac{\text{He}/P}{L_M/G_M}$$

Overall Gas-Side Transfer Units Using the Log-Mean Driving Force: An alternative expression can be used to calculate N_{og} based on the log-mean concentration driving force across the tower. The equations are

$$N_{og} = \frac{y_1 - y_2}{(y - y^*)_{LM}} \tag{11-25}$$

$$(y - y^*)_{LM} = \frac{(y_1 - mx_1) - (y_2 - mx_2)}{\ln [(y_1 - mx_1)/(y_2 - mx_2)]}$$

where y is the bulk concentration and y^* is the interfacial concentration, assuming that all the resistance to mass transfer is in the gas phase; that is, assuming that the tie lines of Fig. 11-2 are vertical. The liquid mole fraction at the bottom of the tower x_1 is determined by material balance around the tower.

Theoretical Stages: An analytical solution is also available for theoretical stages as opposed to transfer units, and the height equivalent to a theoretical plate (HETP). The number of theoretical stages or plates N_P can be determined by counting the steps between the operating line and the equilibrium line as is done with distillation problems (see Example 8-1 in Section 8, Distillation), but a more convenient analytical solution is as follows:

$$Z = (\text{HETP}) N_P$$

$$\text{HETP} = H_{og} \frac{\ln \lambda}{\lambda - 1}$$

$$H_{og} = H_g + \lambda H_l \tag{11-26}$$

$$N_P = \frac{\ln \left[(1-\lambda) \frac{(y_1 - mx_2)}{(y_2 - mx_2)} + \lambda \right]}{\ln \left(\frac{1}{\lambda} \right)}$$

All of these forms are mathematically equivalent and give identical results for the height of the packing.

TABLE 11-2 Values of Variables and Calculated Results for Packing Height

Parameter	Value for 2-in FR	Value for FP2Y
H_g	1.73 ft	1.14 ft
H_l	0.87 ft	0.55 ft
m	1.06	1.06
λ	0.80	0.80
R_{DF}	2.48	2.59
y_1	0.020	0.020
y_2	5.3×10^{-5}	5.3×10^{-5}
x_1	0.0151	0.0151
x_2	0	0
N_g	30.4	30.0
N_{og}	21.67	21.67
N_{og} (log mean)	21.67	21.67
N_P	19.42	19.42
H_{og}	2.43 ft	1.58 ft
HETP	2.71 ft	1.76 ft
Z	52.6 ft	34.2 ft

9. Calculate the height of packing.

Table 11-2 gives the values of the variables and the calculated results for Eqs. (11-23) through (11-26). The various sets of equations all yield the same height of packing. The height for the random packing is 54 percent greater than that of the structured packing. An appropriate safety factor should be added to the height of both packings for this and for subsequent examples.

11-4 Packing Height for Stripping with Straight Equilibrium and Operating Lines

A stream of groundwater flowing at a rate of 700 gal/min and containing 100 ppm of trichloroethylene (TCE) is to be stripped with air to reduce the TCE concentration to 5 ppb (drinking water quality). The tower is to be packed with 2-in polypropylene slotted rings and is to operate at 30 percent of flooding defined at constant liquid rate L, at a gas flow rate four times the minimum. Determine the tower diameter and height. Assume isothermal operation at 50°F and 1 atm. The density of the liquid stream is 62.4 lb/ft^3; that of the gas stream is 0.0778 lb/ft^3.

Calculation Procedure:

1. Construct the equilibrium line.

For dilute solutions, Henry's law will apply. In general, it will be applicable for pressures under 2 atm and liquid mole fractions less than 0.01. For this example, the liquid mole fraction x is (100 lb TCE/10^6 lb soln) (18/132) = 1.37×10^{-5}, where 18 and 132 are the molecular weights of water and trichloroethylene, respectively. Therefore, Henry's law will apply.

The Henry's law constant can be obtained from the literature (e.g., C. Munz and

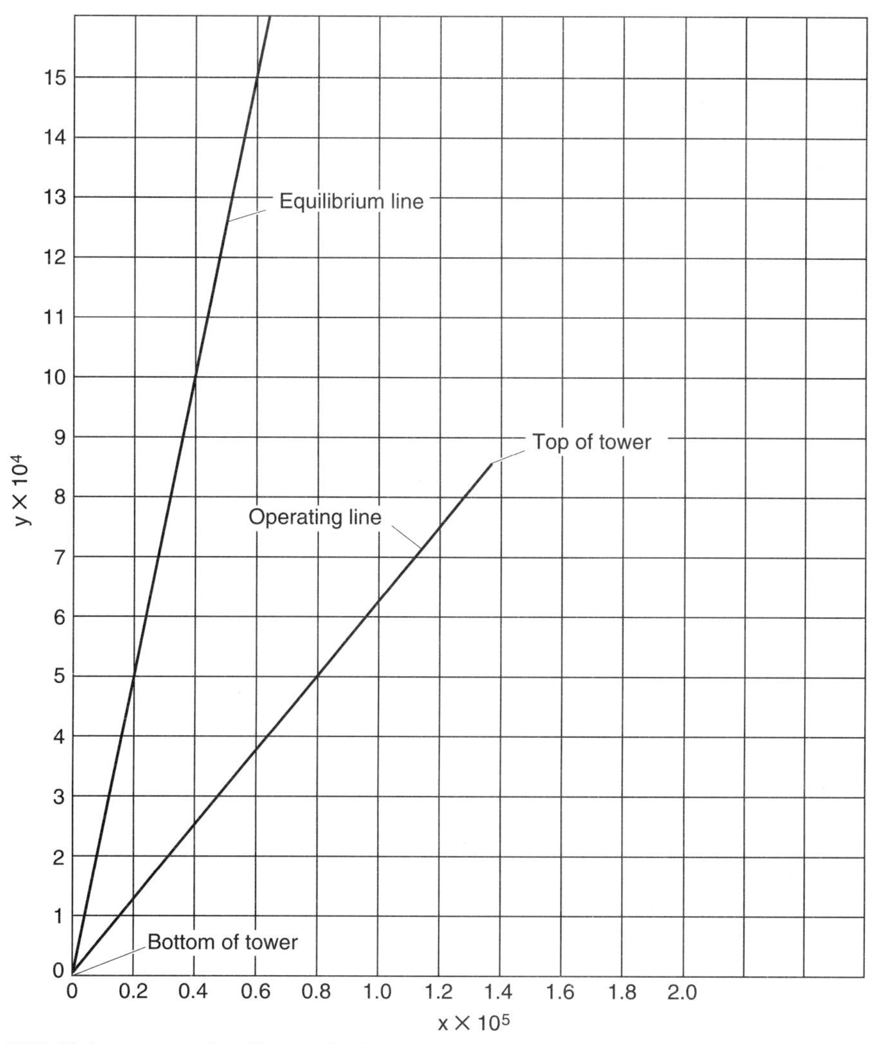

FIG. 11-3 *x-y* operating diagram for Example 11-4.

P. V. Roberts, *Journal of the American Water Works Association,* 79(5), pp. 62–69, 1987). For a temperature of 50°F, a value of 248 atm is calculated.

For operation at 1 atm, accordingly, the equilibrium line is a straight line of slope 248 as shown in Fig. 11-3.

2. Locate the operating line for minimum gas-phase molar flow rate, G_M.

Because this system is dilute, the operating line will be straight. At the bottom of the tower, the concentration of TCE in the air coming into the tower may be assumed to be zero. The liquid mole fraction at 5 ppb, x_1 is $(5 \text{ lb}/10^9 \text{ lb})(18/132)$, or 6.85×10^{-10}. This places the coordinates for the bottom of the operating line very close to the origin of the x-y operating diagram. (For stripping, the operating line lies below the equilibrium line. For absorption, as seen in the preceding example, it lies above the equilibrium line.) The liquid mole fraction at the top of the tower x_2 is 1.37×10^{-5} as calculated in step 1.

The maximum slope of the operating line occurs when one end of it (in this case the top) intersects the equilibrium line. From Eq. (11-10), $y_2 = (248)(1.37 \times 10^{-5})$, or 0.00340. If we use these coordinates in Eq. (11-12), the maximum slope of the operating line is essentially the same as the slope of the equilibrium line, 248. Since the liquid-phase molar flow rate L_M is constant for this example, the maximum operating-line slope defines the minimum value of G_M.

3. Locate the operating line.

The actual gas loading is to be four times the minimum value. Thus, the slope of the operating line is one-fourth that of the operating line of maximum slope, or essentially one-fourth that of the equilibrium line. Thus, $L_M/G_M = 248/4 = 62$. By Eq. (11-12) with $y_1 = 0$, we calculate y_2 as $(62)(1.37 \times 10^{-5} - 6.85 \times 10^{-10})$, or 8.49×10^{-4}. Connect the coordinates at the bottom and the top of the tower (x_1, y_1 and x_2, y_2, respectively) with a straight line to give the operating line as shown in Fig. 11-3.

4. Calculate the gas and liquid loadings and the tower diameter.

Use Eq. (11-1) with $C_{VF} = C_V/f$ and $C_{LF} = C_L$ where f is the fractional approach to flooding, defined at constant liquid loading for this example. Express C_L in terms of C_V using Eq. (11-14). With $L_M/G_M = 62$, $\rho_l = 62.4$ lb/ft³, and $\rho_g = 0.0778$ lb/ft³, Eq. (11-14) gives $C_L/C_V = 611$ (gal · s)/(min)(ft³).

Values of s and c for 2-in slotted rings in polypropylene may be obtained from the packing vendor or may be determined from air-water pressure-drop curves for the packing. The values are 0.042 and 0.72, respectively, in the units of Table 11-1. (Note that these values for polypropylene packing are close to the corresponding values for the same size of stainless steel packing in Table 11-1.) Substituting these values into Eq. (11-1) (as in step 4 of the previous example) gives

$$(C_V/0.3)^{1/2} + 0.042(611C_V)^{1/2} = 0.72$$

from which $C_V = 0.0632$ ft/s. And $C_L = 611C_V = 38.6$ gal/(min)(ft²). From the definition of the capacity factors and F factor, other measures of the gas and liquid throughput may be given as $F_S = 0.499$ (ft/s)(lb/ft³)$^{1/2}$, $U_g = 1.79$ ft/s, and $U_L = 38.6$ gal/(min)(ft²).

Determine the tower diameter from the given flow rate and the liquid loading calculated above. The tower area is (700 gal/min)/[38.6 gal/(min)(ft²)] = 18.1 ft². The tower diameter D is $[(18.1)(4)/\pi]^{1/2} = 4.8$ ft.

5. Determine the HTU values for the liquid and gas resistances.

Determining the values of H_l and H_g to use is usually the most uncertain part of absorption and stripping calculations. Values can often be obtained from the packing vendors. Generalized correlations are available in the literature, but these may not always give reliable results for aqueous systems. Published data on absorption or stripping of highly or sparingly soluble gases such as ammonia and carbon dioxide can be used, with appropriate adjustments. This method is illustrated in Example 11-6, where the following values are calculated for the conditions of this present example: $H_l = 3.48$ ft, and $H_g = 0.54$ ft.

6. *Determine the controlling resistance to mass transfer.*

The controlling resistance is determined from the ratio of the gas to the liquid driving force given by Eq. (11-19). From Eq. (11-17), $\lambda = 248/62 = 4.0$. With the HTUs from step 5, Eq. (11-19) gives $R_{DF} = 0.54/(4.0)(3.48) = 0.039$. Because R_{DF} is less than 1, the major resistance to mass transfer is on the liquid side of the interface. In this case, 96 percent of the resistance is in the liquid. Therefore, the best precision in the calculation of packing height will be obtained by using liquid film transfer units N_l or N_{ol}.

7. *Calculate the number of transfer units.*

For dilute systems in which the equilibrium and operating lines are both straight, the equations for the number of individual liquid-phase transfer units N_l and the number of overall liquid-phase transfer units N_{ol} are

$$N_l = \frac{1 + R_{DF}}{1 - (1/\lambda)} \ln \left\{ \left(1 - \frac{1}{\lambda}\right) \left[\frac{x_2 - (y_1/m)}{x_1 - (y_1/m)} \right] + \frac{1}{\lambda} \right\} \qquad (11\text{-}27)$$

$$N_{ol} = \frac{1}{1 - (1/\lambda)} \ln \left\{ \left(1 - \frac{1}{\lambda}\right) \left[\frac{x_2 - (y_1/m)}{x_1 - (y_1/m)} \right] + \frac{1}{\lambda} \right\}$$

Substituting the appropriate values into these equations gives

$$N_l = \frac{1 + 0.039}{1 - (1/4.0)} \ln \left[\left(1 - \frac{1}{4.0}\right) \left(\frac{1.37 \times 10^{-5} - 0}{6.85 \times 10^{-10} - 0} \right) + \frac{1}{4.0} \right] = 13.32$$

$$N_{ol} = \frac{1}{1 - (1/4.0)} \ln \left[\left(1 - \frac{1}{4.0}\right) \left(\frac{1.37 \times 10^{-5} - 0}{6.85 \times 10^{-10} - 0} \right) + \frac{1}{4.0} \right] = 12.82$$

8. *Calculate the height of packing required.*

Based on individual liquid-transfer units, $Z = H_l N_l = (3.48\text{ ft})(13.32) = 46.3$ ft. Based on overall liquid transfer units, $Z = H_{ol} N_{ol}$. The height of an overall transfer unit based on the liquid resistance is calculated as follows:

$$H_{ol} = H_l + H_g/\lambda \qquad (11\text{-}28)$$

From this equation, $H_{ol} = 3.48 + (0.54/4.0) = 3.615$ ft, and $Z = (3.615\text{ ft})(12.82) = 46.3$ ft.

Related Calculation: When the equilibrium and/or operating lines are curved, Eqs. (11-27) and (11-28) do not apply exactly. In this case, it is necessary to base the design on the individual number of transfer units for the liquid resistance and integrate graphically to determine N_l. This is illustrated in the following example.

11-5 Packed Height for Absorption with Curved Equilibrium and Operating Lines

An air stream containing 30% ammonia is to be contacted with a water stream containing 0.1% ammonia by weight (thus its mole fraction x_2 is 0.00106) in a tower packed with 2-in stainless steel Flexiring (FR) random packing. The air leaving the tower is to contain 1% ammonia. The tower is to be designed to operate at 80% of flood defined at constant liquid-to-gas ratio L/G, and this ratio is to be twice the minimum value. Although significant heat effects would normally occur for this absorption, assume isothermal operation at 80°F and 1 atm total pressure.

Calculation Procedure:

1. Construct the equilibrium line.

Equilibrium data for this system are given in *Perry's Chemical Engineers' Handbook,* 4th ed., p. 14-4, McGraw-Hill, New York (1963). From the data given, calculate values of y at temperatures of 10, 20, 30, and 40°C for a particular range of x values. For a particular x, fit the values of y to a regression equation of the form of Eq. (11-9) [i.e., $\log y = A + (B/T)$], and calculate the y value at 80°F (26.7°C). The equilibrium data interpolated in this way to 80°F are plotted to give the equilibrium line in Fig. 11-4. Henry's law is shown by the short dashed line in this figure. There is good agreement between the equilibrium curve and Henry's law below a liquid mole fraction of about 0.03 but increasing divergence as concentration increases above this value.

2. Construct the limiting operating line.

For concentrated solutions, significant changes will occur in the molar fluxes of gas and liquid. Therefore, base the material balance around the tower on the molar fluxes of the ammonia-free water and the ammonia-free air, L'_M and G'_M, respectively. Typical units are pound-moles of water (or air) per square foot second. The material balance then becomes

$$L'_M\left(\frac{x_1}{1-x_1} - \frac{x_2}{1-x_2}\right) = G'_M\left(\frac{y_1}{1-y_1} - \frac{y_2}{1-y_2}\right) \tag{11-29}$$

The unknowns in this equation are L'_M, G'_M, and x_1. The limiting operating line is that which just touches the equilibrium line at the conditions corresponding to the bottom of the tower. From Fig. 11-4, x_1 is determined to be 0.177 from the equilibrium curve at $y_1 = 0.30$. Equation (11-29) is then solved for L'_M/G'_M, and the values for the liquid and gas mole fractions are entered to give $(L'_M/G'_M)_{\min} = 1.955$.

It is not essential to actually construct the limiting operating line on the operating diagram. If desired, it can be constructed as shown below for the actual operating line, except that $(L'_M/G'_M)_{\min}$ is used rather than L'_M/G'_M.

3. Construct the actual operating line.

The actual L/G is specified to be two times the minimum L/G. Therefore, $L'_M/G'_M = 2(L'_M/G'_M)_{\min} = (2)(1.955) = 3.91$. Since the operating line is curved, corresponding values of x and y are needed at various heights between the top and bottom of the packing. These are determined by making a material balance around the top (or bottom) of the packing, where x and y are the mole fractions leaving the material balance control volume at some point in the bed. Equation (11-29) then becomes

$$\frac{y}{1-y} = \frac{L'_M}{G'_M}\left(\frac{x}{1-x} - \frac{x_2}{1-x_2}\right) + \frac{y_2}{1-y_2} = 3.91\left(\frac{x}{1-x} - 0.00106\right) + 0.0101 \tag{11-30}$$

Select various values of x between x_1 and x_2, substitute into Eq. (11-30), and solve for the

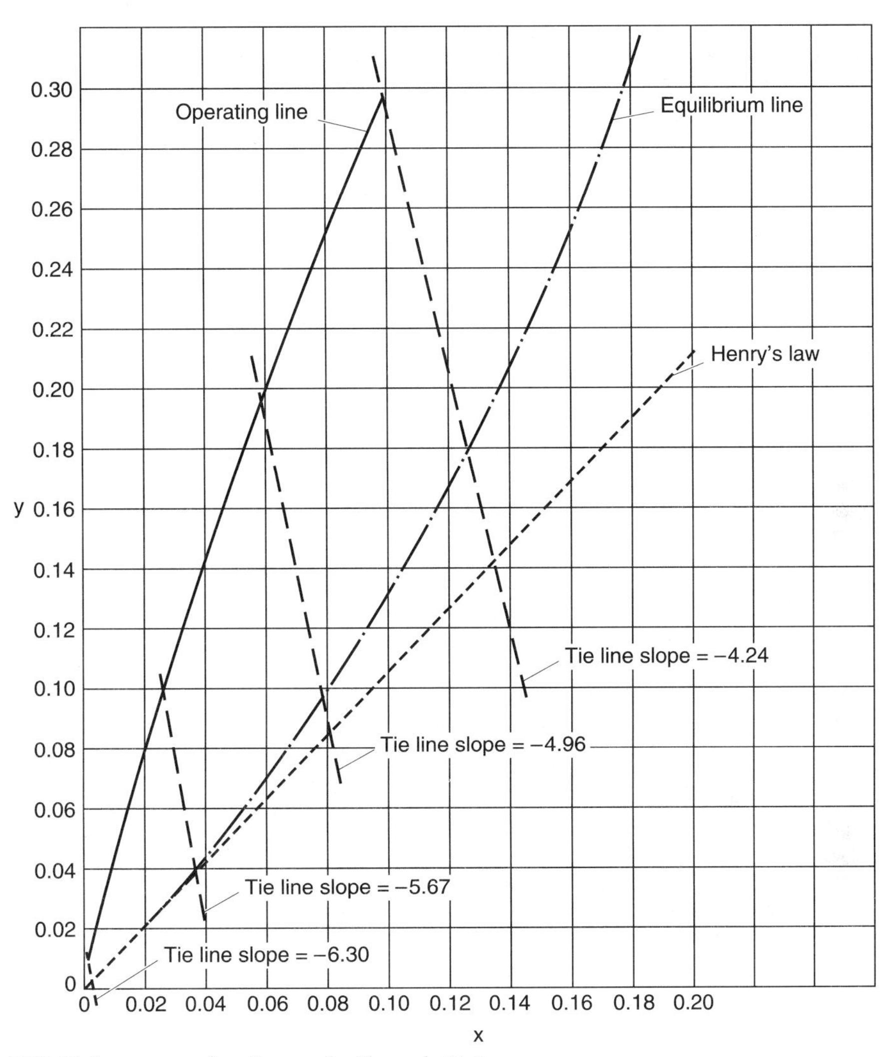

FIG. 11-4 *x-y* operating diagram for Example 11-5.

corresponding values of y. Plot these values on the x-y operating diagram and draw a smooth curve through the points to give the operating line shown in Fig. 11-4. Application of Eq. (11-30) for $y = y_1$ gives $x_1 = 0.0975$ for the mole fraction of ammonia in the liquid leaving the tower. This defines the end of the operating line.

4. Determine the gas and liquid loadings at the bottom and top of the tower.

The maximum loadings will occur at the bottom of the tower, where the approach to flooding is specified to be 80 percent defined at constant L/G. In order to apply Eq. (11-1), the gas and liquid densities must be determined, and the ratio of C_L/C_V must be related to L'_M/G'_M. From the average molecular weight of the gas entering the tower (25.38) and the ideal-gas law, the gas density is 0.0644 lb/ft³. From the concentration of ammonia in the liquid leaving the tower and from published data for specific gravity of aqueous ammonia solutions, the liquid density is 59.9 lb/ft³.

The value of L_M/G_M at the bottom of the tower is given by

$$L_M/G_M = (L'_M/G'_M)(1 - y_1)/(1 - x_1) = (3.91)(1 - 0.30)/(1 - 0.0975) = 3.03.$$

The ratio of C_L to C_V is given by Eq. (11-14), with the average molecular weight of the liquid equal to 17.92 and the average molecular weight of the gas equal to 25.38. Substituting the appropriate values into Eq. (11-14) gives $C_L/C_V = 31.5$ (gal · s)/(min)(ft³).

In Eq. (11-1), $C_{VF} = C_V/f$ and $C_{LF} = C_L/f$. The values of c and s for 2-in Flexiring packing are 0.74 and 0.040, respectively (see Table 11-1). Substituting these values into Eq. (11-1) gives $(C_V/0.80)^{1/2} + 0.040\,(31.5C_V/0.80)^{1/2} = 0.74$. This gives $C_V = 0.292$ ft/s and $C_L = 31.5C_V = 9.20$ gal/(min)(ft²). The loadings at the bottom of the tower in other units are calculated from the defining equations as $F_S = 2.26$ (ft/s)(lb/ft³)$^{1/2}$, $U_g = 8.90$ ft/s, $U_L = 9.20$ gal/(min)(ft²), and $U_l = 0.0205$ ft/s.

Calculate the loadings at the top of the tower from the loadings at the bottom of the tower:

$$U_{gt} = U_{gb}\frac{1 - y_1}{1 - y_2} = 8.90\frac{(1 - 0.3)}{(1 - 0.01)} = 6.29 \text{ ft/s}$$

$$U_{lt} = U_{lb}\frac{\rho_{lb}}{M_{wlb}}\frac{(1 - x_1)}{(1 - x_2)}\frac{M_{wlt}}{\rho_{lt}} = 0.0205\left(\frac{59.9}{17.9}\right)\left(\frac{1 - 0.0975}{1 - 0.00106}\right)\left(\frac{18.02}{62.2}\right) = 0.01796 \text{ ft/s}$$

where the subscripts t and b refer to the top and bottom of the tower, respectively, and the 18.02 and 62.2 are the molecular weight and density of the liquid at the top of the tower. In terms of other units for loading, the F factor at the top is $(6.29)(0.0732)^{1/2} = 1.70$ (ft/s)(lb/ft³)$^{1/2}$ and $U_L = (0.01796)(7.48)(60) = 8.06$ gal/(min)(ft²). (Note that the superficial velocity in feet per second is equivalent to the volumetric flux in cubic feet per square foot-second.)

5. *Determine the individual HTUs for the liquid and gas resistances at the bottom and top of the tower.*

Vendor data for carbon dioxide desorption from water for 2-in FR packing, when corrected to the temperature and system physical properties for this example, give $H_{lb} = 0.83$ ft and $H_{lt} = 0.80$ ft.

Vendor data for ammonia absorption into water for 2-in FR packing, when corrected for temperature, give $H_{gb} = 1.42$ ft and $H_{gt} = 1.25$ ft.

The values of H_l at bottom and top and H_g at bottom and top are close enough to permit the use of average values. The average values are $H_l = 0.82$ ft and $H_g = 1.34$ ft.

6. *Construct the tie lines.*

The tie line slope (TLS) is established using procedures similar to those in Example 3. Equations (11-15) and (11-16) also apply to concentrated systems. For such systems, however, $k_g\alpha$ and $k_l\alpha$ are inversely proportional to y_{BM} and x_{BM}, respectively, which are the log mean mole fractions, between bulk and interface, of the nondiffusing component (air or water, respectively).

For concentrated systems, y_{BM} and x_{BM} can vary considerably across the tower. Therefore, in Eq. (11-21) the numerator and denominator are multiplied by y_{BM}. The first term under the integral is H_g for concentrated systems. This term is theoretically independent of concentration and pressure and is equal to the H_g for dilute systems, for which y_{BM} equals 1 as pointed out in Example 3. The variation of y_{BM} is included in the second term of Eq. (11-21), which integrates to N_g for concentrated systems.

When the loadings at the top and bottom of the tower are such that the value of H_g for dilute systems does not vary appreciably over the tower, as in this example, the first term under the integral of Eq. (11-21) can be considered to be constant at its average value and can be removed from the integral. The equations for a concentrated system are

$$Z = H_g N_g$$

$$H_g = \frac{U_g}{k_g a y_{BM}} \qquad H_l = \frac{U_l}{k_l a x_{BM}}$$

$$N_g = \int_{y_2}^{y_1} \frac{y_{BM}\, dy}{(1-y)(y-y_i)}$$

$$\text{TLS} = -\frac{k_l a \rho_{Ml}}{k_g a (P/RT)} = -\frac{H_g}{H_l}\frac{L_M}{G_M}\frac{y_{BM}}{x_{BM}} = -\frac{H_g}{H_l}\frac{L'_M}{G'_M}\frac{(1-y)}{(1-x)}\frac{y_{BM}}{x_{BM}} \tag{11-31}$$

$$y_{BM} = \frac{(1-y)-(1-y_i)}{\ln\left(\dfrac{1-y}{1-y_i}\right)} \qquad x_{BM} = \frac{(1-x)-(1-x_i)}{\ln\left(\dfrac{1-x}{1-x_i}\right)}$$

In applying these equations, it is helpful to remember that the H_g for concentrated systems is the same as that for dilute systems at the same gas and liquid loadings. It is the $k_g a$ that varies with concentration and requires the introduction of y_{BM} to keep H_g constant.

Tie lines tie together corresponding concentrations on the operating line (bulk concentrations) and the equilibrium line (interfacial concentrations). The tie line slope is determined from the final equality for TLS in Eq. (11-31). The values of H_g and H_l are the average values determined in step 5 (1.34 ft and 0.82 ft, respectively). The value of L'_M/G'_M was determined in step 3 (3.91). Various tie lines are to be constructed over the entire concentration range covered by the operating line. A particular tie line will cross the operating line at x and y, the values of which are used in Eq. (11-31) to calculate y_{BM} and x_{BM}, and in the equation for TLS.

Unfortunately, the equations cannot be solved explicitly, because the equations for y_{BM} and x_{BM} contain the interfacial concentrations, which are unknown until the tie line slope is established. A trial-and-error procedure is required to construct the tie lines. The procedure is illustrated here for the tie line that intersects the operating line at $y = 0.20$:

1. Locate y on the operating line: $y = 0.20$.
2. Determine x from the operating line or from the material balance, Eq. (11-29): $x = 0.059$.
3. Estimate the tie line slope assuming $y_{BM}/x_{BM} = 1$. Thus, $\text{TLS}_1 = -(1.34/0.82)(3.91)(1-0.20)/(1-0.059) = -5.43$.
4. Draw the estimated tie line on the operating diagram.
5. At the intersection of the estimated tie line and the equilibrium line, determine the estimated interfacial concentrations: $y_{i1} = 0.096$ and $x_{i1} = 0.078$.
6. Calculate the estimated values of y_{BM} and x_{BM}. Thus, $y_{BM1} = [(1-0.20)-(1-0.096)]/\ln[(1-0.20)/(1-0.096)] = 0.851$. Similarly, $x_{BM1} = 0.931$.

7. Calculate the second-iteration tie line slope. Thus, $TLS_2 = -(1.34/0.82)(3.91)[(1 - 0.20)/(1 - 0.059)][(0.851)/(0.931)] = -4.96$.
8. Repeat steps 4 through 7 until there is no further change in the tie line slope. For this case, the third iteration gives no further change and accordingly the slope is -4.96.
9. Draw the final tie line on the operating diagram. Slope $= -4.96$ passing through $y = 0.20$, $x = 0.059$. See Fig. 11-4.

This procedure is repeated at various intervals along the operating line. The more tie lines that are constructed, the more accurate the integration to determine N_g. Figure 11-4 shows 4 of the 20 tie lines constructed for the solution of this example.

7. Determine the controlling resistance for mass transfer.

For concentrated solutions, the ratio of the gas-phase driving force to that in the liquid phase in equivalent concentration units is

$$R_{DF} = (H_g/\lambda H_l)(y_{BM}/x_{BM}) \tag{11-32}$$

This differs from Eq. (11-19) for dilute solutions only by the ratio of y_{BM} to x_{BM}. Based on the average values, $H_g/H_l = 1.34/0.82 = 1.634$. Values of the other quantities in Eq. (11-32) vary over the tower. λ is the ratio of the slope of the equilibrium line to that of the operating line. Values of λ can be obtained directly from the operating diagram by determining the slope of the operating line at one end of a particular tie line and the slope of the equilibrium line at the other end of the same tie line. Values of x_{BM} and y_{BM} can be determined with the values of x and x_i and y and y_i at either end of the same tie line.

At the bottom of the tower, the slope of the operating line is 2.48 and the slope of the equilibrium line (determined at the point where it intersects a tie line of slope -4.24) is 1.98, y_{BM} is 0.76, and x_{BM} is 0.89. Accordingly, $R_{DFb} = (1.634)(2.48/1.98)(0.76/0.89) = 1.75$. Using the corresponding values at the top of the tower gives $R_{DFt} = 5.90$. As these values are both greater than 1, the gas phase provides the major resistance to mass transfer throughout the tower. Therefore, using N_g to calculate the number of transfer units in the tower will give the greater precision.

8. Calculate the number of transfer units by graphic integration.

From Eq. (11-31), N_g is equal to the area under a plot of $f(y)$ vs. y between the limits of y_2 and y_1 where $f(y)$ is $y_{BM}/[(1 - y)(y - y_i)]$. For each point of intersection between the operating line and a tie line, $f(y)$ is calculated and plotted against y. The plot is shown in Fig. 11-5. The area under the curve from $y_2 = 0.01$ to $y_1 = 0.30$ is 5.74.

9. Calculate the height of packing.

The height of packing is $Z = H_g N_g = (1.34 \text{ ft})(5.74) = 7.7$ ft.

11-6 Determining HTUs for Air Stripping from Data on Model Systems

Determine the H_l and H_g to be used in Example 11-4 from data published for 2-in Pall rings on carbon dioxide stripping from 23°C water and on ammonia absorption into 20°C water. (See R. Billet and J. Mackowiak, *Chem. Eng. Technol.* 11:213–227, 1988.)

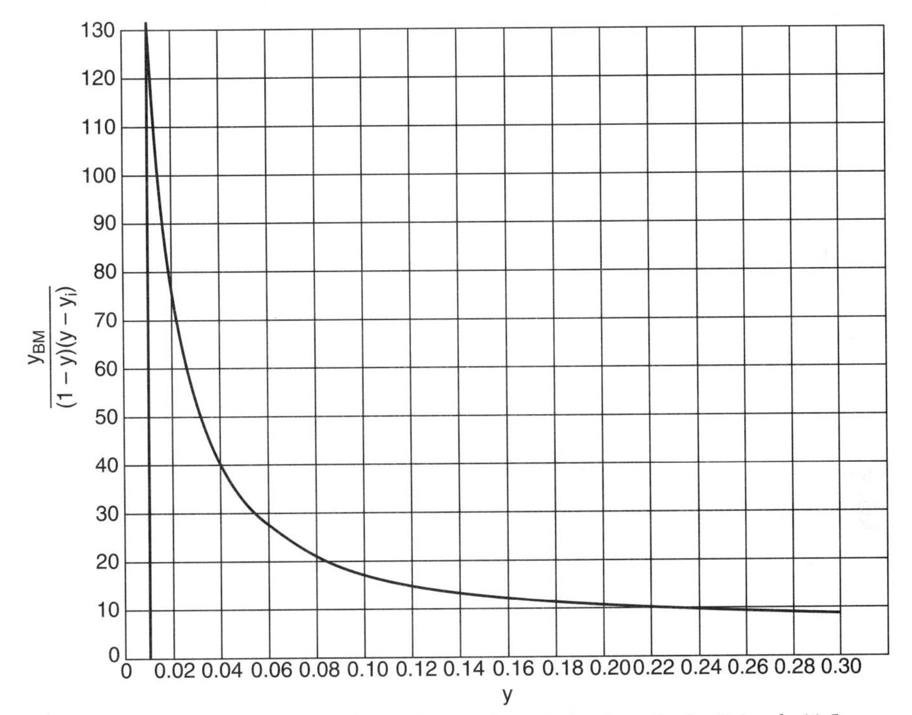

FIG. 11-5 Graphic integration to determine number of transfer units for Example 11-5.

Calculation Procedure:

1. Determine H_l for the model system at an equivalent liquid loading.

Below the load point, H_l is insensitive to the gas loading. Therefore, H_l can be correlated as a function of liquid loading only. For the packing of interest, the published data show a linear log-log relationship between $k_l\alpha$ and the volumetric liquid loading. This relationship can be described by the equation $k_l\alpha = 0.360U_l^{0.81}$, where $k_l\alpha$ is the mass transfer coefficient in s^{-1} and U_l is the superficial liquid velocity in feet per second (or equivalently, the superficial liquid flux in cubic feet per square foot–second). Since $H_l = U_l/k_l\alpha$, the equation for H_l becomes $H_l = 2.78U_l^{0.19}$, where H_l is in feet and U_l is in feet per second. For the liquid loading of Example 11-4, $U_l = 0.086$ ft/s ($U_L = 38.6$ gal/(min)(ft^2) and $H_l = 2.78(0.086)^{0.19} = 1.74$ ft. This is the value of H_l for air stripping carbon dioxide from 23°C water at the given liquid loading.

2. Correct the H_l to the actual system physical properties.

The value of H_l is corrected from the model system to the actual system via the relationship that H_l is proportional to the square root of the Schmidt number ($Sc = \mu/\rho\mathscr{D}$, where μ is viscosity, ρ is density, and $\mathscr{D}$ is diffusion coefficient). The quantities required for the Schmidt number are as follows. For the carbon dioxide–water system (CO_2–H_2O) at 23°C,

$$\mu = 0.936 \text{ cP} \qquad \rho = 62.4 \text{ lb/ft}^3 \qquad \mathscr{D} = 1.83 \times 10^{-5} \text{ cm}^2\text{/s}$$

For the trichloroethylene-water system (TCE–H_2O) at 50°F,

$$\mu = 1.31 \text{ cP} \qquad \rho = 62.4 \text{ lb/ft}^3 \qquad \mathscr{D} = 6.4 \times 10^{-6} \text{ cm}^2\text{/s}$$

$$\frac{(H_l)_{\text{TCE-H}_2\text{O}}}{(H_l)_{\text{CO}_2\text{-H}_2\text{O}}} = \left[\frac{\left(\frac{\mu}{\rho\mathscr{D}}\right)_{\text{TCE-H}_2\text{O}}}{\left(\frac{\mu}{\rho\mathscr{D}}\right)_{\text{CO}_2\text{-H}_2\text{O}}}\right] = \left[\frac{(1.31)}{(0.936)}\,\frac{(1.83 \times 10^{-5})}{(6.4 \times 10^{-6})}\right]^{1/2} = 2.0$$

Therefore, the value of H_l for trichloroethylene-water is (1.74 ft)(2.0) = 3.48 ft.

3. *Determine the value of H_g for the model system at equivalent loadings.*

Determination of H_g is somewhat more complicated, because the cited reference gives H_{og} rather than H_g and these both depend on both the gas and liquid loadings. The data indicate that H_{og} is proportional to $U_l^{-0.48}$ and to $F_S^{0.38}$. For Example 11-4, the F factor is essentially 0.50 (ft/s)(lb/ft^3)$^{1/2}$. For this F factor, the cited reference gives H_{og} as 0.80 ft. This value is for the ammonia-air-water system at 20°C and a liquid loading of 6.14 gal/(min)(ft^2). Correcting this value to the proper liquid loading gives H_{og} = (0.80 ft)(38.6/6.14)$^{-0.48}$ = 0.33 ft.

H_g and H_{og} are related by the equation $H_{og} = H_g + \lambda H_l$, where λ is the ratio of the slope of the equilibrium line to that of the operating line, as discussed in Example 11-3, step 6. For the ammonia-air-water system at 293 K, Eq. (11-9) in Example 11-3 gives He = 0.77 atm. From Eq. (11-11) of Example 11-3, the slope of the equilibrium line m is He/P, i.e., 0.77. From Example 11-4, L_M/G_M = 62. Accordingly, λ = (0.77/62) = 0.0124. The value of H_l determined in step 1 is 1.74 ft for the carbon dioxide–water system at 23°C. This value is corrected to the ammonia-water system at 20°C by using the square root of the Schmidt-number ratio. The values for the Schmidt number for the carbon dioxide–water system are given in step 2. For the ammonia-water system at 20°C, μ = 1.005 cP, ρ = 62.4 lb/ft^3, and $\mathscr{D}$ = 2.06 × 10^{-5} cm^2/s. The value of H_l for the ammonia-water system is [1.74 ft][(1.005/0.936)(1.83 × 10^{-5}/2.06 × 10^{-5})]$^{1/2}$ = 1.70 ft. Then, $H_g = H_{og} - \lambda H_l$ = 0.33 − (0.0124)(1.70) = 0.31 ft.

4. *Correct H_g to the actual system properties.*

The H_g for the model system is corrected to that for the actual system using the relationship that H_g is proportional to the square root of the Schmidt number. The values are as follows. For ammonia-air at 20°C,

$$\mu = 0.018 \text{ cP}$$
$$\rho = 0.0752 \text{ lb/ft}^3$$
$$\mathscr{D} = 0.222 \text{ cm}^2\text{/s}$$

For trichloroethylene-air at 50°F,

$$\mu = 0.018 \text{ cP}$$
$$\rho = 0.0778 \text{ lb/ft}^3$$
$$\mathscr{D} = 0.070 \text{ cm}^2\text{/s}$$

The value of H_g for the actual system is, accordingly,

$$H_g = [0.31 \text{ ft}][(0.0752/0.0778)(0.222/0.070)]^{1/2} = 0.54 \text{ ft}$$

11-7 Packed Height for Absorption with Effectively Instantaneous Irreversible Chemical Reaction

A tower packed with Flexipac Type 2Y structured packing (FP2Y) is to be used to remove 99.5% of the ammonia from an air stream containing an ammonia mole fraction (y_1) of 0.005. The tower is to operate isothermally at 100°F and 1 atm. The ammonia is to be absorbed in an aqueous solution of nitric acid. The tower diameter has been selected to give an F factor of 2.0 (ft/s)(lb/ft^3)$^{1/2}$ and a liquid loading of 15 gal/(min)(ft^2). The density of the air stream is 0.0709 lb/ft^3, and that of the water is 62.0 lb/ft^3. What concentration of acid should be used to ensure that the maximum absorption rate is obtained throughout the entire tower? Compare the height of packing required for absorption with chemical reaction to that for physical absorption.

Calculation Procedure:

1. *Identify the reaction and estimate its rate and reversibility.*

The absorption of ammonia with the liquid phase reaction between dissolved ammonia and a strong acid can be represented by

$$NH_3(g) \leftrightarrow NH_3(aq)$$
$$NH_3(aq) + H^+ \leftrightarrow NH_4^+$$

In general, reactions that consist simply of the transfer of hydrogen ions can be considered to occur "instantaneously." Accordingly, the reaction rate in this example will be considered instantaneous.

All reactions are, to some extent, reversible. This is particularly true of ionic reactions that occur in aqueous solution, as is the case here. The designation of a reaction as "irreversible" (as in this example) is not to say that the reaction cannot proceed in the reverse direction. Rather, it signifies that under the conditions of the absorption, the equilibrium lies strongly in favor of the reaction products.

The criterion for a reaction being effectively irreversible with respect to the absorption of a gas is that the concentration of the unreacted gas in the solution [e.g., $NH_3(aq)$] is so small that its partial pressure at the interface is much less than the partial pressure of the absorbing gas [e.g., $NH_3(g)$] in the gas phase. When this criterion is satisfied, the interfacial mole fraction of the absorbing gas y_i can be considered to be zero. This obviates the need for an equilibrium line or an x-y operating diagram.

The second reaction shown above will be shifted in favor of the product by a large equilibrium constant and by a sufficient concentration of hydrogen ion in solution. For now, the reaction will be assumed to be effectively irreversible, although this assumption will be checked in step 5.

2. *Determine the gas and liquid loadings.*

From the loadings given for this example and the densities of air and water, calculate the molar fluxes. The superficial gas velocity is (see Example 11-1, step 2) $U_g = F_S/(\rho_g)^{1/2} = 2/(0.0709)^{1/2} = 7.51$ ft/s. The molar flux of gas is $G_M = U_g(P/RT) = (7.51)(1)/[(0.73)(560)] = 0.0184$(lb · mol)/(ft^2)(s). The molar flux of liquid is $L_M = U_l\rho_{Ml} = (15)(62.0)/[(7.48)(60)(18)] = 0.115$ (lb · mol)/(ft^2)(s). From these values, the liquid to gas ratio is $L_M/G_M = 0.115/0.0184 = 6.25$.

Note that in the calculation of the liquid molar flux, we assume that the acid concentration is so low that we can use the molecular weight of water as the liquid molecular weight. This assumption later proves to be valid.

3. *Determine the HTUs for the packing at the design conditions.*

When corrected to the design conditions of this example, vendor data on model systems for this packing give $H_g = 0.84$ ft; $H_l = 0.58$ ft.

4. *Calculate the required concentration of absorbent.*

Enough nitric acid must be used to maximize the rate of absorption throughout the tower. As the acid concentration is increased, the reaction plane moves closer and closer to the interface until, at a particular concentration, the reaction plane coincides with the interface. When this condition prevails, the concentration of unreacted $NH_3(aq)$ at the interface becomes zero, the absorption becomes completely limited by the gas film, and the rate of absorption is maximized because y_i is also zero. When this condition prevails throughout the tower, the maximum rate of absorption will be obtained.

The rate of absorption is equal to the rate of mass transfer through the gas film at the interface, which, at steady state, is equal to the rate of mass transfer through the liquid film. (The results of the film theory are used here because of their simplicity and ease of use in engineering calculations.) The rate of absorption is given by

$$\overline{R}a = k_g a(P/RT)(y - y_i) = k_l a \rho_{Ml} x_i [1 + (\mathscr{D}_B / z\mathscr{D}_A)(x_B/x_i)] \tag{11-33}$$

where $\mathscr{D}_B$ = diffusion coefficient of reactant (e.g., H^+) in solution
$\mathscr{D}_A$ = diffusion coefficient of dissolved gas (e.g., NH_3) in solution
ρ_{Ml} = molar density, mol/ft^3
z = number of moles of B (e.g., H^+) reacting with 1 mol of A (e.g., NH_3)
x_B = mole fraction of reactant (e.g., H^+) in bulk liquid.

Other mole fractions in Eq. (11-33) refer to the absorbing gas, as with previous notation. The term in parentheses on the right of Eq. (11-33) is the enhancement factor for chemical reaction in the liquid film. As x_B increases, x_i and y_i decrease until they become effectively zero.

From Eq. (11-10), $x_i = y_i P/\text{He}$. Substituting for x_i in Eq. (11-33) and solving for y_i gives

$$y_i = \frac{k_g a(P/RT)y - k_l a \rho_{Ml} \dfrac{\mathscr{D}_B x_B}{\mathscr{D}_A z}}{k_g a(P/RT) + k_l a \rho_{Ml}(P/\text{He})} \tag{11-34}$$

It is clear from Eq. (11-34) that y_i will be zero if the second term in the numerator is greater than the first. This leads to the desired criterion for x_B:

$$x_B \geq \frac{H_l}{H_g} \frac{G_M}{L_M} \frac{\mathscr{D}_A}{\mathscr{D}_B} zy$$

where Eq. (11-18), Example 11-3, has been used to convert from mass transfer coefficients to HTUs. The liquid-phase diffusion coefficients of NH_3 and H^+ (actually H_3O^+) are approximately equal, and the value of z is 1. The maximum value of y occurs at the bottom of the tower, where $y_1 = 0.005$. Therefore, the minimum value of x_B that will maximize the rate of absorption is $x_B = (0.58/0.84)(0.005/6.25) = 0.000552$.

This is the minimum concentration at the bottom of the tower. The concentration of nitric acid in the liquid feed to the top of the tower must be sufficient to react with all of the ammonia absorbed in addition to providing for x_{B1} to be 0.000552. By material balance, the mole fraction of H^+ in the feed to the tower is $x_{B2} = [(0.005 - 2.5 \times 10^{-5})/6.25] + 0.000552 = 0.00135$. This mole fraction is converted to concentration by multiplying by the liquid molar density [55.10 (g · mol)/L] to give $[B]_2 = 0.0743$ (g · mol)/L, which corresponds to a pH of 1.13. At the bottom of the tower, $[B]_1 = 0.0304$ (g · mol)/L, which gives a pH of 1.52.

5. Check the assumption of irreversibility.

For dissociation of the ammonium ion,

$$NH_4^+ \leftrightarrow H^+ + NH_3(aq)$$

$$K_a = \frac{[H^+][NH_3(aq)]}{[NH_4^+]} = 5.75 \times 10^{-10} \text{ (g · mol)/L} \qquad \text{at } 25°\text{C} \qquad (11\text{-}35)$$

$$[NH_3(aq)] = 5.75 \times 10^{-10} \frac{[NH_4^+]}{[H^+]}$$

The concentrations of NH_4^+ and H^+ must be such that the concentration of ammonia calculated by Eq. (11-35) exerts a partial pressure which is negligible compared to the gas-phase partial pressure of ammonia. The minimum gas-phase partial pressure occurs at the top of the tower. The maximum concentration of NH_4^+ and the minimum concentration of H^+ occur at the bottom of the tower. From the concentrations calculated in step 4, $[H^+] = 0.0304$ (g · mol)/L and $[NH_4^+] = 0.0742 - 0.0304 = 0.0438$ (g · mol)/L, and by Eq. (11-35), $[NH_3(aq)] = 8.28 \times 10^{-10}$ (g · mol)/L. Dividing by the liquid molar density gives the mole fraction $x = 8.28 \times 10^{-10}/55.10 = 1.50 \times 10^{-11}$. From Eq. (11-10), with He = 1.724 atm at 100°F, this mole fraction is equivalent to a partial pressure p of (1.724) (1.50×10^{-11}) or 2.59×10^{-11} atm. This is clearly much less than the partial pressure at either the inlet (0.005 atm) or the outlet (2.5×10^{-5} atm) of the tower. Therefore, the reaction is effectively irreversible.

The above calculation assumes that the liquid makes only a single pass through the tower. In practice, a more effective way of treating the gas would be to recirculate the liquid with a small feed of concentrated acid to the recirculation loop and a small bleed of the recirculation loop contents. This would permit the NH_4^+ concentration to build up in the recirculation loop. However, even if this concentration built up to the solubility limit of ammonium nitrate (NH_4NO_3), about 30 (g · mol)/L, the partial pressure exerted by the dissolved ammonia at the top of the tower would still be nearly two orders of magnitude below the partial pressure of ammonia in the gas at the top of the tower. Therefore, for all practical conditions, the reaction is effectively irreversible, given the acid concentrations calculated in step 4.

6. Calculate the number of gas-phase transfer units.

For dilute systems in which the interfacial mole fraction y_i is kept at zero by virtue of the chemical reaction, the expression for N_g from Eq. (11-22), Example 11-3, becomes

$$N_g = \int_{y_2}^{y_1} \frac{dy}{y} = \ln \frac{y_1}{y_2}$$

Substituting the appropriate values for y gives

$$N_g = \ln (0.005/2.5 \times 10^{-5}) = 5.30$$

7. Calculate the height of packing required.

The height of packing required is $Z = H_g N_g = (0.84 \text{ ft})(5.30) = 4.45$ ft.

8. Calculate the height of packing for physical absorption at the same conditions.

At 100°F, Eq. (11-9), Example 11-3, gives He = 1.72 atm. Equation (11-11) gives $m = 1.72$. By Eq. (11-17), $\lambda = 1.72/6.25 = 0.275$. Equation (11-19) gives $R_{DF} = 0.84/[(0.275)(0.58)] = 5.27$. And Eq. (11-23) for N_g gives

$$N_g = \frac{1 + (1/5.27)}{1 - 0.275} \ln \left[(1 - 0.275) \left(\frac{0.0050 - 0}{2.5 \times 10^{-5} - 0} \right) + 0.275 \right] = 8.17$$

The required height of packing is $Z = H_g N_g = (0.84 \text{ ft})(8.17) = 6.86$ ft. The same result is obtained with Eq. (11-24).

For this example, absorption with chemical reaction gives a 34 percent reduction in the height of the packing relative to absorption without chemical reaction.

12

Liquid Agitation

David S. Dickey, Ph.D.

Manager, Fluid Mixing Technology
Prochem, Div. of Robbins & Myers, Inc.
Springfield, Ohio

REFERENCES: [1] Uhl and Gray, eds.—*Mixing, Theory and Practice,* vol. 1, Academic Press, 1966; vol. 3 Academic Press, 1986; [2] Nagata—*Mixing, Principles and Applications,* Wiley; [3] Bates, Fondy, and Corpstein—*Ind. Eng. Chem. Process Des. Dev. 2*:310, 1963; [4] Gates, Henley, and Fenic—*Chem. Eng.,* Dec. 8, 1975, p. 110; [5] Dickey and Fenic—*Chem. Eng.,* Jan. 5, 1976, p. 139; [6] Dickey and Hicks—*Chem. Eng.,* Feb. 2, 1976, p. 93; [7] Hicks, Morton, and Fenic—*Chem. Eng.,* Apr. 26, 1976, p. 102; [8] Gates, Morton, and Fondy—*Chem. Eng.,* May 24, 1976, p. 144; [9] Hicks and Gates—*Chem. Eng.,* July 19, 1976, p. 141; [10] Hill and Kime—*Chem. Eng.,* Aug. 2, 1976, p. 89; [11] Ramsey and Zoller—*Chem. Eng.,* Aug. 30, 1976, p. 101; [12] Meyer and Kime—*Chem. Eng.,* Sept. 27, 1976, p. 109; [13] Rautzen, Corpstein, and Dickey—*Chem. Eng.,* Oct. 25, 1976, p. 119; [14] Hicks and Dickey—*Chem. Eng.,* Nov. 8, 1976, p. 127; [15] Gates, Hicks, and Dickey—*Chem. Eng.,* Dec. 6, 1976, p. 165; [16] van't Riet—*Ind. Eng. Chem. Process Des. Dev., 18*:357, 1979; [17] Aerstin and Street—*Applied Chemical Process Design,* Plenum; [18] Holland and Chapman—*Liquid Mixing and Processing in Stirred Tanks,* Reinhold; [19] Brodkey, ed.—*Turbulence in Mixing Operations*, Academic Press; [20] Millich and Carraher, eds.—*Interfacial Synthesis,* vol. 1: *Fundamentals,* Marcel Dekker; [21] Oldshue—*Chemical Engineering,* June 13, 1983, p. 82; [22] Coble and Dickey—*AIChE Equipment Testing Procedure, Mixing Equipment (Impeller Type),* AIChE, 1987; [23] Tatterson—*Fluid Mixing and Gas Dispersion in Agitated Tanks,* McGraw-Hill.

12-1 Power Required to Rotate an Agitator Impeller

For a pitched-blade turbine impeller that is 58 in (1.47 m) in diameter and has four 12-in-wide (0.305-m) blades mounted at a 45° angle, determine the power required to operate the impeller at 84 r/min (1.4 r/s) in a liquid with a specific gravity of 1.15 (1150 kg/m^3) and a viscosity of 12,000 cP (12 Pa·s). What size standard electric motor should be used to drive an agitator using this impeller?

Calculation Procedure:

1. Determine the turbulent power number for impeller geometry.

Power number N_P is a dimensionless variable [5] which relates impeller power P to such operating variables as liquid density ρ, agitator rotational speed N, and impeller diameter D as follows:

$$N_P = \frac{P}{\rho N^3 D^5}$$

A conversion factor (see below) is used when working with English engineering units; no factor is necessary for SI metric units. For a given impeller geometry, the power number is a constant for conditions of turbulent agitation. Values of turbulent power numbers for some agitator impellers are shown in Fig. 12-1.

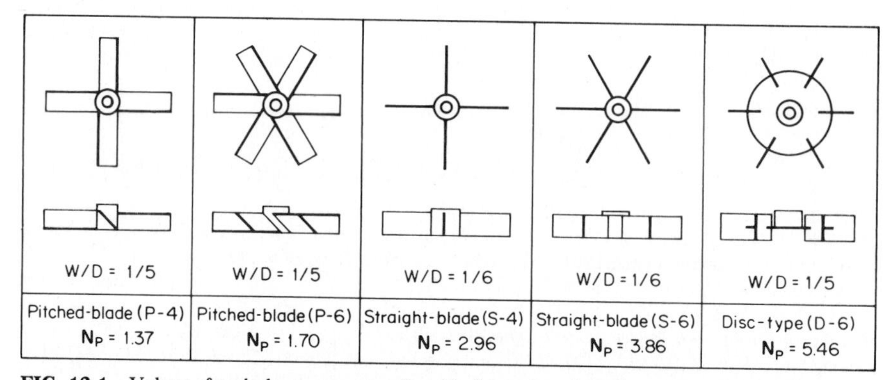

FIG. 12-1 Values of turbulent power number N_P for various impeller geometries. Note: W/D is actual blade-width-to-impeller-diameter ratio.

The pitched-blade impeller described in this example is similar to the four-blade impeller shown in the figure, except that the blade width-to-diameter ratio W/D is not exactly ⅕. To correct for the effect of a nonstandard W/D on a four-blade impeller, a factor of actual W/D to standard W/D raised to the 1.25 power must be applied to the standard turbulent power number $N_P = 1.37$. Thus the turbulent power number for a 58-in-diameter impeller with a 12-in blade width is

$$N_P = 1.37[(12/58)/\tfrac{1}{5}]^{1.25} = 1.37(1.034)^{1.25} = 1.43$$

(The correction factor for nonstandard W/D on a six-blade impeller is the actual W/D to standard W/D raised to the 1.0 power, or simply the ratio actual to standard W/D.)

2. *Determine the power number at process and operating conditions.*

Turbulence in agitation can be quantified with respect to another dimensionless variable, the impeller Reynolds number. Although the Reynolds number used in agitation is analogous to that used in pipe flow, the definition of impeller Reynolds number and the values associated with turbulent and laminar conditions are different from those in pipe flow. Impeller Reynolds number N_{Re} is defined as

$$N_{Re} = \frac{D^2 N \rho}{\mu}$$

where μ is the liquid viscosity. In agitation, turbulent conditions exist for $N_{Re} > 20{,}000$ and laminar conditions exist for $N_{Re} < 10$.

Power number is a function of Reynolds number as well as impeller geometry. A correction factor based on N_{Re} accounts primarily for the effects of viscosity on power. The Reynolds number is computed from the definition found in the previous paragraph and the conditions given in the problem statement; a factor of 10.7 makes the value dimensionless when English engineering units are used:

$$N_{Re} = \frac{10.7 D^2 N \rho}{\mu} = \frac{10.7(58)^2(84)(1.15)}{12{,}000} = 290$$

The viscosity power factor for $N_{Re} = 290$ is found to be 1.2 from Fig. 12-2. The power number for the impeller described in the example is the viscosity factor times the turbulent power number (from the previous step): $N_P = 1.2(1.43) = 1.72$.

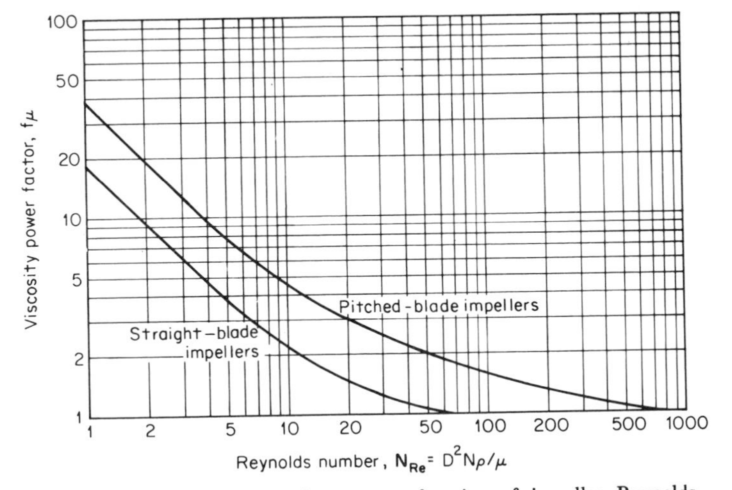

FIG. 12-2 Viscosity power factor as a function of impeller Reynolds number.

3. Compute the shaft horsepower required to rotate the impeller.

Horsepower requirements can be determined by rearranging the definition of power number into $P = N_P \rho N^3 D^5$ and using the value of the power number determined in the previous step. The result must be divided by a factor of 1.524×10^{13} to convert units and give an answer in horsepower: $P = 1.72(1.15)(84)^3(58)^5/(1.524 \times 10^{13}) = 50.5$ hp (37.7 kW).

4. Select a standard motor horsepower.

A typical turbine impeller-type agitator consists of a motor, a specially designed gear reducer, a shaft, and one or more impellers. Although losses through the gear reducer are typically only 3 to 8 percent, slight deviations in actual speed (which enters the power calculation cubed) and fluctuations in process conditions (density and viscosity) make motor loadings in excess of 85 percent of calculated impeller power unadvisable. Therefore, the calculated impeller power of 50.5 hp and a motor loading of 85 percent would indicate a minimum motor power: $P_{motor} = 50.5/0.85 = 59.4$. The next larger commercially available motor is 60 hp (or for metric sizes, 45 kW). Thus, a 60-hp motor should be used on an agitator designed to operate the 58-in-diameter impeller at 84 r/min as described in the example.

Related Calculations: Impeller power requirements are relatively independent of mixing-tank diameter. However, the power numbers shown in Fig. 12-1 assume fully baffled conditions, which for a cylindrical tank would require four equally spaced (at 90°) vertical plate-type baffles. The baffles should extend the full height of the vertical wall (i.e., the straight side) of the tank and should be one-twelfth to one-tenth the tank diameter in width.

For relatively high viscosities, the liquid itself prevents uncontrolled swirling. Therefore, when the liquid viscosity is greater than 5000 cP (5 Pa·s), no baffles are required for most applications. For impellers located less than one impeller diameter from the bottom of the tank, an additional correction factor for the power number may be necessary [3].

The power number provides important design information about the correct motor size necessary to operate an impeller at a given speed. However, these calculations do not give any indication of whether or not the agitation produced is adequate for process requirements. The following example shows a method for determining the horsepower and speed required to achieve a given process result.

12-2 Designing an Agitator to Blend Two Liquids

A process requires the addition of a concentrated aqueous solution with a 1.4 specific gravity (1400 kg/m^3) and a 15-cP (0.015 Pa·s) viscosity to a polymer solution with a 1.0 specific gravity (1000 kg/m^3) and an 18,000-cP (18 Pa·s) viscosity. The two liquids are completely miscible and result in a final solution with a 1.1 specific gravity (1100 kg/m^3) and a 15,000-cP (15 Pa·s) viscosity. The final batch volume will be 8840 gal (33.5 m^3), and the mixing will take place in a 9.5-ft-diameter (2.9-m) flat-bottom tank. Design the agitation system.

Calculation Procedure:

1. *Determine the required agitation intensity.*

Design of most agitation equipment is based on previous experience and a knowledge of the amount of liquid motion produced by a rotating impeller in a given situation. Although absolute rules for agitator design do not exist, good guidelines are available as a starting point for most process applications. The type of liquid motion used for most blending applications is a recirculating-flow pattern with good top-to-bottom motion. Axial-flow impellers, such as the pitched-blade turbines shown in Fig. 12-1, produce the desired liquid motion when operated in a baffled tank. Baffles are required to prevent excessive swirling of low-viscosity liquids (<5000 cP, or 5 Pa·s).

One measure of the amount of liquid motion in an agitated tank is velocity. However, by the very nature of mixing requirements, liquid velocities must be somewhat random in both direction and magnitude. Since actual velocity is difficult to measure and depends on location in the tank, an artificial, defined velocity called "bulk velocity" has been found to be a more practical measure of agitation intensity. "Bulk velocity" is defined as the impeller pumping capacity (volumetric flow rate) divided by the cross-sectional area of the tank. For consistency, the cross-sectional area is based on an "equivalent square batch tank diameter." A "square batch" is one in which the liquid level is equal to the tank diameter.

From previous design experience, the magnitude of bulk velocity can be used as a measure of agitation intensity for most problems involving liquid blending. Bulk velocities in the range from 0.1 to 1.0 ft/s (0.03 to 0.3 m/s) are typical of those found in agitated tanks. An agitator that produces a bulk velocity of 0.1 ft/s is normally the smallest agitator that will move liquid throughout the tank. An agitator capable of producing a bulk velocity of 1.0 ft/s is the largest practical size for most applications. Between these typical limits of bulk velocity, increments of 0.1 ft/s provide 10 levels[1] of agitation intensity that are associated with typical process results, as shown in Table 12-1.

In this example, the two fluids to be mixed have a specific gravity difference of 0.4 and a viscosity ratio of 1200. On the basis of the process capabilities associated with bulk velocities of 0.2 and 0.6 ft/s in Table 12-1, a bulk velocity of 0.4 ft/s (0.12 m/s) should be adequate for this example. Special circumstances, such as a reaction taking place or experience with a similar process, may influence the selection of a bulk velocity.

2. *Compute required impeller pumping capacity.*

To determine the required pumping capacity, the bulk velocity (0.4 ft/s) must be multiplied by the appropriate cross-sectional area. Since a "square batch" is assumed for the design basis of bulk velocity, an equivalent tank diameter T_{eq} is computed by rearranging the formula for the volume of a cylinder with the height equal to the diameter, that is,

$$\frac{\pi}{4} T_{eq}^3 = V$$

For the final batch volume of 8840 gal and the conversion of units,

$$T_{eq} = \left[8840 \text{ gal } (231 \text{ in}^3/\text{gal}) \frac{4}{\pi} \right]^{1/3} = 6.65(8840)^{1/3}$$

$$= 137.5 \text{ in } (3.49 \text{ m})$$

[1] Called ChemScale levels by Chemineer-Kenics.

TABLE 12-1 Agitation Results Associated with Bulk Velocities

Bulk velocity, ft/s (m/s)	Description
0.1 (0.03) ↓ 0.2 (0.06)	Bulk velocities of 0.1 and 0.2 ft/s (0.03 and 0.06 m/s) are characteristic of applications requiring a minimum of liquid motion. Bulk velocity of 0.2 ft/s (0.06 m/s) will • Blend miscible liquids to uniformity if specific gravity differences are less than 0.1 • Blend miscible liquids to uniformity if the viscosity of the most viscous is less than 100 times that of any other • Establish liquid motion throughout the batch • Produce a flat but moving liquid surface
0.3 (0.09) ↓ 0.6 (0.18)	Bulk velocities between 0.3 and 0.6 ft/s (0.09 and 0.18 m/s) are characteristic of most agitation used in chemical processes. Bulk velocity of 0.6 ft/s (0.18 m/s) will • Blend miscible liquids to uniformity if the specific gravity differences are less than 0.6 • Blend miscible liquids to uniformity if the viscosity of the most viscous is less than 10,000 times that of any other • Suspend trace solids (<2%) with settling rates of 2 to 4 ft/min (0.01 to 0.02 m/s) • Produce surface rippling at low viscosities
0.7 (0.21) ↓ 1.0 (0.30)	Bulk velocities between 0.7 and 1.0 ft/s (0.21 and 0.30 m/s) are characteristic of applications requiring a high degree of agitation, such as critical reactors. Bulk velocity of 1.0 ft/s (0.30 m/s) will • Blend miscible liquids to uniformity if the specific gravity differences are less than 1.0 • Blend miscible liquids to uniformity if the viscosity of the most viscous is less than 100,000 times that of any other • Suspend trace solids (<2%) with settling rates of 4 to 6 ft/min (0.02 to 0.03 m/s) • Produce surging surface at low viscosities

Source: From Ref. 7.

A 137.5-in-diameter tank has a cross-sectional area of $(\pi/4)(137.5\text{ in})^2 = 14{,}849\text{ in}^2$ or 103 ft^2, so the required impeller pumping capacity is bulk velocity times cross-sectional area: $(0.4\text{ ft/s})(103\text{ ft}^2) = 41.2\text{ ft}^3/\text{s}$ or 2472 ft^3/min (1.17 m^3/s). Geometry of the actual tank will be taken into consideration by location and number of impellers after the horsepower and speed of the agitator are determined.

3. Select impeller diameter and determine required agitator speed.

The pumping capacity Q for a pitched-blade impeller with four blades ($N_P = 1.37$) can be related to other mixing parameters by the correlation shown in Fig. 12-3. The correlation is between two dimensionless variables: pumping number (Q/ND^3) and Reynolds number ($D^2N\rho/\mu$). Since impeller diameter D and rotational speed N appear in both variables, an iterative solution may be required. A convenient approach to such a solution is as follows:

a. *Select an impeller diameter.* The impeller diameter must be some fraction of the tank diameter, typically between 0.2 and 0.6. For this calculation, an impeller-to-tank-diameter ratio (D/T) of 0.4 will be used. Based on the equivalent tank diameter (137.5 in), an impeller with a 0.4(137.5) = 55 in (1.4 m) diameter will be used. (Blade width will be 11 in, corresponding to a W/D of ⅕.)

b. *Compute initial estimate of impeller Reynolds number.* To compute impeller Reynolds number ($D^2N\rho/\mu$), an initial estimate of rotational speed must be made to begin the iterative solution. Let us assume 100 r/min. Using fluid properties for the final batch, 1.1 specific gravity, and 15,000 cP, the initial estimate of Reynolds number becomes $N_{Re} = 10.7(55)^2(100)(1.1)/15{,}000 = 237$. (The coefficient 10.7 is a conversion factor to make the value dimensionless.)

c. *Determine pumping number and compute speed.* From the correlation for pumping number (Fig. 12-3), at a Reynolds number of 237 and a D/T of 0.4, the pumping number is $N_Q = 0.44$. By rearranging the definition of pumping number $[Q/(ND^3)]$ and using the value obtained from the correlation (0.44), we can calculate a speed for the required pumping capacity of 2472 ft³/min and the impeller diameter of 55 in (4.583 ft):

$$N = \frac{Q}{N_Q D^3} = \frac{2472}{0.44(4.583)^3} = 58.4 \text{ r/min}$$

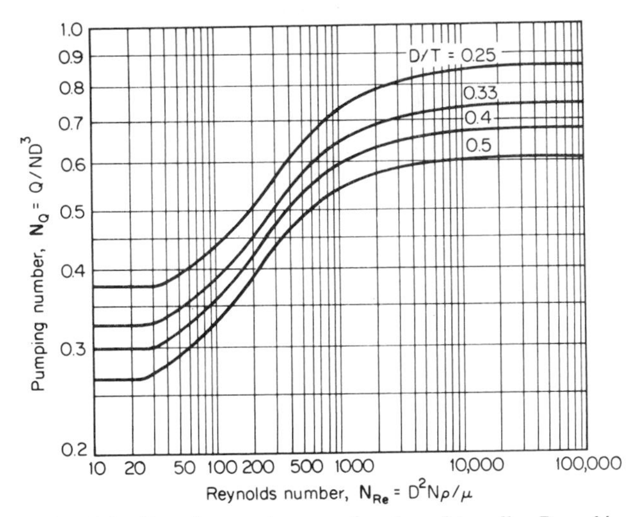

FIG. 12-3 Pumping number as a function of impeller Reynolds number for pitched-blade impeller ($N_P = 1.37$). *(From Chemical Engineering, 1976.)*

The estimated and calculated speeds do not match, and the pumping number is not constant for this Reynolds number range, so an iterative solution for the speed must continue. [In the turbulent range ($N_{Re} > 20{,}000$), where pumping number is constant, no iteration is required and the calculated speed is correct for the design.]

d. Perform an iterative calculation for agitator speed. Successive calculations of Reynolds number (based on the previously estimated speed), pumping number, and agitator speed, similar to steps 3*b* and 3*c*, will converge as follows:

Iteration	Reynolds number	Pumping number	Speed, r/min
2	139	0.38	67.6
3	160	0.40	64.2
4	152	0.39	65.8
5	156	0.395	65.0

Thus a speed of 65 r/min is necessary to provide the pumping capacity of 2472 ft^3/min when using a 55-in-diameter impeller.

4. Select standard speed and motor horsepower.

Although design calculations have determined that an agitator speed of 65 r/min is required, only certain standard output speeds[1] are available with typical industrial gear reducers. The closest standard speed is 68 r/min. If 68 r/min is used instead of the calculated 65 r/min, the bulk velocity will increase to about 0.42 ft/s, a change imperceptible with respect to agitator performance.

The horsepower required to rotate a 55-in-diameter impeller (11-in blade width) at 68 r/min can be computed for the process fluid, using the technique described in Example 12-1. The turbulent power number is 1.37, from Fig. 12-1. The Reynolds number at 68 r/min becomes 161. The viscosity correction factor for this value is 1.35, from Fig. 12-2, which gives a power number N_P of $1.37(1.35) = 1.85$ for the design conditions. From the power number, impeller power can be computed: $P = 1.85(1.1)(68)^3(55)^5/(1.524 \times 10^{13}) = 21.1$ hp. With an 85 percent loading for the motor, a minimum motor horsepower would be $21.1/0.85 = 24.8$ hp, so a 25-hp (18.5-kW) motor would be required. If the next larger standard motor is substantially larger than the minimum motor horsepower, the impeller diameter may be increased by an inch or two to fully utilize the available motor capacity.

5. Specify the number and location of impellers.

The calculations carried out in the previous steps show that a 25-hp agitator operating at 68 r/min will provide sufficient agitation to solve the problem by creating a bulk velocity of 0.4 ft/s. However, these calculations essentially ignore the fact that the process will be carried out in a 9½-ft-diameter tank. The idea behind this final step in the design procedure is that 25 hp at 68 r/min will provide the desired agitation if the number and location of impellers is suitable for the batch height, as related by the ratio of liquid level to tank diameter Z/T.

[1]Standard speeds for common agitator drives are 230, 190, 155, 125, 100, 84, 68, 56, 45, and 37 r/min based on actual speeds of nominal 1800 and 1200 r/min motors and standard gear reductions, which are a geometric progression of the $\sqrt{1.5}$ for enclosed, helical, and spiral bevel gearing (American Gear Manufacturers' Association Standard 420.04, December 1975, p. 29).

TABLE 12-2 Capacity Data for Cylindrical Vessels

Vessel diameter		Volume of cylindrical vessel		Depth and volume of vessel head			
				Standard dished head		ASME flanged and dished head	
ft-in	in	Straight side, gal/in	Square-batch, gal	Depth, in	Volume, gal	Depth, in	Volume, gal
3 ft	36	4.40	159	4.9	11	6.0	16
3 ft 6 in	42	5.99	252	5.7	18	7.2	25
4 ft	48	7.83	376	6.5	27	8.0	37
4 ft 6 in	54	9.91	535	7.3	38	9.0	53
5 ft	60	12.2	734	8.1	52	10	78
5 ft 6 in	66	14.8	977	8.9	70	11	104
6 ft	72	17.6	1269	9.7	90	12	135
6 ft 6 in	78	20.7	1631	11	114	14	170
7 ft	84	24.0	2041	11	142	15	212
7 ft 6 in	90	27.5	2478	12	174	15	261
8 ft	96	31.3	3007	13	212	16	314
8 ft 6 in	102	35.3	3607	14	254	18	375
9 ft	108	39.6	4287	15	301	19	446
9 ft 6 in	114	44.1	5035	15	353	20	524
10 ft	120	48.9	5873	16	414	21	612
10 ft 6 in	126	54	6799	17	480	22	705
11 ft	132	59	7817	18	560	23	806
11 ft 6 in	138	65	8932	20	665	24	926
12 ft	144	70	10148	20	735	25	995

Note: 1.0 ft = 0.3048 m; 1.0 in = 0.0254 m; 1.0 gal/in = 0.149 m^3/m; 1.0 gal = 3.785×10^{-3} m^3.

According to Table 12-2, a 9½-ft-diameter tank holds 44.1 gal/in of liquid level. Therefore, 8840 gal will fill the tank to 8840/44.1 = 200 in. The resulting liquid-level-to-tank-diameter ratio is $Z/T = 200/114 = 1.75$. The following guidelines for number and location of impellers should be applied:

Viscosity, cP (Pa·s)	Maximum level, Z/T	Number of impellers	Impeller clearance	
			Lower	Upper
<25,000 (<25)	1.4	1	$Z/3$	—
<25,000 (<25)	2.1	2	$T/3$	$(2/3)Z$
>25,000 (>25)	0.8	1	$Z/3$	—
>25,000 (>25)	1.6	2	$T/3$	$(2/3)Z$

Since the liquid viscosity is 15,000 cP (<25,000 cP) and the liquid level gives a Z/T of 1.75, two impellers should be used to provide liquid motion throughout the tank.

To properly load the 25-hp motor with a dual impeller system, each impeller should be sized for 25 hp/2 = 12.5 motor hp, or at 85 percent loading, 0.85(12.5 hp) = 10.6 impeller hp. By assuming the same viscosity correction factor (1.35) for the dual impeller size, an initial estimate can be made for power number, $N_P = 1.37(1.35) = 1.85$, and impeller diameter, $D = [1.524 \times 10^{13} P/(N_P \rho N^3)]^{1/5} = \{1.524 \times 10^{13}(10.6)/[1.85(1.1)(68)^3]\}^{1/5} = 47.9$ in. Using this value to compute Reynolds number gives N_{Re}

$= 10.7(47.9)^2(68)(1.1)/15{,}000 = 122$. For $N_{Re} = 122$, the viscosity correction factor is 1.47 from Fig. 12-2, or a power number, $N_P = 1.37(1.47) = 2.01$. Recomputing impeller diameter, that is, $D = \{1.524 \times 10^{13}(10.6)/[2.01(1.1)(68)^3]\}^{1/5} = 47.1$, shows that two 47.1-in-diameter (1.20-m) impellers (with 9.4-in blade width) are equivalent to one 55-in-diameter impeller. The lower impeller should be located $T/3 = 114/3 = 38$ in (0.965 m) off bottom and the upper impeller $(2/3)Z = (2/3)200 = 133$ in (3.39 m) off bottom. Had the tank been 11 or 12 ft in diameter, only one impeller would have been required. Liquid levels that result in $Z/T < 0.4$ are difficult to agitate.

Related Calculations: Repeating the same design calculations but starting with a different impeller diameter will result in other horsepower-speed combinations capable of producing the same bulk velocity (0.4 ft/s). For instance, the following combinations also satisfy the design requirements:

Impeller diameter, in (m)	Motor horsepower	Speed, r/min
48 (1.219)	40	100
50 (1.270)	30	84
58 (1.321)	20	56
62 (1.412)	15	45

As is the case for many agitator design problems, there are several horsepower, speed, and impeller-diameter combinations that solve the problem by producing equivalent results. From the standpoint of energy conservation, large impellers usually require less horsepower to do a given job.

Agitation problems that require other process results, such as the suspension of solids or the dispersion of gas, use other design criteria [8,9].

12-3 Time Required for Uniform Blending

About 150 gal (0.57 m^3) of strong acid must be added to 10,000 gal (37.85 m^3) of slightly caustic waste held in a 12-ft-diameter (3.66-m) tank. The waste has a specific gravity ρ of 1.2 (1200 kg/m^3) and a viscosity μ of 500 cP (0.5 Pa·s). Determine the time required for neutralization if the tank is agitated by a 1-hp (0.75-kW) agitator operating at a rotational speed N of 68 r/min and having a pitched-blade impeller with diameter D of 30 in (0.762 m).

Calculation Procedure:

1. Compute the Reynolds number.

Acid-base neutralizations typically are very fast reactions, so the time required for mixing is usually the limiting factor. Although both liquid motion and molecular diffusion are involved in liquid mixing, the liquid motion dominates the apparent rate of mixing. Impeller agitation creates both large-scale flow patterns and small-scale turbulence, which in combination give efficient and rapid mixing. The effect of turbulent flow patterns is to reduce the distances required for diffusion to almost the molecular scale.

One practical method for quantifying the complicated mixing process in an agitated tank is to measure the time required for a tracer material to blend to uniformity. Such measurements for blend time may use acid-base neutralization with a color-change indicator, a dye tracer, or an ionic salt with an electrode detector. Properly accounting for measurement accuracy, all these methods give essentially the same results for time required to go from extreme segregation to a high degree (>99 percent) of uniformity.

This measured blend time t_b can be expressed as a dimensionless variable by forming a product t_bN with agitator speed. This form of dimensionless blend time, multiplied by the impeller-to-tank-diameter ratio D/T to the 2.3 power, is shown as a function of impeller Reynolds number in Fig. 12-4.

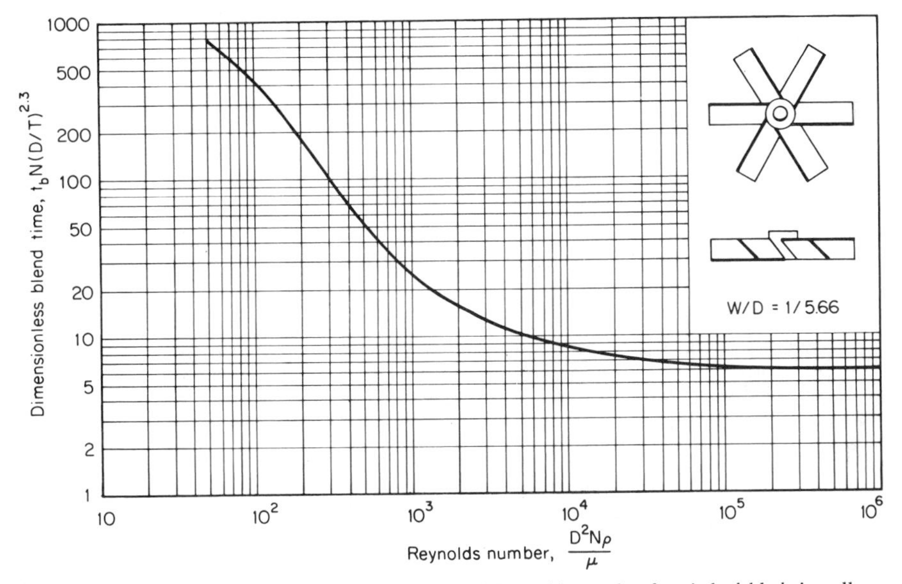

FIG. 12-4 Dimensionless blend time as a function of Reynolds number for pitched-blade impellers.

The independent variable for the correlation is impeller Reynolds number ($D^2N\rho/\mu$), which takes into account the effects of liquid properties on the blend time. To compute the value of the Reynolds number, a coefficient of 10.7 is necessary to put the given units in dimensionless form. Thus, $N_{Re} = 10.7(30)^2(68)(1.2)/500 = 1572$.

2. *Determine dimensionless blend time and D/T.*

A dimensionless blend time $t_bN(D/T)^{2.3} = 18$ is found for a Reynolds number of 1600, from Fig. 12-4. This form of dimensionless blend time takes into account the main geometric effects, as embodied in the impeller-to-tank-diameter ratio, $D/T = 30/144 = 0.208$.

3. *Compute blend time.*

Blend time can be computed directly from rearrangement of the definition and the values for dimensionless blend time, for D/T, and for agitator speed:

$$t_b = \frac{18}{N(D/T)^{2.3}} = \frac{18}{68(0.208)^{2.3}}$$
$$= 9.80 \text{ min (588 s)}$$

The computed blend time is about 10 min, a value that is accurate to ± 10 percent for successive observations of the same process and affected slightly by the location of the acid addition. A reasonable design value for the blend time might be twice the calculated value, or 20 min (1200 s).

Related Calculations: If the waste-neutralization process were continuous and the pH adjustment were relatively small (<3 pH units), a conservative design residence time in the tank might be 10 times the computed blend time, or 100 min.

12-4 Heat Transfer in an Agitated Tank

As part of the final blending operation in a continuous process, it is necessary to cool an oil product from 125°F (325 K) to 100°F (311 K) at the rate of 800 gal/h (0.84×10^{-3} m^3/s).

The oil has the following physical properties at 100°F:

Viscosity μ = 1200 cP (1.2 Pa·s)
Specific gravity ρ = 0.89 (890 kg/m^3)
Heat capacity C_p = 0.52 = Btu/(lb)(°F) [2175 J/(kg)(K)]
Thermal conductivity k = 0.079 Btu/(h)(ft)(°F) [0.137 W/(m)(K)]

The tank diameter T is 9 ft (2.74 m), and the vessel is designed to operate at a 5000-gal (18.9-m^3) capacity. The tank bottom is a standard dished head and the straight side of the tank is fully jacketed. The jacket-side heat-transfer coefficient is estimated to be h_o = 180 Btu/(h)(ft^2)(°F) [1021 W/(m^2)(K)]. The wall thickness and its heat-transfer resistance are assumed to be negligible. If the agitator is 1.5 hp (1.1 kW) [1.15 impeller hp (0.858 kW)] operating at a speed N of 56 r/min, with a 38-in-diameter (D) (0.97-m) impeller, estimate the average cooling water temperature required to cool the oil. Also determine what effect increasing the agitator speed to 100 r/min (assuming appropriate changes were made in the agitator) would have on the temperature of the oil if all other conditions from the first part of the problem remained unchanged.

Calculation Procedure:

1. *Compute the process-side heat-transfer coefficient.*

The correlations for inside (process-side) heat-transfer coefficient in an agitated tank are similar to those for heat transfer in pipe flow, except that the impeller Reynolds number and geometric factors associated with the tank and impeller are used and the coefficients and exponents are different. A typical correlation for the agitated heat-transfer Nusselt number ($N_{Nu} = h_i T/k$) of a jacketed tank is expressed as

$$N_{Nu} = 0.85 N_{Re}^{0.66} N_{Pr}^{0.33} \left(\frac{Z}{T}\right)^{-0.56} \left(\frac{D}{T}\right)^{0.13} \left(\frac{\mu}{\mu_w}\right)^{0.14}$$

All terms in the expression are dimensionless.

The correlation for heat transfer is evaluated with the respective dimensionless groups. With the units stated in the example, the impeller Reynolds number ($N_{Re} = D^2 N\rho/\mu$) requires a conversion factor of 10.7 to make it dimensionless; thus, $N_{Re} = 10.7(38)^2(56)(0.89)/1200 = 642$. The Prandtl number ($N_{Pr} = C_p\mu/k$) requires a conversion factor of 2.42; thus $N_{Pr} = 2.42(0.52)(1200)/0.079 = 19{,}115$. The liquid-level-to-tank-diameter ratio Z/T requires determination of the liquid level for a 5000-gal batch in the tank. A standard dished head holds 301 gal and is 15 in deep, from Table 12-2. The remaining $5000 - 301 = 4699$ gal fills the cylindrical part of the tank at a rate of 39.6 gal/in of height, or to a height of $4699/39.6 = 119$ in. The total liquid level is $Z = 119 + 15 = 134$ in and $Z/T = 134/108 = 1.24$. The impeller-to-tank-diameter ratio D/T is $38/108 = 0.35$. The viscosity ratio will be assumed to be unity ($\mu/\mu_w = 1$) because of lack of data and the very small exponent on the term.

Combining all these values according to the correlation gives a value for the Nusselt number of $N_{Nu} = 0.85(642)^{0.66}(19{,}115)^{0.33}(1.24)^{-0.56}(0.35)^{0.13}(1)^{0.14} = 1212$. The value for the inside heat-transfer coefficient h_i is obtained from the Nusselt number using conductivity and tank diameter (ft); thus, $h_i = N_{Nu}k/T = 1212(0.079)/9 = 10.6$ Btu/(h)(ft²)(°F) [60.1 W/(m²)(K)].

2. *Compute overall heat-transfer coefficient.*

The overall heat-transfer coefficient U_o is simply the series resistance to heat transfer for the inside and outside coefficients. The overall coefficient U_o is $(1/h_i + 1/h_o)^{-1} = (1/10.6 + 1/180)^{-1} = 10.0$ Btu/(h)(ft²)(°F) [56.7 W/(m²)(K)].

3. *Determine total heat load for agitated heat transfer.*

The process requires 25°F cooling of 8000 gal/h of oil. Since the oil has a specific gravity of 0.89, its density is (0.89)(8.337 lb water/gal) = 7.42 lb/gal. The product of volumetric flow, density, heat capacity, and temperature change equals the heat load for cooling the oil: (800 gal/h)(7.42 lb/gal) [0.52 Btu/(lb)(°F)](25°F) = 77,168 Btu/h (22.6 kW). In addition, the power input of the agitator (1.15 hp) also must be dissipated in the form of heat: (1.15 hp)[2545 Btu/(h)(hp)] = 2927 Btu/h. The total heat load q is 77,168 + 2927 = 80,095 Btu/h (23.5 kW).

4. *Compute required coolant temperature.*

The coolant temperature can be determined from heat load and the heat-transfer coefficient because a sufficient temperature difference must exist to drive the heat-transfer rate, that is, $q = U_o A(T_i - T_o)$. The available heat-transfer area A is the jacketed vertical wall in contact with the liquid, since the bottom head is not jacketed; thus, $A = \pi D Z_{ss} = \pi(9\text{ ft})(119/12\text{ ft}) = 280\text{ ft}^2$. The temperature difference is $(T_i - T_o) = q/(U_o A) = 80{,}095/[10.0(280)] = 28.6°\text{F}$. To provide a temperature difference of 28.6°F with respect to the process temperature, the average coolant temperature must be 71.4°F (295 K).

5. Determine the effect of increased agitator speed.

Two effects must be considered when the agitator speed is increased: (1) improved heat-transfer coefficient, and (2) increased power input. The agitator speed enters the heat-transfer correlation in the Reynolds number. For 100 r/min, the Reynolds number N_{Re} becomes $10.7(38)^2(100)(0.89)/1200 = 1146$, which increases the Nusselt number to 1777 and the inside heat-transfer coefficient h_i to $1777(0.079)/9 = 15.6$ Btu/(h)(ft^2)(°F) [88.5 W/(m^2)(K)]. The overall coefficient U_o becomes $(1/15.6 + 1/180)^{-1} = 14.4$ Btu/(h)(ft^2)(°F) [81.7 W/(m^2)(K)], or a 44 percent increase from the lower speed. The horsepower increase associated with increased speed is substantial, because power is roughly proportional to speed cubed (the effect of Reynolds number on power requirement is negligible between $N_{Re} = 600$ and 1200 for a pitched-blade impeller; see Fig. 12-2). The horsepower at 100 r/min is approximately $(1.15 \text{ hp})(100/56)^3 = 6.55$ hp, which results in a heat load of $(6.55 \text{ hp})[2545 \text{ Btu/(h)(hp)}] = 16{,}670$ Btu/h. With the increased power input, the total heat load is $q = 77{,}168 + 16{,}670 = 93{,}838$ Btu/h (27.5 kW).

The combined effects of increased heat transfer and increased heat load can be seen by determining the resultant temperature difference, that is, $(T_i - T_o) = q/(U_oA) = 93{,}838/[14.4(280)] = 23.3$°F. For the same jacket temperature of 71.4°F, the process temperature would be reduced to $71.4 + 23.3 = 94.7$°F (307.8 K) by doubling the agitator speed.

The cost of improved heat transfer by increased speed must be measured against the increased cooling-water requirements and increased operating power and capital cost for the agitator. In general, these increased costs more than offset the benefits of improved heat transfer. Therefore, most agitators designed for heat transfer provide moderate blending (0.2 to 0.3 ft/s bulk velocity) for optimal operation.

Related Calculations: See also Sec. 7 for situations involving both heat transfer and mixing.

12-5 Scale-Up for Agitated Solids Suspension

An agitator must be designed for a solids-suspension operation to be carried out in a 6000-gal (22.7-m^3) tank that is 10 ft (3.05 m) in diameter and has a standard dished bottom. The material to be suspended is insoluble in the liquid and has a particle-size range from 30 to 200 μm with an actual specific gravity ρ of 3.8 (3800 kg/m^3). The liquid is a mineral oil with a specific gravity of 0.89 (890 kg/m^3) and a viscosity of 125 cP (0.125 Pa·s). The suspension is 30 wt% solids and must be sufficiently agitated to give particle uniformity of the large particles to at least three-fourths the liquid level.

Calculation Procedure:

1. Compute suspension density.

To form 1 lb of 30 wt% suspension, 0.3 lb solids must be added to 0.7 lb liquid. The liquid has a density of $0.89(8.337 \text{ lb water/gal}) = 7.42$ lb/gal. Similarly, the solids must displace liquid at 3.8(8.337) = 31.68 lb/gal. Thus, 1 lb of the suspension w. have a volume of 0.7/7.42 + 0.3/31.68 = 0.1038 gal, or a density of (1 lb)/0.1038 gal = 9.63 lb/gal, which is the same as a specific gravity of 9.63/8.337 = 1.16 (1160 kg/m^3).

2. *Determine batch height.*

A 6000-gal batch in a 10-ft-diameter tank with a dished head will put 414 gal in the 16-in-deep dished head (see Table 12-2). The remaining 6000 − 414 = 5586 gal will fill the vertical-wall portion of the tank at a rate of 48.9 gal/in, for a total liquid depth of (5586 gal)/(48.9 gal/in) + 16 in = 130 in (3.3 m).

3. *Use an experimental model to determine required agitation intensity.*

Although physical-property data are available for solid particles, liquid, and suspension, the agitated-suspension characteristics of a relatively wide range of particle sizes in a slightly viscous liquid are almost impossible to predict without making experimental measurements in a small-scale agitated tank. The most direct approach to small-scale testing is to construct a geometrically similar model of the large-scale equipment. Assume that a 1-ft-diameter tank (1/10 scale) is available for such tests. By applying the scale factor (1/10) to the liquid level for the large tank, a (130 in)/10 = 13-in liquid level should be tested in the model.

The testing should determine the intensity of agitation necessary to obtain the desired level of suspension uniformity. Suppose that by visual observation and sample analysis, a pitched-blade impeller with a diameter D of 4 in (with four blades, each 0.8 in wide) operating at a speed N of 465 r/min was found to produce the level of agitation necessary for uniform suspension to three-fourths the total liquid level. With data about impeller diameter and agitator speed in a small tank, it should be possible to scale up performance to the large-scale tank.

4. *Scale-up experimental results for solids suspension.*

To maintain geometric similarity with scale-up, all length dimensions must remain in the same proportion between the small and large equipment. If the large tank diameter is 10 times as large as the small tank, then the impeller diameter should be 10(4 in) = 40 in for the large tank. Similarly, the blade width for the impeller should be 10(0.8 in) = 8 in for a four-blade impeller. Although geometric similarity is not strictly necessary for all agitation scale-up problems, satisfactory results are usually obtained and scale-up relationships are relatively simple.

Since impeller diameter is established by geometric similarity, the agitator speed for the large-scale equipment must be determined to satisfy process requirements. Large-scale speed N_L can be computed by the following relationship:

$$N_L = N_S \left(\frac{D_S}{D_L}\right)^n$$

where N_S is the small-scale speed. The value of the exponent n depends on the type of process result that must be duplicated, since different process results scale up differently. Some typical exponents, their effect on speed, and their scale-up significance are shown in Fig. 12-5.

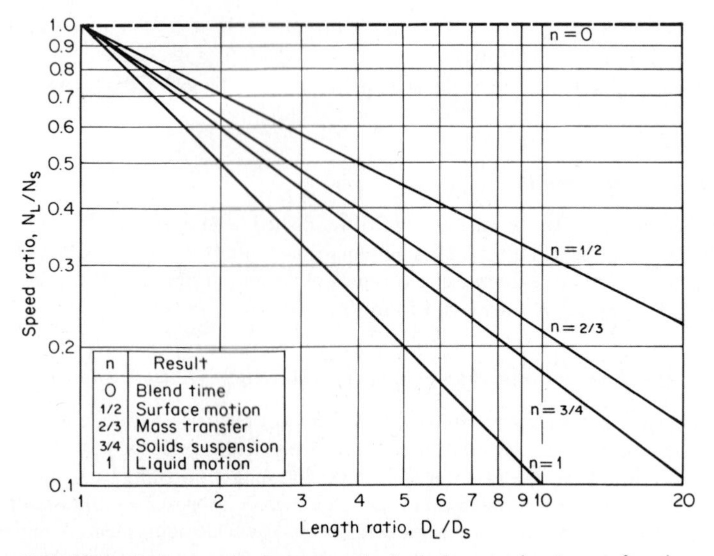

FIG. 12-5 Scale-up rules for geometric similarity; speed ratio as a function of length ratio.

For equivalent solids suspension, an exponent of ¾ will be used to scale up from the small-scale solids suspension speed of 465 r/min:

$$N_L = 465 \left(\frac{4}{40}\right)^{3/4} = 82.7 \text{ r/min}$$

5. *Select standard speed and motor horsepower.*

Although scale-up calculations predict an 82.7 r/min operating speed for the large-scale agitator, only certain standard output speeds are commercially available. The nearest standard speed is 84 r/min. To determine power requirements, use the procedure outlined in Example 12-1. The turbulent-power number is first determined from Fig. 12-1: $N_P = 1.37$. The Reynolds number is then computed for the process conditions: $N_{Re} = 10.7(40)^2(84)(1.16)/125 = 13{,}345$, which is sufficiently turbulent that no correction factor from Fig. 12-2 need be applied to the power number. The calculated impeller power is $1.37(1.16)(84)^3(40)^5/(1.524 \times 10^{13}) = 6.33$ hp. Considering process variations, drive losses, and so forth, an 85 percent motor loading means that a minimum motor horsepower of $(6.33 \text{ hp})/0.85 = 7.45$ hp is required. The next larger standard motor is 7.5 hp (5.5 kW), which is used to drive a 40-in-diameter (1.02-m) impeller at 84 r/min to satisfy the solids-suspension requirements.

Related Calculations: Scale-up problems for agitator design are not always obvious with respect to which scale-up exponent should be used. Problems involving chemical reactions, where kinetics and mixing interact, are often difficult to scale up accurately. Therefore, medium-sized pilot-plant reactors may be necessary to improve the understanding of how mixing influences performance.

12-6 Agitator Design for Gas Dispersion

Pilot-scale testing of an aerobic fermentation process has determined that maximum cell growth rate will consume 16.2 lb O_2 per hour per 1000 gal (5.4×10^{-4} kg O_2 per second per cubic meter) at a temperature of 120°F (322 K), providing the gas rate is sufficient to keep oxygen depletion of the air to less than 10 percent and the O_2 concentration in the broth is at least 2.4 mg/L (2.4×10^{-3} kg/m^3). The overall mass-transfer coefficient $k_La(s^{-1})$ for the process is assumed to behave as in an ionic solution: $k_La = (2.0 \times 10^{-3})(P/V)^{0.7}u_s^{0.2}$, where P/V is power per volume (in watts per cubic meter) and u_s is superficial gas velocity (in meters per second). Design an agitator to carry out the fermentation in a 10,000-gal (37.9-m^3) batch in a 12-ft-diameter (3.66-m) tank. The fermentation broth is initially waterlike and has a specific gravity of 1.0 (1000 kg/m^3) throughout the process.

Calculation Procedure:

1. Determine superficial gas velocity.

The maximum oxygen uptake rate (16.2 lb/h per 1000 gal) will mean that 162 lb O_2 per hour will be required for the 10,000-gal batch. To keep oxygen depletion to less than 10 percent, 10 times the O_2 demand must flow through the tank, that is, $10(162) = 1620$ lb O_2 per hour. Water-saturated air at 120°F contains 21.3 wt% O_2 and has a density of 0.065 lb/ft^3 at 1 atm (*Handbook of Tables for Applied Engineering Science,* CRC Press). To supply 1620 lb O_2 per hour, $1620/0.213 = 7606$ lb air per hour must flow through the tank. This flow rate represents $(7606 \text{ lb/h})/[(0.065 \text{ lb/ft}^3)(60 \text{ min/h})] = 1950 \text{ ft}^3/\text{min}$ (0.92 m^3/s) at 1 atm (101.3 kPa).

To design the agitator, the gas density at the impeller location should be used for computing the superficial gas velocity. Because of liquid head, this necessitates a pressure correction. The total liquid level for 10,000 gal in a 12-ft-diameter tank is roughly 12 ft (see Table 12-2). If the impeller is located one-sixth of the liquid level off bottom, as it should be for gas dispersion, the additional static liquid head is 10 ft of water that must be added to the atmospheric pressure (34 ft of water). The pressure correction makes the actual volumetric flow rate of gas Q_A equal $(1950 \text{ ft}^3/\text{min})(34 \text{ ft of water})/(34 + 10 \text{ ft of water}) = 1507 \text{ ft}^3/\text{min}$. Flow rate divided by cross-sectional area of tank, that is, $\pi(12 \text{ ft})^2/4 = 113 \text{ ft}^2$, gives a superficial gas velocity u_s of $(1507 \text{ ft}^3/\text{min})/113 \text{ ft}^2 = 13.3$ ft/min or 0.22 ft/s (0.067 m/s).

2. Determine overall mass-transfer coefficient.

As a design level for overall mass transfer, the coefficient should be based on the minimum concentration gradient and the maximum transfer rate. The minimum gradient will exist near the liquid surface, where the oxygen saturation concentration in the liquid is a minimum because of the minimum total pressure and the low concentration of oxygen there. The volume (mole) percent oxygen in air with 10 percent depletion is $(18.4 \text{ mol } O_2 - 1.8 \text{ mol } O_2)/(100 \text{ mol gas} - 1.8 \text{ mol gas}) = 0.169$ mol O_2 per mole of gas, which gives an oxygen partial pressure of 0.169 atm. The Henry's law solubility constant H for O_2 at 120°F is 5.88×10^4 (*Handbook of Tables for Applied Engineering Science,* CRC Press) for partial pressure in atmospheres and oxygen concentration in mole fraction dissolved oxygen. Thus the mole fraction oxygen dissolved in the water is $0.169/5.88 \times$

$10^4 = 2.87 \times 10^{-6}$ mol O_2 per mole of liquid. This equals $[2.87 \times 10^{-6}$ mol $O_2/1$ mol $H_2O][(32$ g $O_2/1$ mol $O_2)/(18$ g $H_2O/1$ mol $H_2O)][10^3$ mg $O_2/1$g $O_2][10^3$ g $H_2O/$ 1 L $H_2O] = 5.10$ mg O_2 per liter (5.10×10^{-3} kg O_2 per cubic meter).

The maximum oxygen-transfer rate is 16.2 lb O_2 per hour per 1000 gal (5.4×10^{-4} kg O_2 per second per cubic meter) and the minimum concentration gradient between saturation (5.10×10^{-3} kg O_2 per cubic meter) and bulk concentration (2.4×10^{-3} kg O_2 per cubic meter) is $(5.1 \times 10^{-3}) - (2.4 \times 10^{-3}) = 2.7 \times 10^{-3}$ kg O_2 per cubic meter. The overall mass-transfer coefficient k_La is computed by dividing the transfer rate by the concentration gradient:

$$\frac{5.4 \times 10^{-4}\ \text{kg/(s)(m}^3\text{) O}_2}{2.7 \times 10^{-3}\ \text{kg/m}^3\ \text{O}_2} = 0.2\ \text{s}^{-1}$$

3. *Compute required power per volume for agitation.*

Using the correlation for overall mass-transfer coefficient, that is, $k_La = 2.0 \times 10^{-3}(P/V)^{0.7}u_s^{0.2}$, the design value for mass-transfer coefficient ($0.2\ \text{s}^{-1}$), and the superficial gas velocity (0.067 m/s), the required agitation intensity can be computed: $P/V = [0.2\ \text{s}^{-1}/(0.067\ \text{m/s})^{0.2}(2.0 \times 10^{-3})]^{1/0.7} = 1558\ \text{W/m}^3$ (7.91 hp per 1000 gal).

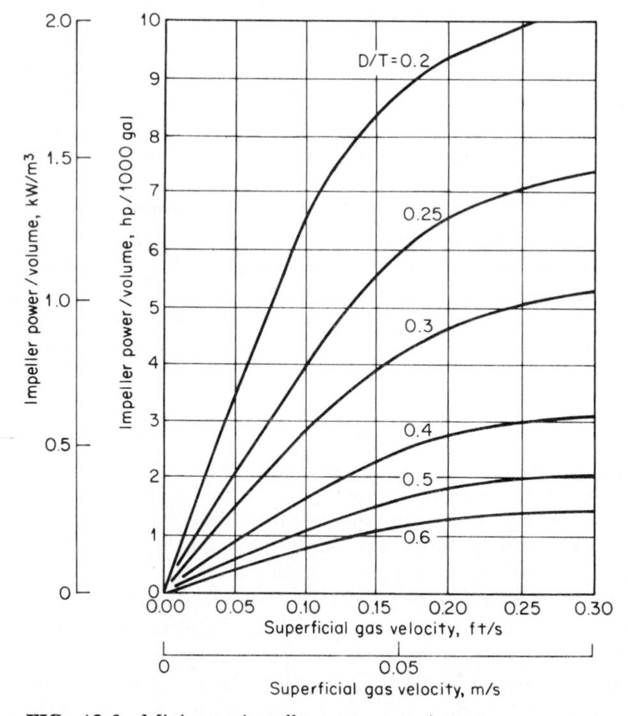

FIG. 12-6 Minimum impeller-power requirement to overcome flooding as a function of superficial gas velocity and D/T. *(From Chemical Engineering, 1976.)*

4. Determine minimum impeller size to prevent flooding.

Flooding in an agitated gas dispersion occurs when the impeller power and pumping capacity are insufficient to control the gas flow rate. A flooding correlation for minimum power per volume and superficial gas velocity is shown in Fig. 12-6 for several impeller-to-tank-diameter ratios (D/T). For a superficial gas velocity of 0.22 ft/s and 7.91 hp per 1000 gal, the minimum D/T is less than 0.25. So any impeller larger than $0.25\ T = 0.25(144\ \text{in}) = 36$ in should produce sufficient agitation to overcome flooding.

5. Determine impeller size for required power input.

The impeller required for the process must draw 7.91 hp per 1000 gal or 79.1 hp for the 10,000-gal batch. This power level must be achieved at the gas flow rate required by the process, which is, $Q_A = 1507\ \text{ft}^3/\text{min}$. The power required to operate an agitator impeller for gas dispersion can be much less than the power required for a liquid without gas. The ratio of power with gas to power without P/P_0 is shown in Fig. 12-7 and is a function of the dimensionless aeration number $N_{Ae} = Q_A/ND^3$.

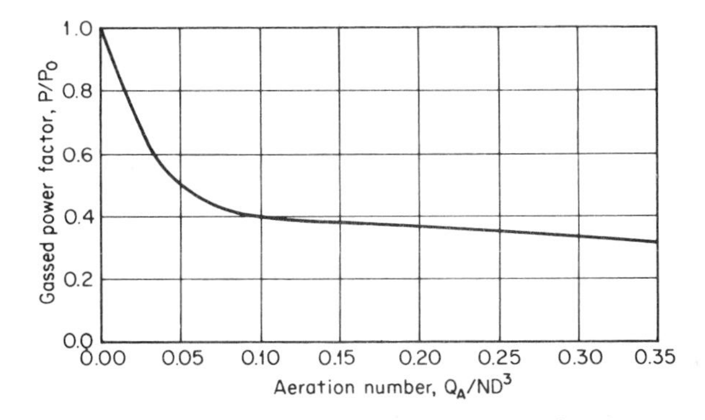

FIG. 12-7 Gassed power factor as a function of aeration number. *(From Chemical Engineering, 1976.)*

For gas-dispersion applications, a radial discharge impeller, such as the straight-blade and disk style impellers in Fig. 12-1, should be used. A straight-blade impeller with a power number N_P of 3.86 will be used for this design. Assume that the operating speed will be 100 r/min.

Since impeller diameter is unknown, aeration number and, therefore, gassed power factor cannot be determined; however, an estimate for power factor P/P_0 of 0.4 is usually a good initial estimate. By rearranging the expression for power number, that is, $N_P = P/(\rho N^3 D^5)$, an expression for impeller diameter can be derived. The factor P/P_0 must be introduced for the effect of gas and a conversion for units makes

$$D = \left[\frac{1.524 \times 10^{13} P}{(P/P_0) N_P N^3}\right]^{1/5} = \left[\frac{1.524 \times 10^{13}(79.1)}{0.4(3.86)(100)^3}\right]^{1/5} = 60.0\ \text{in}$$

With a value for D of 60.0 in (5.0 ft), the aeration number can be computed:

$$N_{Ae} = \frac{1507\ \text{ft}^3/\text{min}}{(100\ \text{r/min})(5.0\ \text{ft})^3} = 0.121$$

The revised estimate for P/P_0 is 0.38 from Fig. 12-7, which gives an impeller diameter of 60.7 in. Therefore, a 60.7-in-diameter (1.54-m) straight-blade impeller with a 10.1-in (0.257-m) blade width operating at 100 r/min should satisfy the process requirements.

6. Select standard size motor.

A motor that is loaded to 85 percent by a 79.1-hp impeller will require a minimum size of (79.1 hp)/0.85 = 93.1 hp, which means a 100-hp (75-kW) motor. This motor and impeller assembly is correctly sized for conditions with the design gas flow. However, because of the gassed power factor, that is, $P/P_0 = 0.38$, should the gas supply be lost for any reason, the impeller power would increase to 78/0.38 = 205 hp and seriously overload the motor. To avoid this problem, some method (typically electrical control) prevents motor operation without the gas supply. When the gas supply is off, the control either stops the agitator motor or, in the case of a two-speed motor, goes to a lower speed.

Related Calculations: The majority of gas-dispersion applications are sized on the basis of power per volume. In aerobic fermentation, levels of 5 to 12 hp per 1000 gal (1 to 2.4 kW/m^3) are typical, while for aerobic waste treatment, levels of 1 to 3 hp per 1000 gal (0.2 to 0.6 kW/m^3) are more common, primarily because of the concentrations and oxygen requirements of the microorganisms.

12-7 Shaft Design for Turbine Agitator

A 15-hp (11.2-kW) agitator operating at 100 r/min has been selected for a process application. The vessel geometry requires two impellers, both of 36-in (0.91 m) diameter, the upper one located 80 in (2.03 m) below the agitator drive and the lower one 130 in (3.30 m) below the drive. Pitched-blade turbine impellers having four blades are to be used. The shaft is to be stainless steel, having an allowable shear stress of 6000 psi (41,370 kPa), an allowable tensile stress of 10,000 psi (68,950 kPa), a modulus of elasticity of 28,000,000 psi (193,000,000 kPa), and a density of 0.29 lb/in^3 (8027 kg/m^3), which represents a weight of $0.228d^2$ per linear inch, where d is shaft diameter in inches. The bearing span for support of the agitator shaft is to be 16 in. What shaft diameter is required for this application?

Calculation Procedure:

1. Determine the hydraulic loads on the shaft.

To rotate the agitator, the shaft must transmit torque from the drive to the impellers. The actual torque required should be found by the sum of the horsepower required by each impeller. However, process conditions can change, so it is better to assume that the full motor power can be outputted from the drive. Maximum torque τ can be found by dividing motor horsepower P_{motor} by shaft speed N. A conversion factor of 63,025 makes the answer come out in English engineering units:

$$\tau = P_{\text{motor}}/N = (63{,}025)(15)/100 = 9454 \text{ in} \cdot \text{lb}$$

The output shaft must be large enough to transmit 9454 in·lb (1068 N·m) at the drive. Only half the torque need be transmitted in the shaft below the first impeller, because the lower impeller should require only half the total power.

If the hydraulic forces on individual impeller blades were always uniformly distributed, torsion considerations would constitute the only significant strength requirement for the shaft. However, real loads on impeller blades fluctuate, due to the shifting flow patterns that contribute to process mixing. For shaft design, using a factor of three-tenths (0.3) approximates the imbalanced force acting at the impeller diameter. This factor (0.3) is typical for pitched-blade turbines with four blades and may be different for other types of impellers. A higher factor should be used if the mixer is subjected to external loads, such as flow from a pump return.

The imbalanced forces result in a bending moment M on the shaft. Such moments must be summed for hydraulic loads at each shaft extension L_n where an impeller is located and must be adjusted for the impeller diameter D_n:

$$M = \sum_n 0.3(P_n/N)(L_n/D_n)$$

In the present example having two equal-size impellers, the motor power is split in half and the moments are calculated at each impeller location:

$$\begin{aligned} M &= 0.3[(63{,}025)(15/2)/100][80/36] + 0.3[(63{,}025)(15/2)/100][130/36] \\ &= 8272 \text{ in} \cdot \text{lb} \end{aligned}$$

Thus, a maximum bending moment of 8272 in·lb (935 N·m) occurs just below the agitator drive, at the top of the shaft extension. The hydraulic loads on the shaft result in a torque and bending moment that the shaft must be strong enough to handle.

2. *Determine the minimum shaft diameter for strength.*

Since the torque and bending moment may act simultaneously on the shaft, these loads must be combined and resolved into shear and tensile stresses on the shaft. The minimum shaft diameter must be the larger of the shaft diameters required by either shear- or tensile-stress limits. The shaft diameter for shear stress d_s can be calculated as follows:

$$d_s = [16(\tau^2 + M^2)^{1/2}/\pi\sigma_s]^{1/3}$$

where σ_s is the allowable shear stress. As already noted, the σ_s value recommended for carbon steel and stainless steel typically used in agitator applications is 6000 psi (41,370 kPa). This stress value is low enough to prevent permanent distortions and to minimize the possibility of fatigue failures. The minimum shaft diameter for shear strength is, accordingly,

$$d_s = [16(9454^2 + 8272^2)^{1/2}/\pi(6000)]^{1/3} = 2.201 \text{ in}$$

The minimum shaft diameter for tensile strength d_t can be calculated using a similar expression:

$$d_t = [16[M + (\tau^2 + M^2)^{1/2}]\pi\sigma_t]^{1/3}$$

where σ_t is the allowable tensile stress. The minimum shaft diameter for tensile strength is, accordingly,

$$d_t = [16(8272 + (9454^2 + 8272^2)^{1/2}]/\pi(10{,}000)]^{1/3} = 2.197 \text{ in}$$

The minimum shaft diameter for shear and tensile limits is 2.201 in (0.056 m). The next larger standard size is 2.5 in (0.0635 m), which provides an adequate initial design.

3. *Calculate the natural frequency of the agitator shaft.*

Shaft strength is not the only limit to agitator shaft design—a long shaft may not be stiff enough to prevent uncontrolled vibrations. A given overhung shaft, of the sort typically used with top-entering agitators, will oscillate at a natural frequency, similar to the vibration of a tuning fork. If the operating speed of the agitator is too close to that frequency, destructive oscillations may occur. Most large agitator shafts are designed with the operating speed less than the first natural frequency, so that even as the agitator is started and stopped, excessive vibrations should not occur.

A typical formula for calculating the first natural frequency (critical speed) of an agitator shaft considers the shaft stiffness, the shaft length, the weights of impellers and shaft, and the rigidity of the shaft mounting:

$$N_c = 37.8d^2(E_y/\rho_m)^{1/2}/LW_e^{1/2}(L + L_b)^{1/2}$$

where N_c = critical speed, r/min
d = shaft diameter, in
L = shaft extension, in
W_e = equivalent weight (lb) of impellers and shaft at shaft extension
L_b = spacing (in) of bearings that support shaft
E_y = modulus of elasticity, lb/in²
ρ_m = density, lb/in³

E_y and ρ are two material properties that characterize the stiffness of the shaft. Substituting the modulus and density values given in the statement of the problem reduces the expression to

$$N_c = 371{,}400d^2/LW_e^{1/2}(L + L_b)^{1/2}$$

For a two-impeller situation, W_e can be calculated as follows:

$$W_e = W_l + W_u(L_u/L)^3 + w_s(L/4)$$

where W_l and W_u are, respectively, the weights of the lower and upper impellers, the upper impeller is located at shaft extension L_u, and w_s is the unit weight of the shaft as given in the statement of the problem. Data on impeller weight can be furnished by the mixer vendor, measured directly by the user, or estimated from dimensions provided by the vendor. For the present case, assume that the impeller hub for a 2.5-in shaft weighs 25 lb (11.4 kg) and a set of blades for a 36-in impeller weighs 34.5 lb (15.7 kg), so each impeller for this agitator weighs 59.5 lb (27.0 kg). Thus,

$$W_e = 59.5 + 59.5(80/130)^3 + [0.228(2.5)^2][130/4] = 120 \text{ lb}$$

and accordingly,

$$N_c = 371{,}400(2.5)^2/[130(120)^{1/2}(130 + 16)^{1/2}] = 135 \text{ r/min}$$

This means that with the proposed 2.5-in shaft, an operating speed near 135 r/min must be avoided. In practice, the design limit for operating speed should be no higher than 65 percent of critical speed. This conservative margin is necessary because of many factors that might reduce the critical speed or increase loads on the shaft. For instance, the critical-speed calculation assumes that the agitator drive support is rigid, but in fact tank nozzles and support structures have some flexibility that reduces the natural frequency. Furthermore, dynamic loads on the impeller, such as those induced by operating near the liquid level, may make the effects of natural frequency more significant.

In the present case, the 100-r/min operating speed given in the statement of the problem is 74 percent of critical speed and therefore is too close to critical speed for safe operation.

4. Redesign the shaft to avoid critical-speed problems.

A larger shaft diameter should overcome critical-speed problems, provided the gear reducer will accept the larger shaft. The larger shaft diameter increases shaft stiffness and thus increases the natural frequency. However, a large shaft also means more weight in the impeller hubs and in the shaft.

For the present case, try a 3.0-in shaft. Assume that vendor information indicates a hub weight of 40 lb (18.2 kg) for such a shaft, thus increasing the impeller weights to 74.5 lb (33.8 kg). Accordingly, the new equivalent weight is

$$W_e = 74.5 + 74.5(80/130)^3 + [0.228(3.0)^2][130/4] = 159 \text{ lb}$$

And the new critical speed is

$$N_c = 371{,}400(3.0)^2/[(130)(159)^{1/2}(130 + 16)^{1/2}] = 169 \text{ r/min}$$

The stated operating speed of 100 r/min is only 59 percent of this critical speed, so the 3.0-in shaft should operate safely.

5. Explore other alternatives for solving the critical-speed problem.

The seemingly obvious answer to a critical-speed problem is to reduce the operating speed of the agitator, and this option should always be checked out. However, it can introduce complications of its own. In particular, the lower speed will change the process performance of the agitator, and accordingly a larger impeller will be required for meeting performance requirements. Because larger impellers weigh more, the critical-speed determination must be made anew.

Apart from its effect on critical speed, such an impeller will affect horsepower requirements. Furthermore, the speed change may require a new gear reducer.

The critical-speed problem may in some cases be solved hand in hand with a common problem related to dynamic loads. One source of dynamic loads on an agitator shaft is the waves and vortices that occur when an impeller operates near the liquid surface, such as when a tank fills or empties. Adding stabilizer fins to the impeller blades will help reduce some of these loads. Such fins also permit the agitator to operate closer to critical speed, perhaps at 80 percent rather than 65 percent.

Suppose, for instance, that the 2.5-in shaft is used and stabilizer fins add 16 lb (7.3 kg) to the weight of the lower impeller. The equivalent weight increases accordingly:

$$W_e = (59.5 + 16) + 59.5(80/130)^3 + [0.228(2.5)^2][130/4] = 136 \text{ lb}$$

and the critical speed decreases:

$$N_c = 371{,}400(2.5)^2/[130(136)^{1/2}(130 + 16)^{1/2}] = 127 \text{ r/min}$$

The operating speed of 100 r/min is now 79 percent of critical, which is just within the aforementioned 80 percent ceiling.

Another alternative to avoid critical-speed problems is the use of a shorter shaft. Reducing the shaft length and impeller extensions by 10 in (0.25 m) reduces equivalent weight to

$$W_e = 59.5 + 59.5(70/120)^3 + [0.228(2.5)^2][120/4] = 114 \text{ lb}$$

At 114 lb (51.8 kg), the critical speed increases to

$$N_c = 371{,}400(2.5)^2/[120(114)^{1/2}(120 + 16)^{1/2}] = 155 \text{ r/min}$$

At 155 r/min, the operating speed is slightly less than 65 percent of critical speed. If the reduced shaft length can be achieved by reducing the mounting height, the impeller location and performance remain unchanged. Otherwise, review process conditions. Other design changes, such as a smaller shaft below the upper impeller, may reduce equivalent weight and increase critical speed.

The additional cost of the material for a larger-diameter shaft is rarely a major factor for carbon or stainless steel, but may be sizeable for special alloys. The additional cost of a larger mechanical seal, for pressurized applications, may add considerable cost regardless of the shaft material.

12-8 Viscosity Determination from Impeller Power

A helical ribbon impeller, 45 in (1.14 m) diameter, is operated in a 47-in (1.19-m) diameter reactor; estimate the fluid viscosity from torque readings. The impeller is a single-turn helix with a 1:1 pitch, so the height of the impeller is the same as the diameter, 45 in (1.14 m). It is a double-flight helix, each blade of which is 4.5 in (0.114 m) wide.

The agitator on the reactor is instrumented with a tachometer and torque meter. In the early stages of a polymerization, the impeller is operated at 37 r/min and the torque reading is 460 in · lb (52 N · m). As the viscosity increases during polymerization, the agitator is slowed to 12 r/min. The reaction is stopped when the torque reaches 27,300 in · lb (3085 N · m). As a final check on the polymer, an additional torque reading of 20,600 in · lb (2328 N · m) is taken at 8 r/min. Assume that the polymer has a specific gravity of 0.92 throughout the polymerization. What is the apparent viscosity of the polymer at the early stage of the process and at the two final conditions?

Calculation Procedure:

1. Estimate viscous power number for the helix impeller.

A helical-ribbon impeller, also called a helix impeller, is used primarily when high-viscosity fluids are being processed. Most of the power data on such impellers have been obtained in the laminar and transitional flow ranges. The effect on power of common geometry factors, i.e., impeller diameter D, tank diameter T, helix pitch P, impeller height H, and helix (blade) width W, can be incorporated into a correlation for a (dimensionless) viscous power number:

$$N_P^* = 96.9[D/(T-D)]^{0.5}[1/P][H/D][W/D]^{0.16}$$

For the impeller described in this problem the power number is

$$N_P^* = 96.9[45/(47-45)]^{0.5}[1/1][45/45][4.5/45]^{0.16} = 318$$

The viscous power number is *defined* in terms of power P, viscosity μ, shaft speed N, and impeller diameter D:

$$N_P^* = P/\mu N^2 D^3 = N_P N_{Re}$$

Thus, viscous power number N_P^* is related to turbulent power number N_P by the factor of the Reynolds number N_{Re}. The viscous power number is chosen as a correlating value because it has a constant value in the viscous, low–Reynolds number range, less than 60 ($N_{Re} < 60$). Using the viscous power number in the laminar range eliminates fluid density from the correlation, which is appropriate.

2. *Estimate viscosity at early stages of polymerization.*

The torque τ measurement, combined with shaft speed, can be converted to power:

$$P = \tau N/63{,}025 = (460)(37)/63{,}025 = 0.27 \text{ hp}$$

In this equation, 63,025 is a conversion factor for dimensional consistency.

At 0.27 hp (0.20 kW), a constant viscous power number can be used to predict an apparent viscosity, with the aid of a 6.11×10^{-15} units-conversion factor:

$$\mu_a = P/N_P^* N^2 D^3 = 0.27/(6.11 \times 10^{-15})(318)(37)^2(45)^3$$
$$= 1114 \text{ cP}$$

Now, a viscosity of 1114 cP is low for viscous (laminar) flow conditions to exist. So, it is prudent to check the magnitude of N_{Re}:

$$N_{\text{Re}} = D^2 N\rho/\mu = (10.7)(45)^2(37)(0.92)/1114 = 662$$

where 10.7 is for dimensional consistency when ρ is specific gravity.

While not fully turbulent, a Reynolds number of 662 is not laminar either. It is in the transitional range, so a correction factor from Fig. 12-8 must be applied to the viscous power number. This is done in step 3.

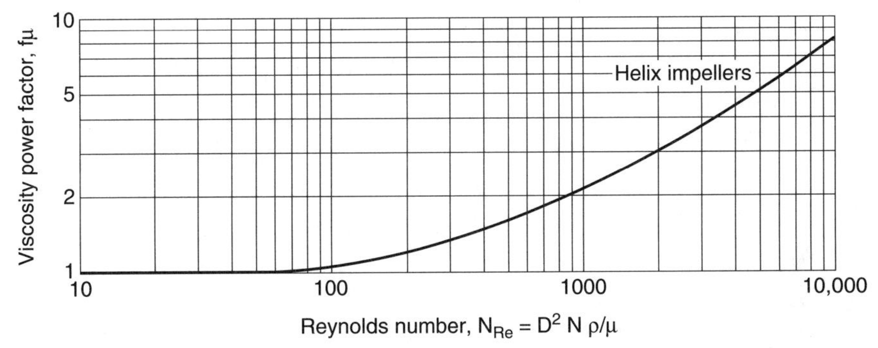

FIG. 12-8 Viscosity power factor for viscous power number of a helix impeller as a function of impeller Reynolds number.

The term "apparent viscosity" refers to a viscosity that has been back-calculated from impeller torque or horsepower. A true-viscosity reading should be measured at a fixed and known shear rate. The effective shear rate developed by a mixing impeller is really a distribution of different shear rates. This distribution is probably most closely related to the shear between the helix blade and the tank wall, but other shear rates may affect power. If viscosity is shear-dependent, as often happens with high-viscosity polymers, the velocity distribution will affect the apparent viscosity. Different impeller types may give different apparent viscosities at the same shaft speed.

Different speeds will provide different shear rates and perhaps different apparent viscosities for many viscous fluids. Apparent viscosities measured by a mixing impeller are more useful for mixer design than those obtained with a viscometer, because the appropriate distribution of shear rates are included in the measurements.

3. Reestimate viscosity for intermediate Reynolds number.

In the intermediate Reynolds number range, $60 < N_{Re} < 20{,}000$ (for a helix impeller), the viscous power number is not constant, nor is the turbulent power number. Figure 12-8, a graph of the viscosity power factor as a function of the Reynolds number, can be used to correct the viscous power number in the transitional range.

At a Reynolds number of 662, the viscosity power factor f_μ is approximately 1.8. Applying this factor to the power number used in the apparent-viscosity calculation of step 2 enables the estimation of another apparent viscosity:

$$\mu_a = 0.27/(6.11 \times 10^{-15})(1.8)(318)(37)^2(45)^3 = 619 \text{ cP}$$

Another estimate of Reynolds number based on a viscosity of 619 cP gives a value of 1191. At a Reynolds number of 1191, the power factor becomes 2.3, which leads to an apparent viscosity of 484. A few more iterations reach an estimated viscosity of 370 cP, based on a correction factor of 3.0 at a Reynolds number of 1994.

Several iterations are required, because power becomes less dependent on viscosity as conditions approach the turbulent range, typically $N_{Re} > 20{,}000$. In the turbulent range, power is independent of viscosity, and impeller power cannot be used to estimate viscosity.

4. Determine apparent viscosities at the end of the process.

As the polymerization proceeds, the viscosity increases. Higher viscosity means higher torque. Reducing the agitator speed is necessary to keep the torque and power within the capabilities of the agitator. At the end of the process, with the agitator turning at 12 r/min, the torque reaches 27,300 in · lb (3085 N · m) and the power required (see step 2) is

$$P = (27{,}300)(12)/63{,}025 = 5.20 \text{ hp}$$

At 5.20 hp (3.88 kW), the apparent viscosity is

$$\mu_a = 5.20/(6.11 \times 10^{-15})(318)(12)^2(45)^3 = 204{,}000 \text{ cP}$$

At 204,000 cP, the Reynolds number is only

$$N_{Re} = (10.7)(45)^2(12)(0.92)/204{,}000 = 1.2$$

which is well into the viscous (laminar) range. At these conditions, power and torque are proportional to viscosity at a set speed and the viscosity power factor is 1.

The same calculations at 8 r/min and 20,600 in · lb (2328 N · m) show the following:

$$P = (20{,}600)(8)/63{,}025 = 2.61 \text{ hp}$$

$$\text{and} \quad \mu_a = 2.61/(6.11 \times 10^{-15})(318)(8)^2(45)^3 = 230{,}000 \text{ cP}$$

Thus, the fluid appears to be more viscous at the lower speed. Another view of the fluid properties is that as agitator speed increases, the shear rate increases and the viscosity decreases. This fluid behavior is called "shear thinning" and is typical of many polymers.

Thus, a properly instrumented reactor may be used as a viscometer for high-viscosity fluids.

13

*Size Reduction**

Ross W. Smith, Ph.D.

Department of Chemical and Metallurgical Engineering
University of Nevada at Reno
Reno, NV

REFERENCES: [1] Herbst and Fuerstenau—*Trans. SME/AIME 241*:538, 1968; [2] Lynch—*Mineral Crushing and Grinding Circuits,* Elsevier; [3] Bond—*Crushing and Grinding Calculations,* Allis-Chalmers; [4] Smith and Lee—*Trans. SME/AIME 241*:91, 1968; [5] Smith—*Mining Engineering,* April 1961.

* Example 13-10 is adapted from an article in *Chemical Engineering* magazine.

13-1 Size Distributions of Crushed or Ground Material

The data in the first three columns of Table 13-1 show the size distributions of a quartz-gold ore obtained by sieving the material. What is the cumulative weight percent passing each sieve size? Also, construct a log-log plot of the cumulative percent versus particle size, and express the resulting relationship as an equation.

TABLE 13-1 Input Data and Calculated Results on Ore Size Distribution (Example 13-1)

Tyler mesh size	Sieve opening, μm	Weight of retained sample, g	Percent retained (by calculation)	Cumulative percent passing (by calculation)
4	4950	0	0	100.0
6	3350	7.20	3.0	97.0
8	2360	27.84	11.6	85.4
10	1700	36.96	15.4	70.0
14	1180	43.68	18.2	51.8
20	850	34.08	14.2	37.6
28	600	27.60	11.5	26.1
35	425	16.08	6.7	19.4
48	300	15.36	6.4	13.0
65	212	8.64	3.6	9.4
100	150	6.48	2.7	6.7
150	106	4.56	1.9	4.8
200	75	3.84	1.6	3.2
270	53	2.16	0.9	2.3
Pan	—	5.52	2.3	—

Calculation Procedure:

1. *Compute the cumulative weight percent passing each sieve size.*

Add up the weights of samples retained on each screen (the weights are shown in the third column of Table 13-1). Express each as a percent of the total, obtaining the percents shown in the fourth column. Sum the cumulative amounts retained and subtract from 100, obtaining the cumulative percents passing. These are shown in the fifth column.

2. *Express the relationship between cumulative percent passing and particle size as an equation.*

Plot cumulative weight percent passing (y) as a function of particle size (x) in microns (as denoted by size of sieve opening) on log-log paper, as in Fig. 13-1.

The straight-line portion of the plot can be expressed as $y = cx^b$. This can be recast into a form that is especially useful for other size-reduction calculations (e.g., see Example 13-2), namely, the Gaudin-Schuhmann equation, that is, $y = 100(x/k)^a$, where a (the distribution modulus) is a constant for a particular size distribution, and k (the size modulus) is the 100 percent size, in microns, of the extrapolated straight-line portion of the plot. By applying least-squares curve fitting to the log-log plot, the values of a and k can be obtained, yielding $y = 100(x/2251)^{1.003}$.

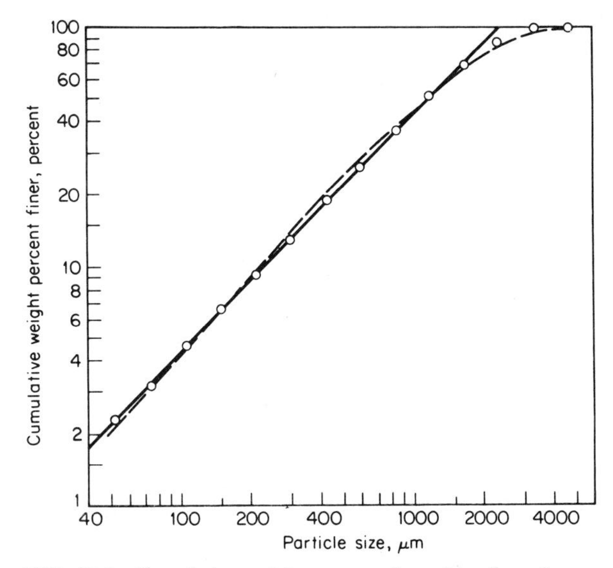

FIG. 13-1 Cumulative weight percent of particles finer than a given size as a function of the particle size (Example 13-1).

Related Calculations: Sometimes the Rosin-Rammler equation is used to represent the size distribution graphically. The dashed line on Fig. 13-1 corresponds to the Rossin-Rammler equation in the form $y = 100 - 100 \exp [-(x/A)^b]$, where, in this case, $A = 1558$ and $b = 1.135$.

13-2 Breakage and Selection Functions from Batch Ball-Mill Tests

Determine the breakage characteristics of a cement rock by using selection and breakage functions, calculating these on the basis of the ball-mill test data in Table 13-2. (A "selection function" is a parameter that represents the resistance of some size fraction to being produced during breakage. The "breakage functions" are related quantities that determine the breakage-product size distribution for material broken in this size fraction.)

TABLE 13-2 Ball-Mill Test Data on Cement Rock (Example 13-2)

Tyler mesh size	Sieve opening, μm	Cumulative mass fraction finer than mesh size after noted grinding time			
		1 min	2 min	4 min	6 min
8	2360	1.000	1.000	1.000	1.000
10	1700	0.450	0.695	0.910	0.966
48	300	0.080	0.175	0.337	0.523
100	150	0.046	0.100	0.196	0.300
200	75	0.030	0.062	0.126	0.195
400	38	0.018	0.046	0.076	0.110

Feed size: −8,+10 mesh (8 × 10 mesh)

Calculation Procedure:

1. Calculate the selection function that pertains to the size fraction of the feed material.

This selection function S_1 is defined by the equation

$$M_1(t) = M_1(0) \exp(-S_1 t)$$

where $M_1(t)$ is the mass fraction of feed remaining after time t. Referring to Table 13-2, note that for this 8 × 10 mesh feedstock, $M_1(t)$ equals 1 minus the cumulative mass fraction finer than no. 10 mesh at time t. Using the data in Table 13-2, we can therefore determine S_1 by plotting log mass fraction of feed versus grinding time, determining the slope of the resulting straight line by least-squares curve fitting and multiplying the slope by -2.303. The resulting value is 0.577 min^{-1}.

2. Determine the production rate constant.

In batch ball milling, the changes in particulate distribution with time can be characterized [1] by

$$\frac{dY_i(t)}{dt} = \overline{F}_i$$

where t is time, $Y_i(t)$ is cumulative mass fraction finer than size x_i at time t for short grinding times, and $\overline{F}_i$ is cumulative zero-order production rate constant for size x_i and the production of fine sizes much smaller than feed size. If, then, a linear plot is made of cumulative mass fraction versus (short) grinding time, the $\overline{F}_i$ constants can be determined for each fine size from the slopes of the curves generated. Figure 13-2 shows such a plot for the present data. From the slopes, Table 13-3 can be assembled.

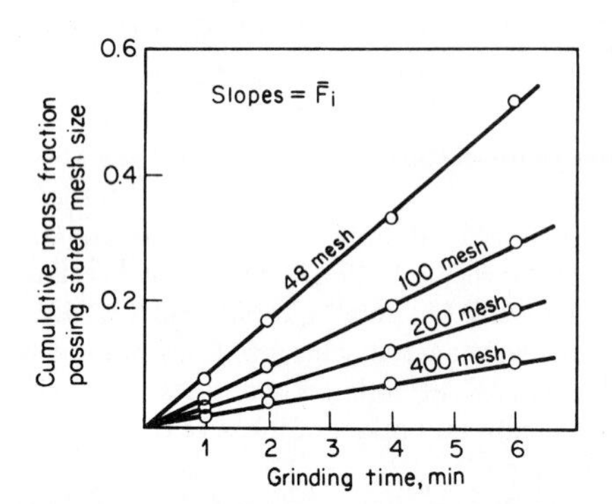

FIG. 13-2 Cumulative mass fraction of ground material as a function of grinding time (Example 13-2).

3. Determine the breakage functions that pertain to the size fraction of the feed material.

Cumulative breakage functions can be calculated [1] from the relation $B_{i,j} = \overline{F}_i/S_j$. Therefore, in the case of the feed-size selection function S_1, the equation is $B_{i,1} = \overline{F}_i/S_1$, and the breakage functions are as follows:

$$B_{48\text{mesh},1} = \frac{\overline{F}_{48\text{mesh}}}{S_1} = \frac{0.0869}{0.577} = 0.151$$

$$B_{100\text{mesh},1} = \frac{0.0500}{0.577} = 0.087$$

$$B_{200\text{mesh},1} = 0.056 \quad \text{and} \quad B_{400\text{mesh},1} = 0.032$$

TABLE 13-3 Calculated Production Rate Constants (Example 13-2)

Particle size x		
Mesh	Microns	Production rate constant $\overline{F}_i$
48	300	0.0869
100	150	0.0500
200	75	0.0325
400	38	0.0183

4. Determine the selection and breakage functions for other size fractions of the same material.

Selection functions for other size intervals may be calculated via the relation

$$S_j = S_1 \left[\frac{(X_j X_{j+1})^{1/2}}{(X_1 X_2)^{1/2}} \right]^a$$

where X_1 and X_2 are the sieve-opening sizes that define the size fraction of the feed material, X_j and X_{j+1} are the sieve openings that define the size fraction whose selection function is now to be calculated, and a is the slope of the log-log plot of the zero-order production rate constants $\overline{F}_i$ against particle size x_i, in microns, from Table 13-3. (That plot is not shown here.) The slope, determined by least-squares curve fitting, is 0.741. This slope, a, happens to be the same as the distribution modulus in the Gaudin-Schuhmann equation (see Example 13-1).

For instance, the selection function for the size fraction, $-10, +14$ mesh (corresponding to sieve openings of 1700 and 1180 μm, respectively) can be calculated thus:

$$S_2 = 0.577 \left\{ \frac{[1700(1180)]^{1/2}}{[2360(1700)]^{1/2}} \right\}^{0.741} = 0.446$$

Similar calculations can be made to find selection functions for other size intervals. Then, cumulative breakage functions can be calculated by the relationship noted in step 3, namely, $B_{i,j} = \overline{F}_i / S_j$.

13-3 Predicting Product Size Distribution from Feedstock Data

Given the feed size distribution, the breakage functions, and the selection functions (probabilities of breakage) for a feedstock to a grinding operation, as shown in Table 13-4, predict the size distribution for the product from the operation.

Calculation Procedure:

1. Predict the weight percentages of broken and unbroken particles within each size range.

In each size range, multiply the feed size percentage F by the corresponding selection function S; this product gives an estimate as to the percentage of feed particles within the

TABLE 13-4 Data on Feedstock to Grinding Operation (Example 13-3)

Size range	Size interval: Tyler mesh	Size interval: Micron size	Feed size distribution F, percent	Breakage matrix B						Selection matrix S
1	−6, +8	−3350,+2360	24	0.18	0	0	0	0	0	1.00
2	−8, +10	−2360,+1700	16	0.22	0.18	0	0	0	0	0.81
3	−10,+14	−1700,+1180	12	0.16	0.22	0.18	0	0	0	0.60
4	−14,+20	−1180,+850	10	0.12	0.16	0.22	0.18	0	0	0.47
5	−20,+28	−850,+600	7	0.10	0.12	0.16	0.22	0.18	0	0.35
6	−28,+35	−600,+425	5	0.08	0.10	0.12	0.16	0.22	0.18	0.24
	−35	−425	26							

range that become broken. Then subtract the product from the feed size percentage; this difference is an estimate as to the particles that remain unbroken. These operations are shown in Table 13-5. In matrix notation (for consistency with subsequent steps), the S's can be considered to form an $n \times n$ diagonal matrix, and the F's an $n \times 1$ matrix.

TABLE 13-5 Estimating the Percentage of Broken and Unbroken Particles (Example 13-3)

Size range	Feed size distribution F, percent	Selection functions S	Particles broken $(S)(F)$, percent	Particles not broken $(F) - (S)(F)$, percent
1	24	1.00	24.00	0
2	16	0.81	12.96	3.04
3	12	0.60	7.20	4.80
4	10	0.47	4.70	5.30
5	7	0.35	2.45	4.55
6	5	0.24	1.20	3.80
−35 mesh	26			

2. ***Predict the size distribution of the product of the breakage of broken particles.***

This product does not exist by itself as a separate entity, of course, because the particles that become broken remain mixed with those which stay unbroken. Even so, the distribution can be calculated by postmultiplying the $n \times n$ lower triangular matrix of breakage functions B in Table 13-4 by the percentage of particles broken, the $n \times 1$ matrix $(S)(F)$, as calculated in step 1. This postmultiplication works out as follows:

Size range	Amount of breakage product in size range
1	0.18(24.00) = 4.32
2	0.22(24.00) + 0.18(12.96) = 7.61
3	0.16(24.00) + 0.22(12.96) + 0.18(7.20) = 7.99
4	0.12(24.00) + 0.16(12.96) + 0.22(7.20) + 0.18(4.70) = 7.38
5	0.10(24.00) + 0.12(12.96) + 0.16(7.20) + 0.22(4.70) + 0.18(2.45) = 6.59
6	0.08(24.00) + 0.10(12.96) + 0.12(7.20) + 0.16(4.70) + 0.22(2.45) + 0.18(1.20) = 5.59

3. Within each size range, sum up the percent of particles that remained unbroken throughout the operation (from step 1) and percent of particles in that range which resulted from breakage (from step 2).

These sums work out as follows. The −35 mesh percentage is found by difference (i.e., it is the residual):

Size interval	Total-product size distribution, percent
1	0 + 4.32 = 4.32
2	3.04 + 7.61 = 10.65
3	4.80 + 7.99 = 12.79
4	5.30 + 7.38 = 12.68
5	4.55 + 6.58 = 11.13
6	3.80 + 5.59 = 9.39
−35 mesh	39.04

Related Calculations: If the grinding system includes a classification step that recycles the larger particles (e.g., those in size ranges 1 and 2) to the mill instead of allowing them to leave with the product, the operation is known as "closed-circuit grinding." In such a case, the preceding sequence can be expanded into an iterative procedure. In essence, steps 1 through 3 are applied anew to the material in size ranges 1 and 2, yielding a "final product" size distribution for this second round of breakage. Steps 1 through 3 are then applied a third time to the material in size ranges 1 and 2 from the second round of breakage; the procedure is repeated until virtually no material remains in those two size ranges.

13-4 Material-Balance Calculations for Closed-Circuit Grinding

In the grinding operation of Fig. 13-3, a ball mill is in closed circuit with a hydrocyclone classifier. The mass flow rates of the classifier feed, oversize, and undersize are denoted by the symbols A, O, and U, respectively. The fineness of classifier feed a, of oversize o, and of undersize u, all expressed as percentages passing a 200 mesh sieve, are 58.5, 48.2, and 96.0 percent, respectively. Undersize is produced at a rate of 20.3 tons/h. What is the percent circulating load? What are the flow rates of classifier feed and oversize?

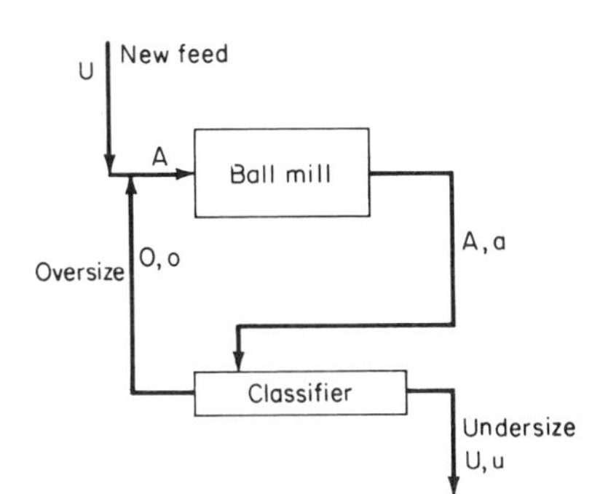

FIG. 13-3 Ball-mill–classifier circuit (Example 13-4).

Calculation Procedure:

1. Find the percent circulating load.

The percent circulating load L is defined by the relationship $L = 100O/U$. By material-balance algebra, then, $L = 100(u - a)/(a - o)$, which is $100(96.0 - 58.5)/(58.5 - 48.2) = 364$ percent.

2. Find the flow rate of oversize.

From step 1, $O = LU/100$, which is $364(20.3)/100 = 73.9$ tons/h.

3. Find the flow rate of classifier feed.

By material-balance algebra, $A = O + U$, which is $73.9 + 20.3 = 94.2$ tons/h.

13-5 Work-Index Calculations

A limestone ore is to be ground in a conventional ball mill to minus 200 mesh. The Bond grindability of the ore for a mesh-of-grind of 200 mesh (75 μm) is determined by laboratory test to be 2.73 g per revolution. The 80 percent passing size of the feed to the test is 1970 μm; the 80 percent passing size of the product from it is 44 μm. Calculate the work index W_i for the material at this mesh-of-grind. (The "work index," defined as the energy needed to reduce ore from infinite size to the state where 80 percent will pass a 100 mesh screen, is a parameter that is useful in calculating size-reduction power requirements; see Example 13-6.)

Calculation Procedure:

1. Employ an empirical formula that yields the work index directly.

The formula [3] is

$$W_i = \frac{44.5}{P_1^{0.23} G^{0.82} \left(\dfrac{10}{P^{0.5}} - \dfrac{10}{F^{0.5}} \right)}$$

where W_i is the work index, in kilowatt-hours per ton, G is grindability, in grams per revolution, P is the 80 percent passing size of the product of the grindability test, in microns, F is the 80 percent passing size of the feed to the grindability test, in microns, and P_1 is the size of the mesh-of-grind of the grindability test, in microns. Thus,

$$W_i = \frac{44.5}{75^{0.23}\, 2.73^{0.82} \left(\dfrac{10}{44^{0.5}} - \dfrac{10}{1970^{0.5}} \right)}$$
$$= 5.64 \text{ kWh/ton}$$

Related Calculations: Although the Bond grindability test is run dry, the work index calculated above is for wet grinding. For dry grinding, the work index must be multiplied by a factor of $4/3$.

A crushability work index can also be empirically calculated. A Bond crushability test (based on striking the specimen with weights) indicates the crushing strength C per unit thickness of material, in foot pounds per inch. This is related to work index W_i, in kilowatthours per ton, by the formula $W_i = 2.59C/S$, where S is the specific gravity.

13-6 Power Consumption in a Grinding Mill as a Function of Work Index

The Bond work index for a mesh-of-grind of 200 mesh for a rock consisting mainly of quartz is 17.5 kWh/ton. How much power is needed to reduce the material in a wet-grinding ball mill from an 80 percent passing size of 1100 μm to an 80 percent passing size of 80 μm?

Calculation Procedure:

1. Employ a formula based on the Bond third theory of comminution.

The formula is

$$W = 10W_i(P^{-0.5} - F^{-0.5})$$

where W is power required, in kilowatthours per ton, W_i is work index, in kilowatthours per ton, P is 80 percent passing size of product, in microns, and F is 80 percent passing size of feed, in microns.

Thus,

$$W = 10(17.5)(80^{-0.5} - 1100^{-0.5})$$
$$= 14.3 \text{ kWh/ton}$$

Related Calculations: The Charles-Holmes equation [4], that is,

$$W = 100^r W_i(P^{-r} - F^{-r})$$

is in principle more accurate than the Bond third theory formula illustrated above. However, it requires determining the parameter r by running grindability tests on the rocks in question at two or more meshes-of-grind. If a single Bond grindability test is run at a mesh-of-grind close to that of the maximum product size from the proposed grinding operation, the Bond third theory equation should be almost as accurate.

13-7 Ball-Mill Operating Parameters

A ball mill having an inside diameter of 12 ft and an inside length of 14 ft is to be used to grind a copper ore. Measurement shows that the distance between the top of the mill and the leveled surface of the ball charge is 6.35 ft. What is the weight of balls in the mill? What is the critical speed of the mill (the speed at which the centrifugal force on a ball in contact with the mill wall at the top of its path equals the force due to gravity)? At what percent of critical speed should the mill operate?

Calculation Procedure:

1. Calculate the volume percent of the mill occupied by the balls.

The volume percent can be calculated from the relationship $V_p = 113 - 126H/D$, where V_p is percent of mill volume occupied by grinding media, H is distance from top to leveled surface, and D is mill inside diameter. Thus, $V_p = 113 - 126(6.35)/12 = 46.3$ percent.

*2. **Calculate the weight of the balls in the mill.***

It can be assumed that loose balls weigh 290 lb/ft^3. (Rods would weigh 390 lb/ft^3, and silica pebbles would weigh 100 lb/ft^3.) Then, weight of balls equals (290)(volume of mill occupied by balls) $= 290\pi(D/2)^2 LV_p/100 = 290(3.14)(12/2)^2(14)(46.3/100) = 213{,}000$ lb.

*3. **Calculate the proper mill speed.***

This can be estimated from the equation $N_o = 57 - 40 \log D$, where N_o is proper speed, in r/min, and D is mill inside diameter, in feet. Thus, $N_o = 57 - 40 \log 12 = 57 - 40(1.079) = 13.8$ r/min. [The relation $N_o = 57 - 40 \log D$ is only approximate. In actual practice, it will be found that short mills ($L < 2D$) often tend to run at slightly higher speeds, and long mills often tend to operate at slightly lower speeds.]

*4. **Calculate the critical mill speed.***

This can be estimated from the equation $N_c = 76.6/D^{1/2}$, where N_c is critical speed, in r/min, and D is mill inside diameter, in ft. Thus, $N_c = 76.6/12^{1/2} = 22.1$ r/min.

*5. **Calculate the percent of critical speed at which the mill should be operated.***

This follows directly from steps 3 and 4. Thus, percent of critical speed equals $100(13.8)/22.1 = 62.4$ percent.

13-8 Maximum Size of Grinding Media

A taconite ore is to be ground wet in a ball mill. The mill has an internal diameter of 13 ft (3.96 m) and is run at 68 percent of critical speed. The work index of the ore is 12.2 kWh/ton, and its specific gravity is 3.3. The 80 percent passing size of the ore is 5600 μm. What is the maximum size of grinding media (maximum diameter of balls) to be used for the operation?

Calculation Procedure:

*1. **Employ an empirical formula that yields the maximum size directly.***

The maximum size grinding media for a ball mill (whether for initial startup or for makeup) may be calculated from the formula

$$M = \left(\frac{F}{K}\right)^{1/2} \left(\frac{SW_i}{100C_s D^{1/2}}\right)^{1/3}$$

where M is maximum size of balls, in inches, F is 80 percent passing size of feed to the mill, in microns, S is specific gravity of the ore, W_i is work index of the ore, in kilowatthours per ton, D is inside diameter of the ball mill, in feet, C_s is fraction of critical speed of the mill, and K is a constant (350 for wet grinding or 330 for dry grinding).

Thus,

$$M = \left(\frac{5600}{350}\right)^{1/2} \left[\frac{3.3(12.2)}{100(0.68)(13)^{1/2}}\right]^{1/3}$$
$$= 2.19 \text{ in} \quad \text{or about } 2\tfrac{1}{4} \text{ in } (0.057 \text{ m})$$

Related Calculation: The maximum-diameter rod to be fed to a rod mill can be calculated from the empirical equation

$$R = \frac{F^{0.75}}{160[W_i S/(C_s D^{1/2})]^{1/2}}$$

where R is maximum diameter of rod, in inches, and the other variables are as in the example.

13-9 Power Drawn by a Grinding Mill

A ball mill with an inside diameter of 12 ft (3.66 m) is charged with 129 tons (117,000 kg) of balls that have a maximum diameter of 3 in (0.076 m) and occupy 46.3 percent of the mill volume. The mill is operated wet at 62.4 percent of critical speed. What is the horsepower needed to drive the mill at this percentage of critical speed?

Calculation Procedure:

1. Employ an empirical formula that yields the horsepower directly.

The formula [5] is

$$\text{hp} = 1.341\,W_b\{D^{0.4}C_s(0.0616 - 0.000575\,V_p) - 0.1(2)^{[(C_s-60)/10]-1}\}$$

where hp is the required horsepower, W_b is the weight of the ball charge, in tons, D is the inside mill diameter, in feet, C_s is the percentage of critical speed at which the mill is operated, and V_p is the percentage of mill volume occupied by balls. Thus,

$$\text{hp} = 1.341(129)\{12^{0.4}(62.4)[0.0616 - 0.000575(46.3)]$$
$$- 0.1(2)^{[(62.4 - 60)/10] - 1}\}$$
$$= 1.341(129)(5.84) = 1010 \text{ hp } (753 \text{ kW})$$

Related Calculation: For large-diameter mills using makeup balls of relatively small maximum size, it is often necessary to subtract a "slump correction" [4] from the braced term of the empirical relation above. The formula for this correction is $[12D - 60B(D - 8)]/240B$, where D is inside mill diameter, in feet, and B is the largest size of makeup ball, in inches. For the present example, the slump correction is $[12(12) - 60(3)(12 - 8]/240(3) = -0.8$. Thus the required horsepower becomes $1.341(129)[5.84 - (-0.8)] = 1150$ hp (858 kW).

13-10 Water Requirements for Closed-Circuit Mill System

A closed-circuit grinding system employs a high-efficiency air classifier wherein the classifier feed (i.e., the mill discharge) is exposed to outside air rather than recirculated air, thus reducing the product-cooling load. The system is shown in Fig. 13-4.

Fresh-feed rate N and mill production rate P are each 200,000 lb/h (90,900 kg/h). The circulating load L is 150 percent (i.e., 1.5). The flow rate of 80°F (300 K) ambient air to the classifier A is 221,000 lb/h (100,500 kg/h). The fresh feed enters the mill at 160°F (344 K). The flow rate of 80°F (300 K) sweep air S to the mill is 53,000 lb/h (24,100 kg/h). Mill power input is 4000 hp.

How much 70°F (294 K) cooling water must be sprayed into the mill to keep the product temperature from exceeding 150°F (339 K)?

Use 0.25 Btu/(lb)(°F) as the specific heat of the air, and 0.19 Btu/(lb)(°F) as the specific heat of the fresh feed, mill discharge D, tailings T, and product. Assume that the fractional heat losses in the mill and classifier circuits are 20 percent and 12 percent, respectively.

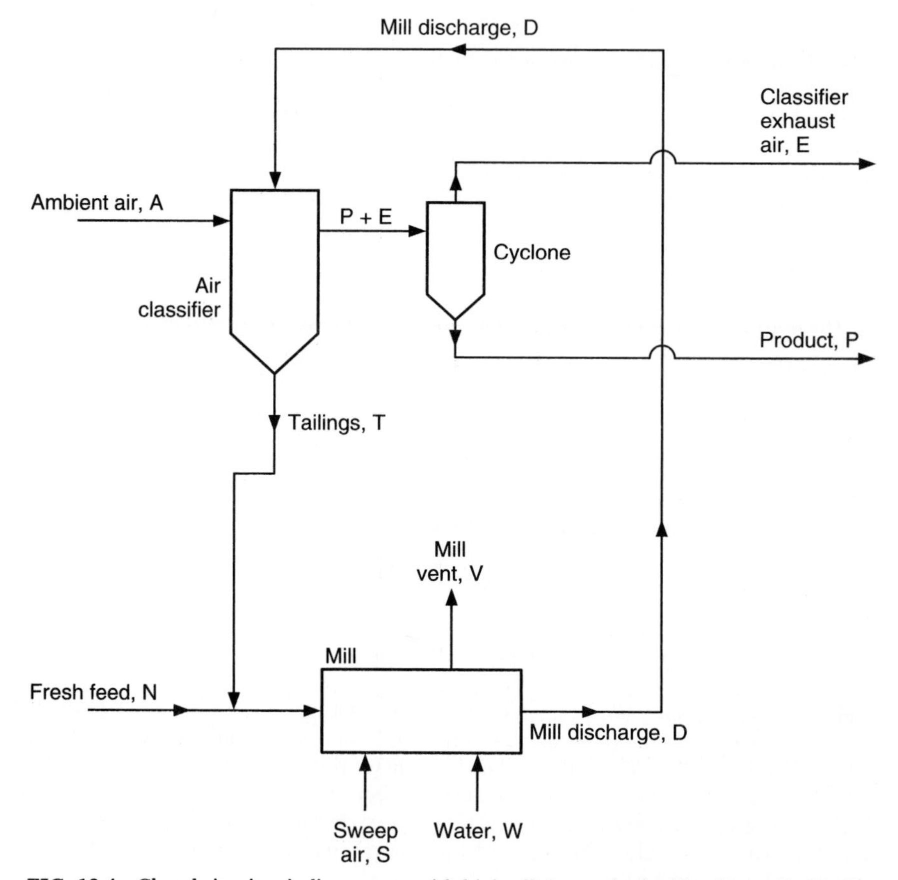

FIG. 13-4 Closed-circuit grinding system with high-efficiency air classifier (Example 13-10).

Assume that the tailings are 15 degrees hotter than the product (i.e., that they are 165°F), that the classifier exhaust E is 2 degrees cooler than the product (i.e., that the exhaust is 148°F), and that the mill vent is 20 degrees cooler than the mill discharge. Assume that the water achieves its cooling via vaporization in the mill, and that amount of water vapor leaving with the mill vent is negligibly small.

Calculation Procedure:

1. *Determine the mill-discharge flow rate D.*

Use the equation $D = P(1 + L)$ where, as noted in the statement of the problem, L is the circulating load. Thus, $D = 200{,}000(1 + 1.5) = 500{,}000$ lb/h.

2. *Calculate the tailings flow rate T.*

The equation is $T = PL$. Thus, $T = 200{,}000(1.5) = 300{,}000$ lb/h.

3. *Determine the enthalpies of the fresh feed N, product P, tailings T, ambient air to classifier A, and classifier exhaust E.*

For all of these streams, use the general formula $H = cm(t - t_o)$, where H is enthalpy in Btu's per hour, c is specific heat as given in the statement of the problem, m is mass flow rate, and t and t_o are, respectively, the stream temperature and reference temperature in degrees fahrenheit. For arithmetical simplicity, use 0°F as the reference temperature throughout. Then

$$H_N = 0.19(200{,}000)(160 - 0) = 6.1 \times 10^6 \text{ Btu/h}$$
$$H_P = 0.19(200{,}000)(150 - 0) = 5.7 \times 10^6 \text{ Btu/h}$$
$$H_T = 0.19(300{,}000)(165 - 0) = 9.4 \times 10^6 \text{ Btu/h}$$
$$H_A = 0.25(221{,}000)(80 - 0) = 4.4 \times 10^6 \text{ Btu/h}$$
$$H_E = 0.25(221{,}000)(148 - 0) = 8.2 \times 10^6 \text{ Btu/h}$$

4. *Estimate the heat loss H_K from the classifier circuit.*

Use the equation $H_K = [p_K/(1 - p_K)](H_E - H_A)$, where p_K is the fractional heat loss (12 percent, or 0.12). Thus,

$$H_K = [0.12/(1 - 0.12)][(8.2 - 4.4) \times 10^6] = 0.5 \times 10^6 \text{ Btu/h}$$

5. *Estimate the heat loss H_M from the mill circuit.*

The equation is $H_M = (p_M)$(power input to mill), where p_M is the fractional heat loss (20 percent, or 0.2). Thus,

$$H_M = (0.2)(4000 \text{ hp})(2545 \text{ Btu/(h)(hp)}) = 2.0 \times 10^6 \text{ Btu/h}$$

6. *Determine H_D, the enthalpy of the mill-discharge stream.*

This is done by making an energy balance around the classifier and solving it for H_D. The energy balance is $H_D + H_A = H_P + H_T + H_E + H_K$. Accordingly, $H_D = (5.7 + 9.4 + 8.2 + 0.5 - 4.4) \times 10^6 = 19.4 \times 10^6$ Btu/h.

7. Calculate the mill-discharge temperature t_D.

Solve the general enthalpy equation (step 3) for t. Thus, $t_D = H_D/cm = (19.4 \times 10^6)/(0.19)(0.5 \times 10^6) = 204°F$, with m being the mill discharge rate determined in step 1.

8. Estimate H_V, the mill-vent enthalpy.

The mill-vent rate equals the sweep-air rate, 53,000 lb/h. From the statement of the problem, the mill-vent temperature is 204 − 20, i.e., 184°F. Accordingly, $H_V = 0.25(53{,}000)(184 - 0) = 2.4 \times 10^6$ Btu/h.

9. Determine H_W, the enthalpy of the water sprayed into the mill.

The water must remove the heat introduced via the fresh feed, the tailings, the sweep air, and the mill power input H_I, less the heat removed via the mill discharge and the mill vent and less the mill heat losses. Thus, $H_W = H_N + H_T + H_S + H_I - H_D - H_V - H_M$. Now, $H_I = (4000 \text{ hp})(2545 \text{ Btu/(h)(hp)}) = 10.2 \times 10^6$ Btu/h, and $H_S = (0.25)(53{,}000)(80 - 0) = 1.1 \times 10^6$ Btu/h. Accordingly, $H_W = (6.1 + 9.4 + 1.1 + 10.2 - 19.4 - 2.4 - 2.0) \times 10^6 = 3.0 \times 10^6$ Btu/h.

10. Calculate the required water rate W.

From enthalpy tables, determine the enthalpy difference between the water entering the mill and the vapor leaving. For the purpose of this example, assume that the difference is 1100 Btu/lb. Then the amount of water needed to satisfy the enthalpy requirement calculated in step 9 is

$$(3.0 \times 10^6 \text{ Btu/h})/(1100 \text{ lb/h}) = 2700 \text{ lb/h } (1225 \text{ kg/h})$$

This is significantly lower cooling duty than would be the case with a conventional closed-circuit grinding system.

Related Calculations: Three related problems are (1) How much will the product temperature change during the hottest part of the year if the water flow rate is substantially raised at the same time? (2) Determine the required water rate assuming that the mill vent V is sent to the classifier instead of being discharged to the atmosphere; (3) Assume that the water rate determined in Problem 2 is at a maximum; how much will the product temperature increase during the hottest period of the year? All three of these call for trial-and-error solution.

Note: This example is adapted from an article by Ivan Klumpar of Badger Engineers, Inc., in *Chemical Engineering,* March 1992.

14

Filtration

Frank M. Tiller, Ph.D.

M.D. Anderson Professor
Department of Chemical Engineering
University of Houston
Houston, TX

Wenfang Leu, Ph.D.

Research Scientist
Department of Chemical Engineering
University of Houston
Houston, TX

REFERENCES: [1] Nelson and Dahlstrom—*Chem. Eng. Progr. 53*:320, 1957; [2] Tiller—*Chem. Engrg. Prog. 51*:282, 1955; [3] Tiller and Crump—*Chem. Engrg. Prog. 73*:65, Oct. 1977; [4] Tiller, Crump, and Ville—*Proceedings of the Second World Filtration Congress (London)*, Sept. 1979.

BASIS FOR FILTRATION CALCULATIONS

MASS BALANCE

An overall filtration material balance based on a unit area is

$$\text{Mass of slurry} = \text{mass of cake} + \text{mass of filtrate}$$

or

$$\frac{w_c}{s} = \frac{w_c}{s_c} + \rho v$$

where w_c is total mass of dry-cake solids per unit area, v is filtrate volume per unit area, s and s_c are, respectively, the mass fraction and average mass fraction of solids in the slurry and cake, and ρ is the density of filtrate. Solving for w_c yields

$$w_c = \frac{\rho s}{1 - s/s_c} v = cv \tag{14-1}$$

where c is concentration expressed by mass of dry cake per unit volume of filtrate.

The value of c can be obtained from Eq. (14-1) (that is, $c = w_c/v$) if it is possible to obtain the mass of solids in the cake. However, draining the slurry can often lead to difficulties in accurate determination of the cake mass. An alternative approach is to consider the cake thickness.

The cake thickness L can be related to the cake mass w_c by

$$w_c = \rho_s(1 - \epsilon_{av})L$$

where ρ_s is the true density of the solid and ϵ_{av} is the average porosity of the cake. Combining the two w_c equations produces

$$L = \frac{c}{\rho_s(1 - \epsilon_{av})} v = c_L v \tag{14-2}$$

where c_L is the ratio of cake thickness L to the per-unit area filtrate volume v. The thickness L is in fact the primary parameter related to filter design. Spacing of leaves, frame thickness, and minimum cake thickness for removal from vacuum drum filters all depend on a knowledge of L.

RATE EQUATIONS

In filtration theory, Darcy's law is used in the form

$$\frac{dp_L}{dw} = \frac{-dp_s}{dw} = \mu\alpha q$$

or

$$\frac{dp_L}{dx} = \frac{-dp_s}{dx} = \mu\rho_s(1 - \epsilon)\alpha q$$

where p_L is hydraulic pressure, p_s is solid compressive or effective pressure, w is mass of cake per unit area in the cake-thickness range from 0 to x, μ is viscosity of the filtrate, α is local specific filtration resistance, and q is superficial velocity of liquid. Solid compressive pressure is defined by $p_s = p - p_L$, where p is filtration pressure.

With respect to the cake cross section, p_s is zero at the cake-slurry interface and reaches its maximum at the cake-medium interface. Conversely, p_L has its maximum value (equal to p) at the cake-slurry interface, whereas at the cake-medium interface it consists solely of p_1, the pressure required to overcome the resistance R_m of the medium.

Integration of the preceding dp_L/dw expression with the assumption that q is constant throughout the cake gives

$$\mu q w_c = \mu c q v = \frac{\Delta p_c}{\alpha_{av}} \tag{14-3}$$

where $\Delta p_c = p - p_1 = p - \mu q R_m$. Substitution and rearrangement give

$$\frac{dv}{dt} = q = \frac{p}{\mu(\alpha_{av} w_c + R_m)} \tag{14-4}$$

The latter equation can be used to calculate v or L as a function of time t, once the filtration mode is specified, e.g., constant-pressure filtration, with p constant; constant-rate filtration, with q constant; or centrifugal-pump filtration, with q as a function of p.

14-1 Constant-Pressure Filtration

The first two columns of Table 14-1 show laboratory data[1] on filtering calcium silicate with an average particle size of 6.5 μm in a 0.04287-m^2 (0.460-ft^2) plate-and-frame filter press operating at a pressure p of 68.9 kPa (10 lbf/in^2) and a slurry-solid mass fraction s of 0.00495. The cake had an average moisture content corresponding to a cake mass fraction of solids s_c of 0.2937. Viscosity of the water μ was 0.001 Pa·s (1.0 cP). The densities of liquid and solid ρ and ρ_s, respectively, were 1000 and 1950 kg/m^3. Calculate the average specific and medium resistances, and set up an equation relating cake thickness L to filtration time t.

Calculation Procedure:

1. Calculate v, the filtrate volume per unit filtration area, and plot v versus t and log v versus log t.

Values of v based on a 0.04287-m^2 area are shown in col. 3 of Table 14-1. The plots are presented in Figs. 14-1 and 14-2. (These figures also include 16 data points that are omitted from the table for simplicity.) The slope of the logarithmic plot can be taken as essentially 0.5 when t exceeds 120 s. Beyond 120 s, the curve of $p/(\mu q_{av})$ versus w_c in step 5 below should be straight enough to give an adequate value of α_{av} at the full applied pressure.

2. Calculate w_c, the total mass of dry-cake solids per unit area.

Use Eq. (14-1). Thus,

$$w_c \frac{1000(0.00495)}{1 - 0.00495/0.2937} v = 5.04v$$

The values of w_c thus calculated are shown in col. 5 of Table 14-1.

[1] M. Hosseini, M.Sc. thesis, University of Manchester, 1977.

TABLE 14-1 Data and Calculated Values for Constant-Pressure Filtration (Example 14-1)

Elapsed time t, s (1)	Filtrate volume V, m^3 (2)	Volume per unit area v, m^3/m^2 (3)	Interpolated v, m^3/m^2 (4)	$w_c = cv$, kg/m^2 (5)	$q = \frac{\Delta v}{\Delta t}$, m/s (6)	$\frac{p}{\mu q_{av}} = \frac{pt}{\mu v}$, 1/m (7)	$\frac{p}{\mu q} = \frac{p\Delta t}{\mu \Delta v}$, 1/m (8)
0	0	0		0		—	
			0.012	0.06	2.56×10^{-3}		2.69×10^{10}
9	10×10^{-4}	0.023		0.12		2.70×10^{10}	
			0.035	0.18	2.40		2.87
19	20	0.047		0.24		2.79	
			0.059	0.30	1.84		3.74
31.5	30	0.070		0.35		3.10	
			0.082	0.41	1.28		5.38
49.5	40	0.093		0.47		3.67	
			0.105	0.53	1.17		5.89
70	50	0.117		0.59		4.12	
			0.129	0.65	1.00		6.89
93	60	0.140		0.71		4.58	
			0.152	0.77	8.52×10^{-4}		8.09
120	70	0.163		0.82		5.07	
			0.175	0.88	7.50		9.19
152	80	0.187		0.94		5.60	
			0.199	1.00	6.57		1.05×10^{11}
187	90	0.210		1.06		6.14	
			0.222	1.12	5.75		1.20
227	100	0.233		1.17		6.71	
			0.245	1.23	5.58		1.23
270	110	0.257		1.30		7.24	

3. Calculate $p/(\mu q_{av})$.

The average rate is simply $q_{av} = v/t$. Then, $p/(\mu q_{av}) = (6.89 \times 10^7)t/v$. Values are listed in col. 7 of Table 14-1.

4. Calculate $p/(\mu q)$.

The instantaneous rate $q = dv/dt$ must be obtained. Inasmuch as v versus t data pertaining to constant-pressure filtration are parabolic, use the following property of parabolas to obtain the slope:

$$\left(\frac{dv}{dt}\right)_{(v_1+v_2)/2} = \frac{\Delta v}{\Delta t}$$

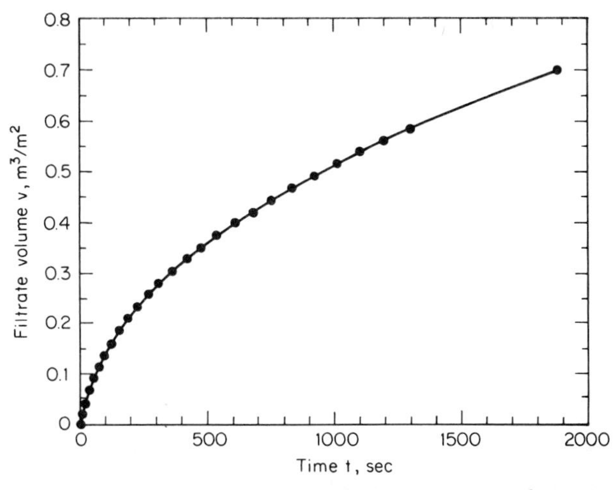

FIG. 14-1 Filtrate volume versus elapsed time (Example 14-1).

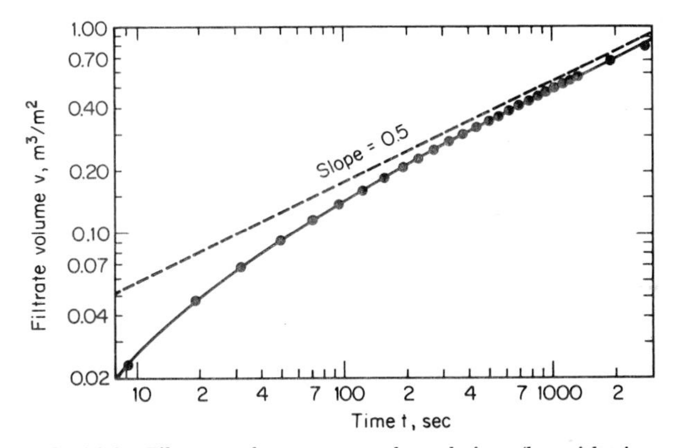

FIG. 14-2 Filtrate volume versus elapsed time (logarithmic scales) (Example 14-1).

This equation states that the angle of the tangent taken at the midpoint $(v + \Delta v/2)$ of a volume interval (not midpoint with respect to time) equals the angle of the secant. The rule is valid regardless of the size of Δv and is generally best applied to smoothed data.

Between 9 and 19 s, $\Delta v/\Delta t = (0.047 - 0.023)/10 = 0.00240$ m/s. This value corresponds to dv/dt at $v = (0.047 + 0.023)/2 = 0.035$ m. Since any size interval can be used, we find that between 93 and 270 s, $\Delta v/\Delta t = (0.257 - 0.140)/177 = 0.00066$ m/s. The value of v to be used is $(0.257 + 0.140)/2 = 0.199$ m.

In Table 14-1, values of $(v_1 + v_2)/2$ and the corresponding derivatives are shown in cols. 4 and 6, respectively. The $p/(\mu q)$ values in col. 8 are obtained from the expression $6.89 \times 10^7/q$.

5. Plot $p/(\mu q)$ and $p/(\mu q_{av})$ versus w_c.

These plots are shown in Fig. 14-3. They enable the calculation of the medium resistance R_m and the specific filtration resistance α_{av} because Eq. (14-4) can be rearranged into the

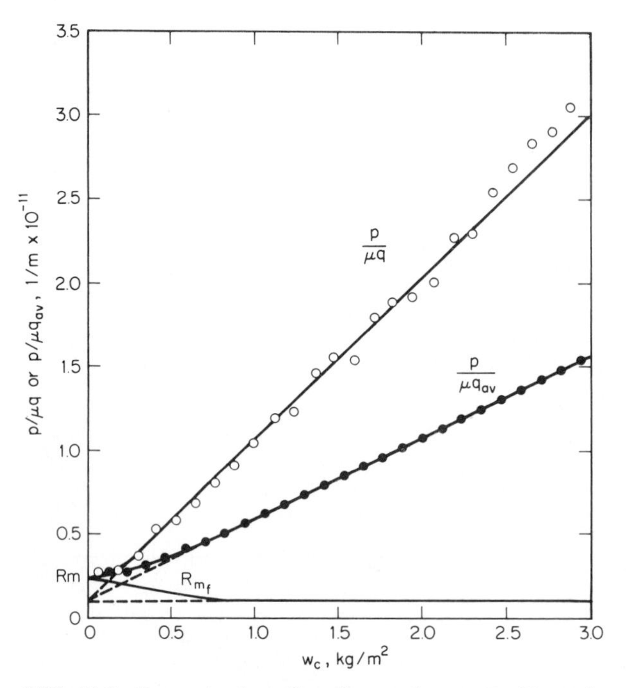

FIG. 14-3 Determination of medium resistance R_m (Example 14-1).

form $p/(\mu q) = \alpha_{av} w_c + R_m$, and there also can be written a similar expression (valid only when α_{av} and R_m are constant): $p/(\mu q_{av}) = (\alpha_{av}/2)w_c + R_m$. Both equations have the same y intercept, namely, R_m. The slope of the first is twice that of the second. It is advisable to plot both lines in order to reach a compromise on the slopes and intercept.

It is important to take enough v and t data to generate the initial curved portions of the plots. The specific filtration resistance is smaller at the start of filtration, and the slopes of both plots have their minimum value when t and w_c equal zero. If the first four points taken during the first 50 s were omitted, only the straight-line portions of the plots would be present; if these were extrapolated along the dotted lines to $w_c = 0$, the resulting value of the y intercept would be false.

6. *Determine the medium resistance R_m.*

R_m is the (true) y intercept in Fig. 14-3, approximately $0.24 \times 10^{11}\ m^{-1}$. (However, use this result with caution, because it is hard to establish operating conditions at $t = 0$.) It frequently happens that the intercept of the $p/(\mu q)$ line is negative. Such a result generally implies (1) migration of fine particles, with subsequent blinding of medium or cake, or (2) sedimentation on a horizontal filter surface facing up.

7. *Calculate α_{av}.*

In this step, do *not* rely on finding a single, constant slope of either of the plots in Fig. 14-3. Such a method would be correct only if the entire plot (including its initial portion) were straight and the false medium resistance were the true medium resistance. Instead, use the first equation in step 5, rearranged into the form $\alpha_{av} = [p/(\mu q) - R_m]/w_c$. The resulting value of α_{av} will vary, with the difference from value to value being greatest when w_c is small. As w_c increases, the value of α_{av} does approach the slope of the $p/(\mu q)$ plot in its straight-line portion, namely, 0.97×10^{11}.

Thus when $w_c = 0.88$ kg/m^2 and $p/(\mu q) = 0.92 \times 10^{11}$ m^{-1}, $\alpha_{av} = [(0.92 \times 10^{11}) - (0.24 \times 10^{11})]/0.88 = 0.77 \times 10^{11}$ m/kg (1.15×10^{11} ft/lb). And when $w_c = 3.0$ kg/m^2, $\alpha_{av} = [(3.03 \times 10^{11}) - (0.24 \times 10^{11})]/3.0 = 0.93 \times 10^{11}$ m/kg.

The pressure drop across the cake is given by the equation $\Delta p_c = p - \mu q R_m$. Thus, at these two points, it equals 50.9 and 63.4 kPa (7.38 and 9.20 lbf/in^2), respectively.

8. *Calculate the average porosity ϵ_{av}.*

Average porosity can be calculated from the equation

$$\epsilon_{av} = \frac{1}{1 + (\rho/\rho_s)s_c/(1 - s_c)} = \frac{1}{1 + (1000/1950)0.2937/(1 - 0.2937)} = 0.824$$

9. *Obtain equations for v versus t and L versus t.*

In Fig. 14-3, the $p/(\mu q_{av})$ plot is an excellent straight line for w_c values greater than about 0.8 kg/m^2, that is, for filtration times greater than about 120 s. Therefore, data *in this range* can be accurately represented by an equation based on the assumption that α_{av} and R_m are constant and are the slope and intercept of the straight-line portion of the plot. Thus, $\alpha_{av} = 0.97 \times 10^{11}$ m/kg, and R_m, found by extending the straight line leftward until it intercepts the vertical axis, is 0.10×10^{11} m^{-1}. Then, substituting $w_c = cv$ and integrating Eq. (14-4) yields the parabola

$$v^2 + \frac{2R_m}{c\alpha_{av}} v = \frac{2p}{\mu c \alpha_{av}} t$$

Noting from step 2 that $c = 5.04$ kg/m^3 and substituting and rearranging, we obtain the relationship $t = 3548v^2 + 145v$. And relating L to v via Eq. (14-2), where $c_L = 5.04/1950(1 - 0.824) = 0.0147$, we obtain $t = (1.64 \times 10^7)L^2 + (9.9 \times 10^3)L$. In these equations, v is in cubic meters per square meter and L is in meters.

As for filtration times of less than 120 s—for instance, with continuous drum or disk filters, where filtration time would normally be less than 60 s—the initial α_{av} of 0.11×10^{11} m/kg and the true R_m of 0.24×10^{11} m^{-1} will yield a reasonable representation of the data. Thus the equations become $t = 402v^2 + 348v$ and $t = (1.86 \times 10^6)L^2 + (2.4 \times 10^4)L$.

Related Calculations: In constant-rate (as opposed to constant-pressure) filtration, $v = qt$ and $w_c = cqt$. Then, from Eq. (14-3), the average specific filtration resistance $\alpha_{av} = (p - p_1)/(\mu c q^2 t)$, where p_1 is the pressure at the interface of the filter medium and the cake. Constant-rate filtrations are usually operated at above 10 lbf/in^2 (69 kPa), and this equation is accurate enough for most purposes. At higher and higher pressures, it becomes more and more acceptable to neglect p_1, which leads to the pressure-time relationship $p = \alpha_{av}\mu c q^2 t$.

Constant-rate filtration is employed sometimes when an improperly used centrifugal pump may break down the slurry particles. In fact, however, centrifugal pumps are most often chosen for filtration operations. The following example shows the relevant calculations.

14-2 Centrifugal-Pump Filtration

A 2% (by weight) aqueous slurry containing solids with an average density of 202.6 lb/ft³ (3244 kg/m³) is to be filtered in a 500-ft² (46.45-m²) filter using a centrifugal pump having the following performance characteristics:

Pressure:										
lbf/in²	15	20	25	30	35	40	45	50	55	60
kPa	103	138	172	207	241	276	310	345	379	414
Flow rate:										
gal/min	500	482	457	420	375	308	232	155	78	0
m³/s	0.0315	0.0304	0.0288	0.0265	0.0237	0.0194	0.0146	0.0098	0.0049	0

The pump has a throttling valve that relates pressure drop to flow rate Q as follows: Δp (throttling) $= 15(Q/500)^2$. The temperature varies from 20 to 27°C (68 to 80.6°F). A series of constant-pressure tests yielded the data on α_{av} and $(1-\epsilon_{av})$ shown in Fig. 14-4, α_{av} being the average specific filtration resistance and ϵ_{av} the average porosity of the cake. Find cake thickness as a function of time.

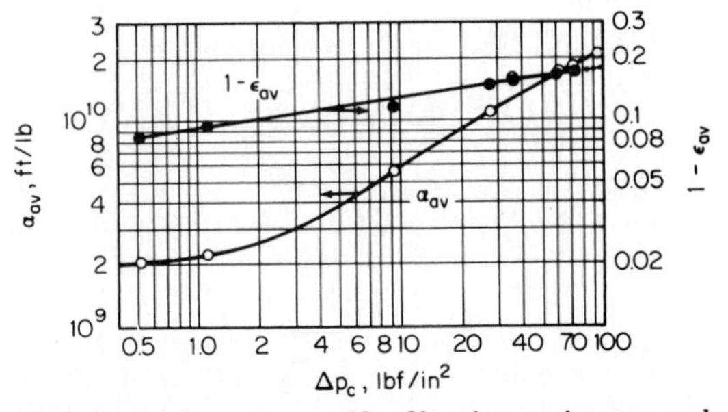

FIG. 14-4 Average specific filtration resistance and average porosity (Example 14-2). (Note: 1 lbf/in² = 6.895 kPa; 1 ft/lb = 6.72 m/kg.)

Calculation Procedure:

1. Select the approach to be used.

If both pressure and filtration rate vary, as in this case, it is necessary to impose the pump characteristics on the filtration equations. No simple formulas can be obtained to relate p to t; instead, a relatively easy numerical integration is used.

Equation (14-3) can be rearranged into the form

$$v = \frac{\Delta p_c}{\mu c q \alpha_{av}}$$

the terms being defined as at the beginning of this section. Now q is a function of p, and α_{av} is a function of Δp_c. Once v has been obtained as a function of p and q, t can be obtained by integration:

$$t = \int_0^v \frac{dv}{q}$$

If data from a series of constant-pressure tests yield values of α_{av} and the average solid fraction s_c, the first equation in this paragraph can be used to find the volume-versus-rate relationship, which can then be used to find the time by integration.

2. *Construct a modified pump curve.*

The characteristic pump curve is plotted as the upper line in Fig. 14-5. Below point U, the pump is unstable and must be throttled so that the pressure does not fall to too low a value. Inasmuch as the throttling pressure is not available to the filter, it must be subtracted from the characteristic curve to give the modified pump curve. From the equation in the statement of the problem, plot the throttling curve. Then, assuming a negligible filter-medium resistance, the pressure drop Δp_c across the cake equals the difference

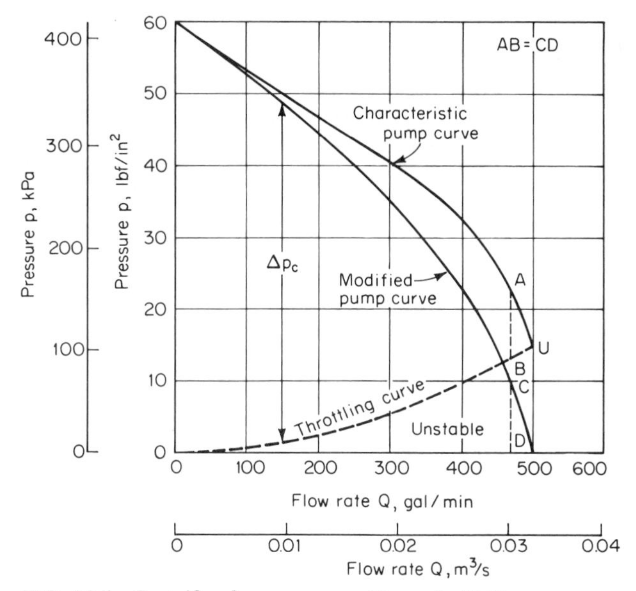

FIG. 14-5 Centrifugal-pump curves (Example 14-2).

between the characteristic and throttling curves. Plot this difference, labeling it the "modified pump curve."

3. *Calculate c, the concentration of cake, that is, the mass of dry cake per unit volume of filtrate.*

As indicated in Eq. (14-1), $c = \rho s/(1 - s/s_c)$, or in the present case, $c = 62.3(0.02)/(1 - 0.02/s_c)$, where 62.3 is the density of water in lbm/ft³. Now, $s_c = \rho_s(1 - \epsilon_{av})/[\rho_s(1 - \epsilon_{av}) + \rho\epsilon_{av}]$, or in the present case, since ρ_s is given as 202.6 lbm/ft³, $s_c = 1/[1 + 0.308\epsilon_{av}/(1 - \epsilon_{av})]$.

Select various values for Δp_c; from Fig. 14-4, determine ϵ_{av}. From the equations in the preceding paragraph, calculate s_c and c. The results are shown in the fourth, fifth, and sixth columns of Table 14-2.

TABLE 14-2 Data and Calculated Results for Centrifugal-Pump Filtration (Example 14-2)

Δp_c, lbf/in²	Q, gal/min	q, ft/s	ϵ_{av}	s_c	c, lbm/ft³	c_L	$\alpha_{av} \times 10^{-9}$, ft/lbm	v, ft³/ft²	L, in	$1/q$, s/ft
0.5	499	0.00224	0.915	0.232	1.364	0.0792	2.1	0.54	0.51	446
1	498	0.00222	0.905	0.254	1.352	0.0702	2.2	1.04	0.88	450
5	487	0.00217	0.880	0.307	1.332	0.0548	3.8	3.14	2.06	461
10	469	0.00209	0.869	0.329	1.327	0.0500	5.7	4.36	2.62	478
15	446	0.00199	0.860	0.346	1.322	0.0466	7.2	5.46	3.05	503
20	418	0.00186	0.853	0.359	1.319	0.0443	9.0	6.24	3.32	538
25	384	0.00171	0.849	0.366	1.318	0.0431	10.0	7.65	3.96	585
30	345	0.00154	0.846	0.371	1.317	0.0422	12.0	8.50	4.30	649

Note: 1 lbf/in² = 6.895 kPa; 1 gal/min = 6.3×10^{-5} m³/s; 1 ft/s = 0.3048 m/s; 1 lbm/ft³ = 16.02 kg/m³; 1 ft/lbm = 0.67 m/kg; 1 ft³/ft² = 0.3048 m³/m²; 1 in = 0.0254 m; 1 s/ft = 3.28 s/m.

4. *Calculate c_L, the ratio of cake thickness L to the filtrate volume per unit filter area v.*

The values of c_L can be calculated from the equation

$$c_L = \frac{c}{\rho_s(1 - \epsilon_{av})}$$

Note that both c and c_L thus vary with Δp_c. The calculated values of c_L appear in Table 14-2.

5. *Calculate q, the superficial velocity (velocity based on unit filter area) corresponding to the values of Δp_c selected in step 3.*

This operation employs Fig. 14-3. The modified pump curve relates Δp_c to Q, the flow rate in gallons per minute. For a given Δp_c, find Q and multiply it by (1 min/60 s)(1 ft³/7.481 gal)(1/500 ft² of filter area) to find q. The calculated values of q appear in Table 14-2.

6. Calculate v, the volume of filtrate per unit of filter area.

Use the first equation in step 1. Thus,

$$v = \frac{\Delta p_c}{\mu c q \alpha_{av}}$$

$$= \frac{(144 \text{ in}^2/\text{ft}^2)[32.17 \text{ lbm}\cdot\text{ft}/(\text{lbf}\cdot\text{s}^2)](\Delta p_c \text{ lbf/in}^2)}{[0.000672 \text{ lbm}/(\text{ft}\cdot\text{s})](c \text{ lbm/ft}^3)(q \text{ ft/s})(\alpha_{av} \text{ ft/lbm})}$$

$$= 6.893 \times 10^6 \frac{\Delta p_c}{c q \alpha_{av}}$$

Use Fig. 14-4 to obtain α_{av} for each value of Δp_c. These values of α_{av} are shown in Table 14-2, and so are the calculated values of v.

7. Calculate L, the cake thickness.

The equation is $L = c_L v$. The values of L thus calculated are shown in Table 14-2, in inches.

8. Plot L, Δp_c, and 1/q against v.

The values are taken from Table 14-2. The resulting smoothed plots appear in Fig. 14-6.

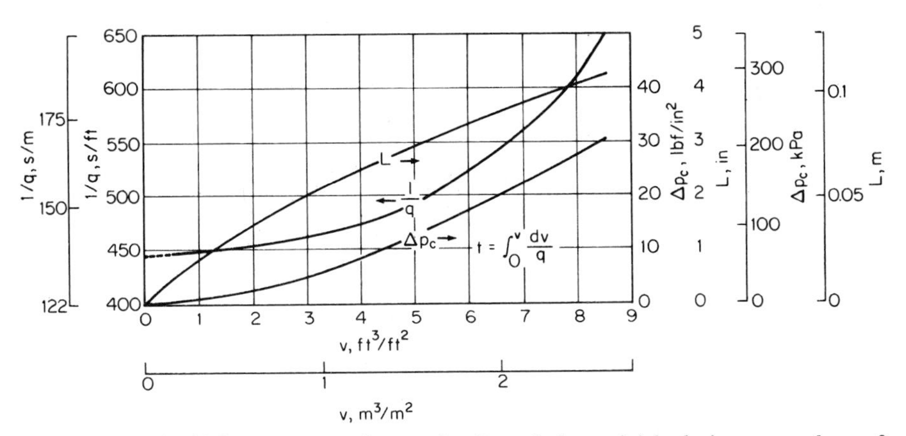

FIG. 14-6 Cake thickness, pressure drop, and reciprocal of superficial velocity versus volume of filtrate; centrifugal-pump filtration (Example 14-2).

9. Find filtration time t.

The time is found by taking the area under the curve of $1/q$ versus v. The result of this integration is shown in the third and fourth columns of Table 14-3. Values of Δp_c and L are repeated in the table for convenience.

TABLE 14-3 Determination of Filtration Time for Centrifugal-Pump Filtration (Example 14-2)

v, ft^3/ft^2	Incremental area, s	t, s	t, min	Δp_c, lbf/in^2	L, in
1	446	446	7.4	0.9	0.80
2	451	897	15.0	2.6	1.47
3	457	1354	22.6	5.0	2.00
4	466	1820	30.3	8.3	2.47
5	482	2302	38.4	12.6	2.90
6	507	2809	46.8	17.2	3.32
7	540	3349	55.8	22.3	3.70
8	582	3931	65.5	27.5	4.10

10. Show cake thickness as a function of time.

This relationship is shown graphically in Fig. 14-7. The figure also shows Δp_c as a function of time.

Related Calculations: Filtrate volume as a function of time can be calculated by dividing each value of L in Fig. 14-7 by c_L and multiplying by the filter area.

Note that this pump proves to be somewhat large for the filter. If the cake thickness is restricted to 2.5 in (0.064 m), the rate would never drop below 95 percent of the initial pump rate.

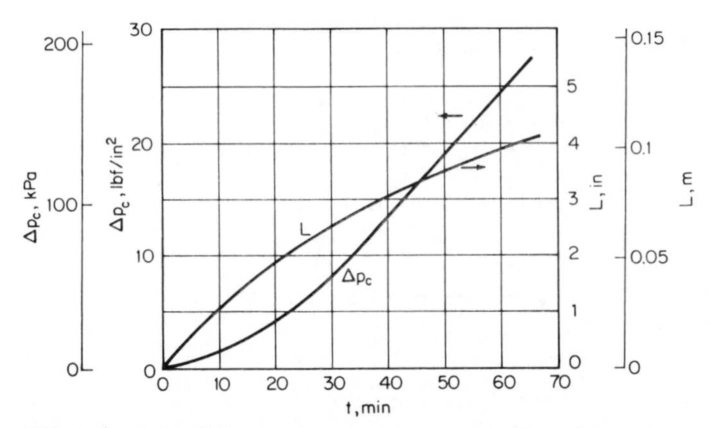

FIG. 14-7 Cake thickness and pressure drop as a function of time; centrifugal-pump filtration (Example 14-2).

14-3 Filter-Cake Washing

A filter cake was washed at a rate of 0.2 gal/(ft^2)(min) [0.0081 m^3/(m^2)(min)] with pure water to remove the soluble salts present in the voids. The cake had a thickness of 2.0 in. (0.051 m) and the following compositions:

	Mass fraction	Density, lbm/ft^3 (kg/m^3)
Inert solids	0.4789	88.6 (1420)
Water	0.4641	62.4 (1000)
Soluble salts	0.0570	85.1 (1363)

Data for instantaneous wash concentration versus time were taken as shown in Table 14-4. At the end of the washing period the cake was analyzed and found to have mass fraction 0.24% of salts on a moisture-free basis. How much water must be used if it is permissible to leave mass fraction 0.67% of soluble material on a moisture-free basis?

TABLE 14-4 Data on Filter-Cake Washing (Example 14-3)

Time t, min	Volume of wash v_w, gal/ft^2	Wash concentration C_w, lbm solute/gal
0	0	0.740
1	0.2	0.739
2	0.4	0.740
3	0.6	0.687
4	0.8	0.480
5	1.0	0.266
6	1.2	0.144
8	1.6	0.0575
10	2.0	0.0313
15	3.0	0.0158
20	4.0	0.0115
40	8.0	0.00624
60	12.0	0.00312
90	18.0	0.00119
120	24.0	0.00070

Note: 1.0 gal/ft^2 = 0.041 m^3/m^2; 1.0 lbm solute/gal = 120.02 kg solute/m^3.

Calculation Procedure:

1. Convert mass fractions into volume fractions and find average porosity.

The volume fraction x_i of component i (inert solid, water, or soluble salts) can be calculated by

$$x_i = \frac{y_i/\rho_i}{\Sigma(y_i/\rho_i)}$$

where y_i and ρ_i are the mass fraction and density of component i, respectively. Then, the initial composition of cake is 40.00% inert solid, 55.04% water, and 4.96% soluble salts by volume. The average porosity ϵ_{av} of the cake is thus also known to be $0.5504 + 0.0496 = 0.6$.

The volume fraction of soluble material on a moisture free-basis ψ_v can be obtained from

$$\psi_v = \frac{\psi_m/\rho_{salt}}{\psi_m/\rho_{salt} + (1 - \psi_m)/\rho_s}$$

where ψ_m is the mass fraction of soluble material on a moisture-free basis and the subscript s refers to the inert solids. Thus the volume fractions of salts on a moisture-free basis are 0.25% at the end of 120 min and 0.70% permissible.

2. *Calculate the average density of the cake, mass of dry inert solid, initial and final mass of soluble salts, and void volume.*

$$\begin{aligned}\text{Average cake density} &= \Sigma(\text{density of component } i)(\text{volume of fraction of component } i) \\ &= 88.6(0.4000) + 62.4(0.5504) + 85.1(0.0496) \\ &= 74.0 \text{ lbm/ft}^3 \ (1186 \text{ kg/m}^3)\end{aligned}$$

$$\begin{aligned}\text{Mass of dry inert solid per unit area of filtration} &= (\text{inert solid density})(\text{cake thickness}) \\ &\quad \times (\text{volume fraction of dry inert solid}) \\ &= 88.6(2/12)(0.4) \\ &= 5.91 \text{ lbm/ft}^2 \ (28.8 \text{ kg/m}^2)\end{aligned}$$

$$\begin{aligned}\text{Initial mass of solute per unit area of filtration} &= (\text{solute density})(\text{cake thickness}) \\ &\quad \times (\text{volume fraction of solute}) \\ &= 85.1(2/12)(0.0496) \\ &= 0.703 \text{ lbm/ft}^2 \ (3.43 \text{ kg/m}^2)\end{aligned}$$

$$\begin{aligned}\text{Final mass of solute per unit area of filtration} &= (\text{mass of dry inert solid})[\psi_m/(1 - \psi_m)] \\ &= 5.91(0.0024)/0.9976 \\ &= 0.0142 \text{ lbm/ft}^2 \ (0.0692 \text{ kg/m}^2)\end{aligned}$$

$$\begin{aligned}\text{Void volume per unit area of filtration} &= (\epsilon_{av})(\text{cake thickness}) \\ &= 0.6(2/12) = 0.1 \text{ ft}^3/\text{ft}^2 \\ &= 0.748 \text{ gal/ft}^2 \ (0.0305 \text{ m}^3/\text{m}^2)\end{aligned}$$

3. *Calculate the average wash concentration $C_{w,av}$.*

The instantaneous wash concentration C_w, shown as the third column in Table 14-4, is plotted against volume of wash (the second column in Table 14-4) in Fig. 14-8. Now, by a material balance for the solute in the cake,

$$\epsilon_{av}L(C_0 - C_{av}) = \int_0^{v_w} C_w dv_w = C_{w,av}v_w$$

where L is cake thickness (and therefore $\epsilon_{av}L$ is void volume per unit area of filtration, calculated in step 2), C_0 is initial concentration of solute in the cake, and v_w is volume of wash per unit area of filtration. Therefore, the total (cumulative) amount of solute removed from the cake may be determined by integrating the instantaneous-concentra-

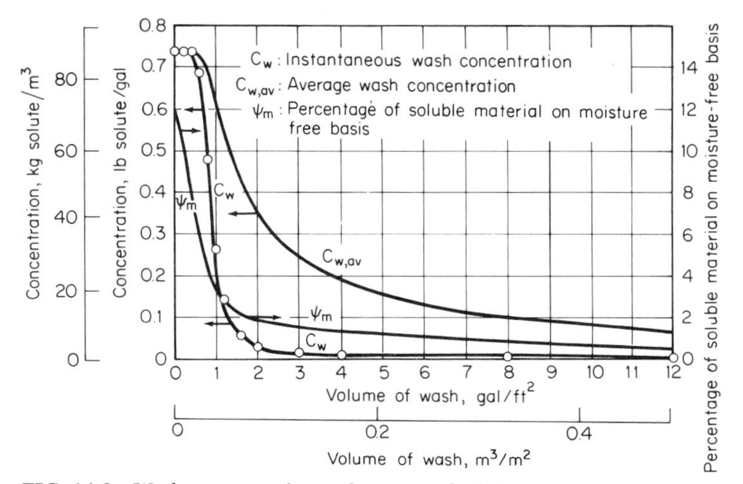

FIG. 14-8 Wash concentration and percent of soluble material versus volume of wash (Example 14-3).

tion-versus-volume data as shown in the first three columns of Table 14-5. Dividing the cumulative amount of solute removed by the cumulative volume of wash used yields the average wash concentration, shown as the fourth column in Table 14-5 and plotted in Fig. 14-8.

4. Calculate the mass of solute remaining in the cake at each moment.

Since 24 gal/ft² (0.98 m³/m²) of wash was used in the run, the total amount of solute removed was 0.8296 lb/ft², as shown in Table 14-5. The amount remaining in the cake is 0.0142 lb/ft², from step 2. Thus the total amount present in the system initially must have been $0.8296 + 0.0142 = 0.8438$ lb/ft² (4.11 kg/m²). (Since step 2 shows that only 0.703 lb/ft² was present in the *cake* initially, the rest must have been in the feed lines to the filter press; it is important to keep this complication in mind when dealing with problems of this kind.)

Then the initial cake concentration C_0 may be calculated from the true initial amount of solute divided by void volume, that is, $0.8438/0.1 = 8.438$ lbm/ft³ (135 kg/m³), and the mass of solute remaining in the cake may be calculated by subtracting the cumulative amount of solute removed from the true initial mass of solute, 0.8438 lbm/ft², as shown in the fifth column of Table 14-5. Since the mass of solid remaining equals $5.91\psi_m/(1 - \psi_m)$, the values in that column can be employed to calculate ψ_m; the results (on a percentage basis) appear as the sixth column in the table.

Thus, if it is permissible to retain 0.67 percent of soluble material, on a moisture-free basis, the table shows that slightly over 10 gal of wash water must be used per square foot (slightly over 0.41 m³/m²).

Related Calculations: The average concentration of solute in the cake consists of mass of solute remaining divided by void volume. These values appear as the final column in Table 14-5.

TABLE 14-5 Results of Cake Washing (Example 14-3)

Range of wash volume, gal/ft^2	Solute removed, lbm/ft^2	Conditions at the end of the interval of wash volume: Cumulative solute removed, lbm/ft^2	Average wash concentration, lbm/gal	Mass of solute remaining, lbm/ft^2	Percent solute on moisture-free basis	Average cake concentration, lbm/gal
0	0	0	—	0.8438	12.49	1.128
0.0–0.1	0.0740	0.0740	0.740	0.7698	11.52	1.029
0.1–0.2	0.0740	0.1480	0.740	0.6958	10.53	0.930
0.2–0.3	0.0740	0.2220	0.740	0.6218	9.52	0.831
0.3–0.4	0.0740	0.2960	0.740	0.5478	8.48	0.732
0.4–0.5	0.0730	0.3690	0.738	0.4748	7.44	0.635
0.5–0.6	0.0705	0.4395	0.733	0.4043	6.40	0.541
0.6–0.7	0.0650	0.5045	0.721	0.3393	5.43	0.454
0.7–0.8	0.0545	0.5590	0.699	0.2848	4.60	0.381
0.8–0.9	0.0420	0.6010	0.669	0.2428	3.95	0.325
0.9–1.0	0.0315	0.6325	0.633	0.2113	3.45	0.282
1.0–1.1	0.0226	0.6551	0.596	0.1887	3.09	0.252
1.1–1.2	0.0163	0.6714	0.560	0.1724	2.83	0.230
1.2–1.3	0.0126	0.6840	0.526	0.1598	2.63	0.214
1.3–1.4	0.0100	0.6940	0.496	0.1498	2.47	0.200
1.4–1.5	0.0079	0.7019	0.468	0.1419	2.34	0.190
1.5–1.6	0.0063	0.7082	0.443	0.1356	2.24	0.181
1.6–1.7	0.0050	0.7132	0.420	0.1306	2.16	0.175
1.7–1.8	0.0042	0.7174	0.399	0.1264	2.09	0.169
1.8–1.9	0.0036	0.7210	0.379	0.1228	2.04	0.164
1.9–2.0	0.0032	0.7242	0.362	0.1196	1.98	0.160
2.0–3.0	0.02190	0.7461	0.249	0.0977	1.63	0.131
3.0–4.0	0.01350	0.7596	0.190	0.0842	1.40	0.113
4.0–5.0	0.01035	0.7700	0.154	0.0738	1.23	0.099
5.0–6.0	0.00888	0.7788	0.130	0.0650	1.09	0.087
6.0–7.0	0.00762	0.7864	0.112	0.0574	0.96	0.077
7.0–8.0	0.00662	0.7931	0.099	0.0507	0.85	0.068
8.0–9.0	0.00575	0.7988	0.089	0.0450	0.76	0.060
9.0–10.0	0.00490	0.8037	0.080	0.0401	0.67	0.054
10.0–11.0	0.00425	0.8080	0.073	0.0358	0.60	0.048
11.0–12.0	0.00350	0.8115	0.068	0.0323	0.54	0.043
12.0–13.0	0.00302	0.8145	0.063	0.0293	0.49	0.039
13.0–14.0	0.00258	0.8171	0.058	0.0267	0.45	0.036
14.0–15.0	0.00222	0.8193	0.055	0.0245	0.41	0.033
15.0–16.0	0.00182	0.8211	0.051	0.0227	0.38	0.030
16.0–17.0	0.00152	0.8226	0.048	0.0212	0.36	0.028
17.0–18.0	0.00132	0.8240	0.046	0.0198	0.33	0.026
18.0–19.0	0.00118	0.8252	0.043	0.0186	0.31	0.025
19.0–20.0	0.00102	0.8262	0.041	0.0176	0.30	0.024
20.0–21.0	0.00100	0.8272	0.039	0.0166	0.28	0.022
21.0–22.0	0.00092	0.8281	0.038	0.0157	0.26	0.021
22.0–23.0	0.00080	0.8289	0.036	0.0149	0.25	0.020
23.0–24.0	0.00075	0.8296	0.035	0.0142	0.24	0.019

Note: 1.0 gal/ft^2 = 0.041 m^3/m^2; 1.0 lbm/ft^2 = 4.88 kg/m^2; 1.0 lbm/gal = 119.8 kg/m^3.

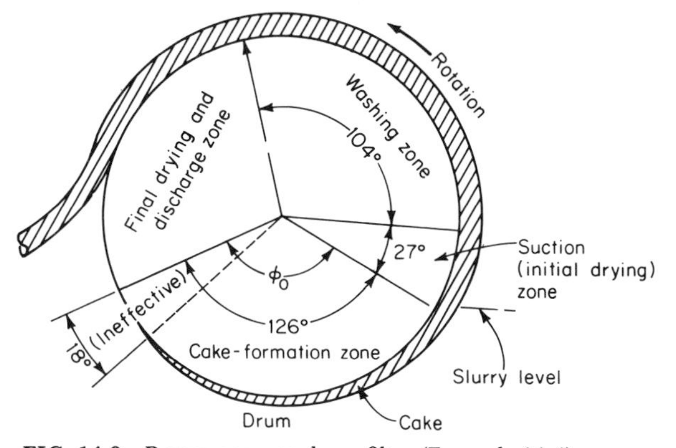

FIG. 14-9 Rotary vacuum drum filter (Example 14-4).

ROTARY-VACUUM-DRUM FILTERS

The salient features of rotary-drum filtration are illustrated in Fig. 14-9, where a cylindrical drum having a permeable surface is revolving counterclockwise partially submerged in a slurry. A pressure differential is usually maintained between the outer and inner surfaces by means of a vacuum pump. However, the drum might be enclosed and operated under pressure. In addition to the vacuum or pressure, each point on the periphery of the drum is subjected to a hydrostatic head of slurry.

Continuous multicompartment drum filters, as illustrated in Fig. 14-9, are normally used on materials that are relatively concentrated and easy to filter. Rates of cake buildup are in the range of 0.05 in/min (0.0013 m/min) to 0.05 in/s (0.0013 m/s). Submergence normally runs from 25 to 75 percent (40 percent being quite common), with rotation speeds from 0.1 to 3 r/min. With these conditions, filtration times could range from 5 s to 7.5 min, the great majority of industrial filtration falling within these limits.

Drum diameters typically are 6 to 12 ft (1.83 to 3.66 m), although larger values may be encountered. With 40 percent submergence and a 12-ft diameter, the hydrostatic head ranges up to 5.7 ft (1.74 m), which is a significant fraction of the driving force in vacuum filtration. Since the slurry will have a density greater than water, the effective head may be as high as 7 ft (2.13 m).

The drum of radius r rotates at an angular velocity of ω rad/s (N r/s). The portion of the drum submerged in the slurry is subtended by an arc ϕ_0. The remaining part of the drum is utilized for washing, drying, and discharge. Filtration through a given portion of the drum is assumed to begin at the instant that that portion enters the slurry; in practice, however, there is a time lag in establishing the full vacuum because of the need to maintain a vacuum seal as each compartment enters the slurry.

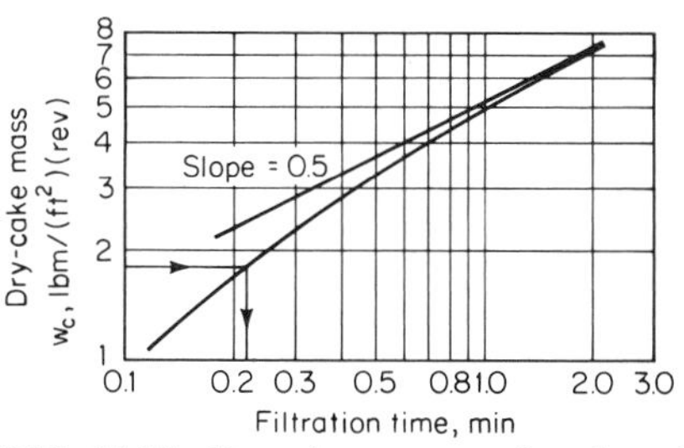

FIG. 14-10 Dry-cake mass as a function of filtration time (Example 14-4). [Note: 1 lbm/(ft^2)(r) = 4.88 kg/(m^2)(r).]

Dry-cake mass per area is shown in Fig. 14-10 as a function of elapsed time during a given revolution of the drum.

CAKE WASHING

Experimental wash curves represented as fraction of solute remaining versus the wash ratio j (ratio of wash to void volume of cake) can be plotted semilogarithmically as in Fig. 14-11 (the solid line). No experimental point will fall on the left of the maximum theoretical curve (the dotted line), which represents perfect displacement.

Cake wash time is the most difficult variable to correlate. Filtration theory suggests three possible correlations: (1) wash time versus $w_c v_w$; where w_c and v_w are total mass of dry solids and volume of wash, each per area of filtration; (2) wash time versus $j w_c$; and (3) wash time per form time versus wash volume per form volume. Fortunately, the easiest correlation (no. 1) usually gives satisfactory results. This curve starts as a straight line, but often falls off as the volume of wash water increases, as in Fig. 14-12.

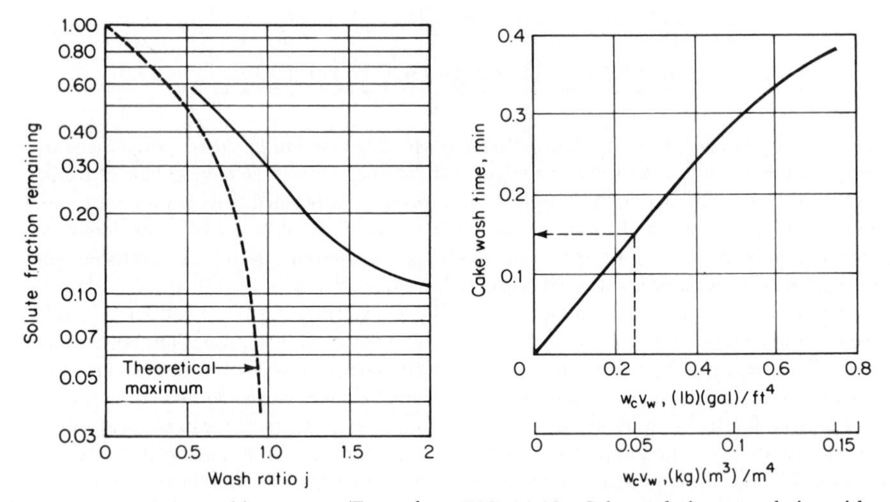

FIG. 14-11 Cake-washing curves (Example 14-4).

FIG. 14-12 Cake-wash time correlation with mass of dry solids w_c and volume of wash v_w per unit of area (Example 14-4).

CAKE MOISTURE CONTENT, AIR RATE

Experience has shown that the following factor is useful in correlating cake-moisture-content data [1]:

$$\text{Correlating factor} = \frac{\text{ft}^3/\text{min}}{\text{ft}^2} \frac{\Delta p_c}{w_c} \frac{t_d}{\mu}$$

where $\text{ft}^3/(\text{min})(\text{ft}^2)$ is air rate through the filter cake, t_d is drying time, Δp_c is pressure drop across the cake, and μ is liquid viscosity.

Figure 14-13 shows the general shape of the curve. The correlating factor chosen for design should be somewhere past the knee of the curve. Values to the left approach an unstable operating range, wherein a small change in operating conditions can result in a relatively drastic change in cake moisture content.

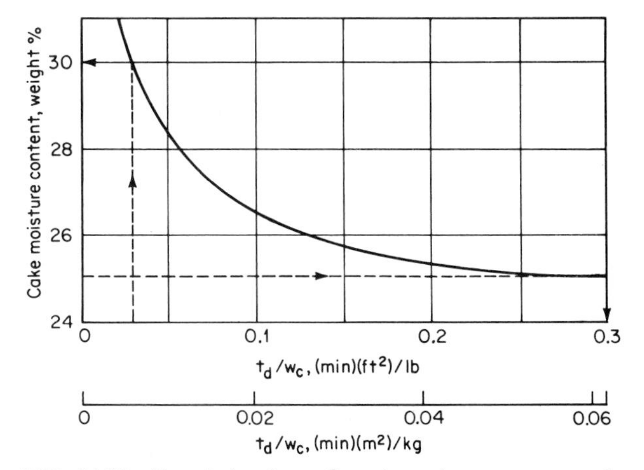

FIG. 14-13 Correlating factor for cake moisture content and air rate (Example 14-4).

If runs are made at constant temperature and vacuum, the pressure drop and viscosity terms can be dropped from the expression. Often air-rate data are not available, but correlations can be obtained without air rates, particularly if the cakes are relatively nonporous. The correlating factor is then reduced to the simplified term t_d/w_c, which involves only drying time and cake weight per unit area per revolution. A substantial degree of data scatter is normally encountered in the moisture-content correlation. Any point selected on the correlation will represent an average operating condition. To ensure that cake moisture content will not exceed a particular value, the correlating factor at the desired minimum should be multiplied by 1.2 before calculating the required drying time.

Air rate through the cake—and thus, vacuum-pump capacity—can be determined from measurements of flow rate as a function of time. Integration of these data over the times involved in the first and second stages of drying in continuous filters yields vacuum-pump-capacity data.

14-4 Design of a Rotary-Vacuum-Drum Filter

A drum filter as illustrated in Fig. 14-9 is to be used for filtering, washing, and drying a cake having the properties given by Figs. 14-10 through 14-13. Air rate through the cake is determined from measurements of flow rate as a function of time with a rotameter as follows:

	Time, min								
	0.05	0.1	0.2	0.3	0.4	0.6	0.8	1.0	1.5
$ft^3/(min)(ft^2)$	2.5	4.2	5.9	6.8	7.45	8.2	8.6	8.75	9.2
$m^3/(min)(m^2)$	0.762	1.28	1.80	2.07	2.27	2.50	2.62	2.67	2.80

These data are also plotted in Fig. 14-14.

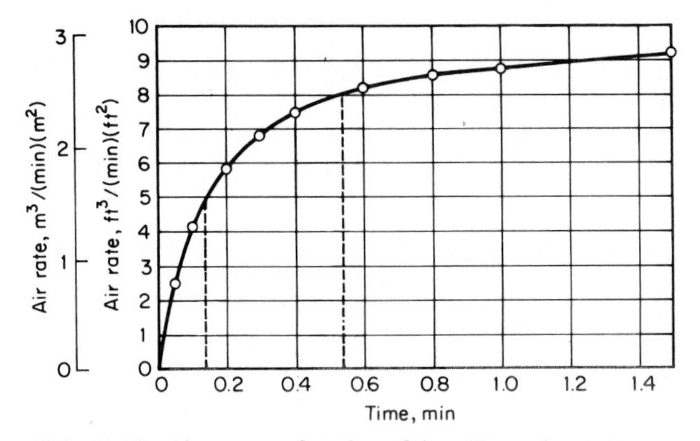

FIG. 14-14 Air rate as a function of time (Example 14-4).

The following conditions and specifications are assumed: slurry contains 40% solids by weight; solute in the liquid is 2%; final cake moisture is 25%; wash ratio (wash volume per void volume) is 1.5; cake mass w_c (in lbm/ft^2) is $7.2L$, where L is cake thickness in inches; maximum submergence is 35 percent or 126°; effective submergence is 30 percent or 108°; maximum washing arc is 29 percent or 104°; suction (initial drying) arc is 7.5 percent or 27°; discharge and resubmergence arc is 25 percent or 90°; and minimum cake thickness is ⅛ in (0.0032 m). Determine the relevant design parameters for the filter.

Calculation Procedure:

1. Calculate the cake mass, find the filtration time for a thickness of 0.25 in (0.0064 m), and determine the minimum cycle needed for cake formation.

The cake mass is given by $w_c = 7.2(0.25) = 1.8$ lbm/ft^2. From Fig. 14-10, filtration time is found to be 0.22 min, and so the filtration rate is (1.8/0.22)(60) = 491 lbm/h per square foot of drum surface. With an effective submergence of 30 percent, the minimum cycle based on cake formation is 0.22/0.3 = 0.73 min/r, which corresponds to 1.37 r/min.

2. Check to see if initial drying or washing can be done within the time available during the minimum cycle from step 1.

Minimum suction time elapses during passage through 27° (7.5 percent) of the perimeter. Therefore, drying time is 0.075(0.73) = 0.06 min, and the correlating factor is t_d/w_c = 0.06/1.8 = 0.033. Based on Fig. 14-13, the dewatered but unwashed (D/u) cake will have a moisture content of 30%. Then with a wash ratio of 1.5, liquid in D/u cake equals (30/70)(1.8) = 0.77 $lbm/(ft^2)(r)$, and quantity of wash equals 1.5(0.77) = 1.155 $lbm/(ft^2)(r)$, or, at 8.33 lbm/gal, 0.14 $gal/(ft^2)(r)$. To calculate wash time, $w_c v_w$ = 1.8(0.14) = 0.25. From Fig. 14-12, the required wash time is 0.15 min. This corresponds to an arc of 0.15/0.73 = 0.21; that is, to 21 percent of the circumference. Since up to 29 percent of the circumference can be used, washing offers no problems.

3. Check the drying time and determine the cycle time.

For a final moisture content of 25%, the simplified correlating factor t_d/w_c from Fig. 14-13 (taking into account the 1.2 factor mentioned above) is approximately $1.2(0.25) = 0.3$ (the 25% moisture content is employed to enter the graph along the ordinate; the 0.25 is read from the graph along its abscissa). With $w_c = 1.8$, $t_d = 0.54$ min and $0.54/0.73 = 0.739$, which takes up nearly three-quarters of the circumference for drying, so a lower speed must be used.

As a first estimate, note that since 25 percent of the arc is needed for discharge and resubmergence, the maximum arc for washing plus final drying is given by 75 − (cake-formation arc) − (suction arc) = 75 − 30 − 7.5 = 37.5 percent. Using the originally calculated washing plus drying times of 0.54 + 0.15 = 0.69 min, then 0.69/0.375 = 1.84 min/r and the washing arc is 0.15/1.84, which is 8.15 percent or 29°.

4. Repeat the calculations of step 2.

Minor adjustments can be made by recalculating each quantity with each change in conditions. Thus initial drying time equals 1.84(0.075) = 0.14 min, so $t_d/w_c = 0.14/1.8 = 0.08$. From Fig. 14-13, D/u moisture is 27%, and accordingly, the liquor in the D/u cake is $(27/73)(1.8) = 0.67$ lbm/(ft^2)(r). The quantity of wash becomes $1.5(0.67) = 1.0$ lbm/(ft^2)(r), or 0.12 gal/(ft^2)(r). Then $w_c v_w = 1.8(0.12) = 0.22$, and from Fig. 14-12, the wash time becomes 0.14 min.

5. Summarize the filtration cycle.

The cycle time is now (0.14 + 0.54)/0.375 = 1.81 min/r, equivalent to 0.55 r/min. The required effective submergence is (0.22/1.81)(100) = 12.2 percent. This is much less than the 30 percent available. The filter valve bridge must delay the start of vacuum or the slurry level can be reduced. If the level is reduced, additional initial drying time will be available, thereby reducing the angle required for washing. The design cycle is as follows:

Operation	Minutes
Form	0.22
Initial dry	0.14
Wash	0.14
Final dry	0.54
Discharge and resubmergence	0.77
Total time	1.81

6. Calculate cake thickness and filtration rate.

Cake thickness is given by L (in inches) = $w_c/7.2$ = 1.8/7.2 = 0.25 in. Taking into account the effective submergence of 12.2 percent, we calculate the filtration rate as 491(0.122) = 59.9 lbm/(h)(ft^2) [0.081 kg/(s)(m^2)]. Experience suggests applying a scaleup factor of 0.8: 0.8(59.9) = 47.9 lbm/(h)(ft^2) [0.065 kg/(s)(m^2)]. This is not intended as a safety factor to allow for increased production; instead, it corrects for deviation owing to the size of the test equipment, to media blinding, and to similar factors.

7. *Calculate the efficiency of solute recovery.*

Assume that Fig. 14-11 applies. With $j = 1.5$, the fraction remaining is 0.145. To be on the safe side, use a value of 0.2. The following calculations are needed:

Solute in feed = (60/40)(0.02) = 0.03 lb solute per pound of feed

Solute in D/u cake = (27/73)(0.02) = 0.0074 lb solute per pound of cake

Solute in washed cake = 0.0074(0.2) = 0.0015 lb solute per pound of washed cake

The fractional recovery, then, equals (0.030 − 0.0015)/0.03 = 0.95. Using 0.145 instead of 0.2 would have yielded a figure of 0.964.

8. *Calculate the air rate.*

The air rate can be calculated based on the data previously presented and shown in Fig. 14-14. During the 0.14 min of initial drying, the average rate is found to be 2.95 $(\text{ft}^3/\text{min})/(\text{ft}^2)(\text{r})$. The average rate during the final 0.54 min of drying is 5.85 $(\text{ft}^3/\text{min})/(\text{ft}^2)(\text{r})$. The total air rate is given by 0.14(2.95) + 0.54(5.85) = 3.57 $(\text{ft}^3/\text{min})/(\text{ft}^2)(\text{r})$. Since there are 1.81 min/r, the air rate is 3.57/1.81 = 1.97 $(\text{ft}^3/\text{min})/\text{ft}^2$ [0.01 $(\text{m}^3/\text{s})/\text{m}^2$].

15

Air Pollution Control

Louis Theodore, Eng.Sc.D.

Professor
Department of Chemical Engineering
Manhattan College
Bronx, NY

REFERENCES: [1] Theodore and Feldman—*Theodore Tutorial: Air Pollution Control Equipment for Particulates,* Research-Cottrell; [2] Theodore, Reynolds, and Richman—*Theodore Tutorial: Air Pollution Control Equipment for Gaseous Pollutants,* Research-Cottrell; [3] Theodore and McGuinn—*U.S. EPA Instructional Problem Workbook: Air Pollution Control Equipment;* [4] *Air Pollution Control Equipment: Selection, Design, Operation and Maintenance,* ETS International (Roanoke, VA); [5] Theodore, Reynolds, and Taylor—*Accident and Emergency Management,* Wiley-Interscience; [6] Theodore and McGuinn—*Pollution Prevention,* Van Nostrand Reinhold; [7] Theodore, personal notes.

15-1 Efficiency of Particulate-settling Chamber

A particulate-settling chamber is installed in a small heating plant that uses a traveling grate stoker. Determine the overall collection efficiency of the chamber, given the following operating conditions, chamber dimensions, and particle-size distribution:

Chamber width: 10.8 ft (3.29 m)

Chamber height: 2.46 ft (0.75 m)

Chamber length: 15.0 ft (4.57 m)

Volumetric flow rate of contaminated air stream: 70.6 std ft³/s (2.00 m³/s)

Flue-gas temperature: 446°F

Viscosity of air stream at 446°F: 1.75×10^{-5} lb/(ft)(s) [2.60×10^{-5}(N)(s)/m²]

Flue-gas pressure: 1 atm (101.3 kPa)

Particle concentration: 0.23 grains/std ft³ (8.13 grains/m³)

Particle specific gravity: 2.65

Standard operating conditions: 32°F, 1 atm (273 K, 101.3 kPa)

Particle-size distribution on the inlet dust:

Particle size range, μm	Average particle diameter, μm	grains/std ft³	grains/m³	Weight %
0–20	10	0.0062	0.219	2.7
20–30	25	0.0159	0.562	6.9
30–40	35	0.0216	0.763	9.4
40–50	45	0.0242	0.855	10.5
50–60	55	0.0242	0.855	10.5
60–70	65	0.0218	0.770	9.5
70–80	75	0.0161	0.569	7.0
80–94	87	0.0218	0.770	9.5
94+	94+	0.0782	2.763	34.0

Assume that the actual terminal settling velocity is one-half of the velocity given by Stokes' law.

Calculation Procedure:

1. *Express the collection efficiency E in terms of the particle diameter d_p by employing the terminal settling velocity for Stokes' law.*

Since the actual terminal settling velocity is assumed to be one-half of the Stokes' law velocity,

$$v_t = g d_p^2 \rho_p / 36\mu$$

and

$$E = v_t BL/q = g\rho_p BL d_p^2 / 36\mu q_a$$

where v_t = terminal velocity, ft/s

g = gravitational constant, 32.2 (lbm)(ft)/(s²)(lbf)

d_p = particle diameter, ft

ρ_p = particle density, lb/ft³

μ = air viscosity, lb/(ft)(s)
B = chamber width, ft
L = chamber length, ft
q = volumetric flow rate for the gas, actual ft^3/s

2. *Determine the particle density.*

As the specific gravity is 2.65, the density is (2.65)(62.4 lb/ft^3) = 165.4 lb/ft^3 (2648 kg/m^3).

3. *Determine the actual volumetric flow rate.*

Use Charles' law to convert from q_s, the flow rate at the standard-conditions temperature T_s, to q_a, the flow rate at the actual temperature T_a:

$$q_a = q_s(T_a/T_s) = 70.6[(446 + 460)/(32 + 460)] = 130 \text{ actual ft}^3/\text{s}$$

4. *Express the collection efficiency in terms of d_p in micrometers.*

The efficiency equation set out in step 1 is for d_p in feet. To adapt it for d_p in micrometers, note that there are 304,800 μm in a foot and accordingly divide the equation by the conversion factor $(304{,}800)^2$. Thus,

$$\begin{aligned} E &= g\rho_p BLd_p^2/36\mu q_a \\ &= (32.2)(165.4)(10.8)(15)d_p^2/(36)(1.75 \times 10^{-5})(130)(304{,}800)^2 \\ &= 1.14 \times 10^{-4} d_p^2 \end{aligned}$$

5. *Calculate the collection efficiency at each average particle size given in the statement of the problem.*

Applying the equation from step 4 and multiplying the answer by 100 (to convert the efficiency from a decimal fraction to a percent) gives the following results:

Average particle diameter, μm	Efficiency, %
10	1.1
25	7.1
35	14
45	23
55	34
65	48
75	64
87	86
94	100

6. *Calculate the overall collection efficiency.*

This is the sum of each of the efficiencies from the previous step multiplied by the weight fraction of the corresponding particle size in the mixture; in other words,

$$E = \Sigma w_i E_i$$

Thus, from the weight fractions given in the statement of the problem,

$$E = (0.027)(1.1) + (0.069)(7.1) + (0.094)(14) + (0.105)(23) + (0.105)(34) + (0.095)(48) + (0.070)(64) + (0.095)(86) + (0.34)(100) = 59.0\%$$

Related Calculations: Instead of following steps 5 and 6 as stated, one can calculate the efficiency at a variety of arbitrary particle diameters, graph the results, and then read the efficiencies for the actual particle diameters from the graph.

15-2 Efficiency of Cyclone Separator

A cyclone separator 2 ft (0.62 m) in diameter, with an inlet width of 0.5 ft (0.15 m) and rated at providing 4.5 effective turns is being considered for removing particulates from offgases from a gravel dryer. Gases to the cyclone have a loading of 0.5 grains/ft^3 (17.7 grains/m^3), with an average particle diameter of 7.5 μm. Specific gravity of the particles is 2.75. Operating temperature is 70°F (294 K) at which the air viscosity is 1.21×10^{-5} lb/(ft)(s) [1.80×10^{-5}(N)(s)/m^2]. Inlet velocity to the cyclone is 50 ft/s (15.2 m/s). The local air-pollution authority requires that the maximum total loading of the cyclone effluent be 0.1 grains/ft^3 (3.53 grains/m^3). Can this cyclone meet that criterion?

Calculation Procedure:

1. Calculate the particle density.

As the specific gravity of the particles is 2.75, the particle density ρ_p is (2.75)(62.4) or 171.6 lb/ft^3 (2747 kg/m^3).

2. Calculate the cut diameter.

The cut diameter d_{pc} for a given cyclone and given gas to be treated is the diameter of the particle that would be collected at 50 percent efficiency by the cyclone. It can be found from the equation

$$d_{pc} = [9\mu B_c/2\pi n_t v_i(\rho_p - \rho)]^{0.5}$$

where
μ = air viscosity, lb/(ft)(s)
B_c = cyclone inlet width, ft
n_t = number of effective turns provided by cyclone
v_i = inlet gas velocity, ft/s
ρ_p = particle density, lb/ft^3
ρ = gas density, lb/ft^3

In this example, the gas density can be assumed negligible in comparison with the particle density, so use ρ_p instead of the density difference.

Thus,

$$d_{pc} = [9(1.21 \times 10^{-5})(0.5)/2\pi(4.5)(50)(171.6)]^{0.5} = 1.5 \times 10^{-5} \text{ ft}$$
$$= 4.57\ \mu\text{m}$$

3. Calculate the ratio of average particle diameter to the cut diameter.

Thus, $d_p/d_{pc} = 7.5/4.57 = 1.64$

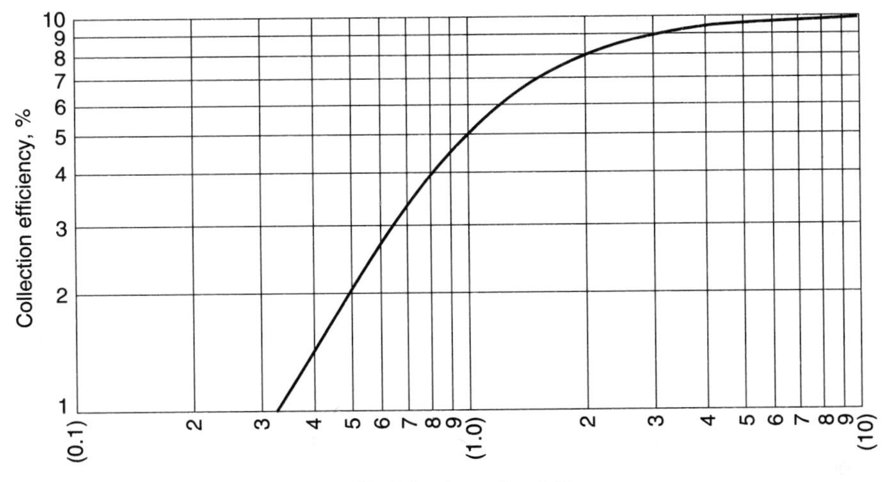

FIG. 15-1 Collection efficiency as a function of particle size ratio.

4. Determine the collection efficiency using Lapple's curve.

Refer to Fig. 15-1. For the particle-size ratio of 1.64 on the abscissa, the curve yields an efficiency of 0.72, or 72 percent, on the ordinate.

5. Determine the collection efficiency required by the air-pollution-control authority.

This efficiency is, simply, (inlet loading − outlet loading)/(inlet loading). In the present example, it equals (0.5 − 0.1)/(0.5), i.e., 0.8, or 80 percent.

6. Does the cyclone meet the requirements of the pollution-control authority?

As an 80 percent efficiency is required whereas the cyclone achieves only a 72 percent efficiency, the cyclone does not meet the requirements.

Related Calculations: If the cyclone is of conventional design and its inlet width is not given, it is relatively safe to assume that the width is one-quarter the cyclone diameter.

Instead of using Lapple's curve, one can solve for cyclone efficiency by using the equation

$$E = 1.0/[1.0 + (d_{pc}/d_p)^2]$$

where d_{pc} is the cut diameter and d_p the average particle diameter.

15-3 Sizing an Electrostatic Precipitator

A duct-type electrostatic precipitator is to be used to clean 100,000 actual ft^3/min (47.2 actual m^3/s) of an industrial gas stream containing particulates. The proposed design of the precipitator consists of three bus sections (fields) arranged in series, each having the same amount of collection surface. The inlet loading has been measured as 17.78 grains/ft^3 (628 grains/m^3), and a maximum outlet loading of 0.08 grains/ft^3 (2.8 grains/m^3) (both volumes corrected to dry standard conditions and 50 percent excess air) is allowed by the local air-pollution regulations. The drift velocity for the particulates has been experimentally determined in a similar installation, with the following results:

First section (inlet): 0.37 ft/s (0.11 m/s)
Second section (middle): 0.35 ft/s (0.107 m/s)
Third section (outlet): 0.33 ft/s (0.10 m/s)

Calculate the total collecting surface required. And find the total mass flow rate of particulates captured in each section.

Calculation Procedure:

1. Calculate the required total collection efficiency E based on the given inlet and outlet loading.

The equation is

$$E = 1 - \text{(outlet loading)}/\text{(inlet loading)}$$

Thus,

$$E = 1 - 0.08/17.78 = 0.9955, \text{ or } 99.55\%$$

2. Calculate the average drift velocity w.

Thus, $w = (0.37 + 0.35 + 0.33)/3 = 0.35$ ft/s (0.107 m/s).

3. Calculate the total surface area required.

Use the Deutsch-Anderson equation:

$$E = 1 - \exp(-wA/q)$$

where E = collection efficiency
w = average drift velocity
A = required surface area
q = gas flow rate

Rearrange the equation as follows:

$$A = \ln(1 - E)/(w/q)$$

For consistency between w and q, convert q from ft^3/min to ft^3/s:

$$100{,}000/60 = 1666.7 \text{ ft}^3\text{/s}$$

Then

$$A = -\ln(1 - 0.9955)/(0.35/1666.7) = 25{,}732 \text{ ft}^2 \text{ (2393 m}^2\text{)}$$

4. Calculate the collection efficiencies of each section.

Use the Deutsch-Anderson equation (from step 3) directly. For the first section, $E_1 = 1 - \exp[-(25{,}732)(0.37)/(3)(1666.7)] = 0.851$. Similarly, E_2 for the second section is found to be 0.835, and E_3 for the third section 0.817.

5. Calculate the mass flow rate $\dot{m}$ of particulates captured by each section.

For the first section, the equation is

$$\dot{m} = (E_1)(\text{inlet loading})(q)$$

Thus,

$$\dot{m} = (0.851)(17.78)(100{,}000) = 1.513 \times 10^6 \text{ grains/min} = 216.1 \text{ lb/min } (1.635 \text{ kg/s})$$

For the second section, the equation is

$$\dot{m} = (1 - E_1)(E_2)(\text{inlet loading})(q)$$

Thus,

$$\dot{m} = (1 - 0.851)(0.835)(17.78)(100{,}000) = 2.212 \times 10^5 \text{ grains/min}$$
$$= 31.6 \text{ lb/min } (0.239 \text{ kg/s})$$

And for the third section, the equation is

$$\dot{m} = (1 - E_1)(1 - E_2)(E_3)(\text{inlet loading})(q)$$

which yields 5.10 lb/min (0.039 kg/s).

The total mass captured is the sum of the amounts captured in each section, i.e., 252.8 lb/min (1.91 kg/s). It is not surprising that a full 85 percent of the mass is captured in the first section.

15-4 Efficiency of a Venturi Scrubbing Operation

A gas stream laden with fly ash is to be cleaned by a venturi scrubber using a liquid-to-gas ratio (q_L/q_G) of 8.5 gal per 1000 actual cubic feet. The fly ash has a particle density of 43.7 lb/ft³ (700 kg/m³). The collection-efficiency k factor equals 200 ft³/gal. The throat velocity is 272 ft/s (82.9 m/s), and the gas viscosity is 1.5×10^{-5} lb/(ft)(s) [2.23×10^{-5} (N)(s)/m²]. The particle-size distribution is as follows:

d_{pi}, μm	Percent by weight, w_i, %
<0.10	0.01
0.1 − 0.5	0.21
0.6 − 1.0	0.78
1.1 − 5.0	13.0
6.0 − 10.0	16.0
11.0 − 15.0	12.0
16.0 − 20.0	8.0
>20.0	50.0

Calculation Procedure:

1. Determine the mean droplet diameter d_o.

Use the Nukiyama-Tanasawa equation:

$$d_o = (16{,}400/v) + 1.45(q_L/q_G)^{1.5}$$

where d_o is the mean droplet diameter in micrometers and v is the throat velocity in feet per second. Thus,

$$d_o = (16{,}400/272) + 1.45(8.5)^{1.5} = 96.23\ \mu\text{m} = 3.16 \times 10^{-4} \text{ ft } (9.63 \times 10^{-5} \text{ m})$$

2. Express the inertial impaction number ψ_1 in terms of particle diameter d_p.

The relationship is

$$\psi_1 = d_p^2 \rho_p v / 9\mu d_o$$

where ρ_p is particle density in lb/ft³ and μ is gas viscosity in lb/(ft)(s). Thus,

$$\psi_1 = d_p^2 (43.7)(272)/(9)(1.5 \times 10^{-5})(3.16 \times 10^{-4})$$
$$= 2.78 \times 10^{11} d_p^2 \qquad \text{with } d_p \text{ in feet}$$

To make the expression suitable for the steps that follow, express it with d_p in micrometers, by dividing by the feet-to-micrometers conversion factor squared (because d_p is squared):

$$2.78 \times 10^{11} d_p^2/(3.048 \times 10^5)^2 = 3.002 d_p^2 \qquad \text{with } d_p \text{ in micrometers}$$

3. Express the individual collection efficiency in terms of d_p.

Use the Johnstone equation:

$$E_i = 1 - \exp(-k q_L \psi_1^{0.5}/q_G)$$
$$= 1 - \exp(-2.94 d_{pi})$$

4. Calculate the overall collection efficiency.

This is done by (a) determining the midpoint size of each of the particle-size ranges in the statement of the problem, (b) using the expression from step 3 to calculate the collection efficiency corresponding to that midpoint size, (c) multiplying each collection efficiency by the percent representation w_i for that size range in the statement of the problem, and (d) summing the results. These steps are embodied in the following table:

d_{pi}, μm	E_i	w_i, %	$E_i w_i$, %
0.05	0.1367	0.01	0.001367
0.30	0.586	0.21	0.123
0.80	0.905	0.78	0.706
3.0	0.9998	13.0	12.998
8.0	0.9999	16.0	16.0
13.0	0.9999	12.0	12.0
18.0	0.9999	8.0	8.0
20.0	0.9999	50.0	50.0

Then $E = \Sigma E_i w_i = 99.83$ percent.

Related Calculations: In situations where the particles are so small that their size approaches the length of the mean free path of the fluid molecules, the fluid can no longer be regarded as a continuum; that is, the particles can "fall between" the molecules. That problem can be offset by applying a factor, the Cunningham correction factor, to the calculation of the inertial-impact-number expression in step 2.

15-5 Selecting a Filter Bag System

It is proposed to install a pulse-jet fabric-filter system to remove particulates from an air stream. Select the most appropriate bag from the four proposed below. The volumetric flow rate of the air stream is 10,000 std ft³/min (4.72 m³/s) (standard conditions being 60°F and 1 atm), the operating temperature is 250°F (394 K), the concentration of pollutants is 4 grains/ft³ (141 grains/m³), the average air-to-cloth ratio is (2.5 ft³/min)/ft², and the required collection efficiency is 99 percent.

Information on the four proposed bags is as follows:

Bag designation	A	B	C	D
Tensile strength	Excellent	Above average	Fair	Excellent
Recommended maximum temperature, °F	260	275	260	220
Cost per bag	$26.00	$38.00	$10.00	$20.00
Standard size	8 in by 16 ft	10 in by 16 ft	1 ft by 16 ft	1 ft by 20 ft

Note: No bag has an advantage from the standpoint of durability.

Calculation Procedure:

1. *Eliminate from consideration any bags that are patently unsatisfactory.*

Bag D is eliminated because its recommended maximum temperature is below the operating temperature for this application. Bag C is also eliminated, because a pulse-jet fabric-filter system requires that the tensile strength of the bag be at least above average.

2. *Convert the given flow rate to actual cubic feet per minute.*

The flow rate as stated corresponds to flow at 60°F, whereas the actual flow q will be at 250°F. Accordingly,

$$q = (10{,}000)(250 + 460)/(60 + 460) = 13{,}654 \text{ actual ft}^3/\text{min} \; (6.44 \text{ actual m}^3/\text{s})$$

3. *Establish the filtering velocity v_f.*

The air-to-cloth ratio is (2.5 ft³/min)/ft². Dimensional simplification converts this to 2.5 ft/min (0.0127 m/s).

4. *Calculate the total filtering area required.*

This equals the actual volumetric flow rate (from step 2) divided by the filtering velocity (from step 3). Thus,

$$\text{Total filtering area} = 13{,}654/2.5 = 5461.6 \text{ ft}^2 \; (507.9 \text{ m}^2)$$

5. *Calculate the filtering area available per bag.*

Consider the operating bag to be in the form of a cylinder, whose wall constitutes the filtering area. This is accordingly calculated from the formula $A = \pi D h$, where A is area, D

is bag diameter, and h is bag height. The two bags still under consideration are Bag A and Bag B. The area of each is as follows:

For Bag A, $A = \pi(8/12)(16) = 33.5\ ft^2\ (3.12\ m^2)$

For Bag B, $A = \pi(10/12)(16) = 41.9\ ft^2\ (3.90\ m^2)$

6. Determine the number of bags required.

The total filtering area required is 5461.6 ft². Accordingly, if Bag A is selected, the number of bags needed is 5461.6/33.5, i.e., 163 bags. If instead Bag B is selected, the number needed is 5461.6/41.9, i.e., 130 bags.

7. Determine the total bag cost.

If Bag A is used, the total cost is (163)($26.00), i.e., $4238. If instead Bag B is used, the total cost is (130)($38.00), i.e., $4940.

8. Select the most appropriate bag.

Since the total cost for Bag A is less than that for Bag B, select Bag A.

15-6 Sizing a Condenser for Odor-carrying Steam

The discharge gases from a meat-rendering plant consist mainly of atmospheric steam, plus a small fraction of noncondensable odor-carrying gases. The stream is to pass through a condenser to remove the steam before the noncondensable gases go to an incinerator or adsorber. Estimate the size of a condenser to treat 60,000 lb/h (7.55 kg/s) of this discharge stream. Assume that the overall heat-transfer coefficient is 135 Btu/(h) (ft²)(°F) [765 W/(m²) (K)], that the enthalpy of vaporization for the steam is 1000 Btu/lb, and that the cooling water enters the condenser at 80°F (300 K) and leaves at 115°F (319 K).

Calculation Procedure:

1. Determine the heat load Q for the condenser.

This equals the flow rate times the enthalpy of vaporization:

$$Q = (60{,}000)(1000) = 6.0 \times 10^7\ \text{Btu/h}\ (17.6 \times 10^6\ \text{W})$$

2. Estimate the log-mean temperature temperature-difference driving force.

The formula is $\text{LMTD} = (t_g - t_l)/\ln(t_g/t_l)$, where LMTD is log-mean temperature difference, t_g is the maximum temperature difference between steam and cooling water, and t_l is the minimum temperature difference between them. The maximum difference is (212 − 80), or 132; the minimum is (212 − 115), or 97. Accordingly, LMTD = (132 − 97)/ln(132/97) = 113.6°F.

3. Calculate the required area of the condenser

The formula is $A = Q/U(\text{LMTD})$, where A is area, Q is heat load, and U is overall heat-transfer coefficient. Thus, $A = (6.0 \times 10^7)/(135)(113.6)$, i.e., 3912 ft² (363.8 m²).

Related Calculations: A comprehensive design procedure for condensers, including several examples of its application, was originally developed by the author and can be found in several of the author's *Theodore Tutorials.* Refer also to Sec. 7 of this handbook, which deals with heat transfer.

Note: This material is original to the author. Some of it has been published elsewhere without authorization.

15-7 Amount of Adsorbent for a VOC Adsorber

Determine the required height of adsorbent in an adsorption column that treats a degreaser-ventilation stream contaminated with trichloroethylene (TCE). Design and operating data are as follows:

Volumetric flow rate of contaminated air: 10,000 std ft^3/min
(4.72 m^3/s), standard conditions being 60°F and 1 atm
Operating temperature: 70°F (294 K)
Operating pressure: 20 psia (138 kPa)
Adsorbent: activated carbon
Bulk density ρ_B of activated carbon: 36 lb/ft^3 (576 kg/m^3)
Working capacity of activated carbon: 28 lb TCE per 100 lb carbon
Inlet concentration of TCE: 2000 ppm (by volume)
Molecular weight of TCE: 131.5

The adsorption column is a vertical cylinder with an inside diameter of 6 ft (1.8 m) and a height of 15 ft (4.57 m). It operates on the following cycle: 4 h in the adsorption mode, 2 h for heating and desorbing, 1 h for cooling, 1 h for standby. An identical column treats the contaminated gas while the first one is not in the adsorption mode. The system is required to recover 99.5 percent of the TCE by weight.

Calculation Procedure:

1. Determine the actual volumetric flow rate of the contaminated gas stream.

The flow rate as stated corresponds to flow at 60°F and 1 atm, whereas the actual flow q is at 70°F and 20 psia. Accordingly,

$$q = 10{,}000[(70 + 460)/(60 + 460)][14.7/20]$$
$$= 7491 \text{ actual ft}^3\text{/min, or } 4.5 \times 10^5 \text{ actual ft}^3\text{/h (3.54 actual m}^3\text{/s)}$$

2. Calculate the volumetric flow rate of TCE.

This flow rate q_{TCE} equals the inlet concentration y_{TCE} of TCE in the gas times the gas flow rate. Thus,

$$q_{\text{TCE}} = (y_{\text{TCE}})(q) = (2000 \times 10^{-6})(4.5 \times 10^5)$$
$$= 900 \text{ actual ft}^3\text{/h (0.007 m}^3\text{/s)}$$

3. *Convert the volumetric flow rate of TCE into mass flow rate.*

For the conversion, rearrange the ideal-gas law, bearing in mnd that mass equals the number of moles times the molecular weight. Thus $\dot{m}$, the mass flow rate in pounds per hour, equals $q_{TCE}(PM/RT)$, where P is the pressure, M the molecular weight, T the absolute temperature, and R the gas constant. Accordingly,

$$\dot{m} = (900)\,(131.5)/(10.73)\,(70 + 460) = 416.2 \text{ lb/h } (0.052 \text{ kg/s})$$

4. *Determine the mass of TCE to be adsorbed during the 4-h period.*

This equals, simply, the required degree of adsorption times the amount of TCE that will pass through the system in 4 h:

$$\text{TCE adsorbed} = (416.2 \text{ lb/h})\,(4 \text{ h})\,(0.995) = 1656.6 \text{ lb}$$

5. *Calculate the volume of activated carbon required.*

To obtain this volume v_{AC}, divide the amount of TCE to be adsorbed by the adsorption capacity of the carbon, and convert from mass to volume by taking into account the bulk density of the carbon:

$$v_{AC} = (\text{TCE to be adsorbed})/(28 \text{ lb TCE}/100 \text{ lb carbon})(\text{bulk density})$$
$$= (1656.6)/(28/100)(36) = 164 \text{ ft}^3 \ (4.64 \text{ m}^3)$$

6. *Find the required height of the carbon in the adsorber.*

This height z equals the volume of carbon divided by the cross-sectional area of the column:

$$z = 164/[\pi(D^2/4)] = 164/[\pi(6^2/4)] = 5.8 \text{ ft } (1.77 \text{ m})$$

Note: This material is original to the author. Some of it has been published elsewhere without authorization.

15-8 Performance of an Afterburner

It is proposed to use a natural-gas–fired, direct-flame afterburner to incinerate toluene in the effluent gases from a lithography plant. The afterburner system is as shown in Fig. 15-2. The flow rate of the 300°F (422 K) effluent is 7000 std ft³/min (3.30 m³/s), standard conditions being 60°F and 1 atm; its toluene content is 30 lb/h (0.0038 kg/s). After passing through the afterburner preheater, the gas enters the afterburner at 738°F (665 K). The afterburner is essentially a horizontal cylinder, 4.2 ft (1.28 m) in diameter and 14 ft (4.27 m) long; it can be assumed to incur heat losses at 10 percent in excess of the calculated heat load. Gases leaving the afterburner are at 1400°F (1033 K).

When reviewing plans for such an installation, the local air-pollution–control agency knows from experience that in order to meet emission standards, the afterburner must operate at 1300 to 1500°F, that the residence time in the vessel must be 0.3 to 0.5 s, and that the velocity within it must be 20 to 40 ft/s. Can this afterburner satisfy those three criteria?

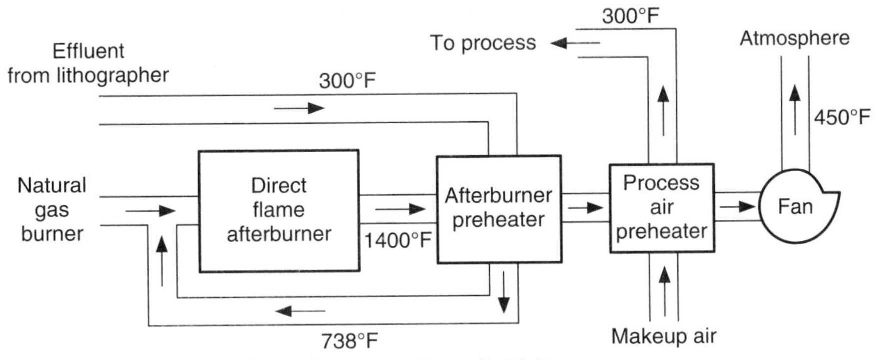

FIG. 15-2 Natural-gas–fired afterburner (Example 15-8).

Use the following data:

Gross heating value of natural gas: 1059 Btu/std ft^3

Volume of combustion products produced per standard cubic foot of natural gas burned: 11.5 std ft^3 (11.5 m^3 per standard cubic meter)

Average available heat from natural gas between 738 and 1400°F: 600 Btu/std ft^3 (22,400 kJ/m^3)

Molecular weight of toluene: 92

Average heat capacity C_{p1} of effluent gases between 0 and 738°F: 7.12 Btu/(lb · mol)(°F)[29.8 kJ/(kg · mol)(K)]

Average heat capacity C_{p2} of effluent gases between 0 and 1400°F: 7.38 Btu/(lb · mol)(°F)[30.9 kJ/(kg · mol)(K)]

Volume of air required to combust natural gas: 10.33 std ft^3 air/std ft^3 natural gas (10.33 m^3/m^3 natural gas)

Calculation Procedure:

1. Convert the gas flow rate to the molar basis.

Since 1 lb-mol of gas at 32°F and 1 atm occupies 359 ft^3, at 60°F it will occupy (359)[(460 + 60)/(460 + 32)], i.e., 379 ft^3. Accordingly, the molar gas flow rate $\dot{n}$ is (7000 actual ft^3/min)/(379 actual ft^3/lb · mol), i.e., 18.47 lb · mol/min (0.139 kg · mol/s).

2. Calculate the total heat load (heating rate) required to raise the gas stream from 738 to 1400°F.

Since the heat capacity data are given with 0°F as a basis, this heat load Q must be based on first cooling the gases from 738 to 0°F, then heating them from 0 to 1400°F. Thus,

$$Q = \dot{n}[C_{p2}(1400 - 0) - C_{p1}(738 - 0)]$$
$$= 18.47[(7.38)(1400) - (7.12)(738)] = 93{,}790 \text{ Btu/min (1648 kW)}$$

3. Determine the actual required heat load, taking into account the 10 percent heat loss.

Thus, actual heat load = 1.1Q = (1.1)(93,790) = 103,169 Btu/min (1813 kW).

4. Find the rate of natural gas needed to satisfy this heat load.

This is, simply, the heat load divided by the available heat. Thus,

Rate of natural gas = 103,169/600 = 171.9 std ft³/min (0.081 m³/s)

5. Determine the volumetric flow rate of the combustion products of the natural gas.

This flow rate q_1 equals the natural gas rate times the volume of combustion products produced per standard cubic foot of natural gas. Thus,

$$q_1 = (171.9)(11.5) = 1976 \text{ std ft}^3/\text{min } (0.932 \text{ m}^3/\text{s})$$

Note that the 11.5 figure already takes into account the air needed to burn the natural gas.

6. Calculate the total volumetric flow rate through the afterburner.

This flow rate q_T equals that of the effluent from the lithography plant plus that of the combustion products from step 5. Thus,

$$q_T = 7000 + 1976 = 8976 \text{ std ft}^3/\text{min } (4.233 \text{ m}^3/\text{s})$$

For the subsequent steps, this figure must be converted to actual cubic feet per minute. Since the afterburner operates at 1400°F, this equals (8976) [(1400 + 460)/(60 + 460)], or 32,106 actual ft³/min (15.14 m³/s).

7. Determine the cross-sectional area of the afterburner.

As the cross section is circular, this area S equals $\pi D^2/4$, where D is the diameter of the afterburner. Thus, $S = \pi(4.2)^2/4 = 13.85$ ft² (4.22 m²).

8. Calculate the velocity through the afterburner.

This residence time t equals the length of the vessel divided by the velocity. Thus, $t = 14/38.6 = 0.363$ s.

9. Find the residence time within the afterburner.

This residence time t equals the length of the vessel divided by the velocity. Thus, $t = 14/38.6 = 0.363$ s.

10. Ascertain whether the afterburner meets the agency's criteria.

The afterburner operates at 1400°F (1033 K), so it meets the temperature criterion. The residence time of 0.363 s meets the residence-time criterion. And the gases pass through the vessel at 38.6 ft/s (11.8 m/s), which satisfies the velocity criterion. Thus, all three criteria are satisfied.

Related Calculations: The determination of the natural gas rate is discussed in more detail in the Wiley-Interscience text *Introduction to Hazardous Waste Incineration* by Theodore and Reynolds.

Note: This material is original to the author. Some of it has been published elsewhere without authorization.

15-9 Preliminary Sizing of an Absorber for Gas Cleanup

Describe how one can make a rough estimate of the required diameter and height for a randomly packed absorption tower to achieve a given degree of gas cleanup without detailed information on the properties of the dirty gas, knowing only that the absorbent is water (or has properties similar to those of water) and that the pollutant has a strong affinity with the absorbent.

Calculation Procedure:

1. Estimate a diameter for the tower.

Use the rule of thumb that superficial gas velocity through the tower (i.e., velocity assuming that the tower is empty) should be about 3 to 6 ft/s (1 to 2 m/s). If we assume, for instance, a velocity of 4 ft/s, then the tower cross section S equals the actual volumetric flow rate of the dirty gas divided by 4. Then the diameter D can be found from the formula $D = (4S/\pi)^{0.5}$.

To illustrate, assume that the volumetric flow rate is 60 actual ft^3/s ($1.7\ m^3/s$). Then a suitable cross section would be about 60/4, i.e., 15 ft^2 ($1.4\ m^2$), and a suitable diameter about $[(4)(15)/\pi]^{0.5}$, i.e., about 4.4 ft (1.3 m). Given the imprecision of the superficial-gas–velocity guideline, it is appropriate to round this figure off to 4 ft.

2. Choose a packing size for the tower.

If D is about 3 ft (1 m), use a packing whose diameter is 1 in. If D is under 3 ft, use smaller packing; if D is greater than 3 ft, use larger.

Continuing the illustration from step 1, since the tower is to be about 4 ft, use packing larger than 1 in, for instance, 1.5 in.

3. Estimate a height for the tower.

The height of an absorption tower equals the product of H_{OG}, the height of a gas transfer unit, and N_{OG}, the number of transfer units needed. It is also prudent to multiply this product by a safety factor of 1.25 to 1.50.

Since equilibrium data are not available, assume that the slope of the equilibrium curve (see Sec. 11 for discussion of this curve) approaches zero. This is not an unreasonable assumption for most solvents that preferentially absorb (or react with) the pollutant. For that condition, the value of N_{OG} approaches $\ln(y_1/y_2)$, where y_1 and y_2 represent the inlet and outlet concentrations, respectively. Accordingly, if for instance the required degree of gas cleanup is 99 percent, then $N_{OG} = \ln[1/(1 - 0.99)] = 4.61$.

As for H_{OG}, since the solvent is either water or similar to it, use the values that are normally encountered for aqueous systems:

Packing diameter, in	H_{OG} for plastic packing, ft	H_{OG} for ceramic packing, ft
1.0	1.0	2.0
1.5	1.25	2.5
2.0	1.5	3.0
3.0	2.25	4.5
3.5	2.75	5.5

Continue the illustration from step 2. Assume that the required cleanup is in fact 90 percent, that 1.5-in ceramic packing is to be used, and that a safety factor of 1.4 is to be used in the height calculation. Then the estimated height equals (safety factor)(N_{OG}) (H_{OG}), i.e., (1.4)(4.61)(2.5), or 16 ft (4.9 m).

Related Calculations: Apart from the rule of thumb concerning superficial velocity (see step 1), be aware of similar guidelines that pertain to mass flow rate through the absorption tower. For plastic packing, the liquid and gas flow rates are both typically around 1500 to 2000 lb/h per square foot of tower cross-sectional area; for ceramic packing, the corresponding range is about 500 to 1000 lb/h.

As a rough estimate of pressure drop for the gas flow through the packing, it is about 0.15 to 0.4 in of water per foot of packing.

For more detail on absorption, see Sec. 11.

Note: This material is original to the author. Some of it has been published elsewhere without authorization.

16

Other Chemical Engineering Calculations*

---

*Unless otherwise indicated, the material in this section is taken from T. G. Hicks, *Standard Handbook of Engineering Calculations*, McGraw-Hill Book Co., Inc.

16-1 Steam Mollier Diagram and Steam-Table Use

(1) Determine from the Mollier diagram for steam: (*a*) the enthalpy of 100 pisa (689.5 kPa) saturated steam; (*b*) the enthalpy of 10 psia (68.9 kPa) steam containing 40% moisture; (*c*) the enthalpy of 100 psia (689.5 kPa) steam at 600°F (315.6°C). (2) Determine from the steam tables: (*a*) the enthalpy, specific volume, and entropy of steam at 145.3 psig (1001.8 kPag); (*b*) the enthalpy and specific volume of superheated steam at 1100 psia (7584.2 kPa) and 600°F (315.6°C); (*c*) the enthalpy and specific volume of high-pressure steam at 7500 psia (51,710.7 kPa) and 1200°F (648.9°C); (*d*) the enthalpy, specific volume, and entropy of 10-psia (68.9-kPa) steam containing 40% moisture.

Calculation Procedure:

1. Use the pressure and saturation (or moisture) lines to find enthalpy.

(a) Enter the Mollier diagram by finding the 100-psia (689.5-kPa) pressure line (Fig. 16-1). In the Mollier diagram for steam, the pressure lines slope upward to the right from the lower left-hand corner. For saturated steam, the enthalpy is read at the intersection of the pressure line with the saturation curve *cef* (Fig. 16-1).

Thus, project along the 100-psia (689.5-kPa) pressure curve until it intersects the saturation curve, point *g*. From here project horizontally to the left-hand scale and read the enthalpy of 100-psia (689.5-kPa) saturated steam as 1187 Btu/lb (2761.0 kJ/kg). (The Mollier diagram in Fig. 16-1 has fewer grid divisions than large-scale diagrams to permit easier location of the major elements of the diagram.)

(b) On a Mollier diagram, the enthalpy of wet steam is found at the intersection of the saturation pressure line with the percent moisture curve corresponding to the amount of moisture in the steam. In a Mollier diagram for steam, the moisture curves slope downward to the right from the saturated liquid line *cd* (Fig. 16-1).

To find the enthalpy of 10-psia (68.9-kPa) steam containing 40% moisture, project along the 10-psia (68.9-kPa) saturation pressure line until the 40% moisture curve is intersected. From here project horizontally to the left-hand scale and read the enthalpy of 10-psia (68.9-kPa) wet steam containing 40% moisture as 750 Btu/lb (1744.5 kJ/kg).

2. Find the steam properties from the steam tables.

(a) Steam tables normally list absolute pressures or temperature in degrees Fahrenheit as one of their arguments. Therefore, when the steam pressure is given in terms of a gage reading, it must be converted to an absolute pressure before the table can be entered. To convert gage pressure to absolute pressure, add 14.7 to the gage pressure, or $p_a = p_g + 14.7$. In this instance, $p_a = 145.3 + 14.7 = 160.0$ psia (1103.2 kPa). Once the absolute pressure is known, enter the saturation pressure table of the steam table at this value and project horizontally to the desired values. For 160-psia (1103.2-kPa) steam, using the ASME or Keenan and Keyes, *Thermodynamic Properties of Steam,* the enthalpy of evaporation $h_{fg} = 859.2$ Btu/lb (1998.5 kJ/kg); enthalpy of saturated vapor $h_g = 1195.1$ Btu/lb (2779.8 kJ/kg), read from the respective columns of the steam tables. The specific volume v_g of the saturated vapor of 160-psia (1103.2-kPa) steam is, from the tables, 2.834 ft^3/lb (0.18 m^3/kg), and the entropy s_g is 1.5640 Btu/(lb)(°F) [6.55 kJ/(kg)(°C)].

(b) Every steam table contains a separate tabulation of properties of superheated steam. To enter the superheated steam table, two arguments are needed—the absolute

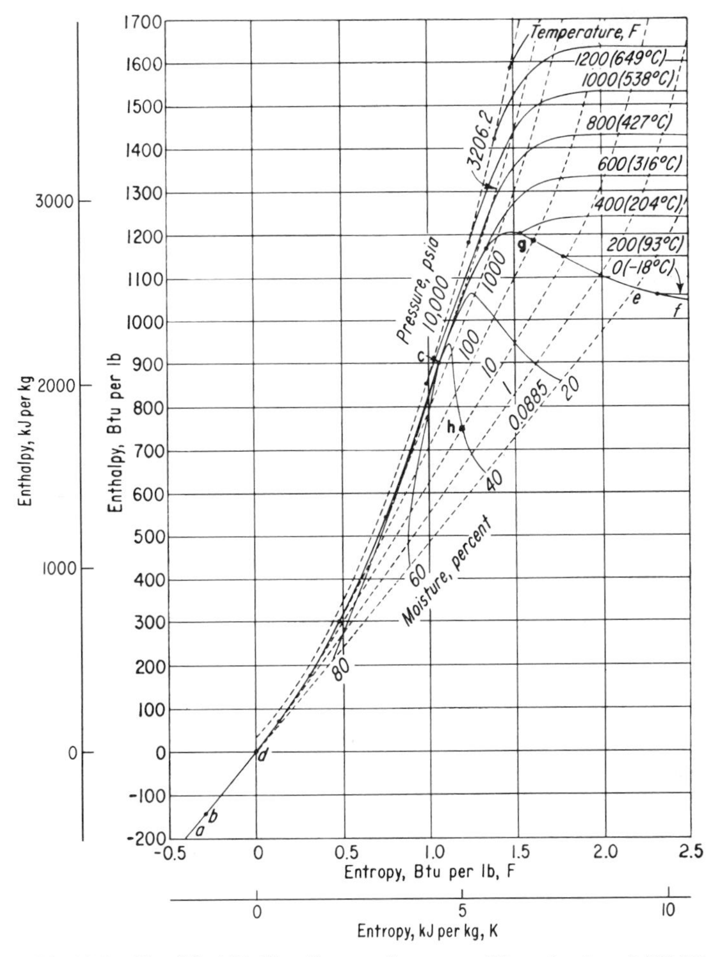

FIG. 16-1 Simplified Mollier diagram for steam. (Note: 1 psia = 6.895 kPa.)

pressure and the temperature of the steam in degrees Fahrenheit. To determine the properties of 1100-psia (7584.5-kPa) steam at 600°F (315.6°C), enter the superheated steam table at the given absolute pressure and project horizontally from this absolute pressure (1100 psia or 7584.5 kPa) to the column corresponding to the superheated temperature (600°F or 315.6°C) to read the enthalpy of the superheated vapor as h = 1236.7 Btu/lb (2876.6 kJ/kg) and the specific volume of the superheated vapor v = 0.4532 ft^3/lb (0.03 m^3/kg).

(c) For high-pressure steam use the ASME *Steam Table,* entering it in the same manner as the superheated steam table. Thus, for 7500-psia (51,712.5 kPa) steam at 1200°F (648.9°C), the enthalpy of the superheated vapor is 1474.9 Btu/lb (3430.6 kJ/kg), and the specific volume of the superheated vapor is 0.1060 ft^3/lb (0.0066 m^3/kg).

(d) To determine the enthalpy, specific volume, and the entropy of wet steam having y percent moisture by using steam tables instead of the Mollier diagram, apply these relations: $h = h_g - yh_{fg}/100$; $v = v_g - yv_{fg}/100$; and $s = s_g - ys_{fg}/100$, where y is percent moisture expressed as a whole number. For 10-psia (68.9-kPa) steam containing 40% moisture, obtain the needed values h_g, h_{fg}, v_g, v_{fg}, s_g, and s_{fg} from the saturation-pressure steam table and substitute in the preceding relations. Thus,

$$h = 1143.3 - \frac{40(982.1)}{100} = 750.5 \text{ Btu/lb } (1745.7 \text{ kJ/kg})$$

$$v = 38.42 - \frac{40(38.40)}{100} = 23.06 \text{ ft}^3/\text{lb } (1.44 \text{ m}^3/\text{kg})$$

$$s = 1.7876 - \frac{40(1.5041)}{100} = 1.1860 \text{ Btu/(lb)(°F) } [4.97 \text{ kJ/(kg)(°C)}]$$

Note that in Keenan and Keyes, *Thermodynamic Properties of Steam*, v_{fg} is not tabulated. Therefore, this value must be obtained by subtraction of the tabulated values, or $v_{fg} = v_g - v_f$. The value v_{fg} thus obtained is used in the relation for the volume of the wet steam. For 10-psia (68.9-kPa) steam containing 40% moisture $v_g = 38.42$ ft^3/lb (2.398 m^3/kg), and $v_f = 0.017$ ft^3/lb (0.0011 m^3/kg). Then, $v_{fg} = 38.42 - 0.017 = 38.403$ ft^3/lb (1.773 m^3/kg).

In some instances, the quality of steam may be given instead of its moisture content in percent. The quality of steam is the percent vapor in the mixture. In the preceding calculation, the quality of the steam is 60% because 40% is moisture. Thus, quality $= 1 - m$, where m is percent moisture, expressed as a decimal.

16-2 Interpolation of Steam-Table Values

(1) Determine the enthalpy, specific volume, entropy, and temperature of saturated steam at 151 psia (1041.1 kPa). (2) Determine the enthalpy, specific volume, entropy, and pressure of saturated steam at 261°F (127.2°C). (3) Determine the pressure of steam at 1000°F (537.8°C) if its specific volume is 2.6150 ft^3/lb (0.16 m^3/kg). (4) Determine the enthalpy, specific volume, and entropy of 300-psia (2068.5-kPa) steam at 567.22°F (297.3°C).

Calculation Procedure:

1. *Use the saturation-pressure steam table.*

Study of the saturation-pressure table shows that there is no pressure value listed for 151 psia (1041.1 kPa). Therefore, it will be necessary to interpolate between the next higher and next lower tabulated pressure values. In this instance, these values are 152 and 150 psia (1048.0 and 1034.3 kPa), respectively. The pressure for which properties are being found (151 psia or 1041.1 kPa) is called the "intermediate pressure." At 152 psia (1048.0 kPa), $h_g = 1194.3$ Btu/lb (2777.5 kJ/kg); $v_g = 2.977$ ft^3/lb (0.19 m^3/kg); $s_g = 1.5683$ Btu/(lb)(°F) [6.57 kJ/(kg)(°C)]; and $t = 359.46$°F (181.9°C). At 150 psia (1034.3 kPa), $h_g = 1194.1$ Btu/lb (2777.5 kJ/kg); $v_g = 3.015$ ft^3/lb (0.19 m^3/kg); $s_g = 1.5694$ Btu/(lb)(°F) [6.57 kJ/(kg)(°C)]; and $t = 358.42$°F (181.3°C).

For the enthalpy, note that as the pressure increases, so does h_g. Therefore, the enthalpy at 151 psia (1041.1 kPa) (the intermediate pressure) will equal the enthalpy at 150 psia (1034.3 kPa) (the lower pressure used in the interpolation) plus the proportional change (difference between the intermediate pressure and the lower pressure) for a 1-psi (6.9-kPa) pressure increase. Or, at any higher pressure, $h_{gi} = h_{gl} + [(p_i - p_l)/(p_h - p_l)](h_h - h_l)$, where h_{gi} is enthalpy at the intermediate pressure, h_{gl} is enthalpy at the lower pressure used in the interpolation; h_h is enthalpy at the higher pressure used in the interpolation, p_i is intermediate pressure, and p_h and p_l are higher and lower pressures used in the interpolation. Thus, using the enthalpy values obtained from the steam table for 150 and 152 psia (1034.3 and 1048.0 kPa), $h_{gi} = 1194.1 + [(151 - 150)/(152 - 150)](1194.3 - 1194.1) = 1194.2$ Btu/lb (2777.7 kJ/kg) at 151 psia (1041.1 kPa) saturated.

Next, study the steam table to determine the direction of change of specific volume between the lower and higher pressures. This study shows that the specific volume decreases as the pressure increases. Therefore, the specific volume at 151 psia (1041.1 kPa) (the intermediate pressure) will equal the specific volume at 150 psia (1034.3 kPa) (the lower pressure used in the interpolation) minus the proportional change (difference between the intermediate pressure and the lower interpolating pressure) for a 1-psi (6.9-kPa) pressure increase. Or, at any pressure, $v_{gi} = v_{gl} - [(p_i - p_l)/(p_h - p_l)](v_l - v_h)$, where the subscripts are the same as above and v is specific volume at the respective pressure. With the volume values obtained from steam tables for 150 and 152 psia (1034.3 and 1048.0 kPa), $v_{gi} = 3.015 - [(151 - 150)/(152 - 150)](3.015 - 2.977) = 2.996$ ft^3/lb (0.19 m^3/kg) at 151 psia (1041.1 kPa) saturated.

Study of the steam table for the direction of entropy change shows that entropy, like specific volume, decreases as the pressure increases. Therefore, the entropy at 151 psia (1041.1 kPa) (the intermediate pressure) will equal the entropy at 150 psia (1034.3 kPa) (the lower pressure used in the interpolation) minus the proportional change (difference between the intermediate pressure and the lower interpolating pressure) for a 1-psi (6.9-kPa) pressure increase. Or, at any higher pressure, $s_{gi} = s_{gl} - [(p_i - p_l)/(p_h - p_l)](s_l - s_h) = 1.5164 - [(151 - 150)/(152 - 150)](1.5694 - 1.5683) = 1.56885$ Btu/(lb)(°F) [6.6 kJ/(kg)(°C)] at 151 psia (1041.1 kPa) saturated.

Study of the steam table for the direction of temperature change shows that the saturation temperature, like enthalpy, increases as the pressure increases. Therefore, the temperature at 151 psia (1041.1 kPa) (the intermediate pressure) will equal the temperature at 150 psia (1034.3 kPa) (the lower pressure used in the interpolation) plus the proportional change (difference between the intermediate pressure and the lower interpolating pressure) for a 1-psi (6.9-kPa) increase. Or, at any higher pressure, $t_{gi} = t_{gl} + [(p_i - p_l)/(p_h - p_l)](t_h - t_l) = 358.42 + [(151 - 150)/(152 - 150)](359.46 - 358.42) = 358.94$°F (181.6°C) at 151 psia (1041.1 kPa) saturated.

2. *Use the saturation-temperature steam table.*

Study of the saturation-temperature table shows that there is no temperature value listed for 261°F (127.2°C). Therefore, it will be necessary to interpolate between the next higher and next lower tabulated temperature values. In this instance, these values are 262 and 260°F (127.8 and 126.7°C), respectively. The temperature for which properties are being found (261°F or 127.2°C) is called the "intermediate temperature."

At 262°F (127.8°C), $h_g = 1168.0$ Btu/lb (2716.8 kJ/kg); $v_g = 11.396$ ft^3/lb (0.71 m^3/kg); $s_g = 1.6833$ Btu/(lb)(°F) [7.05 kJ/(kg)(°C)]; and $p_g = 36.646$ psia (252.7 kPa). At 260°F (126.7°C), $h_g = 1167.3$ Btu/lb (2715.1 kJ/kg); $v_g = 11.763$ ft^3/lb (0.73 m^3/kg); $s_g = 1.6860$ Btu/(lb)(°F) [7.06 kJ/(kg)(°C)]; and $p_g = 35.429$ psia (244.3 kPa).

For enthalpy, note that as the temperature increases, so does h_g. Therefore, the enthalpy at 261°F (127.2°C) (the intermediate temperature) will equal the enthalpy at 260°F (126.7°C) (the lower temperature used in the interpolation) plus the proportional change (difference between the intermediate temperature and the lower temperature) for a 1°F (0.6°C) temperature increase. Or, at any higher temperature, $h_{gi} = h_{gl} + [(t_i - t_l)/(t_h - t_l)](h_h - h_l)$, where h_{gl} is enthalpy at the lower temperature used in the interpolation, h_h is enthalpy at the higher temperature used in the interpolation, t_i is intermediate temperature, and t_h and t_l are higher and lower temperatures used in the interpolation. Thus, using the enthalpy values obtained from the steam table for 260 and 262°F (126.7 and 127.8°C), $h_{gi} = 1167.3 + [(261 - 260)/(262 - 260)](1168.0 - 1167.3) = 1167.65$ Btu/lb (2716.0 kJ/kg) at 261°F (127.2°C) saturated.

Next, study the steam table to determine the direction of change of specific volume between the lower and higher temperatures. This study shows that the specific volume decreases as the pressure increases. Therefore, the specific volume at 261°F (127.2°C) (the intermediate temperature) will equal the specific volume at 260°F (126.7°C) (the lower temperature used in the interpolation) minus the proportional change (difference between the intermediate temperature and the lower interpolating temperature) for a 1°F (0.6°C) temperature increase. Or, at any higher temperature, $v_{gi} = v_{gl} - [(t_i - t_l)/(t_h - t_l)](v_l - v_h) = 11.763 - [(261 - 260)/(262 - 260)](11.763 - 11.396) = 11.5795$ ft^3/lb (0.7 m^3/kg) at 261°F (127.2°C) saturated.

Study of the steam table for the direction of entropy change shows that entropy, like specific volume, decreases as the temperature increases. Therefore, the entropy at 261°F (127.2°C) (the intermediate temperature) will equal the entropy at 260°F (126.7°C) (the lower temperature used in the interpolation) minus the proportional change (difference between the intermediate temperature and the lower temperature) for a 1°F (0.6°C) temperature increase. Or, at any higher temperature, $s_{gi} = s_{gl} - [(t_i - t_l)/(t_h - t_l)](s_l - s_h) = 1.6860 - [(261 - 260)/(262 - 260)](1.6860 - 1.6833) = 1.68465$ Btu/(lb) (°F) [7.1 kJ/(kg)(°C)] at 261°F (127.2°C).

Study of the steam table for the direction of pressure change shows that the saturation pressure, like enthalpy, increases as the temperature increases. Therefore, the pressure at 261°F (127.2°C) (the intermediate temperature) will equal the pressure at 260°F (126.7°C) (the lower temperature used in the interpolation) plus the proportional change (difference between the intermediate temperature and the lower interpolating temperature) for a 1°F (0.6°C) temperature increase. Or, at any higher temperature, $p_{gi} = p_{gl} + [(t_i - t_l)/(t_h - t_l)]\ (p_h - p_l) = 35.429 + [(261 - 260)/(262 - 260)](36.646 - 35.429) = 36.0375$ psia (248.5 kPa) at 261°F (127.2°C) saturated.

3. *Use the superheated-steam table.*

Choose the superheated-steam table for steam at 1000°F (537.9°C) and 2.6150 ft^3/lb (0.16 m^3/kg) because the highest temperature at which saturated steam can exist is 705.4°F (374.1°C). This is also the highest temperature tabulated in some saturated-temperature tables. Therefore, the steam is superheated when at a temperature of 1000°F (537.9°C).

Look down the 1000°F (537.9°C) columns in the superheat table until a specific-volume value of 2.6150 is found. This occurs between 325 psia (2240.9 kPa, $v = 2.636$ or 0.16) and 330 psia (2275.4 kPa, $v = 2.596$ or 0.16). Since there is no volume value exactly equal to 2.6150 tabulated, it will be necessary to interpolate. List the values from the steam table thus: $p = 325$ psia (2240.9 kPa); $t = 1000$°F (537.9°C); $v = 2.636$ ft^3/lb (0.16 m^3/kg); and $p = 330$ psia (2275.4 kPa); $t = 1000$°F (537.9°C); $v = 2.596$ ft^3/lb (0.16 m^3/kg).

Note that as the pressure rises, at constant temperature, the volume decreases. Therefore, the intermediate (or unknown) pressure is found by subtracting from the higher interpolating pressure (330 psia or 2275.4 kPa in this instance) the product of the proportional change in the specific volume and the difference in the pressures used for the interpolation, or, $p_{gi} = p_h - [(v_i - v_h)/(v_l - v_h)](p_h - p_l)$, where the subscripts h, l, and i refer to the high, low, and intermediate (or unknown) pressures, respectively. In this instance, $p_{gi} = 330 - [(2.615 - 2.596)/(2.636 - 2.596)](330 - 325) = 327.62$ psia (2258.9 kPa) at 1000°F (537.9°C) and a specific volume of 2.6150 ft^3/lb (0.16 m^3/kg).

4. Use the superheated-steam table.

When given a steam pressure and temperature, determine, before performing any interpolation, the state of the steam. Do this by entering the saturation-pressure table at the given pressure and noting the saturation temperature. If the given temperature exceeds the saturation temperature, the steam is superheated. In this instance, the saturation-pressure table shows that at 300 psia (2068.5 kPa), the saturation temperature is 417.33°F (214.1°C). Since the given temperature of the steam is 567.22°F (297.3°C), the steam is superheated because its actual temperature is greater than the saturation temperature.

Enter the superheated-steam table at 300 psia (2068.5 kPa) and find the next temperature lower than 567.22°F (297.3°C); this is 560°F (293.3°C). Also find the next higher temperature; this is 580°F (304.4°C). Tabulate the enthalpy, specific volume, and entropy for each of these temperatures thus: $t = 560$°F (293.3°C); $h = 1292.5$ Btu/lb (3006.4 kJ/kg); $v = 1.9128$ ft^3/lb (0.12 m^3/kg); $s = 1.6054$ Btu/(lb)(°F) [6.72 kJ/(kg)(°C)]; and $t = 580$°F (304.4°C); $h = 1303.7$ Btu/lb (3032.4 kJ/kg); $v = 1.9594$ ft^3/lb (0.12 m^3/kg); $s = 1.6163$ Btu/(lb)(°F) [6.77 kJ/(kg)(°C)].

Use the same procedures for each property—enthalpy, specific volume, and entropy—as given in step 2, but change the sign between the lower volume and entropy and the proportional factor (temperature in this instance), because for superheated steam, the volume and entropy increase as the steam temperature increases. Thus,

$$h_{gi} = 1292.5 + \frac{567.22 - 560}{580 - 560}(1303.7 - 1292.5) = 1269.6 \text{ Btu/lb (3015.9 kJ/kg)}$$

$$v_{gi} = 1.9128 + \frac{567.22 - 560}{580 - 560}(1.9594 - 1.9128) = 1.9296 \text{ ft}^3\text{/lb (0.12 m}^3\text{/kg)}$$

$$s_{gi} = 1.6054 + \frac{567.22 - 560}{580 - 560}(1.6163 - 1.6054)$$

$$= 1.6093 \text{ Btu/(lb)(°F) [6.7 kJ/(kg)(°C)]}$$

Note: Also observe the direction of change of a property *before* interpolating. Use a *plus* or *minus* sign between the higher interpolating value and the proportional change depending on whether the tabulated value increases (+) or decreases (−).

16-3 Constant-Pressure Steam Process

Three pounds of wet steam containing 15% moisture and initially at a pressure of 400 psia (2758.0 kPa) expands at constant pressure ($P = C$) to 600°F (315.6°C). Determine the initial temperature T_1, enthalpy H_1, internal energy E_1, volume V_1, entropy S_1, final enthalpy H_2, final internal energy E_2, final volume V_2, final entropy S_2, heat added to the steam Q_1, work output W_2, change in internal energy ΔE, change in specific volume ΔV, and change in entropy ΔS.

Calculation Procedure:

1. *Determine the initial steam temperature from the steam tables.*

Enter the saturation-pressure table at 400 psia (2758.0 kPa) and read the saturation temperature as 444.59°F (229.2°C).

2. *Correct the saturation values for the moisture of the steam in the initial state.*

Sketch the process on a pressure-volume (*P-V*), Mollier (*H-S*), or temperature-entropy (*T-S*) diagram, Fig. 16-2. In state 1, y = moisture content = 15%. Using the appropriate values from the saturation-pressure steam table for 400 psia (2758.0 kPa), correct them for a moisture content of 15%:

$$H_1 = h_g - yh_{fg} = 1204.5 - 0.15(780.5) = 1087.4 \text{ Btu/lb } (2529.3 \text{ kJ/kg})$$

$$E_1 = u_g - yu_{fg} = 1118.5 - 0.15(695.9) = 1015.1 \text{ Btu/lb } (2361.1 \text{ kJ/kg})$$

$$V_1 = v_g - yv_{fg} = 1.1613 - 0.15(1.1420) = 0.990 \text{ ft}^3/\text{lb } (0.06 \text{ m}^3/\text{kg})$$

$$S_1 = s_g - ys_{fg} = 1.4844 - 0.15(0.8630) = 1.2945 \text{ Btu/(lb)(°F) } [5.4 \text{ kJ/(kg)(°C)}]$$

3. *Determine the steam properties in the final state.*

Since this is a constant-pressure process, the pressure in state 2 is 400 psia (2758.0 kPa), the same as state 1. The final temperature is given as 600°F (315.6°C). This is greater than the saturation temperature of 444.59°F (229.2°C). Hence, the steam is superheated when in state 2. Use the superheated-steam tables, entering at 400 psia (2758.8 kPa) and 600°F (315.6°C). At this condition, $H_2 = 1306.9$ Btu/lb (3039.8 kJ/kg), and $V_2 = 1.477$ ft³/lb (0.09 m³/kg). Then, $E_2 = h_{2g} - P_2V_2/J = 1306.9 - 400(144)(1.477)/778 = 1197.5$ Btu/lb (2785.4 kJ/kg). In this equation, the constant 144 converts psia to psfa, and J is the mechanical equivalent of heat and equals 778 ft·lb/Btu (1 N·m/J). From the steam tables, $S_2 = 1.5894$ Btu/(lb)(°F) [6.7 kJ/(kg)(°C)].

4. *Compute the process inputs, outputs, and changes.*

$W_2 = (P_1/J)(V_2 - V_1)m = [400(144)/778](1.4770 - 0.9900)(3) = 108.1$ Btu (114.1 kJ). In this equation, m is the weight of steam used in the process, which is 3 lb (1.4 kg). Then,

$$Q_1 = (H_2 - H_1)m = (1306.9 - 1087.4)(3) = 658.5 \text{ Btu } (694.4 \text{ kJ})$$

$$\Delta E = (E_2 - E_1)m = (1197.5 - 1014.1)(3) = 550.2 \text{ Btu } (580.2 \text{ kJ})$$

$$\Delta V = (V_2 - V_1)m = (1.4770 - 0.9900)(3) = 1.461 \text{ ft}^3 (0.041 \text{ m}^3)$$

$$\Delta S = (S_2 - S_1)m = (1.5894 - 1.2945)(3) = 0.8847 \text{ Btu/°F } (1.680 \text{ kJ/°C})$$

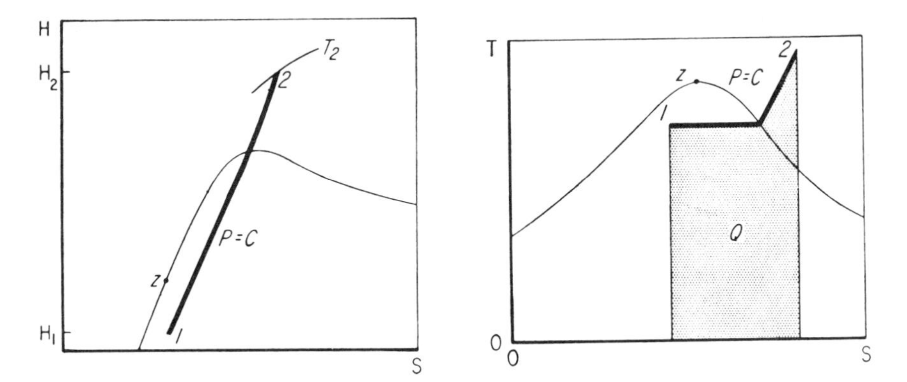

FIG. 16-2 Constant-pressure process (Example 16-3).

5. Check the computations.

The work output W_2 should equal the change in internal energy plus the heat input, or $W_2 = E_1 - E_2 + Q_1 = -550.2 + 658.5 = 108.3$ Btu (114.3 kJ). This value very nearly equals the computed value of $W_2 = 108.1$ Btu (114.1 kJ) and is close enough for all normal engineering computations. The difference can be traced to calculator input errors. In computing the work output, the internal-energy change has a negative sign because there is a decrease in E during the process.

Related Calculations: Use this procedure for all constant-pressure steam processes.

16-4 Constant-Volume Steam Process

Five pounds (2.3 kg) of wet steam initially at 120 psia (827.4 kPa) with 30% moisture is heated at constant volume ($V = C$) to a final temperature of 1000°F (537.8°C). Determine the initial temperature T_1, enthalpy H_1, internal energy E_1, volume V_1, final pressure P_2, final enthalpy H_2, final internal energy E_2, final volume V_2, heat added Q_1, work output W, change in internal energy ΔE, change in volume ΔV, and change in entropy ΔS.

Calculation Procedure:

1. Determine the initial steam temperature from the steam tables.

Enter the saturation-pressure table at 120 psia (827.4 kPa), the initial pressure, and read the saturation temperature: $T_1 = 341.25$°F (171.8°C).

2. Correct the saturation values for the moisture in the steam in the initial state.

Sketch the process on *P-V*, *H-S*, or *T-S* diagrams (Fig. 16-3). Using the appropriate values from the saturation-pressure table for 120 psia (827.4 kPa), correct them for a moisture content of 30%:

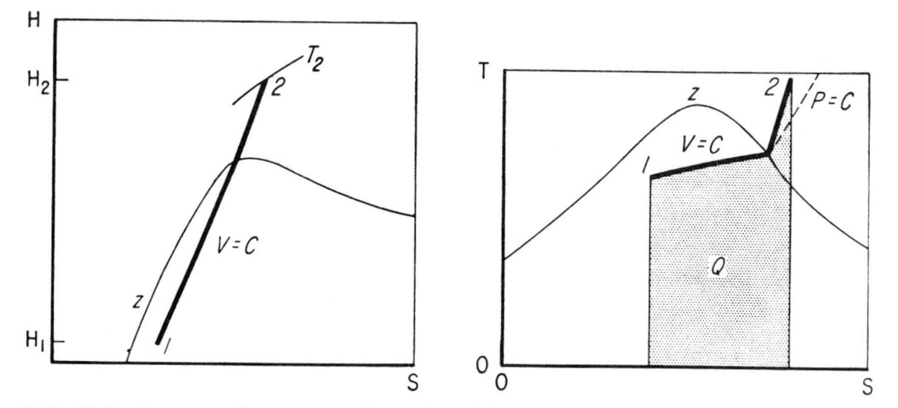

FIG. 16-3 Constant-volume process (Example 16-4).

$$H_1 = h_g - yh_{fg} = 1190.4 - 0.3(877.9) = 927.0 \text{ Btu/lb } (2156.2 \text{ kJ/kg})$$

$$E_1 = u_g - yu_{fg} = 1107.6 - 0.3\,(795.6) = 868.9 \text{ Btu/lb } (2021.1 \text{ kJ/kg})$$

$$V_1 = v_g - yv_{fg} = 3.7280 - 0.3(3.7101) = 2.6150 \text{ ft}^3\text{/lb } (0.16 \text{ m}^3\text{/kg})$$

$$S_1 = s_g - ys_{fg} = 1.5878 - 0.3(1.0962) = 1.2589 \text{ Btu/(lb)(°F) } [5.3 \text{ kJ/(kg)(°C)}]$$

3. Determine the steam volume in the final state.

$T_2 = 1000°\text{F}$ (537.8°C), given. Since this is a constant-volume process, $V_2 = V_1 =$ 2.6150 ft^3/lb (0.16 m^3/kg). The total volume of the vapor equals the product of the specific volume and the number of pounds of vapor used in the process, or total volume equals $2.6150(5) = 13.075$ ft^3 (0.37 m^3).

4. Determine the final steam pressure.

The final steam temperature (1000°F or 537.8°C) and the final steam volume (2.6150 ft^3/lb or 0.16 m^3/kg) are known. To determine the final steam pressure, find in the steam tables the state corresponding to the preceding temperature and specific volume. Since a temperature of 1000°F (537.8°C) is higher than any saturation temperature (705.4°F or 374.1°C is the highest saturation temperature for saturated steam), the steam in state 2 must be superheated. Therefore, the superheated-steam tables must be used to determine P_2.

Enter the 1000°F (537.8°C) column in the steam table and look for a superheated-vapor specific volume of 2.6150 ft^3/lb (0.16 m^3/kg). At a pressure of 325 psia (2240.9 kPa), $v = 2.636$ ft^3/lb (0.16 m^3/kg); $h = 1542.5$ Btu/lb (3587.9 kJ/kg); and $s =$ 1.7863 Btu/(lb)(°F) [7.48 kJ/(kg)(°C)]; and at a pressure of 330 psia (2275.4 kPa), $v = 2.596$ ft^3/lb (0.16 m^3/kg); $h = 1524.4$ Btu/lb (3545.8 kJ/kg); and $s = 1.7845$ Btu/(lb)(°F) [7.47 kJ/(kg)(°C)]. Thus, 2.6150 lies between 325 and 330 psia (2240.9 and 2275.4 kPa). To determine the pressure corresponding to the final volume, it is necessary to interpolate between the specific-volume values, or $P_2 = 330 - [(2.615 - 2.596)/(2.636 - 2.596)](330 - 325) = 327.62$ psia (2258.9 kPa). In this equation, the volume values correspond to the upper (330 psia or 2275.4 kPa), lower (325 psia or 2240.9 kPa), and unknown pressures.

5. Determine the final enthalpy, entropy, and internal energy.

The final enthalpy can be interpolated in the same manner, using the enthalpy at each volume instead of the pressure. Thus $H_2 = 1524.5 - [(2.615 - 2.596)/(2.636 - 2.596)](1524.5 - 1524.4) = 1524.45$ Btu/lb (3545.8 kJ/kg). Since the difference in enthalpy between the two pressures is only 0.1 Btu/lb (0.23 kJ/kg) (= 1524.5 − 1524.4), the enthalpy at 327.62 psia could have been assumed equal to the enthalpy at the lower pressure (325 psia or 2240.9 kPa), or 1524.4 Btu/lb (3545.8 kJ/kg), and the error would have been only 0.05 Btu/lb (0.12 kJ/kg), which is negligible. However, where the enthalpy values vary by more than 1.0 Btu/lb (2.3 kJ/kg), interpolate as shown if accurate results are desired.

Find S_2 by interpolating between pressures, or

$$S_2 = 1.7863 - \frac{327.62 - 325}{330 - 325}(1.7863 - 1.7845)$$

$$= 1.7854 \text{ Btu/(lb)(°F) [7.5 kJ/(kg)(°C)]}$$

$$E_2 = H_2 - P_2V_2/J = 1524.4 - \frac{327.62(144)(2.615)}{778}$$

$$= 1365.9 \text{ Btu/lb (3177.1 kJ/kg)}$$

6. Compute the changes resulting from the process.

$Q_1 = (E_2 - E_1)m = (1365.9 - 868.9)(5) = 2485$ Btu (2621.8 kJ); $\Delta S = (S_2 - S_1)m = (1.7854 - 1.2589)(5) = 2.6325$ Btu/°F (5.0 kJ/°C).

By definition, $W = 0$; $\Delta V = 0$; $\Delta E = Q_1$. Note that the curvatures of the constant-volume line on the T-S chart (Fig. 16-3) are different from the constant-pressure line (Fig. 16-2). Adding heat Q_1 to a constant-volume process affects only the internal energy. The total entropy change must take into account the total steam mass $m = 5$ lb (2.3 kg).

Related Calculations: Use this general procedure for all constant-volume steam processes.

16-5 Constant-Temperature Steam Process

Six pounds (2.7 kg) of wet steam initially at 1200 psia (8274.0 kPa) and 50% moisture expands at constant temperature ($T = C$) to 300 psia (2068.5 kPa). Determine the initial temperature T_1, enthalpy H_1, internal energy E_1, specific volume V_1, entropy S_1, final temperature T_2, final enthalpy H_2, final internal energy E_2, final volume V_2, final entropy S_2, heat added Q_1, work output W_2, change in internal energy ΔE, change in volume ΔV, and change in entropy ΔS.

Calculation Procedure:

1. Determine the initial steam temperature from the steam tables.

Enter the saturation-pressure table at 1200 psia (8274.0 kPa), and read the saturation temperature: $T_1 = 567.22$°F (297.3°C).

2. *Correct the saturation values for the moisture in the steam in the initial state.*

Sketch the process on *P-V*, *H-S*, or *T-S* diagrams (Fig. 16-4). Using the appropriate values from the saturation-pressure table for 1200 psia (8274.0 kPa), correct them for the moisture content of 50%.

$$H_1 = h_g - y_1 h_{fg} = 1183.4 - 0.5(611.7) = 877.5 \text{ Btu/lb (2041.1 kJ/kg)}$$

$$E_1 = u_g - y_1 u_{fg} = 1103.0 - 0.5(536.3) = 834.8 \text{ Btu/lb (1941.7 kJ/kg)}$$

$$V_1 = v_g - y_1 v_{fg} = 0.3619 - 0.5(0.3396) = 0.1921 \text{ ft}^3\text{/lb (0.012 m}^3\text{/kg)}$$

$$S_1 = s_g - y_1 s_{fg} = 1.3667 - 0.5(0.5956) = 1.0689 \text{ Btu/(lb)(°F) [4.5 kJ/(kg)(°C)]}$$

3. *Determine the steam properties in the final state.*

Since this is a constant-temperature process, $T_2 = T_1 = 567.22°\text{F}$ (297.3°C); $P_2 =$ 300 psia (2068.5 kPa), given. The saturation temperature of 300 psia (2068.5 kPa) is 417.33°F (214.1°C). Therefore, the steam is superheated in the final state because 567.22°F (297.3°C) > 417.33°F (214.1°C), the saturation temperature.

To determine the final enthalpy, entropy, and specific volume, it is necessary to interpolate between the known final temperature and the nearest tabulated temperatures greater and less than the final temperature. Or, at $T = 560°\text{F}$ (293.3°C), $v = 1.9128$ ft³/lb (0.12 m³/kg); $h = 1292.5$ Btu/lb (3006.4 kJ/kg); and $s = 1.6054$ Btu/(lb)(°F) [6.72 kJ/(kg)(°C)]. At $T = 580°\text{F}$ (304.4°C), $v = 1.9594$ ft³/lb (0.12 m³/kg); $h = 1303.7$ Btu/lb (3032.4 kJ/kg); and $s = 1.6163$ Btu/(lb)(°F) (6.76 kJ/(kg)(°C)]. Then,

$$H_2 = 1292.5 + \frac{567.22 - 560}{580 - 560}(1303.7 - 1292.5) = 1296.5 \text{ Btu/lb (3015.7 kJ/kg)}$$

$$S_2 = 1.6054 + \frac{567.22 - 560}{580 - 560}(1.6163 - 1.6054)$$

$$= 1.6093 \text{ Btu/(lb)(°F) [6.7 kJ/(kg)(°C)]}$$

$$V_2 = 1.9128 + \frac{567.22 - 560}{580 - 560}(1.9594 - 1.9128) = 1.9296 \text{ ft}^3\text{/lb (0.12 m}^3\text{/kg)}$$

$$E_2 = H_2 - P_2 V_2 / J = 1296.5 - 300(144)(1.9296)/778$$

$$= 1109.3 \text{ Btu/lb (2580.2 kJ/kg)}$$

4. *Compute the process changes.*

$Q_1 = T(S_2 - S_1)m$, where T_1 is absolute initial temperature. Thus, $Q_1 = (567.22 + 460)(1.6093 - 1.0689)(6) = 3330$ Btu (3513.3 kJ), or 555 Btu/lb (1291 kJ/kg). Then,

$$\Delta E = E_2 - E_1 = 1109.3 - 834.8 = 274.5 \text{ Btu/lb (638.5 kJ/kg)}$$

$$\Delta H = H_2 - H_1 = 1296.5 - 877.5 = 419.0 \text{ Btu/lb (974.6 kJ/kg)}$$

$$W_2 = (Q_1 - \Delta E)m = (555 - 274.5)(6) = 1683 \text{ Btu (1776 kJ)}$$

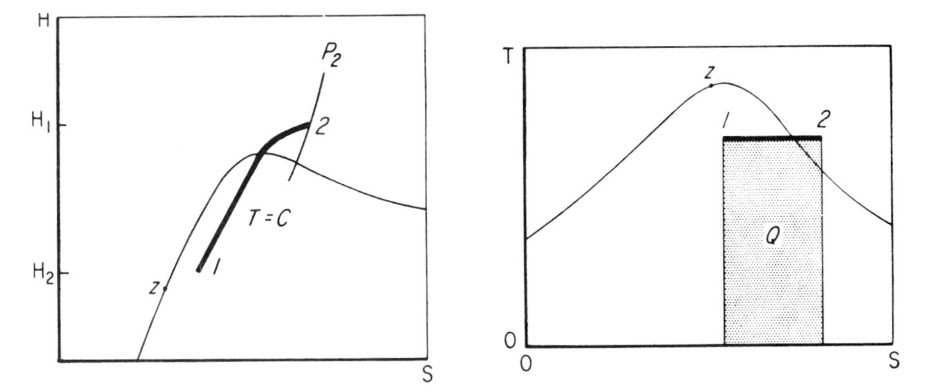

FIG. 16-4 Constant-temperature process (Example 16-5).

$$\Delta S = S_2 - S_1 = 1.6093 - 1.0689 = 0.5404 \text{ Btu/(lb)(°F) [2.3 kJ/(kg)(°C)]}$$

$$\Delta V = V_2 - V_1 = 1.9296 - 0.1921 = 1.7375 \text{ ft}^3\text{/lb (0.11 m}^3\text{/kg)}$$

Related Calculations: Use this procedure for any constant-temperature steam process.

16-6 Constant-Entropy Steam Process

Ten pounds (4.5 kg) of steam expands under two conditions—nonflow and steady flow—at constant entropy ($S = C$) from an initial pressure of 2000 psia (13,790.0 kPa) and a temperature of 800°F (426.7°C) to a final pressure of 2 psia (13.8 kPa). In the steady-flow process, assume that the initial kinetic energy E_{k1} equals the final kinetic energy E_{k2}. Determine the initial enthalpy H_1, initial internal energy E_1, initial volume V_1, initial entropy S_1, final temperature T_2, percent moisture y, final enthalpy H_2, final internal energy E_2, final volume V_2, final entropy S_2, change in internal energy ΔE, change in enthalpy ΔH, change in entropy ΔS, change in volume ΔV, heat added Q_1, and work output W_2.

Calculation Procedure:

1. Determine the initial enthalpy, volume, and entropy from the steam tables.

Enter the superheated-vapor table at 2000 psia (13,790.0 kPa) and 800°F (427.6°C) and read H_1 = 1335.5 Btu/lb (3106.4 kJ/kg); V_1 = 0.3074 ft^3/lb (0.019 m^3/kg); and S_1 = 1.4576 Btu/(lb)(°F) [6.1 kJ/(kg)(°C)].

2. Compute the initial energy.

$$E_1 = H_1 - \frac{P_1 V_1}{J} = 1335.5 - \frac{2000(144)(0.3074)}{778} = 1221.6 \text{ Btu/lb (2841.1 kJ/kg)}$$

3. Determine the vapor properties on the final state.

Sketch the process on *P-V*, *H-S*, or *T-S* diagrams (Fig. 16-5). Note that the expanded steam is wet in the final state because the 2-psia (13.8-kPa) pressure line is under the saturation curve on the *H-S* and *T-S* diagrams. Therefore, the vapor properties in the final state must be corrected for the moisture content. Read, from the saturation-pressure steam table, the liquid and vapor properties at 2 psia (13.8 kPa). Tabulate these properties thus: s_f = 0.1749 Btu/(lb)(°F) [0.73 kJ/(kg)(°C)]; s_{fg} = 1.7451 Btu/(lb)(°F) [7.31 kJ/(kg)(°C)]; s_g = 1.9200 Btu/(lb)(°F) [8.04 kJ/(kg)(°C)]; h_f = 93.99 Btu/lb (218.6 kJ/kg); h_{fg} = 1.022.2 Btu/lb (2377.6 kJ/kg); h_g = 1.116.3 Btu/lb (2596.5 kJ/kg); u_f = 93.98 Btu/lb (218.6 kJ/kg); u_{fg} = 957.9 Btu/lb (2228.1 kJ/kg); u_g = 1051.9 Btu/lb (2446.7 kJ/kg); v_f = 0.016 ft^3/lb (0.00010 m^3/kg); v_{fg} = 173.71 ft^3/lb (10.8 m^3/kg); v_g = 173.73 ft^3/lb (10.8 m^3/kg).

Since this is a constant-entropy process, $S_2 = S_1 = s_g - y_2 s_{fg}$. Solve for y_2, the percent moisture in the final state. Thus, $y_2 = (s_g - S_1)/s_{fg} = (1.9200 - 1.4576)/1.7451 = 0.265$, or 26.5 percent.

Then,

$$H_2 = h_g - y_2 h_{fg} = 1116.2 - 0.265(1022.2) = 845.3 \text{ Btu/lb } (1966.2 \text{ kJ/kg})$$

$$E_2 = u_g - y_2 u_{fg} = 1051.9 - 0.265(957.9) = 798.0 \text{ Btu/lb } (1856.1 \text{ kJ/kg})$$

$$V_2 = v_g - y_2 v_{fg} = 173.73 - 0.265(173.71) = 127.7 \text{ ft}^3/\text{lb } (8.0 \text{ m}^3/\text{kg})$$

4. Compute the changes resulting from the process.

The total change in properties is for 10 lb (4.5 kg) of steam, the quantity used in this process. Thus,

$$\Delta E = (E_1 - E_2)m = (1221.6 - 798.0)(10) = 4236 \text{ Btu } (4469.2 \text{ kJ})$$

$$\Delta H = (H_1 - H_2)m = (1335.5 - 845.3)(10) = 4902 \text{ Btu } (5171.9 \text{ kJ})$$

$$\Delta S = (S_1 - S_2)m = (1.4576 - 1.4576)(10) = 0 \text{ Btu/°F } (0 \text{ kJ/°C})$$

$$\Delta V = (V_1 - V_2)m = (0.3074 - 127.7)(10) = -1274 \text{ ft}^3 \ (-36.1 \text{ m}^3)$$

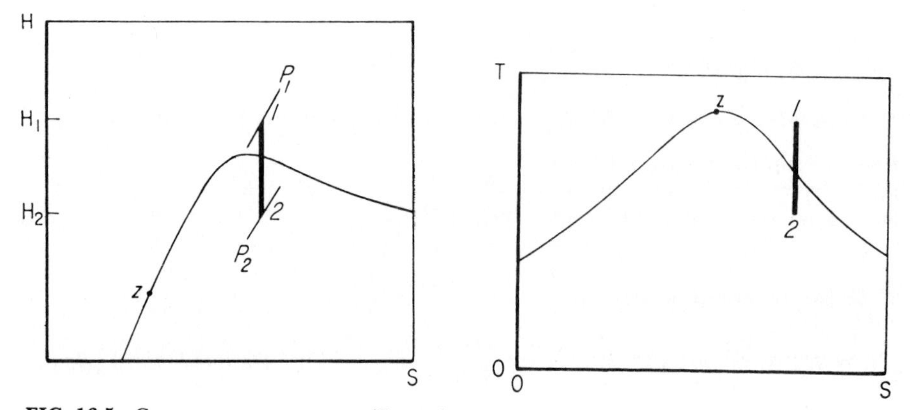

FIG. 16-5 Constant-entropy process (Example 16-6).

Q_1 = 0 Btu. (By definition, there is no transfer of heat in a constant-entropy process.) Nonflow $W_2 = \Delta E$ = 4236 Btu (4469.2 kJ). Steady flow $W_2 = \Delta H$ = 4902 Btu (5171.9 kJ). *Note:* In a constant-entropy process, the nonflow work depends on the change in internal energy. The steady-flow work depends on the change in enthalpy and is larger than the nonflow work by the amount of the change in the flow work.

16-7 Irreversible Adiabatic Expansion of Steam

Ten pounds (4.5 kg) of steam undergoes a steady-flow expansion from an initial pressure of 2000 psia (13,790.0 kPa) and a temperature 800°F (426.7°C) to a final pressure of 2 psia (13.9 kPa) at an expansion efficiency of 75 percent. In this steady flow, assume there is no kinetic-energy change. Determine ΔE, ΔH, ΔS, ΔV, Q, and W_2.

Calculation Procedure:

1. *Determine the initial vapor properties from the steam tables.*

Enter the superheated-vapor tables at 2000 psia (13,790.0 kPa) and 800°F (426.7°C) and read H_1 = 1335.5 Btu/lb (3106.4 kJ/kg); V_1 = 0.3074 ft^3/lb (0.019 m^3/kg);. E_1 = 1221.6 Btu/lb (2840.7 kJ/kg); and S_1 = 1.4576 Btu/(lb)(°F) [6.1 kJ/(kg)(°C)].

2. *Determine the vapor properties in the final state.*

Sketch the process on *P-V*, *H-S*, or *T-S* diagrams (Fig. 16-6). Note that the expanded steam is wet in the final state because the 2-psia (13.9-kPa) pressure line is under the saturation curve on the *H-S* and *T-S* diagrams. Therefore, the vapor properties in the final state must be corrected for the moisture content. However, the actual final enthalpy cannot be determined until after the expansion efficiency $(H_1 - H_2)/(H_1 - H_{2s})$ is evaluated.

To determine the final enthalpy H_2, another enthalpy H_{2s} must first be computed by assuming a constant-entropy expansion to 2 psia (13.8 kPa) and a temperature of 126.08°F (52.3°C). Enthalpy H_{2s} will then be that corresponding to a constant-entropy

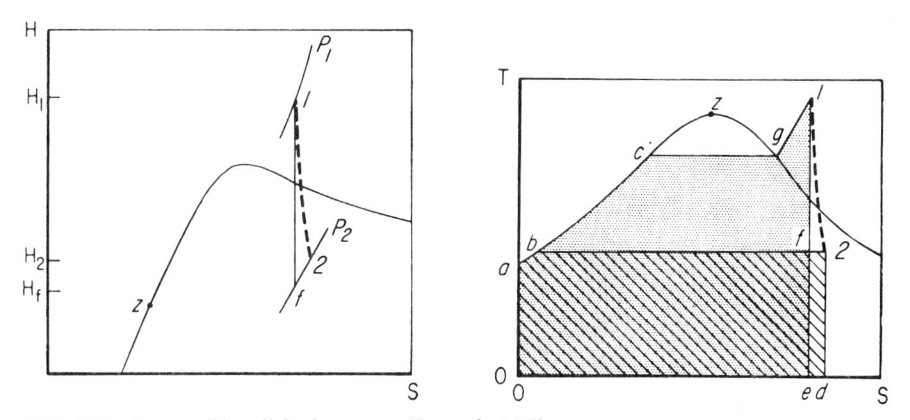

FIG. 16-6 Irreversible adiabatic process (Example 16-7).

expansion into the wet region and the percent moisture will be that corresponding to the final state. This percentage is determined by finding the ratio of $s_g - S_1$ to s_{fg}, or $y_{2s} = (s_g - S_1)/s_{fg} = (1.9200 - 1.4576)/1.7451 = 0.265$, where s_g and s_{fg} are entropies at 2 psia (13.8 kPa). Then, $H_{2s} = h_g - y_{2s}h_{fg} = 1116.2 - 0.265(1022.2) = 845.3$ Btu/lb (1966.2 kJ/kg). In this relation, h_g and h_{fg} are enthalpies at 2 psia (13.8 kPa).

The expansion efficiency, given as 0.75, then is $(H_1 - H_2)/(H_1 - H_{2s})$ = actual work/ideal work = 0.75 = $1335.5 - H_2/(1335.5 - 845.3)$. Solve for $H_2 = 967.9$ Btu/lb (2251.3 kJ/kg).

Next, read from the saturation-pressure steam table the liquid and vapor properties at 2 psia (13.8 kPa). Tabulate these properties thus: $h_f = 93.99$ Btu/lb (218.6 kJ/kg); $h_{fg} = 1022.2$ Btu/lb (2377.6 kJ/kg); $h_g = 1116.2$ Btu/lb (2596.3 kJ/kg); $s_f = 0.1749$ Btu/(lb)(°F) [0.73 kJ/(kg)(°C)]; $s_{fg} = 1.7451$ Btu/(lb)(°F) [7.31 kJ/(kg)(°C)]; $s_g = 1.9200$ Btu/(lb)(°F) [8.04 kJ/(kg)(°C)]; $u_f = 93.98$ Btu/lb (218.60 kJ/kg); $u_{fg} = 957.9$ Btu/lb (2228.1 kJ/kg); $u_g = 1051.9$ Btu/lb (2446.7 kJ/kg); $v_f = 0.016$ ft^3/lb (0.0010 m^3/kg); $v_{fg} = 173.71$ ft^3/lb (10.84 m^3/kg); and $v_g = 173.73$ ft^3/lb (10.85 m^3/kg).

Since the actual final enthalpy H_2 is different from H_{2s}, the final actual moisture y_2 must be computed using H_2. Thus, $y_2 = (h_g - H_2)/h_{fg} = (1116.2 - 967.9)/1022.2 = 0.1451$. Then,

$$E_2 = u_g - y_2u_{fg} = 1051.9 - 0.1451(957.9) = 912.9 \text{ Btu/lb (2123.4 kJ/kg)}$$

$$V_2 = v_g - y_2v_{fg} = 173.73 - 0.1451(173.71) = 148.5 \text{ ft}^3\text{/lb (9.3 m}^3\text{/kg)}$$

$$S_2 = s_g - y_2s_{fg} = 1.9200 - 0.1451(1.7451)$$

$$= 1.6668 \text{ Btu/(lb)(°F) [7.0 kJ/(kg)(°C)]}$$

3. Compute the changes resulting from the process.

The total change in properties is for 10 lb (4.5 kg) of steam, the quantity used in this process. Thus,

$$\Delta E = (E_1 - E_2)m = (1221.6 - 912.9)(10) = 3087 \text{ Btu (3257.0 kJ)}$$

$$\Delta H = (H_1 - H_2)m = (1335.5 - 967.9)(10) = 3676 \text{ Btu (3878.4 kJ)}$$

$$\Delta S = (S_2 - S_1)m = (1.6668 - 1.4576)(10) = 2.092 \text{ Btu/°F (4.0 kJ/°C)}$$

$$\Delta V = (V_2 - V_1)m = (148.5 - 0.3074)(10) = 1482 \text{ ft}^3 \text{ (42.0 m}^3\text{)}$$

$Q = 0$. By definition, $W_2 = \Delta H = 3676$ Btu (3878.4 kJ) for the steady-flow process.

16-8 Irreversible Adiabatic Steam Compression

Two pounds (0.9 kg) of saturated steam at 120 psia (827.4 kPa) with 80 percent quality undergoes nonflow adiabatic compression to a final pressure of 1700 psia (11,721.5 kPa) at 75 percent compression efficiency. Determine the final steam temperature T_2, change in internal energy ΔE, change in entropy ΔS, work input W, and heat input Q.

Calculation Procedure:

1. *Determine the vapor properties in the initial state.*

From the saturation-pressure steam tables, $T_1 = 341.25°F$ (171.8°C) at a pressure of 120 psia (827.4 kPa) saturated. With $x_1 = 0.8$, $E_1 = u_f + x_1 u_{fg} = 312.05 + 0.8(795.6) = 948.5$ Btu/lb (2206.5 kJ/kg), using internal-energy values from the steam tables. The initial entropy S_1 equals $s_f + x_1 s_{fg} = 0.4916 + 0.8(1.0962) = 1.3686$ Btu/(lb)(°F) [5.73 kJ/(kg)(°C)].

2. *Determine the vapor properties in the final state.*

Sketch a *T-S* diagram of the process (Fig. 16-7). Assume a constant-entropy compression from the initial to the final state. Then, $S_{2s} = S_1 = 1.3686$ Btu/(lb)(°F) [5.7 kJ/(kg)(°C)].

The final pressure, 1700 psia (11,721.5 kPa), is known, as is the final entropy, 1.3686 Btu/(lb)(°F) [5.7 kJ/(kg)(°C)], with constant-entropy expansion. The *T-S* diagram (Fig. 16-7) shows that the steam is superheated in the final state. Enter the superheated steam table at 1700 psia (11,721.5 kPa), project across to an entropy of 1.3686, and read the final steam temperature as 650°F (343.3°C). (In most cases, the final entropy would not exactly equal a tabulated value and it would be necessary to interpolate between tabulated entropy values to determine the intermediate pressure value.)

From the same table, at 1700 psia (11,721.5 kPa) and 650°F (343.3°C), $H_{2s} = 1214.4$ Btu/lb (2827.4 kJ/kg), and $V_{2s} = 0.2755$ ft^3/lb (0.017 m^3/lb). Then, $E_{2s} = H_{2s} - P_2V_{2s}/J = 1214.4 - 1700(144)(0.2755)/788 = 1127.8$ Btu/lb (2623.3 kJ/kg). Since E_1 and E_{2s} are known, the ideal work W can be computed. Thus, $W = E_{2s} - E_1 = 1127.8 - 948.5 = 179.3$ Btu/lb (417.1 kJ/kg).

3. *Compute the vapor properties of the actual compression.*

Since the compression efficiency is known, the actual final internal energy can be found from compression efficiency = ideal W/actual $W = (E_{2s} - E_1)/(E_2 - E_1)$, or $0.75 = 1127.8 - 948.5/(E_2 - 948.5)$, and $E_2 = 1187.6$ Btu/lb (2762.4 kJ/kg). Then, $E = (E_2 - E_1)m = (1187.6 - 948.5)(2) = 478.2$ Btu (504.5 kJ) for 2 lb (0.9 kg) of steam. The actual work input W equals $\Delta E = 478.2$ Btu (504.5 kJ). By definition, $Q = 0$.

Finally, the actual final temperature and entropy must be computed. The final actual internal energy $E_2 = 1187.6$ Btu/lb (2762.4 kJ/kg) is known. Also, the *T-S* diagram (Fig. 16-7) shows that the steam is superheated. However, the superheated-steam tables do not list the internal energy of the steam. Therefore, it is necessary to assume a final temperature for the steam and then compute its internal energy. The

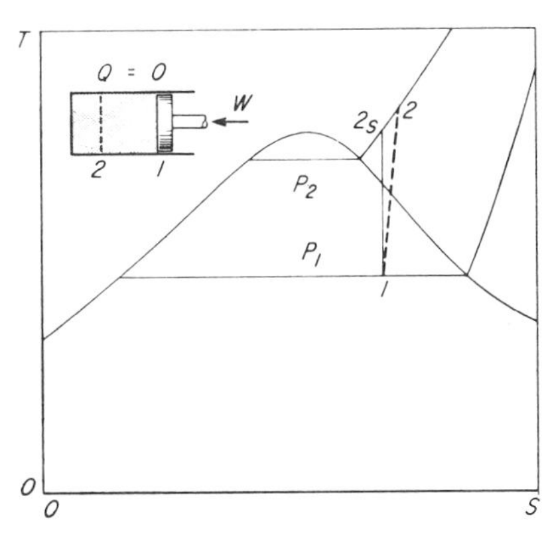

FIG. 16-7 Irreversible adiabatic compression process (Example 16-8).

computed value is compared with the known internal energy and the next assumption is adjusted as necessary. Therefore, assume a final temperature of 720°F (382.2°C). This assumption is higher than the ideal final temperature of 650°F (343.3°C) because the T-S diagram (Fig. 16-7) shows that the actual final temperature is higher than the ideal final temperature. Using values from the superheated-steam table for 1700 psia (11,721.5 kPa) and 720°F (382.2°C),

$$E = H - \frac{PV}{J} = 1288.4 - \frac{1700(144)(0.3283)}{778} = 1185.1 \text{ Btu/lb } (2756.5 \text{ kJ/kg})$$

This value is less than the actual internal energy of 1187.6 Btu/lb (2762.4 kJ/kg). Therefore, the actual temperature must be higher than 720°F (382.2°C), since the internal energy increases with temperature. To obtain a higher value for the internal energy to permit interpolation between the lower, actual, and higher values, assume a higher final temperature—in this case the next temperature listed in the steam table, or 740°F (393.3°C). Then, for 1700 psia (11,721.5 kPa) and 740°F (393.3°C),

$$E = 1305.8 - \frac{1700(144)(0.3410)}{778} = 1198.5 \text{ Btu/lb } (2757.7 \text{ kJ/kg})$$

This value is greater than the actual internal energy of 1187.6 Btu/lb (2762.4 kJ/kg). Therefore, the actual final temperature of the steam lies somewhere between 720 and 740°F (382.2 and 393.3°C). Interpolate between the known internal energies to determine the final steam temperature and final entropy. Thus,

$$T_2 = 720 + \frac{1178.6 - 1185.1}{1198.5 - 1185.1}(740 - 720) = 723.7°\text{F } (384.3°\text{C})$$

$$S_2 = 1.4333 + \frac{1187.6 - 1185.1}{1198.5 - 1185.1}(1.4480 - 1.4333)$$

$$= 1.4360 \text{ Btu/(lb)(°F) } [6.0 \text{ kJ/(kg)(°C)}]$$

$$\Delta S = (S_2 - S_1)m = (1.4360 - 1.3686)(2) = 0.1348 \text{ Btu/°F } (0.26 \text{ kJ/°C})$$

Note that the final actual steam temperature is 73.7°F (40.9°C) higher than that (650°F or 343.3°C) for the ideal compression.

Related Calculations: Use this procedure for any irreversible adiabatic steam process.

16-9 Throttling Processes for Steam and Water

A throttling process begins at 500 psia (3447.5 kPa) and ends at 14.7 psia (101.4 kPa) with (1) steam at 500 psia (3447.5 kPa) and 500°F (260.0°C), (2) steam at 500 psia (3447.5 kPa) and 4% moisture, (3) steam at 500 psia (3447.5 kPa) and 50% moisture, and (4) saturated water at 500 psia (3447.5 kPa). Determine the final enthalpy H_2, temperature T_2, and moisture content y_2 for each process.

Calculation Procedure:

1. *Compute the final-state conditions of the superheated steam.*

From the superheated-steam table for 500 psia (3447.5 kPa) and 500°F (260.0°C), H_1 = 1231.3 Btu/lb (2864.0 kJ/kg). By definition of a throttling process, $H_1 = H_2$ = 1231.3 Btu/lb (2864.0 kJ/kg). Sketch the *T-S* diagram for a throttling process (Fig. 16-8).

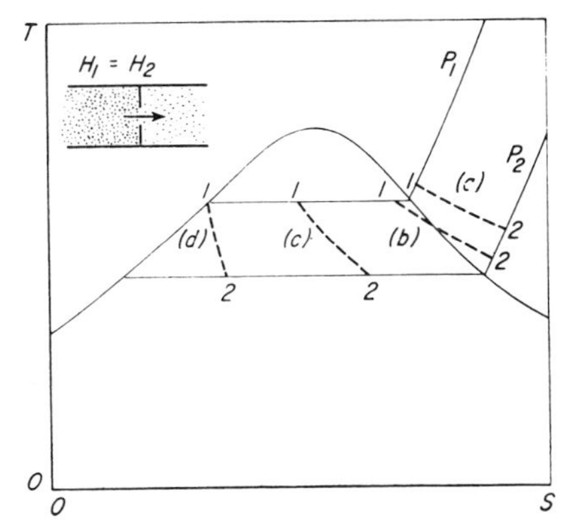

FIG. 16-8 Throttling process for steam (Example 16-9).

To determine the final temperature, enter the superheated-steam table at 14.7 psia (101.4 kPa), the final pressure, and project across to an enthalpy value equal to or less than the known enthalpy, 1231.3 Btu/lb (2864.0 kJ/kg). [The superheated steam table is used because the *T-S* diagram (Fig. 16-8) shows that the steam is superheated in the final state.] At 14.7 psia (101.4 kPa) there is no tabulated enthalpy value that exactly equals 1231.3 Btu/lb (2864.0 kJ/kg). The next lower value is 1230 Btu/lb (2861.0 kJ/kg) at T = 380°F (193.3°C). The next higher value at 14.7 psia (101.4 kPa) is 1239.9 Btu/lb (2884.0 kJ/kg) at T = 400°F (204.4°C). Interpolate between these enthalpy values to find the final steam temperature. Thus,

$$T_2 = 380 + \frac{1231.3 - 1230.5}{1239.9 - 1230.5}(400 - 380) = 381.7\text{°F } (194.3\text{°C})$$

The steam does not contain any moisture in the final state because it is superheated.

2. *Compute the final-state conditions of the slightly wet steam.*

Determine the enthalpy of 500-psia (3447.5-kPa) saturated steam from the saturation-pressure steam table: h_g = 1204.4 Btu/lb (2801.4 kJ/kg), and h_{fg} = 755.0 Btu/lb (1756.1 kJ/kg). Correct the enthalpy for moisture: $H_1 = h_g - y_1 h_{fg}$ = 1204.4 − 0.04(755.0) = 1174.2 Btu/lb (2731.2 kJ/kg). Then, by definition, $H_2 = H_1$ = 1174.2 Btu/lb.

Determine the final condition of the throttled steam (wet, saturated, or superheated) by studying the *T-S* diagram. If a diagram was not drawn, enter the saturation-pressure steam table at 14.7 psia (101.4 kPa), the final pressure, and check the tabulated h_g. If the tabulated h_g is less than H_1, the throttled steam is superheated. If the tabulated h_g is greater than H_1, the throttled steam is saturated. Examination of the saturation-pressure steam table shows that the throttled steam is superheated because $H_1 > h_g$.

Next, enter the superheated-steam table to find an enthalpy value H_1 at 14.7 psia (101.4 kPa). There is no value equal to 1174.2 Btu/lb (2731.2 kJ/kg). The next lower value is 1173.8 Btu/lb (2730.3 kJ/kg) at T = 260°F (126.7°C). The next higher value at 14.7 psia (101.4 kPa) is 1183.3 Btu/lb (2752.4 kJ/kg) at T = 280°F (137.8°C). Interpolate between these enthalpy values to find the final steam temperature. Thus,

$$T_2 = 260 + \frac{1174.2 - 1173.8}{1183.3 - 1173.8}(280 - 260) = 260.8°\text{F } (127.1°\text{C})$$

This is higher than the temperature of saturated steam at 14.7 psia (101.4 kPa)—212°F (100°C)—giving further proof that the throttled steam is superheated. The throttled steam therefore does not contain any moisture.

3. *Compute the final-state conditions of the very wet steam.*

Determine the enthalpy of 500-psia (3447.5-kPa) saturated steam from the saturation-pressure steam table. Thus, $h_g = 1204.4$ Btu/lb (2801.4 kJ/kg), and $h_{fg} = 755.0$ Btu/lb (1756.1 kJ/kg). Correct the enthalpy for moisture: $H_1 = H_2 = h_g - y_1 h_{fg} = 1204.4 - 0.5(755.0) = 826.9$ Btu/lb (1923.4 kJ/kg). Then, by definition, $H_2 = H_1 = 826.9$ Btu/lb (1923.4 kJ/kg).

Compare the final enthalpy, that is, $H_2 = 826.9$ Btu/lb (1923.4 kJ/kg), with the enthalpy of saturated steam at 14.7 psia (101.4 kPa), that is, 1150.4 Btu/lb (2675.8 kJ/kg). Since the final enthalpy is less than the enthalpy of saturated steam at the same pressure, the throttled steam is wet. Since $H_1 = h_g - y_2 h_{fg}$, $y_2 = (h_g - H_1)/h_{fg}$. With a final pressure of 14.7 psia, use h_g and h_{fg} values at this pressure. Thus,

$$y_2 = \frac{1150.4 - 826.9}{970.3} = 0.3335 = 33.35 \text{ percent}$$

The final temperature T_2 of the steam is the same as the saturation temperature at the final pressure of 14.7 psia (101.4 kPa), or $T_2 = 212°\text{F}$ (100°C).

4. *Compute the final-state conditions of saturated water.*

Determine the enthalpy of 500-psia (3447.5-kPa) saturated water from the saturation-pressure steam table at 500 psia (3447.5 kPa); $H_1 = h_f = 449.4$ Btu/lb (1045.3 kJ/kg) $= H_2$, by definition. The *T-S* diagram (Fig. 16-8) shows that the throttled water contains some steam vapor. Or, comparing the final enthalpy of 449.4 Btu/lb (1045.3 kJ/kg) with the enthalpy of saturated liquid, 180.07 Btu/lb (418.8 kJ/kg), at the final pressure, 14.7 psia (101.4 kPa), shows that the liquid contains some vapor in the final state because its enthalpy is greater.

Since $H_1 = H_2 = H_g - y_2 h_{fg}$, $y_2 = (h_g - H_1)/h_{fg}$. Using enthalpies at 14.7 psia (101.4 kPa) of $h_g = 1150.4$ Btu/lb (2675.8 kJ/kg) and $h_{fg} = 970.3$ Btu/lb (2256.9 kJ/kg) from the saturation-pressure steam table, $y_2 = (1150.4 - 449.4)/970.3 = 0.723$.

The final temperature of the steam is the same as the saturation temperature at the final pressure of 14.7 psia (101.4 kPa), or $T_2 = 212°\text{F}$ (100°C).

Note: Calculation 2 shows that when starting with slightly wet steam, it can be throttled (expanded) through a large enough pressure range to produce superheated steam. This procedure is often used in a throttling calorimeter to determine the initial quality of the steam in a pipe. When very wet steam is throttled (calculation 3), the net effect may be to produce drier steam at a lower pressure. Throttling saturated water (calculation 4) can produce partial or complete flashing of the water to steam. All these processes find many applications in power-generation and process-steam plants.

16-10 Use of a Humidity Chart[1]

The temperature and dew point of the air entering a certain dryer are 130 and 60°F (328 and 289 K), respectively. Using a humidity chart (Fig. 16-9), find the following properties of the air: its humidity, its percentage humidity, its adiabatic-saturation temperature, its humidity at adiabatic saturation, its humid heat, and its humid volume.

Calculation Procedure:

1. *Find the humidity.*

In Fig. 16-9, the humidity is the ordinate (along the right side of the graph) of the point on the saturation line (the 100 percent humidity line) that corresponds to the dew point, the latter being read from the abscissa along the bottom. In the present case, the humidity is found to be 0.011 lb water per pound of dry air (0.011 kg water per kilogram of dry air).

2. *Find the percentage humidity.*

Find the dry-bulb temperature, that is, 130°F, along the abscissa, erect a perpendicular to intersect the 0.011-lb humidity line (at point A in Fig. 16-9), and find the percentage-humidity line (interpolating a line if necessary) that passes through that intersection. In this case, the 10 percent line passes through, so the percentage humidity is 10 percent.

3. *Find the adiabatic-saturation temperature.*

Find the adiabatic-cooling line (these are the straight lines having negative slope) that passes through point A, interpolating a line if necessary, and read the abscissa of the point (point B) where this line intersects the 100 percent humidity line. This abscissa is the adiabatic-saturation temperature. In the present case, it is 80°F (300 K).

4. *Find the humidity at adiabatic saturation.*

The humidity at adiabatic saturation is the ordinate, along the right side of the graph, of point B. Its value is 0.022 lb water per pound of dry air (0.022 kg water per kilogram of dry air).

5. *Find the humid heat.*

Find the intersection (point C) of the 0.011-lb-humidity line with the humid-heat-versus-humidity line, and read the humid heat as the abscissa of point C along the top of the graph. This abscissa is 0.245 Btu/(°F)(lb dry air) [or 1024 J/(K)(kg dry air)].

6. *Find the humid volume.*

Erect a perpendicular through the abscissa (along the bottom of the graph) that corresponds to 130°F, the dry bulb temperature. Label the intersection of this perpendicular

[1]Adapted from McCabe and Smith—*Unit Operations of Chemical Engineering,* 3d ed., McGraw-Hill, Inc.

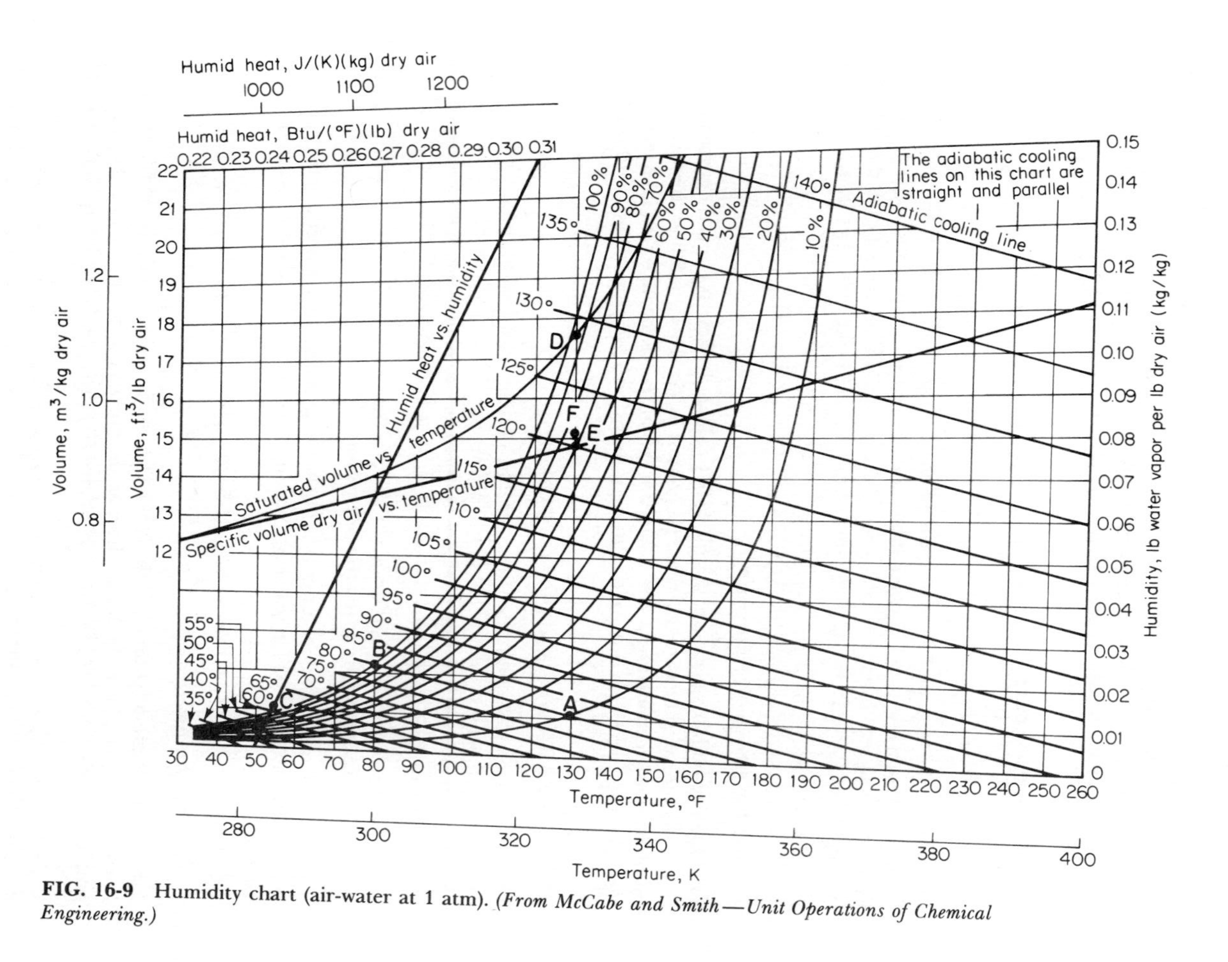

FIG. 16-9 Humidity chart (air-water at 1 atm). *(From McCabe and Smith—Unit Operations of Chemical Engineering.)*

with the saturated-volume-versus-temperature line as point D, and the intersection of the perpendicular with the specific-volume-dry-air-versus-temperature line as point E. Then, along line ED, find point F by moving upward from point E by a distance equal to

$$\overline{ED}\ [(\text{percentage humidity})/100]$$

or, in the present case $(\overline{ED})(10/100)$, where $\overline{ED}$ is the length of line segment $\overline{ED}$. The humid volume is the ordinate of point F as read along the left side of the graph. In this case, the humid volume is 15.1 ft^3/lb dry air (0.943 m^3/kg dry air).

Related Calculations: Do not confuse percentage humidity with relative humidity. "Relative humidity" is the ratio of the partial pressure of the water vapor to the vapor pressure of water at the temperature of the air, this ratio usually being expressed as a percent. "Percentage humidity" is the ratio of the actual humidity to the saturation humidity that corresponds to the gas temperature, which is also usually expressed as a percent. At all humidities other than 0 or 100 percent, the percentage humidity is less than the relative humidity.

16-11 Blowdown and Makeup Requirements for Cooling Towers[1]

A cooling tower handles 1000 gal/min (0.063 m^3/s) of circulating water that is cooled from 110 to 80°F (316 to 300 K). How much blowdown and makeup are required if the concentration of dissolved solids is allowed to reach three times the concentration in the makeup?

Calculation Procedure:

1. Set out material-balance equations for the cooling tower.

When the system is at equilibrium, the makeup must equal the losses, so, by definition,

$$M = E + B + W \tag{16-1}$$

where M is makeup, E is evaporation loss, B is blowdown, and W is windage loss, all being expressed as percent of circulation.

Since the evaporation water will be essentially free of dissolved solids, all solids introduced with the makeup water must be removed by the blowdown plus windage loss, or

$$Mp_m = (B + W)p_c$$

where p_m is concentration of solids in the makeup and p_c is concentration of solids in the circulating water, both in parts per million.

For cooling towers, the concentration in the recirculating water is arbitrarily defined as "cycles of concentration" C, namely, C = (concentration in cooling water)/(concentration in makeup water). Thus,

[1]Adapted from *Chemical Engineering,* June 21, 1976.

$$M = \frac{(B + W)p_c}{p_m} = (B + W)C \qquad (16\text{-}2)$$

2. *Make appropriate assumptions about windage and evaporation losses and set out and solve an equation for blowdown.*

Windage losses will be about 1.0 to 5.0 percent for spray ponds, 0.3 to 1.0 percent for atmospheric cooling towers, and 0.1 to 0.3 percent for forced-draft cooling towers; for modern forced-draft towers (as in this example), 0.1 percent can be assumed. As for evaporation losses, they are 0.85 to 1.25 percent of the circulation for each 10-degree drop in Fahrenheit temperature across the tower; it is usually safe to assume 1.0 percent, so $E = \Delta T/10$, where ΔT is the temperature drop across the tower. Therefore, in the present case,

$$M = \frac{\Delta T}{10} + B + 0.1 \qquad (16\text{-}3)$$

Combining Eqs. (16-2) and (16-3) gives

$$B = \frac{\Delta T}{10(C - 1)}$$

In the present case, then $B = (110 - 80)/[10(3 - 1)] = 1.5$ percent. Thus the blowdown requirement is 1.5 percent of 1000 gal/min, or 15 gal/min (9.45×10^{-4} m^3/s).

3. *Find the makeup requirement.*

From Eq. (16-1), $M = (110 - 80)/10 + 1.5 + 0.1 = 4.6$ percent, or 46 gal/min (2.9×10^{-3} m^3/s).

16-12 Water-Softener Selection and Analysis[1]

Select a water softener that will treat 100 gal/min (22.7 m^3/h) of water at 60°F (289 K) that has the following dissolved components (with concentrations in parts per million as $CaCO_3$): calcium, 300; sodium, 100; magnesium, 100; and total cations, 500. The unit must produce water with no more than 2 ppm hardness and must operate for 8 h between regenerations. Manufacturer's data on the water-softening resin to be used include the following: (1) a regeneration level of 4 lb NaCl per cubic foot (64 kg/m^3) will result in 2 ppm hardness leakage and a capacity of 16 kgr/ft^3 (36 kg/m^3), assuming standard cocurrent operation; (2) pressure drop per linear foot of bed (per 0.305 m of bed) for 60°F water at a linear velocity equivalent to 7.1 gal/(min)(ft^2) [17.3 $m^3/(h)(m^2)$] is 0.6 lb/in^2 (4.14 kPa); (3) a flow rate of 6.4 gal/(min)(ft^2) [15.6 $m^3/(h)(m^2)$] will bring about a bed expansion of 60 percent; and (4) rinse requirements are 25 to 50 gal/ft^3 (3.34 to 6.68 m^3/m^3). Determine the resin volume needed, the pressure drop, the backwash requirement, the regenerant requirement, and the required volume of rinse water.

[1]Courtesy of Rohm & Haas Co.

Calculation Procedure:

1. *Determine the amount of water to be treated per cycle and the amount of hardness to be removed.*

Softening of water requires use of a cation-exchange resin operated in sodium form to exchange divalent hardness cations for sodium regenerated with aqueous sodium chloride solution. Total amount of water to be treated is (100 gal/min)(60 min/h)(8 h/cycle) = 48,000 gal/cycle (182 m^3/cycle).

In determining the quantity of hardness to be removed, neglect the 2 ppm allowable hardness in the effluent (this is a conservative simplification) and assume complete hardness removal. Since the influent hardness is expressed as parts per million (equivalents as $CaCO_3$), it is necessary to convert to units consistent with resin manufacturers' capacity data, usually expressed as kilograins (as $CaCO_3$) per cubic foot of resin. A total of 400 ppm hardness (calcium plus magnesium) is to be removed. Convert this to kilograins as $CaCO_3$. Thus, (400 ppm as $CaCO_3$)(48,000 gal/cycle)/(1000 gr/kgr)(17.1 ppm per grain per gallon) = 1120 kgr (73 kg) as $CaCO_3$ per cycle.

2. *Establish regeneration level and resin capacity.*

An optimal level of regeneration exists for each softening application. This relates level of regeneration (pounds of regenerant per cubic foot of softening resin), leakage (ions not exchanged and thus appearing in the effluent), and operating capacity. In the present case, the desired information is given (based on information from the resin manufacturer) in the statement of the example: The optimal regeneration level is 4 lb NaCl per cubic foot (64 kg NaCl per cubic meter).

3. *Determine volume of softening resin needed.*

The hardness load per cycle is 1120 kgr, from step 1, and the resin capacity is given as 16 kgr/ft^3. So the amount of resin needed is 1120 kgr/(16 kgr/ft^3) = 70 ft^3 (1.98 m^3).

However, if water production must be continuous, two softening units must be obtained, so that one can be regenerated while the other is in service. The alternative is to supply storage facilities for several hours' production of water at 100 gal/min.

4. *Determine column dimensions, pressure drop, and backwash requirements.*

In conventional water softening, an acceptable space velocity is usually between 1 and 5 gal/(min)(ft^3) [8 and 40 m^3/(h)(m^3)]. In the present case, space velocity is (100 gal/min)/70 ft^3 = 1.43 gal/(min)(ft^3), which is within the normal range and thus is acceptable.

Normal *linear* velocity in a softening unit is equivalent to the range 4 to 10 gal/(min)(ft^2) [9.75 to 24.4 m^3/(h)(m^2)]. If the velocity is too high, the pressure drop is excessive; too low a velocity can cause poor distribution of flow through the unit. As for bed depth, it should normally be 3 to 6 ft (0.9 to 1.8 m).

Given these norms, determination of column dimensions is usually done by trial and error. Thus assume a bed depth of 5 ft (1.5 m). Then, cross-sectional area will be 70 ft^3/5 ft = 14 ft^2, and linear velocity will be equivalent to (100 gal/min)/14 ft^2 = 7.1 gal/(min)(ft^2), which is acceptable because it falls in the normal range.

The column diameter is $(\text{area} \times 4/\pi)^{1/2} = (14 \times 4/\pi)^{1/2} = 4.2$ ft (1.28 m). In establishing the column height, allow adequate head space, or freeboard, to permit backwashing. A good allowance is 100 percent of the bed height. Thus the column height is twice the bed height, or 10 ft (3.05 m).

The pressure drop per foot of bed depth is given in the statement of the example as 0.6 lb/in^2. Thus total pressure drop for the resin bed is $[0.6 \text{ lb/(in}^2)(\text{ft})](5 \text{ ft}) = 3.0$ lb/in^2 (21 kPa). This excludes the pressure drop due to the liquid distributors and collectors in the column, as well as that due to auxiliary fittings and valves.

Backwashing is necessary to keep the bed in a hydraulically classified condition, to minimize pressure drop, and to remove resin fines and suspended solids that have been filtered out of the influent water. Normal practice is to backwash at the end of each run for about 15 min, so as to obtain about 50 to 75 percent bed expansion. The flow rate required to achieve this expansion is obtained from the manufacturers' data. As noted in the statement of the example, an appropriate flow rate in this case is 6.4 gal/(min)(ft^2). The total backwash rate is thus $[6.4 \text{ gal/(min)(ft}^2)](14 \text{ ft}^2) = 90$ gal/min. The total water requirement, then, is $(90 \text{ gal/min})(15 \text{ min}) = 1350$ gal (5.11 m^3).

5. Determine regenerant requirement and flow rate.

The sodium chloride regeneration level necessary to hold leakage to 2 ppm is 4 lb NaCl per cubic foot, as noted in step 2. The salt (100% basis) requirement, then, is $(4 \text{ lb/ft}^3)(70 \text{ ft}^3) = 280$ lb NaCl per cycle. Now salt is typically administered as a 10% solution at a rate of 1 gal/(min)(ft^3). The density of such a solution is 8.94 lb/gal. Thus the volumetric requirement is (280 lb NaCl per cycle)/(0.10 lb NaCl per pound solution)(8.94 lb solution per gallon), or about 310 gallons per cycle. This should be fed at a flow rate of $[1 \text{ gal/(min)(ft}^3)](70 \text{ ft}^3) = 70$ gal/min (15.9 m^3/h).

6. Determine required volume of rinse water.

The resin bed must be rinsed with water following regeneration with salt. Rinse-water requirements are obtained from manufacturers' literature; in the present case, they are 25 to 50 gal/ft^3. At 35 gal/ft^3, the total rinse required is $(35 \text{ gal/ft}^3)(70 \text{ ft}^3) = 2450$ gallons per cycle. Normal practice is to rinse at 1 gal/(min)(ft^3) (in this case, 70 gal/min, or 15.9 m^3/h) for the first 10 to 15 min and then at 2 gal/(min)(ft^3) (here, 140 gal/min, or 31.8 m^3/h) for the remainder.

Related Calculations: This example assumes standard cocurrent operation of the column with downflow feed and downflow regeneration. Countercurrent operation is a special case that is best handled by a manufacturer of ion-exchange equipment.

16-13 Complete Deionization of Water[1]

Select a deionization system to treat 250 gal/min (56.8 m^3/h) of water at 60°F (289 K) that has the following dissolved components (concentrations in parts per million as $CaCO_3$ equivalent): calcium, 75; sodium, 50; magnesium, 25; chloride, 30; sulfate, 80;

[1]Courtesy of Rohm & Haas Co.

bicarbonate, 40; and silica, 10 (as SiO_2). Maximum tolerable sodium leakage is 2 ppm; silica leakage is to be under 0.05 ppm. Service-cycle length must be at least 12 h. The system is to use sodium hydroxide (available at 120°F) and sulfuric acid for regeneration.

Manufacturers' data on the resins to be used include the following: For the cation-exchange resin: (1) a regeneration level of 6 lb H_2SO_4 per cubic foot (96 kg H_2SO_4 per cubic meter) will result in a sodium leakage of 2.0 ppm and an operating capacity of 15.6 kgr/ft^3 (35 kg /m^3); (2) pressure drop per foot (per 0.305 m) of bed depth for 60°F water at a linear velocity equivalent to 8.6 gal/(min)(ft^2) [21 m^3/(h)(m^2)] is 0.75 lb/in^2 (5.2 kPa); (3) a flow rate of 6.4 gal/(min)(ft^2) [15.6 m^3/(h)(m^2)] will bring about a bed expansion of 60 percent; and (4) rinse requirements are 25 to 50 gal/ft^3 (3.34 to 6.68 m^3/m^3) using deionized rinse water. For the anion-exchange resin: (1) a regeneration level of 4 lb NaOH per cubic foot (64 kg NaOH per cubic meter) will result in a silica leakage of 0.05 ppm and an operating capacity of 15.3 kgr/ft^3 (35 kg/m^3); (2) pressure drop per foot (per 0.305 m) of bed depth for 60°F water at a linear velocity equivalent to 8.5 gal/(min)(ft^2) [20.8 m^3/(h)(m^2)] is 0.85 lb/in^2 (5.9 kPa); (3) a flow rate of 2.6 gal/(min)(ft^2) [6.34 m^3/(h)(m^2)] will bring about a bed expansion of 60 percent; and (4) rinse requirements are 40 to 90 gal/ft^3 (5.34 to 12.0 m^3/m^3) using deionized water. Determine the resin volumes, the pressure drops, the backwash, regenerant and rinse-water requirements, and overall operating conditions.

Calculation Procedure:

1. *Decide on the ion-exchange system to be used.*

Deionization requires replacement of all cations by the hydrogen ion, accomplished by use of a cation-exchange resin in the hydrogen form, as well as replacement of all anions by the hydroxide ion, accomplished by use of an anion exchanger in the hydroxide form. Since complete removal of all anions, including carbon dioxide and silica, is required, it will be necessary to use a strongly basic anion exchanger, regenerated with sodium hydroxide. The simplest system, a strongly acidic cation exchanger followed by a strongly basic anion exchanger, will be employed here. More elaborate and, in some cases, more efficient systems involving use of degassing equipment, stratified beds of strong- and weak-electrolyte resins, or mixed-bed units are beyond the scope of this handbook.

2. *Specify the cation-exchange column.*

a. *Determine quantity of water to be treated per cycle and quantity of cations to be removed.* The amount of water is (250 gal/min)(60 min/h)(12 h/cycle) = 180,000 gal (681 m^3). To determine the cation load, neglect the 2 ppm of sodium leakage and assume complete removal of all cations. Since the influent cation load is expressed as parts per million (equivalents as $CaCO_3$), it is necessary to convert to units consistent with resin manufacturers' capacity data, usually expressed as kilograins (as $CaCO_3$) per cubic foot of resin. Total cation load in this case is 75 + 50 + 25 ppm. Converting, (150 ppm)(180,000 gal/cycle)/(1000 gr/kgr)(17.1 ppm per grain per gallon) = 1580 kgr (102 kg) as $CaCO_3$ per cycle.

b. *Establish regeneration level and resin capacity.* Using manufacturers' data will determine the least amount of regenerant that will produce water of acceptable quality.

In the present case, the desired information is given in the statement of the example: The optimal regeneration level is 6 lb H_2SO_4 per cubic foot (96 kg H_2SO_4 per cubic meter). *Note:* Care must be taken in regenerating cation-exchange resins with sulfuric acid to avoid precipitation of calcium sulfate.

c. Determine the volume of cation-exchange resin needed. The cation load per cycle is 1580 kgr, from step 2a, and the resin capacity is given as 15.6 kgr/ft^3. So the amount of resin needed is 1580 kgr/(15.6 kgr/ft^3) = 101 ft^3 (2.86 m^3). However, if water production must be continuous, two units are needed, so that one can be regenerated while the other is in service. The alternative is to provide storage for several hours of production of water at 250 gal/min.

d. Determine column dimensions, pressure drop, and backwash requirement. In conventional water treatment, an acceptable space velocity is usually between 1 and 5 gal/(min)(ft^3) [8 and 40 m^3/(h)(m^3)]. In the present case, space velocity is (250 gal/min)/101 ft^3 = 2.5 gal/(min)(ft^3), which is thus acceptable.

Normal *linear* velocity is equivalent to the range 4 to 10 gal/(min)(ft^2) [9.75 to 24.4 m^3/(h)(m^2)]. Given this norm, determination of column dimensions is usually trial-and-error. Thus, assume a bed depth of 3.5 ft (1.07 m). Then the cross-sectional area will be 101 ft^3/3.5 ft = 28.9 ft^2, and linear velocity will be equivalent to (250 gal/min)/28.9 ft^2 = 8.6 gal/(min)(ft^2), which is acceptable.

If either space velocity or linear velocity had been considerably greater than the normal ranges, it would have been necessary to assign more resin.

The column diameter is $(\text{area} \times 4/\pi)^{1/2} = (28.9 \times 4/\pi)^{1/2} = 6.1$ ft (1.86 m). In establishing the column height, allow adequate head space, or freeboard, to permit backwashing. A good allowance is 100 percent of the bed height. Thus the column height is twice the bed height, or 7 ft (2.13 m).

The pressure drop per foot of bed depth is given in the statement of the example as 0.75 lb/in^2. Thus total pressure drop for the cation-resin bed is [0.75 lb/(in^2)(ft)](3.5 ft) = 2.6 lb/in^2 (17.9 kPa). This excludes the pressure drop due to valves, fittings, or liquid distributors or collectors.

Backwashing is necessary to keep the bed in a hydraulically classified condition, to minimize pressure drop and provide for proper flow distribution, as well as to remove resin fines and suspended solids that have filtered out of the water. Normal practice is to backwash at the end of each run to achieve 50 to 75 percent bed expansion. The flow rate required for this expansion is given in the statement of the example as 6.4 gal/(min)(ft^2). The total backwash rate is, thus, [6.4 gal/(min)(ft^2)](28.9 ft^2) = 185 gal/min (42 m^3/h).

e. Determine regenerant requirement and flow rate. The sulfuric acid regeneration level to hold sodium leakage to 2 ppm is 6 lb/ft^3, as noted earlier. The total acid requirement, then, is (6 lb/ft^3)(101 ft^3) = 606 lb (275 kg) per cycle.

A typical technique to avoid precipitation of calcium sulfate is to administer half the regenerant as a 2% solution and then the rest at 4%. Thus, in this case, the first step would require (½ × 606 lb H_2SO)/(8.43 lb solution per gallon)(0.02 lb H_2SO_4 per pound solution), or about 1800 gal (6.8 m^3) of 2% acid solution. The second stage requires (½ × 606)/[8.54(0.04)], or about 890 gal (3.37 m^3) of 4% acid solution. (The 8.43 and 8.54 lb/gal are densities of the acid solutions.) Each stage should be fed at a rate of 1 to 1.5 gal/(min)(ft^3), or in this case about 100 to 150 gal/min (23 to 34 m^3/h).

f. Determine the required volume of rinse water. The column must be rinsed with water after regeneration. Rinse-water requirements, as noted earlier, are 25 to 50 gal/ft^3. In the present case, for 101 ft^3, the requirement is about 2500 to 5000 gal (9.5 to 19 m^3). The first portion should be administered at 1 gal/(min)(ft^3) [8 m^3/(h)(m^3)], and the rest at 1.5 gal/(min)(ft^3) [12 m^3/(h)(m^3)].

3. *Specify the anion-exchange column.*

a. *Determine quantity of water to be treated and quantity of anions to be removed.* The amount of water, from step 2a, is 180,000 gal. Total anion load is $30 + 80 + 40$ ppm. Converting, (150 ppm)(180,000 gal/cycle)/(1000 gr/kgr)(17.1 ppm per grain per gallon) = 1580 kgr (102 kg) as $CaCO_3$ per cycle.

b. *Establish regeneration level and resin capacity.* As given in the statement of the problem, the optimal regeneration level is 4 lb NaOH per cubic foot (4 kg NaOH per cubic meter), associated with an operating capacity of 15.3 kgr/ft^3.

c. *Determine the volume of anion-exchange resin needed.* The anion load per cycle is 1580 kgr, from step 3*a*, and the resin capacity is 15.3 kgr/ft^3. So the amount of resin needed is $1580/15.3 = 103$ ft^3 (2.91 m^3). However, if water production must be continuous, it is necessary to either install a second anion-exchange column in parallel, so that one can be regenerated while the other is in service, or else provide for water storage.

d. *Determine column dimensions, pressure drop, and backwash requirement.* Space velocity is (250 gal/min)/103 ft^3 = 2.4 gal/(min)(ft^3), which falls within the acceptable range (see step 2*d*). As in step 2*d*, assume a bed depth of 3.5 ft (1.07 m). Then the cross-sectional area of the bed will be $103/3.5 = 29.4$ ft^2, and linear velocity will be equivalent to (250 gal/min)/29.4 ft^2 = 8.5 gal/(min)(ft^2), which is also acceptable.

The column diameter is $(\text{area} \times 4/\pi)^{1/2} = (29.4 \times 4/\pi)^{1/2} = 6.12$ ft (1.86 m). Allowing 100 percent head space for backwashing, the column height is twice the bed height, or 7 ft (2.13 m).

The pressure drop per foot of bed depth is given in the statement of the example as 0.85 lb/in^2. Thus total pressure drop for the anion-exchange bed is $(0.85)(3.5) = 3$ lb/in^2 (20.7 kPa). This excludes the pressure drop due to valves, fittings, or liquid distributors or collectors.

As for the backwash requirement, as discussed in step 2*d*, the flow rate required is [2.6 gal/(min)(ft^2)](29.4 ft^2) = 76 gal/min (17.3 m^3/h).

e. *Determine the regenerant requirement and flow rate.* The sodium hydroxide regeneration level is 4 lb NaOH per cubic foot, as noted earlier. Total hydroxide requirement, then, is (4 lb/ft^3)(103 ft^3) = 412 lb (187 kg) NaOH per cycle. Regenerant concentration is typically 4% NaOH solution having a density of 8.68 lb/gal. Total regenerant-solution requirement, then, is (412 lb NaOH per cycle)/(0.04 lb NaOH per pound of solution)(8.68 lb solution per gallon), or 1190 gal per cycle (4.5 m^3 per cycle). This should be applied at about 0.5 gal/(min)(ft^3) [4 m^3/(h)(m^3)].

f. Determine the required volume of rinse water. Rinse-water requirements, as noted earlier, are 40 to 90 gal/ft^3. In the present case, for 103 ft^3, the requirement is about 4000 to 9000 gal (15.1 to 34.1 m^3). The first bed volume (i.e., first 103 $ft^3 \times 7.48$ gal/ft^3, or 750 gal) should be applied at about 50 gal/min (11.4 m^3/h), and the remainder should be applied at about 150 gal/min (34.1 m^3/h).

16-14 Cooling-Pond Size for a Known Heat Load

How many spray nozzles and what surface area is needed to cool 10,000 gal/min (630.8 L/s) of water from 120 to 90°F (48.9 to 32.2°C) in a spray-type cooling pond if the average wet-bulb temperature is 60°F (15.6°C)? What would the approximate dimensions of the cooling pond be? Determine the total pumping head if the static head is 10 ft (29.9 kPa), the pipe friction is 35 ft of water (104.6 kPa), and the nozzle pressure is 8 lb/in^2 (55.2 kPa).

Calculation Procedure:

1. Compute the number of nozzles required.

Assume a water flow of 50 gal/min (3.2 L/s) per nozzle; this is a typical flow rate for usual cooling-pond nozzles. Then, the number of nozzles required equals 10,000 gal/(min)/(50 gal/min per nozzle) = 200 nozzles. If six nozzles are used in each spray group in a series of crossed arms, with each arm containing one or more nozzles, then 200 nozzles divided by 6 nozzles per spray group means that 33⅓ spray groups will be needed. Since a partial spray group is seldom used, 34 spray groups would be chosen.

2. Determine the surface area required.

Usual design practice is to provide 1 ft^2 (0.09 m^2) of pond area per 250 lb (113.4 kg) of water cooled for water quantities exceeding 1000 gal/min (63.1 L/s). Thus, in this pond, the weight of water cooled equals (10,000 gal/min)(8.33 lb/gal)(60 min/h) = 4,998,000, say, 5,000,000 lb/h (630.0 kg/s). Then, the area required, using 1 ft^2 of pond area per 250 lb of water (0.82 m^2 per 1000 kg) cooled, is 5,000,000/250 = 20,000 ft^2 (1858.0 m^2).

As a cross-check, use another commonly accepted area value: 125 Btu/(ft^2)(°F) [2555.2 kJ/(m^2)(°C)], based on the temperature difference between the air wet-bulb temperature and the warm entering-water temperature. This is the equivalent of (120 − 60)(125) = 7500 Btu/ft^2 (85,174 kJ/m^2) in this spray pond, because the air wet-bulb temperature is 60°F (15.6°C) and the warm-water temperature is 120°F (48.9°C). The heat removed from the water is (pounds per hour of water)(temperature decrease, in °F)(specific heat of water) = (5,000,000)(120 − 90)(1.0) = 150,000,000 Btu/h (43,960.7 kW). Then, area required equals (heat removed, in Btu/h)/(heat removal, in Btu/ft^2) = 150,000,000/7,500 = 20,000 ft^2 (1858.0 m^2). This checks the previously obtained area value.

3. Determine the spray-pond dimensions.

Spray groups on the same header or pipe main are usually arranged on about 12-ft (3.7-m) centers with the headers or pipe mains spaced on about 25-ft (7.6-m) centers (Fig. 14-10). Assume that 34 spray groups are used, instead of the required 33⅓, to provide an equal number of groups in two headers and a small extra capacity.

Sketch the spray pond and headers (Fig. 16-10). This shows that the length of each header will be about 204 ft (62.2 m), because there are seventeen 12-ft (3.7-m) spaces between spray groups in each header. Allowing 3 ft (0.9 m) at each end of a header for fittings and cleanouts gives an overall header length of 210 ft (64.0 m). The distance

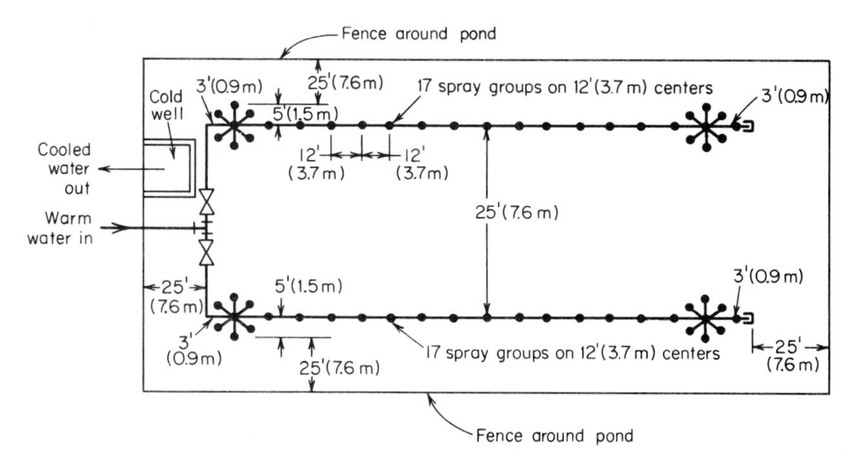

FIG. 16-10 Spray-pond nozzle and piping layout (Example 16-14).

between headers is 25 ft (7.6 m). Allow 25 ft (7.6 m) between the outer sprays and the edge of the pond. This gives an overall width of 85 ft (25.9 m) for the pond, assuming the width of each arm in a spray group is 10 ft (3.0 m). The overall length will then be $210 + 25 + 25 = 260$ ft (79.2 m). A cold well for the pump suction and suitable valving for control of the incoming water must be provided, as shown in Fig. 16-10. The water depth in the pond should be 2 to 3 ft (0.6 to 0.9 m).

4. Compute the total pumping head.

The total head, expressed in feet of water, equals static head + friction head + required nozzle head $= 10 + 35 + 8(0.434) = 48.5$ ft of water (145.0 kPa). A pump having a total head of at least 50 ft of water (15.2 m) would be chosen for this spray pond. If future expansion of the pond is anticipated, compute the probable total head required at a future date and choose a pump to deliver that head. Until the pond is expanded, the pump would operate with a throttled discharge. Normal nozzle inlet pressures range from about 6 to 10 lb/in^2 (41.4 to 69.0 kPa). Higher pressures should not be used, because there will be excessive spray loss and rapid wear of the nozzles.

Related Calculations: Unsprayed cooling ponds cool 4 to 6 lb (1.8 to 2.7 kg) of water from 100 to 70°F per square foot (598.0 to 418.6° C/m^2) of water surface. An alternative design rule is to assume that the pond will dissipate 3.5 Btu/(h)(ft^2) [19.9 J/(m^2)(°C)(s)] of water surface per degree difference between the wet-bulb temperature of the air and the entering warm water.

16-15 Process Temperature-Control Analysis

A water storage tank (Fig. 16-11) contains 500 lb (226.8 kg) of water at 150°F (65.6°C) when full. Water is supplied to the tank at 50°F (10.0°C) and is withdrawn at the rate of 25 lb/min (0.19 kg/s). Determine the process time constant and the zero-frequency process gain if the thermal sensing pipe contains 15 lb (6.8 kg) of water between the tank

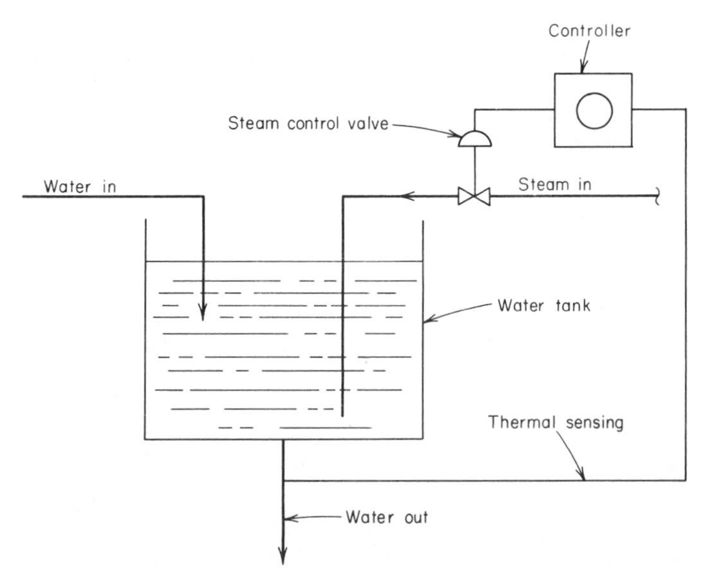

FIG. 16-11 Temperature control of a simple process (Example 16-15).

and thermal bulb and the maximum steam flow to the tank is 8 lb/min (0.060 kg/s). The steam flow to the tank is controlled by a standard linear regulating valve whose flow range is 0 to 10 lb/min (0 to 0.076 kg/s) when the valve operator pressure changes from 5 to 30 lb/in^2 (34.5 to 206.9 kPa).

Calculation Procedure:

1. Compute the distance-velocity lag.

The time in minutes needed for the thermal element to detect a change in temperature in the storage tank is the "distance-velocity lag," which is also called the "transportation lag" or "dead time." For this process, the distance-velocity lag d is the ratio of the quantity of water in the pipe between the tank and the thermal bulb, that is, 15 lb (6.8 kg), and the rate of flow of water out of the tank, that is, 25 lb/min (0.114 kg/s), or $d = 15/25 = 0.667$ min.

2. Compute the energy input to the tank.

This is a "transient control process"—i.e., the conditions in the process are undergoing constant change instead of remaining fixed, as in steady-state conditions. For transient process conditions, the heat balance is $H_{in} = H_{out} + H_{stor}$, where H_{in} is heat input, in Btu/min; H_{out} is heat output, in Btu/min; H_{stor} is heat stored, in Btu/min.

The heat input to this process is the enthalpy of vaporization h_{fg}, in Btu/(lb)(min), of the steam supplied to the process. Since the regulating valve is linear, its sensitivity s is the (flow-rate change, in lb/min)/(pressure change, in lb/in^2). Or, using the known valve characteristics, $s = (10 - 0)/(30 - 5) = 0.4$ lb/(min)(psi) [0.00044 kg/(kPa)(s)].

With a change in steam pressure of p lb/in^2 (p' kPa) in the valve operator, the change in the rate of energy supply to the process is $H_{in} = 0.4$ lb/(min)(psi) $\times p \times h_{fg}$. Taking h_{fg} as 938 Btu/lb (2181 kJ/kg), $H_{in} = 375p$ Btu/min ($6.6p'$ kW).

3. Compute the energy output from the system.

The energy output H_{out} equals pounds per minute of liquid outflow times liquid specific heat, Btu/(lb)(°F), $\times$ (T_a − 150°F), where T_a is tank temperature, in °F, at any time. When the system is in a state of equilibrium, the temperature of the liquid in the tank is the same as that leaving the tank or, in this instance, 150°F (65.6°C). But when steam is supplied to the tank under equilibrium conditions, the liquid temperature will rise to 150 + T_r, where T_r is temperature rise, in °F (T_r' in °C), produced by introducing steam into the water. Thus, the preceding equation becomes $H_{out} = 25$ lb/min $\times$ 1.0 Btu/(lb)(°F) $\times T_r = 25T_r$ Btu/min ($0.44T_r'$ kW).

4. Compute the energy stored in the system.

With rapid mixing of the steam and water, H_{stor} = liquid storage, in pounds, times liquid specific heat, in Btu(lb)(°F), times $T_rq = 500 \times 1.0 \times T_rq$, where q is derivative of the tank outlet temperature with respect to time.

5. Determine the time constant and process gain.

Write the process heat balance, substituting the computed values in $H_{in} = H_{out} + H_{stor}$, or $375p = 25T_r + 500T_rq$. Solving, $T_r/p = 375/(25 + 500q) = 15/(1 + 20q)$.

The denominator of this linear first-order differential equation gives the process system time constant of 20 min in the expression $1 + 20q$. Likewise, the numerator gives the zero-frequency process gain of 15°F/(lb)(in^2).

Related Calculations: This general procedure is valid for any liquid using any gaseous heating medium for temperature control with a single linear lag. Likewise, this general procedure is also valid for temperature control with a double linear lag and pressure control with a single linear lag.

16-16 Control-Valve Selection for Process Control

Select a steam control valve for a heat exchanger requiring a flow of 1500 lb/h (0.19 kg/s) of saturated steam at 80 psig (551.6 kPag) at full load and 300 lb/h (0.038 kg/s) at 40 psig (275.8 kPag) at minimum load. Steam at 100 psig (689.5 kPag) is available for heating.

Calculation Procedure:

1. Compute the valve flow coefficient.

The valve flow coefficient C_v is a function of the maximum steam flow rate through the valve and the pressure drop that occurs at this flow rate. When choosing a control valve for a process control system, the usual procedure is to assume a maximum flow rate for the valve based on a considered judgment of the overload the system may carry. Usual overloads do not exceed 25 percent of the maximum rated capacity of the system. Using this overload range as a guide, assume that the valve must handle a 20 percent overload, or 0.20(1500) = 300 lb/h (0.038 kg/s). Hence, the rated capacity of this valve should be 1500 + 300 = 1800 lb/h (0.23 kg/s).

The pressure drop across a steam control valve is a function of the valve design, size, and flow rate. The most accurate pressure-drop estimate that is usually available is that given in the valve manufacturer's engineering data for a specific valve size, type, and steam flow rate. Without such data, assume a pressure drop of 5 to 15 percent across the valve as a first approximation. This means that the pressure loss across this valve, assuming a 10 percent drop at the maximum steam flow rate, would be $0.10 \times 80 = 8$ psig (55.2 kPag).

With these data available, compute the valve flow coefficient from $C_v = WK/3(\Delta p P_2)^{0.5}$, where W is steam flow rate, in lb/h, K equals 1 + (0.0007 × °F superheat of the steam), p is pressure drop across the valve at the maximum steam flow rate, in lb/in^2, and P_2 is control-valve outlet pressure at maximum steam flow rate, in psia. Since the steam is saturated, it is not superheated, and $K = 1$. Then, $C_v = 1500/3(8 \times 94.7)^{0.5} = 18.1$

2. *Compute the low-load steam flow rate.*

Use the relation $W = 3(C_v \Delta p P_2)^{0.5}/K$, where all the symbols are as before. Thus, with a 40-psig (275.5-kPag) low-load heater inlet pressure, the valve pressure drop is $80 - 40 = 40$ psig (275.8 kPag). The flow rate through the valve is then $W = 3(18.1 \times 40 \times 54.7)^{0.5}/1 = 598$ lb/h (0.75 kg/s).

Since the heater requires 300 lb/h (0.038 kg/s) of steam at the minimum load, the valve is suitable. Had the flow rate of the valve been insufficient for the minimum flow rate, a different pressure drop, i.e., a larger valve, would have to be assumed and the calculation repeated until a flow rate of at least 300 lb/h (0.038 kg/s) was obtained.

Related Calculations: The flow coefficient C_v of the usual 1-in-diameter (2.5-cm) double-seated control valve is 10. For any other size valve, the approximate C_v valve can be found from the product $10 \times d^2$, where d is nominal body diameter of the control valve, in inches. Thus, for a 2-in-diameter (5.1-cm) valve, $C_v = 10 \times 2^2 = 40$. Using this relation and solving for d, the nominal diameter of the valve analyzed in steps 1 and 2 is $d = (C_v/10)^{0.5} = (18.1/10)^{0.5} = 1.35$ in (3.4 cm); use a 1.5-in (3.8-cm) valve because the next smaller standard control valve size, 1.25 in (3.2 cm), is too small. Standard double-seated control-valve sizes are: ¾, 1, 1¼, 1½, 2, 2½, 3, 4, 6, 8, 10, and 12 in. Figure 16-12 shows typical flow-lift characteristics of popular types of control valves.

To size control valves for liquids, use a similar procedure and the relation $C_v = V(G/\Delta p)$, where V is flow rate through the valve, in gal/min, Δp is pressure drop across the valve at maximum flow rate, in lb/in^2, and G is specific gravity of the liquid. When a liquid has a specific gravity of 100 SSU or less, the effect of viscosity on the control action is negligible.

To size control valves for gases, use the relation $C_v = Q(GT_a)^{0.5}/[1360(\Delta p P_2)^{0.5}]$, where Q is gas flow rate, in ft^3/h at 14.7 psia (101.4 kPa) and 60°F (15.6°C), T_a is temperature of the flowing gas, in °F abs = 460 + F; other symbols as before. When the valve outlet pressure P_2 is less than $0.5P_1$, where P_1 is valve inlet pressure, in psia, use the value of $P_1/2$ in place of $(\Delta p P_2)^{0.5}$ in the denominator of the relation.

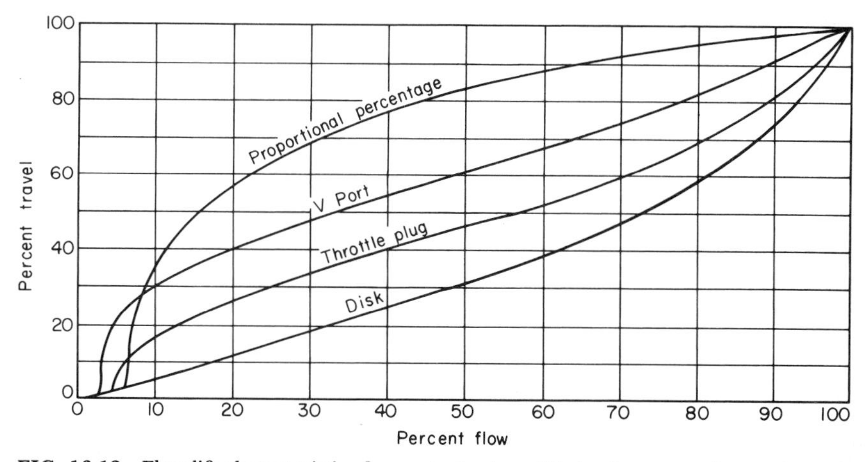

FIG. 16-12 Flow-lift characteristics for control valves. *(Taylor Instrument Process Control Division of Sybron Corporation.)*

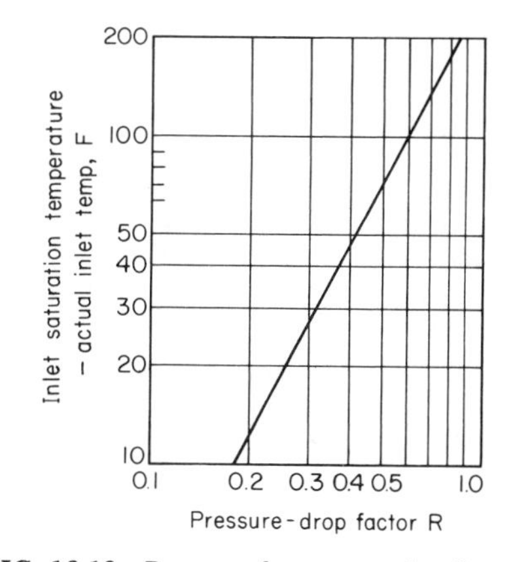

FIG. 16-13 Pressure-drop correction factor for water in the liquid state. *(International Engineering Associates.)*

To size control valves for vapors other than steam, use the relation $C_v = W(v_2/\Delta p)^{0.5}/63.4$, where W is vapor flow rate, in lb/h, v_2 is specific volume of the vapor at the outlet pressure P_2, in ft^3/lb; other symbols as before. When P_2 is less than $0.5P_1$, use the value of $P_1/2$ in place of Δp and use the corresponding value of v_2 at $P_1/2$.

When the control valve handles a flashing mixture of water and steam, compute C_v using the relation for liquids given earlier after determining which pressure drop to use in the equation. Use the *actual* pressure drop or the *allowable* pressure drop, whichever is smaller. Find the allowable pressure drop by taking the product of the supply pressure, in psia, and the correction factor R, where R is obtained from Fig. 16-13. For a further discussion of control-valve sizing, see Considine—*Process Instruments and Controls Handbook,* McGraw-Hill, and G. F. Brockett and C. F. King—"Sizing Control Valves Handling Flashing Liquids," Texas A & M Symposium.

16-17 Control-Valve Characteristics and Rangeability

A flow control valve will be installed in a process system in which the flow may vary from 100 to 20 percent while the pressure drop in the system rises from 5 to 80 percent. What is the required rangeability of the control valve? What type of control-valve characteristic should be used? Show how the effective characteristic is related to the pressure drop the valve should handle.

Calculation Procedure:

1. Compute the required valve rangeability.

Use the relation $R = (Q_1/Q_2)(\Delta P_2/\Delta P_1)^{0.5}$, where R is valve rangeability, Q_1 is valve initial flow, in percent of total flow, Q_2 is valve final flow, in percent of total flow, P_1 is initial pressure drop across the valve, in percent of total pressure drop, and P_2 is percent final pressure drop across the valve. Substituting, $R = (100/20)(80/5)^{0.5} = 20$.

2. Select the type of valve characteristic to use.

Table 16-1 lists the typical characteristics of various control valves. Study of Table 16-1 shows that an equal-percentage valve must be used if a rangeability of 20 is required. Such a valve has equal stem movements for equal-percentage changes in flow at a constant pressure drop based on the flow occurring just before the change is made.[1] The equal-percentage valve finds use where large rangeability is desired and where equal-percentage characteristics are necessary to match the process characteristics.

3. Show how the valve effective characteristic is related to pressure drop.

Figure 16-14 shows the inherent and effective characteristics of typical linear, equal-percentage, and on-off control valves. The inherent characteristic is the theoretical performance of the valve.[1] If a valve is to operate at a constant load without changes in the

[1] E. Ross Forman—"Fundamentals of Process Control," *Chemical Engineering,* June 21, 1965.

TABLE 16-1 Control-Valve Characteristics

Valve type	Typical flow rangeability	Stem movement
Linear	12–1	Equal stem movement for equal flow change
Equal percentage	30–1 to 50–1	Equal stem movement for equal-percentage flow change*
On-off	Linear for first 25 percent of travel; on-off thereafter	Same as linear up to on-off range

*At constant pressure drop.

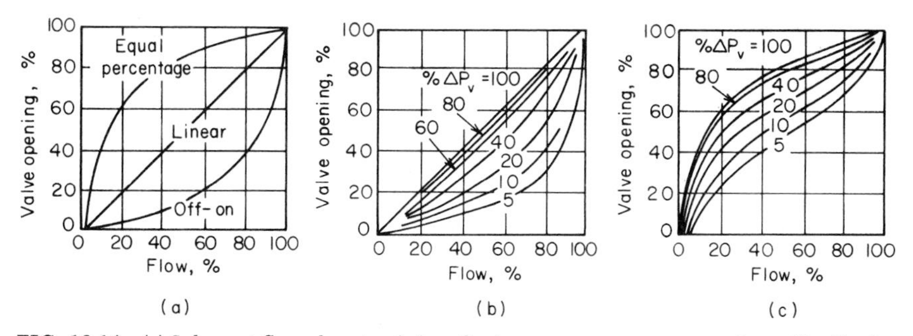

FIG. 16-14 (*a*) Inherent flow characteristics of valves at constant pressure drop; (*b*) effective characteristics of a linear valve; (*c*) effective characteristics of a 50:1 equal-percentage valve.

flow rate, the characteristic of the valve is not important, since only one operating point of the valve is used.

Figure 16-14*b* and *c* give definite criteria for the amount of pressure drop the control valve should handle in the system. This pressure drop is not an arbitrary value, such as 5 lb/in^2, but rather a percent of the total dynamic drop. The control valve should take at least 33 percent of the total dynamic system pressure drop if an equal-percentage valve is used and is to retain its inherent characteristics. A linear valve should not take less than a 50 percent pressure drop if its linear properties are desired.

There is an economic compromise in the selection of every control valve. Where possible, the valve pressure drop should be as high as needed to give good control. If experience or an economic study dictates that the requirement of additional horsepower to provide the needed pressure is not worth the investment in additional pumping or compressor capacity, the valve should take less pressure drop with the resulting poorer control.

16-18 Cavitation, Subcritical, and Critical-Flow Considerations in Controller Selection

Using the sizing formulas of the Fluid Controls Institute, Inc., size control valves for the cavitation, subcritical, and critical flow situations described below. Show how accurate the FCI formulas are.

Cavitation: Select a control valve for a situation where cavitation may occur. The fluid is steam condensate; inlet pressure P_1 is 167 psia (1151.5 kPa); ΔP is 105 lb/in^2 (724.0 kPa); inlet temperature T_1 is 180°F (82.2°C); vapor pressure P_v is 7.5 psia (51.7 kPa).

Subcritical gas flow: Determine the valve capacity required at these conditions; fluid is air; flow Q_g is 160,000 std ft^3/h (1.3 std m^3/s); inlet pressure P_1 is 275 psia (1896 kPa); ΔP is 90 lb/in^2 (620.4 kPa); gas temperature T_1 is 60°F (15.6°C).

Critical vapor flow: A heavy-duty angle valve is suggested for a steam pressure-reducing application. Determine the capacity required and compare an alternate valve type. The fluid is saturated steam; flow W is 78,000 lb/h (9.8 kg/s); inlet pressure P_1 is 1260 psia (8688 kPa); and outlet pressure P_2 is 300 psia (2068.5 kPa).

TABLE 16-2 Critical Flow Factors for Control Valves at 100 percent Lift

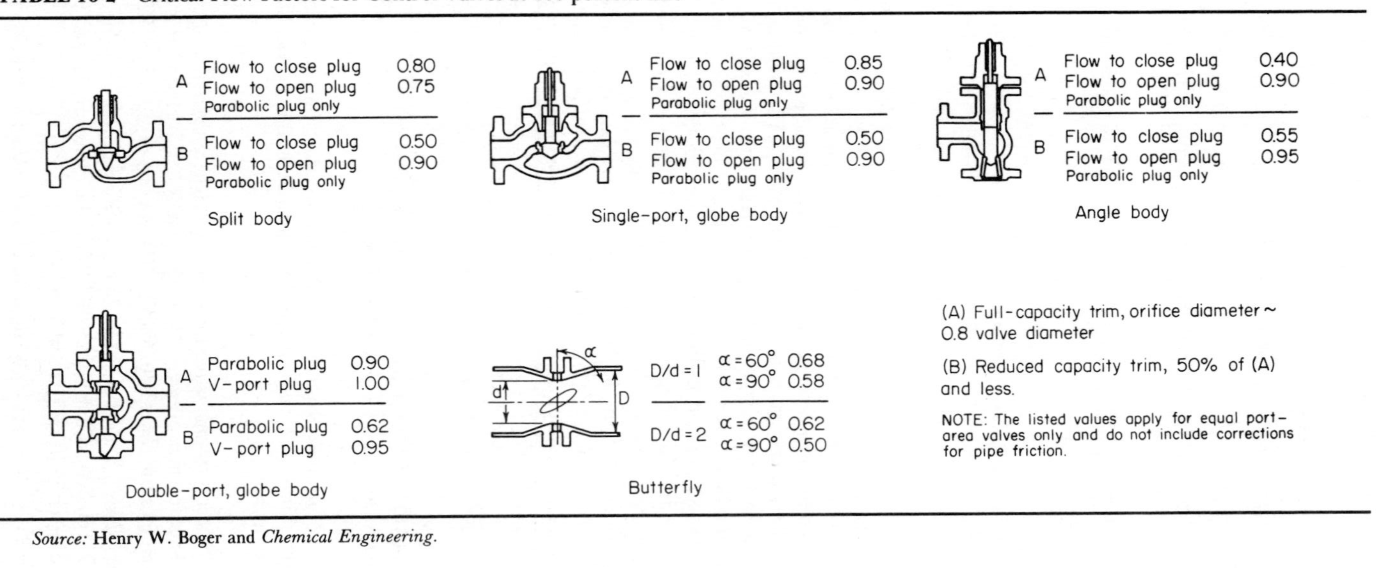

Split body

Trim	Flow direction	Factor
A	Flow to close plug	0.80
A	Flow to open plug	0.75
	Parabolic plug only	
B	Flow to close plug	0.50
B	Flow to open plug	0.90
	Parabolic plug only	

Single-port, globe body

Trim	Flow direction	Factor
A	Flow to close plug	0.85
A	Flow to open plug	0.90
	Parabolic plug only	
B	Flow to close plug	0.50
B	Flow to open plug	0.90
	Parabolic plug only	

Angle body

Trim	Flow direction	Factor
A	Flow to close plug	0.40
A	Flow to open plug	0.90
	Parabolic plug only	
B	Flow to close plug	0.55
B	Flow to open plug	0.95
	Parabolic plug only	

Double-port, globe body

Trim	Plug	Factor
A	Parabolic plug	0.90
A	V-port plug	1.00
B	Parabolic plug	0.62
B	V-port plug	0.95

Butterfly

D/d	α	Factor
D/d = 1	$\alpha = 60^\circ$	0.68
D/d = 1	$\alpha = 90^\circ$	0.58
D/d = 2	$\alpha = 60^\circ$	0.62
D/d = 2	$\alpha = 90^\circ$	0.50

(A) Full-capacity trim, orifice diameter ~ 0.8 valve diameter

(B) Reduced capacity trim, 50% of (A) and less.

NOTE: The listed values apply for equal port-area valves only and do not include corrections for pipe friction.

Source: Henry W. Boger and *Chemical Engineering.*

Calculation Procedure:

1. Choose the valve type and determine its critical-flow factor for the cavitation situation.

If otherwise suitable (i.e., with respect to size, materials, and space considerations), a butterfly control valve is acceptable on a steam-condensate application. Find, from Table 16-2, the value of the critical flow factor $C_f = 0.68$ for a butterfly valve with 60° operation.

2. Compute the maximum allowable pressure differential for the valve.

Use the relation $\Delta P_m = C_f^2(P_1 - P_v)$, where ΔP_m is maximum allowable pressure differential, in lb/in^2, P_1 is inlet pressure, in psia, and P_v is vapor pressure, in psia. Substituting, $\Delta P_m = (0.68)^2(167 - 7.5) = 74$ lb/in^2 (510.2 kPa). Since the actual pressure drop, 105 lb/in^2 (724.0 kPa), exceeds the allowable drop, 74 lb/in^2 (510.2 kPa), cavitation *will* occur.

3. Select another valve and repeat the cavitation calculation.

For a single-port top-guided valve with flow to open plug, find $C_f = 0.90$ from Table 16-2. Then, $\Delta P_m = (0.90)^2(167 - 7.5) = 129$ lb/in^2 (889.5 kPa).

In the case of the single-port top-guided valve, the allowable pressure drop, 129 lb/in^2 (889.5 kPa), exceeds the actual pressure drop, 105 lb/in^2 (724.0 kPa), by a comfortable margin. This valve is a better selection because cavitation will be avoided. A double-port valve might also be used, but the single-port valve offers lower seat leakage. However, the double-port valve offers the possibility of a more economical actuator, especially in larger valve sizes. This concludes the steps for choosing the valve where cavitation conditions apply.

4. Apply the FCI formula for subcritical flow.

The FCI formula for subcritical gas flow is $C_v = Q_g/[1360(\Delta P/GT)^{0.5}][(P_1 + P_2)/2]^{0.5}$, where C_v is valve flow coefficient, Q_g is gas flow, in std ft^3/h, ΔP is pressure differential, in lb/in^2, G is specific gravity of gas at 14.7 psia (101.4 kPa) and 60°F (15.6°C), and T is absolute temperature of the gas, in R; other symbols as given earlier. Substituting, $C_v = 160{,}000/[1360(90/520)^{0.5}][(275 + 185)/2]^{0.5} = 18.6$.

5. Compute C_v using the unified gas-sizing formula.

For greater accuracy, many engineers use the unified gas-sizing formula. Assuming a single-port top-guided valve installed open to flow, Table 16-2 shows $C_f = 0.90$. Then, $Y = (1.63/C_f)(\Delta P/P_1)^{0.5}$, where Y is defined by the equation and the other symbols are as given earlier. Substituting, $Y = (1.63/0.90)(90/275)^{0.5} = 1.04$. Figure 16-15 shows the flow correlation established from actual test data for many valve configurations at a maximum valve opening and relates Y and the fraction of the critical flow rate.

Find from Fig. 16-16 the value of $Y - 0.148Y^3 = 0.87$. Compute $C_v = Q_g(GT)^{0.5}/[834C_f(Y - 0.148Y^3)]$, where all the symbols are as given earlier. Or, $C_v = 160{,}000(520)^{0.5}/[834(0.90)(275)(0.87)] = 20.4$. This value represents an error of approximately 10 percent in the use of the FCI formula.

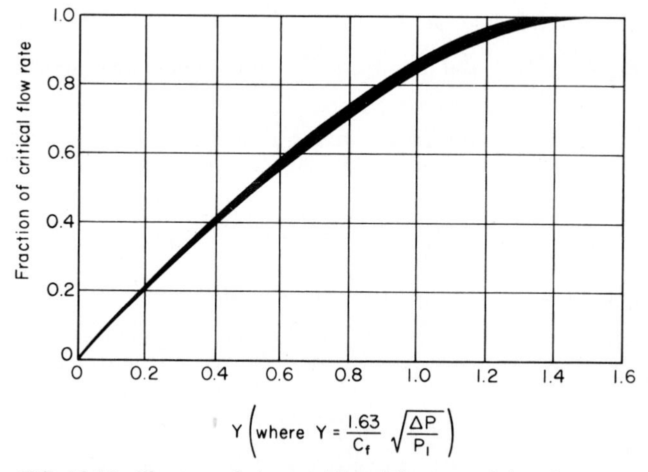

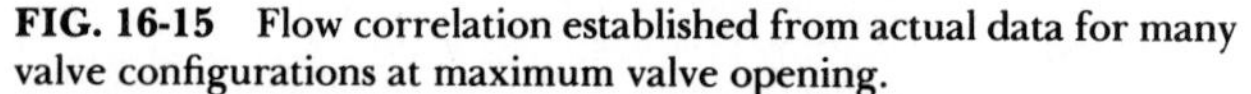

FIG. 16-15 Flow correlation established from actual data for many valve configurations at maximum valve opening.

6. Determine C_f for critical vapor flow.

Assuming reduced valve trim for a heavy-duty angle valve, $C_f = 0.55$ from Table 16-2.

7. Compute the critical pressure drop in the valve.

Use $\Delta P_c = 0.5(C_f)^2 P_1$, where P_c is critical pressure drop, in lb/in^2; other symbols are as given earlier. Substituting, $\Delta P_c = 0.5(0.55)^2(1260) = 191$ lb/in^2 (1316.9 kPa).

8. Determine the value of C_y.

Use the relation $C_v = W/[1.83 C_f P_1]$, where the symbols are as given earlier. Substituting, $C_v = 78{,}000/[1.83(0.55)(1260)] = 61.5$. A lower C_v could be attained by using

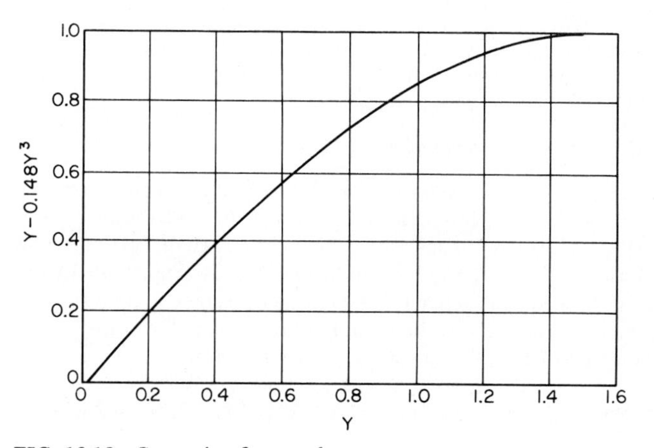

FIG. 16-16 Correction-factor values.

the valve flow to open, but a more economical choice is a single-port top-guided valve installed open to flow.

For a single-port top-guided valve flow to open, $C_f = 0.90$ from Table 16-2. Hence, $C_v = 78{,}000/[1.83(0.90)(1260)] = 37.6$.

A lower capacity is required at critical flow for a valve with less pressure recovery. Although this may not lead to a smaller body size because of velocity and stability considerations, the choice of a more economical body type and a smaller actuator requirement is attractive. The heavy-duty angle valve finds its application generally on flashing-hydrocarbon liquid service with a coking tendency.

This Calculation Procedure is the work of Henry W. Boger, Engineering Technical Group Manager, Worthington Controls Co.

16-19 Indirect Drying of Solids

An indirect dryer consisting of a heating section, a constant-rate–drying section, and a falling-rate–drying section in series is to lower the water content of 1000 lb/h (454 kg/h) of feed from 20 to 5 percent. The feed temperature is 60°F (289 K), and the product leaves the dryer at 260°F (400 K). The specific heat of the solid is 0.4 Btu/(lb)(°F) [1.67 kJ/(kg)(K)]; that of water is 1.0 Btu/(lb)(°F) [4.19 kJ/(kg)(K)]. The heating medium is 338°F (443 K) steam. The heat-transfer rates in the heating, constant-rate–drying and falling-rate–drying sections are 25, 40, and 15 Btu/(h)(ft^2)(°F), respectively. The surface loading in the three sections is 100, 80, and 60 percent, respectively. Thermal data for the moisture and solids are as follows:

Evaporation enthalpy:	970.3 Btu/lb (2257 kJ/kg)
Water enthalpy at 212°F:	180.2 Btu/lb (419 kJ/kg)
Water-vapor enthalpy, averaged over 212 to 260°F	1159.0 Btu/lb (2696 kJ/kg)
Solids temperature during constant-rate drying	212°F (373 K)
Product moisture at start of falling-rate drying	10 percent

Determine the heat load for each section of the dryer, as well as the area required for each.

Calculation Procedure:

1. Calculate the flow rate of dry solids W_S.

As the feed moisture is 20 percent and the total flow rate is 1000 lb/h, $W_S = 1000(1 - 0.20) = 800$ lb/h (363 kg/h).

2. Determine the product flow rate W_P.

Since W_S is 800 lb/h and the product moisture is 5 percent, $W_P = 800/(1 - 0.05) = 842.1$ lb/h (382 kg/h).

3. Calculate the total amount of liquid to be removed.

Because the feed rate is 1000 lb/h and the product rate 842.1 lb/h, the liquid removed (i.e., evaporated) is 1000 − 842.1 = 157.9 lb/h (71.7 kg/h).

4. *Determine the amount of moisture in the in-process material as it passes from the constant-rate section into the falling-rate section.*

As the moisture content at this point is 10 percent and the dry-solids rate is 800 lb/h, the flow rate for the in-process material at this point is 800/(1 − 0.10) lb/h. The moisture in the material is, then, [800/(1 − 0.10)][0.10], or 88.9 lb/h (40.4 kg/h).

5. *Calculate the moisture leaving in the final product.*

The product rate is 842.1 lb/h and its moisture content 5 percent, so the moisture leaving in the final product is (842.1)(0.05), or 42.1 lb/h (19.1 kg/h).

6. *Determine the moisture removed in the falling-rate zone.*

Because the moisture entering this zone is 88.9 lb/h and the amount leaving in the product is 42.1 lb/h, the amount removed from the in-process material in this zone is 88.9 − 42.1, or 46.8 lb/h (21.2 kg/h).

7. *Find the amount of moisture removed in the constant-rate zone.*

The total removed in the dryer is 157.9 lb/h and the amount removed in the falling-rate zone is 46.8 lb/h, so the amount removed in the constant-rate zone is 157.9 − 46.8, or 111.1 lb/h (50.4 kg/h).

8. *Calculate Q_{HS}, the heat load for the solid in the heating zone.*

Now, $Q_{HS} = W_S C_S \Delta T$, where W_S is the solid flow rate, C_S the specific heat of the solid, and ΔT the temperature rise for the solid in this zone. Thus, $Q_{HS} = (800)(0.4)(212 - 60) = 48{,}640$ Btu/h (14.25 kW).

9. *Calculate Q_{HL}, the heat load for the liquid in this zone.*

Use the same procedure as in step 8, but with the liquid flow rate and specific heat. Thus, $Q_{HL} = [(0.20)(1000)][1.0][212 - 60] = 30{,}400$ Btu/h (8.91 kW).

10. *Find Q_H, the total heat load for this zone.*

It is the sum of the solid and liquid heat loads: $Q_H = 48{,}640 + 30{,}400 = 79{,}040$ Btu/h (23.16 kW).

11. *Determine Q_C, the heat load in the constant-rate–drying zone.*

The heat in this zone serves solely for evaporation, with no sensible heating. The amount of water evaporated in this zone was found in step 7. Then, $Q_C = (970.3)(111.1) = 107{,}800$ Btu/h (31.58 kW).

12. *Calculate Q_{FS}, the heat load for the solid in the falling-rate zone.*

As in step 8, $Q_{FS} = (800)(0.4)(260 - 212) = 15{,}360$ Btu/h (4.5 kW).

13. *Find Q_{FE}, the heat load for evaporation in the falling-rate zone.*

The amount of water removed in this zone was found in step 6. For the heat load per pound of evaporated water, use the difference in enthalpy between that for water at 212°F and the averaged value for water vapor between 212 and 260°F, as stated at the beginning of the problem. Thus, $Q_{FE} = (46.8)(1159 - 180.2) = 45{,}808$ Btu/h (13.4 kW).

14. *Calculate Q_{FL}, the heat load for unevaporated liquid in the falling-rate zone.*

This is the liquid that remains as moisture in the final product. The amount was calculated in step 5. In the falling-rate zone, it becomes heated from 212 to 260°F. As in step 8, $Q_{FL} = (42.1)(1.0)(260 - 212) = 2021$ Btu/h (592 W).

15. *Calculate Q_F, the total heat load in the falling-rate zone.*

Sum the quantities calculated in the previous three steps. Thus, $Q_F = 15{,}360 + 45{,}808 + 2021 = 63{,}189$ Btu/h (18.51 kW).

16. *Calculate ΔT_{mH}, the log-mean temperature difference in the heating zone.*

Let T_i and T_o be the heating-medium (i.e., steam) temperature at the zone inlet and outlet, respectively, and t_i and t_o correspondingly be the temperature of the in-process material at the zone inlet and outlet. Then

$$\Delta T_{mH} = [(T_i - t_o) - (T_o - t_i)]/\ln[(T_i - t_o)/(T_o - t_i)]$$
$$= [(338 - 212) - (338 - 60)]/\ln[(338 - 212)/(338 - 60)] = 192.1 \text{ Fahrenheit degrees}$$

17. *Determine A_H, the surface area required in the heating zone.*

Use the equation $A = Q/U\Delta TL$, where A is the required surface area, Q the heat load (from step 10), U the overall heat-transfer coefficient, ΔT the log mean temperature difference, and L the surface loading. Thus, $A = 79{,}040/(25)(192.1)(1.0) = 16.46$ ft^2 (1.53 m^2).

18. *Calculate A_C, the surface area required in the constant-rate–drying zone.*

In this case, the temperature difference between the 338°F steam and the 212°F in-process material is constant. Thus, via the equation from step 17, $A = 107{,}800/(40)(338 - 212)(0.8) = 26.74$ ft^2 (2.49 m^2).

19. Calculate ΔT_{mF}, the log-mean temperature difference in the falling-rate zone.

As in step 16, $\Delta T_{mF} = [(338 - 260) - (338 - 212)]/\ln[(338 - 260)/(338 - 212)] = 100.1$ Fahrenheit degrees.

20. Calculate A_F, the surface area required for the falling-rate zone.

As in step 18, $A = 63{,}189/(15)(100.1)(0.6) = 70.14$ ft^2 (6.52 m^2).

21. Determine the total surface area required for the dryer.

From steps 17, 18, and 20, the total is $16.46 + 26.74 + 70.14 = 113.3$ ft^2 (10.54 m^2).

22. Find the total heat load.

From steps 10, 11, and 15, the total is $79{,}040 + 107{,}800 + 63{,}189 = 250{,}029$ Btu/h (73.3 kW).

Related Calculations: This example is adapted from *Process Drying Practice* by Cook and DuMont, published in 1991 by McGraw-Hill. More details are available in that source. Similar calculation for direct dryers is far more complex, involving the psychrometric relations of moist air. Trial-and-error loops are required, and manual calculation is not only time-consuming but also error-prone. A sequence of equations suitable for setting into a computer program can be found in the aforementioned Cook and DuMont.

16-20 Vacuum Drying of Solids

A material having a wet bulk density of 40 lb/ft^3 (640 kg/m^3) and containing 30% water is to be fully dried in a rotary vacuum batch dryer. The dryer is 5 ft (1.5 m) in diameter and 20 ft (6.1 m) long, and has a working volume of 196 ft^3 (5.5 m^3) and a wetted surface of 206 ft^2 (19.1 m^2). The maximum product temperature is 125°F (325 K). Cooling water at 85°F (302 K) is available to condense the water vapor removed, in a shell-and-tube surface condenser. The condenser is to maintain a vacuum of 85 torr, at which level the vapors will condense at 115°F (319 K). Pilot studies indicate that the dryer will dry the material at an effective rate of 1.0 lb/(h)(ft^2) [4.9 kg/(h)(m^2)] and a peak rate of 2.0 lb/(h)(ft^2) [9.8 kg/(h)(m^2)]. On average, how many pounds per hour of dry product can it produce? How much condenser surface is required?

Calculation Procedure:

1. Determine the charge that the dryer can handle.

Multiply the working volume of the dryer by the wet bulk density of the feed. Thus, (196 ft^3)(40 lb/ft^3) = 7840 lb (3559 kg).

2. *Determine the rate at which the dryer can remove vapor.*

Multiply the wetted surface by the effective drying rate. Thus, $(206\ \text{ft}^2)[1.0\ \text{lb}/(\text{h})(\text{ft}^2)] =$ 206 lb/h (93.5 kg/h).

3. *Determine the amount of water to be removed per batch.*

Multiply the amount of material charged by its water content. Thus, $(0.30)(7840\ \text{lb}) =$ 2352 lb (1068 kg) of water.

4. *Calculate the drying time per batch.*

Divide the water to be removed by the rate at which the dryer can remove it. Thus, $2352\ \text{lb}/(206\ \text{lb/h}) = 11.4\ \text{h}$.

5. *Determine the amount of dry product produced per batch.*

Subtract the amount of water removed (see step 3) from the amount of material charged (step 1). Thus, $7840\ \text{lb} - 2352\ \text{lb} = 5488\ \text{lb}$ (2492 kg) dry product.

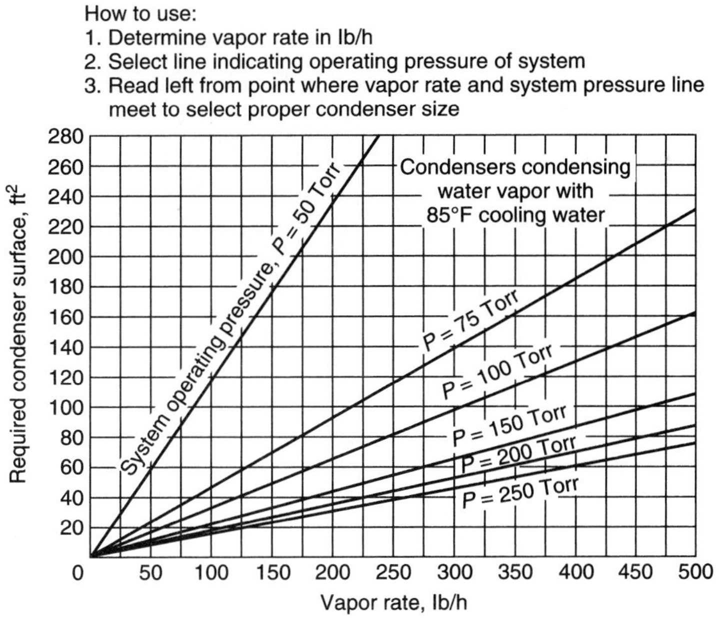

FIG. 16-17 Shell-and-tube condenser surface as a function of vapor rate and system operating pressure (cooling water at 85°F). *(From Chemical Engineering, January 17, 1977, copyright 1977 by McGraw-Hill, Inc., New York. Reprinted by special permission.)*

6. Determine the average production rate in pounds per hour.

Divide the amount of dry product per batch by the drying time per batch. Thus, 5488 lb/11.4 h = 481 lb/h (218 kg/h).

7. Determine the amount of water removed under peak drying conditions.

Multiply the wetted surface area by the peak drying rate. Thus, $(206\ ft^2)[2.0\ lb/(h)(ft^2)] = 412$ lb/h (187 kg/h).

8. Determine the required condenser surface.

Enter the graph in Fig. 16-17 with 412 lb/h peak water removal as abscissa and system operating pressure of 75 torr as the parameter. From the ordinate, read 190 ft^2 (17.7 m^2) as the required condenser surface.

Related Calculations: Figure 16-17 is valid for condensers of shell-and-tube design, employing cooling water at 85°F. This example is adapted from "Vacuum Dryers," *Chem. Eng.*, January 17, 1977.

16-21 Estimating Thermodynamic and Transport Properties of Water

Estimate the following properties of liquid water at 80°F: (1) vapor pressure, (2) density, (3) latent heat of vaporization, (4) viscosity, (5) thermal conductivity. Also, estimate the following properties for saturated water vapor at 200°F: (6) density, (7) specific heat, (8) viscosity, (9) thermal conductivity. And calculate the boiling point of water at 30 psia.

Calculation Procedure: For each of the estimates, the procedure consists of using correlation equations that have been derived by regression analysis of the properties of saturated steam, as discussed in more detail under "Related Calculations." The temperatures are to be entered into the equations in degrees Fahrenheit and the pressure (for the boiling-point example) in pounds per square inch absolute. The results are likewise in English units, as indicated below. These correlations are valid only over the range 32 to 440°F.

1. Estimate the vapor pressure.

The equation is

$$P = \exp[10.9955 - 9.6866 \ln T + 1.9779(\ln T)^2 - 0.085738(\ln T)^3]$$

Thus, $P = \exp[10.9955 - 9.6466(\ln 80) + 1.9779(\ln 80)^2 - 0.085738(\ln 80)^3] = 0.504$ psia.

2. Estimate the liquid density.

The equation is

$$\rho_L = 62.7538 - 3.5347 \times 10^{-3}T - 4.8193 \times 10^{-5}T^2$$

Thus, $\rho_L = 62.7538 - 3.5347 \times 10^{-3}(80) - 4.8193 \times 10^{-5}(80)^2 = 62.16\ lb/ft^3$.

3. Estimate the latent heat of vaporization.

The equation is

$$\Delta H_{vap} = 1087.54 - 0.43110T - 5.5440 \times 10^{-4}T^2$$

Thus, $\Delta H_{vap} = 1087.54 - 0.43110(80) - 5.5440 \times 10^{-4}(80)^2 = 1049.5$ Btu/lb.

4. Estimate the liquid viscosity.

The equation is

$$\mu_L = -0.23535 + 208.65/T - 2074.8/T^2$$

Thus, $\mu_L = -0.23535 + 208.65/80 - 2074.8/(80)^2 = 2.05$ lb/(h)(ft).

5. Estimate the liquid thermal conductivity.

The equation is

$$k_L = 0.31171 + 6.2278 \times 10^{-4}T - 1.1159 \times 10^{-6}T^2$$

Thus, $k_L = 0.31171 + 6.2278 \times 10^{-4}(80) - 1.1159 \times 10^{-6}(80)^2 = 0.369$ Btu/(h)(ft)(°F).

6. Estimate the vapor density.

The equation is

$$\rho_V = \exp(-9.3239 + 4.1055 \times 10^{-2}T - 7.1159 \times 10^{-5}T^2 + 5.7039 \times 10^{-8}T^3)$$

Thus, $\rho_V = \exp[-9.3239 + 4.1055 \times 10^{-2}(200) - 7.1159 \times 10^{-5}(200)^2 + 5.7039 \times 10^{-8}(200)^3] = 0.0301$ lb/ft³.

7. Estimate the specific heat of the vapor.

The equation is

$$C_P = 0.43827 + 1.3348 \times 10^{-4}T - 5.9590 \times 10^{-7}T^2 + 4.6614 \times 10^{-9}T^3$$

Thus, $C_P = 0.43827 + 1.3348 \times 10^{-4}(200) - 5.9590 \times 10^{-7}(200)^2 + 4.6614 \times 10^{-9}(200)^3 = 0.478$ Btu/(lb)(°F).

8. Estimate the vapor viscosity.

The equation is

$$\mu_V = 0.017493 + 5.7455 \times 10^{-5}T - 1.3717 \times 10^{-8}T^2$$

Thus, $\mu_V = 0.017493 + 5.7455 \times 10^{-5}(200) - 1.3717 \times 10^{-8}(200)^2 = 0.028435$ lb/(h)(ft).

9. Estimate the thermal conductivity of the vapor.

The equation is

$$k_V = 0.0097982 + 2.2503 \times 10^{-5}T - 3.3841 \times 10^{-8}T^2 + 1.3153 \times 10^{-10}T^3$$

Thus, $k_V = 0.0097982 + 2.2503 \times 10^{-5}(200) - 3.3814 \times 10^{-8}(200)^2 + 1.3153 \times 10^{-10}(200)^3 = 0.01400$ Btu/(h)(ft)(°F).

10. Estimate the boiling point.

The equation is

$$T_B = \exp[4.6215 + 0.34977 \ln P - 0.03727(\ln P)^2 + 0.0034492(\ln P)^3]$$

Thus, $T_B = \exp[4.6215 + 0.34977(\ln 30) - 0.03727(\ln 30)^2 + 0.0034492(\ln 30)^3] =$ 248.6°F.

Related Calculations: These equations are presented in Dickey, D.S.,* Practical Formulas Calculate Water Properties, Parts 1 and 2, *Chem. Eng.*, Sept. 1991, pp. 207, 208 and November 1991, pp. 235, 236. That author developed the equations via regression analysis of the properties of saturated steam as presented in *Perry's Chemical Engineers' Handbook,* 6th ed., McGraw-Hill. The maximum error (compared to tabulated values) and sample standard deviation of the correlations are as follows:

Property correlation	Maximum error, %	Sample standard deviation
Vapor pressure	3.64	1.84
Liquid density	0.26	0.10
Latent heat	0.63	0.19
Liquid viscosity	20.23	6.13
Liquid thermal conductivity	0.43	0.20
Vapor density	7.81	1.99
Vapor specific heat	0.49	0.13
Vapor viscosity	0.37	0.19
Vapor thermal conductivity	1.09	0.29
Boiling point	2.66	1.43

The original reference also includes correlations for liquid and vapor specific volume, liquid thermal expansion coefficient, liquid and vapor enthalpy, liquid specific heat, and liquid and vapor Prandtl numbers.

* Dickey is the author of the section on liquid agitation in this present handbook.

Index

ABOUT THE EDITOR

Nicholas P. Chopey is executive editor at *Chemical Engineering* magazine with more than 35 years of experience in publishing and industry. A former process engineer with Esso Standard Oil Company (now Exxon), he received his bachelor's degree in chemical engineering from the University of Virginia and his master's degree in economics from New York University. Mr. Chopey is a member of the American Institute of Chemical Engineers and the American Society for Engineering Education.